Classical Topics in Mathematics

Mathematics is the queen of sciences. She is pure, noble and attractive, and also has a distinct character in comparison with subjects in sciences such as physics: its permanent relevance and eternal validness of its theories and theorems. Whatever was once proved will stay true forever.

Mathematics is a vast subject, and many new concepts, theories and results spring up like mushrooms after spring rain. Similarly, there is also a large number of new mathematics books appearing in libraries and on bookshelves. Probably due to the usefulness of mathematics and its foundational nature, there seems to be more books in mathematics than in other subjects. On the other hand, only a limited number, or even a few, of them stand out and are appreciated and used by many people. The best test on the quality of books is the test of time.

In this series of books *Classical Topics in Mathematics*, we have selected books written by leading experts on topics which are well-tested by time. We believe that in spite of the passage of time, their power and value have not diminished, and hence they bear the trademark of the classical mathematics.

The first volumes of this series consist of an annotated version of Klein's masterpiece *Lectures on the icosahedron and the solution of equations of the fifth degree*, and the first English translation of Klein and Fricke's four big volumes on modular functions and automorphic functions. For this series, we have tried to pick books which share or reflect Klein's vision of the grand unity of mathematics.

The publication of this series of books is consistent with the motto of the Higher Education Press: to provide high quality books on the essential mathematics to the world mathematics community at an affordable price.

Classical Topics in Mathematics

(*Series Editor*: *Lizhen Ji*)

1. Lectures on the Theory of Elliptic Modular Functions, First Volume
Felix Klein, Robert Fricke (Authors), Arthur M. DuPre (Translator)

2. Lectures on the Theory of Elliptic Modular Functions, Second Volume
Felix Klein, Robert Fricke (Authors), Arthur M. DuPre (Translator)

3. Lectures on the Theory of Automorphic Functions, First Volume
Robert Fricke, Felix Klein (Authors), Arthur M. DuPre (Translator)

4. Lectures on the Theory of Automorphic Functions, Second Volume
Robert Fricke, Felix Klein (Authors), Arthur M. DuPre (Translator)

5. Lectures on the Icosahedron and the Solution of Equations of the Fifth Degree
(With a new introduction and commentary)
Felix Klein

6. The Bochner Technique in Differential Geometry
Hung-Hsi Wu

7. Lectures on Differential Equations and Differential Geometry
Louis Nirenberg

CTM 4
Classical Topics in Mathematics

Robert Fricke
Felix Klein

Lectures on the Theory of Automorphic Functions

Second Volume

自守函数理论讲义

第二卷

Translated by Arthur M. DuPre

HIGHER EDUCATION PRESS

Lectures on the Theory of Automorphic Functions, Second Volume by Robert Fricke, Felix Klein, translated by Arthur M. DuPre from *VORLESUNGEN ÜBER DIE THEORIE DER AUTOMORPHEN FUNCTIONEN, ZWEITER BAND*, first published 2017 by Higher Education Press

4 Dewai Dajie, Beijing 100120, P.R. China

Cataloging in Publication Data
图书在版编目（CIP）数据
自守函数理论讲义. 第二卷 = Lectures on the Theory of Automorphic Functions, Second Volume: 英文/(德) 弗里克（Robert Fricke）, (德) 克莱因（Felix Klein）著; (美) 迪普雷（Art DuPre）译. — 北京: 高等教育出版社, 2017.10
(数学经典论题 / 季理真主编)
ISBN 978-7-04-047839-6

Ⅰ. ①自… Ⅱ. ①弗… ②克… ③迪… Ⅲ. ①自守函数—数学理论—英文Ⅳ. ①O174.54

中国版本图书馆CIP数据核字 (2017) 第 194800 号

Copy Editor: Liping Wang
Cover Design: Lingbo Wang
Print Editor: Yimin Zhao

787mm×1092mm 1/16
37.5 Printed Sheets

Printed and Bound in P.R. China by Beijing Shengtong Printing Co., Ltd.
ISBN: 978-7-04-047839-6
1 2017

Why should one open and read Klein-Fricke and Fricke-Klein?

Much of the most important developments of number theory in the 20th and early 21st century, especially those related to algebraic number theory, modular forms, and the connexion between special values of L-functions and arithmetic, grew out of the great discoveries made by German mathematicians in the 19th century. Klein was one of the leading figures in 19th century German mathematics, and the books by Klein and Fricke give the first systematic account of the theory of modular forms.

John Coates

1 They are classics

There is no question that almost every mathematician has heard of Felix Klein, and almost every student and mathematician in geometry has also heard of the famous Erlangen program proposed by him when he became a full professor at the age of 23.

In collaboration with his former student Robert Fricke, Klein wrote two classic books: *Vorlesungen über die Theorie der elliptischen Modulfunctionen* (Lectures on the Theory of Elliptic Modular Functions), and *Vorlesungen über die Theorie der automorphen Functionen* (Lectures on the Theory of Automorphic Functions). They will be referred to as Klein-Fricke and Fricke-Klein in the following discussion in view of the different orders of their names on the book covers.

Probably many people working in number theory, automorphic forms and discrete subgroups of Lie groups have heard both Klein-Fricke and Fricke-Klein books. On the other hand, most people working in Teichmüller theory have only heard of the book Fricke-Klein in connection with Fricke spaces.

Klein-Fricke and Fricke-Klein are classic books and have exerted their influence on mathematics. On the other hand, they unfortunately fit one definition of a classic book by Mark Twain in 1900: it is "something that everybody wants to have read and nobody wants to read". Maybe a variant describes them better: A classic book is "a book which people have heard but have never read".

There are other definitions of classic books. For example, one definition is that "a classic is a book that has never finished saying what it has to say". Another says: "The idea of a classic implies something that has continuance and consistence, and which produces unity and tradition, fashions and transmits itself, and endures."

We hope that the reader will eventually agree that the books Klein-Fricke and Fricke-Klein fit these definitions and also a definition by Goethe: "Ancient works are classical not because they are old, but because they are powerful, fresh, and healthy."

Though more people in the previous generations had cited Klein-Fricke and Fricke-Klein, recent citations and explanations of Fricke-Klein also do occur. For example, one recent major exposition of a part of the book Fricke-Klein is given in a large book [MSW].

Fricke-Klein was the first systematic exposition of the uniformization theorem for Riemann surfaces and related moduli spaces and Teichmüller spaces. Lars Ahlfors referred to the Fricke-Klein in a commentary in his collected works [Ah, p. 122] and wrote:

> An incredibly patient reader of Fricke-Klein, a two volume, 1300-page book, might have been able to discern that Fricke had anticipated Teichmüller's idea of a space $\mathscr{T}_g$ of marked Riemann surfaces of genus g with a $3g-3$ dimensional complex structure.

Lipman Bers also cited it in his papers when he discussed Teichmüller space and mapping class groups (see for example [Ber] and [BG]).

When I was working with Armand Borel on compactifications of locally symmetric spaces, he told me that the idea of adding rational boundary points (or filling in cusp points) and putting a Satake topology on the partial compactification of the upper half-plane must be in Fricke-Klein, though I did not and could not check due to the language barrier.

To many people in number theory and group theory, the famous congruence subgroup problem is natural and important. Once I asked an expert what was the exact motivation for raising and studying this problem, he said that it must be explained in Klein-Fricke. Various papers and books also cited Klein-Fricke as the first place where a non-congruent finite index subgroup of $\mathrm{SL}(2,\mathbb{Z})$ appeared.

When I became interested in moduli space and Teichmüller space, the name of Fricke space appears often in books and papers (for example, [BG] [Ber] [FM] [Bes]), and some people said that Teichmüller space should be called Fricke space. I asked experts for more detail, and they always said that it must be in Fricke-Klein.

On the other hand, these books of Fricke and Klein were written in German and never translated, unlike many classical books written in German which were translated into English shortly after publication. Since the German dominance in mathematics ended with the World War II, fewer people can read German outside Germany now.

Now both Klein-Fricke and Fricke-Klein are available for the first time in English translation. As Kazhdan said, "I definitely agree that it is important to know such important works as lectures by Fricke and Klein." There are many good reasons to read these classic books. One of the best reasons is the curiosity: To see what is really inside them, to unearth unknown treasure, and to understand the origin or history of many modern topics.

For example, Kleinian groups are central in the low dimensional topology and one dimensional complex dynamics.

The question then arises: *How did it start? How was it related to the famous Felix Klein?*

Though it is not easy to name specific major results in the work of Klein, most people will agree that he was a great mathematician and one of the most influential mathematicians in the nineteenth century.

Then another question arises: *What kind of mathematics did Klein do?*

Many people have heard the sad story that Klein competed with Poincaré, and this competition ruined Klein's ability to do any more research.

Then the question is: *What did Klein and Poincaré compete exactly? Was it really important mathematics? Are they still relevant now?*

To me and hopefully to some other people, these questions and the above quote from Ahlfors give enough motivation to *open and read Fricke-Klein and Klein-Fricke.*

If these questions are not enough or irrelevant to some readers, probably a really good reason is to follow a Chinese saying: "*It is always beneficial to open any book!*"

When I was a student, I heard Shing-Tung Yau giving this advice many times, and he practiced it. (He read books rows after rows at the mathematics library at University of California at Berkeley during his graduate student days and is stilling reading all kinds of books now.)

Therefore, it is simply rewarding to open these heavy volumes written by two great mathematicians and see what is inside them. And it is also simply fun to trace out the history of some parts of mathematics we practice and love.

Some people like to say "History is now". So it is perhaps also important to heed the warning of Weyl in the preface to his famous book on Riemann surfaces [We, p. IX]:

> The younger generation is only too inclined to forget the connection of the new with the old!

In a similar spirit, Chern wrote in the foreword to the Chinese edition of Atiyah's collected works:

> No matter how refined or improved a new account is, the original papers on a subject are usually more direct and to the point. When I was young, I was benefited by the advice to read Henri Poincaré, David Hilbert, Felix Klein, Adolf Hurwitz, etc. I did better with Wihelm Blaschke, Elie Cartan and Heinz Hopf. This has also been in the Chinese tradition, when we were told to read Confucius, Han Yu in prose, and Tu Fu in poetry.

See more discussion about reading classics in the commentary by Toledo near the end of this book.

Unity of and connections between different subjects are essential characteristics of beautiful and deep mathematics. With the rapid development of mathematics, there are very few, if any, who could command all subjects of mathematics. People often say that Hilbert, Poincaré and Weyl were the last universalists. Klein should belong there too, but he was also more special since he wrote extensively and carefully with a grand vision in mind. His writing can show his unified view of mathematics. See more discussion about this in the commentary by Mazur in this book.

The biggest exposition project related to Klein was Klein's encyclopedia: A German mathematical encyclopedia published in six volumes from 1898 to 1933 with a total length of 20,000 pages.

But the writings which showed most clearly the trademark of Klein are these four volumes of Klein-Fricke and Fricke-Klein together with Klein's book on the icosahedron. Therefore, if you want to see mathematics through the eyes of Klein, open and read these books!

2 The main work of Klein and his competition with Poincaré

To understand better the books Klein-Fricke and Fricke-Klein, it is probably important to understand better the work of Klein which led to these books, and how the appearance of Poincaré influenced him and led to their competition.

No one knows the work of Klein better than himself. Let us quote from Klein in the late stage of his career when he wrote his famous book on the history of mathematics in the 19th century [Kl1, p. 325]:

> We now turn to the theory of *automorphic functions!* This is the main area of my own work, so that I am much concerned to illuminate the exposition with personal memories and to bring the report up to the time, 1882/1883, when I was prevented from working further because of illness. Since I will often have to refer to them, perhaps I may first name the three works on this subject that were composed or instigated by me:
>
> 1. Klein: *Vorlesungen über das Ikosaeder und die Auflösung der Gleichungen vom fünften Grade* [Lectures on the Icosahedron and the Solution of Equations of the Fifth Degree] 1884.
> 2. Klein-Fricke: *Vorlesungen über die Theorie der elliptischen Modulfunctionen* [Lectures on the Theory of Elliptic Modular Functions] Vol. 1, 1890; Vol. 2, 1892.
> 3. Fricke-Klein: *Vorlesungen die Theorie der automorphen Functionen* [Lectures on the Theory of Automorphic Functions] Vol. 1, 1897; Vol. 2, 1901, 1911, 1912.
>
> Item 1 is essentially intended as a textbook; Item 2 and Item 3 have the character of monographs.

Later Klein described his interaction with Poincaré [Kl1, p. 355]:

> It is now time for me to tell of the appearance of H. Poincaré and of the personal relations which developed between us and which laid the foundations for the further development of the whole subject [of automorphic forms].
> First I must make some remarks on the terminology. When Poincaré began, his knowledge of the German work was very faulty, and what he called the groups with limit-circles "groupes fuchsiens", an undeserved name. When I made him aware of the general functions, he called them "kleinéennes". Thus a great historical confusion arose here.

Klein continued [Kl1, p. 359]:

> Now I may ... tell of my own work on this subject. ... My first note on this, which was published at New Year's 1882 .... In this paper I published that

fundamental theorem on the limit-circle case, which I above called the *Central Theorem* ... because of its special simplicity; namely, that one can, in one and only essentially only one way, uniformize any Riemann surface with $p \geq 2$ without branch points by means of automorphic functions with a limit-circle. This theorem ... was brought forth under the most difficult external circumstances....

The proof was in fact very difficult. I used the so-called *continuity method*, which compares the manifold of Riemann surfaces with a given p to the corresponding manifold of automorphic groups with limit-circles. I never doubted that the method of proof was correct, but everywhere I ran into gaps in my knowledge of function theory or in function theory itself. I could only postulate the resolution of these difficulties, which were in fact completely resolved only 30 years later (in 1912) by Koebe.

This could not stop me from stating still more general theorems ... As imperfect and unresolved as much of that paper is, the structure of the train of thought as a whole has remained, and was not even displaced by Poincaré's subsequent papers in Volumes 1, 3, 4, 5 of the then founded *Acta Mathematica*.

In fact, I was again able to precede Poincaré by a little, for my offprints were sent off at the end of November 1882; while the first issue of the *Acta*, which contains Poincaré's first paper, appeared at the beginning of December....

The price I had to pay for my work was extraordinarily high—my health completely collapsed. In the following years I had to take long leaves and to renounce all productive activity. Things did not go well again until the autumn 1884, but I have never regained my earlier level of productivity. I never returned to elaborate my earlier ideas. And later, when I was at Göttingen, I turned to extending the domain of my work and to general tasks of organizing our science. Thus one can understand that since then I have only occasionally touched on automorphic functions. My real productive activity in theoretical mathematics perished in 1882. All that has followed, insofar as it is not been purely expository, has been merely a matter of working out details.

Thus Poincaré had a free field and, until 1884, went on to publish his five great papers on the new functions.

3 Vision of Klein and his grand book project

Klein was a mathematician of vision. His Erlangen program has had changed people's perspective of geometry and of the role of transformation groups. One of the fundamental contributions of Klein and Lie is their emphasis on the important role played by group theory and their initiation of such a point of view. See [Ya] and [JP] and references there. These books by Klein and Fricke fit into his grand book project which formed his lasting contribution to mathematics.

Felix Klein's famous Erlangen program made the theory of group actions into a central part of mathematics. In the spirit of this program, Klein set out to write a grand series of books which unified many different subjects of mathematics including number theory, geometry, complex analysis, discrete subgroups, complex analysis etc.

The first book on the icosahedron and the solution of equations of the fifth degree showed close relations between three seemingly different subjects: the symmetries of the icosahedron, the solution to fifth degree algebraic equations, and the differential equation of hypergeometric functions. (See the commentary by Jeremy Gray in this book for more detail). It was translated into English in 1888, four years after its original German version was published in 1884. It was followed by two volumes on elliptic modular functions by Klein and Fricke, and two more volumes on automorphic functions by Fricke and Klein. These four volumes are vast generalizations of the first book and contain the highly original works of Poincaré and Klein on automorphic forms. They have been very influential in the development of discrete subgroups of Lie groups and automorphic forms.

In the preface to [Kl2], Klein explained relations between these books:

> The theory of Icosahedron has during the last few years obtained a place of such importance for nearly all departments of modern analysis, that it seemed expedient to publish a systematic exposition of the same. Should this prove acceptable, I propose to continue in the same course and to treat in a similar manner the subject of Elliptic Modular Functions, and the general investigations newly made of Single-valued Functions, with linear transformations into themselves. Thus a treatise of several volumes would grow, in which I should expect to promote science, at least in so far as it might introduce many to realms of modern mathematics rich in far-stretching vistas.

These five books contain many original ideas, striking examples, explicit computations and details which are not available at other places. They will be very valuable references for people at all levels and allow the reader to see the unity of mathematics through the eyes of one of the most influential mathematicians, Felix Klein. Indeed, by reading them, the reader can learn more about the unity of mathematics in a concrete situation.

For example, it is still very valuable to read his book on the icosahedron. Most modern literature concentrate on its part on singularities. Arnold[1] promoted strongly to understand better the singularities first studied by Klein in this book:

> Unexpectedly enough, the classification of the simplest singularities in all these problems turns out to be related to the Lie, Coxeter, and Weyl groups A_k, D_k, E_h to Artin's and Brieskorn's braid groups, and to the classification of the platonics in Euclidean three space.
>
> The occurrence of the diagrams A, D, E and of Coxeter groups in such different situations as the simple Lie algebra theory, the classification of simple categories of linear spaces (Gabriel, Gel'fand-Ponomarev, Roiter-Nasarova), the

[1] Quoted from page 20 of the paper *Critical Points of Smooth Functions* by Arnold in *Proceedings of the International Congress of Mathematicians*, Vancouver, 1974. See also Problem VIII posed by Arnold on page 46 of the book *Mathematical Developments Arising from Hilbert Problems*. Proceedings of Symposia in Pure Mathematics, Vol. XXVIII. American Mathematical Society, Providence, 1976.

> Kodaira classification of elliptic curves degenerations, the theory of platonics, and the theory of simple singularities gives an impression of a wonderful chain of coincidences of the results of independent classifications (certain relations between some of them being known, others suspected). As we will now see, the classification of more complex singularities provides new wonderful coincidences, where Lobatchevski triangles and automorphic functions take part.

But the action of finite groups on sphere and fundamental domains discussed in this book of Klein also have had a far reaching consequence in the contemporary mathematics. Indeed, as Klein wrote [Kl1, p. 315]:

> The special subject of *group theory* extends through all of modern mathematics. As an ordering and clarifying principle, it intervenes in the most various domains.... When Lie and I then undertook to work out the significance of group theory for various domains of mathematics, ...

In the books [Kl2], Klein-Fricke and Fricke-Klein, discrete groups play an essential role. Klein made this clear on page 2 of the preface of [Kl2]:

> My indebtedness to Professor Lie dates back to the years 1869–1870, when we were spending that last period of our student-life in Berlin and Paris together in intimate comradeship. At that time we jointly conceived the scheme of investigating geometric or analytic forms susceptible of transformation by means of groups of changes. This purpose has been of directing influence in our subsequent labours, though these may have appeared to lie asunder. Whilst I primarily directed my attention to groups of discrete operations, and was thus led to the investigation of regular solids and their relations to the theory of equations, Professor Lie attacked the more recondite theory of continued groups of transformations, and therewith of differential equations.

Among these books planned by Klein, only the first one [Kl2] was written by Klein. The other two books Klein-Fricke and Fricke-Klein were written by Fricke with substantial contributions from Klein, as the reader can see from the prefaces of these books. It should be emphasized that contributions from both Klein and Fricke were substantial.

Given the grand scale and rich contents of these books, one natural question is: *What were people's reactions to these books?* Let us take a look at some reviews about them.

In a Math Review of the paper [Sl] which tried to explain the book [Kl2], Dieudonnie wrote:

> Klein's Lectures on the icosahedron, published in 1884, soon became one of the most famous books on mathematics for a half-century; no other book so forcefully put in the limelight the fundamental unity of mathematics. Klein showed that three apparently disjoint theories: the symmetries of the icosahedron (geometry), the resolution of fifth degree equations (algebra) and the differential equation of hypergeometric functions (analysis) were in fact dominated by the structure of a single object, the simple group A_5 of 60 elements. It is the group of rotations leaving the icosahedron invariant, the monodromy

group of a hypergeometric equation, and the commutator subgroup of the Galois group of the "general" fifth degree equation.

There have been many modern expositions on the book on the icosahedron, especially in many writings of Peter Slodowy.[2]

In his review of Klein-Fricke, Cole wrote [Co, p. 105]:

> The theory of the modular functions and the allied branches has been one of the chief series of investigations to which Professor Klein has devoted himself in the period of some twenty years over which his scientific activity now extends. It is characteristic of these investigations that they are not included as a subordinate part in any of the great mathematical theories heretofore commonly so recognized. Their distinctive tendency is in the direction of the combination and unification of the latter into a broader method of research. This idea has been developed by Klein to an extent and with an elaboration which has long since entitled it to recognition as an independent, and in the highest degree productive mathematical point of view.

He continued [Co, p. 109]:

> The value to the student of this breadth of treatment is simply inestimable. No one can study long under Klein without obtaining an intelligent comprehension of most of the great tendencies in mathematics. This is at present particularly desirable in view of the extreme degree of specialization which has come to prevail among mathematicians. Of all the great services which Klein has rendered to mathematics, there is none more valuable than his successful unification of its heretofore rapidly diverging branches.

In his review of the Klein-Fricke and an earlier version of the Fricke-Klein, Hutchinson wrote [Hu]:

> Among the various classes of special functions which have hitherto engaged the attention of mathematicians, that of most recent origin, and of by far the largest content (at least potentially) is the automorphic functions. This vast subject, the growth of the past quarter of a century, owes its rapid development to the genius and assiduity of the two eminent mathematicians, Klein and Poincaré, as well as to the comparatively high state of perfection of other mathematical disciplines which have been forced to contribute their assistance to this new field. Klein and Poincaré have each brought to this subject a breadth of knowledge and a corresponding wealth of ideas truly remarkable at the present day when multiplicity of interests hardly permits the investigator any other choice than to specialize within limits more or less narrow. Klein in particular has by precept and example urged the importance of a closer unity among all departments of mathematical thought, and the great advantage to

[2] For example, P. Slodowy, Platonic solids, Kleinian singularities, and Lie groups. *Algebraic geometry* pp. 102–138, Lecture Notes in Math., 1008, Springer, Berlin, 1983. And Groups and special singularities. *Singularity theory* (Trieste, 1991), 731–799, World Sci. Publ., River Edge, 1995.

> be derived from bringing to the aid of any one field the combined resources of all the others. It is in his work on the automorphic functions that he has given the most brilliant illustration of this mode of treatment. In this work he has been ably seconded by the efforts of the pupils and co-laborers he has been so fortunate as to gather about him at the Göttingen school, among whom Robert Fricke is especially deserving of the highest praise for the ability and industry which he has brought to the arduous labor of preparing for publication the volumes now before us, and for the wealth of material which he himself has contributed to the subject.
> These volumes form the sequel to, and final elaboration of the ideas contained in the volumes previously published, namely, the Ikosaeder (by Klein), and the Elliptische Modulfunctionen (by Klein and Fricke). The former of these works deals exhaustively with the finite groups of transformations on a single variable, and the functions associated with them. The two large volumes of the latter treat in a very elaborate manner the modular group, and the general ideas necessary to be followed out in the study of all such groups and their functions. One of the principal objects of this fullness of discussion in the Modulfunctionen is to pave the way for a subsequent discussion of the general theory of automorphic functions without the restraint of burdensome details. The advantage of thus disposing of preliminary details, and familiarizing the reader with many of the fundamental ideas as applied to a concrete example is clearly perceived when we observe how completely are the various elements in the present work fused into an organic whole, each part in strong and vital contact with every other part.

Among many reasons to read these classical books Klein-Fricke and Fricke-Klein, one reason is that some old ways can help people to understand better some key results, since the first intuitions are often the most direct and reasonable. For example, one of the main results in Fricke-Klein is a proof of the uniformization of Riemann surfaces. There are now many different and rigorous proofs. Klein and Poincaré used the method of continuity to prove it, which gives an intuitive and convincing argument for the validity of the theorem. More importantly, they explained motivations better than many modern books. Specifically, the moduli space of compact Riemann surfaces of genus g has the same dimension as the moduli space of compact hyperbolic surfaces of genus g. An injective map from the latter to the former suggests that every Riemann surface should come from a hyperbolic surface. This is reasonable since it is open and hence the map from any one point can be extended to nearby points. Furthermore, the fact that the proof for a result of individual Riemann surfaces involves deformation and the totality of Riemann surfaces is also appealing. This is probably the first important application of moduli spaces.

It is perhaps appropriate to end this section with a quote from Arnold [Li, p. 435] about Klein's books:

> One-half of the mathematics I know comes from the book of F. Klein *Development of Mathematics in the 19th Century.*

4 Sources of many contemporary things and unearthed treasure

Both books Klein-Fricke and Fricke-Klein contain a wealth of information. Since they were related to research of Klein and some of his contemporaries, they contain first systematic expositions of various topics. We mention some of them here.

Klein-Fricke gave the first comprehensive study of the modular group $SL(2, \mathbb{Z})$, its congruence subgroups and more general finite index subgroups. For example, it gave the first example of finite index subgroup of $SL(2, \mathbb{Z})$ which is not congruence subgroup. In other words, the congruence subgroup problem has a negative answer for $SL(2, \mathbb{Z})$. It also contains very detailed discussion of the elliptic modular function. See the commentary by Borcherds for many striking properties of the j-invariant of elliptic curves.

It seems that there are treasures to be found. As Deligne wrote, "I would be glad if someone could extract what is buried in it, especially the modular meaning of the functions Fricke and Klein considered."

In some sense, the earlier book Klein-Fricke dealt with more mature theories. But the latter book Fricke-Klein was a research monograph describing some new and ongoing works. It gave the first detailed discussion of relations between Riemann surfaces and hyperbolic metrics. For example, the main problem they tried to solve is the uniformization theorem of Riemann surfaces. To do this, they needed to develop theories of Fuchsian groups, moduli space of hyperbolic polygons (not identical to but related to the Teichmüller space of hyperbolic surfaces), moduli space of Fuchsian groups, which is also essentially the character variety of surface groups.

Consequently, the idea of Teichmüller space and mapping class group were there, but they were not clearly defined, and some proofs were not carried out in a well-organized way either.

To show that these quotients of the Poincaré disks are Riemann surfaces, automorphic functions were also developed.

Though they need moduli spaces of Riemann surfaces in order to apply the continuity method to prove the uniformization theorem, they did not really study moduli spaces of Riemann surfaces and hyperbolic surfaces for their own sake, but only used them as step-stones to prove the uniformization theorem of Riemann surfaces.

As most people working in Teichmüller theory know, Fenchel-Nielson coordinates are convenient for many purposes. For example, they give a clear and quick proof that the Teichmüller space $\mathcal{T}_g$ of Riemann surfaces of genus $g \geq 2$ is homeomorphic to the Euclidean space $\mathbb{R}^{6g-6}$. But it is not well-known that this idea of decomposing a general hyperbolic surface into simple pieces including pairs of pants, and hence essentially the Fenchel-Nielsen coordinates, was in Fricke-Klein. See the commentary by William Harvey about this and other results in Fricke-Klein and, topics in discrete subgroups which are still actively pursued now. There are also strong assertions about similar facts related to Teichmüller (or Fricke) spaces in books [FM] [ACG] without references. With this translation of the book Fricke-Klein, you can read and check for yourself.

It is perhaps helpful to put all five books [Kl2], Klein-Fricke, Fricke-Klein together. To solve quintic equations by using icosahedron invariants, Klein developed actions of discrete groups, fundamental domains for discrete groups. More precisely, in Remark

1 in [Kl1, pp. 321-322], Klein explained fundamental domains of the symmetry group of regular solids and then wrote [Kl1, p. 322]:

> Remark 1 on p. 321 leads to the consideration of other discontinuous but infinite groups of linear substitutions of one variable, and from this grows the general theory of single-valued automorphic functions.

The book Klein-Fricke gave a comprehensive study of the modular group $SL(2,\mathbb{Z})$ and elliptic modular functions which serves as a model for Fuchsian groups and automorphic functions. See the commentary by Borcherds for a comprehensive summary of results on modular forms in Klein-Fricke.

It is fitting to mention that two important generalizations of $SL(2,\mathbb{Z})$ were started in Fricke-Klein. The first consists of discrete subgroups of $SL(2,\mathbb{R})$, $SL(2,\mathbb{C})$, and more generally discrete subgroups of Lie groups, especially lattices and arithmetic subgroups of Lie groups. The currently active topic of discrete subgroups of semisimple Lie groups and locally symmetric spaces can be traced to Fricke-Klein. The second consists of mapping class groups of surfaces and their actions on Teichmüller spaces. Both generalizations have had huge impact on the geometric group theory.

The analogy between moduli spaces of Riemann surfaces and locally symmetric spaces has stimulated a lot of work in the past several decades. It culminated in work of Maryam Mirzakhan and her Fields Medal in 2014.

5 Why should one care about modular forms and automorphic forms

These books of Fricke and Klein gave systematic expositions of modular and automorphic forms for the first time, and their impact on the later development is huge.

To people not working in these subjects, one natural question is why it is important to study modular forms and automorphic forms?

The distinguished physicist C.N. Yang[3] asked me such a question on December 8, 2016: "Perhaps you can explain to me why number theorists are interested in modular functions."

It is impressive that a person of 95 years old is still so interested in one of the key subjects of the contemporary mathematics. The commentary by Borcherds gives a comprehensive summary of various special properties of modular functions, especially the j-function. For the convenience of the reader, I include my answer to Prof. Yang. Hope that a general answer from an amateur in number theory to an old lover of mathematics might be interesting to some readers outside the field of number theory.

Briefly, there are three kinds of reasons for the importance and application of modular forms and automorphic forms:

1. It has a good origin starting from problems in number theory studied by Lagrange, Gauss, Klein, Poincaré et al.

[3] Symmetry and group theory is an integral part of Prof. Yang's academic life. He received the Nobel Prize for his work with T.-D. Lee on violation of the law of parity conservation. He is also famous for the Yang-Mills theory and the Yang-Baxter equation.

2. The basic modular function, j-function, gives a natural explanation of certain number fields and plays a similar role like the exponential function for the field of rational numbers.
   Understanding all number fields was and still is the major problem in number theory, and it is basically the problem of solving algebraic equations.
3. Each modular form or more a general automorphic form gives a sequence of numbers which can be related to other problems in number theory and arithmetic algebraic geometry.
   Basically, the fact that a sequence of numbers comes this way is the essence of many problems, such as the solution by Wiles of Fermat's last theorem (conjecture), the Langlands program, and Borcherds' solution of the Monster Mooshine. These sequences of numbers can be incorporated into L-functions.

The theory of modular and automorphic forms has a distinguished and interesting (or complicated) history. In his book *Disquisitiones Arithmeticae*, Gauss studied binary quadratic forms and equivalence classes of binary quadratic forms, and related them to the action of $\mathrm{SL}(2,\mathbb{Z})$ on the upper half plane $\mathbb{H}^2$: The quotient space $\mathrm{SL}(2,\mathbb{Z})\backslash\mathbb{H}^2$ classifies the equivalence classes of binary positive definite quadratic forms of determinant 1. This space is also the moduli space of compact Riemann surfaces of genus 1, and equivalently the moduli space of elliptic curves over $\mathbb{C}$. Consequently, people are interested in functions on the upper half plane $\mathbb{H}^2$ which are invariant under $\mathrm{SL}(2,\mathbb{Z})$.

Due to this modular interpretation, $\mathrm{SL}(2,\mathbb{Z})$ is called the *modular group*, and such functions are called *modular functions*. The first important such function is the j-invariant of elliptic curves. This gives a bijection between the above moduli space (or the quotient $\mathrm{SL}(2,\mathbb{Z})\backslash\mathbb{H}^2$) and the complex plane $\mathbb{C}$.

But the invariance under $\mathrm{SL}(2,\mathbb{Z})$ is too strong, and people impose some other conditions under the action by adding automorphy factors. Then the theory of general modular forms for $\mathrm{SL}(2,\mathbb{Z})$ was born. One can also replace $\mathrm{SL}(2,\mathbb{Z})$ by a subgroup of finite index, in particular principal congruence subgroups, and Klein initiated a systematic study of them and detailed discussion was given in the book Klein-Fricke.

Slightly later, Poincaré and also Klein initiated the more general theory of automorphic forms, which made Poincaré first famous (Poincaré made connection with the hyperbolic geometry and applied it to the uniformization theorem of Riemann surfaces), and which also led to the competition between them.

One major result of modular forms and automorphic functions in number theory is the Kronecker-Weber theorem: every abelian extension of $\mathbb{Q}$ is contained within some cyclotomic field. People have tried to understand abelian extensions of other number fields. For imaginary quadratic fields, their extensions can be obtained by special values of the j-function. This is called Kronecker's Jugendtraum ("dream of youth"). See [Vl] for a historical and systematic discussion. One major problem in number theory now is to generalize this to other number fields.

Many problems in number theory such as the number of ways an integer can be represented as sums of squares can be solved via modular forms (or rather theta functions). They are related to coefficients of modular/automorphic forms.

The basic point is that the coefficients of modular forms (and more general automorphic forms) are special and satisfy certain symmetry. This is reflected by the fact that L-functions associated with them satisfy functional equations. (The fact that Riemann zeta function satisfies the functional equation follows from the Poisson summation formula, or a modular property of theta function.)

There is the known Hecke and Weil's converse theorem: gives sufficient conditions for a Dirichlet series to be the Mellin transform of a modular form, i.e., coming from a modular form as above.

This also explains the power of the Langlands program: many sequences of numbers arising in number theory are encoded in L-functions, and they should come from automorphic forms (or representations) in the sense that their L-functions are related.

In the vaguest sense, the modular property of L-function of a sequence of number catches the special property of the sequence. The recent work of Borcherds on the Monstrous Moonshine about the Monster group (the largest sporadic finite simple group) is one such instance also.

Besides these applications in number theory and related subjects, the work of Poincaré and Klein on Fuchsian groups and their actions on the hyperbolic plane has had a huge impact on moduli spaces of compact Riemann surfaces and the low dimensional topology such as the work and conjecture of Thurston.

6 Preparatory reading for Klein-Fricke and Fricke-Klein

Reading classics may not be easy, and this happens to be the case with these books of Klein and Fricke. Both books Klein-Fricke and Fricke-Klein are not easy to read, and even the famous and more expository book [Kl2] is not easy to many people. For example, there have been several recent attempts to explain some ideas of [Kl2] in modern languages, such as [To, §1.6] [Sh] [Na] [Gr]. But these expositions do not seem to convey completely all the major ideas and results of [Kl2] or its connections with other work of Klein, in particular Klein-Fricke and Fricke-Klein. Therefore, it is very valuable to read the original writing.

On the one hand, to show that it is probably not easy for many people to understand [Kl2], it might be helpful to quote from the first page of a letter by Serre [Se]:

> I am sorry to have been so slow in answering your query about Klein's Icosahedron book. I have been looking at it, off and on, for the past weeks without being able to write anything.

On the other hand, some good overview of and introduction to [Kl2] were given by Klein himself, for example, in Lectures IX and X of his book *Lectures on Mathematics* [Kl3]. Lecture IX dealt with solutions of higher algebraic equations, and Klein wrote [Kl3, pp. 67-68]:

> ...since the year 1858, a very different point of view should have been adopted. This is the year in which Hermite and Kronecker, together with Brioschi, found the solution of the equation of the fifth degree, at least in its fundamental ideas. This solution of the quintic equation is often referred to as a "solution by elliptic functions"; but this expression is not accurate, at least not as a counterpart to the "solution by radicals". Indeed, the elliptic functions enter unto the solution of the equation of the fifth degree, as logarithms might be said to enter

> into the solution of an equation by radicals, because the radicals can be computed by means of logarithms. *The solution of an equation will,* in the present lecture, *be regarded as consisting in its reduction to certain normal equations.*
> ...
> I have simplified the solution of the quintic, and think that I have reduced it to the simplest form, by introducing the *icosahedron equation* as the proper normal equation. In other words, the icosahedron equation determines the typical irrationality to which the solution of the equation of the fifth degree can be reduced.

Lecture X dealt with hyperelliptic and Abelian functions, and Klein concluded with [Kl3, p. 84]:

> I emphasize these special problems because the Abelian functions have always been regarded as one of the most interesting achievements of modern mathematics, so that every advance we make in this theory gives a standard by which we can measure our own efficiency.

The chapter [Kl1, Chapter VIII] gives a more systematic discussion of the books [Kl2], Klein-Fricke and Fricke-Klein.

Reading these original books will help one understand better theories in them. But there is another important reason as well. In the book Fricke-Klein, one can see how important mathematics was created, problems were met and how methods and new concepts were introduced. Though results and solutions may not be final or polished, they give people a sense of how great people did mathematics.

To help the reader, we have also asked some experts in the fields to provide some commentaries on these books, which might serve as guides and introductions to the books Fricke-Klein and Klein-Fricke. They all appear near the end of the book.

7 Why does Fricke-Klein have a reputation of being difficult and not clear

On the one hand the book Klein-Fricke was considered well-written. For example, in a book review, Cole wrote [Co, p. 109]:

> Turning to the Modulfunctionen, one cannot but admire the simplicity and perfect proportion with which Dr. Fricke has developed the subject. So far as possible, everything is traced from first principles. The network of interwoven theories is constructed with a painstaking elaboration of details and a rare geniality of method. A clearer and more scholarly presentation than that before us could hardly be imagined.

On the other hand, the later book Fricke-Klein is considered difficult and even badly written by some people. This is in sharp contrast with the fact that many people have enjoyed some writings of Klein on elementary mathematics, for example, his books on elementary from advanced points of view such as [Kl3] [Kl4] [Kl5] [Kl6], which have been translated into many languages and are still in print.

Similarly, Fricke also wrote several other books after his joint books with Klein, and was famous for being a clear writer, at least according to reviews of some more elementary books by him.

For example, Dickson wrote in a review of a book titled *Lecture on Algebra* by Fricke [Di]:

> Professor Fricke is an experienced writer of unusual clearness. In addition to his recent book on elliptic functions, he was a joint author with Klein of extended treatises on elliptic modular functions and automorphic functions.

In an email to the author, Serre commented:

> When I need things from their period [Fricke and Klein], I rather look at books by Fricke alone, such as the three volumes of his 'Algebra' . The third volume is especially interesting: It contains a lot of information on modular forms which are difficult to find elsewhere; this volume might be worth translating.

In a review of another book by Fricke [Do], Dowling wrote:

> Dr. Fricke's style is well-known the world over and needs no comment.

White wrote in a review of yet another book by Fricke about his style [Wh]:

> In attractive, inductive sequence, mainly in geometric style, are presented linear functions and inversion with respect to a circle, stereographic projection, typical rational functions, exponential and hyperbolic functions, then the general concept of analytic functions of the complex variable. The style is that of the lecturer rather than that of the critic, and the requirements of rigor are met by making definitions and hypotheses amply inclusive.

Therefore, one curious question arises: Why did such excellent and clear writers Klein and Fricke write such a difficult or confusing book as Fricke-Klein?

One possible explanation is that those books by each of Klein and Fricke or both of them receiving positive reviews and comments were about topics which were well-developed or more elementary, but for the book Fricke-Klein, they were writing something about research in progress. They were struggling themselves to understand the topics. Some results needed were not developed yet. Therefore, the exposition cannot be too organized or clear.

Fricke was aware of this and wrote in the preface:

> the author of a coherent work on a mathematical subject, on which publications or lectures have already appeared, will have many opportunities to flesh out concepts, fill in gaps, and correct errors. Publications about researches concerning a subject in the full flow of development will show traces of the preliminary and the incomplete. Therefore, lectures, as far as they are to be lively and effective, only seldom have all the details of a completed textbook.

As several people commented, the nonstandard style of the book also caused additional difficulty for the modern reader.

However, such a book as Fricke-Klein about fresh result has a feel of freshness.

8 A few facts about Robert Fricke

Fricke is now probably best known as the co-author of the books Klein-Fricke and Fricke-Klein. But he was a major mathematician on his own, and wrote several other books as we saw in the previous section. According to [MSW, p. 3],

> Fricke was highly respected both as a mathematician and personally, working closely with Klein to develop much of the theory of what are now called Kleinian groups .... He played a leading role in the university administration, being Rektor from 1904–1906 and again from 1921–1923. His activity extended to state educational affairs where his experience as a school teacher was valued and he held several official posts. These many duties account for the long delay between the appearance of the first and second volumes (in 1897 and 1912 respectively) of the second opus 'Vorlesungenuber die Theorie der automorphen Functionen', of which Fricke was really the author, although with much input from Klein. In the final volume, Fricke took the opportunity to use new developments like Cantor's set theory and Brouwer's theory of dimension to solve a number of problems which had been unresolved in the past.

Fricke was also one of the editors of *Dedekind's Complete Works* published in 1930 and one of the editors of *Felix Klein's Complete Works* published in 1921–1923. He was President of the German Mathematical Society in 1920.

9 Special features of this translation and acknowledgments

The story of how this project on four books by Klein-Fricke and Fricke-Klein is a bit unusual, and the editing process is also complicated. Consequently, it is perhaps helpful to explain it. In 2014, while doing a research project on the history of Teichmüller space with Norbert A'Campo and Athanase Papadopoulos, I was puzzled by several questions including the following ones: (1) whether Fricke defined the so-called Fricke space and studied its basic properties such as the Fenchel-Nielsen coordinates as the following paper of Bers and Gardiner says, "Fricke spaces" Adv. in Math. 62 (1986), no. 3, 249—284, (2) whether Fricke defined the mapping class group and its action on the Fricke space.

Most people refereed to the two volumes by Fricke and Klein without precise page references or concrete statements. But I could not read German and could not check the books by Fricke-Klein. These are fundamental concepts in Teichmüller theory. Since Fricke lived before Teichmüller, positive answers would imply that they were defined ahead of Teichmüller. Therefore, these questions are important, and I asked several people about these questions but could not get any clear answer, since very few people I asked have read them. Linda Keen told me that Art Dupre translated these books in the 1970s and also gave me his email address. I contacted Art in March 2014 and started a series of email correspondence and phone calls with him. Art sent me a part of his handwritten translation manuscript. It was neatly written but it was not easy to read due to the style of writing and the mathematical content. After talking to Liping Wang of the Higher Education Press and obtaining her strong support for the idea of publishing these books, I proposed this publishing project to Art. Art was surprised and said that many big publishers had proposed this idea but could not do it, and he expressed doubt at my proposal. I assured Art that the Higher Education Press can do it since I knew that Liping Wang agreed to do it. It turned out to be a huge and time consuming project.

The first nontrivial step involved turning over 2200 pages of handwritten manuscripts into Latex. Though Art's writing was neat, but it was still difficult at places. Liping Wang recruited over several people in China to carry out this step, they are Jiale Huang, Fanlin Meng, Zhimeng Fan, Lei Qian, Chuliang Song, and it still took a while in spite of splitting of the project. Among them, Jiale Huang carried out a substantial part of the work and persisted till the end of the project.

The second step involved checking translation, English writing and mathematical formulas, and redrawing many pictures in Latex. This turned out to be much more difficult than the first big step. We tried to ask many people to help, but very few people were willing to take on the difficult proofreading project. One obvious reason is that very few people could read German, especially due to the difficult style (to the modern reader) and the mathematical content. Pat Boland kindly agreed to proofread Volume 1 of the Fricke-Klein books on automorphic forms since he is a native English speaker, but could only finish the first half of Volume 2 of the Fricke-Klein books since it became too demanding.

Liping Wang and other editors of the Higher Education Press: Peng Li, Huaying Li and Xiaoli Wu each copy-edited all four volumes of Klein-Fricke, and Fricke-Klein multiple times. Though both the languages and mathematics were difficult for them, they persisted. For example, they checked all mathematical equations carefully. To make the correspondence with the original German easier, they hand-inserted the original page numbers on the margin of this English version. All these steps took a long time, and more than three years have passed before the book project reached the final step. In the meantime, we also asked many experts to comment on various aspects of these books and to make suggestions for the young mathematicians and students. Due to the size of these four books and the nature of the topics, it is reasonable that there is room for improvement.

Reference

[Ah] L. Ahlfors, Collected papers. Vol. 2. 1954–1979. Contemporary Mathematicians. Birkhäuser, Boston, Mass., 1982. xix+515 pp.

[ACG] E. Arbarello, M. Cornalba, P. Griffiths, Geometry of algebraic curves. Vol. II. With a contribution by Joseph Daniel Harris. Grundlehren der Mathematischen Wissenschaften, 268. Springer, Heidelberg, 2011. xxx+963 pp.

[BG] L. Bers, F. Gardiner, Fricke spaces. Adv. in Math. 62 (1986), no. 3, 249–284.

[Ber] L. Bers, Quasiconformal mappings, with applications to differential equations, function theory and topology. Bull. Amer. Math. Soc. 83 (1977), no. 6, 1083–1100,

[Bes] M. Bestvina, A primer on mapping class groups [book review of the book [FM] by Farb and Margalit]. Bull. Amer. Math. Soc. (N.S.) 51 (2014), no. 4, 691–700.

[Co] F.N. Cole, Book Review: Vorlesungen über die Theorie der elliptischen Modulfunctionen. Bull. Amer. Math. Soc. 1 (1892), no. 5, 105–120.

[Di] L. Dickson, Book Review: Lehrbuch der Algebra, verfasst mit Benutzung von Heinrich Webers gleichnamigem Buche. Bull. Amer. Math. Soc. 31 (1925), no. 7, 372–373.

[Do] L. Dowling, Book Review: Die elliptischen Funktionen und ihre Anwendungen. Erster Teil. Bull. Amer. Math. Soc. 23 (1917), no. 7, 319–322.

[FM] B. Farb, D. Margalit, A primer on mapping class groups. Princeton Mathematical Series, 49. Princeton University Press, Princeton, NJ, 2012. xiv+472 pp.

[Gr] M. Green, On the analytic solution of the equation of fifth degree. Compositio Math. 37 (1978), no. 3, 233–241.

[Hu] J. Hutchinson, Book Review: Vorlesungen über die Theorie der Automorphen Functionen. Bull. Amer. Math. Soc. 9 (1903), no. 9, 470–492.

[JP] L. Ji, A. Papadopoulos. Sophus Lie and Felix Klein: the Erlangen program and its impact in mathematics and physics. IRMA Lectures in Mathematics and Theoretical Physics, 23. European Mathematical Society (EMS), Zürich, 2015. xviii+330 pp.

[Kl1] F. Klein, Development of mathematics in the 19th century. Translated from the German by M. Ackerman. Lie Groups: History, Frontiers and Applications, IX. Math Sci Press, Brookline, Mass., 1979. ix+630 pp.

[Kl2] F. Klein, Lectures on the icosahedron and the solution of equations of the fifth degree. Translated into English by George Gavin Morrice. Second and revised edition. Dover Publications, Inc., New York, N.Y., 1956. xvi+289 pp.

[Kl3] F. Klein, Lectures on Mathematics. AMS Chelsea Publishing, American Mathematical Society, 2000, 109 pages.

[Kl4] F. Klein, Elementary mathematics from an advanced standpoint. Arithmetic, algebra, analysis. Reprint of the 1932 translation. Dover Publications, Inc., Mineola, NY, 2004. x+274 pp.

[Kl5] F. Klein, Elementary mathematics from an advanced standpoint. Geometry. Reprint of the 1949 translation. Dover Publications, Inc., Mineola, NY, 2004. x+214 pp.

[Kl6] F. Klein, Famous problems of elementary geometry. The duplication of the cube, the trisection of an angle, the quadrature of the circle. Dover Publications, Inc., New York, 1956. xi+92 pp.

[Li] S.T. Lui, An interview with Vladimir Arnol'd. Notices Amer. Math. Soc. 44 (1997), no. 4, 432–438.

[MSW] D. Mumford, C. Series, D. Wright, Indra's pearls. The vision of Felix Klein. Cambridge University Press, New York, 2002. xx+396 pp.

[Na] O. Nash, On Klein's icosahedral solution of the quintic. Expo. Math. 32 (2014), no. 2, 99–120.

[Se] J.-P. Serre, Extensions icosaédriques. Seminar on Number Theory, 1979–1980, Exp. No. 19, 7 pp., Univ. Bordeaux I, Talence, 1980.

[Sh] J. Shurman, Geometry of the quintic. A Wiley-Interscience Publication. John Wiley & Sons, Inc., New York, 1997. xii+200 pp.

[Sl] P. Slodowy, Das Ikosaeder und die Gleichungen fünften Grades. [The icosahedron and quintic equations] Arithmetik und Geometrie, 71–113, Math. Miniaturen, 3, Birkhäuser, Basel, 1986.

[To] G. Toth, Finite Möbius groups, minimal immersions of spheres, and moduli. Universitext. Springer-Verlag, New York, 2002. xvi+317 pp.

[Vl] S.G. Vladut, Kronecker's Jugendtraum and modular functions. Studies in the Development of Modern Mathematics, 2. Gordon and Breach Science Publishers, New York, 1991. x+411 pp.

[Ya] I.M. Yaglom, Felix Klein and Sophus Lie. Evolution of the idea of symmetry in the nineteenth century. Birkhäuser Boston, Inc., Boston, MA, 1988. xii+237 pp.

[We] H. Weyl, The concept of a Riemann surface. ADIWES International Series in Mathematics Addison-Wesley Publishing Co., Inc., Reading, Mass.-London 1964 xi+191 pp.

[Wh] H.S. White, Review: Robert Fricke, Kurzgefasste Vorlesungen über verschiedene Gebiete der höheren Mathematik, mit Berüchsichtigung der Anwendungen, Bull. Amer. Math. Soc. 7 (7) (1901), 316–318.

February 15, 2017

Lizhen Ji
University of Michigan

Preface

Almost 15 years have passed since the first volume of this work was published; and there needs to be an explanation of why, after so much time, the second volume is appearing. As part of the general preparations for the second volume, I had, immediately after completing the first volume, appearing in 1901, given a lecture, in which I gave a detailed treatment of the existence of automorphic functions, the theory of automorphic forms, and Poincaré series. But in the years following, after several repeated attempts to bring this to a conclusion, I was always interrupted by time-consuming official obligations; and I have for years entertained the fear that Klein's full publication plan, beginning with "Lectures on the Icosahedron", would fail to complete its last stage.

In the meantime, this turned out to be a blessing. In the middle of the following decade, the detailed investigations of P. Koebe made possible a new development in the treatment of the fundamental theorem of the existence of single-valued polymorphic functions on Riemann surfaces. The last unfinished part of this project will be devoted to this. The lively interest which this development generated, gave me the incentive to complete the task, all the more, since I was now in a position, because of Koebe's contributions, to provide this work with a satisfying conclusion, after a decade.

In the last period of development of the present work, Klein's own researches of 30 years ago on automorphic functions were once again applied decisively. The larger part of those researches, in which Klein had obtained the detailed theory of elliptic modular functions, and had enriched the theory of elliptic functions with a picture of the many relations, lay behind this. In further natural development, Klein was led to the general functions which are transformed by linear fractional transformations into themselves (automorphic functions) at the same time as Poincaré, stimulated by Hermite and Fuchs' work on differential equations, and both arrived at the same results. By means of personal letters between these two researchers, which were highly fruitful and lively, both soon arrived at the fundamental existence theorem of polymorphic functions; and Klein succeeded in proving the reentrant cut theorem, the boundary circle theorem for arbitrary genus p, as well as, finally, the general fundamental theorem, in section IV, published in volume 21 of the Math. Annalen on p. 141ff. in Fall 1882, which was the high point of the development at that time.

But with the definitive proving of these theorems, there was still something unsatisfactory. Certainly Klein and Poincaré had, independently of one another, developed the approach of continuity observations; Klein had used this approach to sketch a proof of his general theorem, and Poincaré had, barely a year later, at least in the simplest case of a boundary circle, remarkably deepened this. But even in this restricted circumstance, one can hardly speak of a satisfactory solution; and by means of further theorems of Klein, no attempt was made to really carry through the continuity theorem.

For Klein himself, the hindrance arose, that his health suffered from his being extremely overworked, so that in 1882, his investigations came to an abrupt and complete halt. But there is no doubt that the conditions were not yet ripe to finish an exact proof of the fundamental theorem. Imagine at that time how striking and remarkable was Klein's thinking. Arising from a vivid intuition, his way of thinking had borne the fruit of the working out of the inner relationship of many separate disciplines, and led to a whole series of results, which were delineated through many relations, and gleaming results. I mean, not on account of the final outer form, but on account of the inner procession of ideas, which required the creation of the aesthetic of present-day mathematics, I believe that this direction had its most beautiful flowering in Klein's work of the transformation of the seventh degree of elliptic functions (Math. Annalen vol 14, 1878). There was a barrier at the time to the setting up of the fundamental theorems, and obtaining exact proofs for these lay just on the other side of this barrier.

But what was not possible at that time became so with the passage of time. A new period of development began, which expressed at first its principal critical character by its tendency to obtain principles. Weierstrass' tradition gave a new impetus to the work of French mathematicians, and here as well as there, Cantor's results on set theory had an increasingly important influence on the further development of mathematics. A series of detailed works of E. Picard, and a large publication of Poincaré, published in the 1890s, gave the first direct and complete proof of the boundary circle theorem, and gave a further proof, belonging to the new period in 1907. Above all, however, the younger generation, imbued with fresh strength, in this new period, have taken the fundamental theorem as the inspiration for their life's work. First and foremost, P. Koebe, in a work lasting years, has proved all theorems of Klein in a more unified manner, by use of the so-called covering spaces; and more recently, the labors of Koebe and, through his set-theoretic works, L.E,J. Brouwer, has secured the original proof on the basis of continuity considerations.

These investigations of Brouwer and Koebe are described fully in last year's Naturforscherversammlung in Karlsruhe. I have not found it necessary to wait for the conclusion of these investigations, but I believe, since now, much attention is now directed toward these developments, the right moment has arrived to stop and present the aforementioned works. The position which I have taken with regard to the fundamental theorems is the following: True to the program which I have first presented (p. 45ff), I have treated the proof of the boundary circle, the principal circle, and the reentrant cut theorems in connection to Koebe's investigations on the basis of covering spaces, with a sketch of his "iteration procedure", and its use in the proof of the most general theorems of Klein. As opposed to this, I have, in the chapter on the continuity proof, from the beginning, foresworn a complete generality, and restricted myself, (on the basis of my own investigations) and given a detailed descriptive treatment of the group and surface continua, from which, the correctness of the continuity proof appears perspicuous by the direct observation of the simpler, more approachable cases.

Therefore, to the chapter on the continuity proof, belongs the way of thinking just described above, in which this entire work is rooted. Also, in the detailed developments in the appendix, I wanted to return to this way of thinking. I treat here, to a certain extent, using an old work of Wiman's, a special chapter of the transformation theory of automorphic functions, which appears in an interesting way to penetrate the general theory of the equation of the sixth degree. Besides, in applying the fundamental theorems of the theory of automorphic functions to a large part of the Riemann theory of algebraic functions, I want to draw attention to these applications. They lead us back to the tradition of "Lectures on the Icosahedron and the Solution of Equations of the Fifth Degree", and offer new perspectives for understanding transcendental function theory in the challenging problems of algebra.

The purpose of the preceding lines is now also, to give expression to my feeling of appreciation for support and inspiration. First and foremost, my advisor Klein earns my heartfelt thanks. As concerns the origin of the entire work, which bears Klein and my name, I have already spoken in the preface to the first volume, and need not repeat that. It is indeed only natural, that during such a work, extending over several years, I have again and again been inspired to fashion it into a self-contained whole. But Klein has, in spite of this, pursued many other branches of this work with undiminished vigor. At the same time Klein has, apart from the chapter on continuity methods, shared with me many corrections and a large number of valuable comments. Especially for this second volume, Klein's influence, apart from his following the works beginning in the 1880s, is principally in two directions.

First, in the 1890s, the prematurely deceased E. Ritter, tying in to the lectures of Klein and, under his direction, set the theory of automorphic forms and Poincaré's series onto a new basis, which was definitive for the first draft of this volume, appearing in 1901. Second, Koebe, during his stay at Gottingen, attended for several semesters Klein's seminar lectures, and through this, as well as through stimulating conversations with Klein, obtained all the tools necessary for the proof of the fundamental theorem. I have already characterized the importance of this development for the present volume. The now completed work brings to a conclusion the entire publication plan, in which I, following a request of Klein's in Fall 1887, worked out the theory of modular functions. I am pleased that I have succeeded in effecting this plan almost a quarter of a century later, with the sincere friendship and close scientific collaboration with Klein.

After Klein, I am indebted for my inspiration to the brilliant works and ideas of H. Poincarè. I will not repeat here, which I have spoken of in the first volume, the wonderful results of the genius of the French mathematicians. In any case, the reader of this book will recognize the name of Poincaré from the many citations. But I will, in addition, give thanks of a more personal nature. My own results have depended heavily on the arithmetic group law; and I believe that the reader will get the impression that, especially in the last chapter, beautiful results arise in this subject. Poincaré's work in 1887 on the groups belonging to the indefinite ternary quadratic forms, has inspired me to move in this direction for about 20 years.

L.E.J. Brouwer was gracious enough to briefly share with me, before publication, his groundbreaking theorem on the invariance of dimension. Brouwer has spoken of the importance of this theorem for carrying through the continuity proof in a highly interesting note appearing in this year's Gottingen Nachrichten. As I was occupied in Fall 1910 with corrections to the chapter on the continuity method, I was, through the many friendly letters of Brouwer's concerning his theorem, in the position, to point out an extremely important application to the further development of the continuity method (p. 413).

It was with the liveliest interest and sincerest of appreciation, that I received many letters and personal notifications, through which Prof. Koebe kept me abreast of the rapid progress of his detailed investigations. In particular, however, I must acknowledge that Prof. Koebe, on the occasion of my last visit to Leipzig, on the centennial celebration of our highly esteemed publishing firm B.G. Teubner, at a time when I was occupied with Koebe's works, spoke with me in a conversation on many levels about many points concerning which I wanted to discuss.

Since my advisor Klein was otherwise occupied, and could not make corrections to the chapter on the continuity method, I was overjoyed to be notified by letter that Dr. Bieberbach would help correct that chapter. According to my wishes then, Bieberbach, Klein and myself corrected the two following chapters on the principal circle and boundary circle theorems, as well as the reentrant cut theorem. I am most heartfully obligated to Dr. Bieberbach for the hard work he has contributed. Because of his detailed knowledge of the latest function theory, and his thoroughly careful corrections, I thank him for a long series of valuable remarks.

And now, finally, I have to share a very special thanks. It was certainly a bit much, for this volume to have been published in a series of parts with long interruptions. But our highly respected publication firm did not let the book suffer; with known liberality, it always supported my wishes in a most accommodating way, and gave the book a technically polished form. So may this book now, under the proven leadership of the firm B.G. Teubner, make its way into the world, and may it do its part in advancing and broadening our beautiful science, and may it find readers, in which is reflected the same joy and devotion, which we have experienced in writing it.

Brunswick, February 20, 1912

Robert Fricke

Contents

Commentaries

Second Volume

Theory and application of the unique automorphic function of one variable [1]

[1] In the first volume of the present lectures, the concept of the *properly discontinuous groups of linear substitutions of are complex variable* ζ is initially subjected to investigation geometrically, group-theoretically and arithmetically. We have hereby recognized, in the polygon resp. polyhedron nets of the individual group, an important attribute of it, in order, first, to understand the character of the individual group, as well as to obtain a suitable overview of all existing groups of our kind.

The consideration of the function-theoretic significance of our groups is now to lead us to the central principle which rules our further discussions. If an individual group Γ is proposed, then we ask for *functions* $z = \varphi(\zeta)$*, which do not alter their value, in case one exercises a substitution of that group on the argument* ζ. For these functions $\varphi(\zeta)$,the name *"linear-automorphic"* or briefly *"automorphic" functions*,has become current.

The function $\zeta = f(z)$, inverse to the individual function $z = \varphi(\zeta)$, will correspondingly have the property, *that be closed (encirclements) upon the variable z, it goes over into linear functions of itself*:

$$\zeta' = \frac{\alpha\zeta + \beta}{\gamma\zeta + \delta}.$$

Such a function $\zeta = f(z)$ is accordingly to be designated as a "*linearly-polymorphic*" or briefly "*polymorphic*" function.

The first part will occupy itself with the theory of the automorphic functions for a given group Γ. Hereby as already in the "Lectures on Modular Functions", we shall make extended use of *form-theoretic* considerations. Moreover, by attaching to the developed theory of analytic, especially algebraic, functions in the development of the [2] concept of automorphic functions, we shall be required to restrict our considerations to *polygon groups,* i.e. to exclude those groups which are first properly discontinuous in the interior of the ζ-halfspace (and not already in the ζ-plane).

The further parts will treat applications of the theory of automorphic functions to algebraic functions, as well as to other functions and function systems coming into consideration for Riemann surfaces. In this regard, several *fundamental theorems on the existence of linearly-polymorphic functions on given Riemann surfaces* will be basic. The questions, whose answering is furnished by those theorems, is to occupy us very soon (cited in §1.11). These existence theorems will effect for us an interesting as well as extended broadening of our function-theoretic knowledge; and they will, on the other hand, permit us to essentially work out further the theory of algebraic models in various directions, by evaluation of the geometric-function-theoretic perceptions and tools, which were developed in Vol. I and in the first part of the present volume.

[1] The page numbers in the book margin refer to the page numbers of the original German version of this book, some of which are cited or referenced in the book.

In both the German and the English versions, each page number appears at the beginning of the corresponding page.

Part I Narrower theory of the single-valued automorphic functions of one variable

1

Concept, existence and fundamental properties of the automorphic functions

The concept of the automorphic functions of a given group is, first of all, to be delimited somewhat more precisely, and above all, we shall concern ourselves with the proof of the existence of such functions. We base hereby on the principles explained in "M" I on p. 508ff., by means of which *Schwarz* and *Neumann* established the general *Riemannian Existence theorems*. Thereby, it will yield itself, at the same time, that the various automorphic functions, which we will obtain for one and the same group, are connected with one another by algebraic relations. In such a way, we find, for each group, an associated *algebraic model*, a circumstance, which will be fundamental for building up our theory. All these fundamental developments, moreover, base themselves again on the use of a *discontinuity domain* of the group; this object, utilized in extended fashion in the first volume, will consequently, here too, be one of the principal tools for the function-theoretic investigations. [3]

1.1 Definition of the automorphic functions

The regular division of the ζ-plane, which belongs to a proposed polygon group of the first kind, consists either of one, two or of infinitely many polygon nets N [cf. I p. 164 [1]]. Let us pick out an individual one of these nets; let it consist of the polygons [4]
$P_0, P_1, P_2, \ldots$. The substitutions, through which P_0 is carried over into $P_0, P_1, P_2, \ldots$, let us call $V_0 = 1, V_1, V_2, \ldots$. The latter, in their totality, form a group Γ, which is either identical with the proposed polygon group Γ or represents a proper subgroup of it.

If the latter is the case, then in the following consideration, in place of the original group, we take as a basis the group Γ of the substitutions $V_0, V_1, V_2, \ldots$ contained in it. The restriction lying herein, as we shall recognize further below, is only an *inessential one.*

In contrast to this, it signifies an *essential* restriction, if we assume from now on, *that the individual polygon P_0 has only "finitely" many sides*, or, what amounts to the same thing, *that the group Γ can be produced by "finitely" many generators.* This restriction appears warranted, in consideration of our function-theoretic tools. Moreover,

[1] Citations to the first volume of the present work are to be given in the manner followed in the text; references with mere page numbers refer themselves to the present second volume.

also in volume I, we had let the same restriction enter for each of the more detailed study of the groups.

Because of the shape of the polygon P_0 and the associated net N, all principal theorems from I are assumed. We may also assume, that the sides of P_0 are exclusively full circles resp. circular arcs. Let P_0 have n cycles of "fixed" vertices, and let the closed surface F, arising from P_0 by bending together boundary curves coordinated to one another, have the genus p. As in volume I, we then assign the polygon P_0, and with this, the group Γ, the "character" (p,n). The polygon P_0 can, as such, even if p is arbitrarily large, be represented by a simply connected domain; in particular, this will always hold, in case the associated net N has one and only one boundary curve.

The initial polygon P_0 is either identical with the discontinuity domain of Γ or represents a part of it. Accordingly, the name "discontinuity domain", which had a group-theoretic meaning in volume I, is no longer fitting in each case. In order to utilize a notation free of objection, and in order to characterize the fundamental significance of P_0 for the following function-theoretic developments, we take up, from now on, the name "*fundamental domain*" for P_0, and furthermore, for the sake of brevity, leave off the index 0 in the notation P_0 from now on, for the most part[2].

[5] After these preliminary remarks let us go to our proper subject. If any group of transformations of one or several variables is given, then it is a well known mathematical development of ideas, that one seek out *invariants of this group*, i.e., such functions of the variables, which do not alter their value, in case one exercises a transformation of the group on the variables.

In the application to our group Γ, we work out these ideas further in the following manner, by forming the concept of an *"analytic" function $\varphi(\zeta)$ of the complex variable ζ, which, for the exercise of each individual substitution $V_k = \begin{pmatrix}\alpha_k & \beta_k\\ \gamma_k & \delta_k\end{pmatrix}$ of the group Γ on its argument ζ, does not alter its value*:

$$\varphi\left(\frac{\alpha_k\zeta+\beta_k}{\gamma_k\zeta+\delta_k}\right)=\varphi(\zeta), \qquad k=0,1,2,\dots \tag{1}$$

This is to signify, that $\varphi(\zeta)$, continued analytically from any point of the initial polygon to the homologous point of the polygon P_k on a suitable path, is to again reach the initial value at the end.

In order to fashion this concept into one not too circumstantial, and to further make it more precise, let us also explicitly prescribe, for the function $\varphi(\zeta)$, the properties:

$\varphi(\zeta)$ is to possess an essential singularity at no point of the fundamental domain;

$\varphi(\zeta)$ is to be everywhere unbranched and single-valued in the fundamental domain.

By reserving the further explanation of these conditions, we define, first of all, *that each function $\varphi(\zeta)$, outfitted with the required properties, is to be called an automorphic function belonging to the group Γ resp. to the fundamental domain P.*

---

[2] Klein has introduced the conception of the "fundamental domain" in his first investigations on modular functions in Vol. 14 of the Mathem. Annalen on p. 133 (1878) and defined it in complete generality later on in Vol. 21 of the same journal on p. 141 ff. (1882).

The significance of the last requirements, however, we shall be able to present in this fashion, by exhibiting the *form of the power series development*, which $\varphi(\zeta)$ is to permit in a neighborhood of the individual place ζ_0 of the fundamental domain.

If ζ_0, now, does not furnish a boundary point of P, then let us set:

$$t = \zeta - \zeta_0, \qquad \text{resp.} \qquad t = \frac{1}{\zeta}, \tag{2}$$

according as ζ_0 is finite or ∞. Then, in the neighborhood of this place, there is a power series converging there:

$$\varphi(\zeta) = t^m (a_0 + a_1 t + a_2 t^2 + a_3 t^3 + \cdots), \tag{3}$$

where $a_0 \neq 0$, and m signifies a finite whole positive or negative number or zero. For $m \geqslant 0$, $\varphi(\zeta)$ obviously behaves regularly, while for $m < 0$, the function $\varphi(\zeta)$ has a polar [6]
discontinuity there.

This approach will remain valid without change, in case ζ_0 represents a boundary point of the fundamental domain different from a fixed vertex. What, in this case, we have to understand by the "neighborhood of the place ζ_0 of the fundamental domain", will emerge from the representations of the next paragraph. Meanwhile, one wants only to remind oneself, that each boundary point of the kind in question can be changed into an interior point of this domain by "allowed modification".

Finally, as far as the fixed vertices of the polygon P go, we may imagine P so chosen, according to I on p. 128, that these are exclusively *elliptic* or *parabolic*. For the sake of simple notation, we have fashioned P in detail, so that, for the individual such vertex, the two polygon sides reaching up to it are coordinated to one another by the associated generator V; i.e., this vertex is not to belong together in a cycle with other vertices different from it. Then, by means of an orbit curve of the substitution V, we cut off a small sector of the polygon P at the vertex under discussion, as this is pictured in the parabolic case by the appended figure 1. *This sector S we designate as the "neighborhood of the fixed vertex ζ_0 in the fundamental domain".*

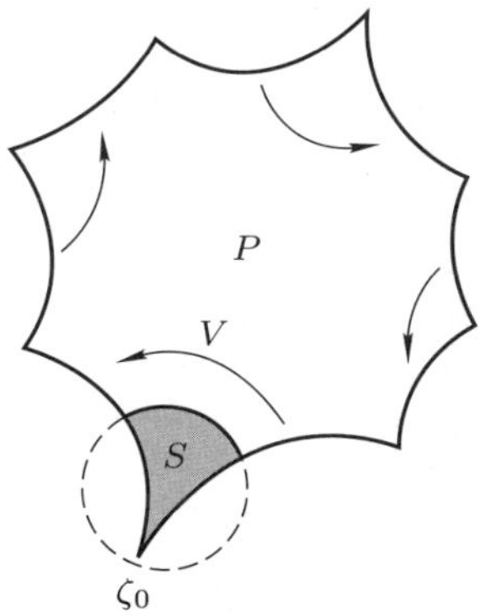

Fig. 1.

In order to characterize the representations of $\varphi(\zeta)$ in this neighborhood, we treat here, at the same time, the *mapping of the neighborhood in question onto the surface of a simply and completely covered circle.* If there is an elliptic vertex ζ_0, then one brings the associated substitution $\zeta' = V(\zeta)$ into the form:

$$\frac{\zeta' - \zeta_0}{\zeta' - \zeta_0'} = e^{\frac{2\pi i}{l}} \frac{\zeta - \zeta_0}{\zeta - \zeta_0'}, \tag{4}$$

where ζ_0' is the second fix-point and l is the period of V. For a parabolic vertex ζ_0, we correspondingly have for V the form:

$$\frac{1}{\zeta' - \zeta_0} = \frac{1}{\zeta - \zeta_0} + \gamma; \tag{5}$$

[7] here γ will be identical with the third coefficient in the original form $V = \begin{pmatrix} \alpha & \beta \\ \gamma & \delta \end{pmatrix}$ of the substitution, in case one chooses the latter in agreement with $\alpha + \delta = 2$. The neighborhood under consideration of the fixed vertex is now simply mapped by:

$$t = \left(\frac{\zeta - \zeta_0}{\zeta - \zeta_0'} \right)^l \qquad \text{resp.} \qquad t = e^{\frac{2i\pi}{\gamma} \cdot \frac{1}{\zeta - \zeta_0}} \tag{6}$$

into the t-plane onto the surface of a circle about the origin $t = 0$[3].

The automorphic function $\varphi(\zeta)$, in the neighborhood of that vertex, now permits a development of the form (3) by means of the t given in (6), where again $a_0 \neq 0$ and m is finite and integral. Our requirement therefore simply amounts to this, that $\varphi(\zeta)$, considered as a function of t, is regular or only polarly discontinuous in the neighborhood of $t = 0$.

Through the requirement of the developability into a series (3) at the individual place ζ_0 of the fundamental domain, the two properties last placed on $\varphi(\zeta)$ are brought completely to expression, as long as P represents a *simply* connected domain. If this is not the case, then the function $\varphi(\zeta)$, although it is unbranched, can nevertheless be multiple valued inside P; for the continuation over such closed paths running in
[8] P, which do not let themselves be drawn together to points in this domain, need not necessarily reproduce the initial function values. By our determination, however, such kinds of unbranched, but multiple-valued, functions are expressly excluded.

[3] Loxodromic and hyperbolic vertices, as was already remarked in the text, do not come up for a clever choice of the polygon P. One has not always made use of this circumstance, rather, initially also has admitted polygons with tips of the kind cited. Hereby then, as one knows today, the theorem is fundamental, that indeed the neighborhood of a loxodromic vertex is mappable onto a circular disc, but not a hyperbolic one. This exceptional place of the hyperbolic tip, however, was not sufficiently attended to at the beginning, which was occasionally the source of incorrect conceptions and proofs with gaps; cf. Comples rendus, Vol. 93, p. 582 (1881) and Mathem. Annalen Vol. 19, p. 558 ff. (1881). Concerning this, see the detailed representations of Klein in the paper, "*Über den Begriff des functionen-theoretischen fundamentalbereichs*", Math. Ann. Vol. 130 (1891) or also the investigation in I on p.143ff.

1.2 Production of an elementary potential of the second kind belonging to the fundamental domain

Poincaré[4] provides the existence proof of the automorphic functions, by forming them analytically; and indeed, by *quotients of certain convergent series,* with which we have to occupy ourselves in detail below in the third and fourth chapter. Klein[5] founds his parallel considerations on the general *Riemannian existence theorems* where then, in the precise proof of this theorem, the methods of Schwartz and Neumann come into force, which we already considered in "M" I on p. 508 ff.[6]. In "M" I, these considerations referred themselves to a closed Riemann surface. Here, in its place, we have our fundamental domain P. It will reveal itself, that this distinction is an inessential one.

We next investigate complex functions of $\zeta = \xi + i\eta$, separate its real components, which, taken by themselves, represent *logarithmic potentials.* For these potentials, we use in general, the notation $u(\xi,\eta)$. These possess the fundamental properties represented in "M" I on p. 504 ff.. In particular, they are single-valued and uniquely continuable inside those domains, for which we defined them, as well as continuous everywhere there in general. However, we shall, at the same time, also considered potentials, which become polarly discontinuous at a prescribed place ζ_0 as the real component of $\frac{1}{\zeta-\zeta_0}$.

The production of suitable potentials $u(\xi,\eta)$ bases itself, first of all, on the solution of the so-called *boundary-value problem* ("M" I on pp. 510 and 513), which is carried through directly only for circular domains, namely, by means of the Poisson integral. Hereby, we have yet to first expand the representation of "M" I in a certain respect. [9]

First of all, namely, the fixed vertices of the fundamental domain P play a similar exceptional role, as the branch-points of a multiple sheeted Riemann surface in "M" I on p. 510. For the individual fixed vertex, we again bring in a triangular sector S, as we cut it off in the previous paragraph by means of an orbit curve of the associated substitution V (cf. figure 1). By the function t, defined in (6) on p. 7, this sector S is mapped onto a circular disc of the t-plane. After we have solved the boundary-value problem for this circular disc, we carry over the values of the potential obtained backwards onto the sector S. For the sector S, the boundary-value problem is thus solvable in the following sense: on the side of S furnished by the orbit, a continuous series of values is assigned, in such a manner, that the same values occur at the two endpoints of this side. *Then, on S, there is one and only one everywhere continuous potential* $u(\xi,\eta)$ *of our kind, which goes over continuously into the assigned boundary values, has equal values at corresponding points along the two sides of S running out from the fixed polygon vertex, and, upon continuation over one of these sides into the corresponding sector*

---

4 See, in particular, the papers "*Mémoire sur les fonctions fuchsiennes*" and "*Mémoire sur les groups Kleinén's*" Acta Math. Vol. 1., p. 193 and Vol. 3, p. 49 (1882 and 83), as well as the preceding notes of Poincaré's in Vols. 93 and 94 of the Comptes rendus (1881 and 82).

5 Compare the essay "*Neue Beiträge zur Riemann'schen Functionentheorie*", Math. Ann. Vol. 21 p. 141 (1882).

6 E. Ritter treats the existence proof of the automorphic functions tying into Klein and indeed, for the genus $p = 0$ in the essay "*Die eindeutigen automorphen Formen vom Geschlechte null*" Math. Ann. Vol. 41 p. 1 (1892), for arbitrary p in the paper "*Die multiplicative Formen auf algebraischen Gebilden, etc*". Math. Ann. Vol. 44 p. 349 (1893).

P_1 belonging to the neighboring polygon P_1, exhibits the equivalent distribution of values of that in S.

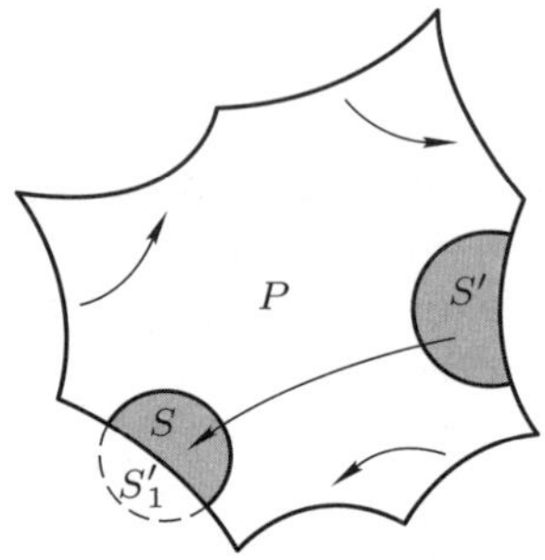

Fig. 2.

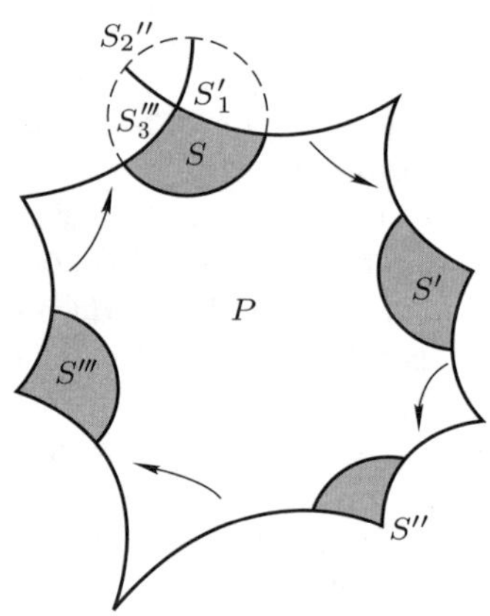

Fig. 3.

[10] We consider, secondly, a circular disc, which, without reaching a vertex of P, reaches over a side of this polygon into a neighboring polygon P_1. The circular disc is divided by this polygon side into two triangular sectors S and S'_1, the latter of which let lie in P_1 and be equivalent with the sector S' of the initial polygon P. In figure 2, such a sector pair S, S' is drawn[7].

We also add, at the same time, the third case, that a circular disc surrounds an accidental vertex of P. If this vertex is surrounded in the polygon net by ν polygons all together, then one will immediately spread the circle in question out into ν triangular sectors $S, S', \dots, S^{(\nu-1)}$ of the fundamental domain. This is depicted in figure 3, where $\nu = 4$.

[7] The neighborhood "belonging to the fundamental domain" of a boundary point of P different from a vertex will always consist two sectors S, S' corresponding to one another in this manner.

We now immediately gather both cases together, by taking $\nu \geqslant 2$. On those ν sector sides, which originate from the original circular periphery, assign continuous sequences of boundary values, in such a manner, that equal values are present at two corresponding endpoints of the particular side. *Then in the domain consisting of the* ν *separated pieces* $S, S', \ldots, S^{(\nu-1)}$, *there is one and only one everywhere continuous potential* $u(\xi,\eta)$, *which goes over continuously into the prescribed boundary values, assumes the same value at each of the two coordinated points of such sector sides, which originate from sides of the domain P, and, upon continuation over a side of this kind into the adjacent sector* $S_k^{(i)}$ *of the neighboring domain* P_k, *assumes the same distribution of values, as in the sector* $S^{(i)}$ *of P equivalent to* $S_k^{(i)}$.

Now turn to the "*method of passage to the limit by the procedure of alternation*" (cf. "M" I on pp. 511 & 514. If two partially intersecting domains are given, and if the boundary-value problem is solvable for each of the domains individually, then, on the basis of that method, the solution of this problem is also successful for the new domain arising from the two given domains by union. the sectors resp. cycles of sectors are hereby usable in the given fashion as domains, just as the circular discs used in "M" I on p. 512 ff.

In detail, one will proceed, say, in the composition of the domains, by first completely covering the boundary of P with sectors. In figure 4, this is carried out for a canonical polygon of character (0, 4). First of all, sectors are situated at the four fixed vertices. Apart from these vertices, we still have five further cycles of sectors; the associated sectors are provided, in each case, with the same numeral. For this wreath-like domain, the boundary values would be assigned along the inner, heavily marked boundary. At every two corresponding outer boundary points, then, the same value $u(\xi,\eta)$ occurs, and upon continuation over this boundary, $u(\xi,\eta)$ shows the behavior already expressed many times, of exhibiting a distribution of values equivalent with P in the neighboring polygons. If the domain P, as such, is multiple connected, then naturally, one will have need of several chains of sectors for covering the entire boundary. [11]

Finally, for the continuation of the alternating procedure, we also cover the entire interior of P intricately with circular discs, and shall always be successful in this process with *finitely* many circular discs and sectors.

Furthermore, exactly as in "M" I on p. 511, we must take care, the potential finally appearing becomes infinite at an arbitrarily choosable place ζ_0 of the fundamental domain, as the real component of $(\zeta-\zeta_0)^{-1}$. Here we assume the place ζ_0 finite and different from a fixed vertex, otherwise, we would have to replace $(\zeta-\zeta_0)^{-1}$ by ζ resp. by t^{-1}, where t is given by the first or second formula (6) on p. 7. We attain the goal mentioned, by always choosing, for all circular discs resp. cycles of sectors containing ζ_0, in place of the everywhere finite potential, that one which becomes discontinuous at ζ_0 in the prescribed manner. In fact, the potential $u(\xi,\eta)$ arising at the end, has itself the desired discontinuity point at the place ζ_0 in this manner; through this, however, the required guarantee is procured for the further development, that $u(\xi,\eta)$ is not identical with a constant. [12]

If we now exercise on P and $u(\xi,\eta)$ all substitution of the group Γ, then we obtain uniquely defined potentials for all polygons $P, P_1, P_2, \ldots$ of the net. Thereby, it is exactly a consequence of the procedure followed, that the analytic continuation of the poten-

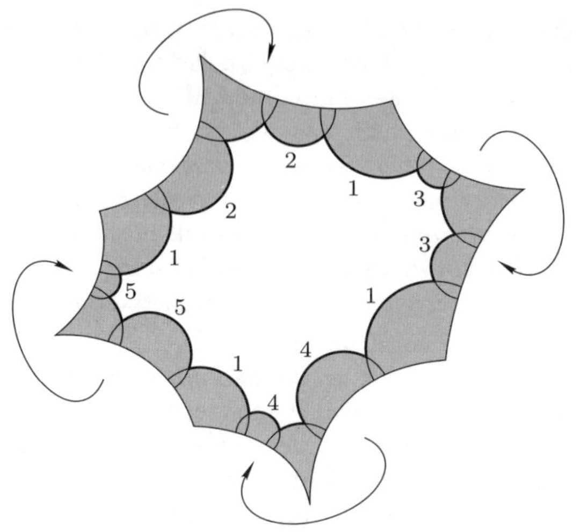

Fig. 4.

tial prescribed in the individual polygon over the boundary into a neighboring polygon loads by itself to the potential defined there. *The potential $u(\xi,\eta)$ is uniquely defined for the entire domain of the connected net of the polygons $P, P_1, P_2, \ldots$, and exhibits the principal property of the automorphic functions, of possessing the same value at equivalent points of the polygons $P, P_1, P_2, \ldots$; $u(\xi,\eta)$ is an "automorphic potential".*

If we designate the potential $u(\xi,\eta)$ thus obtained as an "*elementary potential of the second kind belonging to the fundamental domain P*" then this terminology is justified by "M" I on p. 506, to the extent that $u(\xi,\eta)$ becomes discontinuous of one and only one place of the fundamental domain in the manner described above, and therefore it behaves exactly as an elementary potential serving for the production of an integral of the second kind on the Riemann surface.

1.3 Production of automorphic functions of the group Γ

It is now easy matter, from elementary potentials $u(\xi,\eta)$ of the second kind, to obtain automorphic functions themselves for our group Γ. We form, first of all, the *potential conjugate* to $u(\xi,\eta)$ (cf. "M" I p. 505):

$$v(\xi,\eta) = \int_{(\xi_0,\eta_0)}^{(\xi,\eta)} \left(\frac{\partial u}{\partial \xi} d\eta - \frac{\partial u}{\partial \eta} d\xi\right), \tag{1}$$

where we choose the fixed lower limit (ξ_0,η_0) inside P, say, and let the integration path run entirely in the interior of the net.

From the automorphic character of $u(\xi,\eta)$ and the conformality of the mapping of the various polygons of the net to one another, it only follows at first, that the differential of v:

$$dv(\xi,\eta) = \frac{\partial u}{\partial \xi} d\eta - \frac{\partial u}{\partial \eta} d\xi$$

has automorphic character. For $v(\xi,\eta)$, this character by no means yields itself in general, indeed in case the net N represents a multiple connected domain, we are not immediately able to conclude the uniqueness of v. *Rather, upon continuation from any place to an equivalent place or (for multiple connectivity of the net) back to the initial place, the potential $v(\xi,\eta)$, general speaking, will only first reproduce itself up to a real additive constant.* [13]

For the continuation of the investigation, the use of the closed surface F belonging to the fundamental domain P is expedient, whose mapping onto P and the polygon net N we already explained in I on p.178 ff. The paths in the polygon net, and similarly, the paths between equivalent points in N gave closed circuits on F; and conversely, the individual such circuit on F furnishes a uniquely determined path, from, say, a place coming into consideration of the initial polygon P, of the kinds characterized in N, if one will only avoid those n places on F, which correspond to the fixed vertices of P. For these places, moreover, we again take up the notation $e_1, e_2, \ldots, e_n$ used in the first volume.

We now convince ourselves easily, *that a closed circuit on F, which lets itself be drawn together into a point, always furnishes a path in the polygon net with vanishing integral value $\int dv(\xi,\eta)$.* A continuous deformation of the circuit on F, for which we avoid the places $e_2, e_2, \ldots, e_n$, furnishes, namely, a uniquely determined continuous deformation of the corresponding path in the polygon net, for which the associated integral $\int dv$ remains unaltered. One wants only to pay attention, that, in case it is a question, in the polygon net of an open path, the initial and endpoint in their respective polygons describe equivalent paths whereby, then, the invariability of the integral is evident from the automorphic character of $dv(\xi,\eta)$. On the strength of this, we draw the circuit together on F into a series of arbitrarily small circles about the points e[8]. For the individual such circle, we then simply have $\int dv = 0$; for one wants only to notice that the neighborhood of the individual point e, resp., the corresponding sector of P, maps itself onto a circular disc in the t-plane by means of the functions t defined above, in which disc $v(\xi,\eta)$ is an everywhere continuous single-valued potential.

If now, the genus p of the group Γ is equal to zero, then we arrive at a result of the highest simplicity. In this case, namely, each circuit on F can be continuously drawn together into a point; now, therefore, we also have in $v(\xi,\eta)$ a potential with automorphic character. *If the genus p of the group Γ is equal to zero, we have:* [14]

$$\varphi(\zeta) = u(\xi,\eta) + i\,v(\xi,\eta), \tag{2}$$

and obtain an expression provided with all the characteristics of an automorphic function; moreover, this automorphic function $\varphi(\zeta)$ of Γ even has the property of having a pole of the first order at an arbitrarily chosen place of the fundamental domain P, but, except for this, of being everywhere continuous in P.

If the genus $p > 0$, then, as is known, there are always circuits on F, which do not let themselves be drawn together into points; and each such circuit lets itself be produced

[8] Since, in the text, we are concerned with potentials of the *second* kind, we may, according to known theorems, push the integration path over the discontinuity place of our potential without alteration of the value of the integral. For the sake of simplicity, the terminology is so chosen in the immediately following objectives, that that discontinuity point does not lie exactly at a fixed polygon vertex.

from $2p$ elementary circuits. Corresponding to this, the expression:

$$Z = u(\xi,\eta) + iv(\xi,\eta) \tag{3}$$

exhibits, on the closed surface F, the character of an "*elementary integral of the second kind*", which has only one pole, at a place of F to be chosen arbitrarily, and even one such of the first order, and furthermore, exhibits purely imaginary periods on the surface.

From here, however, we arrive at automorphic functions in the same manner, as we produced, in "M" I on p. 540, from the integrals of the second kind of a Riemann surface, its algebraic functions. We pick out, first of all, any μ poles on F, and construct the μ associated functions (3), which we call $Z_1, Z_2, \dots, Z_\mu$. Next, with constants c, we write the expression:

$$e_1 Z_1 + e_2 Z_2 + \dots + e_\mu Z_\mu, \tag{4}$$

and will so determine these constants e, that the expression (4), continued over any closed path on F, reproduces itself. If suffices to set up this requirement for the $2p$ elementary circuits considered above. With this, however, there yield themselves, in all, $2p$ linear homogeneous equations with real coefficients for the $e_1, e_2, \dots, e_\mu$. According to this, we shall, in any case, not be able to give usable systems of number e not all vanishing, if we choose μ sufficiently large, e.g., $\mu > 2p$. *For each such system, we recognize in the expression (4) an automorphic function $\varphi(\zeta)$ of our group, which has at least two poles in the fundamental domain (insofar as at least two coefficients e are different from 0), and which, consequently is certainly not identical with a constant.*

With this, we have carried out *the proof of the existence of associated automorphic functions for each of our groups.*

1.4 Mapping of the fundamental domain P onto a closed Riemann surface

[15]

An individual one of the automorphic function $\varphi(\zeta)$ constructed so far has only simple poles in the fundamental domain P. The number of the latter we will call μ; and we again assume, for the sake of simple terminology, that none of these is situated at a fixed polygon vertex. We set the problem, *of determining the number ν of the zeros of $\varphi(\zeta)$ in the fundamental domain P.*

For this purpose, we utilize the well-known rule:

$$\nu - \mu = \frac{1}{2i\pi} \int_{(P)} \frac{d\varphi(\zeta)}{\varphi(\zeta)}, \tag{1}$$

where the integral provided with the index (P) is to be extended in the positive sense over the entire boundary of the domain P. If P is multiple connected, then, to be sure, one would first have to produce simple connectivity by bringing in suitable crosscuts. However, the integral carried out on both banks of the individual such crosscut would mutually cancel itself, so that we are also able to ignore these crosscuts from the beginning.

One must take care, the $\varphi(\zeta)$ never vanishes or becomes infinite along the integration path. In this regard, only the fixed vertices of P cause difficulty; for we are able, moreover, if necessary by allowed modification of P, to attain, that the polygon boundary runs through neither a pole nor a zero of $\varphi(\zeta)$.

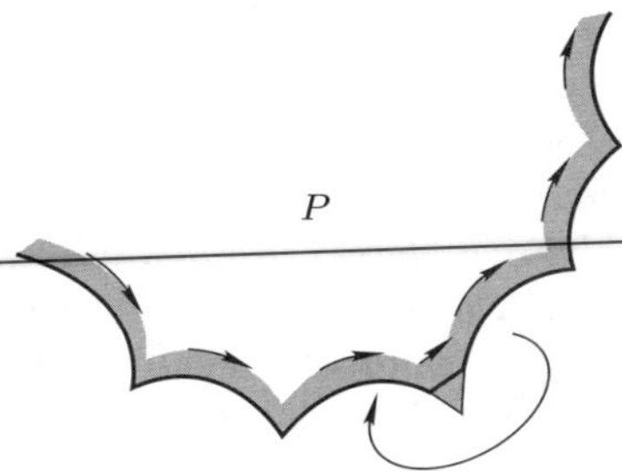

Fig. 5.

But if a zero of order ν' lies at an individual fixed vertex, then we again cut off a sector there by means of an associated orbit curve, and choose this sector so small, that apart from the vertex itself, no further zero or pole of $\varphi(\zeta)$ occurs on it. We now carry the integration, as figure 5 shows by means of the heavily marked path, immediately over the orbit curve cutting off the sector, avoiding the vertex in question.

At the same time, we determine the share, which this orbit curve belonging to the sector cut off furnishes for our integral. To this end, we transform the latter into the form:

$$\frac{1}{2i\pi}\int \frac{d\log\varphi(\zeta)}{dt}\,dt; \tag{2}$$

by t we hereby understand the function (6) on p. 7 belonging to our vertex, by which [16]
the sector is mapped onto a circular disc. Since the integration in (2) refers itself to the boundary of the circular disc in the negative sense, then we have, as value of the integral, $-\nu'$, understanding by ν', as determined the order of the zero of $\varphi(\zeta)$ at the vertex in question.

This consideration moreover remains unrestrictedly valid, in case $\nu' = 0$, i.e., in case $\varphi(\zeta)$ has no zero at the fixed vertex considered.

Now let zeros of the orders $\nu_1, \nu_2, \ldots, \nu_n$ of $\varphi(\zeta)$ be situated at the n fixed vertices, whereby, naturally, ν_k is to be taken $= 0$, in case $\varphi(\zeta)$ does not vanish at the kth vertex. Apart from these vertices, however, let ν_0 zeros be present all together. We then cut off sectors in the manner described from all fixed vertices, and call the remaining domain P'. In place of (1), under these circumstances, we get the more precise formula:

$$\nu_0 - \mu = \frac{1}{2i\pi}\int_{(P')} \frac{d\varphi(\zeta)}{\varphi(\zeta)} \tag{3}$$

where the integral is to be extended over the boundary of P' in the positive sense of circuit.

The n orbit curves lying at the fixed vertices furnish, as we say, the value $(-\nu_1 - \nu_2 - \cdots - \nu_n)$ for the right side of (3). The remaining part of the boundary of P' puts itself together from pairs of boundary curves, each equivalent with one another. Thereby, one of the two boundary curves of the individual pair is always to be run through in the direction opposite to the other. The consequence is, that the values of the integrals on the curves of the individual pair exactly cancel; the integration in (3) on the right side thus leads to:

$$\nu_0 - \mu = -\nu_1 - \nu_2 - \cdots - \nu_n$$

so that the total number ν of all zeros is equal to total number μ of all poles. In known fashion, we immediately extend this result to the following theorem: *Our automorphic function $\varphi(\zeta)$, which has μ simple poles in the fundamental domain, assumes each arbitrarily prescribed complex value at μ places of the domain P all together, which naturally may coincide in any manner.* We will designate this number μ from now on, characteristic for $\varphi(\zeta)$, as the "*valence*" of the automorphic function $\varphi(\zeta)$.

[17] The result obtained we will now clothe in a geometric expression form, highly important for the sequel. We write

$$\varphi(\zeta) = z \tag{4}$$

and will conformably map the fundamental domain P onto the plane of the variable z defined by this. Since corresponding boundary points of P furnish the same value z, we arrive at a closed image, which is uniquely mapped onto P, and which, according to the theorem just proved, covers the z-plane μ-fold everywhere. At $z = \infty$, the μ sheets run separately. At a place z_0 situated in the finite part of the plane, a branching enters then and only then, if in the development:

$$z - z_0 = t^m(a_0 + a_1 t + a_2 t^2 + \cdots), \tag{5}$$

to be understood in the sense of (3) on p. 5, the exponent $m > 1$. Such branching places, however, only occur in finite number. Namely, if there were infinitely many, then those would accumulate themselves at least at one place z; but this contradicts, in known fashion, the fact, that z admits a convergent development (5) in the neighborhood of *each* place of the fundamental domain, and therefore exhibits regular behavior.

By recapitulating, we have arrived at the following fundamental theorem: *the fundamental domain P is mapped by the automorphic function $z = \varphi(\zeta)$ onto an ordinary μ-sheeted Riemann surface of the genus p; this mapping is a conformal one, if one ignores the n fixed polygon vertices and the branch-points coming up in finite number.* The surface obtained can, in each respect, replace the closed surface F used so far, which we had only first mapped into the fundamental domain P continuously in the sense of analysis situs, but not yet conformably. On this account we shall denote the Riemann surface itself from now on by F, or if we will indicate its number of sheets, by F_μ.

1.5 The totality of all automorphic functions belonging to a group Γ and their principal properties

On the basis of the proved mappability of the polygon P resp. the polygon net N onto a Riemann surface F, it is easy to make a series of fundamental assertions concerning the

existence and character of all automorphic functions, belonging to the group Γ resp. to the fundamental domain P. We will hereby call the function $z = \varphi(\zeta)$, just utilized for the mapping, $z(\zeta)$ for short, in order to be able to utilize the symbol $\varphi(\zeta)$ for an arbi- [18]
trary one of our automorphic functions; the function inverse to $z(\zeta)$ we shall denote by $\zeta(z)$. In order, moreover, to emphasize the single-valuedness of the mapping of P onto F especially clearly, one could imagine the surface provided with that crosscut system, which correspond to the boundary of the polygon P.

Now simply utilize the known mapping between the surface F and the net N, according to which all the closed circuits on F correspond precisely to all the paths in N joining equivalent points. Now consider an arbitrary automorphic function $\varphi(\zeta)$ of our fundamental domain P as a function $\varphi(\zeta(z))$ on the Riemann surface F. Since $\varphi(z)$ is single-valued and without essential singularities in P, then the same thing holds for $\varphi(\zeta(z))$ on F. Since $\varphi(\zeta)$ reproduces itself upon continuation over paths between equivalent points, then $\varphi(\zeta(z))$ shows itself to be single-valued for arbitrarily circuits on the uncut surface. According to a well-known function-theoretic fundamental theorem (cf. "M" I on p. 499), it follows: $\varphi(\zeta(z))$ *is an algebraic function belonging to the Riemann surface F.*

If, conversely, $w(z)$ is an arbitrary algebraic function of F, then consider $w(z(\zeta))$ as a function of ζ. This is obviously single-valued and free of essential singularities inside P; upon continuation over a path between equivalent points, however, it will reproduce itself. *Thus,* $w(z(\zeta))$ *is an automorphic function of P in the sense of the definition fixed by us.*

We have arrived, in this manner, at the following "general existence theorem of automorphic function": *The totality of all automorphic functions* $\varphi(\zeta)$ *belonging to the group* Γ *resp. to the fundamental domain P is exactly furnished by the totality of all algebraic functions* $w(z)$ *belonging to the Riemann surface F, by considering the latter as functions* $w(z(\zeta))$ *of* ζ *by mediation of* $z(\zeta)$. Corresponding to this, and tying into the notion of "algebraic model", we will call the totality of the automorphic functions of the group in question an "*automorphic model*". Later on, we shall also have to consider "forms" and "integrals" as belonging to the automorphic model.

The result obtained is fundamental in two respects. First, it is especially interesting, *that by giving the fundamental domain P (exactly as well as by giving a closed Riemann surface), we have the definition of a uniquely determined associated algebraic model.* The fundamental domain, which, by means of the association of its boundary curves, [19]
can be converted as a closed model, is, so to speak, a new form of the Riemann surface. On the other hand, the known fundamental theorems on the functions of an algebraic model carry themselves over without further ado to the functions of the automorphic model, i.e., to the automorphic functions of our fundamental domain resp. the entire polygon net.

By pursuing the previous approach, we recognize, first of all, *that each of our automorphic functions* $\varphi(\zeta)$ *receives a finite "valence"* μ, *which gives, at how many places of the fundamental domain* $\varphi(\zeta)$ *become equal to an arbitrarily prescribed complex value.* If, in particular, the genus $p = 0$, then there are "*single-valued*" functions, which we will also designate as "*principal functions*". Concerning these principal functions, the investigations in "M" I on p. 534 furnish the theorem: *if the genus p of the fundamental domain P is equal to zero, then there are* ∞^3 *principal functions, which represent* *themselves in one of them, $\varphi(\zeta)$, as linear fractional functions*:

$$\frac{a\varphi(\zeta)+b}{c\varphi(\zeta)+d}, \tag{1}$$

with non-vanishing determinants, $ad-bc\neq 0$.

If $p>0$, then only for $\mu\geqslant 2$ are there automorphic functions, and indeed, the lower bound $\mu=2$ is always reached for $p=1$ and $p=2$, but for $p>2$ only then, if the algebraic model defined by P is hyperelliptic. Generally, however, the rules developed in "M" I on p. 556 hold for the minimal values of μ.

If $\varphi_1(\zeta)$ and $\varphi_2(\zeta)$ are any two automorphic functions of our group Γ, and μ_1 and μ_2 are the values of these, then the following theorems holds: *the functions φ_1, φ_2 are bound to one another by an algebraic relation*:

$$G(\varphi_1,\varphi_2)=0, \tag{2}$$

which attains the degree μ_2 in φ_1 and the degree μ_1 in φ_2. This relation is either irreducible and then of the genus p, or $G(\varphi_1,\varphi_2)$ represents a power higher than the first of an irreducible function.

For chosen φ_1, one can always easily so determine the functions φ_2, that the equation (2) becomes irreducible. *But if $G(\varphi_1,\varphi_2)=0$ is irreducible, then each automorphic function of our group Γ lets itself be represented as a rational function $R(\varphi_1,\varphi_2)$ of the two particular functions φ_1,φ_2*. Conversely, each function $R(\varphi_1,\varphi_2)$ self-evidently furnishes an automorphic function of our model. In the lowest case, in the place of this
[20] theorem, one can put the simpler one: *if the genus p of the fundamental domain P is equal to zero, then each function of our automorphic model lets itself be represented as a rational function $R(\varphi)$ of a principal function.*

Further remarks refer themselves to the dependency of the automorphic functions of ζ. In the immediate neighborhood of each boundary point of the polygon net N, the polygons P_k press themselves together in infinite number. Since an automorphic function $\varphi(\zeta)$ assumes a prescribed complex value μ times in an individual polygon, $\varphi(\zeta)$ will assume an arbitrarily chosen value infinitely often in the immediate vicinity of the boundary point. It follows: *each boundary point of the polygon net N is an essential singular point of the individual automorphic function $\varphi(\zeta)$.*

If the net N has one or infinitely many boundary curves, then each of them is covered everywhere densely with essential singular points of $\varphi(\zeta)$. In analytic continuation of $\varphi(\zeta)$ we are indeed able to come arbitrarily close to each individual boundary curve; but we are not able to reach, much less pass over it. Rather, the convergence radii of the power-series developments for $\varphi(\zeta)$, upon approaching the boundary curve, will approach the value 0. Following "M" I on p. 110, we express this briefly thus: *the boundary curve resp. all infinitely many boundary curves of the net N represent a "natural boundary" for each associated automorphic function $\varphi(\zeta)$.*

The parabolic polygon cusps play a special role, as was already discussed in "M" on p. 235 ff. The individual parabolic cusp belongs to the boundary points; nevertheless, if we approach the parabolic point not exactly tangentially to the associated boundary curve, we obtain a determinate boundary value of the function[9].

[9] If the polygon net spreads out over the entire ζ-plane (up to isolatedly situated boundary points), then this always reaches up to the individual parabolic point from two opposite di-

In the interior of the net N, $\varphi(\zeta)$ is unboundedly analytically continuable and, as an automorphic function, has, in each polygon P_k, the equivalent distribution of values as in P. From this follows: *not only in P, but in the entire net N, i.e., in the entire domain of definition, the individual one of our automorphic functions $\varphi(\zeta)$ is a single-valued function of its argument ζ*. We recognize the remarkable theorem, to be pursued even further later: *the algebraic functions of the Riemann surface F, which in general are* [21]
"multiple" valued in z, in case one makes ζ the independent variable, become "single valued" functions of this argument ζ. The known theorem of elliptic functions, that the algebraic functions of a Riemann surface of genus 1 are single-valued functions of the associated everywhere finite integral, hereby subsumes itself as a special case, as we soon pursue even further. Here we already gauge the wide-ranging significance, which the automorphic functions have for the theory of algebraic models.

Finally, we return have once more to the general case of a group Γ, for which an individual net N is one of two or infinitely many equivalent ones N, N' resp. $N, N', N'', \ldots$. If we have constructed $\varphi(\zeta)$ for N, then there are indeed, in all further nets $N', N'', \ldots$, functions with equivalent distribution of values. If (which would certainly contradict the fundamental concept of an analytic function determined by Weierstrass) we consider the totality of all these function values as *one* function $\varphi(\zeta)$, then this would be invariant with respect to all substitutions of Γ.

One is tempted, say by mediation of the polyhedra of the ζ-halfspace, to look for an analytic connection between the function values in the different nets. The detailed investigations concerning this, however, do not yet exist.

1.6 Classification and closer study of the elementary automorphic functions

The classification of all groups Γ in I on p. 164 ff. was effected according to group-theoretic-geometric principles. Next to this, for the classification of the automorphic functions, function theoretic viewpoints will naturally also be the determining factor, so that the two classifications will not run entirely parallel to one another. We next treat here first only the "elementary" automorphic functions. By these are to be understood not only for automorphic functions, which are rational and elementarily transcendent in the ordinary sense; but we reckon here also those automorphic functions, which represent doubly periodic functions or functions related to these. In fact, indeed, the theory of the doubly periodic functions, so far as we have to utilize it here, is scrutinized from all angles.

[22]

I. Cyclic functions

In this category, we count all automorphic functions $\varphi(\zeta)$ which belong to cyclic groups Γ.

1. Functions of cyclic groups of elliptic or parabolic substitutions

rections. The boundary values of an individual function to be reached on the two sides, will only then have to be identical, if the parabolic polygon cusps lying opposite one another are equivalent with respect to Γ.

The functions belonging here we will easily treat. One gives the fundamental domain of the individual group of this kind, according to "M" I on pp. 188 and 190, the shape of a circular crescent, whose cusps, in particular in the parabolic case reach up to one and the same point ζ_0 from opposite sides. The genus p is 0 is both cases. We have already become acquainted with the associated *principal functions* above (p. 7) in the form:

$$\varphi(\zeta) = \left(\frac{\zeta-\zeta_0}{\zeta-\zeta_0'}\right)^l \qquad \text{resp.} \qquad \varphi(\zeta) = e^{\frac{2i\pi}{\gamma}\cdot\frac{1}{\zeta-\zeta_0}}, \tag{1}$$

as we mapped individual tips of the circular crescents (namely elliptic or parabolic polygon vertices) onto full circles by means of these functions.

2. Functions of cyclic groups of hyperbolic or loxodromic substitutions

Consider the generating substitution V of a loxodromic group in its normal form:

$$\frac{\zeta'-\zeta_0}{\zeta'-\zeta_0'} = \kappa e^{\vartheta i}\frac{\zeta-\zeta_0}{\zeta-\zeta_0'} \tag{2}$$

and count the hyperbolic case here, at the same time, as the special case $\vartheta = 0$. According to I on p. 66, we regard the fundamental domain with the shape of a circular ring, whose boundaries are to belong to that pencil of circles, which has the two fix-points ζ_0, ζ_0' of V as boundary points. In the hyperbolic case, these circles are the level curves of V. One sees, that in the present case, we have the genus $p = 1$.

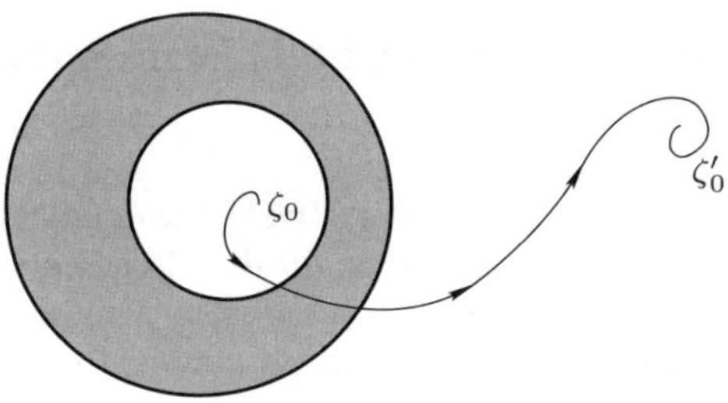

Fig. 6.

In order to form the associated functions, we cut the ζ-plane along a double-
[23] spiraled resp. circular orbit curve of V (cf. figure 6), and map the cut plane, by means of the function:

$$u = \log\left(\frac{\zeta-\zeta_0}{\zeta-\zeta_0'}\right), \tag{3}$$

onto a strip of the u-plane bounded by two parallel lines, as is represented by figure 7. The cut ring-formed fundamental domain maps itself onto a rectilinear parallelogram with the vertices: u_0, $u_0+2i\pi$, $u_0+\log\kappa+i\vartheta$, $u_0+\log\kappa+i(\vartheta+2\pi)$, where $\log\kappa$ is to be taken real. The opposite sides of the parallelogram are mapped to one another by the substitutions:

$$u' = u+2i\pi, \quad u' = u+\log\kappa+i\vartheta, \tag{4}$$

of which the second corresponds to the generator V and the first a circuit about ζ_0. We argue from this, without further ado, to the following result: the automorphic functions $\varphi(\zeta)$ belonging to the cyclic groups under discussion are the single-valued doubly periodic functions of the argument u and periods ω_1, ω_2, which are given by:

$$u = \log\left(\frac{\zeta-\zeta_0}{\zeta-\zeta_0'}\right), \quad \omega_1 = 2i\pi, \quad \omega_2 = \log\kappa + i\vartheta.$$

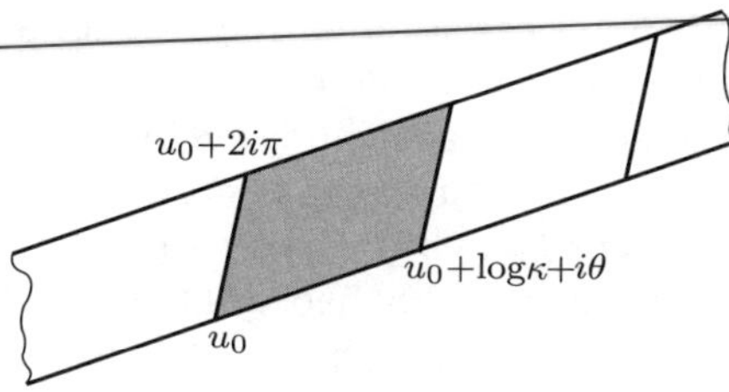

Fig. 7.

The *hyperbolic* special case is thereby distinguished by a *purely imaginary period ratio*[10].

II. Functions of the regular bodies [24]

The *elliptic rotation groups* or *groups of the regular bodies*, as is known, are the only non-cyclic groups of finite order. We have here, first of all, the groups of the dihedral type, then the tetrahedral group, the octahedral group and icosahedral group. The structure of these groups and the divisions of the ζ-plane into nets of finitely many circular-arc triangles belonging to them are sufficiently discussed in "Icos" and "M" I.

---

[10] The following parenthetical discussion refers itself to he loxodromic case. If one bounds off a domain B by two double-spiraled level curves mapped onto one another by means of V, then this is mapped in the u-plane, into a stripe bounded by two parallel lines, which crosses over the stripe of figure 7 perpendicularly, and whose opposite sides correspond with one another by the second parabolic substitution (4). According to this, the domain B is mapped onto the full plane t by means of the function:

$$t = e^{\frac{2i\pi u}{\log\kappa + i\vartheta}} = e^{\frac{2i\pi}{\log\kappa+i\vartheta}\log\left(\frac{\zeta-\zeta_0}{\zeta-\zeta_0'}\right)}.$$

If we now cut off from B, by means of an orbit curve of V, a loxodromic tip reaching up to ζ_0, then the latter will map itself onto a circular disc in the t-plane. We already mentioned the possibility of this mapping above (p. 7) for a loxodromic (non-hyperbolic) tip.

If, in this situation, we shall designate $t(\zeta)$ as a principal function of the fundamental domain B of genus 0, then, nevertheless, one may not see in $t(\zeta)$ a single-valued automorphic function. B is not a group-theoretic discontinuity domain at all, and correspondingly, $t(\zeta)$ is an infinitely multiple-valued function of ζ. Compare the investigations on multiple-valued automorphic functions at the close of the chapter.

The automorphic functions belonging to these groups we call, for short, "*functions of the regular bodies*"; they represent *national functions of* ζ*, which, in "Icos.", section I, are subjected to a multifaceted algebraic-function-theoretic investigation.*

III Doubly periodic functions

1. Functions with additive periods

Under this rubric belong the automorphic functions of the *parabolic rotation groups* or *doubly periodic groups.* Each such group has one single boundary point. If ζ is so chosen, that this boundary point lies at $\zeta = \infty$, then the fundamental domain has the shape of a "period parallelogram", and on the other hand, each period parallelogram furnishes a suitable fundamental domain of the kind belonging here (cf. figure 8). *Consequently, the ordinary doubly periodic functions subsume themselves here as a special kind of elementary automorphic functions.* The considerations of §2 ff., with this, also obtain a fundamental significance for the theory of the doubly periodic functions, insofar as they contain the existence proof of these functions for given periods; in the relevant textbooks one is accustomed to carrying out this proof ordinarily by formation of the theta series or the product development of the sigma functions. The variable ζ, for the individual elliptic model, naturally becomes an *everywhere finite integral,* in which
[25] the algebraic functions of the model are single-valued, as well, too, the integrals, which will emerge even more clearly in connection with our later developments.

The particular parabolic rotation groups with elliptic substitutions discussed in I on p. 222 ff. All arise by extensions of the highest simplicity from groups of the kind just mentioned, which have parallelogrammatic discontinuity domains. We shall consequently be able to consider here as disposed of without further ado also the maximal parabolic rotation groups, which moreover all belong to the genus $p = 0$.

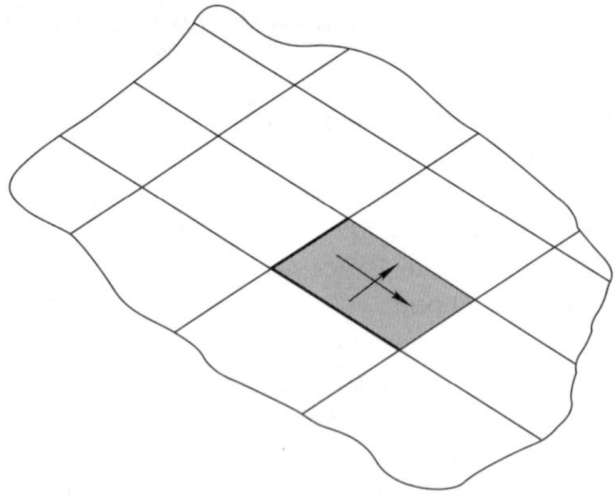

Fig. 8.

2. Functions with multiplicative periods

Here, finally, belong the automorphic functions of the polygon groups with *two* boundary points, considered in I on p. 236. If we situate the latter at $\zeta = 0$ and $\zeta = \infty$, then the two generators of the groups have the form $\zeta' = \alpha_1\zeta$, $\zeta' = \alpha_2\zeta$. In this spirit, we call the associated functions $\varphi(\zeta)$ *multiplicitively periodic*[11]. In addition, we by no means have to do here with essentially new functions; it is simply a question of ordinary additively doubly periodic functions of the argument $u = \log\zeta$ and the periods $\omega_1 = \log\alpha_1, \omega_2 = \log\alpha_2$.

1.7 Preparations for the classification of the higher automorphic functions

[26]

In order not to have to interrupt, later on, the classification of the automorphic functions yet remaining, we present an investigation, which concerns the *hyperbolic rotation groups* or *principal circle groups.*

The detailed theory of these groups, which one finds in I on p. 284 ff., we have based on the canonical polygons situated in the projective plane. The existence proof of the functions $\varphi(\zeta)$ attached to the polygons of the ζ-plane. We must consequently first consider somewhat more closely the relation of these latter polygons to those of the projective plane.

The individual polygon situated in the projective plane, which, as such, always represents a simply-connected domain, receives a certain genus p by means of its coordination of sides, and it possesses a determinate number n of fixed vertices. We then assigned the polygon the symbol (p, n) as "character". From now on, in place of these notations we would rather use $p', n', (p', n')$, and call the polygon itself P', in order to reserve the notations $p, n, (p, n)$, P for the corresponding fundamental domain of the ζ-plane, we may furthermore assume P' as a rectilinear polygon of $(2n' + 4p')$ sides, whose more precise shape is discussed in I on p. 310 ff.

Among the n' fixed vertices, let μ be hyperbolic, which reach out into the exterior of the absolute ellipse. The remaining $\nu = n' - \mu$ vertices will then be elliptic or parabolic.

For the ζ-plane, in any case, only the part of P' will come into consideration, which lies in the interior of the ellipse. But we must imagine this interior as double-sided, and let both sides be connected along the ellipse. By letting the principal circle coincide, in the ζ-plane, with the real ζ-axis, the upper side of the interior of the ellipse will correspond, say, to the positive ζ-halfplane, the lower to the negative. On both sides, we imagine copies of the polygon P' positioned exactly over one another; obviously these furnish two images symmetric to one another along the real ζ-axis.

It now amounts to a radical distinction, whether the number $\mu = 0$ or > 0.

In the first case, i.e., if no hyperbolic vertices come up on P', then the two images just mentioned are symmetric to one another with respect to the principal circle (the real ζ-axis), are *not* connected, and *one* of them, say that one in the positive halfplane, [27]
furnishes the fundamental domain P. *If* $\mu = 0$, *i.e., if there are no hyperbolic vertices in*

[11] These functions have been such studied. Compare, e.g., the sixth section in O. Rauschberger's "*Lehrbuchder Theorie der periodischen Functionsen einer Variablen u.s.w.*" (Leipzig, 1884).

P', then the fundamental domain P is uniquely mapped onto the projective polygon P', and has the same character $p = p'$, $n = n'$; the principal circle, in this case, is a boundary circle in its entire extent.

If $\mu > 0$, then the polygons P' laid out on both sides of the interior of the ellipse are connected in μ finitely extended segments of the ellipse. The corresponding thing will hold for the two images in the ζ-plane, which now, taken together, represent the fundamental domain P. *If the polygon P' has $\mu > 0$ hyperbolic vertices, then the projective polygon P' is mapped 1-2-valuedly onto the fundamental domain P; the latter then represents a μ-fold connected domain, symmetric to itself along the real ζ-axis.* Figure 9 illustrates the case $p' = 1$, $n' = 2$, $\mu = 2$. The Riemann surface F belonging to P, corresponding to the transition from ζ to the conjugate complex value, admits a symmetric transformation into itself, for which a symmetry line comes up consisting of μ separated lines.

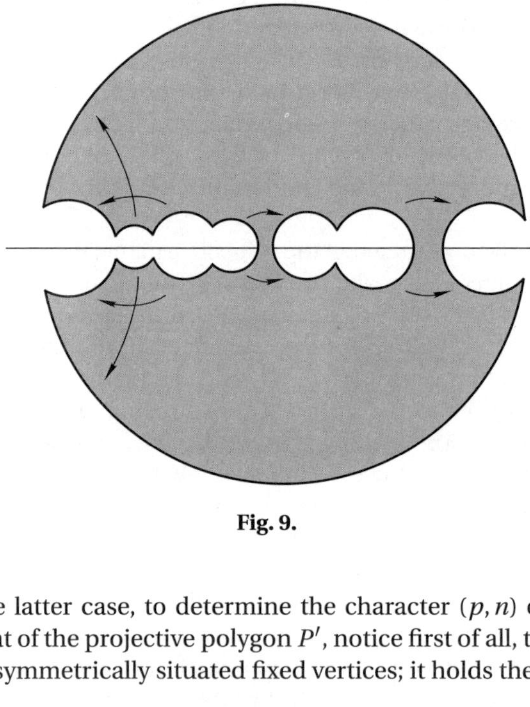

Fig. 9.

[28] In order, in the latter case, to determine the character (p, n) of the fundamental domain P from that of the projective polygon P', notice first of all, that P obviously has $\nu = n' - \mu$ *pairs* of symmetrically situated fixed vertices; it holds therefore:

$$n = 2(n' - \mu) = 2\nu. \tag{1}$$

For the determination of p, we attach to the shape of P', which represents a polygon of $(2n' + 4p')$ sides with *one* cycle of accidental vertices lying in the interior of the ellipse. Correspondingly, P has *two* such vertex cycles, to which correspond, on the closed surface F belonging to P, two symmetrical points, to be denoted, say, by A and B. We imagine the points A, B marked as "boundary points" of the surface F, but besides these, also the 2ν places of F corresponding to the fixed vertices of P. If we initially ignore all further crosscuts, then the surface "bounded" in such a way represents a $(2p+2v+2)$-fold connected domain. By addition of those crosscuts, which correspond to the boundary of P, the degree of the connectivity is decreased by μ. There are the following crosscuts:

1. p' pairs of conjugate crosscuts a_k, b_k, each individual one of which begins and ends at A;
2. p' pairs of crosscuts, which lie symmetrically with respect to the ones just cited, and each individual one of which, just on this account, begins and ends at B;
3. v crosscuts from A to places, which correspond to fixed polygon vertices;
4. v crosscuts from B conjugate to those;
5. μ crosscuts from A to B.

A known fundamental equation of analysis situs thus furnishes:

$$(2p+2v+2)-4p'-2v-\mu=\mu,$$

from which we conclude for the genus p:

$$p=2p'+\mu-1. \tag{2}$$

We note after this, as a result: *if P' has $\mu>0$ hyperbolic vertices, then the character of the fundamental domain P is given by:*

$$(2p'+\mu-1, 2n'-2\mu),$$

in case (p', n') is that of P'.

A further remark refers to all polygon groups Γ resp. the auromorphic models belonging to them. [29]

Let two groups Γ, Γ' be proposed, of which one is transformed into the other by the substitution $S(\zeta)=\frac{a\zeta+b}{c\zeta+d}$, $\Gamma'=S^{-1}\Gamma S$; we then say, that Γ and Γ' belong to the same "class". Thereby, we obtain, from the functions $\varphi(\zeta)$ of Γ, those of Γ', simply in $\varphi[S(\zeta)]$. *All groups of one and the same class furnish automorphic models not essentially different; in particular, one and the same algebraic model corresponds to them.*

For the classification of the function $\varphi(\zeta)$, moreover, also here again, those of the groups in I on p. 165 are not the final determining factors. Thus, e.g., the l.c. division into two types carried out for the principal-circle groups is left aside as not especially essential for the function-theoretic investigation. If we have a group Γ, which also contains substitutions, which interchange the two ζ-halfplanes, then there is in Γ a subgroup of index 2, which always only transforms the individual halfplanes into themselves again. One will then easily pass over from the functions of Γ_2 to those of the group Γ[12].

[12] This is naturally not to prevent us from also occasionally evaluating groups Γ of such a kind as examples. Thus, e.g., at the close of the first paragraph in the next chapter, we shall consider the groups of the kind discussed belonging to the genus zero.

1.8 Classification and closer study of the higher automorphic functions

On the basis of the preparations presented, we now move to the classification of the higher automorphic functions themselves.

I. Principal-circle functions.

1. Functions with circular boundary

In this place the functions of the hyperbolic rotation groups with $\mu = 0$ arrange themselves. Further division principles for this kind of automorphic functions are:

a) division into "*genera*" according to the number p.

b) division of all automorphic models of the individual genus into "*species*" according to the number n of the fixed vertices. As in I on p. 261 ff., (p, n) is the "*species character*". We remember, that for $p = 0$, the number $n \geqslant 3$ must hold, and correspondingly, for $p = 1$, $n \geqslant 1$, while for $p \geqslant 2$, the number n can be any non-negative whole number.

[30] c) division of the automorphic models of the individual species into "*families*", each individual one of which receives a determinate "*signature*" $(p, n; l_1, l_2, \ldots, l_n)$. The l are whole numbers > 1 or ∞; they give the periods of the generating substitutions belonging to the fixed vertices. Two combinations of numbers l, which agree except for the sequence of the l, furnish one and the same family. According to I on p. 389, the following important theorem holds: *the individual family represents a "continuum" of* $\infty^{2n+6p-6}$ *essentially different automorphic models.* Next to this, there is the theorem: *between the different families, there is no continuous transition*[13].

We arrange here, at the same time, next to these, the discussion of a few important special cases.

α) The functions with circular boundary of genus zero

As already mentioned, $n \geqslant 3$ holds in this case. The individual family $(0, n; l_1, l_2, \ldots, l_n)$ includes ∞^{2n-6} models. In the lowest case $n = 3$, the dimension of this manifold is zero; we know from I, *that the family* $(0, 3; l_1, l_2, l_3)$ *reduce to be a single group class and with this, essentially to one model.* According to I on p. 287, a group of this kind is always capable of extension by reflections. In this way, we arrive at the well-known *regular-symmetric nets of circular-arc triangles with boundary circle* (cf. "M" I on p. 93); the functions $\varphi(\zeta)$ under discussion we accordingly also call "*functions of the circular triangles*" or "*triangle functions*".

For $n \geqslant 4$, the possibility of the extension by a reflection $\overline{V}$ always represents a particularity. If it takes place, then, at the same time, we will determine the effect of the transformation $\overline{V}$ on an associated principal function $z(\zeta)$. In $z'(\zeta) = z[V(\zeta)]$ we have before us a complex variable depending on ζ, which indeed has equal values at all equivalent points ζ. But first, if we replace z' by its conjugate value $\overline{z}'$, then we obtain in $\overline{z}'(\zeta)$ an automorphic function in the ordinary sense[14]. Obviously, $\overline{z}'$ is again a principal function and, as such, is linear in z; but then z' will be linear in $\overline{z}$, where $\overline{z}$

[13] At least inside the the domain of the *single-valued* automorphic functions, alone important for us.

[14] $z'(\zeta)$ could effect a conformal mapping "with reversal of the angle".

signifies the value complex conjugate to z. Therefore it follows: *corresponding to the reflections $\overline{V}$, the principal function z itself experiences a linear substitution of the second kind*: [31]

$$z' = \frac{a\bar{z}+b}{c\bar{z}+d}. \tag{1}$$

This substitution (corresponding to the points of the symmetry circle of $\overline{V}$) has infinitely many fix-points; according to "M" I on p. 198 it thus itself represents a *reflection with a real symmetry circle.*

Further on, one has two cases to distinguish: either the n places of the z-plane corresponding to the fixed polygon vertices all lie on that symmetry circle, or these places are partially or thoroughly symmetrically situated in pairs with respect to that circle. The first case furnishes an especially elegant result: here, the interior (and naturally also the exterior) of the symmetry circle of the z-plane is mapped onto a circular-arc n-angle of the ζ-halfplane with the angles $\frac{\pi}{l_1}, \frac{\pi}{l_2}, \cdots, \frac{\pi}{l_n}$, and the entire fundamental domain P can be built up from two symmetric polygons of this kind[15].

β) The family $(p,0)$ of the functions with circular boundary

The species $(p,0)$ entering for $p \geqslant 2$ contains only a single family and represents a continuum of ∞^{6p-6} automorphic models. The special thing here is, *that now the individual fundamental domain P furnishes a Riemann surface, for which the neighborhood of each place*[16] *is mapped "without exception" one-to-one onto the ζ-halfplane.* One can also state this by saying, *that ζ represents an "unbranched" linearly-polymorphic function on that Riemann surface F.*

For the lowest case $p = 2$, the underlying canonical octangle P' and its transformation theory is developed in detail in I on p. 333.

2. Principal-circle functions with isolatedly situated boundary points

For all principal-circle functions not yet cited, the projective polygon has $\mu > 0$ hyperbolic vertices. These functions $\varphi(\zeta)$ are then analytically continuable from one ζ-halfplane into the other over the real axis (the principal circle); *they exist, therefore, in the entire ζ-plane, except they will have infinitely many isolatedly situated boundary points on the principal circle.* In the further classification, the consideration of the function-theoretic fundamental domain P will again come to the forefront; we have, analogously, as under 1, the following classification principles: [32]

a) Division into "*genera*" according to the number p of the fundamental domain P.

b) Division of all automorphic models of the individual genus into "*species*", according to the number $n = 2\nu$ of the fixed vertices; let (p,n) or $(p,2\nu)$ again be called the "*character*" of the species. Hereby, ν is a non-negative whole number, and therefore n is an even number. Except, for $p = 0$, ν must always be > 1, and for $p = 1$, ν must correspondingly be taken > 0; for the combinations hereby excluded would furnish the values $p' = 0$, $n' < 3$ for the projective polygon P', which are to be excluded.

c) The models of the individual species $(p,2\nu)$, according to the numerical value of p', fall into finitely many, namely $1+[\frac{p}{2}]$, divisions, which we will designate as "phyla".

---

15 See, moreover, the furthergoing investigations in §1 of the next chapter.

16 The neighborhood of a branch-point is self-evidently to be thought of a multiple-sheeted.

In fact, in consequence of (2) on p. 28, only the values $0,1,2,\ldots,[\frac{p}{2}]$ are admissible for p', where by the symbol $[A]$, we understand the greatest whole number contained in A[17]. The individual phylum then receives the "*character*" $(p,2\nu,p')$ or (p,n,p'). The degree μ of the connectivity of P and the number n' of vertices of P' now determine themselves to be:

$$\mu = 1+p-2p', \quad n' = 1+\frac{1}{2}n+p-2p'. \tag{2}$$

[33] d) Division of all automorphic models of the individual polygon into "*families*", each individual one of which receives a "*signature*" $(p,2\nu,p';l_1,l_2,\ldots,l_\nu)$. Hereby, the numbers $l_1,l_2,\ldots,l_\nu$ are used in exactly the same meaning, as for the functions with boundary circle.

For the extent of the individual family, the theorem from I on p. 389 is again the thing to use. The number $(3n'-\nu+6p'-6)$, with consideration of (2), express itself thus:

$$3n'-\nu+6p'-6 = n+3p-3.$$

In consequence of this, we make note of the theorem: *the individual family represents a continuum of* ∞^{n+3p-3} *different automorphic models. Between the various families there is no continuous connection.*

The fundamental domains do not have the shape of canonical polygons anywhere here. However, according to the rules developed in I on p. 182 ff., one can again obtain this shape, if one attaches no particular value to the circular form of the polygon sides.

As an especially important special case, we note the polygon with $p'=0, n=0$, including only one single family. entering for each $p \geqslant 2$. In this case, we have $n'=p+1$, and moreover, to these conditionally corresponds a continuum of ∞^{3p-3} automorphic models. In I on p. 305, we gave a canonical polygon P' of character $(0,n')$ the shape of a rectilinear $2n'$-angle with all concave angles; the figure given l.c. is reproduced here as figure 10. In the present case, the fixed vertices $e_1,e_2,\ldots,e_{n'}$ are all hyperbolic and therefore lie outside the ellipse. The movable vertex E we indeed imagine, as usual, in the interior of the ellipse; but nothing prevents us from letting the point E move, *in the interior* of the polygon N_0' of the figure, out over beyond the ellipse, e.g., into the tip reaching up to $e'_{n'}$. Visualize, under these circumstances, the position of the $2n'$-angle marked heavily in the figure. No single vertex any longer lies on or inside the ellipse; the two sides running out from $e_{n'}$ lie entirely outside the ellipse and are therefore lost
[34] all together for the ζ-plane, but in contrast to this, the remaining $(2n'-2)$ sides each have a middle portion running in the interior of the ellipse. It yields itself, *that the associated fundamental domain P is bounded by* $2n'-2=2p$ *full circles, running separately*

[17] The division of the individual species into $1+[\frac{p}{2}]$ phyla corresponds precisely to the division of all "orthosymmetric" Riemann surfaces of the given genus p into $1+[\frac{p}{2}]$ kinds, according to the number μ of the transition or symmetry lies. Compare the following pamphlets and papers of Klein "*Über Riemann's Theory der algebraischen Functionen und ihrer integrale*" p. 72 (Leipzig, 1882), "*Über conforme Abbildung von Flächen*", Math. Ann. Vol. 19 (1891), "*Über die Realitätsverhältnisse bei der einen beliebigen Geschlechte zugehörigen Normaleurre der* φ", Mathem. Ann. Vol. 42 (1892), as well as the investigation of G. Weichold "*Über symmetrische Riemann'sche Flächen und die Periodicitäts moduln der zugehörigen Abel'schen Normalintegrate erster Gattung*", Zeitschrift f. Math. u. Phys. Vol. 23, p. 321 (1883).

and coordinated to one another pairwise, whereby two full circles corresponding to one another are always neighboring. Figure 11 illustrates the case $p = 3$.

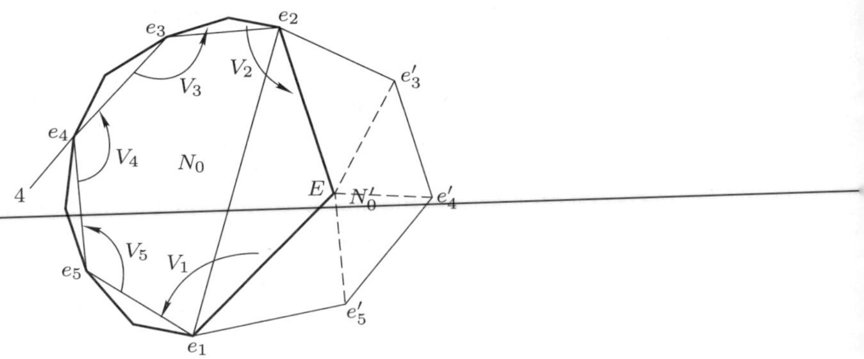

Fig. 10.

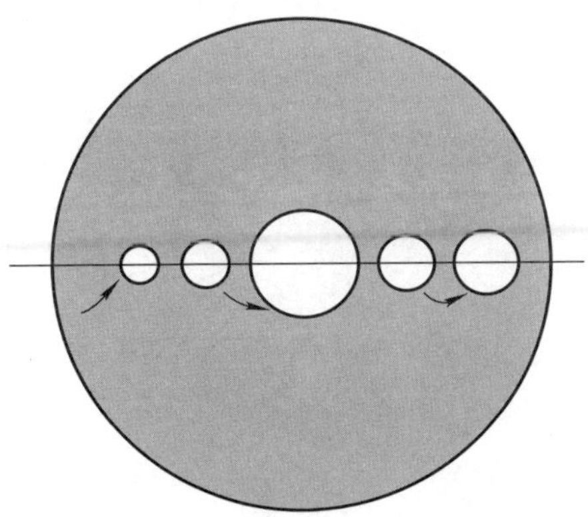

Fig. 11.

II. Higher automorphic functions without principal circle

In the more precise division of the functions without principal circle, the lack of a detailed knowledge of the non-rotation groups acts as a hindrance. We must here satisfy ourselves with a few indications, and with consideration of the boundary set, we initially, make the following division:

1. Functions with infinitely many isolated boundary points;
2. Functions with one (non-analytic) boundary curve;
3. Functions with infinitely many (circular or non-analytic) boundary curves.

Inside each of these categories, one will then divide further, according to genus p and the number n of fixed vertices, into "*species*", and the species further into "*families*".

[35] We draw attention here to two especially important *special cases*:

α) Automorphic models of genus 0

The automorphic models belonging here we again arrange together into families, whereby the individual family is assigned a signature $(0, n; l_1, l_2, \dots, l_n)$, as for the principal-circle functions. The fundamental domain P is bounded in the canonical form by $2n$ boundary curves pairwise coordinated to one another[18], and has n fixed, as well as n accidental, vertices belonging to a cycle. The number n must hereby be > 3, nevertheless, we already have non-rotation groups for $n = 4$.

A special case is again the one, that the fundamental domain lets itself be built up from two circular-arc n-angle, symmetric to one another, with the angles

$$\frac{\pi}{l_1}, \frac{\pi}{l_2}, \dots, \frac{\pi}{l_n}.$$

Regular-symmetric quadrangle nets of this kind were studied in detail in I on p. 399 ff. The developments at that time, in particular, gave us the theorem, *that such automorphic models furnished by regular-symmetric quadrangle nets are able to occur in all these categories, distinguished in* 1, 2 *and* 3, *according to the nature of the boundary manifold.*

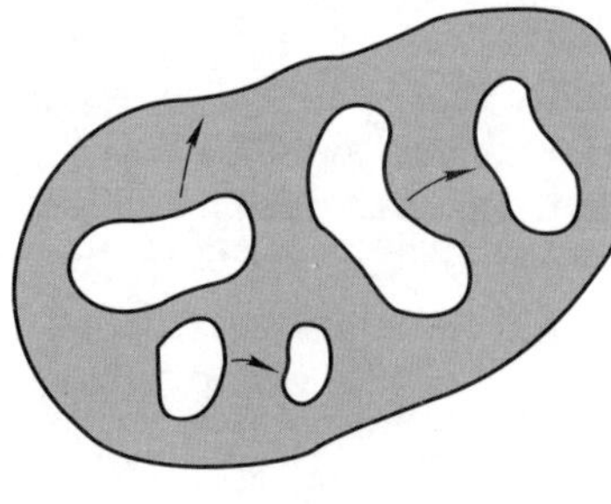

Fig. 12.

It needs scarcely to be remarked, that the principal-circle functions with $p = 0$, cited under α on p. 30 subsume themselves here as a special case.

β) Automorphic models of fundamental domains with $2p$ closed boundary curves running separately

The models intended here represent the generalization of those models discussed on p. 34, whose domains P were bounded by p pairs of full circles coordinated to one another, along with a principal circle. Here belong the automorphic models of such fundamental domains, for which there need be no principal circle, and which, in place
[36] of circles, are bounded by $2p$ separately running, closed, pairwise equivalent curves (cf.

[18] Whether these always let themselves be fixed as $2n$ circular arcs, will remain to be seen. The normal polygons (cf. I on p. 123 ff.) for the groups under discussion are indeed bounded exclusively by circular arcs; however, the number of the sides is, in general $> 2n$ for them.

figure 12), but otherwise shaped in any fashion[19]. *Thus, it is a question of models of the species* $(p, 0)$, *and indeed, these belong consistently in the first of the three categories just distinguished, i.e., we have functions with infinitely many isolatedly situated boundary points.*

As a special case, we next have to consider the one, that the p generating substitutions are exclusively *hyperbolic*. Here then, in particular, again belonging as special cases those groups, *which let themselves be generated by* $(p+1)$ *reflections in isolatedly situated full circles.* In this case, the fundamental domain P lets itself be built up from two domains symmetric to one another, the individual one of which is bounded by those $(p+1)$ full circles. Naturally, here too, principal-circle groups again subsume themselves here as special cases, for which the $(p+1)$ full circles have a common orthogonal circle.

1.9 The integrals of the automorphic models

In the theory of elliptic functions, there is the important theorem, that an arbitrary integral of the second kind is a *single-valued* function of associated integrals of the first kind. This is only a special case of a general theorem referring to automorphic models, which is now to be derived.

For any proposed automorphic model, let us imagine the fundamental domain P, at the same time, put into the canonical form, in order to be able to utilize the associated system of the generating operations in the form agreed upon in I on p. 186. On the associated Riemann surface F, the algebraic or Abelian integrals of the model are given generally by the expression:

$$J = \int R(w, z) dz \tag{1}$$

(cf. "M" I on p. 501). We now want to conceive the *individual such integral as a function of* ζ on the basis of the mapping of the Riemann surface onto the polygon net. We denote this, in this spirit, by $J(\zeta)$ and call it "*an integral belonging to the proposed automorphic model*". It is clear from the very beginning, *that according to the Weierstrassian* [37]
definition of analytic function, such an integral $J(\zeta)$ *represents a function defined only for the domain of the polygon net* N, *and has each boundary point of the net as an essential singular point.*

In order, now, at the same time, to discuss the question of the multiple valuedness of the function $J(\zeta)$, we initially assume, that the polygon net N has *one* boundary curve, and with this, represents a *simply* connected domain[20] under these circumstances, the domain P is also simply connected; let it have the character (p, n). Here

[19] The boundary curves may self-evidently also represent full circles; only we must again let it remain to be seen, whether, in each case, one can give the $2p$ boundary curves circular form by allowed modification. If, according to I on p. 123ff., one forms an associated normal polygon, then the boundary curves are indeed exclusively circular arcs, but the occurrence of accidental vertices does not then appear to be excluded.

[20] Also, the case of the doubly periodic functions subsumes itself here. The associated polygon net N has a single boundary point, and thus covers a simply connected domain.

then, the following fundamental relations are present: each closed path in N can be drawn together into a point inside N, accordingly, to it will correspond a closed circuit on the surface F, which likewise can be drawn together into a point. To each proper period-path on F (which cannot be drawn together into a point) must conversely correspond an open path in the net between two equivalent points.

The application of these remarks to the integrals of the first and second kinds of the automorphic model is effected immediately. The continuation of the individual such integral $J(\zeta)$ over a closed circuit in the net always leads back to the initial value $J(\zeta)$, since to this circuit these do not correspond a proper periodpath on the surface. We obtain the theorem: *if the polygon net N has one and only one boundary curve, then the automorphic integrals of the first and second kinds are "single-valued" functions of* ζ[21]. *These remain completely unaltered with respect to the n elliptic and parabolic generators $V_1, V_2, \dots, V_n$; upon exercise of one of the remaining $2p$ generators V_{a_k}, V_{b_k}, the individual integral $J(\zeta)$ takes on an additive period increase.* The exercise of an elliptic or parabolic substitution, in fact, amounts to a simple circuit about one of the points $e_1, e_2, \dots, e_n$ on the surface F; we return once more below to the effect of the V_{a_k}, V_{b_k}.

For the integrals of the third kind, the relations lie somewhat more complicatedly. Such an integral has logarithmic points of discontinuity on the surface F, and the en-
[38] circlement of an individual such point furnishes a period increase. If, however, it happens, in particular, that to such a discontinuity point corresponds a *parabolic* vertex of the domain P, then all arbitrary multiple circuits about that point of F will furnish open paths in the net. One obtains the theorem: *under the condition of a simply connected net N, there are furnished, from the integral $J(\zeta)$ of the third kind, those, but also only those, single-valued functions of ζ, whose logarithmic discontinuity points lie exclusively at parabolic vertices. With respect to the associated parabolic generator V_i, the individual such function $J(\zeta)$ will alter itself by an additive constant, but except for this, will take part in the properties cited before of the integrals of the first and second kinds.*

If the net N is multiply connected, then these are always closed paths in N, which do not draw themselves together into points on F and consequently represent proper period-paths. *Now, accordingly, the integrals $J(\zeta)$, speaking generally, will no longer furnish single-valued, but infinitely multiple valued functions of ζ.*

But there are exceptions to this general rule in special cases. Let us turn ourselves e.g., to an automorphic model of the kind discussed under β on p. 35 with a domain P, which is bounded by $2p$ closed boundary curves running separately. These $2p$ boundary curves furnish p reentrant cuts on the surface F, which we choose as $a_1, a_2, \dots, a_p$. The general theory of generators (cf. I on p. 184 ff.) so fashions itself here in particular that the substitutions $V_{b_1}, \dots, V_{b_p}$ all become equal to 1; as generators of the group there remain only the substitutions $V_{a_1}, \dots, V_{a_p}$, between which a relation does *not* exist.

Now, according to "M" I on p. 531, there are normal integrals of the second kind, which upon continuation over the p paths a_k, experience *no* period growth different from 0. Therefore, with consideration of the non-existence of a relation between the V_{a_k}, it follows the theorem: *for a fundamental domain P with $2p$ separately run-*

[21] This theorem has been given by Klein in the paper "*Neue Beiträge zur Riemann'schen Functionentheorie*" Abschn II, Math. Ann. Vol. 21 (1882).

ning boundary curves, there are always "single-valued" integrals $J(\zeta)$ of the second kind. From the further remarks made concerning these integrals in "M" I on p. 531, it also follows: *in particular, there are integrals of this kind, which only become discontinuous with $(\zeta-\zeta_0)^{-1}$ at one arbitrarily choosable place ζ_0 of the fundamental domain; the individual such integral is uniquely determined up to an additive constant by its place of discontinuity.* In the course of the fourth chapter, we shall represent such integrals as functions of ζ by convergent series and investigate them in more detail.

We now return once more to the case first considered, of a model with one boundary curve, but restrict ourselves to the associated integrals $J(\zeta)$ of the first and second [39]
kinds, which are single-valued under all circumstances. Let an individual integral $J(\zeta)$ have the periods $\omega_{11},\dots,\omega_{1p},\omega_{21},\dots,\omega_{2p}$ corresponding to the generators $V_{a_1},\dots,V_{a_p}$, $V_{b_1},\cdots,V_{b_p}$. The individual substitution V of the group Γ then carries J over into:

$$J'=J+m_1\omega_{11}+\cdots+m_p\omega_{1p}+n_1\omega_{21}+\cdots+n_p\omega_{2p}. \tag{2}$$

The meaning of the whole numbers m,n hereby coming up yields itself, if one writes the substitution V, under consideration, symbolically as a product:

$$V=\prod(V_1,\dots,V_n;V_{a_1},\dots,V_{b_p}). \tag{3}$$

then m_k is obviously the sum of the exponents of V_{a_k} coming up in (3) on the right side, and the $n_1,\dots,n_p$ are to be defined correspondly.

For $p=0$, $J(\zeta)$ is simply an automorphic function of the group Γ. If, however, one has $p>0$, then infinitely many transformations (2) come up, *which, in their totality, obviously form an Abelian group with parabolic substitutions permutable with one another.* From the considerations presented before, the theorem then immediately arises: *the Abelian group G stands in the relation of $1-\infty$-valued homomorphism to the group Γ*[22]. Hereby, the identical substitution inside G corresponds to a distinguished subgroup Γ_∞ of index ∞ in Γ. *This Γ_∞ includes all substitutions V, which leave invariant all the integrals of the first and second kinds; but these substitutions are furnished by all those symbolic products (3), for which the sum of the exponents of the individual generators V_{a_k}, V_{b_k} is equal to zero in each product*[23].

We also draw a remarkable conclusion for the case of an automorphic model with boundary circle. The group Γ of the character (p,n), in this case (cf. I on p. 313 ff.), we had produced by composition from p groups of the character $(1,1)$ and one group of the character $(0,n+p)$, whereby these characters refer themselves to the projective [40]
polygon. The latter group had the $(n+p)$ generators:

$$V_1,\dots,V_n,V_{a_1}^{-1}V_{b_1}V_{a_1}V_{b_1}^{-1},\dots,V_{a_p}^{-1}V_{b_p}V_{a_p}V_{b_p}^{-1}.$$

All these substitutions now obviously belong to the group Γ_∞ just considered; therefore follows: *the "single-valued" integrals $J(\zeta)$ belonging to Γ remain unaltered with respect to all substitutions of the group of character $(0,n+p)$ considered.*

---

[22] For the names "homomorphism" and "isomorphism" of groups, compare the note in I on p. 194.

[23] The approaches of the text we carried further for a special case belonging to the theory of the modular functions in the paper of the author "*Über ausgezeichnete Untergruppen in der Gruppe der elliptischen Modulfunctionen*" (Mathem. Ann. Vol. 31, 1887).

Moreover, the latter group furnishes, in the ζ-plane, a fundamental domain of the genus $p-1$, which is intersected by the principal circle in p separated pieces and is decomposed in such a way into two mutually symmetric halves. On the Riemann surface belonging to this fundamental domain, the functions considered are indeed single-valued, but they possess the system of the p symmetry lines as a natural boundary.

1.10 General single-valuedness theorem. Application to linear differential equations

The theorems concerning single-valuedness of the integrals $J(\zeta)$ in ζ are only special cases of the more general single-valuedness theorem. In order, in the formulation of the latter, to arrive at simple relations, we now restrict ourselves simply to the *case of an automorphic model with "one" boundary curve.* Under this assumption, the following theorem holds: *a function, which, apart from the n points corresponding to the fixed polygon vertices, is everywhere regular or only polarly discontinuous on the Riemann surface F, and which preserves these properties upon unrestricted continuation over the surface, is a single-valued function in dependency on ζ inside the polygon net N, in case it only satisfies the one requirement, to reproduce itself for an l_i-fold circuit about each such place of the surface, which corresponds to an elliptic polygon vertex of the period l_i.* Notice, that no kind of restrictive assumptions are made here concerning the multiple valuedness of the function with respect to period-paths on the surface, as well, too, as with respect to the circuits about the points corresponding to the parabolic polygon vertices.

For the proof of our theorem, we carry out an arbitrary closed circuit in the net N and imagine the corresponding path constructed on F. That circuit and its image on the surface F are then to be drawn together; however, pushing it over fixed polygon ver-
[41] tices is hereby to be forbidden. Under these circumstances, the closed path lets itself be drawn together into a series of looplike circuits about elliptic polygon vertices. A single such loop, however, on the surface F, becomes an l_i-fold circuit about the corresponding point. In consequence of our assumption, the proposed function will reproduce itself for this circuit, and therefore, in general, for the originally chosen closed path, i.e., it is single-valued in ζ[24].

One now knows, besides the integrals $J(\zeta)$ of our automorphic model, also an extended class of functions, to which the single-valuedness theorem set up finds application. These are the *solutions* of such *linear homogeneous differential equations* of arbitrary order m, whose coefficients are algebraic functions of the surface F. Thereby, to be sure, on account of the singular points of the differential equations, certain assumptions are made, to be immediately formulated even more precisely.

[24] The single-valuedness theorem treated in the text has already played a prominent role in the first investigation of Poincaré and Klein on automorphic functions. In the more general form, this was stated by Klein in the paper "*Neue Beiträge zur Riemann'schen Functionen theorie*", Mathem. Annlen Vol. 21 on p. 214 (1882). Poincaré has also treated. Namely, in his first note in the Comples rendus Vol. 92(1881) and Vol. 93 (1882), the example of the linear differential equations to be considered immediately in the text.

By arranging F as a Riemann surface over the z-plane (cf. p. 17), we give the differential equation in question the form:

$$\frac{d^m Z}{dz^m} + R_1(w,z)\frac{d^{m-1}Z}{dz^{m-1}} + R_2(w,z)\frac{d^{m-2}Z}{dz^{m-2}} + \cdots + R_m(w,z)Z = 0. \quad (1)$$

Let a "*fundamental system*" of solutions of this equation be given by $Z_1, Z_2, \ldots, Z_m$[25]. This function system then has the basic property, *of substituting itself linearly homogeneously for arbitrary closed circuits on F*. Therefore, if for an individual circuit, Z_k goes over into Z'_k, then, for the $Z'_1, Z'_2, \ldots, Z'_m$, one has the representation:

$$\begin{cases} Z'_1 = \alpha_{11}Z_1 + \alpha_{12}Z_2 + \cdots + \alpha_{1m}Z_m, \\ Z'_2 = \alpha_{21}Z_1 + \alpha_{22}Z_2 + \cdots + \alpha_{2m}Z_m, \\ \cdots\cdots\cdots\cdots \\ Z'_m = \alpha_{m1}Z_1 + \alpha_{m2}Z_2 + \cdots + \alpha_{mm}Z_m, \end{cases} \quad (2)$$

with m^2 coefficients α_{ik} of non-vanishing determinant. For the closed circuits, how- [42]
ever, not only the proper period paths on F thereby come into consideration, but also circuits about individual "*singular points*" of the differential equation, which we assume is the number ν. The substitutions (2) belonging to these singular points we call symbolically, $S_1, S_2, \ldots, S_\nu$, while to the $2p$ elementary period paths a_k, b_k are to correspond the substitutions S_{a_k}, S_{b_k} of our fundamental system.

By returning from the Riemann surface to the fundamental domain P, we have the conditions to formulate concerning the position of the singular points: *the singular points, in P, are to occur exclusively at fixed vertices.* Thus, the number of the singular places ν must be $\leq n$. If, however, in particular, a singular place of the differential equation lies at the individual elliptic vertex with the period l_i, then we further assume, *that the associated substitution S_i, for its part, is also of a finite period λ_i, and that λ_i is a divisor of the number l_i.*

Under these circumstances, the preconditions for the validity of our single-valuedness theorem are obviously fulfilled. We obtain the theorem: *each integral Z of the differential equation (1) represents a "single-valued" function $Z(\zeta)$ inside the net N; thereby, the single-valued functions $Z_1(\zeta), Z_2(\zeta), \ldots, Z_m(\zeta)$, furnished by the fundamental system of the equation (1), substitute themselves linearly-homogeneously with respect to the substitutions V of the group Γ.*

The totality of these substitutions of the $Z_1, \ldots, Z_m$ will again form a group G, which has as generators the substitutions denoted above by $S_1, \ldots, S_\nu, S_{a_1}, \ldots, S_{b_p}$. This group stands in a relation to the group Γ which is single-valued in one direction. In fact, to each V there corresponds a uniquely determined S; thus, the substitutions S_{a_k}, S_{b_k} correspond to the V_{a_k}, V_{b_k}, and let the $S_1, S_2, \ldots, S_\nu$ correspond to the $V_1, V_2, \ldots, V_\nu$, while for $n > \nu$, the identical substitutions $= 1$ belongs simultaneously to the $V_{\nu+1}, \ldots, V_n$. Also, conversely, an S can be uniquely coordinated to a V; necessary conditions for this are that $\nu = n$, and that the periods denoted above by λ_i are exactly equal to the l_i. We

[25] The elementary theorems to be utilized here from the theory of the linear differential equations one finds treated in detail in L. Schlesinger "*Handbueh der theorie der linearen Differentialglcichungen*", Vol. 1 (Leipzlg, 1895). See also L. Heffter "*Einleitung in die Theorie der linearen Differentialgleichungen*" (Leipzig 1894) or the modern textbooks on higher analysis.

thus make note: *the groups* Γ *and G, in each case, stands in the relation of one-sided, single-valued "homomorphism"; however, in special cases, "isomorphism" can also be present.*

We yet return in detail to these approaches in a later section of the present work, and shall have to occupy ourselves, in particular, also with systems of convergent se-
[43] ries, which represent such functions families $Z_1(\zeta), Z_2(\zeta), \ldots, Z_m(\zeta)$.

1.11 ζ as a linearly polymorphic function. The fundamental problem

Having just changed the functions of the Riemann surface F around into functions of ζ, we now want, conversely, to carry over onto the closed Riemann surface F the stock of values of the variable ζ restricted to the net N. We thereby imagine F as a multiple sheeted surface over the plane of an automorphic function $z = \varphi(\zeta)$ belonging to the proposed group Γ. We obtain thus, the *function* $\zeta = f(z)$, *inverse to* the automorphic function $z = \varphi(\zeta)$, which we would designates, following the language agreed to on p.1, as a "*linearly polymorphic*" or briefly, *"polymorphic" function on the Riemann surface* F. In fact, the function $\zeta = f(z)$ is indeed *multiple valued* on F, and apart from a few elementary cases, even *infinitely multiple valued; and each branch* ζ' *of this function to be reached by an arbitrary circuit on the surface is a linear function of the initial branch* ζ:

$$\zeta' = \frac{\alpha\zeta + \beta}{\gamma\zeta + \delta}. \tag{1}$$

Above (p. 29), we designated automorphic models, which were transformable into one another, and therefore belonged to one and the same "class", as not essentially different. If we exercise on ζ the substitution denoted at that time by S:

$$\zeta' = S(\zeta) = \frac{a\zeta + b}{c\zeta + d}, \tag{2}$$

with arbitrary complex coefficients of non-vanishing determinant $(ad - bc)$, then we obtain a new polymorphic function $\zeta' = S(\zeta)$, which we designate as *belonging to the same "class" as* ζ. Since the coefficients a, b, c, d of S only come into force by means of their quotients, *then the individual class thus defined includes* ∞^6 *polymorphic functions* ζ (insofar as the domain of values of a complex variable has to count as doubly infinite).

In order to represent in more detail the behavior of a particular function ζ of the class at an individual place of the surface, we must reach back to the remarks on p. 5 ff. We now place a corresponding quantity τ opposite the development quantity t there, which we define for a place z_0 of the surface, situated in the finite part of the plane, by:

$$\tau = z - z_0, \qquad \text{or} \qquad \tau = \sqrt[m]{z - z_0}, \tag{3}$$

[44] According as it is a question of an ordinary place or an m-sheeted branch point z_0; for the infinitely distant point of the Riemann surface, there enters, correspondingly:

$$\tau = \frac{1}{z}, \qquad \text{resp.} \qquad \tau = \sqrt[m]{\frac{1}{z}}. \tag{4}$$

Then, at each place of the surface, without exception, we have a development:

$$t = \tau(\alpha_0 + \alpha_1\tau + \alpha_2\tau^2 + \cdots) \qquad \text{with} \quad \alpha_0 \neq 0$$

for the quantity t defined on p. 5 resp. 7. In this general scheme, the following special representations are contained.

1. For a point of the surface, which does not correspond to an elliptic or parabolic polygon vertex, there is one particular function ζ of the class, which admits the representation:

$$\zeta = \tau(1 + \alpha_1\tau + \alpha_2\tau^2 + \alpha_3\tau^3 + \cdots). \tag{5}$$

2. For the elliptic vertex with the angle $\frac{2\pi}{l}$, there is, at the corresponding place of the surface, a particular function ζ in the class with the development:

$$\zeta = \tau^{\frac{1}{l}}(1 + \beta_1\tau + \beta_2\tau^2 + \beta_3\tau^3 + \cdots). \tag{6}$$

3. Finally, for such a place of the surface, which corresponds to a parabolic vertex, one has a special ζ with:

$$\zeta = \log\tau + (\gamma_1\tau + \gamma_2\tau^2 + \gamma_3\tau^3 + \cdots). \tag{7}$$

The remaining polymorphic functions of the class are then, in each case, according to (2), expressable linearly in the ζ represented. Beyond this, it is naturally by no means to be asserted, that there is one and the same function ζ, to which the representations (5), (6), and (7) refer.

We finally take note of the theorem: *the polymorphic functions are only branched at those n places of the surface F, which correspond to the fixed polygon vertices; the elliptic vertices furnish finitely-sheeted, the parabolic ones ∞-sheeted branchpoints.* We will again denote these n branch places in more detail by $z = e_1, z = e_2, \ldots, z = e_n$; naturally, the individual branch-points only lie in a determinate sheet of the Riemann surface at $z = e_k$.

The classification of the polymorphic functions is immediately given along with that of the automorphic models; we consequently need not spend time here in detail.

As especially approachable, and just on this account, exceedingly important, we have, in any case, to designate the "*polymorphic functions with principal circle*". Naturally, for this category of polymorphic functions, there again occurs the fundamental case distinction, whether the principal circle is a boundary circle or not. Here, then, there are always particular functions $\zeta = f(z)$, for which *all* the *substitutions* (1) entering for analytic continuation exhibit *"real" coefficients* $\alpha, \beta, \gamma, \delta$. [45]

We mention further, as a remarkable special case, when fixed polygon vertices are lacking; here we arrive at the concept of the "*unbranched polymorphic function*". Thereby, in particular, there may be a boundary circle, or also a principal circle which is not a boundary circle. Also, a function of this kind can map the cut surface onto a polygon without principal circle. In the latter case, the special case is again especially important, that the polygon to be mapped is bounded by $2p$ closed curves, which are mapped onto one another by p loxodromic, or perhaps, in particular, hyperbolic, substitutions. The great advantage accrues to these special cases, that the mapping of the Riemann surface onto the polygon is especially transparent: the $2p$ boundary curves of the polygon correspond to the banks of a system of p reentrant cuts on the surface F.

The introduction of our polymorphic functions $\zeta = f(z)$ in the Riemann surface F appears especially important, in consideration of the single-valuedness theorem treated on p. 37 ff., according to which, not only the algebraic functions, but also even numerous additional functions existing on the surface, are single-valued in ζ. Then too, the investigations of Klein and Poincaré , right at the very beginning, have turned themselves to the fundamental question, inverse in a certain measure to our line of development so far, whether perhaps there might exist, on *each* Riemann surface, polymorphic functions of suitable character. In this regard, again preferring the principal-circle case, we already want to formulate here the following questions, which we call the "*fundamental problems*", and whose solutions will furnish the "*fundamental theorems*" to be treated later.

I. *If, on an arbitrary given Riemann surface of the genus p over the z-plane, n randomly chosen branch places* $e_1, e_2, \ldots, e_n$ *are marked, then we ask for the existence of a polymorphic function* $\zeta = f(z)$, *branched at these places, which maps the canonical cut surface (cf. I on p. 182 ff.) onto a polygon of our kind with boundary circle and of the signature* $(p, n; l_1, l_2, \ldots, l_n)$, *understanding by the* l_i *arbitrary whole number* > 1 *or* ∞.

[46] II. *Let there be given an orthosymmetric Riemann surface of arbitrary genus p over the z-plane, whose symmetry lines consists of* μ *closed lines; then* μ *is one of the number* $p+1, p-1, p-3, \ldots$, *and therefore* $p' = \frac{1}{2}(p - \mu + 1)$ *is a whole number from the series* $0, 1, 2, \ldots, [\frac{p}{2}]$. *If we now mark, at random,* $n = 2\nu$ *pairwise symmetric places* $e_1, e_1', \ldots, e_\nu, e_\nu'$, *does there then always exist a polymorphic function* $\zeta = f(z)$, *branched at these places, which maps the suitably cut surface onto a principal-circle polygon of the kind discussed on p. 27 and of the signature* $(p, 2\nu, p'; l_1, l_2, \ldots, l_\nu)$, *which is intersected in* μ *segments by the principal circle and is thereby decomposed into two symmetric halves? Naturally, the whole numbers* l_i *are thereby again to be chosen* > 1 *or* ∞, *but otherwise arbitrarily.*

III *Again, if an arbitrary Riemann surface is given, whose genus p, however, is to be* > 0, *and if this surface is cut along p reentrant cuts not meeting one another, then we shall ask for the existence of an unbranched polymorphic function* $\zeta = f(z)$, *which maps the cut surface onto a domain bounded by* $2p$ *closed boundary curves, pairwise coordinated to one anther, of the kind considered under* β *on p. 35.*[26]

Naturally, it may also be a question of unbranched polymorphic functions in the cases I and II, for one has simply to take the number denoted by n equal to zero. Fur-

[26] See, for the existence theorem for polymorphic functions, the two equally titled notes of Klein "*Über indeutige Functionen mit linearen Transformations in sich*", Mathem. Ann. Vol. 19, p. 565, Vol. 20, p. 49 (1882), as well as the recapitulatory representation in second IV of the paper "*Neue Beiträge zur Riemann'schen Functionentheorie*" Math. Ann. Vol. 20, p. 206 (1882). For the models with $p = 0$, Poincaré had already set up the existence theorem in question somewhat earlier, concerning which one wants to compare his announcements in Comples Rendus of April and August 1881, as well as the paper "*Sur les fonctions uniformes qui se reproduisent par des substitutions linéaires*" Math. Ann. Vol. 19, p. 553 (1882). Poincaré then let the generalizations follow, touching in many points with the brief preceding investigations of Klein, in the Comptes rendus of April 1882, as well as, in the most detailed manner, in the paper "*Sur les groupes des équations linéaires*", Acta. Mathem. Vol. 4 p. 201 (1883).

thermore, self-evidently, the underlying surfaces in the problems I and III may also be orthosymmetric or dissymmetric.

Only the fundamental problem III refers to the polymorphic functions without principal circle. One could, in this regard, go essentially further, and, in particular, Klein, in the third of his works just cited, has already taken up a far more general stand- [47]
point. We have, in I on p. 443 ff., produced a discontinuity domain by the process of composition from ν polygons with boundary circles of the signatures:

$$(p_i, n_i; l_1^{(i)}, l_2^{(i)}, \dots, l_{n_i}^{(i)}), \quad i = 1, 2, \cdots, \nu, \tag{8}$$

which as such (i.e., without addition of crosscuts) represents a ν-fold connected domain. The associated Riemann surface is of the genus $p = \sum_{i=1}^{\nu} p_i$, and it carries, corresponding to the boundary of the discontinuity domain, ν different canonical cut systems of the kind discussed in I on p. 183, whereby, however, its connectivity still remains a ν-fold one.

If we now imagine a Riemann surface of the genus p with a decomposition of this kind chosen at random and if we prescribe arbitrary whole numbers $l > 1$ or ∞ for the individual signatures, then it is a question of the existence of a polymorphic function on this Riemann surface, which accomplishes a mapping of the surface onto a discontinuity domain of the kind described above. Klein, l.e., also allows, besides the principal-circle groups, any cyclic non-loxodromic groups, as well as elliptic and parabolic rotation groups to take part in the composition; in this way, he arrives at a somewhat more general fundamental problem, to which the studies in chapter IV of the work cited are devoted.

The further treatment of the fundamental problems set up is here, for the moment, to be pushed aside. We have to occupy ourselves in the most detail with these in the second section; their full disposal is one of the principal goals of the present work.

1.12 Differential equations of the third order for the polymorphic functions

We return here, albeit cursorily to the known "*Schwartzian differential expression*"

$$[\zeta]_z = \frac{\dfrac{d^3\zeta}{dz^3}}{\dfrac{d\zeta}{dz}} - \frac{3}{2}\left[\frac{\dfrac{d^2\zeta}{dz^2}}{\dfrac{d\zeta}{dz}}\right]^2 \tag{1}$$

of the third order, which was already studied in "Icos." on p. 74 and "M" I on pp. 97 and 63.[27]. The basic property of this expression is, that it remains unaltered, if one replaces [48]
ζ by an arbitrary linear fractional function of itself:

$$\left[\frac{a\zeta + b}{c\zeta + d}\right]_z = [\zeta]_z. \tag{2}$$

[27] For the literature of the expression (1), see the remarks in "Icos." on p. 74.

Consequently, if we understand by ζ a polymorphic function of our kind on a Riemann surface over the z-plane, then $[\zeta]_z$ will represent a single-valued function on the surface.

In order to draw into investigation the behavior of this single-valued function at any place z_0 of the surface, we may substitute for ζ an arbitrary particular one of the ∞^3 functions of the class. In order to be able to make use of the series developments (5) ff. on p. 44, let us calculate out the relation (cf. "Icos." p. 75):

$$[\zeta]_z = [\zeta]_\tau \left(\frac{d\tau}{dz}\right)^2 + [\tau]_z. \tag{3}$$

On the other hand, from the validity of the development (5) on p. 44, we determine the following development for $[\zeta]_\tau$:

$$[\zeta]_\tau = 6(\alpha_2 - \alpha_1^2) + 24(\alpha_3 - 2\alpha_2\alpha_1 + \alpha_1^3)\tau + \cdots. \tag{4}$$

Now consider, first of all, such places of the surface, at which ζ is not branched. If z_0 is a place of this kind situated in the finite part of the plane, which, at the same time, does not represent a branch-point of the surface, then it yields itself immediately from (4), that $[\zeta]_\tau$ is regular in the neighborhood of this place.

But if we have an m-sheeted branch-point of the surface lying in the finite part of the plane, then we use τ in the sense of the second formula (3) on p. 43 and find:

$$[\zeta]_z = \frac{m^2-1}{2m^2}\tau^{-2m} + \frac{1}{m^2}\tau^{-2m+2}[\zeta]_\tau. \tag{5}$$

at this place of the surface, $[\zeta]_\tau$ thus has a pole of order $2m$.

At $z = \infty$ we may assume the surface as unbranched; then from (3) it immediately emerges, that $[\zeta]_z$ has a zero of fourth order at $z = \infty$ in the individual sheet.

By now passing on to the investigation of the elliptic and parabolic branch places of the surface, we assume (which poses no difficulty) the latter so chosen, that none of these places coincides with an infinitely distant point or a branch-point of the surface, with use of the developments (6) resp. (7) on p. 44, we shall thus have to take $\tau = z - z_0$

[49] and therefore $[\zeta]_z = [\zeta]_\tau$. In case (6) on p. 44 however:

$$[\zeta]_\tau = \frac{1}{2}\left(1 - \frac{1}{l^2}\right)\tau^{-2} + a_{-1}\tau^{-1} + a_0 + a_1\tau + \cdots, \tag{6}$$

and this formula also holds, at the same time, in the parabolic case (7) on p. 44, if one substitutes $l = \infty$. We fixed a pole of the second order for $[\zeta]_z$ at each of the places in question.

The preceding considerations establish the result: *the polymorphic functions of the proposed class are integrals of a differential equation of third order of the following form:*

$$\frac{\frac{d^3\zeta}{dz^3}}{\frac{d\zeta}{dz}} - \frac{3}{2}\left[\frac{\frac{d^2\zeta}{dz^2}}{\frac{d\zeta}{dz}}\right]^2 = R(w, z), \tag{7}$$

which right side is a determinate algebraic function of the surface; and one can say, that $\frac{a\zeta+b}{c\zeta+d}$ *is the general integral of this differential equation, in case* ζ *signifies a particular integral and* a, b, c, d *are arbitrary complex constants.*

Concerning the function $R(w, z)$, the following remarks let themselves be made:

1. *It has, in this specific situation, a pole of order* $2m$ *at an* m*-sheeted branch-point of the surface,*
2. *It exhibits a pole of the second order at each of the n branch places* $e_1, e_2, \dots, e_n$ *of the polymorphic functions.*
3. *Except for this,* $R(w, z)$ *is everywhere regular on the surface and, in particular, at the place* ∞ *of each sheet, possesses a zero of fourth order.*

A few further remarks still let themselves be made concerning the series developments of $R(w, z)$ at the places $z = \infty$. But even if we add these remarks, $R(w, z)$ is still by no means determined in general from the properties enumerated. We pursue these relations somewhat further for the case of the genus $p = 0$, where we take as a starting point the principal function $z = \varphi(\zeta)$ associated to the plane. In this case, in place of $R(w, z)$, we get a rational function of z, which in this specific case has poles of second order at the n places e_i and except for this, is everywhere regular. Since, besides this, the rational function in question has a zero of fourth order at $z = \infty$, then it is, at the same time, the quotient of a rational integral function of degree $(2n-4)$ and the product $\prod_{i=1}^{n}(z-e_i)^2$.

In order to recognize more precisely the form of the function under discussion, we [50]
notice, that $[\zeta]_z \prod\limits_{i=1}^{n}(z-e_i)$ is equal to the quotient of an entire function of the $(2n-4)^{\text{th}}$ degree and the function of the nth degree:

$$(z-e_1)(z-e_2)\cdots(z-e_n).$$

If one applies the partial fraction decomposition to this quotient, there then arises an equation, to which we give the form:

$$[\zeta]_z \prod_{i=1}^{n}(z-e_i) = 2G_{n-4}(z) + \sum_{i=1}^{n}\frac{A_i}{z-e_i},$$

where $G_{n-4}(z)$ is an entire function of z of degree $(n-4)$. For the determination of the constants A_i, we develop the left side of this equation, in the neighborhood of $z = e_i$, into a series according to $\tau = z - e_i$ and find immediately, on the basis of (6):

$$A_i = \frac{1}{2}\left(1-\frac{1}{l_i^2}\right)(e_i-e_1)\cdots(e_i-e_{i-1})(e_i-e_{i+1})\cdots(e_i-e_n).$$

We arrive, in this way, at the result: *if, in the case* $p = 0$*, one starts from a principal function* $z = \varphi(\zeta)$*, then the polymorphic function* $\zeta = f(z)$*, inverse to it, is an integral of a differential equation of third order of the following form:*

$$[\zeta]_z = \frac{2}{\prod\limits_{i=1}^{n}(z-e_i)}\left(G_{n-4}(z) + \sum_{i=1}^{n}\frac{\frac{1}{4}\left(1-\frac{1}{l_i^2}\right)}{z-e_i}(e_i-e_1)\cdots(e_i-e_{i-1})(e_i-e_{i+1})\cdots(e_i-e_n)\right). \tag{8}$$

The $(n-3)$ *coefficients of the integral function* $G_{n-4}(z)$ *are initially yet unknown complex constants.* In the lowest case, $n = 3$, no function G comes up; here therefore, the right side of the equation (8) is uniquely determined. In the higher cases $n > 3$, we shall designate the $(n-3)$ constants initially remaining unknown as the *accessary parameters* of the differential equation; we return to these in the treatment of the fundamental theorem.

The differential equation given in "M" I on p. 63 under (5) for $n = 3$, moreover, does not directly subsume itself under the equation (8) just obtained, since, for the
[51] equation at that time, one of the three points e falls at $z = \infty$. If, corresponding to this assumption, we introduce $z' = (z - e_3)^{-1}$, then since obviously $[z]_{z'} = 0$, formula (3) furnishes:

$$[\zeta]_{z'} = [\zeta]_z \left(\frac{dz}{dz'}\right)^2 = [\zeta]_z (z-e_3)^4,$$

with consideration of (6), one thus finds, as the development of $[\zeta]_{z'}$ at $z = e_3$ resp. $z' = \infty$:

$$[\zeta]_{z'} = \frac{1}{2}\left(1 - \frac{1}{l_3^2}\right)(z-e_3)^2 + \cdots = \frac{1}{2}\left(1 - \frac{1}{l_3^2}\right) z'^{-2} + \cdots,$$

so that, at this place, there is a zero of second order. If, from now on, we again leave off the index in z' and, besides this, situate the two branch places e_1 and e_2 remaining in the finite part of the plane to $z = 0$ and $z = 1$, then

$$[\zeta]_z z^2 (z-1)^2 = A(z-1)^2 + Bz^2 + Cz(z-1).$$

The initial terms of the series developments of $[\zeta]_z$ for the neighborhoods of $z = 0, 1$ and ∞ serve for the determination of the coefficients A, B, C; we find, as a special from of the differential equation of third order for this case:

$$[\zeta]_z = \frac{1 - \frac{1}{l_1^2}}{2z^2} + \frac{1 - \frac{1}{l_2^2}}{2(z-1)^2} + \frac{\frac{1}{l_1^2} + \frac{1}{l_2^2} - \frac{1}{l_3^2} - 1}{2z(z-1)}, \tag{9}$$

which agrees with equation (5) in "M" I on p. 63.[28]

1.13 Generalization of the concept of automorphic functions

The single-valuedness of our automorphic functions is a basic property of theirs, which will be of fundamental significance, especially for the analytic representations of these functions to be considered in the third chapter. In contrast to this, it is worth mentioning, that most of the definitions set up so far, as well as the methods utilized, also preserve their validity for the treatment of *multiple-valued* functions.

We will, in this regard, follow a completely general approach. Let there be deter-
[52] mined in the ζ-plane, by finitely many boundary curves, an arbitrary simply or multiple

[28] Several invariant-theoretic investigations have been presented concerning the form of the differential equation (8) and the homogeneous linear differential equation of the second order connected with it (which we study later on). We cite the relevant papers below in a connected manner.

connected domain P, which may also reach over itself in any multiple sheeted fashion and thereby exhibit individual branch-points in its interior. The boundary curves of P are to be coordinated to one another in pairs, in such a way, that P, exactly as in polygons considered above, can be changed around into a closed surface by the bending together of boundary curves coordinated to one another. The coordination of the boundary curves is to be determined by analytic transformations $\zeta' = T_1(\zeta), \zeta' = T_2(\zeta), \ldots$.

The question now will be, *whether the domain P, thus defined, can be considered as a "fundamental domain" for a model of analytic functions of ζ, which would then be designated as automorphic functions of the transformation group to be generated from the $T_1, T_2, \ldots$.* Here it now reveals itself, that the method utilized on p. 8 for the existence proof of single-valued automorphic functions, in fact, remains usable. *If it is possible to cover the domain P, as at that time (p. 11), imbricately with finitely many circular discs and circular sectors, and if one can solve the boundary-value problem for the individual covering pieces, then the existence proof of the functions in question lets itself be disposed of as earlier.* Hereby, obviously, only those sectors resp. cycles of sectors are to be drawn into closer investigation, which lie at the "fixed" vertices of P, where the question of the mappability of such sectors onto circular discs would have to be treated. All remaining circular discs situated in the interior of P (also those, which include interior branch-points), as well as all circular sectors lying on the boundary, but not at fixed vertices, offer no difficulty.

This approach, except for the discontinuity domain of our groups Γ, has only first been considered for the generalization lying closest at hand. This generalization consists in this, that all transformations T, now as before, are to be linear substitutions, and that P possesses, without exception, elliptic or parabolic fixed vertices. But the angles at the elliptic vertices may be entirely arbitrary, and the domain P may reach over itself, as well over the polygon net N of the ζ-plane arising from P, entirely or partially, arbitrarily in any manner at all, indeed, it may cover it infinitely many times. We recognize immediately, *that the domain P considered defines, as a fundamental domain, a model of associated multiple-valued "linear"-automorphic functions.*

The individual one of these functions $z = \varphi(\zeta)$ again maps the fundamental domain onto an ordinary Riemann surface over the z-plane; and on this, the function $\zeta = f(z)$, inverse to $\varphi(\zeta)$, has the character of a linearly-polymorphic function, since, [53]
in particular, they satisfy a certain differential equation $[\zeta]_z = R(w, z)$ of third order. Just from the point of view of these differential equations, as well as certain homogeneous linear differential equations of the second order, which we shall soon find closely bound to those equations of the third order, one has approached the study of the polymorphic functions under discussion and their inverse, infinitely-valued, automorphic functions. Thereby, the lowest case, namely, that the fundamental domain P is of genus zero and has three fixed vertices, already offered a most extensive field of investigation. We refer, in this regard, to the handwritten lectures of Klein "*Über die hypergeometrische function*" from the winter semester 1893/1894[29], as well as the paper of Eschilling "*Beiträge zur geometrischen Theorie der Schwarz'ern s-function*"[30].

[29] Elaborated by E. Ritter, by commission from B.G Teubner, Leipzig.

[30] Göttinger Dissertation of 1893, printed in the Mathem. Anna. Vol. 44 on p. 162.

2

Form-theoretic discussions for the automorphic models of genus zero

We now turn ourselves to the detailed consideration of the automorphic models of genus zero. Hereby, after a brief glance at the forms of the associated fundamental domains, we give the investigation a turn, which, after the discoveries accumulated in the field of the modular functions, promises particular success. In fact, the use of the *modular forms* was of fundamental importance for the discovery of the simplest functions of a proposed fundamental domain and the algebraic connections subsisting between them, as well as in the study of analytic representations in the form of infinite series or products. In the former regard, compare, e.g., the form-theoretic developments in "M" II on p. 431 ff., which there furnish the foundation for the transformation of the eleventh order of the elliptic functions. In the latter respect, notice, that the numerous infinite series and products deduced from the theory of doubly periodic functions are not directly modular functions, but first directly represent modular forms of non-vanishing dimension. [54]

These relations emerging for the modular functions are typical for the entire domain of the single-valued automorphic functions. First by splitting the variable ζ into the quotient of the *binary variable* ζ_1, ζ_2, as it was already done in I on p. 194 ff., and by introduction of the homogeneous "*automorphic forms*" $\varphi(\zeta_1, \zeta_2)$ in ζ_1, ζ_2, we do arrive at the simplest quantities of the individual automorphic model. On the other hand, it is the automorphic *forms*, which are directly represented by the series to be studied in detail below, and which were discovered by Poincaré and named after him(see the introduction and the first paragraph in the next chapter).

Poincaré himself based his developments concerning automorphic functions resp. forms on the series just mentioned. In contrast to this, the concept of the automorphic [55]
forms is to be built up here, without any wide-ranging use of series developments, by means of a "*differentiation process*" from the automorphic functions obtained in the previous chapter. Thereby, by taking up the form-theoretic viewpoint, our developments gain extraordinarily in richness of relations, so that the restriction to the genus *zero*, initially taken into prospect, will appear justified.

The form-theoretic standpoint will give us cause, besides the polymorphic function ζ, to now consider "*polymorphic forms*" ζ_1, ζ_2 on the Riemann surface as well; these stand in an interesting relation to certain homogeneous linear differential equations of the second order, which we shall consider below.

Historically, the study of the polymorphic *forms* in Riemann's principal work on the P-function[1] and in Schwarz's investigations on hypergeometric series[2] has been in existence even earlier than that of the automorphic *functions*. Thereby, it is certainly to be remarked, that in the works cited, the form-theoretic terminology does not yet come explicitly into utilization; we shall have to refer in detail to the matter in question below.

Poincaré, too, still avoids the homogeneous method of consideration. In contrast with this, the form-theoretic standpoint was consistently carried through by Klein, as earlier for the icosahedron and the modular functions so too in his function-theoretic lectures from the beginning of the ninetieth year.[3] The distinguished investigations of Ritter's, which attach themselves to these lectures [4] attests in the most detailed manner
[56] to the essential advantage of the *form-theoretic methods*.

2.1 Shapes of the fundamental domains for the models of genus zero

Our further arguments we attach judiciously to the "*canonical shape*" of the fundamental domain, which according to I on p. 182, is attainable in *each* case. We thereby have to do with a polygon P of $2n$ sides, if we have the character $(0, n)$; figure 13 schematically illustrates the relations for $n = 5$. The n *fixed* vertices denoted in the figure by $\varepsilon_1, \varepsilon_2, \dots, \varepsilon_n$ are the fixed points of the generators $V_1, V_2, \dots, V_n$. The n accidental vertices $E_1, E_2, \dots, E_n$ form one cycle; to them corresponds the one relation subsisting here:

$$V_1 \cdot V_2 \cdots V_n = 1 \tag{1}$$

between the generators. Along with this equation, we naturally also get the relations:

$$V_k^{l_k} = 1,$$

corresponding to the periods of the elliptic generators as well, too, under certain circumstances, as secondary relations.

In the plane of an associated principal function $z = \varphi(\zeta)$, let the places $e_1, e_2, \dots, e_n$ correspond to the fixed vertices; *one* point E corresponds to the n accidental vertices. The "canonical cross-cut system", which leads to the polygon of figure 13, then consists, in known fashion, of n cuts running form E to the places $e_1, e_2, \dots, e_n$.

The *polygon net N arising from the proposed polygon P can as well have a boundary circle or a non-analytic boundary curve as it covers the entire ζ-plane up to isolatedly situated boundary points.* In the case of a boundary circle, one can always assume the

---

1 "*Beitrüge zur Theorie der durch die Gauss'sche Reihe darstellbaren Functionen*", Abhandl. der Göttinger Gesell. d. Wiss. of 1857 or collected works on p. 62.

2 "*Über diejenigen Fälle, in welchen die Gauss'sche hypergeometrische Reiheeine algebraische Function ihres vierten Argumentes ist*", Journ. Mathem. Vol.75 (1872).

3 Comp. the compilation in I, preface on p. VI.

4 Here, Ritter's dissertation comes mainly into consideration "*Die eindeutigen automorphen Formen vom Geschlechte null, eine Revision und Erweiterung der Poincaréschen Sätze*", Mathem. Ann. Vol. 41, p. 1(1892).

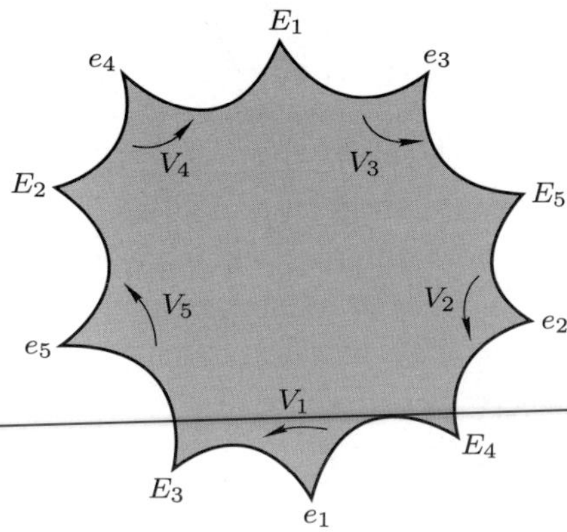

Fig. 13.

$2n$ sides of the polygon as $2n$ circular arcs. Whether this is also always possible in the remaining cases, must remain undecided in I.

As examples of cases *without* principal circle, let the two placed together in figure 14 count. The polygon P drawn on the left side, consisting of two quadrangles symmetric to one another(which one will easily change around into the canonical shape), has the signature$(0,4;\infty,\infty,\infty,\infty)$; and is cut out of figure 156 in I, where one of the two symmetric quadrangles carries the notation $\overline{P}'$. As one wants to see there, we arrive here at a net N *with a non-analytic boundary curve.* The second polygon drawn in figure 14, of the signature$(0,4;2,2,3,3)$, arises from the net represented in figure 153 in I; but the latter covers *the entire ζ-plane up to isolatedly situated boundary points.* The transformation of this domain, too, into the canonical shape is easily carried out. Choose, say, the four points $\varepsilon_1,\varepsilon_2,\varepsilon_3,\varepsilon_4$ as fixed vertices of a canonical domain. Of the four generators V_1,V_2,V_3,V_4 belonging to the canonical domains thus considered, the first two are the substitutions denoted in the figure by V_1 and V_2. For the substitution V_3' of the figure, we get: [57]

$$V_3' = V_2V_3 = V_1^{-1}V_4^{-1}.$$

We next shift our attention to *polygons with principal circle,* which, however, is *not a boundary circle.* We have considered polygons of this category on p. 26 ff. and must, in order to arrive at genus zero, substitute in the formulas there:

$$p = p' = 0,\ \mu = 1,\ n = 2(n'-1).$$

The segment of the principal circle belonging to the fundamental domain is a symmetry line of the *orthosymmetric fundamental domain* present here; we are able to so choose the principal function z, that this segment of the principal circle maps onto the real z-axis.

The next thing to do in the present case, is to choose the place E *on* the real z-axis. The canonical polygon then has a shape, such as the example of figure 15 represents for $n=6$. The characteristic thing is, that $(n-2)$ of the n accidental vertices coincide in pairs on the principal circle. [58]

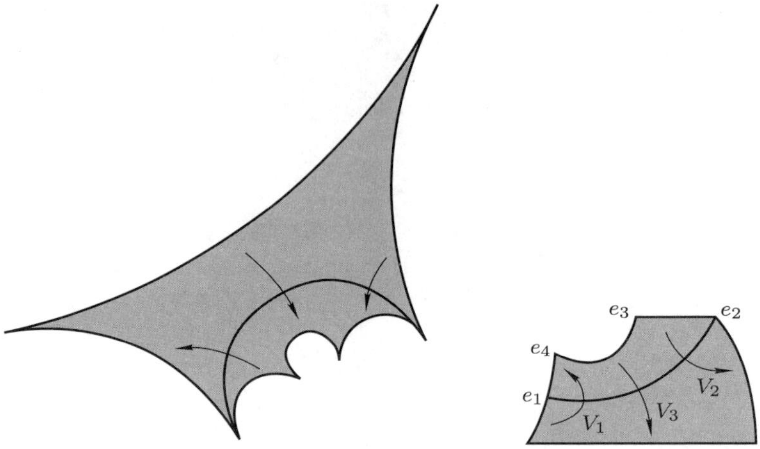

Fig. 14.

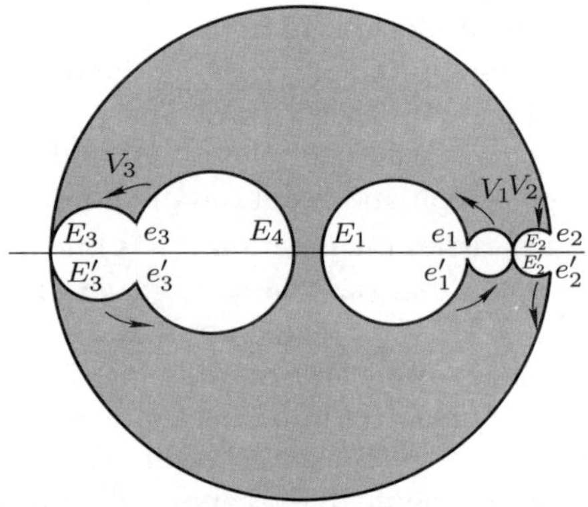

Fig. 15.

The shape of this polygon, by means of the considerations carried out in the projective plane in I on p. 299 ff., is especially easily understood. We have, in this plane, a polygon of character $(0, n')$, which has *one* fixed (hyperbolic) vertex outside the ellipse. We then meet exactly with the relations represented in figure 15, if, in the spirit of the developments just cited, we choose the point denoted there by E *on* the ellipse. If we then move this even more out over the ellipse, then the polygon just considered changes into that of figure 16. We recognize, *that our group lets itself be generated by shuffling into one another $n' - 1 = \frac{1}{2}n$ cyclic elliptic or parabolic groups* (cf. I on p. 191). The generators $V_1, V_2, \ldots, V_{\frac{1}{2}n}$ are independent of one another; the relation (1) is satisfied by holding for the rest of the generators:

$$V_{\frac{1}{2}n+1} = V_{\frac{1}{2}n}^{-1}, \ldots, V_{n-1} = V_2^{-1}, V_n = V_1^{-1}. \tag{2}$$

If, on the other hand, the point E of the projective plane, wanders into the interior of the ellipse, then one arrives at the shape of the fundamental domain drawn in figure

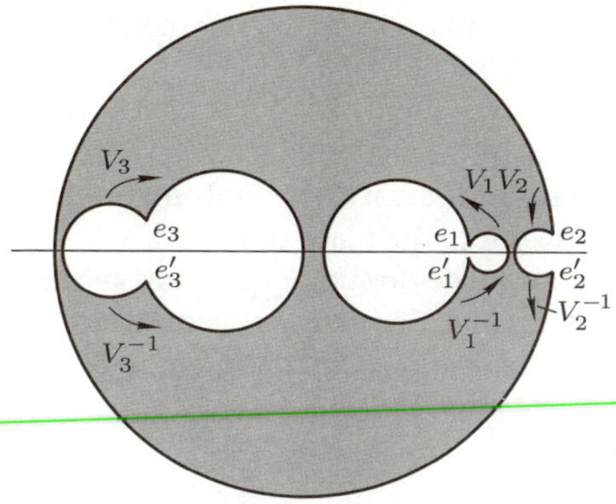

Fig. 16.

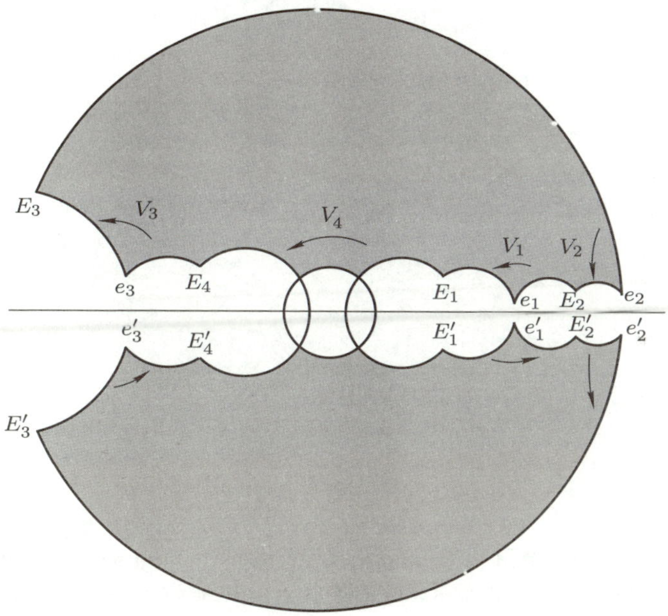

Fig. 17.

17. Here, E has even wandered over the polar of the hyperbolic cusp; this polar furnished the dotted circle in figure 17. One does not want to omit visualizing to oneself [59]
the cutting of the planes of the principal function corresponding to the two shapes of the fundamental domain last given.

In the two cases just considered, the points e in the z-plane were situated symmetrically to one another with respect to the symmetry circle (the real z-axis). Naturally, the case can also enter, that the points e, partially or all together lie on the symmetry circle.[5] Figure 18 furnishes an orthosymmetric fundamental domain of character $(0,7)$,

[5] Compare, moreover, the statements on p. 31 concerning these polygons.

for which three of the seven points e lie on the symmetry circle of the z-plane. The associated cutting of the z-plane, which is not directly the canonical one, one also wants to produce for oneself here. The symmetry circle of the z-plane maps itself onto the heavily marked circular arcs of the figure; these will represent N symmetry circles in the polygon net. The generating system deviates from the canonical one, insofar as the substitution $V_2' = V_1 V_2$ moves into the place of V_2.

[60] In order to obtain a knowledge developed as multi-facetly as possible of the most important types of fundamental domains of genus zero, we finally consider such groups Γ, which contain substitutions, which interchange *the two sides on the principal circle*[6].

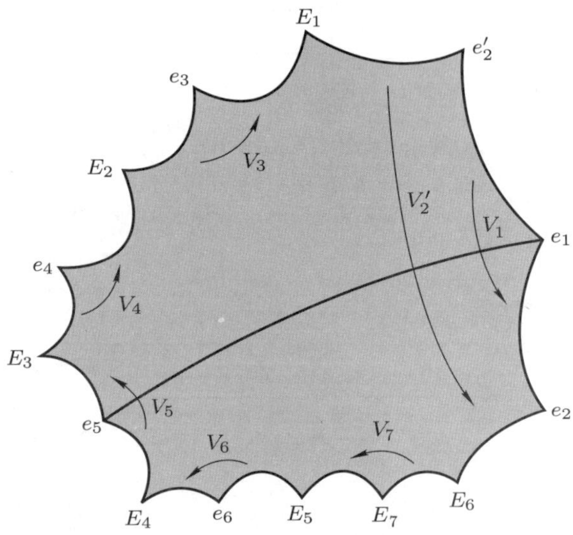

Fig. 18.

An individual group Γ of this category contains a distinguished subgroup Γ_2 of index 2, whose substitutions always transform the individual sides of the principal circle only into itself again. In the ζ-plane, this Γ_2 has an orthosymmetric polygon, which is intersected by the principal circle in μ segments. It is again a question of a polygon of the character (p, n) of that category, which was discussed above on p. 31 under 2., and to which we also just attached; if the character of the associated projective polygon is (p', n'), the relations (4) on p. 32 hold for both characters and the number μ. In the special case lying before us now, the orthosymmetric polygon resp. the polygon net N of the Γ_2 now admits, besides the symmetric transformation in the principal circle, one more transformation of the *first* kind of period two, which interchanges the two symmetric polygon halves, and whose addition leads to the group Γ. The individual half thus furnishes a discontinuity domain for Γ, whereby the μ principal circle segments
[61] are either assigned to one another in pairs, or are also able to be mapped onto them-

[6] These groups, by the classification in I on p. 164 ff., are named under III c 2 β.

selves individually. The immediate perception now informs us, that this discontinuity domain *is always and only then of genus 0, if first, the genus p' of the projective polygon is equal to 0, and secondly, each of the μ segments is coordinated to itself.*

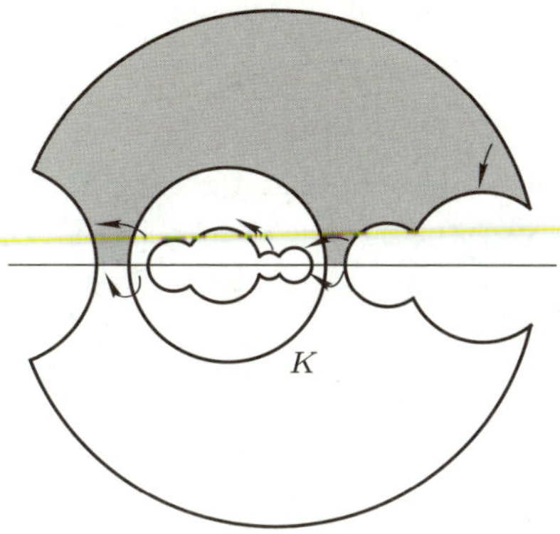

Fig. 19.

Let the polygon shaded in figure 19 furnish us a first example. Here $\mu = 2$, and the circle denoted in the figure by K arranges itself next to the principal circle as a second symmetry circle. The transformation of the first kind of period 2 named above is that elliptic substitution of this period, which has, as fixed points, the intersection points of the circle K with the principal circle.

An even more interesting example is the following: We start out from a discontinuity domain of the second kind, which is bounded by $(p+1)$ separately running, full circles cutting the principal circle orthogonally (in figure 20, the circles in question are called K_1, K_2, K_3, K_4, and $p = 3$).

The reflections $\overline{V}_1, \overline{V}_2, \ldots, \overline{V}_{p+1}$ in these $(p+1)$ full circles are the generators of the associated group, which we call $\overline{\Gamma}_2$. This group permits the extension by addition of the reflection $\overline{V_0}$ in the principal circle and furnishes in such a way, the group $\overline{\Gamma}$, as whose [62]
discontinuity domain we choose, say, the half of the previous domain shaded in figure 20. The most comprehensive subgroup Γ of the first kind contained in $\overline{\Gamma}$ we are able to generate from:

$$\overline{V}_0\overline{V}_{p+1}, \overline{V}_1\overline{V}_{p+1}, \ldots, \overline{V}_p\overline{V}_{p+1}.$$

The boundary of the associated fundamental domain of the first kind is marked heavily in figure 20; we recognize immediately, *that we have here a principal circle group of genus zero with the signature:*

$$(0, 2p+2; 2, 2, \ldots, 2),$$

of whose generators, the first namely $\overline{V}_0\overline{V}_{p+1}$, interchange the two sides of the principal circle.[7]

[7] In order, moreover, to invest the fundamental domain of the group Γ utilized in the text with the canonical shape, one will proceed according to the prescription of I on p. 305. One will

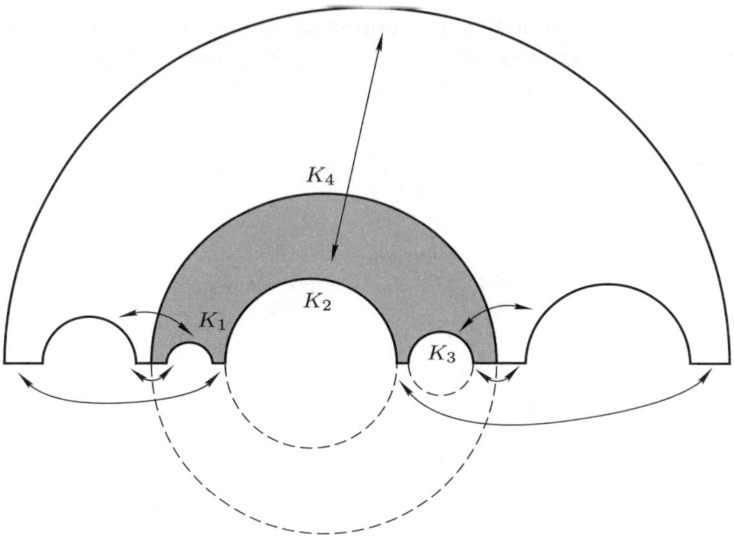

Fig. 20.

If we leave out $\overline{V}_0\overline{V}_{p+1}$, then $\overline{V}_1\overline{V}_{p+1},\ldots,\overline{V}_p\overline{V}_{p+1}$ form the p generators of the maximal subgroup of the *first* kind contained in $\overline{\Gamma}_2$. Let the latter be called Γ_2; we know, that it exhibits the genus p. Obviously, as a distinguished subgroup of index 2, Γ_2 is
[63] contained in the group Γ of genus 0. We therefore make note for later the result: *To the group* Γ_2 *of genus p belongs an automorphic model of "hyperelliptic" character.*

Furthermore, this entire development also lets itself be generalized to the case that the symmetry circles, placed at the forefront which bound discontinuity domains of the second kind, touch one another or even intersect; naturally, the intersection angles thereby coming up must be aliquot parts of π. The conclusion of the hyperelliptic character of the associated group Γ also remains subsisting here.

For many of the groups considered, there are, moreover, besides the primary relations between the generators, also even *secondary* relations. Thus, in the case of the domain standing on the right in figure 14, one has the secondary relation:

$$(V_2V_1)^3 = 1, \tag{3}$$

whose significance emerges from the polygon net drawn in I on p. 435. For the groups illustrated in figures 15 ff., there subsist even $(n'-1)$ secondary relations; for the equations (2) on p. 58 are to be conceived as such. Completely generally, one also has secondary relations for the groups last discussed and illustrated by figures 19 and 20. In order, e.g., to spend time with the case illustrated by figure 20, here are $(p+1)$ secondary relations, corresponding to the $(p+1)$-fold connectivity of the discontinuity domain of the group denoted by $\overline{\Gamma}_2$. The situation is this, that the $(2p+2)$ fixed vertices of a canonical polygon of the group Γ in question, coordinated in pairs, furnish the fixed points

choose a point E in one of the symmetric halves of the previous domain, carry out from here $n = 2p+2$ cuts to the vertices of this half-domain and hang the n pieces of this half arising onto the other half-domain in the correct order.

of $(p+1)$ elliptic substitutions of period 2. The $(2p+2)$ generators are simply identical in pairs here, and in these $(p+1)$ identities, one possesses the secondary relations in question. As one sees, these relations never cause particular difficulties in the concrete cases, where they might occur. Only these concrete cases are so variously fashioned among one another, that it appears difficult, to make general statements.

2.2 Recapitulation of homogeneous variables, substitutions and groups

The developments in I on p. 194 ff. were denoted to the introduction of the homogeneous variables ζ_1, ζ_2 and to the discussion of the homogeneous form of the substitutions and groups. The points of the development at that time most essential for the investigation following next are to be briefly gathered together here.

The *homogeneous variables* ζ_1, ζ_2, whose quotient is $\frac{\zeta_1}{\zeta_2} = \zeta$, are to be so restricted initially, that ζ appears *restricted to the polygon net* N of a proposed group Γ. Except for this, the determinations hold, which one will always make when operating with homogeneous variables: *Neither* ζ_1 *nor* ζ_2 *is able to become infinite,* and also, a *simultaneous vanishing of both variables* is to be *excluded.* The value pairs ζ_1, ζ_2, admissible after this,we call from now on, for short, "*allowed value pairs*". We are able to satisfy these prescriptions, e.g., in this manner, by setting: [64]

$$\zeta_1 = \frac{\zeta\tau}{1+|\zeta|}, \quad \zeta_2 = \frac{\tau}{1+|\zeta|}, \tag{1}$$

understanding by τ an arbitrary complex number, which however, is to be consistently different from 0 and ∞; thereby, in the usual way, by $|\zeta|$ is meant the the absolute value of ζ, and ζ is to be restricted to the net N. Furthermore, we have only mentioned the scheme (1) here in passing.

The *homogeneous* ζ*-substitutions:*

$$\zeta_1' = \alpha\zeta_1 + \beta\zeta_2, \quad \zeta_2' = \gamma\zeta_1 + \delta\zeta_2, \tag{2}$$

which, as in I on p. 194, we denote symbolically by U, are to be formed *unimodularly,* i.e., have the determinant $\alpha\delta - \beta\gamma = 1$. To the individual non-homogeneous substitution V then always corresponds two unimodular homogeneous U, which go over into one another by simultaneous sign change of the four coefficients. We expressed this l.c. thus, *that the non-homogeneous group* Γ *is 1-2-valued homomorphic to the associated unimodular homogeneous group.* The theorem holds, that admissible value systems ζ_1, ζ_2 are always carried over into just such value system again by the substitutions of the homogeneous group.

The form of the *generators* of the homogeneous group is determined in I on p. 198. If it is a question of an *elliptic* substitution, which belongs to the polygon vertex ε_k, then we also imagine this particular value $\zeta = \varepsilon_k$ represented as a quotient of $\varepsilon_k^{(1)}$ and $\varepsilon_k^{(2)}$, and understand by (ζ, ε_k) the expression:

$$(\zeta, \varepsilon_k) = \zeta_1\varepsilon_k^{(2)} - \zeta_2\varepsilon_k^{(1)}. \tag{3}$$

For the second fixed point ε'_k of the substitution in question, we correspondingly set $\varepsilon'_k = \varepsilon'^{(1)}_k : \varepsilon'^{(2)}_k$ and form, by analogy with (3), the expression (ζ, ε'_k). Then we have for the elliptic generator U_k:

$$(\zeta', \varepsilon_k) = e^{\frac{\pi i}{l_k}} (\zeta, \varepsilon_k), \quad (\zeta', \varepsilon'_k) = e^{\frac{-\pi i}{l_k}} (\zeta, \varepsilon'_k), \tag{4}$$

[65] if l_k is the period of V_k. In this form, U_k is unimodular and furnishes, recalculated into the form (2), $\alpha + \delta = 2\cos\frac{\pi}{l_k}$.

The limiting case $l_k = \infty$ leads to the homogeneous form of a parabolic generator U_k, for which, following I on p. 198, we would in fact assume $\alpha + \delta = 2$. In agreement with this, in case ε_k is the parabolic polygon vertex, one will initially obtain, from the form (5) on p. 6, for V_k, as a homogeneous substitution U_k:

$$\zeta'_1 - \varepsilon_k \zeta'_2 = \zeta_1 - \varepsilon_k \zeta_2, \quad \zeta'_2 = \gamma \zeta_1 + (1 - \varepsilon_k \gamma) \zeta_2.$$

If here, besides ε_k, one arbitrarily introduces yet a second fixed point $\zeta = \varepsilon'_k$, different from ε_k, then one can also calculate U_k around into:

$$\zeta'_1 - \varepsilon_k \zeta'_2 = \zeta_1 - \varepsilon_k \zeta_2, \quad \zeta'_1 - \varepsilon'_k \zeta'_2 = (\varepsilon_k - \varepsilon'_k)\gamma(\zeta_1 - \varepsilon_k \zeta_2) + (\zeta_1 - \varepsilon'_k \zeta_2).$$

We now split ε_k arbitrarily into the quotient $\varepsilon^{(1)}_k : \varepsilon^{(2)}_k$, but will then control a corresponding splitting of ε'_k in this way, by setting:

$$\varepsilon'^{(2)}_k = \frac{2i\pi \varepsilon^{(2)}_k}{\gamma(\varepsilon_k - \varepsilon'_k)}, \quad \varepsilon'^{(1)}_k = \varepsilon'_k \cdot \varepsilon'^{(2)}_k.$$

Then, as one easily calculates, it follows, as the homogeneous form of the parabolic generator U_k:

$$(\zeta', \varepsilon_k) = (\zeta, \varepsilon_k), \quad (\zeta', \varepsilon'_k) = 2i\pi(\zeta, \varepsilon_k) + (\zeta, \varepsilon'_k). \tag{5}$$

We shall, as a rule, make use, further on, of this representation of the parabolic substitution.

The *primary relation* (1) on p. 56 between the $V_1, V_2, \ldots, V_n$, according to I on p. 202, has the homogeneous form:

$$U_1 U_2 U_3 \cdots U_n = (-1)^n, \tag{6}$$

whereby the homogeneous substitution $\zeta'_1 = -\zeta_1, \zeta'_2 = -\zeta_2$ is symmetrically denoted by -1. The nature of possible *secondary* relations between the generators, as we discussed at the close of the first paragraph for the examples discussed there, is clarified in general in I on p. 173 ff. For a group with only one boundary circle or a single non-analytic boundary curve, as is known, such relations do not occur. If we have a group with secondary relations, then we shall easily recalculate this into the homogeneous form(cf. below p. 95), and we note for ourselves, for this form, the schema:

$$\begin{cases} \Pi_1(U_1, U_2, \ldots, U_n) = (-1)^{m_1}, \\ \cdots\cdots\cdots\cdots \\ \Pi_s(U_1, U_2, \ldots, U_n) = (-1)^{m_s}, \end{cases} \tag{7}$$

whereby the values on the right side are able to be determined without trouble for each concrete group. [8] [66]

2.3 General definition of the automorphic forms

After the preparatory developments of the last two paragraphs, we go on to the *definition of the automorphic forms*. Moreover, we are thereby able, if nothing is said to the contrary, to consider as proposed, *at the same time an automorphic model of arbitrary genus*, in order not to have to repeat the same definitions later for $p > 0$.

Among the properties, which we want to consider as characteristic for the automorphic forms, we propose the following: *An automorphic form $\varphi(\zeta_1, \zeta_2)$ is to be a homogeneous function of integral or rational fractional positive or negative dimension; thereby, $\varphi(\zeta_1, \zeta_2)$ is to be defined for the entire domain of allowed value pairs of the variables ζ_1, ζ_2, and be everywhere unbranched in this domain.* We call the dimension of the form $\varphi(\zeta_1, \zeta_2)$ in ζ_1, ζ_2, d for short, and also denote the form more precisely by $\varphi_d(\zeta_1, \zeta_2)$. However, we shall fix the concept of an "unbranched" form as follows: Let $\zeta_1^{(0)}, \zeta_2^{(0)}$ be two particular values with a quotient $\zeta^{(0)} = \zeta_1^{(0)} : \zeta_2^{(0)}$ which furnishes a point in the interior of the underlying polygon net. Further, let α_1, α_2 be two complex constants, for which:

$$\alpha_1 : \alpha_2 \neq \zeta_1^{(0)} : \zeta_2^{(0)}$$

holds; the determinant $(\zeta, \alpha) = \zeta_1\alpha_2 - \zeta_2\alpha_1$ will then not vanish for the value pair $\zeta_1^{(0)}, \zeta_2^{(0)}$. The requirement, that $\varphi_d(\zeta_1, \zeta_2)$ be unbranched at the place $\zeta_1^{(0)}, \zeta_2^{(0)}$, is then simply to be defined in this way, that:

$$(\zeta, \alpha)^{-d}\varphi_d(\zeta_1, \zeta_2) \tag{1}$$

is a *function* of ζ unbranched at $\zeta = \zeta^{(0)}$.

Now describe, for any initial pair ζ_1, ζ_2, inside the domain of admissible values ζ_1, ζ_2, a continuous path to a value pair equivalent with the initial pair with respect to the homogeneous unimodular group: [67]

$$\zeta_1' = \alpha\zeta_1 + \beta\zeta_2, \quad \zeta_2' = \gamma\zeta_1 + \delta\zeta_2.$$

Along this path, $\varphi(\zeta_1, \zeta_2)$ is uniquely continuable, and indeed, the automorphic form is to again assume, at the end of the path, its initial value up to a multiplicative constant:

$$\varphi(\alpha\zeta_1 + \beta\zeta_2, \gamma\zeta_1 + \delta\zeta_2) = \mu \cdot \varphi(\zeta_1, \zeta_2). \tag{2}$$

---

[8] There is now a special case, in which the $U_1, U_2, \ldots, U_n$ do not yet generate the entire homogeneous group. If, namely, n is even and the $U_1, U_2, \ldots, U_n$ are parabolic throughout, and if, besides this, either there are no secondary relations (7) at all or only such, whose right sides consistently read $+1$, then it is in no way possible, from the $U_1, U_2, \ldots, U_n$, by repetition or combination, to produce the homogeneous substitution -1. If, in this case, one will form the entire homogeneous group, then it is necessary, to yet expressly add the substitution -1 to the n homogeneous generators $U_1, U_2, \ldots, U_n$.

Herein, vis-a-vis the determinations followed for the automorphic functions, there lies a generalization in two directions. First, namely, while the automorphic functions considered so far are absolutely invariant with respect to the substitutions of the underlying group Γ, for the automorphic forms, we also allow multiplicators μ different from 1 which, moreover, we shall make even more precise soon. *Above all, however, we secondly completely ignore, initially, the requirement of the single valuedness of the forms $\varphi(\zeta_1,\zeta_2)$ in ζ_1,ζ_2.* If, in the value domain of the variable pair, we describe a closed circuit, then, upon continuation over this, $\varphi(\zeta_1,\zeta_2)$ may go over into itself, multiplied by a multiplicator μ, different from 1.

It is important to consider, that the requirement set on the automorphic forms, to be unbranched, does not contradict the multiple-valuedness of the $\varphi(\zeta_1,\zeta_2)$ at all. The theorem even holds, *that an automorphic form of fractional dimension d is multiple-valued under all circumstances.* Namely, if we substitute, for ζ_1 and ζ_2, their expressions (1) on p. 64 into:

$$\varphi(\zeta_1,\zeta_2) = \tau^d \varphi\left(\frac{\zeta}{1+|\zeta|}, \frac{1}{1+|\zeta|}\right), \tag{3}$$

then for ζ held fixed, we have to do with a function of τ alone. But τ is now an arbitrary complex variable with the exclusion of the two values $\tau = 0$ and $\tau = \infty$; τ is therefore restricted to a *two fold* connected domain. In the interior of the latter, τ^d is indeed unbranched for fractional d, but not single-valued, rather it assumes the factor $e^{2i\pi d}$, different from 1, upon a circuit about the point $\tau = 0$.

Conceived in this generality, the automorphic forms shall accomplish for us the same thing, as the modular forms inside the theory of the modular functions. We effected the transition to the functions at that time, by producing quotients of modular forms of the same dimension. Reserving further determinations concerning the forms $\varphi(\zeta_1,\zeta_2)$ we state here, at the same time, the analogous theorem: *The quotient of two automorphic forms of our proposed model, which have the same dimension, and which, for equal paths from ζ_1,ζ_2 to the equivalent value pairs ζ_1',ζ_2', assume the same multi-*
[68] *plicator μ, is an automorphic function of the model.*

We had also assigned the automorphic functions the property, that they were not to be essentially singular at any place of the fundamental domain. We brought this requirement to expression above(p. 5 ff.), by demanding the developability into a series of the forms (3) on p. 5, for an automorphic function $\varphi(\zeta)$, at each individual place ζ_0 of the fundamental domain, where the development quantity t is more closely defined there.

We will now proceed in completely corresponding fashion for the automorphic forms.

If we first consider any place of the fundamental domain, *which is different from an elliptic or parabolic vertex,* then we set:

$$t = \zeta - \zeta_0 \quad \text{or} \quad t = \frac{1}{\zeta}, \tag{4}$$

according as the chosen place lies in the finite part of the place at ζ_0 or infinitely distant. We then form the product set down before under (1) according to the prescription given there. This is only dependent on the quotient $\zeta_1 : \zeta_2 = \zeta$, and is to be either regular or only polarly discontinuous at the place considered. It follows: *An automorphic* *form $\varphi_d(\zeta_1,\zeta_2)$ is to permit, at each place of the fundamental domain different from a fixed vertex, a representation:*

$$\varphi_d(\zeta_1,\zeta_2) = (\zeta,\alpha)^d t^m (a_0 + a_1 t + a_2 t^2 + \cdots), \tag{5}$$

convergent in the neighborhood of this place, where a_0 is different from zero, and m signifies a finite whole number. If $m \neq 0$, then we say that $\varphi(\zeta_1,\zeta_2)$ has a zero of mth order or a pole of $(-m)^{\text{th}}$ order at the place considered, according as $m > 0$ or $m < 0$.

If, secondly, we have an *elliptic* vertex ε, then again, let ε' be the second fixed point of the associated substitution. By using the symbol (ζ,ε') in the sense on p. 64, we now form the product, depending only on the quotient $\zeta_1 : \zeta_2 = \zeta$:

$$(\zeta,\varepsilon')^{-d}\varphi_d(\zeta_1,\zeta_2).$$

This is to represent a function of ζ unbranched at the point ε; there will accordingly exist a development, convergent in the neighborhood of $\zeta = \varepsilon$:

$$\varphi_d(\zeta_1,\zeta_2) = (\zeta,\varepsilon')^d \left[a_0 \left(\frac{\zeta-\varepsilon}{\zeta-\varepsilon'}\right)^m + a_1' \left(\frac{\zeta-\varepsilon}{\zeta-\varepsilon'}\right)^{m+1} + a_2' \left(\frac{\zeta-\varepsilon}{\zeta-\varepsilon'}\right)^{m+2} + \cdots \right],$$

where, $a_0 \neq 0$, and m signifies a finite whole number. Now, upon exercise of the elliptic substitution coming into question: [69]

$$(\zeta',\varepsilon) = e^{\frac{\pi i}{l}}(\zeta,\varepsilon), \quad (\zeta',\varepsilon') = e^{\frac{-\pi i}{l}}(\zeta,\varepsilon'), \tag{6}$$

$\varphi_d(\zeta_1,\zeta_2)$ is now to be reproduced up to a multiplicative constant, μ. Consequently, $(\zeta,\varepsilon')^{-d}\varphi_d(\zeta_1,\zeta_2)$ will also behave itself similarly. Now, however, on account of $a_0 \neq 0$, the behavior of $(\zeta,\varepsilon')^{-d}\varphi_d$ upon exercise of (6), is easy to deduce directly from the initial term of the series development written down; one finds, *that $(\zeta,\varepsilon')^{-d}\varphi_d$ assumes the root of unity $e^{\frac{2i\pi m}{l}}$ as a factor with respect to (6).* It follows, that in the series development put down for φ_d, only the exponents $m, m+l, m+2l, \ldots,$ will make their appearance, and we arrive at the result: *In the neighborhood of an elliptic vertex ε, an automorphic form $\varphi_d(\zeta_1,\zeta_2)$ permits a convergent development of the form:*

$$\varphi_d(\zeta_1,\zeta_2) = (\zeta,\varepsilon')^d \left(\frac{\zeta-\varepsilon}{\zeta-\varepsilon'}\right)^m \left[a_0 + a_1 \left(\frac{\zeta-\varepsilon}{\zeta-\varepsilon'}\right)^l + a_2 \left(\frac{\zeta-\varepsilon}{\zeta-\varepsilon'}\right)^{2l} + \cdots \right], \tag{7}$$

where a_0 is different from zero and m signifies a finite whole number. Upon utilization of the development quantity:

$$t = \left(\frac{\zeta-\varepsilon}{\zeta-\varepsilon'}\right)^l, \tag{8}$$

formula (7) lets itself be written thus:

$$\varphi_d(\zeta_1,\zeta_2) = (\zeta,\varepsilon')^d t^{\frac{m}{l}} (a_0 + a_1 t + a_2 t^2 + \cdots). \tag{9}$$

We shall ascribe $\varphi_d(\zeta_1,\zeta_2)$ "in the fundamental domain", at the corresponding place e, a zero of order $(\frac{m}{l})$ resp. a pole of order $(\frac{-m}{l})$, in case m is not equal to zero, but signifies a positive resp. negative number. we return below to the precise determination of the multiplicator of the form $\varphi_d(\zeta_1,\zeta_2)$ corresponding to the substitution (6). That, furthermore, zeros of "fractional" order(measured in the fundamental domain) come up here, ought not be surprising. A zero of order $\frac{m}{l}$, measured in ζ, receives the integral order m.

The relations are somewhat different for a *parabolic vertex* ε. The requirement of the non-branchedness of φ_d refers itself only to the places in the interior of the polygon net, and therefore does not come into consideration for parabolic vertices. For
[70] the formation of a function of $\zeta_1:\zeta_2=\zeta$ alone, we shall accordingly have to deviate here from the rule followed in the production of the product (1), to the extent, that we utilize the determinant (ζ,ε), which vanishes at the place ε. In consequence of (5), $(\zeta,\varepsilon)^{-d}\varphi_d(\zeta_1,\zeta_2)$ assumes any multiplicator at all, upon exercise of the parabolic substitution. Accordingly, if, tying in to (5) on p. 65 and with inessential deviation from (6), we put down the development quantity t in the form:

$$t=e^{\frac{(\zeta,\varepsilon')}{(\zeta,\varepsilon)}}=e^{\frac{2i\pi}{\gamma(\varepsilon-\varepsilon')}}\cdot e^{\frac{2i\pi}{\gamma}\frac{1}{\zeta-\varepsilon}}, \tag{10}$$

then $(\zeta,\varepsilon)^{-d}\varphi_d(\zeta_1,\zeta_2)$ will be a function of t, which assumes a factor μ upon a circuit about $t=0$. *Here, we shall best prepare for future developments, if we give the restrictive prescription, that a multiplicator μ, coming up for a parabolic substitution, is always to be a root of unity:*

$$\mu=e^{\frac{2i\pi m}{l}}, \tag{11}$$

whose finite degree l is to be subjected here, initially, to no further restriction. To this determination corresponds the series development:

$$\varphi_d(\zeta_1,\zeta_2)=(\zeta,\varepsilon)^d\,t^{\frac{m}{l}}(a_0+a_1t+a_2t^2+\cdots)$$

or, explicitly:

$$\varphi_d(\zeta_1,\zeta_2)=(\zeta,\varepsilon)^d\left(e^{\frac{(\zeta,\varepsilon')}{(\zeta,\varepsilon)}}\right)^{\frac{m}{l}}\left[a_0+a_1e^{\frac{(\zeta,\varepsilon')}{(\zeta,\varepsilon)}}+a_2e^{\frac{2(\zeta,\varepsilon')}{(\zeta,\varepsilon)}}+\cdots\right], \tag{12}$$

where, repeating ourselves, l is a positive whole number not yet more closely determined, m signifies an arbitrary whole number and a_0 is different from zero. If m is not equal to zero, then we say, that φ_d, at the place of the fundamental domain which corresponds to the vertex ε, has a zero of order $\frac{m}{l}$ or a pole of order $(\frac{-m}{l})$, according as $m>0$ or <0. Notice, that here, for the enumeration of the order of the zero resp. infinity, we have ignored the vanishing of the factor (ζ,ε), which is given at $t=0$, approximately by:

$$(\zeta,\varepsilon)=\frac{(\varepsilon,\varepsilon')}{\log t}.$$

This circumstance, in the calculations following next, as well as later on, for a convergence consideration in the theory of the Poincaré series, obtains its particular significance.

2.4 The differentiation process and the principal forms of the models of genus zero

In the actual construction of automorphic forms, the method followed in "M" I on p. 117 for the production of modular forms is to be exemplary for us. From the modular function J, we produced the fundamental modular forms g_2, g_3, Δ l.e. by a "differentiation process". One can hereby designate as a basic fact, that the *"differentiation form" of second dimension:* [71]

$$(\zeta, d\zeta) = \zeta_1 d\zeta_2 - \zeta_2 d\zeta_1 = -\zeta_2^2 d\zeta, \tag{1}$$

upon exercise of our substitution, goes over into

$$(\zeta', d\zeta') = (\alpha\delta - \beta\gamma)(\zeta, d\zeta),$$

and is *therefore absolutely invariant with respect to the "unimodular" substitutions of our homogeneous group* Γ.

From this, there arises a "*differentiation process*", by means of which we produce automorphic forms from the functions $\varphi(\zeta)$. If $\varphi(\zeta)$ is any automorphic function of the proposed model, then we obtain, in:

$$\varphi_{-2}(\zeta_1, \zeta_2) = \frac{d\varphi(\zeta)}{(\zeta, d\zeta)} = \frac{1}{-\zeta_2^2}\frac{d\varphi(\zeta)}{d\zeta}, \tag{2}$$

an automorphic form of the $(-2)^{\text{nd}}$ *dimension of this model, which also has the particular property of being absolutely invariant with respect to the homogeneous group.*

Up until now, the development, since the beginning of the third paragraph is valid for automorphic models of each genus. Since we now return *again exclusively to genus zero,* let us apply the differentiation process to a *principal function* $z = \varphi(\zeta)$ of the model. Let the arising form:

$$\varphi_{-2}(\zeta_1, \zeta_2) = \frac{dz}{(\zeta, d\zeta)} = \frac{1}{-\zeta_2^2}\frac{dz}{d\zeta} \tag{3}$$

be designated as a *principal form* of the model.

In order to represent the behavior of this principal form more precisely in the fundamental domain P, we introduce, for the individual place of this domain, the associated development quantity t and put the principal form into the form:

$$\varphi_{-2}(\zeta_1, \zeta_2) = \frac{-1}{\zeta_2^2}\frac{dz}{dt}\frac{dt}{d\zeta}. \tag{4}$$

the factor $\frac{dz}{dt}$ has a pole of the second order at the place $z = \infty$, but is otherwise finite and different from zero in the entire fundamental domain, including the fixed vertices. For the sake of simplicity suppose the pole of z not to lie right at a fixed vertex of the domain P. [72]

The factor $\frac{dt}{d\zeta}$, apart from the fixed elliptic or parabolic vertices and the point $\zeta = \infty$ possibly belonging to the domain, is always finite and different from zero. But at the place $\zeta = \infty$, one easily recognizes:

$$-\frac{dt}{\zeta_2^2 d\zeta} = \frac{dt}{(\zeta, d\zeta)}$$

as a finite non-vanishing form. There remain, therefore, only the fixed polygon vertices to investigate. For an elliptic vertex ε, we have:

$$\frac{dt}{d\zeta} = l\left(\frac{\zeta-\varepsilon}{\zeta-\varepsilon'}\right)^{l-1}\frac{\varepsilon-\varepsilon'}{(\zeta-\varepsilon')^2} = \frac{l(\varepsilon-\varepsilon')}{(\zeta-\varepsilon')^2}t^{(1-\frac{1}{l})},$$

while for a parabolic vertex ε, one is correspondingly led to the equation:

$$\frac{dt}{d\zeta} = -\frac{2i\pi}{\gamma}\frac{1}{(\zeta-\varepsilon)^2}\cdot e^{\frac{2i\pi}{\gamma(\varepsilon-\varepsilon')}}\cdot e^{\frac{2i\pi}{\gamma}\frac{1}{\zeta-\varepsilon}} = -\frac{2i\pi}{\gamma}\cdot\frac{1}{(\zeta-\varepsilon)^2}t.$$

Now in the parabolic case, in the estimation of the order of the zero, the factor $(\zeta-\varepsilon)^2$ in the denominator does not come into consideration(comp. the close of the previous paragraph). By conceiving both cases together, we are thus able to state, that $\frac{dt}{d\zeta}$ receives a zero of order $(1-\frac{1}{l})$ at the individual fixed vertex, whereby, for a parabolic vertex, l is to be taken to ∞.

By recapitulating, we have obtained the theorem: *The individual principal form $\varphi_{-2}(\zeta_1,\zeta_2)$ of our model of genus zero has n zeros at the n fixed vertices of the fundamental domain P, and indeed, the zero has order* $(1-\frac{1}{l_k})$ *at the vertex l_k; further, $\varphi_{-2}(\zeta_1,\zeta_2)$ has a pole of order* 2, *which coincides with the pole of the underlying principal function $z=\varphi(\zeta)$. Except for this, however, our principal form is finite and different from zero at all places, of the fundamental domain.*

2.5 The family of prime forms and the ground forms for automorphic models with $p=0$

[73] If, as earlier, we denote the value of the principal function z at the vertex ε_k by e_k, then the function

$$\prod_{k=1}^{n}(z-e_k)^{1-\frac{1}{l_k}}$$

will obviously have precisely the same zeros as the principal form $\varphi_{-2}(\zeta_1,\zeta_2)$, and also agree with the latter in the order of each individual zero, the product:

$$\varphi_{-2}(\zeta_1,\zeta_2)\cdot\prod(z-e_k)^{-(1-\frac{1}{l_k})} \tag{1}$$

will consequently be everywhere finite and different from zero in the fundamental domain, apart from the one pole of the principal function z.

But in order to determine the behavior of the product (1) at the discontinuity place of z, we have to form the expression:

$$2-\sum_{k=1}^{n}\left(1-\frac{1}{l_k}\right)=\frac{2}{\nu}. \tag{2}$$

The number ν defined by the right side of this equation is just that one, which was denoted by N in "Icos." on p. 117 ff., and which had the values $2n, 12, 24$ and 60 for the four groups of the regular solids. For those automorphic models of genus zero, which are furnished by parabolic rotation groups, $\nu=\infty$(cf. I on p. 223 ff); this case however, counts as excluded for the next considerations. *For the higher automorphic models with infinitely many boundary points, interesting us first and foremost, ν is always a negative whole or rational number.*

The product (1), at the place named, now obviously has a pole of order $\frac{2}{\nu}$ resp. a zero of order $(-\frac{2}{\nu})$. The $(-\frac{\nu}{2})^{\text{th}}$ power of the product will consequently have a *zero of first order* there. Since, besides this, neither zero, nor poles come up, *we shall have obtained, in that power:*

$$\left\{\frac{dz}{(\zeta,d\zeta)}\prod_{k=1}^{n}(z-e_k)^{-(1-\frac{1}{l_k})}\right\}^{-\frac{\nu}{2}}, \tag{3}$$

an unbranched automorphic form of dimension ν, which is free from poles in the entire fundamental domain P, and exhibits only a single zero of the first order, namely, at the discontinuity place of z. We denote the form (3) so obtained by $z_2(\zeta_1,\zeta_2)$ from now on. [74]

A form, which is everywhere finite in the fundamental domain and has a single zero there of the first order, is to be designated as a "*prime form*". We have a first such prime form in $z_2(\zeta_1,\zeta_2)$. Obviously, however, we also have a prime form in:

$$z_1(\zeta_1,\zeta_2)=z(\zeta)\cdot z_2(\zeta_1,\zeta_2), \tag{4}$$

whose zero coincides with the zero of the principal function z, and which behaves itself exactly as $z_2(\zeta_1,\zeta_2)$ with respect to the operations of the homogeneous group. *On the strength of this, however, we arrive, at the same time, in:*

$$az_1(\zeta_1,\zeta_2)+bz_2(\zeta_1,\zeta_2)=(az+b)\cdot z_2, \tag{5}$$

at a binary family of prime forms of the dimension ν, which all behave themselves as $z_2(\zeta_1,\zeta_2)$ with respect to the homogeneous substitutions $U_1,U_2,\dots$, and whose zero is freely movable in the fundamental domain with the variable parameter $a:b$. The quotient of any two forms of the family, which have different zeros, then conversely, again furnishes a principal function of the model.

Corresponding to the decomposition of z into the quotient $z_1:z_2$, the value e_k of the principal function z at the individual fixed vertex is now also to be split into $e_k^{(1)}:e_k^{(2)}$. We then have, in:

$$(z,e_k)=z_1e_k^{(2)}-z_2e_k^{(1)}, \tag{6}$$

a form of the binary family (5), which vanishes simply at the vertex ε_k in the fundamental domain P. If it is a question of an elliptic vertex, then (z,e_k), measured in the polygon net, will have a zero of order l_k at this place. The consequence is, *that even:*

$$Z_k(\zeta_1,\zeta_2)=\sqrt[l_k]{(z,e_k)} \tag{7}$$

is also an unbranched form of our automorphic model. We will designate this as the "*ground form*" belonging to the elliptic vertex ε_k. In the case of a parabolic vertex ε_k, $(z,e_k)^{\lambda_k}$ is unbranched with *arbitrary real rational or irrational exponents* λ_k. To the ground form (7) of an elliptic vertex would properly correspond here the function [75]

$\log(z, e_k)$, which is similarly unbranched in the value domain of the ζ_1, ζ_2. However, we remain in agreement with our determinations on p. 70, *if we also define a ground form $Z_k(\zeta_1, \zeta_2)$ for the individual parabolic vertex ε_k as in (7), by thereby understanding by l_k a positive whole number, concerning which we dispose of at pleasure from case to case.*

Between each three forms of the binary family (5), there obviously subsists a linear relation. But now, in case Z_h, Z_i, Z_k are any three of our n ground forms, $Z_h^{l_h}, Z_i^{l_i}, Z_k^{l_k}$ are three forms of the family named. In fact, one then also shows immediately, *that between Z_h, Z_i, Z_k, the relation subsists:*

$$(e_i, e_k) Z_h^{l_h} + (e_k, e_h) Z_i^{l_i} + (e_h, e_i) Z_k^{l_k} = 0. \tag{8}$$

2.6 Behavior of the automorphic forms $\varphi_d(\zeta_1, \zeta_2)$ with respect to the group generators

For the further development, it will now be of fundamental importance, that we determine the *behavior of the automorphic forms $\varphi_d(\zeta_1, \zeta_2)$, especially the prime and ground forms obtained, with respect to the unimodular substitutions $U_1, U_2, \ldots, U_n$.* Thereby, since it is a question in general of multiple-valued forms, we have always to also give the *path*, on which the original value pair ζ_1, ζ_2 is to go over into the equivalent one (ζ_1', ζ_2'). In this regard, the following consideration is to be presented.

Since an automorphic form $\varphi_d(\zeta_1, \zeta_2)$ is presentable as a product of ζ_2^d and a function of ζ alone then for the individual generator U, it will be sufficient, that we determine the transition from ζ to ζ' and simultaneously, that from ζ_2 to ζ_2'. One can imagine both paths carried out either simultaneously or one after the other. We begin, by describing a particular transition from ζ, ζ_2 to ζ', ζ_2', which we then generalize afterwards in required fashion.

First, if the point ζ in the initial domain P in the neighborhood of the vertex ε belonging to U, then ζ' will similarly lie close to ε, in the neighboring polygon P'; *we then move ourselves directly along an orbit curve of U from ζ to ζ'.* We assume, for the sake of simplicity, that the values $\varepsilon, \varepsilon'$ are both finite for U[cf. formula (4) on p. 64 and (5) on p. 65]. The initial value ζ_2 will then be $\neq 0$; and we then conclude, from the formulas just cited, for an elliptic substitution:

$$\zeta_2' = e^{\frac{-\pi i}{l}} \frac{\zeta - \varepsilon'}{\zeta' - \varepsilon'} \zeta_2, \tag{1}$$

[76] while, for a parabolic substitution, there correspondingly holds:

$$\zeta_2' = 2i\pi\zeta_2 \frac{\varepsilon^{(2)}}{\varepsilon'^{(2)}} \frac{\zeta - \varepsilon}{\zeta' - \varepsilon'} + \frac{\zeta - \varepsilon'}{\zeta' - \varepsilon'} \zeta_2. \tag{2}$$

Since now, ζ and ζ' are only slightly different from ε, and moreover, $\varepsilon \neq \varepsilon'$, one has approximately, in the first case, $\zeta_2' = e^{-\frac{\pi i}{l}} \zeta_2$, in the second $\zeta_2' = \zeta_2$. *On the strength of this, we simply prescribe the rectilinear path from ζ_2 to ζ_2', which then, in particular, for a parabolic generator, will all the more draw itself together into a point, the closer ζ is chosen to the parabolic vertex.*

In order now to furnish our prescription in the necessary generality, let us set, in detail:

$$\varphi_d(\zeta_1,\zeta_2) = \zeta_2^d \psi(\zeta). \tag{3}$$

The function $\psi(\zeta)$ then, according to p. 66, is unbranched everywhere in the interior of the polygon net N, apart only, say, from the place $\zeta = \infty$ (i.e., from the value $\zeta_2 = 0$), as far as the latter belongs to the interior of the net N at all. If we thus continuously deform the path from ζ to ζ', without pushing it across $\zeta = \infty$ or over a boundary point of the group, and if we similarly modify the path from ζ_2 to ζ_2' continuously, without letting it slide over $\zeta_2 = 0$,[9] then the multiplicator of the form $\varphi_d(\zeta_1,\zeta_2)$, belonging to the initial path, will hereby remain maintained unaltered. The same thing will hold, if we also undertake continuous alterations of ζ and ζ_2 as well, whereby ζ' and ζ_2' experience equivalent alterations. In these alterations, one will initially again avoid only $\zeta = \infty$ resp. $\zeta_2 = 0$.

Moreover, the effect of pushing the path over $\zeta = \infty$, in case this point belongs to the net N, is very easy to follow. Namely, one has, in:

$$\psi'(\zeta) = \zeta_1^{-d}\varphi_d(\zeta_1,\zeta_2) = \zeta^{-d}\cdot\psi(\zeta), \tag{4}$$

a function unbranched at $\zeta = \infty$. Thus, $\psi(\zeta)$ is branched exactly as ζ^d is at this place. It follows, that the multiplicator of the form $\varphi_d(\zeta_1,\zeta_2)$ again remains unaltered, in case, upon pushing the path of ζ over ∞, the path of ζ_2 simultaneously slides over the origin $\zeta_2 = 0$.[10] We will call all the transitions between the equivalent value pairs ζ,ζ_2 and [77]
ζ',ζ_2' resp. ζ,ζ_2 and ζ',ζ_2', arising by means of the alterations discussed so far of the original path from ζ_1,ζ_2 to ζ_1',ζ_2', as "*primitive paths*".

In the case of a simply connected net N, we arrive at the most general paths from the primitive paths, by hanging onto the path of the point ζ_2, yet *an arbitrary number of circuits about the zero $\zeta_2 = 0$ in the possible sense.*[11] Hereby, only the first factor of $\zeta_2^d\psi(\zeta)$ experiences an alteration, which one gives immediately. If now, N is of infinitely high connectivity, then we are also able to add to the path of ζ *circuits about a set of boundary points or families of boundary curves.* However, we ignore the generalization lying herein; for, in fact, we shall be able to determine the effect of such a closed circuit, by combining, according to the direction of the corresponding "*secondary relation*", the primitive paths of the generators taking part in this latter.[12]

Our task now amounts essentially to determining the effect of the path from ζ,ζ_2 to ζ',ζ_2', determined at the beginning of the paragraph, on a proposed form $\varphi_d(\zeta_1,\zeta_2)$. If it is a question of an elliptic substitution U[cf. formula (6) on p. 69], then, in the neighborhood of the associated vertex ε, utilize the representation (7) on p. 69 for $\varphi_d(\zeta_1,\zeta_2)$. An easy investigation shows, that $\varphi_d(\zeta_1,\zeta_2)$, after carrying out the path, has assumed the factor $e^{\frac{-\pi i d}{l}+\frac{2i\pi m}{l}}$. For a parabolic substitution[cf. formula (5) on p. 65], the analytic representation (12) on p. 70 is to be utilized; thereby, then, the multiplicator $e^{\frac{2i\pi m}{l}}$

---

9 Remember, that the value ∞ is always forbidden for ζ_2.

10 that, moreover, the pushing over $\zeta = \infty$ and $\zeta_2 = 0$ always have to enter simultaneously, is already grounded on our original determinations, insofar as the value ∞ is inadmissible for ζ_1.

11 Circuits in the negative sense will be indicated by negative values of σ.

12 Do not neglect, to make quite clear in detail, these relations, by using the figures 16,19 and 20.

comes up. From this, there then arises immediately as a principal result: *If we effect an individual generator U by means of a transition, which arises from a primitive path by addition of σ circuits of the value ζ_2 about its origin, then, according as U is elliptic or parabolic, the individual automorphic form $\varphi_d(\zeta_1, \zeta_2)$ alters itself by the multiplicator:*

$$\mu = e^{\frac{-\pi i d}{l} + \frac{2\pi i m}{l} + 2\pi i d\sigma} \quad \text{resp} \quad \mu = e^{\frac{2i\pi m}{l} + 2i\pi d\sigma} \tag{5}$$

[78] *provided that the form, measured in the fundamental domain, exhibits a zero of order $\frac{m}{l}$ (pole of order $\frac{-m}{l}$) at the associated vertex ε.*[13]

The application of this general theorem to the prime- and ground forms now offers no kind of further difficulty thus, e.g., for the ground form $Z_k(\zeta_1, \zeta_2)$ belonging to the vertex ε_k, we have to substitute the value $\frac{\nu}{l_k}$ for the dimension d; the number m is to be taken equal to 1, in case the substitution U_k itself is exercised, otherwise, one must set $m = 0$, if one of the $(n-1)$ remaining generators is being considered in its effect on $Z_k(\zeta_1, \zeta_2)$. For the family of the prime forms (5) on p. 74, one can always set $m = 0$.

2.7 The ground forms for the groups of the circular-arc triangles

For the illustration of the preceding general investigations, the forms of the circular-arc triangles are to serve, i.e., the automorphic forms of those models, which are to be characterized by the signatures$(0,3;l_1, l_2, l_3)$. We will hereby also consider in passing the models, which, in the language used on p. 21, consist of "elementary" automorphic functions.

I. Groups of the regular solids

The interest, which is taken here in the "forms" of the regular solids, bases itself on the circumstance, that the theorems on the dimensions of the associated ground forms Z_1, Z_2, Z_3, and their behavior with respect to the group generators, find a common expression in the general theorems of the previous paragraph.

First, as far as the *dimensions* go, we always denote these by d_1, d_2, d_3 for the three ground forms, while, as above, ν is the dimension of the prime form. With the aid of
[79] formulas (2) and (7) on p. 73 ff., we then easily calculate ourselves the following table:

Table 2.1.

	Signature	ν	d_1	d_2	d_3
dihedron	$(0,3;2,2,n)$	$2n$	n	n	2
tetrahedron	$(0,3;2,3,3)$	12	6	4	4
octahedron	$(0,3;2,3,4)$	24	12	8	6
icosahedron	$(0,3;2,3,5)$	60	30	20	12

These are now, in fact, the degrees of the *integral rational* forms Z_1, Z_2, Z_3, as they are calculated in "Icos" on p. 47 ff; thus, e.g., in the case of the octahedron, we have:

[13] For forms of *integral* dimension, consequently, arbitrary circuits of the argument ζ_2 about its origin remain without effect, which is also clear at the very beginning.

$$Z_1(\zeta_1,\zeta_2) = \zeta_1^{12} - 33\zeta_1^8\zeta_2^4 - 33\zeta_1^4\zeta_2^8 + \zeta_2^{12},$$
$$Z_2(\zeta_1,\zeta_2) = \zeta_1^8 + 14\zeta_1^4\zeta_2^4 + \zeta_2^8,$$
$$Z_3(\zeta_1,\zeta_2) = \zeta_1\zeta_2(\zeta_1^4 - \zeta_2^4),$$

whereby, the general relation brought into account under (8) on p. 75, according to "Icos" on p. 55, reads here in particular thus:

$$Z_1^2 - Z_2^3 + 108Z_3^4 = 0.$$

The individual ground form $Z_k(\zeta_1,\zeta_2)$, upon *exercise of the unimodular generators* U, now assumes the multiplicator:

$$\mu = e^{\frac{\pi i(2-d_k)}{l}} \quad resp. \quad \mu = e^{-\frac{\pi i d_k}{l}}, \tag{1}$$

according as it is a question of U_k or one of the other two other generators. We thus find, e.g., that the three icosahedral ground forms are absolutely invariant with respect to all generators. Of the octahedral forms, however, only Z_2 is absolutely invariant for all three U; Z_1 experiences a sign change for U_1 and U_3, Z_3 similarly for U_1 and U_3. In "Icos" on pp. 55 and 57 these theorems are obtained by direct calculation.

The ground forms of the regular solids are set up exhaustively by Klein in the paper "*Über binäre Formen mit linearen Transformationen in sich*"[14]. It is thereby a question of the same forms, which were designated and investigated as "prime forms" in the developments by Fuchs on algebraically integrable linear differential equations of the second order[15]).

II. Parabolic rotation groups of genus zero

[80]

For the sake of completeness, let us also briefly pursue, how our general investigations fashion themselves in the case of the three *parabolic rotation groups of the signatures* $(0,3;2,3,6), (0,3;2,4,4), (0,3;3,3,3)$. We had initially excluded these models, since $\nu = \infty$ for them; but in fact, we have to do here with simple relations, as we briefly discuss, say, for *the example of the first case* $(0,3;2,3,6)$.

By placing the boundary point of the group at $\zeta = \infty$, we are able to put the unimodular homogeneous substitutions into the following form:

$$\zeta_1' = e^{\frac{\pi i}{l}}\zeta_1 + \beta\zeta_2, \quad \zeta_2' = e^{-\frac{\pi i}{l}}\zeta_2. \tag{2}$$

The consequence is, that the differentiation process(cf. p. 71) now furnishes, in $\frac{dz}{d\zeta}$, a "*function*", which remains unaltered by the operations of the "non-homogeneous" group, apart from a multiplicative sixth root of unity. The zeros and poles of this function in the fundamental domain(cf. p. 71 ff.), however, show that this is identical up to a constant factor, with:

---

14 Mathem. Annalen. Vol. 9 (1875).

15 Compare the work "*Über die linearen Differentialgleichungen zweiter ordnung, welche algebraische integrale besitzen, und eine neue Anwendung der Invariantentheorie*", Journ. Math., Vol. 81 (1875).

$$(z-e_1)^{\frac{1}{2}}(z-e_2)^{\frac{2}{3}}(z-e_3)^{\frac{5}{6}},$$

whereby, we again assume for the sake of simplicity, that all three values e are finite. We are able to imagine z chosen straight away, so that the equation subsists:

$$\frac{dz}{d\zeta}=(z-e_1)^{\frac{1}{2}}(z-e_2)^{\frac{2}{3}}(z-e_3)^{\frac{5}{6}}. \tag{3}$$

Conversely, ζ then represents itself as an *integral of the first kind of the underlying elliptic model here thus:*[16]

$$\zeta=\int\frac{dz}{\sqrt[6]{(z-e_1)^3(z-e_2)^4(z-e_3)^5}}. \tag{4}$$

It may also be mentioned, that the function:

$$t(\zeta)=\sqrt[3]{\frac{z-e_2}{z-e_3}}, \tag{5}$$

[81] up to a constant factor, represents the $\wp$ function of the elliptic model; in fact, the integral (4) then also easily transforms itself into the form:

$$\zeta=\frac{3}{\sqrt{e_2-e_3}}\int\frac{dt}{\sqrt{e_2-e_1-t^3(e_3-e_1)}},$$

which is no longer essentially different from the normal form of the integral of the first kind in the equianharmonic case.

Groups of the circular-arc triangles with principal circle

In the models to be discussed now, which consist of *higher automorphic functions* in the sense on p. 29 ff., the dimension ν of the prime forms is, in general, a fractional negative number. If we set $\nu=-\nu'$, then, according to (2) on p. 73, one has:

$$\frac{1}{l_1}+\frac{1}{l_2}+\frac{1}{l_3}+\frac{2}{\nu'}=1. \tag{6}$$

From this equation here arises, first of all, the requirement, that there are only *finitely* many cases with integral values ν'; in fact, one finds by direct discussion of equation (6), *that here are 98 models all together of the signature* $(0,3;l_1,l_2,l_3)$, *whose prime forms have integral negative dimensions* ν.

If we also set the further requirement, that *the three ground forms are also to be of integral negative dimension,* then we have to separate out those signatures $(0,3;l_1,l_2,l_3)$, for which the least common multiple of l_1,l_2,l_3 goes into ν'. However, in case one or more of the numbers l_1,l_2,l_3 are equal to ∞, then hereby, the smallest common multiple of the finite numbers l is to be meant.

Since we want to study the models thus considered, more closely here, then for the sake of simplicity, of two groups, which we immediately recognize as commeasurable,

[16] This elliptic model corresponds to the automorphic model of that subgroup, which consists of all parabolic substitutions of the total group considered in the text.

only one is to be allowed. The transition to the forms of the other group, namely, is always to be effected hereby through simple algebraic operations. We still arrive thus, all together, at *ten* models, which we want to collect here, arranged according to a decreasing number ν of dimensions; we thereby give, at the same time, in Table 2.2, also the numbers d_1, d_2, d_3 of the dimensions of the associated ground forms

Table 2.2.

signature	$-\nu$	$-d_1$	$-d_2$	$-d_3$
(0,3;2,3,7)	84	42	28	12
(0,3;2,3,8)	48	24	16	6
(0,3;2,4,5)	40	20	10	8
(0,3;2,3,9)	36	18	12	4
(0,3;2,3,10)	30	15	10	3
(0,3;2,3,12)	24	12	8	2
(0,3;2,4,6)	24	12	6	4
(0,3;2,3,18)	18	9	6	1
(0,3;2,3,∞)	12	6	4	12
(0,3;2,4,12)	12	6	3	1

In the case of the modular functions, which is arranged here in the next to last place, [82] as the third ground form $z_3(\zeta_1,\zeta_2)$, the prime form vanishing at the vertex in question of the fundamental domain was chosen.

For the determination of the multiplicators μ, which the individual ground form assumes with respect to the unimodular generators U_1, U_2, U_3, the prescriptions made at the close of §6 on p. 78 hold again. In the case of the modular group, the three ground forms of the dimensions $-6, -4, -12$ are the well-known modular forms, denoted in "M" I on p. 118 ff. by $g_3(\omega_1,\omega_2), g_2(\omega_1,\omega_2), \Delta(\omega_1,\omega_2)$; these are absolutely invariant with respect to the unimodular generators U_1, U_2, U_3(cf. "M" I on p. 118). Besides this, however, one wants to convince oneself of the remarkable theorem: *Except for the three modular forms g_2, g_3, Δ, only the three ground forms of the model of the signature* $(0,3;2,3,7)$ *are absolutely invariant with respect to all three generating substitutions U_1, U_2, U_3.* Invariance up to sign change occurs in the three cases $(0,3;2,3,8), (0,3;2,4,5), (0,3;2,4,6)$.

We attach yet a pair of historical notes. Schwarz, in his investigations, indeed knew very well the triangle nets with principal circle, but his analytic developments connected with setting up the ground forms do not yet refer themselves to these triangle nets. The first one, who has set up the ground forms of the models in question(even [83] if not yet in their simplest form, entering upon the use of binary variables (ζ_1,ζ_2), has been Halphen[17]. Immediately thereupon, Poincaré[18] has then developed, at the same time, a theory of ground forms for arbitrary models of genus zero, on the basis of his series. However, these developments are also given in non-homogeneous form, as was already remarked on p. 55. As was likewise mentioned on p. 55, Klein and Ritter have

[17] "*Sur les fonctions, qui proviennent de la série de Gauss*", Comples rendus, Vol. 92(1881).
[18] In the paper "*Mémorie sur les fonctions fuchsiennes*", Acta Mathematica, Vol. 1(1882).

consistently carried out the use of the binary variables ζ_1,ζ_2 and the homogeneous forms.

2.8 The single-valued automorphic forms and their multiplicator systems

The prime- and ground forms, as we have seen, e.g., in the case of the groups of the circular-arc triangles, are of integral dimension in only a few cases. These forms, although they are always unbranched, are consequently by no means always single-valued automorphic forms(cf. p. 67). Nevertheless, the prime- and ground forms will provide the foundation for the study of the *single-valued automorphic forms*, to which we now turn ourselves.

The following definition is to be placed at the forefront: *An unbranched automorphic form $\varphi_d(\zeta_1,\zeta_2)$ of the integral dimension d, belonging to a proposed group Γ of the genus $p=0$, is to be designated as a"single-valued" automorphic form, in case the multiplicator μ belonging to any individual substitution of the group is always one and the same, whatever path, lying inside the domain of admissible valued ζ_1,ζ_2, the arguments ζ_1,ζ_2 may describe upon exercise of the substitution.*

The next question would be, whether perhaps, *each* automorphic form, which is unbranched and of *integral* positive or negative dimension, is single-valued. This is in fact the case, if the polygon net N has *one* boundary curve and is therefore *simply* connected. We convince ourselves of this easily, by a consideration, such as we already had to carry out on p. 75 ff. Namely, if we understand by $\alpha=\alpha_1:\alpha_2$ any particular value, then we have, in:

$$(\zeta_1\alpha_2-\zeta_2\alpha_1)^{-d}\varphi_d(\zeta_1,\zeta_2)=\varphi(\zeta),$$

[84] a function of ζ, which is unbranched in the simply connected domain N which never reaches over itself, and is therefore single-valued. But since the factor $(\zeta_1\alpha_2-\zeta_2\alpha_1)^{-d}$, on account of the integral value d, is single-valued, then, in fact, there subsists the theorem: *For a group Γ with simply connected net N, each unbranched automorphic form of "integral" dimension d is a "single-valued" form.*

In the further development of the concept of the single-valued automorphic forms, as long as nothing is expressly said to the contrary, we shall now initially draw into consideration only the *case of a simply connected net N*. The consideration of "secondary" relations between the generators $U_1,U_2,\dots,U_n$, which enter for multiple connected nets N, will certainly offer no kind of difficulty in the concrete cases, which we shall consider below. But for taking the secondary relations into consideration "in general", we are lacking a systematic theory of the discontinuity polyhedra, which in fact was avoided in volume I.

We now consider the multiplicators of the forms $\varphi_d(\zeta_1,\zeta_2)$ for arbitrary substitutions of the unimodular homogeneous group, as this was first done in general form by E. Ritter in the paper named on p. 55. As a first theorem we have: For a proposed single-valued form $\varphi_d(\zeta_1,\zeta_2)$ of the underlying group Γ, let the multiplicators μ and μ' correspond to the two operations U and U'; then the substitution $U\cdot U'$, arising by combination from U and U', will obviously belong to the multiplicator $\mu \cdot \mu'$. *With respect to the combination of the operations of the homogeneous group* Γ, *the associated multiplicators multiple themselves.*

Now let the multiplicators $\mu_1, \mu_2, \ldots, \mu_n$ correspond to the n generators $U_1, U_2, \ldots, U_n$. The multiplicator:

$$\mu_1^{\alpha_1} \cdot \mu_2^{\alpha_2} \cdot \mu_3^{\alpha_3} \ldots \mu_n^{\alpha_n} \tag{1}$$

would then belong to an arbitrary substitution of the group, where the exponents α represent any positive or whole numbers or 0. But now the multiplicator 1 must belong to the substitution 1 and the multiplicator $(-1)^d$ to the substitution -1. As long, therefore, as we are able to produce one of the substitutions ± 1 from the generators $U_1, U_2, \ldots, U_n$, the corresponding multiplicator (1) must be equal to $(\pm 1)^d$. The essentially different ways of generating one of the substitutions ± 1 from the $U_1, U_2, \ldots, U_n$, however, are furnished by the relations: [85]

$$U_1^{l_1} = -1, U_2^{l_2} = -1, \ldots, U_n^{l_n} = -1, U_1 \cdot U_2 \cdots U_n = (-1)^n \tag{2}$$

(cf. p. 64 ff.), where, however, among the first n equations, all those are to disappear, which have parabolic U. If we denote $\mu_1, \mu_2, \ldots, \mu_n$ as the "*multiplicator system*" of the form $\varphi_d(\zeta_1, \zeta_2)$, then we have the important theorem: *Each of the multiplicator systems coming up for the single-valued automorphic forms of dimension d satisfy the relations:*

$$\mu_1^{l_1} = (-1)^d, \mu_2^{l_2} = (-1)^d, \ldots, \mu_n = (-1)^d, \mu_1 \cdot \mu_2 \cdots \mu_n^{l_n} = (-1)^{nd}, \tag{3}$$

where, naturally, among the first n relations, those are again to be left out, which contain multiplicators of parabolic substitutions.

Concerning the number d, the only thing which comes into consideration here, obviously, is whether it is even or odd. If we accordingly distinguish "*multiplicator systems for even resp. odd dimensions*", then we have the result: *The multiplicator systems for even dimensions satisfy the relations:*

$$\mu_1^{l_1} = 1,\ \mu_2^{l_2} = 1,\ \ldots,\ \mu_n^{l_n} = 1,\ \mu_1 \cdot \mu_2 \cdots \mu_n = 1. \tag{4}$$

For those of odd dimensions, however, we have the following:

$$\mu_1^{l_1} = -1,\ \mu_2^{l_2} = -1,\ \ldots, \mu_n^{l_n} = -1,\ \mu_1 \mu_2 \cdots \mu_n = (-1)^n. \tag{5}$$

If l_k is a finite number, i.e., if U_k is elliptic, then according to this, μ_k is a root of unity of the degree l_k resp. $2l_k$:

$$\mu_k = e^{2i\pi \frac{\nu_k}{l_k}} \qquad \text{for} \quad d \equiv 0 \pmod 2, \tag{6}$$

$$\mu_k = e^{i\pi \frac{2\nu_k+1}{l_k}} \qquad \text{for} \quad d \equiv 1 \pmod 2, \tag{7}$$

whereby, in each case, ν_k can be assumed as a whole number from the series $0, 1, 2, \ldots, (l_k - 1)$. If $l_k = \infty$, i.e., if we have a parabolic substitution U_k, then, at first, no power of μ_k needs to be identical with $+1$ or -1. However, we remain in agreement with our determinations on p. 70, by stipulating, *that, from now on, the multiplicator* μ_k, *for a parabolic generator* U_k, *is also to be a root of unity:*

$$\mu_k = e^{2i\pi\frac{\nu_k}{l'_k}}, \tag{8}$$

hereby, l'_k is to signify any finite whole number, and ν_k will be a number from the series $0, 1, 2, \ldots, (l'_k - 1)$.

Naturally, the $\mu_1, \mu_2, \ldots, \mu_n$ must also still satisfy the last relation (4) resp. (5), to which we return at once.

[86] Any system $\mu_1, \mu_2, \ldots, \mu_n$, which satisfies the relations (4) resp. (5), as well as, if need be, corresponds to the determinations (8), is to be designated as a "*theoretically possible*" *multiplicator system,* whereby, the question, whether to this system there might also belong forms $\varphi_d(\zeta_1, \zeta_2)$, remains undecided for the moment. The principal question will next be that for the number of all theoretically possible multiplicator systems for a given group Γ. For the answering of this question, the following argument will be basic. If, in $\mu_1, \mu_2, \ldots, \mu_n$ and $\mu'_1, \mu'_2, \ldots, \mu'_n$, one has any two theoretically possible multiplicator systems, then, in:

$$\mu''_1 = \mu_1\mu'_1, \quad \mu''_2 = \mu_2\mu'_2, \ldots, \mu''_n = \mu_n\mu'_n, \tag{9}$$

one again has such a multiplicator system. If we denote the two given systems symbolically by M and M′, then let the system (9), arising by multiplication, be written symbolically as a product M″ = MM′ . Naturally, M′M = MM′; and one has, in M″, a system for an even dimension, in case M and M′ either both belong to an even dimension or both to an odd dimension, while in the other cases, M″ represents a multiplicator system for an odd dimension. We notice, finally, that the particular multiplicator system:

$$\mu_1 = 1, \mu_2 = 1, \ldots, \mu_n = 1 \tag{10}$$

comes up in each case as belonging to an even dimension d.

The results just obtained now permit the following formulation in the language of group theory: *All the theoretically possible multiplicator systems* $M_0, M_1, M_2, \ldots$ *for our group* Γ, *form an Abelian group with respect to multiplication; thereby, either it is a question of all systems belonging to an even dimension d, or the multiplicator systems for even d form a subgroup of index two in the total Abelian group.* Thereby, understand by M_0 the system (10), which one will denote symbolically by $M_0 = 1$.

The number of all systems $1, M_1, M_2, \ldots$, and with this, the order of the Abelian group in question is to be determined in the following paragraph. It will thereby be a matter of an arithmetic investigation, which is of great significance for extending the theory of the single-valued automorphic forms.

2.9 The number of all multiplicator systems M for a given group Γ

[87]

In the determination of the number of all systems M, which are theoretically possible for our underlying group Γ, one has three cases to distinguish, of which the first two are very easily disposed.

I. Among the generators $U_1, U_2, \ldots, U_n$, there are at least two parabolic substitutions

Among the n substitutions $U_1, U_2, \dots, U_n$, let there be $n' \geq 2$ parabolic ones all together. The remaining $(n - n')$ substitutions U_k will then be elliptic, and for the individual one of these, μ_k is restricted by (6) resp. (7) on p. 85 to l_k different possibilities. Now choose, according as $d \equiv 0$ or $\equiv 1 \pmod 2$, according to (6) resp. (7) on p. 85, an arbitrary system of multiplicators for the elliptic generators hereby, one has the choice between a finite number of combinations. The n' multiplicators of the parabolic generators still remaining must then be chosen as any roots of unity in agreement with the last relation (4) resp. (5) on p. 85. Obviously, one may arbitrarily determine $(n' - 1)$ of these roots of unity, whereupon the last is uniquely calculable from the relation in question. *If, among the n generators $U_1, U_2, \dots, U_n$, $n' \geq 2$ parabolic substitutions are contained all together, then there are $\infty^{n'-1}$ multiplicator systems for even as well as for odd dimension.*

II. Among the generators $U_1, U_2, \dots, U_n$, there is contained only one parabolic substitution

Let $U_1, U_2, \dots, U_{n-1}$, be elliptic, while U_n is the one parabolic substitution contained in the present generating system. Here we always have, first of all, for even as well as for odd d, $l_1 l_2 \cdots l_{n-1}$ systems $\mu_1, \mu_2, \dots, \mu_{n-1}$. The multiplicator μ_n, however, is in each case uniquely calculable from the $\mu_1, \mu_2, \dots, \mu_{n-1}$ by means of the last relation (4) resp. (5) on p. 85. *If, among the generators $U_1, U_2, \dots, U_n$ the last, U_n is the only parabolic one, then there are, all together, $2 l_1 l_2 \cdots l_{n-1}$ different multiplicator systems, of which one half belong to even, the other half to odd d.*

III. The generators $U_1, U_2, \dots, U_n$ are all elliptic [88]

Here, somewhat further-going developments are required; we consider initially only the case $d \equiv 0 \pmod 2$.

1. Number of the multiplicator system for even dimension d

The group-theoretic language at the close of the previous paragraph is now to be brought into utilization. We denote the Abelian group of all theoretically possible multiplicator systems with even d symbolically by $[l_1, l_2, \dots, l_n]$; the order of the group $[l_1, l_2, \dots, l_n]$ we have to determine. The multiplicators, in any case, have the form:

$$\mu_1 = e^{\frac{2i\pi\nu_1}{l_1}}, \quad \mu_2 = e^{\frac{2i\pi\nu_2}{l_2}}, \dots, \mu_n = e^{\frac{2i\pi\nu_n}{l_n}}. \tag{1}$$

In order, however, to satisfy the last relation (4) on p. 85, we have to choose the whole numbers $\nu_1, \nu_2, \dots, \nu_n$ in such a manner, that:

$$\frac{\nu_1}{l_1} + \frac{\nu_2}{l_2} + \cdots + \frac{\nu_n}{l_n} = G,$$

where G is any whole number. An alteration of ν_k by an integral multiple of l_k has no alteration of the multiplicator system (1) as consequence. Accordingly, it will be no restriction of the generality, if we take $G = 0$. On the strength of this, the requirement still to be placed on the ν_k assumes the form:

$$\frac{\nu_1}{l_1} + \frac{\nu_2}{l_2} + \cdots + \frac{\nu_n}{l_n} = 0. \tag{2}$$

Now, we have to set up the number of the solutions $\nu_1, \nu_2, \ldots, \nu_n$ of this Diophantine equation. Thereby, two such solutions $\nu_1, \nu_2, \ldots, \nu_n$ and $\nu'_1, \nu'_2, \ldots, \nu'_n$ will not count as distinct, for which:

$$\nu'_1 \equiv \nu_1 \pmod{l_1}, \quad \nu'_2 \equiv \nu_2 \pmod{l_2}, \ldots, \quad \nu'_n \equiv \nu_n \pmod{l_n}$$

occurs.

Let the least common multiple of $l_1, l_2, \ldots, l_{n-1}$ be called L_n. Call s_n the greatest common divisor of l_n and L_n, and set $l_n = s_n t_n$, where t_n will then be relatively prime to $L_n s_n^{-1}$. From (2) follows:

$$\left(\frac{L_n}{l_1}\right)\nu_1 + \left(\frac{L_n}{l_2}\right)\nu_2 + \cdots + \left(\frac{L_n}{l_{n-1}}\right)\nu_{n-1} + \frac{L_n s_n^{-1} \nu_n}{t_n} = 0;$$

and since here the first $(n-1)$ terms have integral values, then ν_n is divisible by t_n. Because of this, we write $\nu_n = u t_n$, understanding by u a whole number.

[89] In particular, there now exist solution systems $\nu_1, \nu_2, \ldots,$ with $u = 1$. In fact, the equation:

$$\left(\frac{L_n}{l_1}\right)\nu_1 + \left(\frac{L_n}{l_1}\right)\nu_2 + \cdots + \left(\frac{L_n}{l_{n-1}}\right)\nu_{n-1} = -L_n s_n^{-1}$$

is solvable by whole numbers $\nu_1, \nu_2, \ldots, \nu_{n-1}$, since the $(n-1)$ expressions standing in the parentheses on the left represent whole numbers without a common factor > 1.

According to this, let us make note as a first theorem: *For the number system* $\nu_1, \nu_2, \ldots, \nu_n$ *belonging to the multiplicator systems* M *of our group* $[l_1, l_2, \ldots, l_n]$, ν_n *is always divisible by* t_n*; in particular, we are able to produce a number system:*

$$\nu_1^{(0)}, \nu_2^{(0)}, \ldots, \nu_{n-1}^{(0)}, t_n \tag{3}$$

whose multiplicator system we let be called $\mathrm{M}^{(0)}$.

The combination of two systems M and M′ will now be effected, by joining the number systems $\nu_1, \nu_2, \ldots, \nu_n$ and $\nu'_1, \nu'_2, \ldots, \nu'_n$ additively into $\nu_1 + \nu'_1, \nu_2 + \nu'_2, \ldots, \nu_n + \nu'_n$. Thus, the multiplicator system $(\mathrm{M}^{(0)})^{s_n}$ has a ν_n divisible by l_n, and will consequently belong to the group $[l_1, l_2, \ldots, l_{n-1}]$. If, for an arbitrary M, the number $\nu_n = u t_n$, then $\mathrm{M}(\mathrm{M}^{(0)})^{-u}$ belongs to the same group $[l_1, l_2, \ldots, l_{n-1}]$. Hereby, an inductive path is sketched for the determination of the order of the groups. According to known theorems on Abelian groups, we initially have the result: *The order of the Abelian group* $[l_1, l_2, \ldots, l_n]$ *is equal to the* s_n*-fold of the order of the group* $[l_1, l_2, \ldots, l_{n-1}]$.

We now agree upon the following notation. Let the greatest common divisor of the two numbers l_i, l_k be called δ_{ik}; call δ_{ikm} the greatest whole number simultaneously going into l_i, l_k, l_m and also utilize the corresponding notation for four, five, etc. numbers l.

The order of the group $[l_1]$ is equal to 1. For the group $[l_1, l_2]$, s_2 is identical with the greatest common divisor δ_{12} of l_1 and l_2; *the order of the latter group is thus* δ_{12}. In order to determine, for the group $[l_1, l_2, l_3]$, the greatest common divisor s_3 of the number l_3 and the least common multiple L_3 of l_1 and l_2, one proceeds, according to the "method of inclusion and exclusion" in the following manner: the greatest common divisor of l_3 and l_1 is δ_{13}, that of l_3 and l_2 is δ_{23}. In the product $\delta_{13}\delta_{23}$, the common divisor δ_{123} of all three numbers l_1, l_2, l_3 is counted twice. If we remove it once by division, then we have, in $\frac{\delta_{13}\delta_{23}}{\delta_{123}}$, the greatest common divisor s_3 of l_3 and L_3 sought. It follows: *the order of the group* $[l_1, l_2, l_3]$ *is equal to:* [90]

$$\frac{\delta_{12}\delta_{13}\delta_{23}}{\delta_{123}}. \tag{4}$$

It is easy to generalize the rule herein coming to light. The greatest common divisor s_n of l_n and the least common multiple L_n of the numbers $l_1, l_2, \ldots, l_{n-1}$ represents itself in general for $n = 4, 5, \ldots$, thus:

$$s_4 = \frac{\delta_{14}\delta_{24}\delta_{34}\delta_{1234}}{\delta_{124}\delta_{134}\delta_{234}},$$

$$s_5 = \frac{\delta_{15}\delta_{25}\delta_{35}\delta_{45}\delta_{1245}\delta_{1345}\delta_{2345}}{\delta_{125}\delta_{135}\delta_{145}\delta_{235}\delta_{245}\delta_{345}\delta_{12345}},$$

$$\cdots\cdots\cdots\cdots$$

$$s_n = \frac{\delta_{1n}\ldots\delta_{n-1,n}\delta_{123n}\ldots\delta_{n-3,n-2,n-1,n}\delta_{12345n}\cdots}{\delta_{12n}\ldots\delta_{n-2,n-1,n}\delta_{1234n}\ldots\delta_{n-4,n-3,n-2,n-1,n}\cdots}. \tag{5}$$

But instead of proceeding syntactically according to the method of inclusion and exclusion in the construction of these formulas, the following direct line of proof is far more convincing.

The factors δ standing above and below the fraction line in the expression of s_n are composed exclusively of such prime factors, which also go into l_n. Let q be an individual one of these prime factors, and q^α the highest power of this going into l_n, the expression given for s_n depends symmetrically on the numbers $l_1, l_2, \ldots, l_{n-1}$, and will thus remain unaltered for any other arrangements of these numbers. We make use of this circumstance in the following manner. Let q^{α_k} be the highest power of q going into l_k, where k is to be taken $1, 2, 3, \ldots, n-1$; we then imagine the $(n-1)$ numbers l so arranged, that

$$\alpha_1 \geq \alpha_2 \geq \alpha_3 \geq \cdots \geq \alpha_{n-1}$$

occurs. The consequence is, *that the highest power of* q *contained in* $\delta_{\ldots,k,n}$ *will be independent of how many and which indices precede the last two* k, n *in* $\delta_{\ldots,k,n}$. We call this highest power of q, say, q^{β_k}; obviously, one has $\beta_k = \alpha_k$, in case $\alpha_k \leq \alpha$, and $\beta_k = \alpha$, in case $\alpha_k \geq \alpha$ occurs.

Now write the expression for s_n thus:

$$s_n = \delta_{1n}\frac{\delta_{1n}\ldots\delta_{n-1,n}\delta_{123n}\ldots\delta_{n-3,n-2,n-1,n}\cdots}{\delta_{12n}\ldots\delta_{n-2,n-1,n}\ldots\delta_{1234n}}. \tag{6}$$

Here, next count off, above the fraction line, those factors which contain q in the highest power $q^{\beta_{k+1}}$. It is therefore a question of the number of factors $\delta_{\ldots,k+1,n}$, whereby k is a determinate number of the series $1, 2, \ldots, n-2$. Among the [91]

$$\delta_{2n}, \ldots, \delta_{n-1,n}$$

there occurs one such factor, but among the $\delta_{123n},\ldots,\delta_{n-3,n-2,n-1,n}$ $\binom{k}{2}$ etc.; the total number of the factors sought is:

$$1+\binom{k}{2}+\binom{k}{4}+\binom{k}{6}+\cdots \tag{7}$$

Analogously, one finds, as the number of the factors with the highest power $q^{\beta_{k+1}}$ in the denominator of the right side of (6):

$$\binom{k}{1}+\binom{k}{3}+\binom{k}{5}+\binom{k}{7}+\cdots.$$

This number, however, is identical with the number given in (7). If, one after the other, we take $k=1,2,\ldots,n-2$, it then follows, that in (6) on the right side, all factors q let themselves be completely removed from the fraction multiplied by δ_{1n}.

Now notice, that q^{α_1} is also exactly the highest power of q contained in L_n. The first factor δ_{1n} on the right side of (6) thus furnishes just the correct power q^{β_1} for s_n.

Therefore, the consideration holds for each arbitrary prime factor q of l_n, the correctness of the expression of s_n given in (5) is actually verified.

With this, we are immediately at our goal: *the number of the theoretically possible multiplicator systems for even dimension d, in the case of all generators elliptic, given for n = 3,4,... as well as generally is given, by:*

$$\begin{gathered}\frac{\delta_{12}\delta_{13}\delta_{23}}{\delta_{123}},\ \frac{\delta_{12}\delta_{13}\delta_{14}\delta_{23}\delta_{24}\delta_{1234}}{\delta_{123}\delta_{124}\delta_{134}\delta_{234}},\\ \cdots\cdots\\ \frac{\delta_{12}\delta_{13}\ldots\delta_{n-1,n}\delta_{1234}\ldots\delta_{n-3,n-2,n-1,n}\delta_{123456}\cdots}{\delta_{123}\delta_{124}\ldots\delta_{n-2,n-1,n}\delta_{12345}\ldots\delta_{n-4,n-3,n-2,n-1,n}\cdots}.\end{gathered} \tag{8}$$[19]

2. Number of the multiplicator systems for odd dimension d

The number of the theoretically possible multiplicator systems for odd dimension d is either identical with the number determined in (8), or there are no multiplicator systems at all for odd d. To decide concerning this alternative, let us consult the forms
[92] of the multiplicators $\mu_1,\ldots,\mu_n$ given in (7) on p. 85. The whole numbers $\nu_1,\nu_2\ldots,\nu_n$ hereby utilized must, for the satisfaction of the last relation (5) on p. 85, in:

$$\frac{2\nu_1+1}{l_1}+\frac{2\nu_2+1}{l_2}+\cdots+\frac{2\nu_2+1}{l_n}=G, \tag{9}$$

furnish a whole number G congruent modulo 2 with n.

If the numbers $l_1,\ldots,l_n$ are all odd, then we satisfy this requirement, in fact, by:

[19] The determination given in the text of the number of all multiplicator systems for even d has first most recently been carried through by the author and is given here for the first time.

$$\nu_1 = \frac{l_1 - 1}{2}, \quad \nu_2 = \frac{l_2 - 1}{2}, \quad \ldots, \quad \nu_n = \frac{l_n - 1}{2}. \tag{10}$$

If at least one number, say l_k is even, then the alteration of ν_k by $\frac{1}{2}l_k$ has an alteration of G by 1 as a consequence. The condition $G \equiv n \pmod 2$ therefore presents no difficulty here; and it is no restriction of the generality, if we set $G = 0$ in (9). If we then multiply equation (9) by the smallest common multiple L of $l_1, l_2, \ldots, l_n$, then it assumes the form:

$$2\left(\frac{L}{l_1}\nu_1 + \frac{L}{l_2}\nu_2 + \cdots + \frac{L}{l_n}\nu_n\right) + \left(\frac{L}{l_1} + \frac{L}{l_2} + \cdots + \frac{L}{l_n}\right) = 0. \tag{11}$$

For each admissible system $\nu_1, \ldots, \nu_n$, an integral value of the sum would have to stand here in the first as well as in the second parentheses; therefore, the condition:

$$\frac{L}{l_1} + \frac{L}{l_2} + \cdots + \frac{L}{l_n} \equiv 0 \pmod 2 \tag{12}$$

must be satisfied. If this is the case, then equation (11) also lets itself always be solved by whole numbers $\nu_1, \ldots, \nu_n$, since the $Ll_1^{-1}, \ldots, Ll_n^{-1}$ have no common factor > 1. If, however, the condition (12) is not fulfilled, then equation (11) is not satisfied at all by integral ν.

In order now to discuss the congruence (12), then bring in the highest power 2^λ, which is contained in one of the numbers $l_1, \ldots, l_n$. Only the l divisible by 2^λ furnish terms of odd numerical value in (12); all remaining terms represent even numbers. Therefore, the theorem yields itself: *If all numbers l are odd or if an "even" number of the numbers $l_1, l_2, \ldots, l_n$ are divisible by the highest power 2^λ of 2 coming up in any of them, then the number of the multiplicator systems for odd d is given in (8); on the other hand, there are no multiplicator systems for odd d, in case the $l_1, l_2, \ldots, l_n$ are divisible in odd number by the highest power 2^λ with $\lambda > 0$.*[20]

2.10 Example for the determination of the number of the multiplicator systems M, the effect of secondary relations

The general number of determinations in the foregoing paragraph are now to be illustration in a few special cases and examples. We shall hereby principally consider such cases, for which parabolic substitutions do not come up. [93]

1. *If no two of the numbers $l_1, l_2, \ldots, l_n$ have a common divisor different from 1,* then there is only the one multiplicator system $\mu_1 = \mu_2 = \cdots = \mu_n = 1$ for even d; and only in case all numbers $l_1, l_2, \ldots, l_n$ are odd, do we get a system (10) for odd d. For the first ten groups Γ, to which the table on p. 82 refers, there is consequently only a single system M.

One will also suitably consider further groups of that table as examples here. Thus, for the group of the signature $(0,3;2,3,8)$ one finds two systems M, both of which belong to $d \equiv 0 \pmod 2$. The group $(0,3;2,3,18)$ has twelve multiplicator systems, half of which belong to even d.

[20] This theorem was first set up by Ritter in §6 of his work named on p. 55.

2. *If a common divisor different from* 1 *of two numbers l is never contained in a third,* then the number of the systems M for even d is given by $\delta_{12}\delta_{13}\cdots\delta_{n-1,n}$. Thus, for the signature $(0,3;2,3,12)$, we have six systems M, which belong to even d.

3. *If all numbers $l_1,\ldots,l_n$ are equal to one another* and equal to l, then all the factors δ utilized in the expression (8) on p. 91 have the common value l. But the number of the factors δ in that expression above the fraction line is:

$$\binom{n}{2}+\binom{n}{4}+\binom{n}{6}+\cdots,$$

while below the fraction line, there stand:

$$\binom{n}{3}+\binom{n}{5}+\binom{n}{7}+\cdots$$

The first of these numbers exceeds the latter by $(n-1)$. With this, l^{n-1} multiplicator systems exist for even dimension d; and we get just as many systems M for odd d, in case l is odd or for even l the number n is similarly even.

4. In the case of the modular group $(0,3;2,3,\infty)$, we have one parabolic generator. According to the rule developed in II on p. 87, we shall obtain twelve multiplicator systems here, of which six belong to odd dimension.

[94] In the modular group, there are still contained a few *subgroups with only one parabolic generator each.* Here, in particular, belong those three congruence subgroups of the indices 5,7 and 11, which furnish the well-known resolvents of the $5^{\text{th}}, 7^{\text{th}}$ and 11th degree of the modular equations of $6^{\text{th}}, 8^{\text{th}}$ and 12th degree. The discontinuity domains of these groups are given in "M" I on pp. 647 and 753 as well as in "M" II on p. 420. One has, e.g., for the second group, the domain reproduced adjacently in figure 21. In consequence of this figure, we are concerned with a group of the signature $(0,5;2,2,2,3,\infty)$. The number of all systems M in 48 here; half of these belong to odd dimension d.

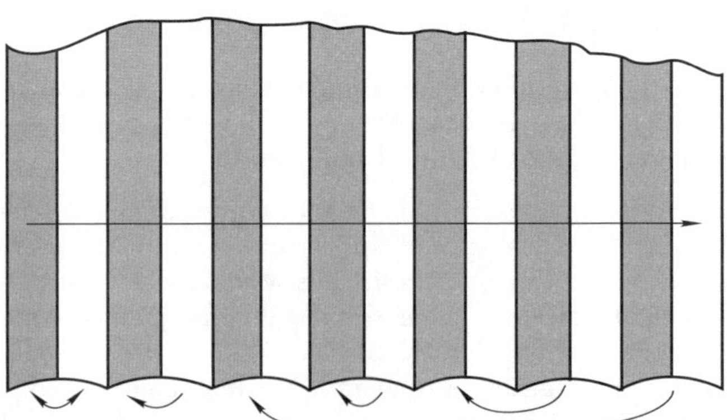

Fig. 21.

This will be the place, for us to also illustrate, by a few examples, the significance and effect of possible *secondary relations* between the generators. In the case, s secondary relations are present, we had already noted for ourselves the schema (7) on p. 65 for these. For the multiplicators $\mu_1, \mu_2, \ldots, \mu_n$ for even dimension d, there arise from these, just as many new equations:

$$\prod\nolimits_1(\mu_1, \mu_2, \ldots, \mu_n) = 1, \ldots, \prod\nolimits_s(\mu_1, \mu_2, \ldots, \mu_n) = 1, \tag{1}$$

while for odd d, beside the respective relations (5) on p. 85, the s additional ones arrange themselves:

$$\prod\nolimits_1(\mu_1, \mu_2, \ldots, \mu_n) = (-1)^{m_1}, \ldots, \prod\nolimits_s(\mu_1, \mu_2, \ldots, \mu_n) = (-1)^{m_s}. \tag{2}$$

The relations (1) are naturally always satisfied by the multiplicator system $M_0 = 1$. *In general, these relations define a subgroup inside the Abelian group of all* M *considered in the previous paragraph, for even d; this subgroup now represents for us the totality of all theoretically possible systems* M *of even dimension.* For odd dimension, there are then either just as many systems M or none at all, concerning which, one has to decide according to the rules of the previous paragraph by addition of the relations (2). [95]

We now return immediately to the example on p. 57 ff.

1. To the domain given on the right side in figure 14, there belonged a group of the signature $(0,4;2,3,2,3)$, which is generatable by shuffling two tetrahedral groups into one another. The associated discontinuity polyhedron has *one* edge running in the ζ-half space, which does *not* reach the ζ-plane. To it corresponds the secondary relation $(V_1V_2)^3 = 1$ already named on p. 63.

In order to determine the homogeneous form of this relation, consider, that V_1 and V_2 generate a tetrahedral group by themselves. If, for the latter, we form the generating system U_1, U_2, U_3' according to the general prescriptions on p. 64, then $U_1U_2U_3' = -1$; and since $U_3'^3 = -1$, then obviously:

$$(U_1U_2)^3 = 1 \tag{3}$$

follows as the homogeneous form of the secondary relation.

The rules of the previous paragraph how initially furnish us with *six* systems M for even d, and just as many for odd dimension. For even d, namely, one has:

$$\mu_1 = \mu_3 = \pm 1, \mu_2 = \mu_4^{-1} = e^{2i\pi\frac{k}{3}} \qquad (k = 0, 1, 2);$$

for odd dimension, it suffices to add the one system:

$$\mu_1 = i, \mu_2 = -1, \mu_3 = i^3, \mu_4 = -1,$$

from which all six systems arise by combination with the six systems given before.

But now, corresponding to the relation (3), we have to require $(\mu_1\mu_2)^3 = 1$: *there remain only three systems with even d and none for odd dimension.*

2. If we take the example of the figures 15 ff, and set, in particular, say:

$$l_1 = l_6 = 6, l_2 = l_5 = 10, l_3 = l_4 = 15,$$

then the rules of the previous paragraph furnish 27000 systems M each for even and odd dimension.

Here we have *three* secondary relations, namely:

$$V_1 V_6 = 1, V_2 V_5 = 1, V_3 V_4 = 1.$$

[96] One must return to the forms (6) on p. 64 of the homogeneous generators, in order to recognize, as the homogeneous form of these relations:

$$U_1 U_6 = 1,\ U_2 U_5 = 1,\ U_3 U_4 = 1. \tag{4}$$

There follows $\mu_6 = \mu_1^{-1}$, $\mu_5 = \mu_2^{-1}$, $\mu_4 = \mu_3^{-1}$; and since, with this, the relation $\mu_1 \cdot \mu_2 \cdots \mu_6 = 1$ is satisfied spontaneously, the remaining relations (4), resp. (5) on p. 85 furnish the result, *that of all* $2 \cdot 27000$ *systems* M, *only* 900 *remain theoretically possible here for each kind of dimension.*

3. Finally, we also consider the case of figure 20, but leave arbitrary the number p, which has the value 3 in figure 20.

The group, in non-homogeneous form, has $(2p+2)$ elliptic substitutions of period 2 as generators, which are identical in pairs; in the $(p+1)$ equations thus arising, we have just as many secondary relations, which exhaust all relations of this kind subsisting in the present case.

In the homogeneous form, these substitutions become of the period 4; and it turns out, that they are inverse to one another in pairs, which we bring to expression by:

$$U_1 U_2 = 1,\ U_3 U_4 = 1,\ \ldots,\ U_{2p+1} \cdot U_{2p+2} = 1. \tag{5}$$

According to the example 3 at the beginning of the present paragraph, the rules of the preceding paragraph would furnish 2^{2p+1} systems M here for even dimension and just as many for odd d. The effect of the secondary relations is this, *that the number of the theoretically possible multiplicator systems now reduces to* 2^{p+1} *for each of the two kinds of dimension.*

2.11 Representation of all unbranched automorphic forms of a model with $p = 0$ by the prime and ground forms

The main question of the next development will be whether for each theoretically possible multiplicator system, there exist associated single-valued automorphic forms or not. This question we shall be able to answer, by treating here, beforehand, the *representability in general of all unbranched automorphic forms of the group* Γ *by means of the prime forms* z_1, z_2 *introduced on p. 73 ff., resp., the ground forms*:

$$Z_k = (z, e_k)^{\frac{1}{l_k}}.$$

[97] The next thing is to develop a fundamental theorem concerning the number of the zeros and poles of a form $\varphi_d(\zeta_1, \zeta_2)$.

Let $\varphi_d(\zeta_1,\zeta_2)$ be any unbranched automorphic form of our group Γ; the dimension d of this form represents a whole number or a rational fraction. The prime form $z_2(\zeta_1,\zeta_2)$ has the dimension v, which, according to p. 73 is calculated from:

$$\frac{2}{v}=2-\sum_{k=1}^{n}\left(1-\frac{1}{l_k}\right). \tag{1}$$

The quotient of φ_d^v and z_2^d is of dimension zero. Let it be called $\psi(\zeta)$; then we have:

$$[\varphi_d(\zeta_1,\zeta_2)]^v=\psi(\zeta)[z_2(\zeta_1,\zeta_2)]^d. \tag{2}$$

We now bring in the zeros and poles of the unbranched form $\varphi_d(\zeta_1,\zeta_2)$ "lying in the fundamental domain" and remember, that, at the fixed vertices of the domain, zeros and poles of certain "fractional" orders are also able to occur (cf. p. 69). Since $\psi(\zeta)$ can alter itself, with respect to the generators, only by multiplicative roots of unity of finite degree (cf. p. 77 ff.), there are n powers of ψ with a finite exponent, which is absolutely invariant with respect to all generators, and therefore represents an automorphic *function* of Γ. Accordingly, for $\psi(\zeta)$ too, the total order of the zeros will be equal to the total order of the poles in the fundamental domain. The prime form z_2 is pole-free and has only simple zeros in the fundamental domain. On the strength of this, the formula (2) immediately furnishes the important fundamental theorem: *For an unbranched automorphic form of dimension d of our group Γ, the total order of the zeros exceeds that of the poles by an amount $\frac{d}{v}$ in the fundamental domain.*

If we ignore the elementary group Γ, then v is a negative number (cf. p. 73). For all non-elementary groups, therefore, for the individual form, the total order of the zeros will be greater or smaller than that of the poles, according as the dimension d is negative or positive.

Should a form $\varphi_d(\zeta_1,\zeta_2)$ be free of poles, then the total order of the vanishing is $\frac{d}{v}$. It follows: *A form of non-vanishing dimension d cannot be everywhere finite and simultaneously different from zero in the fundamental domain.*

Two unbranched forms, which have precisely (in position and order) the same zeros and poles, have the same dimension. Their quotient, raised to a suitable power with [98]
finite exponent, will represent an everywhere finite automorphic function of Γ and is therefore identical with a constant. *An unbranched automorphic form of Γ is consequently uniquely determined up to a constant factor by its zeros and poles.*

On the basis of this last theorem, it is now easy, to express an arbitrary unbranched form $\varphi_d(\zeta_1,\zeta_2)$ by the ground forms and prime forms of our group Γ. Suppose, at the kth fixed polygon vertex, there is situated a zero of the order $\frac{m_k}{l_k}$ (pole of the order $-\frac{m_k}{l_k}$) of $\varphi_d(\zeta_1,\zeta_2)$, as far as this vertex is elliptic; and for $l_k=\infty$, i.e., in the parabolic case, let the corresponding order be given by the rational fraction $\frac{m_k'}{l_k'}$, not to be initially restricted further. All further zeros and poles are of integral order; let the former be situated at $z=x_1,x_2,\ldots,x_\tau$, the latter at $z=y_1,y_2,\ldots,y_\sigma$, where each place is entered as often as the multiplicity of the associated zero resp. pole requires.

In order then to accomplish the representation taken into prospect of φ_d, we split the individual x into the quotient $x':x''$, y correspondingly into $y':y''$ and let:

$$(z,x_1)=z_1x_1''-z_2x_1',\ldots,(z,y_1)=z_1y_1''-z_2y_1',\ldots, \tag{3}$$

understanding by $z_1 = z_1(\zeta_1,\zeta_2)$ and $z_2 = z_2(\zeta_1,\zeta_2)$ our two prime forms obtained on p. 74.

Besides them, we also have to bring in the ground forms considered on p. 74, and by means of the theorem last set up, immediately arrive at the result: *The arbitrarily proposed unbranched automorphic form $\varphi_d(\zeta_1,\zeta_2)$ of our group, with the aid of a constant C remaining undetermined here, lets itself be represented in the prime forms and ground forms thus:*

$$\varphi_d(\zeta_1,\zeta_2) = C\frac{(z,x_1)(z,x_2)\cdots(z,x_\tau)}{(z,y_1)(z,y_2)\cdots(z,y_\sigma)}\prod_{k=1}^{n}{}'(z,e_k)^{\frac{m_k}{l_k}}, \tag{4}$$

where, by the upper index on the product sign, it is to be indicated, that in the the case of a parabolic vertex, the exponent $\frac{m_k}{l_k}$ is to be replaced by $\frac{m'_k}{l'_k}$.

By separating out, from the individual factor of the product $\prod$ standing on the right, a suitable power of (z,e_k) with *integral* exponent, and adding it to the factors standing before the product sign, one can attain, that the n exponents standing under the product sign signify positive proper fractions or are equal to zero. We may consequently
[99] assume, at the same time, of the expression (4) of our automorphic form $\varphi_d(\zeta_1,\zeta_2)$, that the conditions:

$$0 \leqslant m_k < l_k \quad \text{resp.} \quad 0 \leqslant m'_k < l'_k \tag{5}$$

are fulfilled. Before the product sign on the right side of the equation (4), there stands an expression rational and homogeneous in z_1, z_2, whose dimension in z_1, z_2 is obviously $\tau-\sigma$ and may be denoted by m. On the strength of this, we are also able to write the representation (4) of the form $\varphi_d(\zeta_1,\zeta_2)$, in abbreviation, thus:

$$\varphi_d(\zeta_1,\zeta_2) = R_m(z_1,z_2)\prod_{k=1}^{n}{}'(z,e_k)^{\frac{m_k}{l_k}}. \tag{6}$$

The total order of the zero and that of the poles are to be read off immediately on the right sides of the formulas (4) resp. (6). The excess of the former number over that of the latter, however, is equal to $\frac{d}{\nu}$, according to a theorem set up above. We have: *Between the dimension d of the form $\varphi_d(\zeta_1,\zeta_2)$ and the numbers $m, m_1, m_2, \ldots$ coming into force for the representation* (6), *here subsists the relation:*

$$d = \nu\left(m + \sum_{k=1}^{n}{}'\frac{m_k}{l_k}\right), \tag{7}$$

where, by the index on the summation sign, is again to be indicated, that for the parabolic vertices, $\frac{m_k}{l_k}$ is to be replaces by $\frac{m'_k}{l'_k}$.[21]

2.12 Existence theorem for single-valued forms $\varphi_d(\zeta_1,\zeta_2)$ for given multiplicator system M

We now specialize the preceding developments for the *single-valued* forms $\varphi_d(\zeta_1,\zeta_2)$ and thereby return to the question raised at the beginning of the previous paragraph,

[21] Compare, for the present and following developments, §8 of the work of Ritter cited on p. 55, to which the presentation of the text attaches itself quite closely.

whether, for each theoretically possible multiplicator system $\mu_1, \mu_2, \dots, \mu_n$, associated forms exist or not.

For this purpose, first of all, we determine the factor μ_k according to the rule (5) on p. 78, which the form $\varphi_d(\zeta_1,\zeta_2)$ represented in (4) resp. (6) in §2.11 assumes, in case we effect the generating substitution U_k. We have, since d is integral:

$$\mu_k = e^{2i\pi\left(\frac{m_k}{l_k} - \frac{d}{2l_k}\right)} \quad \text{resp.} \quad \mu_k = e^{2i\pi\frac{m'_k}{l'_k}} \tag{1}$$

according as U_k is elliptic or parabolic. [100]

On the other hand, independently of the preceding developments, we now imagine any theoretically possible multiplicator system $\mu_1, \dots, \mu_n$ proposed. In order thereby not to have to always distinguish the cases of even and odd dimension d, we write the multiplicator μ_k in the form:

$$\mu_k = e^{2i\pi\frac{\lambda_k}{2l_k}}, \tag{2}$$

with the determination, that the whole number λ_k is to satisfy the congruence:

$$\lambda_k \equiv d \pmod 2 \tag{3}$$

[cf. (6) and (7) on p. 85]. Naturally, if one will, one can prescribe the condition $0 \leqslant \lambda_k < 2l_k$ for the whole number λ_k. But, in consideration of the further calculations to follow, it is more expedient, to ignore this condition. The formula (2) refers itself initially only to the case of an elliptic substitution U_k; for a parabolic one, the multiplicator one would have:

$$\mu_k = e^{2i\pi\frac{\lambda'_k}{l'_k}}. \tag{4}$$

However, it is more advantageous here, to subsume formula (4) under (2); we simply have to stipulate, *that for $l_k = \infty$, λ_k is also to be infinitely large, and indeed, in such a manner, that $\frac{\lambda_k}{2l_k}$ becomes equal to the finite rational fraction $\frac{\lambda'_k}{l'_k}$.*

For our arbitrarily proposed multiplicator system M, now choose, according as $d \equiv 0$ or $\equiv 1 \pmod 2$, an even or odd dimension d *arbitrarily* and assume, that the form $\varphi_d(\zeta_1,\zeta_2)$ represented under (6) on p. 99, actually belongs to this dimension d and this multiplicator system. Then the multiplicator μ_k, just determined from the representation (6) on p. 99 in the form (1), must be identical with that prescribed in (2). We satisfy it and the inequality (5) on p. 99 simultaneously, *by determining the exponent m_k of the kth ground form in the representation* (6) *on p. 99 from:*

$$\frac{m_k}{l_k} = \frac{d+\lambda_k}{2l_k} - E\left[\frac{d+\lambda_k}{2l_k}\right]. \tag{5}$$

Hereby, in known fashion, $E[\alpha]$ signifies the greatest whole number not exceeding the [101]
real number x. The congruence (3) guarantees an *integral* value for the m_k thus determined. In the parabolic case, thanks to our determination just made, the equation (5) assumes the form:

$$\frac{m'_k}{l'_k} = \frac{\lambda'_k}{l_k} - E\left[\frac{\lambda'_k}{l'_k}\right] \tag{6}$$

The equation (5), with the utilization of the relation (7) of the previous paragraph, now furnish:

$$d = \nu\Big\{m + \sum_{k=1}^{n}\Big(\frac{d+\lambda_k}{2l_k}\Big) - \sum_{k=1}^{n} E\Big[\frac{d+\lambda_k}{2l_k}\Big]\Big\},$$

$$d\Big(\frac{2}{\nu} - \sum_{k=1}^{n}\frac{1}{l_k}\Big) = 2m + \sum_{k=1}^{n}\frac{\lambda_k}{l_k} - 2\sum_{k=1}^{n} E\Big[\frac{d+\lambda_k}{2l_k}\Big].$$

In consequence of (1) on p. 97, the difference standing in the parentheses on the left has the value $(2-n)$, and therefore, follows the representation for m:

$$2m = d(2-n) - \sum_{k=1}^{n}\frac{\lambda_k}{l_k} + 2\sum_{k=1}^{n} E\Big[\frac{d+\lambda_k}{2l_k}\Big]. \tag{7}$$

The sum standing in the first place here on the right side of the equation, according to the last relation (4) resp. (5) on p. 85, has an integral value congruent modulo 2 with $d\cdot n$. The consequence is, that an *integral* value calculates itself from (7) for m. Now, for the construction of the right side of formula (6) on p. 99, one has yet only to choose at random a homogeneous rational function $R_m(z_1, z_2)$ of the dimension m just determined, to then actually passes a form $\varphi_d(\zeta_1,\zeta_2)$ of the desired kind.

With this, the fundamental theorem is proved: *For each theoretically possible multiplicator system, and each dimension d chosen in agreement with* (3), *there are associated single-valued automorphic forms; in the individual case, the n numbers m_k resp. m'_k are uniquely determined from* (5) *resp.* (6), *and similarly, the number m yields itself integrally and uniquely from* (7); *on the other hand, $R_m(z_1, z_2)$ remains freely choosable as a rational homogeneous function of this dimension m of the z_1, z_2.*

The multiplicator system $\mathrm{M}_0 = 1$, i.e., written explicitly,

$$\mu_1 = 1,\ \mu_2 = 1,\ \ldots,\ \mu_n = 1,$$

exists in each case, and required $d \equiv 0 \pmod 2$. The associated forms, which are abso-
[102] lutely invariant with respect to all substitutions of the homogeneous group Γ, are to be designated as "properly" automorphic forms of the group Γ. *For each group Γ and for each integral dimension d, there exist properly automorphic forms.*

2.13 Relations between multiplicator systems inverse to one another.

Inside the Abelian group of all multiplicator systems, to the individual system M, according to a well-known coordination, always belongs an inverse system, which is to be denoted by $\overline{\mathrm{M}} = \mathrm{M}^{-1}$.

For two inverse systems M and $\overline{\mathrm{M}} = \mathrm{M}^{-1}$, two corresponding multiplicator μ_k and $\overline{\mu}_k$ will always be *reciprocal to one another*:

$$\mu_k \cdot \overline{\mu}_k = 1. \tag{1}$$

A system M is accordingly always and only then inverse to itself, in case all multiplicators in M are equal to $+1$ or -1, as, e.g., this occurs for the multiplicator system

$\mathrm{M}_0 = 1$. Obviously, two systems M and $\overline{\mathrm{M}}$, inverse to one another, either both belong to even or both to odd dimension d.

For the preparation of later developments, which concern the fundamental theorems on representation of the automorphic forms in the form of Poincaré series, we bring in any two dimensions d, $\overline{d}$ for M and $\overline{\mathrm{M}}$, which satisfy the equation:

$$d + \overline{d} + 2 = 0. \tag{2}$$

It is a question, under the assumption (2), of setting up a relation for the two systems M, $\overline{\mathrm{M}}$ inverse to one another, by means of an elementary arithmetic consideration, which relation will come into utilization for the investigation just mentioned.

If we write the multiplicators in the form (2) on p. 100, then we satisfy the condition (1), by setting:

$$\frac{\overline{\lambda}_k}{2l_k} = 1 - \frac{\lambda_k}{2l_k},$$

which, according to the agreement made on p. 100, holds without further ado for the parabolic U_k coming up, say upon utilization of (2), one obtains:

$$\frac{\overline{d}+\overline{\lambda}_k}{2l_k} = \frac{-d-2+2l_k-\lambda_k}{2l_k} = -\frac{d+\lambda_k}{2l_k} + \left(1-\frac{1}{l_k}\right),$$

$$E\left[\frac{\overline{d}+\overline{\lambda}_k}{2l_k}\right] = E\left[-\frac{d+\lambda_k}{2l_k} + \frac{l_k-1}{l_k}\right]. \tag{3}$$

The right side of this equation lets itself be written more simply. Notice, first of all, for the case of a finite l_k that on account of $\lambda_k \equiv d \pmod 2$, the factor 2 in the denominator of the first term disappears. If now, $-\frac{d+\lambda_k}{2l_k}$ has an integral value, then the right side of (3) is equal to this value, which we are also able to write $-E\left[\frac{d+\lambda_k}{2l_k}\right]$. If $-\frac{d+\lambda_k}{2l_k}$ is not integral, but equal to a rational fraction at the denominator l_k, then of the two numbers: [103]

$$-\frac{d+\lambda_k}{2l_k} \quad \text{and} \quad -\frac{d+\lambda_k}{2l_k} + \frac{l_k-1}{l_k},$$

differing from the *proper* fraction $\frac{l_k-1}{l_k}$, either the second is equal to a whole number, or there lies a whole number between them. The *whole number intended* here in both cases is $-E\left[\frac{d+\lambda_k}{2l_k}\right]$:

$$-\frac{d+\lambda_k}{2l_k} < -E\left[\frac{d+\lambda_k}{2l_k}\right] \leqslant -\frac{d+\lambda_k}{2l_k} + \frac{l_k-1}{l_k}.$$

Now since the middle term is less than 1 from the right side of this inequality, then it follows:

$$E\left[-\frac{d+\lambda_k}{2l_k} + \frac{l_k-1}{l_k}\right] = -E\left[\frac{d+\lambda_k}{2l_k}\right].$$

Therefore, in the elliptic case, we are able to write the equation (3) thus:

$$E\left[\frac{\overline{d}+\overline{\lambda}_k}{2l_k}\right] = -E\left[\frac{d+\lambda_k}{2l_k}\right]. \tag{4}$$

If $l_k = \infty$, i.e., if we have a parabolic substitution U_k, then the equation (3) reads, in particular:

$$E\Big[\frac{\overline{\lambda}'_k}{l'_k}\Big] = E\Big[-\frac{\lambda'_k}{l'_k} + 1\Big].$$

Here it makes a difference, whether $\frac{\lambda'_k}{l'_k}$ is integral or not; namely, according as the former or latter occurs, it obviously holds:

$$E\Big[\frac{\overline{\lambda}'_k}{l'_k}\Big] = -E\Big[\frac{\lambda'_k}{l'_k}\Big] + 1 \quad \text{resp.} \quad E\Big[\frac{\overline{\lambda}'_k}{l'_k}\Big] = -E\Big[\frac{\lambda'_k}{l'_k}\Big]. \tag{5}$$

The second of these two formulas subsumes itself under the rule (4).

Now one wants to calculate the number systems $m, m_1, \ldots, m_n$ and $\overline{m}, \overline{m}_1, \ldots, \overline{m}_n$ for the forms belonging to M and d on the one hand, as well as $\overline{\mathrm{M}}$ and $\overline{d}$ on the other.
[104] For the number m_k, resp. , the number m'_k coming into its place in the parabolic case, there holds the equation (5) on p. 100 resp. (6) on p. 101, the equation for $\overline{m}_k$:

$$\frac{\overline{m}_k}{l_k} = \frac{\overline{d} + \overline{\lambda}_k}{2l_k} - E\Big[\frac{\overline{d} + \overline{\lambda}_k}{2l_k}\Big].$$

One now rewrites, on the basis of the calculations just developed, this as:

$$\frac{\overline{m}_k}{l_k} = -\frac{d + \lambda_k}{2l_k} + \Big(1 - \frac{1}{l_k}\Big) + E\Big[\frac{d + \lambda_k}{2l_k}\Big]. \tag{6}$$

In its place, however, in the parabolic case, we get:

$$\frac{\overline{m}'_k}{l'_k} = -\frac{\lambda'_k}{l'_k} + 1 + E\Big[\frac{\lambda'_k}{l'_k}\Big] - \varepsilon_k, \tag{7}$$

where ε_k is equal to 0 in general and only then has the value 1, if the multiplicator $\mu_k = 1$ is present.

These equations (6) and (7) one wants to combine with (5) resp. (6) on p. 100; and one finds:

$$\frac{m_k}{l_k} + \frac{\overline{m}_k}{l_k} = 1 - \frac{1}{l_k} \quad \text{resp.} \quad \frac{m'_k}{l'_k} + \frac{\overline{m}'_k}{l'_k} = 1 - \varepsilon_k. \tag{8}$$

For the numbers m and $\overline{m}$, in consequence of (7) on p. 99, the following hold:

$$m = \frac{d}{\nu} - \sum_{k=1}^{n}{}' \frac{m_k}{l_k}, \qquad \overline{m} = \frac{\overline{d}}{\nu} - \sum_{k=1}^{n}{}' \frac{\overline{m}_k}{l_k}.$$

On the basis of (8), with utilization of the abbreviation terminology $\sum \varepsilon_k = q$, one finds from this:

$$m + \overline{m} = \frac{d + \overline{d}}{\nu} - \sum_{k=1}^{n} \Big(1 - \frac{1}{l_k}\Big) + q,$$

whereby q is the number of the parabolic multiplicators 1, *which occur in the system* $\mu_1, \mu_2, \ldots, \mu_n$. On the basis of the relations (1) on p. 97 and (2) on p. 102, the last equation goes over into:

$$m + \overline{m} + 2 = q. \tag{9}$$

We make note of this result in the form of the theorem: *For two multiplicator systems* M *and* $\overline{\mathrm{M}}$, *inverse to one another, and two dimensions d and $\overline{d}$ bound to one another by the relation* (2) *on p. 102, the associated whole numbers m, m_k resp. m'_k, $\overline{m}$, $\overline{m}_k$ resp. $\overline{m}'_k$ are joined by the relations* (8) *and* (9), *where ε_k and q have the given meaning.* [105]

2.14 Integral forms and forms with prescribed poles

For the number of poles, which an automorphic form of given multiplicator system and given dimension might have, the number m is, above all, the determining factor.

From the determinations (5) on p. 99 follows:

$$0 \leqslant \frac{m_k}{l_k} \leqslant 1 - \frac{1}{l_k} \quad \text{resp.} \quad 0 \leqslant \frac{m'_k}{l'_k} < 1,$$

$$0 \leqslant \sum_{k=1}^{n}{}' \frac{m_k}{l_k} \leqslant \sum_{k=1}^{n} \left(1 - \frac{1}{l_k}\right),$$

where the second equality sign in the last inequality can only hold, if all U_k are elliptic. With utilization of (7) on p. 99, it now follows after a short intermediate calculation: *for the number m, we have:*

$$\frac{d}{\nu} \geqslant m \geqslant \frac{d}{\nu} - \sum_{k=1}^{n} \left(1 - \frac{1}{l_k}\right), \tag{1}$$

where again, the second equality sign can only hold, if all U_k are elliptic.

In the application of these conditions, we may assume $\nu < 0$, by considering the elementary automorphic models (for which $\nu > 0$ resp. $\nu = \infty$) as excluded.

If we now assume, that the dimension $d > 0$, then $m \leqslant -1$ immediately follows from (1). For a form $\varphi_d(\zeta_1, \zeta_2)$ of positive dimension d, the dimension m of the rational form $R_m(z_1, z_2)$ coming into force in the representation (6) on p. 99 is always $\leqslant -1$.

For the negative dimension d, we want to draw a few further conclusions under the assumption $d \leqslant -2$. These dimensions will become especially important in the theory of Poincaré series. On account of $\nu < 0$, it follows:

$$\frac{d}{\nu} - \sum_{k=1}^{n} \left(1 - \frac{1}{l_k}\right) \geqslant -\frac{2}{\nu} - \sum_{k=1}^{n} \left(1 - \frac{1}{l_k}\right) = -2.$$

If one now has $d < -2$, then $m > -2$ follows. *The forms of a negative dimension $d < -2$ have $m \geqslant -1$.*

For $d = -2$, we have $m \geqslant -2$. Should the equation $m = -2$ subsist here, then parabolic U_k may not come up, and $m_k = l_k - 1$ must hold for all $k = 1, 2, \ldots, n$. Then, however, in consideration of $d = -2$, it follows from (5) on p. 100, that for each of the n indices k, the number λ_k is divisible by $2l_k$, i.e., that the multiplicator system M = 1 is present. Upon occurrence of the latter condition, conversely, $m = -2$ always follows [106]
from $d = -2$ when U_k are elliptic. *The properly automorphic forms of the dimension -2 have $m = -2$ or $m \geqslant -1$, according as all U_k are elliptic or not; for the improperly automorphic forms of the dimension $d = -2$, $m \geqslant -1$ holds.*

After these preparatory remarks concerning the number m, we introduce the following notations. If a form $\varphi_d(\zeta_1,\zeta_2)$ is everywhere finite in the fundamental domain, then we call it "*pole-free*" or "*integral*"; otherwise, we denote it as "*one-poled*", "*two-poled*" etc., according to the number of poles. The form represented in (4) on p. 98 would be called σ*-poled.* It is expedient, to hold fast to this enumeration, even if a pole moves into a vertex of the fundamental domain, at which a zero of fractional order is situated (whereby, naturally, a partial compensation of the poles and the zeros takes place); *i.e., we will not reduce the order of the pole in such a case and correspondingly naturally retain the zero unshortened.*

If a multiplicator system M and dimension d are given in advance, and if one will, by means of the scheme (4) on p. 98, construct a form with the least possible number of zeros and poles, then one must take either $\tau = m$, $\sigma = 0$ (namely, for $m \geqslant 0$) or $\tau = 0$, $\sigma = -m$ (for $m < 0$). Should $\sigma = 0$, then m must be $\geqslant 0$ and therefore $d < 0$ occurs. *The integral automorphic forms all have negative dimension d in ζ_1, ζ_2.* From the remarks above concerning m follows further: *The forms of a dimension $d \leqslant -2$ with the "minimal number" of zeros and poles are either one-poled or integral; only the properly automorphic forms of dimension -2 without parabolic generator U_k are at least two-poled.*

For the illustration of these statements, first of all let the *group* Γ *of the signature* (0,3;2,3,7) serve. As we already determined on p. 93, there is only the *one* multiplicator system M = 1 for this group, so that all single-valued forms are properly automorphic and of even dimension. For the dimensions $d = -2, -4, \ldots$, the forms with the minimal number of poles and zeros are the following here:

$$\varphi_{-2}(\zeta_1,\zeta_2) = \frac{(z,e_1)^{1/2}(z,e_2)^{2/3}(z,e_3)^{6/7}}{G_2(z_1,z_2)},$$

$$\varphi_{-4}(\zeta_1,\zeta_2) = \frac{(z,e_2)^{1/3}(z,e_3)^{5/7}}{G_1(z_1,z_2)},$$

$$\varphi_{-6}(\zeta_1,\zeta_2) = \frac{(z,e_1)^{1/2}(z,e_3)^{4/7}}{G_1'(z_1,z_2)},$$

$$\ldots\ldots,$$

[107] where the G_2, G_1, G_1' are integral forms of the second resp. first dimension in z_1, z_2, in which m in each case can be chosen arbitrarily. The first integral form enters for $d = -12$, namely, the ground form $(z,e_3)^{1/7}$. The forms following, for $d = -14, -16, \ldots$, are then again first one-poled.

In the *modular group* Γ, one has twelve multiplicator systems, which in the sense of p. 86, form a cyclic group. One can produce the latter from the multiplicator system belonging to odd d:

$$\mu_1 = e^{\frac{\pi i}{2}},\ \mu_2 = e^{\frac{\pi i}{3}}, \mu_3 = e^{\frac{\pi i}{6}}. \tag{2}$$

For this multiplicator system, we have for the dimension $d = -1$, from (5) on p. 100, the values $m_1 = 0$, $m_2 = 0$, while $m_3' = 1$ is to be taken equal to 1. The equation (7) on p. 99 gives, further, $m = 0$, so that, apart from a multiplicative constant, we obtain only the *one* integral form $\sqrt[12]{\Delta}$, understanding by Δ the modular form of first level (discriminant) thus denoted. For the further dimensions following, $d = -3, -5, -7, -9, \ldots$, the formulas (5) ff. on p. 100 furnish, as the forms with the minimal number of zeros and poles belonging to the multiplicator system (2):

$$\frac{g_3 g_2^2 \sqrt[12]{\Delta}}{G_1(g_2^3, g_3^2)}, \; g_2 \sqrt[12]{\Delta}, \; g_3 \sqrt[12]{\Delta}, g_2^2 \sqrt[12]{\Delta}, \ldots;$$

Here, g_2 and g_3 are the known modular forms of the $(-4)^{\text{th}}$ and $(-6)^{\text{th}}$ dimensions. Only for $d = -3$ is there a one-poled form, while all remaining forms are integral. For the properly automorphic forms of the $(-2)^{\text{nd}}$ dimension, we have $m = -1$ here; the one-poled forms of this kind are:

$$\frac{g_3 g_2^2}{G_1(g_2^3, g_3^2)}.$$

With consideration of the application in the following chapter, we fix here one more pair of statements concerning forms with σ poles.

If we utilize the representation (4) resp. (6) on p. 98 for the individual such form $\varphi_d(\zeta_1,\zeta_2)$, then in the numerator of $R_m(z_1, z_2)$ there stands a rational integral form of the dimension $\tau = m + \sigma$. Since this has $(m + \sigma + 1)$ terms, the theorem holds: *If the multiplicator system* M *and the dimension d are given, and σ poles (of the first order) are arbitrarily prescribed, there are only then associated forms, if $m + \sigma \geqslant 0$; and indeed, the most general form then lets itself be built up linearly and homogeneously with constant coefficients from $(m + \sigma + 1)$ particular ones.* Naturally the poles may coincide in any manner and furnish poles of higher order in such a way; also, the case of the pole free forms with $\sigma = 0$ is to be included. [108]

Thc following discussions have a more incidental character. If, of the σ poles just prescribed, σ_0 coincide as the place given by $z = y_1$, then we write the representation (4) on p. 98 for $\varphi_d(\zeta_1,\zeta_2)$ thus:

$$\varphi_d(\zeta_1,\zeta_2) = (z, y_1)^{-\sigma_0}(A_0 z_1^\tau + A_1 z_z^{\tau-1} z_2 + \cdots + A_\tau z_2^\tau) \frac{\prod{}'(z, e_k)^{\frac{m_k}{l_k}}}{(z, y_2)(z, y_3)\cdots}. \tag{3}$$

If, for the neighborhood of the pole in question, we now carry out a development of $\varphi_d(\zeta_1,\zeta_2)$ according to powers of the associated development quantity t, which, according to the position of the poles, is given by (4) on p. 68 resp. (8) on p. 69 or (10) on p. 70, then one easily makes clear to oneself with the aid of the equation just given, that the coefficients of the powers of t thereby coming up will be built up linearly and homogeneously in the yet unknown coefficients $A_0, A_1, \ldots, A_\tau$ on the right side of (12). If now, in the series development under discussion of the $\varphi_d(\zeta_1,\zeta_2)$, the σ_0 coefficients of the powers of t with negative exponents are given in advance, then we shall say, $\varphi_d(\zeta_1,\zeta_2)$ becomes infinite as the place in question *"in prescribed fashion"*. For the $A_0, A_1, \ldots, A_\tau$, then, σ_0 linear equations are given. With this follows: *There are, for given* M *and d, still $(m+1)$ linearly independent forms, which have prescribed poles and become infinite at the individual one of these in prescribed fashion.*

Here, initially, $m + 1$ is considered > 0. For $m + 1 = 0$, there are exactly $\sigma = \tau + 1$ equations for the $(\tau + 1)$ coefficients A. *For $m + 1 = 0$, $\varphi_d(\zeta_1,\zeta_2)$ is uniquely determined by the way it becomes infinite.* Correspondingly, it follows further: *If $m + 1 < 0$, then for prescribed manner of becoming infinite, there can only then be an associated form,* *if the prescribed coefficients of the powers of t with negative exponents satisfy* $-(m+1)$ *relations, determined in the individual case.*

For the properly automorphic forms of dimension -2, we have arrived back as our initial point for the formation of automorphic forms, namely to the differentiation process (cf. p. 71). If we now, conversely, multiple all the properly automorphic forms of $(-2)^{\text{nd}}$ dimension with $(\zeta, d\zeta)$ and integrate, we would not only obtain all the automorphic functions, but also even the *integrals* of an automorphic model. This development, however, first has a further interest for the automorphic models of the genus
[109] $p > 0$. By accordingly not failing to deviate out attention from natural examples, as, say, the representations belonging to the theory of modular functions:

$$J = \frac{9i}{\pi}\int g_2^3 g_3 \Delta^{-1}\cdot(\zeta, d\zeta),\ \log J = \frac{9i}{\pi}\int g_2^{-1} g_3\cdot(\zeta, d\zeta),\ldots,$$

we shall push aside all detailed investigations on the "integrations" of the form of $(-2)^{\text{nd}}$ dimension until the discussion of the automorphic models of an arbitrary genus p.

The theory of single-valued automorphic forms, hereby coming to a preliminary conclusions, has familiarized us with the basic properties of these forms in the fundamental domain and with their behavior with respect to the substitutions of the group, and last of all, has also informed us on the manifold of the forms with given d and M. The following chapter will bring us, essentially new viewpoints, that we obtain, in the Poincaré series, analytic representations for our forms, for which the automorphic character comes immediately into appearance. Before we go over to this, however, the theory of the polymorphic functions laid out on p. 43 ff. is yet to be developed further form-theoretically in the case $p = 0$.

2.15 The ζ_1, ζ_2 as linearly-polymorphic forms of the z_1, z_2

The inversion of an isolated automorphic function $z = \varphi(\zeta)$ led us above (p. 43) to the definition of associated "*linearly-polymorphic*" or briefly "*polymorphic*" functions $\zeta = f(z)$, which substituted itself linearly for circuits on the underlying Riemann surface. Since we will now take up the corresponding form-theoretic developments, we naturally restrict ourselves, here again, initially, only to models of the genus $p = 0$. For a proposed model of this kind, let $z = \varphi(\zeta)$ be an associated principal function. The inverse function $\zeta = f(z)$ is then a polymorphic function, which is branched at the n places $z = e_1, e_2, \ldots, e_n$ in known fashion, but is single-valued in the neighborhood of each other place z.

Now, by setting $z = z_1 : z_2$, we introduce, in place of z, the "binary variables" z_1, z_2 and prescribe in the usual manner, *that* z_1, z_2 *is always to be finite and, at the same time, may never vanish.* The value systems z_1, z_2 then still admissible are the "*allowed*
[110] *value pairs*". The first question is to be, *now we are able to split the polymorphic function* $\zeta = f(z)$ *into the quotient* $\zeta_1 : \zeta_2$ *of two forms:*

$$\zeta_1 = f_1(z_1, z_2),\ \zeta_2 = f_2(z_1, z_2), \tag{1}$$

in such a way, that to allowed value pairs of the arguments z_1, z_2, again, always correspond only allowed value pairs ζ_1, ζ_2.

This splitting, as emerges from the discussions on p. 73, is now already given by the expression (3) on p. 73 of the prime form z_2. By replacing z by $z_1 : z_2$ in this expression, as well as, on the other hand, putting $-\zeta_2^2 d\zeta$ for $(\zeta, d\zeta)$ then after easy intermediate calculation, the equations arise:

$$\left\{\begin{aligned} \zeta_1 &= f_1(z_1, z_2) = C\frac{\zeta}{\sqrt{\frac{d\zeta}{dz}}} z_2 \prod_{k=1}^{n} (z, e_k)^{-\frac{1}{2}\left(1-\frac{1}{l_k}\right)}, \\ \zeta_2 &= f_2(z_1, z_2) = C\frac{1}{\sqrt{\frac{d\zeta}{dz}}} z_2 \prod_{k=1}^{n} (z, e_k)^{-\frac{1}{2}\left(1-\frac{1}{l_k}\right)}, \end{aligned}\right. \tag{2}$$

where C signifies a certain constant. The expression (3) on p. 73 having served as the definition of the prime form z_2 for given ζ_1, ζ_2, then, conversely, we are able, by the formulas (2) just given, and indeed, with arbitrary finite, non-vanishing constant C, for given z_1, z_2, to define the forms ζ_1, ζ_2 sought in the most general manner. These forms ζ_1, ζ_2 are of the dimension:

$$\lambda = \nu^{-1} = 1 - \frac{1}{2}\sum_{k=1}^{n}\left(1 - \frac{1}{l_k}\right) \tag{3}$$

and, in fact, fulfill the requirement, of always again assuming only allowed value systems for allowed value pairs z_1, z_2.

For the investigation of the forms ζ_1, ζ_2 in the neighborhood of an individual place z_1, z_2, similar arguments are basic, as above (p. 68 ff.) for the automorphic forms: *The forms ζ_1, ζ_2 are branched in the z-plane at the n places $e_1, e_2, \ldots, e_n$; further branch-points do not occur.* If the dimension λ is integral, then the ζ_1, ζ_2 are single-valued at each of the places z different from the e_k. If one has fractional dimension λ, then the ζ_1, ζ_2, for z held constant, already show themselves to be multiple-valued with respect to circuits of the variable z_2 about its origin (cf. the corresponding considerations on p. 67 with respect to the automorphic forms).

For the dimension λ, the inequality obviously holds:

$$1 - \frac{n}{2} \leqslant \lambda \leqslant 1 - \frac{n}{4}. \tag{4}$$

An integral dimension λ does not come up at all for $n = 3$. For $n = 4$, one only has [111]
an integral dimension in the one case $l_1 = l_2 = l_3 = l_4 = \infty$, namely $\lambda = -1$. But already in the next following case $n = 5$ one counts off 156 cases with integral λ, namely $\lambda = -1$; examples are:

$$l_1 = 2,\ l_2 = 3,\ l_3 = 7,\ l_4 = 43, l_5 = 1806,$$
$$l_1 = 2,\ l_2 = 5,\ l_3 = 6,\ l_4 = 12, l_5 = 20,$$
$$l_1 = 3,\ l_2 = 4,\ l_3 = 5,\ l_4 = 6, l_5 = 20,$$
$$\ldots\ldots$$

In the elementary cases, where the principal function $z=\varphi(\zeta)$ is rational or elliptic, λ becomes positive resp. equal to zero. The majority of the following developments remain in force in these cases; nevertheless, these are to count as excluded[22].

The principal property of the ζ_1, ζ_2 consists in their *behavior with respect to closed circuits of the variable pair* z_1, z_2 in their value domains. In the investigation of this behavior of the ζ_1, ζ_2, we shall all the less need to spend time, since we were very detailed above (p. 75 ff.) in the corresponding consideration of the automorphic forms. Obviously, we now have to do with the reversal of development at that time. We next note the theorem:

If, for fixed z, the z_1, z_2 alter themselves by a common multiplicator μ, then correspondingly, as forms of dimension λ, ζ_1, ζ_2 assume a common factor μ^λ, which is uniquely determined in each case by the path described by z_2. On the strength of this, we easily conclude from the developments on p. 75 ff. just cited:

For a closed circuit by the variable pair z_1, z_2 in its value domain, the forms ζ_1, ζ_2 go over into linear homogeneous functions of themselves:

$$\zeta_1'=\alpha\zeta_1+\beta\zeta_2,\ \zeta_2'=\gamma\zeta_1+\delta\zeta_2; \tag{5}$$

this linear substitution, in case the circuit does not run right through a branch place e_k, is uniquely determined in each case, and, in particular, the determinant $(\alpha\delta-\beta\gamma)$ is equal to an arbitrary root of unity.

In order to bring to expression the fundamental property of the forms ζ_1, ζ_2 hereby coming to light, we will designate these as "*linearly-polymorphic forms*"; in abbrevia-
[112] tion, we use the name "*polymorphic forms*".

In order to explicitly give the substitutions (5) for simple circuits about the branch-points e_k, we first produce, from the ζ_1, ζ_2, with arbitrary finite and non-simultaneously vanishing parameters A, B, the "*linear family of forms*":

$$A\zeta_1+B\zeta_2=Af_1(z_1,z_2)+Bf_2(z_1,z_2). \tag{6}$$

Any two forms chosen from this family, whose quotient is not constant, are obviously able to be utilized in place of ζ_1, ζ_2 for the representation of the entire family.

We now turn ourselves to an individual branch-point e_k, to which we let correspond an *elliptic* polygon vertex. For the neighborhood of this point e_k, we take, from the family (6), the two particular forms:

$$(\zeta,\varepsilon_k)=\zeta_1\varepsilon_k^{(2)}-\zeta_2\varepsilon_k^{(1)},\ (\zeta,\varepsilon_k')=\zeta_1\varepsilon_k'^{(2)}-\zeta_2\varepsilon_k'^{(1)}.$$

According to the developments on p. 69, a representation holds for $(z,e_k)=z_1e_k^{(2)}-z_2e_k^{(1)}$:

$$(z,e_k)=(\zeta,\varepsilon_k')^\nu\left(\frac{\zeta-\varepsilon_k}{\zeta-\varepsilon_k'}\right)^{l_k}\left\{a_0+a_1\left(\frac{\zeta-\varepsilon_k}{\zeta-\varepsilon_k'}\right)^{l_k}+\cdots\right\},$$

where $a_0\neq 0$. From this follows, conversely:

$$(\zeta,\varepsilon_k')^\nu=e_k^{(2)}z_2[a_0'+a_1'(z-e)+\cdots],$$

22 See, besides this, "Icos" on p. 71 ff.

again with $a_0' \neq 0$, so that (ζ, ε_k') is itself a finite and unbranched form in the neighborhood of e_k.[23] On the strength of this, one easily also obtains a corresponding representation for (ζ, ε_k). *The two special forms (ζ, ε_k) and (ζ, ε_k') permit, in the neighborhood of e_k, the representations:*

$$\begin{cases} (\zeta, \varepsilon_k) = (z, e_k)^{\frac{1}{l_k}} F_1(z_1, z_2), \\ (\zeta, \varepsilon_k') = F_2(z_1, z_2), \end{cases} \tag{7}$$

where F_1 and F_2 are forms which are finite, different from zero and unbranched in the neighborhood of e_k mentioned.

The form $F_2(z_1, z_2)$ can now be written as $z_2^{\lambda} F_2(z, 1)$, where $F_2(z, 1)$ is a single-valued function in the neighborhood of e_k. If we therefore describe a small closed curve with z about the branch-point e_k in the positive sense of the circuit, while z_2 remains fixed, then the form $F_2(z_1, z_2)$ will remain unaltered. By reaching back to the known behavior of the quotient $(\zeta, \varepsilon_k) : (\zeta, \varepsilon_k')$ for that circuit of z, we find as a result: if, for fixed z_2, z circles the branch-point e_k of elliptic character in the positive sense, then the two forms (ζ, ε_k) and (ζ, ε_k'), suitable for the representation of the entire family (6), go over into: [113]

$$(\zeta', \varepsilon_k) = e^{\frac{2i\pi}{l_k}} (\zeta, \varepsilon_k), \; (\zeta', \varepsilon_k') = (\zeta, \varepsilon_k'). \tag{8}$$

If, to the branch-point e_k, there corresponds a *parabolic* polygon vertex, then we refer the entire family (6) to those two forms, which are furnished by the expressions (ζ, ε_k), (ζ, ε_k') defined on p. 65 for this case. For these forms, one calculates from the developments on p. 70 as *representations for the neighborhood of the parabolic branch-point e_k:*

$$\begin{cases} (\zeta, \varepsilon_k) = F_1(z_1, z_2), \\ (\zeta, \varepsilon_k') = \log(z - e_k) F_1(z_1, z_2) + F_2(z_1, z_2), \end{cases} \tag{9}$$

where again, $F_1(z_1, z_2)$ and $F_2(z_1, z_2)$ signify forms, which are finite, non-vanishing and unbranched in the neighborhood in question.

On the basis of these representations, we now immediately find, further: *If, for fixed z_2, z circles the parabolic branchpoint e_k in the positive sense, then the forms (ζ, ε_k) and (ζ, ε_k') go over into:*

$$(\zeta', \varepsilon_k) = (\zeta, \varepsilon_k), (\zeta', \varepsilon_k') = 2i\pi(\zeta, \varepsilon_k) + (\zeta, \varepsilon_k'). \tag{10}$$

Here, therefore, we come back directly to the substitution (5) on p. 65.

We would now properly have to add one more investigation, which has as its purpose the generalization of the "primitive" circuits by the z_1, z_2 initially described here. But since the investigation is precisely analogous to that on p. 75ff., we leave its further effectuation to the reader.

---

[23] For the sake of brevity, it is assumed in the calculations of the text, that $e_k^{(2)}$ does not vanish, or (in other words) that the branch-point under consideration is not situated at $z = \infty$ $(z_2 = 0)$ However, the result given in (1) in the text also remains subsisting for $e_k^{(2)} = 0$.

2.16 Other forms of the polymorphic forms. History

For the introduction of the polymorphic forms resp. the splitting of the polymorphic functions $\zeta = f(z)$ into two binary quantities, one often lets the two expressions of zero
[114] dimension:

$$Z_1 = \frac{\zeta}{\sqrt{\frac{d\zeta}{dz}}}, \quad Z_2 = \frac{1}{\sqrt{\frac{d\zeta}{dz}}} \tag{1}$$

come to the forefront of consideration, in place of ζ_1, ζ_2.[24] In this way, e.g., Riemann engineered the splitting of ζ in the substantial fragment "*Gleichgewicht der Electricität auf Cylindern mit Kreisförmigem Querschnitt und parallelen Axen*"[25]. One shows without trouble by direct calculation, *that these "polymorphic forms of zero dimension" substitute themselves "unimodularly" in the form:*

$$Z_1' = \alpha Z_1 + \beta Z_2, \; Z_2' = \gamma Z_1 + \delta Z_2, \tag{2}$$

in this case ζ goes over into $\frac{\alpha\zeta+\beta}{\gamma\zeta+\delta}$ upon a circuit by z. Naturally, in (2), in each case, one has before oneself a determinate one of the two homogeneous unimodular substitutions, which correspond to the non-homogeneous ζ substitution.

In more detail, the developments given by Riemann at the place cited belong to such automorphic models, which we placed under β on p. 35; in particular, it is a question of groups with semipolygons, which are bounded by $(p+1)$ separately running symmetry circles. However, the essential viewpoints of the Riemann development in question also retain their validity for the remaining automorphic models.

Our forms ζ_1, ζ_2 above are connected with the Z_1, Z_2 by the relations:

$$\begin{cases} CZ_1 = \zeta_1 \cdot z_2^{-1} \prod_{k=1}^{n} (z, e_k)^{\frac{1}{2}\left(1-\frac{1}{l_k}\right)}, \\ CZ_2 = \zeta_2 \cdot z_2^{-1} \prod_{k=1}^{n} (z, e_k)^{\frac{1}{2}\left(1-\frac{1}{l_k}\right)}. \end{cases} \tag{3}$$

We find: *The Z_1, Z_2 no longer satisfy the condition of assuming exclusively "allowed value systems" for "allowed" value pairs z_1, z_2; rather, both quantities Z_1, Z_2 have a pole of the first order at $z = \infty$ and a zero of the order $\frac{1}{2}\left(1-\frac{1}{l_k}\right)$ at each individual branch-point.* The Z_1, Z_2 are distinguished, insofar as they are of zero dimension in the z_1, z_2;
[115] but, for this, one must lose, in the bargain, the unnecessary property just named.

In place of (7) on p. 112, we get the representations for an elliptic branch-point e_k:

$$\begin{cases} (Z, \varepsilon_k) = (z-e_k)^{\frac{1}{2}\left(1+\frac{1}{l_k}\right)} [a_0 + a_1(z-e) + a_2(z-e)^2 + \cdots], \\ (Z, \varepsilon_k') = (z-e_k)^{\frac{1}{2}\left(1-\frac{1}{l_k}\right)} [a_0' + a_1'(z-e) + a_2'(z-e)^2 + \cdots], \end{cases} \tag{4}$$

whereby $a_0 \neq 0$, $a_0' \neq 0$ holds. For a parabolic point e_k, we have, correspondingly:

---

[24] Above (p. 74 ff.), we have denoted the ground forms belonging to the n fixed polygon vertices by $Z_1, Z_2, Z_3, \ldots$. A confusion of the expressions denoted here by Z_1, Z_2 with those ground forms is scarcely to be feared.

[25] Remarked by H.Weber from bequeathed indications and formulas; see Riemann's gesammelte Werke on p. 413.

$$\begin{cases}(Z,\varepsilon_k)=(z-e_k)^{\frac{1}{2}}[a_0+a_1(z-e)+a_2(z-e)^2+\cdots],\\(Z,\varepsilon_k')=(z-e_k)^{\frac{1}{2}}\{\log(z-e_k)[a_0+a_1(z-e)+\cdots]+[a_0'+a_1'(z-e)+\cdots]\},\end{cases}\tag{5}$$

again with $a_0\neq 0$, $a_0'\neq 0$. The consequence is, that in place of the formulas (8) and (10) on p. 113, we get the following:

$$(Z',\varepsilon_k)=-e^{\frac{\pi i}{l_k}}(Z,\varepsilon_k),\ (Z',\varepsilon_k')=-e^{-\frac{\pi i}{l_k}}(Z,\varepsilon_k');\tag{6}$$

$$(Z',\varepsilon_k)=-(Z,\varepsilon_k),\ (Z',\varepsilon_k')=-2i\pi(Z,\varepsilon_k)-(Z,\varepsilon').\tag{7}$$

Here in fact, as one sees, unimodular substitutions are present both times.

With the Z_1, Z_2 and the linear family to be produced from these quantities:

$$AZ_1+BZ_2,\tag{8}$$

we have now also attained a direct connection with the "*Riemannian P-function*". The Riemannian developments in the paper "*Beiträge zur Theorie der durch die Gauss'sche Reihe $F(\alpha,\beta,\gamma;x)$ darstellbaren Functionen*"[26], only first refer themselves to the case of three branch-points; however, approaches for the generalization already find themselves in the bequeathed paper, stemming from the year 1857, "*Zwei allgemeine Sütze über lineare Differentialgleichungen mit algebraischen Coefficienten*"[27]. If we utilize the terminology used by Riemann, then we would denote any form from the linear family (8) by the symbol:

$$P\begin{Bmatrix}e_1, & e_2, & \ldots, & e_n, & \\ \frac{1}{2}\left(1+\frac{1}{l_1}\right), & \frac{1}{2}\left(1+\frac{1}{l_2}\right), & \ldots, & \frac{1}{2}\left(1+\frac{1}{l_n}\right), & z\\ \frac{1}{2}\left(1-\frac{1}{l_1}\right), & \frac{1}{2}\left(1-\frac{1}{l_2}\right), & \ldots, & \frac{1}{2}\left(1-\frac{1}{l_n}\right), & \end{Bmatrix}.\tag{9}$$

Hereby, under the notation e_k of the individual branch-point are placed the two associated "exponents" [cf. (4) and (5)]. [116]

Klein has first representated the form-theoretic development of the Riemannian P-function in the paper "*Über Normierung der linearen Differentialgleichungen zweiter Ordnung*"[28] and repeatedly explained it in lectures[29]. The standpoint there is even somewhat more general than that developed here so far. We choose a series of real exponents $\lambda_1'',\lambda_2'',\ldots,\lambda_n''$ arbitrarily, and write:

$$\lambda_1''+\frac{1}{l_1}=\lambda_1',\ \lambda_2''+\frac{1}{l_2}=\lambda_2',\ldots,\lambda_n''+\frac{1}{l_n}=\lambda_n',\tag{10}$$

so that the exponent pairs λ_k', λ_k'' have prescribed differences $\frac{1}{l_k}$. In consequence of (7) on p. 112 and (9) on p. 113, we have, in:

---

26 In the 7th volume of the Abhandlungen der Gesellsch. d. Wiss. zu Göttingen (1857); cf. also Riemann's gesammelte Werke on p. 62.

27 Riemann's gesammelte Werke on p. 357.

28 Mathem. Annalen Vol. 38 on p. 144 (1890).

29 It is a question of the lectures on linear differential equations cited in the foreword to the first volume on page VI, as well as lectures on the hyper geometric function.

$$\zeta_1 \prod_{k=1}^{n} (z, e_k)^{\lambda''_k}, \ \zeta_2 \prod_{k=1}^{n} (z, e_k)^{\lambda''_k}, \tag{11}$$

forms of these exponents; and we denote an arbitrary form of the associated linear family by:

$$P\left\{\begin{matrix} e_1, \ e_2, \ \dots, \ e_n, \\ \lambda'_1, \ \lambda'_2, \ \dots, \ \lambda'_n; \ z_1, \ z_2 \\ \lambda''_1, \ \lambda''_2, \ \dots, \ \lambda''_n, \end{matrix}\right\}. \tag{12}$$

A form of our original family (6) on p. 112 would, in this sense, be denoted symbolically by:

$$P\left\{\begin{matrix} e_1, \ e_2, \ \dots, \ e_n, \\ \frac{1}{l_1}, \ \frac{1}{l_2}, \ \dots, \ \frac{1}{l_n}; \ z_1, \ z_2 \\ 0, \ 0, \ \dots, \ 0, \end{matrix}\right\}. \tag{13}$$

The dimension of the forms (12) in the z_1, z_2 determines itself to be:

$$1 + \sum_{k=1}^{n} \frac{\lambda'_k + \lambda''_k - 1}{2}. \tag{14}$$

[117] All these families of forms furnish representations of ζ as quotients of two linearly-polymorphic forms. But only the forms ζ_1, ζ_2 considered originally satisfy, as we emphasize once more, the condition, of always being finite and never vanishing, in case this also holds for the arguments z_1, z_2. Corresponding to this, we shall also have to expect, that for the following developments concerning the differential equations of the second order of the polymorphic forms, the ζ_1, ζ_2 will lead to especially elegant and transparent results.

2.17 Differential equations of second order for the polymorphic forms of zero dimension

An individual polymorphic *function* $\zeta = f(z)$ satisfies a certain differential equation of third order, which, according to p. 50, in the case of the genus $p = 0$, has the form:

$$\frac{\dfrac{d^3\zeta}{dz^3}}{\dfrac{d\zeta}{dz}} - \frac{3}{2} \frac{\left(\dfrac{d^2\zeta}{dz^2}\right)^2}{\left(\dfrac{d\zeta}{dz}\right)^2} = 2R(z), \tag{1}$$

understanding by $R(z)$ the following rational function:

$$R(z) = \frac{1}{\prod_{k=1}^{n}(z - e_k)} \cdot \left\{ G_{n-4}(z) + \sum_{k=1}^{n} \frac{\frac{1}{4}\left(1 - \frac{1}{l_k^2}\right)}{z - e_k} (e_k - e_1) \cdots (e_k - e_{k-1})(e_k - e_{k+1}) \cdots (e_k - e_n) \right\}. \tag{2}$$

For $n = 3$, the first term G_{n-4} in the parenthesis falls out; for $n > 3$, $G_{n-4}(z)$ signifies an integral rational function of the $(n-4)^{\text{th}}$ degree, whose $(n-3)$ coefficients are uniquely determined in each case, i.e., for individual automorphic models and models with a chosen principal function z, but are certainly unknown initially.

Corresponding to this, the polymorphic *forms* of the individual linear family satisfy a *linear differential equation of second order*, to whose closer investigation we will now turn ourselves. In order thereby to attach to older and more known investigations, we consider, in the present paragraph, initially, the linear differential equation of second order belonging to the *polymorphic forms of "zero dimension"*:

$$Z_1 = \frac{\zeta}{\sqrt{\frac{d\zeta}{dz}}}, \ Z_2 = \frac{1}{\sqrt{\frac{d\zeta}{dz}}}. \tag{3}$$

For this purpose, we understand by: [118]

$$Z = AZ_1 + BZ_2 \tag{4}$$

an arbitrary "form" of the linear family arising from the Z_1, Z_2. Next to equation (4), arrange those two equations, which arise by twice differentiating according to z, eliminate the two constants A and B between these three equations. We are able to invest the result of elimination with the form:

$$D_1 \frac{d^2Z}{dz^2} + D_2 \frac{dZ}{dz} + D_3 Z = 0,$$

where the coefficients D have the meaning:

$$\begin{aligned}
D_1 &= Z_1 \frac{dZ_2}{dz} - Z_2 \frac{dZ_1}{dz}, \\
D_2 &= Z_2 \frac{d^2Z_1}{dz^2} - Z_1 \frac{d^2Z_2}{dz^2}, \\
D_3 &= \frac{dZ_1}{dz}\frac{d^2Z_2}{dz^2} - \frac{dZ_2}{dz}\frac{d^2Z_1}{dz^2}.
\end{aligned}$$

For the more precise calculation of these coefficients D, we differentiate the equation $Z_2 : Z_1 = \zeta^{-1}$ according to z and find:

$$D_1 = -Z_1^2 \zeta^{-2} \frac{d\zeta}{dz} = -1,$$

from which it immediately follows:

$$D_2 = -\frac{dD_1}{dz} = 0.$$

Our differential equation of the second order for Z thus assumes the preliminary form:

$$\frac{d^2Z}{dz^2} = D_3 Z. \tag{5}$$

For the determination of D_3, differentiate $\zeta = Z_1 : Z_2$ according to z:

$$\frac{d\zeta}{dz} = -\frac{D_1}{Z_2^2} = \frac{1}{Z_2^2}.$$

By logarithmic differentiation of this equation follows:

$$\frac{\frac{d^2\zeta}{dz^2}}{\frac{d\zeta}{dz}} = -2\frac{\frac{dZ_2}{dz}}{Z_2},$$

from which, by one more differentiation, one obtains:

$$\frac{\frac{d^3\zeta}{dz^3}}{\frac{d\zeta}{dz}} - \frac{\left(\frac{d^2\zeta}{dz^2}\right)^2}{\left(\frac{d\zeta}{dz}\right)^2} = -2\frac{\frac{d^2Z_2}{dz^2}}{Z_2} + 2\frac{\left(\frac{dZ_2}{dz}\right)^2}{Z_2^2},$$

[119] If one utilizes the preceding equation here for the elimination of the last term on the right, and replaces the quotient of $\frac{d^2Z_2}{dz^2}$ and Z_2 in the first term on the right by D_3 on the basis of (5), then it follows:

$$\frac{\frac{d^3\zeta}{dz^3}}{\frac{d\zeta}{dz}} - \frac{3}{2}\frac{\left(\frac{d^2\zeta}{dz^2}\right)^2}{\left(\frac{d\zeta}{dz}\right)^2} = -2D_3.$$

The comparison with (1) furnishes, with this, $D_3 = -R(z)$.

In this way, we have arrived at the result: *The polymorphic "forms of zero dimension" Z_1, Z_2, as well as, generally each form Z from the linear family to be produced from it, satisfy the linear homogeneous differential equation of second order:*

$$\frac{d^2Z}{dz^2} + R(z)\cdot Z = 0, \tag{6}$$

where $R(z)$ is the rational function of z represented in equation (2) *and even more closely characterized there.* Hereby one has, in $(AZ_1 + BZ_2)$, in case A and B are considered as arbitrary constants here, the "general" integral of this differential equation.

We will also add an extension, not inessential for the sequel, for the case $n = 3$. Here, above (p. 51), we had also pursued the assumption, that the three values e_1, e_2, e_3 here, in particular, were identical with 0, 1, ∞. The differential equation of third order for $\zeta = f(z)$ then had the form (9) on p. 51. We argue from this immediately to the theorem: *If, in the case $n = 3$, one places the branch-points e_k at $z = 0, 1, \infty$, then the polymorphic forms Z_1, Z_2 of zero dimension satisfy the differential equation of second order*

$$\frac{d^2Z}{dz^2} + \left(\frac{1-\frac{1}{l_1^2}}{4z^2} + \frac{1-\frac{1}{l_2^2}}{4(z-1)^2} + \frac{\frac{1}{l_1^2}+\frac{1}{l_2^2}-\frac{1}{l_3^2}-1}{4z(z-1)}\right)Z = 0. \tag{7}$$

In the differential equations (6) and (7), the terms with the first derivative of z are lacking. According to known theorems of the theory of differential equations, it follows from this, that the two forms Z_1, Z_2 substitute themselves *unimodularly* for circuits by the z, which was also already determined above (p. 115). If, in the differential equation (6) or (7), we introduce the function $H = Z \cdot \Phi(z)$ in place of Z, understanding by $\Phi(z)$ a given function of z, then, in the transformed equation, in general, a term with the first derivative of H will come up; and the substitutions corresponding to the circuits will only then still be unimodular, if Φ^2 is a single-valued function of z. [120]

In particular, we will transform the differential equation (7) by means of:

$$Z = H \cdot z^{\frac{1}{2}\left(1-\frac{1}{l_1}\right)}(z-1)^{\frac{1}{2}\left(1-\frac{1}{l_2}\right)}. \tag{8}$$

Hereby, we introduce, in place of the l_1, l_2, l_3, three quantities α, β, γ by:

$$\begin{cases} \alpha = \dfrac{1}{2}\left(1 - \dfrac{1}{l_1} - \dfrac{1}{l_2} - \dfrac{1}{l_3}\right), \\ \beta = \dfrac{1}{2}\left(1 - \dfrac{1}{l_1} - \dfrac{1}{l_2} + \dfrac{1}{l_3}\right), \\ \gamma = 1 - \dfrac{1}{l_1}, \end{cases} \tag{9}$$

equations, which immediately invert into:

$$\frac{1}{l_1} = 1 - \gamma, \ \frac{1}{l_2} = \gamma - \alpha - \beta, \ \frac{1}{l_3} = \beta - \alpha. \tag{10}$$

As the form of the transformed equation, it follows, after brief intermediate calculation:

$$z(z-1)\frac{d^2H}{dz^2} + [(\alpha+\beta+1)z - \gamma]\frac{dH}{dz} + \alpha\beta H = 0; \tag{11}$$

but this is the *differential equation of the hypergeometric seties.* We are led thus to the result: *If, in the case* $n = 3$, *one places the branch-points* e_k *as* $0, 1, \infty$ *and splits the polymorphic function* ζ *into the quotient of the two quantities:*

$$\begin{cases} H_1 = \dfrac{\zeta}{\sqrt{\frac{d\zeta}{dz}}} z^{-\frac{1}{2}\left(1-\frac{1}{l_1}\right)}(z-1)^{\frac{1}{2}\left(1-\frac{1}{l_2}\right)}, \\ H_2 = \dfrac{1}{\sqrt{\frac{d\zeta}{dz}}} z^{-\frac{1}{2}\left(1-\frac{1}{l_1}\right)}(z-1)^{\frac{1}{2}\left(1-\frac{1}{l_2}\right)}, \end{cases} \tag{12}$$

then one has, in these H_1, H_2, *obtained integrals of the hypergeometric differential equation* (11). We have, with this, arrived back at the starting point of the oft-cited investigations of H. A. Schwarz[30].

2.18 Invariant form of the differential equation for the polymorphic forms ζ_1, ζ_2

[121]

We now return to the polymorphic forms $\zeta_1 = f_1(z_1, z_2)$, $\zeta_2 = f_2(z_1, z_2)$, which we had placed at the forefront above (p. 110). An arbitrary form of the associated family

30 "*Über diejenigen Fälle, in welchen die Gauss'sche hypergeometrische Reihe eine algebraische Function ihres rierten Argumentes ist.*" Journ. Math. Vol. 75 (1872).

$A\zeta_1 + B\zeta_2$ we call $f(z_1, z_2)$. Its dimension λ in z_1, z_2 is represented in (3) on p. 110. The forms $f(z_1, z_2)$ distinguish themselves from the forms Z of zero dimension of the previous paragraph by a factor, which we already determined on p. 114; according to the formulas (3) at that time, we have to let:

$$f = z_2^{\lambda}\Phi(z)Z,\ \Phi(z) = \prod_{k=1}^{n}(z-e_k)^{-\frac{1}{2}\left(1-\frac{1}{l_k}\right)}. \tag{1}$$

In order now to obtain the *differential equation of second order of the polymorphic forms* $f(z_1, z_2)$, we have to calculate, in terms of f, the equation (6) on p. 119 set up for Z, by means of the relation just given between Z and f. Hereby, our investigations peaks in the obtention of an *invariant* representation of the differential equation in question for the forms f. In this respect, notice beforehand, that the exercise of the transformation:

$$z_1' = az_1 + bz_2,\ z_2' = cz_1 + dz_2, \tag{2}$$

in the sense of our original developments, amounts to another choice of the principal function of our automorphic model. The conjugacy of all these principal functions has as consequence here, conversely, *that the theory of polymorphic forms will have the character of invariance with respect to the transformation* (2). This situation will, in fact, now come immediately into expression in the form of the differential equation for the forms f, insofar as we shall build up the differential equation by means of *covariant* constructions, namely from transvectants.

We have thereby to define the transvectant precess for arbitrary algebraic or transcendental forms. If $\varphi(z_1, z_2)$ and $\psi(z_1, z_2)$ are any two forms, then we denote the νth transvectant of φ and ψ symbolically by $(\varphi, \psi)_\nu$ and define this explicitly by[31]:

$$(\varphi,\psi)_\nu = \frac{\partial^\nu\varphi}{\partial z_1^\nu}\cdot\frac{\partial^\nu\psi}{\partial z_2^\nu} - \binom{\nu}{1}\frac{\partial^\nu\varphi}{\partial z_1^{\nu-1}\partial z_2}\cdot\frac{\partial^\nu\psi}{\partial z_2^{\nu-1}\partial z_1} + \cdots + (-1)^\nu\frac{\partial^\nu\varphi}{\partial z_2^\nu}\cdot\frac{\partial^\nu\psi}{\partial z_1^\nu}.$$

[122] Further on, only the first three cases come onto consideration:

$$\begin{cases} (\varphi,\psi)_0 = \varphi\cdot\psi, \\ (\varphi,\psi)_1 = \dfrac{\partial\varphi}{\partial z_1}\dfrac{\partial\psi}{\partial z_2} - \dfrac{\partial\varphi}{\partial z_2}\dfrac{\partial\psi}{\partial z_1}, \\ (\varphi,\psi)_2 = \dfrac{\partial^2\varphi}{\partial z_1^2}\dfrac{\partial^2\psi}{\partial z_2^2} - 2\dfrac{\partial^2\varphi}{\partial z_1\partial z_2}\dfrac{\partial^2\psi}{\partial z_1\partial z_2} + \dfrac{\partial^2\varphi}{\partial z_2^2}\dfrac{\partial^2\psi}{\partial z_1^2}, \end{cases} \tag{3}$$

where, in particular, for $\nu = 0$, we have the product of both forms φ, ψ before us and for $\nu = 1$, their functional determinant.

In order, furthermore, not to have to present invariant-theoretic considerations going too far, we attach the derivation of the desired differential equation for the polymorphic forms $f(\zeta_1, \zeta_2)$ as closely as possible to the non-homogeneous terminology.

---

[31] See, e.g., Gordan-Kerschensteiner "*Vorlesungen über Invariantentheorie*" (Leipzig, 1887) Vol. 2 on p. 36. Moreover, the expression $(\varphi, \psi)_\nu$ of the text deviates by a factor from the definition of the νth transvectant given l.e.. This modification was undertaken for the purpose of simplifying the formulas to be developed later on.

Let the product of all n expressions (z, e_k) belonging to the branch-points be called:

$$u(z_1, z_2) = \prod_{k=1}^{n} (z, e_k) = \prod_{k=1}^{n} (z_1 e_k^{(2)} - z_2 e_k^{(1)}). \tag{4}$$

We have defined, in $u(z_1, z_2)$, a first rational form important for setting-up the differential equation in question. The transition to the non-homogeneous terminology is effected for this product $u(z_1, z_2)$ by:

$$u(z_1, z_2) = z_2^n \cdot u_0(z); \tag{5}$$

therefore, one has:

$$u_0(z) = \prod_{k=1}^{n} e_k^{(2)} \cdot \prod_{k=1}^{n} (z - e_k) = \prod_{k=1}^{n} e_k^{(2)} \left[z^n - z^{n-1} \sum_{k=1}^{n} e_k + \cdots \right]. \tag{6}$$

The differential equation, the obtaining of which is our goal, now has the form given below as equation (12), in which the left side represents the sum of three transvectants.

With this, according to the prescription of (3), first form the second transvectant $(f, u)_2$ of our polymorphic form f of dimension λ and the rational integral form u of dimension n. If one puts the expressions (1) and (5) in for f and u, then after an elementary intermediate calculation, one finds: [123]

$$(f, u)_2 = z_2^{\lambda+n-4} \left[n(n-1) u_0 \frac{d^2(\Phi Z)}{dz^2} - 2(n-1)(\lambda-1) \frac{du_0}{dz} \frac{d(\Phi Z)}{dz} + \lambda(\lambda-1) \frac{d^2 u_0}{dz^2} (\Phi Z) \right].$$

In the further development of this expression, for the sake of brevity, we will denote the derivatives of u_0, Φ and Z according to z by upper indices and, furthermore, replace Z'', according to (6) on p. 119, by $-ZR(z)$, we obtain in this way:

$$\begin{aligned} (f, u)_2 = z_2^{\lambda+n-4} \{ & 2(n-1) Z' [n u_0 \Phi' - (\lambda-1) u_0' \Phi] \\ & + Z[n(n-1) u_0 \Phi'' - 2(n-1)(\lambda-1) u_0' \Phi' + \lambda(\lambda-1) u_0'' \Phi - n(n-1) u_0' \Phi R(z)] \} \cdots \end{aligned} \tag{7}$$

We understand, further, by $v(z_1, z_2)$ a rational integral form of $(n-2)^{\text{nd}}$ degree to be immediately determined more closely, and set:

$$v(z_1, z_2) = z_2^{n-2} \cdot v_0(z).$$

For the first transvectant of the forms f and v, it follows:

$$(f, v)_1 = z_2^{\lambda+n-4} \{ (n-2) v_0 \Phi Z' + Z[(n-2) v_0 \Phi' - \lambda v_0' \Phi] \}. \tag{8}$$

Now add equations (7) and (8) and determine $v_0(z)$ in such a manner, that on the right side of the equation arising, the expression multiplied by Z' vanishes identically. We follow this, if we define v_0 by:

$$v_0(z) = 2 \frac{n-1}{n-2} \left[(\lambda-1) u_0' - n u_0 \frac{d \log \Phi}{dz} \right]. \tag{9}$$

One immediately recognizes in v_0 a rational *integral* function, whose degree is $\leqslant n-1$. But if we write the expression of v_0 explicitly:

$$2\frac{n-1}{n-2}\prod_{k=1}^{n} e_k^{(2)}\cdot\left[(\lambda-1)\frac{d\prod(z-e_k)}{dz}+\frac{1}{2}n\prod_{k=1}^{n}(z-e_k)\cdot\sum_{k=1}^{n}\frac{1-\frac{1}{l_k}}{z-e_k}\right],$$

then it follows; as the coefficient of the power z^{n-1}:

$$2\frac{n-1}{n-2}\prod_{k=1}^{n} e_k^{(2)}\left[n(\lambda-1)+\frac{1}{2}n\sum_{k=1}^{n}\left(1-\frac{1}{l_k}\right)\right].$$

But, in consequence of (3) on p. 110, this vanishes; *thus, the rational integral function* $v_0(z)$ *given by* (9) *is of degree* $(n-2)$, *and, as was already assumed, we in fact have, in*

$$v(z_1,z_2)=z_2^{n-2}v_0(z),$$

a rational integral form of $(n-2)^{nd}$ *dimension.*

[124] Now, utilizating the formulas (1) and (9), we write the sum of the equations (7) and (8) in the following form:

$$(f,u)_2+(f,v)_1=-f\cdot z_2^{n-4}w_0(z), \tag{10}$$

where $w_0(z)$ has the following meaning:

$$w_0(z)=n(n-1)u_0\left[R+2\left(\frac{\Phi'}{\Phi}\right)^2+\frac{\Phi''}{\Phi}\right]+\lambda v_0'-\lambda(\lambda-1)u_0''.$$

In order now, finally, to study the function $w_0(z)$ hereby defined more closely, we deduce from the second equation (1) by logarithmic differentiation:

$$\frac{\Phi'}{\Phi}=-\frac{1}{2}\sum_{k=1}^{n}\frac{1-\frac{1}{l_k}}{z-e_k}.$$

If we differentiate once more with respect to z and combine the arising equation with the one just given, then we obtain:

$$2\left(\frac{\Phi'}{\Phi}\right)^2-\frac{\Phi''}{\Phi}=\frac{1}{4}\left(\sum_{k=1}^{n}\frac{1-\frac{1}{l_k}}{z-e_k}\right)^2-\frac{1}{2}\sum_{k=1}^{n}\frac{1-\frac{1}{l_k}}{(z-e_k)^2}.$$

The product of this expression with $\prod(z-e_k)$ obviously becomes $(n-2)$-fold infinite and simply infinite at each of the n points e_k. In detail, one easily determines the kind of infinity at each of the latter places and arrives in such a way at the equation:

$$\begin{aligned}&2\left(\frac{\Phi'}{\Phi}\right)^2-\frac{\Phi''}{\Phi}\\&=\frac{1}{\prod\limits_{k=1}^{n}(z-e_k)}\left\{G_{n-2}(z)-\frac{1}{4}\sum_{k=1}^{n}\frac{1-\frac{1}{l_k^2}}{z-e_k}(e_k-e_1)\cdots(e_k-e_{k-1})(e_k-e_{k+1})\cdots(e_k-e_n)\right\}.\end{aligned}$$

Comparison with the expression of $R(z)$ represented in (2) on p. 117 yields, that

$$u_0(z)\left[R(z)+2\left(\frac{\Phi'}{\Phi}\right)^2-\frac{\Phi''}{\Phi}\right]$$

represents an integral rational function of $(n-2)^{\text{nd}}$ degree of z, and the same accordingly holds for $w_0(z)$. But if one calculates the coefficients, which the powers z^{n-2} and z^{n-3} have in the developed expression of $w_0(z)$, then one finds, that both vanish. *According to this,* $w_0(z)$ *is identical to* 0 *for* $n=3$ *and for* $n>3$, *represents a rational integral function of* $(n-4)^{th}$ *degree.* If we therefore set:

$$z_2^{n-4}w_0(z)=w(z_1,z_2),$$

then $w(z_1,z_2)$ is a rational integral form of $(n-4)^{\text{th}}$ degree resp. identical with 0 for $n=3$. [125]

Moreover, one also arrives at the same result from the equation (10). Namely, for the form $w(z_1,z_2)$ just defined, we calculate the form:

$$w(z_1,z_2)=-\frac{(f,u)_2+(f,v)_1}{f}. \tag{11}$$

Now, for each individual place z, the branch-points e_k as well as $z=\infty$ not excluded, we are able to give a particular form f satisfying the last equation, which is finite and unbranched here (cf. p. 112 ff.). We accordingly recognize from (11) that $w(z_1,z_2)$ never becomes infinite; now since $w(z_1,z_2)$ is a rational form, then it is consequently even a rational integral form. Its dimension in $z_1:z_2$ determines itself from the the right side of (11) to be $(n-4)$, whereby we have again obtained the result already mentioned.

According to the first formula (3), we may write the product of the forms f and w, for the sake of formal uniformity, as the transvectant $(f,w)_0$, so that equation (10) obtains the form :

$$(f,u)_2+(f,v)_1+(f,w)_0=0. \tag{12}$$

With this, we have now actually obtained the invariant form of the differential equation of second order. *The polymorphic forms f of our family* $(A\zeta_1+B\zeta_2)$ *are characterized, by the sum, represented in* (12) *of the transvectants of a particular form f with the rational integral forms u, v, w of the dimensions n, $n-2$, $n-4$, vanishing.*

The thoroughly simple and transparent construction of the differential equation obtained, gains yet a heightened interest, if we recapitulate the following concerning the rational forms coming up in the individual forms: *The form $u(z_1,z_2)$, by means of its zeros, furnishes the n branch-points $e_1,e_2,\dots,e_n$; the form $v(z_1,z_2)$, in case the e_k count as calculated, further yields the numbers $l_1,l_2,\dots,l_n$, which determine the signature $(0,n;l_1,l_2,\dots,l_n)$; the form $w(z_1,z_2)$ finally, if the e_k and l_k are known, permits the calculation of the $(n-3)$ accessory parameters of the differential equation (cf. p. 50), i.e., the $(n-3)$ coefficients of the integral function $G_{n-4}(z)$ coming up in the differential equation* (6) *on p. 119.* For further going investigations, it will probably be advisable to directly designate the coefficients of w as accessory parameters.

We finally return here, in particular, to the lowest case $n=3$. Here w vanished identically, u is a form of the third dimension and v a linear form. The invariant form of the differential equation of the second order therefore becomes here: [126]

$$(f,u)_2 + (f,v)_1 = 0. \tag{13}$$

Now let there follow a few more historical remarks on the invariant representation of our differential equation.

The first development lying in this direction originates from Hilbert. In his paper '*Über eine Darstellungsweise der invarianten Gebilde im binären Formengebiete*"[32], he has given the hypergeometric differential equation just that form, to which we were just led in (13). Hilbert's developments, to be sure, only refer themselves to the case of algebraic integrals of the hypergeometric differential equation; however, as we have seen, Hilbert's equation also remains directly valid for the cases of transcendental solutions as well.

The investigations of Pick "*Über eine Normalform gewisser Differentialgleichungen zweiter und dritter Ordnung*"[33] concern arbitrary polymorphic functions and forms on surfaces of genus $p = 0$, and with this, exactly the differential equations initially coming into consideration for us here. It is determined there, to what extent one has to ascribe the property of invariants to these differential equations, and, in particular, an invariant way of writing the function denoted above (p. 117) by $R(z)$ is developed.

The paper of Klein, cited on p. 116, gives, in essence, the definition of the polymorphic forms, which we called ζ_1, ζ_2, but on the other hand, only touches in passing the invariant form of the differential equations.

Finally, however, in the paper "*Zur Geometrie der linearen algebraischen Differentialgleichungen und binären Formen*"[34], the general form (12) of our differential equation has be set up by E. Waelsch. To be sure, the investigations of Waelsch also only first refer themselves to differential equations with rational solutions; however as we saw, Waelsch's equation remains valid directly for our transcendental forms ζ_1, ζ_2.

2.19 Series representation of the polymorphic forms in the case $n = 3$

[127] We are able to see the significance of the linear differential equations of the second order obtained, in this, that, starting from them, we are able to calculate, without difficulty, *power series development according to z* for our polymorphic forms, namely, as recurrent series of second order. Thereby, naturally, besides the e_k and l_k, we must also consider as known the accessory parameters of the differential equation. In the lowest case $n = 3$, accessory parameters do not yet come up. Let it therefore be allowed, to briefly go into this case, and with this, into the *doctrine of the hypergeometric series.* In fact, our considerations are to hold for the polymorphic forms H_1, H_2, satisfying the hypergeometric differential equation and defined in (12) on p. 120. For the obtention of series for other forms, as, e.g., for the Z_1, Z_2, one would then have to go back to their relation to the H_1, H_2.

Furthermore, one wants to conceive the following developments more only as a parenthetical excursion into an important neighboring field, which our investigations directly touch. A more exhaustive treatment of the hypergeometric series, which, above

[32] Mathem. Annalen Vol. 30, p. 15 (1887).

[33] Mathem. Annalen, Vol. 38, p. 139 (1890).

[34] Schriften der Deutschen Prager mathematischen Gesellschaft of 1892 on p. 78 ff.

all, would also have to have as their content the questions of the convergence and the more precise function-theoretic character of these series, would deviate us too far from our own goals.

The hypergeometric series:

$$F(\alpha,\beta,\gamma;z)=1+\frac{\alpha\cdot\beta}{1\cdot\gamma}z+\frac{\alpha(\alpha+1)\beta(\beta+1)}{1\cdot 2\cdot\gamma(\gamma+1)}z^2+\cdots, \tag{1}$$

provided with the notation $F(\alpha,\beta,\gamma;z)$ by Gauss[35], as is known[36], furnishes particular integrals of the hypergeometric differential equation obtained on p. 120:

$$z(z-1)\frac{d^2H}{dz^2}+[(\alpha+\beta+1)z-\gamma]\frac{dH}{dz}+\alpha\beta H=0 \tag{2}$$

to be given in manifold ways. For the neighborhood of the singular point $z=0$, we have the two integrals:

$$F(\alpha,\beta,\gamma;z),\; z^{1-\gamma}F(\alpha-\gamma+1,\beta-\gamma+1,2-\gamma;z); \tag{3}$$

next to these, for the neighborhood of $z=1$, we have the integrals: [128]

$$F(\alpha,\beta,\alpha+\beta-\gamma+1;1-z),\; (1-z)^{\gamma-\alpha-\beta}F(\gamma-\beta,\gamma-\alpha,\gamma-\alpha-\beta+1;1-z), \tag{4}$$

as well as, finally, for the neighborhood of $z=\infty$:

$$z^{-\alpha}F\left(\alpha,\alpha-\gamma+1,\alpha-\beta+1;\frac{1}{z}\right),\; z^{-\beta}F\left(\beta,\beta-\gamma+1,\beta-\alpha+1;\frac{1}{z}\right). \tag{5}$$

For the sake of simplicity, we assume that the l_1, l_2, l_3 are all three finite; then the integrals of the three given pairs will be linearly independent of one another.

We only mention in passing, that according to a theory first carried through completely by Kummer[37], each of the particular integrals set up can always be represented in *four* ways, all together, by the hypergeometric series. This theorem bases itself on the circumstance, that the original series (1) also lets itself be written in the following three ways:

$$\begin{gathered}(1-z)^{\gamma-\alpha-\beta}F(\gamma-\alpha,\gamma-\beta,\gamma;z),\\ (1-z)^{-\alpha}F\left(\alpha,\gamma-\beta,\gamma;\frac{z}{z-1}\right),\\ (1-z)^{-\beta}F\left(\beta,\gamma-\alpha,\gamma;\frac{z}{z-1}\right).\end{gathered}$$

35 "*Disquisitiones generates circa seriem infinitam* $1+\frac{\alpha\cdot\beta}{1\cdot\gamma}x+\cdots$", Abhandl. der. Gött. Gesell. d. Wiss. of 1813 or Gauss' Werke, Vol. 3 on p. 125. "*Determinatio seriei nostrae per aequationem differentialem secundi ordinis*", Gauss' Werke, Vol. 3 on p. 207.

36 See, e.g., the lecture of Klein's "*Über die hypergeometrischen Function*" of the winter semester 1893/94 on p. 63 ff. or Schlesinger's "*Handbuch der Theorie der linearen Differentialgleichungen*" Vol. 1, p. 265.

37 "*Über die hypergeometrische Riehe*", Journ. Math. Vol. 15 (1836).

One consequently has 24 series, all together, at one's disposal for the integration of the equation (2).

We now consider, first of all, the *neighborhood of the branch-point* $z=0$ and for this purpose construct the forms:

$$(H,\varepsilon_1)=H_1-\varepsilon_1 H_2,\ (H,\varepsilon_1')=H_1-\varepsilon_1' H_2.$$

According to the developments on p. 112 ff., (H,ε_1), for a circuit about the point $z=0$, will behave itself as $z^{\frac{1}{l_1}}$, while (H,ε_1') is single-valued here. Accordingly, (H,ε_1') and (H,ε_1), up to constant factors, will be equal to the two functions of z given in (3), so that we obtain the scheme:

$$\begin{cases}(H,\varepsilon_1)=C_1 z^{1-\gamma}F(\alpha-\gamma+1,\beta-\gamma+1,2-\gamma;z),\\ (H,\varepsilon_1')=C_2 F(\alpha,\beta,\gamma;z).\end{cases} \tag{6}$$

[129] For the determination of the constants one has to present somewhat circumstantial calculations. First form the quotient of these two equations and call the quotient of C_1 and C_2, C say:

$$\frac{\zeta-\varepsilon_1}{\zeta-\varepsilon_1'}=C\cdot z^{1-\gamma}\frac{F(\alpha-\gamma+1,\beta-\gamma+1,2-\gamma;z)}{F(\alpha,\beta,\gamma;z)}. \tag{7}$$

The next thing to do then, is to determine this quotient C. One lets z grow from 0 to 1, and reaches the boundary of convergence at $z=1$, for both hypergeometric series, which occur in (7). However, Gauss has determined, that our series still converge at the place $z=1$ in question, and we calculated, as the value of the series, and with this, that of the function $F(\alpha,\beta,\gamma;z)$, at $z=1$:

$$F(\alpha,\beta,\gamma;1)=\frac{\prod(\gamma-1)\prod(\gamma-\alpha-\beta-1)}{\prod(\gamma-\alpha-1)\prod(\gamma-\beta-1)}. \tag{8}$$

Hereby, $\prod(x)$ signifies that so denoted by Gauss and investigated more closely l.e.[38]. The variable z is now to wander on the real axis from 0 to 1; and we prescribe, that $z^{1-\gamma}=z^{\frac{1}{l_1}}$ is to hereby be taken real and positive. For this alteration z, ζ will wander on a side of the associated circular-arc triangle from ε_1 to ε_2. We accordingly find, by substituting $z=1$ into (7) and utilizing the rule (8), *the value for the constant* C:

$$C=\frac{\varepsilon_2-\varepsilon_1}{\varepsilon_2-\varepsilon_1'}\frac{\prod(-\alpha)\prod(-\beta)\prod(\gamma-1)}{(1-\gamma)\prod(\gamma-\alpha-1)\prod(\gamma-\beta-1)}. \tag{9}$$

In order now, further, to calculate C_1 and C_2, we draw the conclusion from (7), that at $z=0$, ζ is represented, in a first approximation by

$$\zeta=\varepsilon_1+C(\varepsilon_1-\varepsilon_1')z^{1-\gamma}.$$

[130] If one calculates $\frac{d\zeta}{dz}$ from this, and introduces the value obtained into (12) on p. 120,

[38] We note furthermore, that the requirement $\gamma-\alpha-\beta>0$ to be set for the validity of the statements just made (see Gauss' Werke, Vol. 3 on p. 143) is always fulfilled for the developments in the text, in consequence of the second equation (10) on p. 120.

then it follows the value of H_2^2 at $z=0$:

$$H_2^2 = \frac{(-1)^{\gamma-\alpha-\beta-1}}{C(\varepsilon_1-\varepsilon_1')(1-\gamma)}.$$

But for $z=0$, $H_1=\varepsilon_1 H_2$, and therefore the second formula (6), likewise specialized for $z=0$:

$$(H,\varepsilon_1') = (\varepsilon_1-\varepsilon_1')H_2 = C_2.$$

Accordingly, one finds for the constant C_2:

$$C_2 = \pm(-1)^{\frac{\gamma-\alpha-\beta-1}{2}}\sqrt{\frac{\varepsilon_1-\varepsilon_1'}{C(1-\gamma)}}. \tag{10}$$

The unique determination of the root of unity standing here on the right before the radical sign poses no difficulty, as soon as one has uniquely defined the functions of z, standing on the right in the formulas (12) on p. 120, for the neighborhood of $z=0$ coming into question here. With C and C_2, $C_1=C\cdot C_2$ is now also directly known.

Entirely analogous considerations hold for the *neighborhoods of the other two branch-points $z=1$ and $z=\infty$*. However, as we will show, e.g., for $z=1$, one can also proceed in a somewhat different way, with utilization of the results so far. For $z=1$, there hold the schemata corresponding to the formulas (6):

$$\begin{aligned}(H,\varepsilon_2) &= C_1'(1-z)^{\gamma-\alpha-\beta}F(\gamma-\beta,\gamma-\alpha,\gamma-\alpha-\beta+1;1-z),\\ (H,\varepsilon_2') &= C_2'F(\alpha,\beta,\alpha+\beta-\gamma+1;1-z).\end{aligned}$$

If one sets $z=0$, and utilizes the rule (8), then it follows:

$$\begin{aligned}H_1-\varepsilon_2H_2 &= C_1'\frac{\prod(\gamma-\alpha-\beta)\prod(-\gamma)}{\prod(-\alpha)\prod(-\beta)},\\ H_1-\varepsilon_2'H_2 &= C_2'\frac{\prod(\alpha+\beta-\gamma)\prod(-\gamma)}{\prod(\beta-\gamma)\prod(\alpha-\gamma)}.\end{aligned}$$

But the values of H_1 and H_2 were just determined precisely for $z=0$; if one puts these into the last two equations, then C_1', C_2' let themselves be calculated immediately.

We illustrate the preceding calculations for the *example* $l_1=2$, $l_2=4$, $l_3=6$, which already occupied us in I on p. 554. It is again to be a question here of that ζ, which is pictured by figure 176 l.c.. The values denoted here by ε_1 and ε_2 we shall then identify with:

$$\varepsilon_1 = \frac{i\sqrt{2}}{-1+\sqrt{3}},\ \varepsilon_2 = i,$$

and ε_1', ε_2' are simply complex conjugate to ε_1 resp. ε_2. One is led to the following results for the neighborhood of $z=0$: [131]

$$\frac{(-1+\sqrt{3})\zeta-i\sqrt{2}}{(-1+\sqrt{3})\zeta+i\sqrt{2}} = 2(\sqrt{2}-\sqrt{3})\frac{\prod(-\frac{1}{24})\prod(-\frac{5}{24})}{\prod(-\frac{13}{24})\prod(-\frac{17}{24})}\sqrt{z}\cdot\frac{F(\frac{13}{24},\frac{17}{24},\frac{3}{2};z)}{F(\frac{1}{24},\frac{5}{24},\frac{1}{2};z)},$$

and for the forms H_1 and H_2:

$$H_1 - \frac{i\sqrt{2}}{-1+\sqrt{3}} H_2$$

$$= 2e^{\frac{\pi i}{8}} \sqrt[4]{2}\sqrt{1+\sqrt{3}}\sqrt{-\sqrt{2}+\sqrt{3}}\sqrt{\frac{\Pi(-\frac{1}{24})\Pi(-\frac{5}{24})}{\Pi(-\frac{13}{24})\Pi(-\frac{17}{24})}}\sqrt{z}F\Big(\frac{13}{24}, \frac{17}{24}, \frac{3}{2}; z\Big),$$

$$H_1 + \frac{i\sqrt{2}}{-1+\sqrt{3}} H_2$$

$$= -e^{\frac{\pi i}{8}} \sqrt[4]{2}\sqrt{1+\sqrt{3}}\sqrt{\sqrt{2}+\sqrt{3}}\sqrt{\frac{\Pi(-\frac{13}{24})\Pi(-\frac{17}{24})}{\Pi(-\frac{1}{24})\Pi(-\frac{5}{24})}}F\Big(\frac{1}{24}, \frac{5}{24}, \frac{1}{2}; z\Big).$$

Upon calculating out these formulas, the rule of the $\prod$-function coming to expression by:

$$\prod(x+1) = (x+1)\cdot\prod(x)$$

was made use of.

2.20 Representation of the polymorphic forms in the case $n = 3$ by definite integrals

For the solutions of the hypergeometric differential equation, one has known, for a long time, a *representation by definite integrals*. We also take up this representation here, even if only incidentally, so to speak, since we are able to attach to this, as it appears, a promising approach for the arithmetic representation of the triangle group, as well as for the development of the associated automorphic forms in the form of theta series.

By means of the complex variable w, we put down the following integral:

$$\int w^{\beta-1}(1-w)^{\gamma-\beta-1}(1-zw)^{-\alpha}dw. \tag{1}$$

[132] Hereby, α, β, γ are to be understood in the previous sense (9) on p. 120, and the "parameter" z of the integral is any fixed value of the variable so denoted previously, fixed during the integration with respect to w. The α, β, γ are always rational fractions here, so that an algebraic function of w stands under the integral (1), which is branched at the four places $w = 0, 1, z^{-1}, \infty$, in a manner easily given in each case. To the algebraic function belongs a certain Riemann surface over the w-plane, which we call F_w.

From the values of α, β, γ given on p. 120, and the inequality:

$$0 \leqslant \frac{1}{l_1} + \frac{1}{l_2} + \frac{1}{l_3} < 1,$$

one easily reads off, that in (1), we have before us an *integral of the first kind* of the surface F_w. If we therefore bring in *any closed curve on* F_w, in order to extend the integral (1) over this, then we obtain, as the value of the integral (1), the *period of the integral of the first kind under discussion* belonging to that circuit.

Now, as is already remarked at the introduction to the paragraph, the very remarkable theorem subsists, *that the integral in* (1) *in question, in dependency on the parameter z, satisfies the hypergeometric differential equation* (11) *on p. 120.* For the proof of this assertion, we refer to the literature of the subject, concerning which let a few notes be inserted here.

Euler[39] was already led to the consideration of our integral (1); however, for him it was a question only of real values of w, and the integral was extended between the limits 0 and 1. If the exponents resp. the quantities α, β, γ, are not chosen in the special manner present here, but arbitrarily at first, then, to be sure, under certain circumstances, the integral can be discontinuous at the lower limit 0 or as the upper limit 1 or also at both. Consequently, in order that the Euler approach thus remain directly usable, the inequalities $\beta > 0$, $\gamma > \beta$ have to be considered as fulfilled. But then, one can easily show, that the integral considered by Euler, assuming $|z| < 1$, is equal to $F(\alpha,\beta,\gamma;z)$ up to a constant factor. This constant factor, however, is recognized im- [133]
mediately as the Euler integral of the first kind:

$$B(\beta,\gamma-\beta) = \int_0^1 w^{\beta-1}(1-w)^{\gamma-\beta-1}dw.$$

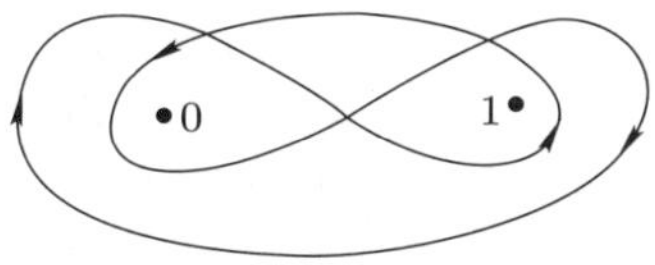

Fig. 22.

If the integral (1) is not of the first kind, then one will simply avoid the difficulties, which result from the infinities of the integral, by adopting Riemann's method of conceiving the integral (1), taken as a departure point above. Now w is complex, and one will simply have to avoid the four branch-points with the closed integration curve on the surface F_w, in order to always have a finite and determinate integral value. How one has to choose the closed curve, will naturally depend completely on the structure of the surface F_w. But it is remarkable, that in the so-called "*double circuits*" about two of the branch-points, as they were first utilized by C. Jordan[40], L. Pochhammer[41] and P. A. Nekrasoff[42], one always has closed paths of the kind desired. A double circuit about the branch-points 0 and 1 is represented in figure 22. As one sees, each of the two branch-points are circled first in one direction, then in the opposite one just on this account, there results a closed path on F_w.

The theorem above, that the integral (1), as a function of z, satisfies the hypergeometric differential equation, one can now show simply by substituting the function of z given by (1) into the differential equation and carrying out the differentiations according to the "parameter" z under the integral sign. For the further effectuation of this

39 See the "*Institutiones caleuli integralis*", Vol. 2. Abschn. 1 Kaps. 10 and 11 (Petersburg 1769)

40 "*Cours d'analyse de l'école polytechnique*", t. 3, p. 241 ff. (1887).

41 See his papers "*Über ein Integral mit doppeltem Umlauf*", Mathem. Ann. Vol. 35 (1890) and "*Über eine Klasse von Integralen mit geschlossener Integrationscurve*", ibid. Vol. 37 (1890).

42 "*Über lineare Differentialgleichungen, welche mittelst bestimmter Integrale integriert werden*" Mathem. Ann. Vol. 38 (1891).

approach see, besides the papers just cited above, the oft-cited lectures of Klein on the "*hypergeometric Function*" (cf. p. 53). Upon the instigation of Klein, C. Schellenberg
[134] has given a "*Neue Behandlung der hypergeometrischen Function auf Grund ihrer Definition durch das bestimmte Integral*"[43], where the emphasis is laid on the symmetric utilization of all elements of equal standing, as well as, on the other hand, on fashioning the hypergeometric function as *an integral transcendental function of the exponents* α, β, γ.

In order now to present the significance, which the integral (1) has for our theory of polymorphic forms, we turn ourselves first of all, to the *special case of the modular functions.* For this purpose, we set $l_1 = l_2 = l_3 = \infty$, and find, as the corresponding values α, β, γ:

$$\alpha = \frac{1}{2}, \ \beta = \frac{1}{2}, \ \gamma = 1.$$

Our integral (1) assumes the form:

$$\int \frac{dw}{\sqrt{w(1-w)(1-zw)}}, \tag{2}$$

i.e., we arrive at the *elliptic integral of the first kind of "modulus" z*. The surface F_w is of genus $p = 1$, and we come back here directly to the well known conception, *that the two functions* H_1, H_2 *of* z *represent the periods of the elliptic integral* (2) *of the first kind in dependency on the modulus* z. This tie-in to the theory of the elliptic functions, however, was fundamental for the development of the modular functions, insofar as one obtained immediately usable approaches for the representation and calculation of the modular functions and modular forms in the series of developments of the doubly periodic functions and, above all, of the theta functions.

We now obtain here the interesting realization, that these relations, known and basic in the theory of the modular functions, are capable of generalization to each group $(0,3;l_1,l_2,l_3)$ of our kind. *In place of the elliptic functions, we get the Abelian functions of an algebraic model defined by the function of the complex variable* w*:*

$$w^{\beta-1}(1-w)^{\gamma-\beta-1}(1-zw)^{-\alpha},$$

with a "modulus" z to be thought of as variable. In particular, in place of the elliptic integral (2), here is, in general, be Abelian integral (1) of the first kind; let the periods of this, referred to any cut system, be $\omega_1, \omega_2, \ldots, \omega_{2p}$, understanding by p the genus of the
[135] algebraic model. The theorem then holds: *the periods* $\omega_1, \omega_2, \ldots, \omega_{2p}$ *are representable as functions of* z *by two linearly independent integrals* H_1, H_2 *of the hypergeometric differential equation* (11) *on p. 120 in the form:*

$$\omega_i = a_i H_1 + b_i H_2 \quad (i = 1, 2, \ldots, 2p) \tag{3}$$

with $4p$ *constant coefficients a, b*[44].

43 Inaugural-Dissertation (Göttingen, 1892).

43 Compare principally "M" II, where the approaches mentioned in the text are carried through in the most extended fashion.

44 J. Wellstein (in an announcement in a letter to the author) has been able to directly carry out this reduction of all $2p$ periods to two quantities in an investigation not yet published. Well-

The formulas (3) establish a connection between our originally underlying homogeneous group Γ of the signature $(0,3;l_1,l_2,l_3)$ and a certain *group of integral unimodular substitutions of the periods* $\omega_1,\omega_2,\ldots,\omega_{2p}$, which comes up in the theory of linear transformation of the Abelian functions. In fact, each to each *closed circuit of the movable branch-point* $w = z^{-1}$ corresponds a periods transformation:

$$\omega_i' = \alpha_{i1}\omega_1 + \alpha_{i2}\omega_2 + \cdots + \alpha_{i,2p}\omega_{2p} \tag{4}$$

of the kind just mentioned. *The total group of all substitutions* (4) *reachable by "monodromy" of the branch-point* $w = z^{-1}$ *is directly isomorphic with our group* Γ *of the signature* $(0,3;l_1,l_2,l_3)$. *In the case of the modular functions, both groups are also formally identical; in general, they will be transformable into one another by the* $2p$ *equations* (3). In the group of the substitutions (4) and the $2p$ relations (3) we have accordingly obtained a new approach to an arithmetic representation of our triangle group Γ of the signature $(0,3;l_1,l_2,l_3)$.

For the representation of the automorphic functions and forms of the model $(0,3;l_1,l_2,l_3)$, one would now have to bring in the *theta series* of the associated algebraic model of genus p, following the relations for the elliptic modular functions. One will thereby, first and foremost, evaluate the "*zeros*" of the even theta functions, which only depend on the periods. In the two special cases $(0,3;4,4,\infty)$ and $(0,3;3,\infty,\infty)$, which both belong to the genus $p = 2$, this approach has actually been carried through by H. Burkhardt in the paper "*Über die Darstellung einiger Fälle der automorphen Primformen durch speciette Thetareihen*"[45]. Unfortunately, however, just the two cases have not been able to furnish new results, since the two groups $(0,3;4,4,\infty)$ and $(0,3;3,\infty,\infty)$ are commensurable with the modular group. The representations of the [136]
prime forms given by Burkhardt then also belong in fact, to the known series of developments of elliptic modular functions[46].

In this state of affairs, one can only designate it as desirable, to carry through the approach developed, in other promising cases, and indeed, especially in such cases, for which commensurability with the modular group is avoided at the very beginning.

---

stein operates hereby with especially apposite "reduced" out systems and furthermore this investigation refers itself not only to the case of integrals with three fixed branch places $0,1,\infty$ coming into question in the text, but to integrals with n fixed branch places $e_1,e_2,\ldots,e_n$.

[45] Math. Annalen, Vol. 42 p. 185 (1892).

[46] Compare, for this, paragraph 15 in the paper of A. Hurwitz "*Über endliehe Gruppen linearer Substitutionen, welche in der Theorie der elliptischen Transcendenten auftreten*", Math. Ann. Vol. 27 (1885). There, in general, even analytic representations of elliptic modular forms are derived from theta series of higher p, whereby Burkhardt's formulas subsume themselves as special cases.

3

Theory of Poincaré series with special discussions for the models of genus zero

Before we carry the form theory of automorphic models of an arbitrary genus further, we will become acquainted with the series of developments, already cited often, which Poincaré has introduced and made into the foundation of his entire analytic theory of automorphic functions. We shall hereby be able to attach immediately to an automorphic model of *arbitrary genus p*. The special discussions, which we shall present, lying in with our general representations for the particular genus $p = 0$, place us in the position, of bringing our investigations on the models of the lowest genus to a definitive conclusion. The Poincaré procedure, which in natural fashion, puts down analytic expressions which are invariant with respect to all the substitutions of a proposed group Γ, has already been advanced by Poincaré in his first note in the Comptes Rendus, cited on p. 41, and carefully established in the paper "Mémoire sur les functions [1]", as well as carried over, in the later paper "Mémoire sur les groups Kleinéans"[2], to models without principal circle. The studies, which Poincaré carries out for the proof of the convergence of the series in question, belong to the most important developments of the entire theory of automorphic functions. [137]

After the statements on p. 55, one will expect, that Poincaré, in the formation of the series in question does *not* avail himself of homogeneous variables ζ_1, ζ_2, but operates, rather, with non-homogeneous ones. This is in fact the case, and has the consequence, that Poincaré series are *not absolutely invariant* with respect to the substitutions of Γ, but, upon exercise of the individual substitution, *assumes a factor independent of* ζ. The [138]
series, accordingly, show a behavior similar to that of the theta series upon increase of their argument periods. This circumstance caused Poincaré, to designate his series, as "Fuchsion" resp. "Kleinian" "Theta series".

Now Ritter, in his dissertation, cited on p. 55, tying into Klein's lectures, has established the series developments under discussion by taking the homogeneous variables ζ_1, ζ_2 as a point of departure. Furthermore, it is hereby a question of only a formal reworking of Poincaré's series.

Ritter has undertaken, from yet another point of view, an essential extension of the Poincaré approach. Namely while Poincaré's own series (put into homogenous form)

---

[1] Acta mathematica, Vol.1, p. 193 (1882).

[2] Acta mathematica, Vol. 3, p. 49 (1883).

refer themselves to single-valued forms of the multiplicator system $M_0 = 1$, and therefore, only to properly automorphic forms, Ritter's developments extend themselves to *single-valued forms of arbitrary multiplicator system* M.

Since, furthermore, the introduction of the series in question is Poincaré's most original accomplishment, Ritter has departed from the names utilized by Poincaré himself and has taken up, in their place, the quite justified personal name "*Poincaré series*", which we shall consistently utilize from now on.

3.1 The approach to the Poincaré series

Let an arbitrary group Γ, with the substitutions V_0, V_1, ⋯, initially written non-homogenously, be proposed, all of which transform an associated polygon net N into itself; let the "initial polygon" or the "fundamental domain" P_0 be transformed by V_k into the polygon P_k. The corresponding homogeneous Γ, in case we write its substitutions "unimodularly", is either isomorphic or 2-1-valued homomorphic with the non-homogeneous group. The substitutions of the homogeneous group we denote in general in the usual manner, by U and write U_k explicitly thus:

$$\zeta_1^{(k)} = \alpha_k\zeta_1 + \beta_k\zeta_2, \qquad \zeta_2^{(k)} = \gamma_k\zeta_1 + \delta_k\zeta_2. \tag{1}$$

Let the genus p of the fundamental domain P_0, and with this, that of the associated automorphic model, be a completely arbitrary one until further notice.

The concept of the unbranched automorphic forms was determined on p. 66 for groups Γ of arbitrary genus p. The *single-valued* automorphic forms, which represent a special class of unbranched forms and always have *integral dimension,* we shall de-
[139] fine for our groups Γ here exactly as above for $p = 0$, namely, by the requirement: *the multiplicator* μ*, which a single-valued form* $\varphi_d(\zeta_1,\zeta_2)$ *assumes for the exercise of a substitution* U *of the homogeneous* Γ, *is always to be one and the same, whatever path the arguments* ζ_1,ζ_2 *might describe inside the domain of admissible values* ζ_1,ζ_2 *upon exercise of the substitution.* The admissible value systems ζ_1,ζ_2 are thereby to be defined as on p. 64. We immediately deduce, from the requirement formulated, the theorem: *if* μ_i *and* μ_k *are the multiplicators of the single-valued form* $\varphi_d(\zeta_1,\zeta_2)$, *which correspond to the arbitrary operations* U_i, U_k*, then the multiplicator* $\mu_i \cdot \mu_k$ *belongs as well to* U_iU_k *as to* U_kU_i.

If we give the canonical shape to the domain P_0, then this furnishes $(n+2p)$ generators $U_1,\ldots,U_n$, $U_{a_1},U_{b_1},\ldots,U_{a_p},U_{b_p}$, as far as (p,n) is the "character" of Γ. The individual form, correspondingly, possess a system of $(n+2p)$ generating multiplicators, from which, according to the theorem given, all remaining multiplicators of this form are to be produced. For the groups of genus $p = 0$, only multiplicators came into consideration, which were *roots of unity.* For the multiplicators corresponding to the parabolic generators U_i occurring, say, as well as to the hyperbolic U_{a_k},U_{b_k}, it is not yet necessarily, from the nature of the group and the concept of the single-valued forms, that they represent roots of unity all the same, we determine, *that the* $(n+2p)$ *generating multiplicators, and with these, generally, all multiplicators without exception, are to be of absolute value* 1. We come back to the significance of this determination, to the relations subsisting between the generating multiplicators, say, etc. in the next chapter during the detailed form theory of the automorphic models, with $p > 0$. Meanwhile, the previous determinations are sufficient, in order to lay out the theory of the Poincaré series in the extension chosen by Ritter, at the same time, for models of arbitrary p.

We now try to obtain, for an individual given multiplicator system, analytic expressions for associated single-valued automorphic forms of prescribed integral dimension d. To this purpose, we go back to the way of writing the substitutions of Γ introduced in (1), and denote, with μ_k, the multiplicator belonging to U_k. We understand by $H_d(\zeta_1,\zeta_2)$ *any homogeneous rational function of the* ζ_1,ζ_2 *of dimension d* and, with the aid of this, we put down the following series:

$$\varphi_d(\zeta_1,\zeta_2) = \sum_k \mu_k^{-1} H_d(\zeta_1^{(k)},\zeta_2^{(k)}), \tag{2}$$

or, written in more detail:

$$\varphi_d(\zeta_1,\zeta_2) = \sum_k \mu_k^{-1} H_d(\alpha_k\zeta_1+\beta_k\zeta_2, \gamma_k\zeta_1+\delta_k\zeta_2), \tag{3}$$

summed over $k = 0,1,2,3,\ldots$, i.e., over all operations of the homogeneous group Γ. For [140]
the present, one wants to consider $\varphi_d(\zeta_1,\zeta_2)$ as a symbol for the series put down on the right, since nothing is determined yet concerning its convergence and function-theoretic character, we will designate each such series (2) and (3), from now on, as a "*Poincaré series*" of our group Γ.

If $H_d(\zeta_1,\zeta_2)$ is not an integral form, then this has poles in the ζ-plane. If one of these poles coincides with a boundary point of the group Γ, then, where we also pick out ζ_0 in the polygon net, in each neighborhood, chosen ever so small, of that boundary point, there will still be situated unboundedly many points $V_k(\zeta_0)$ equivalent with ζ_0. The behavior of the corresponding series terms in (2) and their sum is then more difficult to determine. Certainly, we come back below (in §3.3) to the still approachable case, that a pole of sufficiently low order occurs at a parabolic cusp. But in order to avoid the difficulty under discussion here, in setting up the theory of the Poincaré series, one has, at the very beginning, placed the following prescription first and foremost: *of the poles of the form* $H_d(\zeta_1,\zeta_2)$, *none is to coincide with a boundary point of the group.* We shall always hold fast to this determination, except for the development in §3 already mentioned.

As a *rational* form, $H_d(\zeta_1,\zeta_2)$ is either pole-free or has a *finite* number of poles in the net N. In the latter case, mark all the places equivalent with these poles in the polygon P_0, as well as in all remaining polygons of the net N. At the individual such place, one term of the series (2) becomes *polarly* discontinuous, or several in *finite* number, in case, namely, several poles of $H_d(\zeta_1,\zeta_2)$ occur at points, which are equivalent with respect to Γ. In the latter case, it is possible, that the poles of these terms, in the formation of the sum (2), mutually destroy one another, an occurrence, which will yet occupy us below.

We now state the following theorem: *if we are able to prove, that the series (2) is convergent absolutely and uniformly in the neighborhood of an "arbitrary" place* ζ_0 *of the net N, except for finitely many terms, which are polarly discontinuous at* ζ_0, *then this furnishes a single-valued automorphic form* $\varphi_d(\zeta_1,\zeta_2)$ *with the prescribed multiplicator system and the given dimension d.* Namely, if we write the series (3) symbolically, somewhat more briefly, in the form:

$$\sum_k \mu_k^{-1} H_d[U_k(\zeta_1,\zeta_2)].$$

[141] Applying the U_i leads to:

$$\sum_k \mu_k^{-1} H_d[U_k U_i(\zeta_1,\zeta_2)] = \mu_i \sum_k (\mu_i\mu_k)^{-1} H_d[U_k U_i(\zeta_1,\zeta_2)].$$

Now $U_k U_i$, with $k = 0,1,2,\ldots$, also runs through the group Γ, and $\mu_i\mu_k$ is the associated multiplicator. However, because of the absolute, and, just on this account, unconditional, convergence, the series on the right side of the last equation multiplied with μ_i has the same summation value as the series (3). If we yet add, that in the neighborhood of the arbitrary place ζ_0 of the net, the "uniform" convergence of the series (3) was assumed, then this series will furnish a single-valued homogeneous analytic function of ζ_1,ζ_2 in the net N, which assumes the factor μ_i with respect to U_i. The stated theorem is, with this, actually proved.

If the homogeneous group is not mapped isomorphically onto the nonhomogeneous one, but 2-1-valued homeomorphically, the former contains the substitution -1. For this homogeneous substitution, each system M of even dimension furnishes the multiplicator $\mu = 1$, and each one of odd dimension, $\mu = -1$. The consequence is, that the terms of the series (2) now become equal to one another in pairs. *According to this, it will be sufficient, if while forming the series (2), we always admit only one substitution from each two homogeneous U corresponding to the same substitution V.*

Furthermore, in the sequel, we shall mostly utilize the *series (2) in non-homogeneous form.* We hereby set:

$$H_d(\zeta_1,\zeta_2) = \zeta_2^d H_d(\zeta,1) = \zeta_2^d H(\zeta), \tag{4}$$

and obtain, with this, the following representation of our series:

$$\varphi_d(\zeta_1,\zeta_2) = \zeta_2^d \sum_k \mu_k^{-1} H\left(\frac{\alpha_k\zeta+\beta_k}{\gamma_k\zeta+\delta_k}\right)(\gamma_k\zeta+\delta_k)^d, \tag{5}$$

where the series terms are now uniquely coordinated to the substitutions V_k of the *non-homogeneous* group.

If we leave out the factor ζ_2^d and specialize the formula (5) for even dimensions $d = -2m$ and for the multiplicator system $M_0 = 1$, then, in:

$$\theta(\zeta) = \sum_k H\left(\frac{\alpha_k\zeta+\beta_k}{\gamma_k\zeta+\delta_k}\right)(\gamma_k\zeta+\delta_k)^{-2m}, \tag{6}$$

[142] we arrive at the "theta function" placed by Poincaré at the forefront. The equation:

$$\varphi(\zeta_1^{(i)},\zeta_2^{(i)}) = \varphi(\zeta_1,\zeta_2),$$

then furnishes, for the θ-function represented in (6), as behavior with respect to the substitution V_i:

$$\theta\left(\frac{\alpha_i\zeta+\beta_i}{\gamma_i\zeta+\delta_i}\right) = (\gamma_i\zeta+\delta_i)^{2m}\theta(\zeta).$$

The $\theta(\zeta)$ does not behave itself absolutely invariantly with respect to V_i, rather assumes, *a factor dependent on* ζ, was already stated above (p. 137).

Everything now depends on actually carrying through the convergence investigation of the Poincaré series.

3.2 First convergence study of the Poincaré series

Poincaré, in the first of the two papers cited on p. 137, has carried through the convergence investigation of his series in the case of the principal-circle groups according to two different methods, of which the first preserves its power of proof in the case of arbitrary polygon groups. We consequently enter here, first of all, into this first Poincaré convergence study. A pair of preliminary arguments are thereby to be presented in advance.

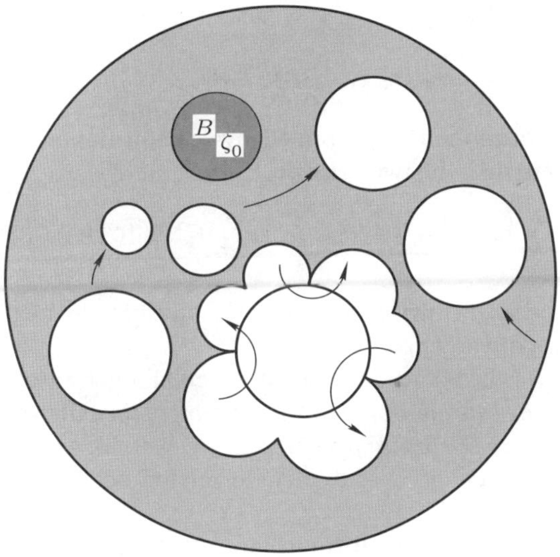

Fig. 23.

First of all, in the course of the proof, it will be requisite, that we have a net N with one or infinitely many boundary curves. But this is no restriction of the general validity of our results. Namely, the theorem holds: *if the series of the dimension d, for the groups with boundary curves, are absolutely and uniformly convergent everywhere in the interior of the nets under consideration, then the same convergence also takes place for the series of just this dimension d for the groups with only isolatedly situated bounding points.* Namely, are we to investigate the convergence of a proposed series [143]
of the dimension d for a group Γ as the last kind for the neighborhood B of a place ζ_0 of the associated net, then compose Γ, according to the direction of figure 23, with a principal circle group of the signature (0, 3; 0, 0, 0) say, which we so choose, that the domain B also lies inside the net as the group arising by composition. In the figure,

the polygon boundary of this principal-circle group and the principal circle itself belonging to it are marked heavily. By means of the composition, we have group with infinitely many boundary curves. One now has only to consider, that each component of an absolutely and uniformly convergent series again represents a similarly convergent series, in order to recognize the correctness of the assertion stated for the Poincaré series of the original group.

We will secondly prove: *the convergence of the individual Poincaré series is independent of an arbitrary linear transformation, which we might exercise on* ζ *resp* ζ_1 *and* ζ_2. If namely, by means of a unimodularly written substitution say:

$$\zeta_1 = A\zeta_1' + B\zeta_2', \qquad \zeta_2 = C\zeta_1' + D\zeta_2'. \tag{1}$$

We go over to the new variables ζ_1' and ζ_2', and set:

$$H_d(A\zeta_1' + B\zeta_2', C\zeta_1' + D\zeta_2') = H_d'(\zeta_1, \zeta_2).$$

Then, written symbolically:

$$H_d[U_k(\zeta_1, \zeta_2)] = H_d'[U_k'(\zeta_1', \zeta_2')],$$

where U_k' is the substitution arising from U_k by transformation by means of the operation (1). According to this, the two series:

$$\sum_k \mu_k^{-1} H_d[U_k(\zeta_1, \zeta_2)], \qquad \sum_k \mu_k^{-1} H_d'[U_k'(\zeta_1', \zeta_2')]$$

are equal to one another termwise, and each in fact, can take the place of the other for consideration of the convergence.

We make use of this last consideration as we go along, by assuming, *that the point* $\zeta = \infty$ *lies neither in the interior nor on the boundary of the polygon net N, which consequently will lie entirely in the finite part of the plane.* Were this not to take place initially, then we could immediately go over to a circle-related net, which fulfills our requirement.

We now write our series in the form (5) on p. 141 and also leave out the factor ζ_2^d
[144] (which does not interest us in the convergence study); therefore, we have to investigate the convergence of the following series:

$$\sum_k \mu_k^{-1} H\left(\frac{\alpha_k\zeta + \beta_k}{\gamma_k\zeta + \delta_k}\right)(\gamma_k\zeta + \delta_k)^d.$$

If, as above, we utilize the notation $\zeta^{(k)} = V_k(\zeta)$ for the unimodularly written nonhomogeneous substitutions, then the series in question can also be written thus:

$$\sum_k \mu_k^{-1} H[V_k(\zeta)] \left(\frac{dV_k(\zeta)}{d\zeta}\right)^{-\frac{d}{2}}. \tag{2}$$

Now let an "arbitrary" point ζ_0 be picked out in the "interior" of the polygon net and surrounded with a small domain B, which is never to come near the boundary of the net N. In this domain B, we will investigate the convergence of the series (2) and accordingly restrict ζ to the domain B.

Let the domains equivalent with B be called $B_0 = B, B_1, B_2, \ldots$ say, We assume B chosen in such a way, that the domains B_k do not reach over one another, in the case of an elliptic fixpoint ζ_0, one will thus cut off B from a polygon corner whose vertex is at ζ_0. While ζ describes the domain B, $V_k(\zeta)$ will obviously run through the equivalent domain B_k.

If there are poles of the rational function $H(\zeta)$ in the net, then surround each of these with a small circle running in the interior of the net. These circles, insofar as they were to lie in the "interior" of the net, will only have points in common with *finitely* many domains B_k. The series terms in (2) belonging to these last domains, we imagine as left out; because of their finite number, they play no role in the question of convergence. Let the domains B still remaining be called $B'_0, B'_1, B'_2, \ldots$; let the upper index on the summation sign remind us of the shortening of the series thus introduced.

Since now, the domains B'_k are everywhere finitely distant from the poles of the rational function $H(\zeta)$, then the absolute values of the function values $H[V_k(\zeta)]$ coming into consideration for us will all not exceed a *givable finite* quantity G. If we consequently pass from the shorted series (2) to the *series of the absolute values* and utilize, that $|\mu_k| = 1$, then it follows:

$$\sum_k{}' |\mu_k^{-1} H[V_k(\zeta)]| \cdot \left|\frac{dV_k(\zeta)}{d\zeta}\right|^{-\frac{d}{2}} \leq G \cdot \sum_k{}' \left|\frac{dV_k(\zeta)}{d\zeta}\right|^{-\frac{d}{2}}.$$

The investigation is hereby reduced to the question of the convergence of the series: [145]

$$\sum_k{}' \left|\frac{dV_k(\zeta)}{d\zeta}\right|^{-\frac{d}{2}}. \tag{3}$$

Now for the individual point ζ of the domain B, the geometric meaning of the absolute value $|\frac{dV_k(\zeta)}{d\zeta}|$ is this, that this furnishes the "ratio of magnification" or the *"modulus" of the conformal mapping of the immediate neighborhood of the point ζ onto the domain B_k*. This mapping modulus decreases to the limit 0, if we approach the boundary of the net N more and more. With this, in order that the series (3) is convergent, d *must be a "negative" whole number*; if we therefore set $-d = d'$, then d' is to be taken ≥ 1, and it is a question of the convergence of the series:

$$\sum_k{}' \left|\frac{dV_k(\zeta)}{d\zeta}\right|^{\frac{d'}{2}}. \tag{4}$$

Now draw into consideration the magnitude of the *oscillation, which the mapping modulus* $\left|\frac{dV_k(\zeta)}{d\zeta}\right|$ *experiences, if ζ describes the domain B*. For this purpose, we write

$$\left|\frac{dV_k(\zeta)}{d\zeta}\right| = \frac{1}{|\gamma_k|^2|\zeta + \frac{\delta_k}{\gamma_k}|^2},$$

whereby, it is to be noticed, that the coefficients γ_k of all substitutions of the group Γ, apart from the identity $V_0 = 1$, may be assumed as different from 0[3]. If we denote the greatest value of a variable quantity in the domain B by appending the index M, and the smallest, corresponding, by appending m, then it follows from the last equation:

$$\frac{\left|\frac{dV_k(\zeta)}{d\zeta}\right|_M}{\left|\frac{dV_k(\zeta)}{d\zeta}\right|_m} = \left(\frac{\left|\zeta + \frac{\delta_k}{\gamma_k}\right|_M}{\left|\zeta + \frac{\delta_k}{\gamma_k}\right|_m}\right)^2 .$$

Now notice, that $V_k^{-1}(\infty) = -\frac{\delta_k}{\gamma_k}$. The point $-\frac{\delta_k}{\gamma_k}$ is thus equivalent with ∞ for each k
[146] and therefore situated outside the net N. Now $|\zeta + \frac{\delta_k}{\gamma_k}|$ is simply equal to the distances of the two points ζ and $-\frac{\delta_k}{\gamma_k}$. According to this, $|\zeta + \frac{\delta_k}{\gamma_k}|_m$ is greater than the smallest distance m, which one can discover between a boundary point of B and one such of N. Since B remains finitely distinct everywhere from the boundary of the net N, then m is a determinate positive number different from 0 (cf. figure 24)[4] Since, further, all points $-\frac{\delta_k}{\gamma_k}$ for $k = 1, 2, 3, \ldots$ lie in the finite part of the plane, then one can place, around ζ_0 say, a circle with determinate finite radius, which includes all points $-\frac{\delta_k}{\gamma_k}$. But then, $|\zeta + \frac{\delta_k}{\gamma_k}|_M$ is obviously smaller than the greatest distance M, which exists between a boundary point of B and a point of the periphery of the circle just mentioned. Hereby, M is again a finite number (cf. figure 24). If we now set the quotient $\frac{M^2}{m^2} = C$, then the right side of the last equation is $< C$, and therefore the inequality yields itself:

$$\left|\frac{dV_k(\zeta)}{d\zeta}\right|_M < C \left|\frac{dV_k(\zeta)}{d\zeta}\right|_m , \tag{5}$$

i.e., the quotient of the greatest value of the mapping modulus in question in the domain B and the smallest value entering there is smaller than C, where C is a positive independent of k.

We now understand by J_0 the area of the surface of the domain B and correspondingly denote the area of B_k by J_k. Then we have in each case:

$$\left|\frac{dV_k(\zeta)}{d\zeta}\right|_m^2 J_0 \leq J_k \leq \left|\frac{dV_k(\zeta)}{d\zeta}\right|_M^2 J_0 . \tag{6}$$

[147] From this and from the inequality (5) follows, for an arbitrary point ζ in the domain B:

$$\left|\frac{dV_k(\zeta)}{d\zeta}\right|^2 \leq \left|\frac{dV_k(\zeta)}{d\zeta}\right|_M^2 < C^2 \left|\frac{dV_k(\zeta)}{d\zeta}\right|_m^2 \leq C^2 \frac{J_k}{J_0} \tag{7}$$

---

[3] If, namely, $\gamma_k = 0$, then V_k has a fixpoint $\zeta = \infty$. All the non-elliptic fixpoints, however, lie on the boundary of N and therefore not at $\zeta = \infty$. Possible elliptic fixpoints lie isolated outside; and one can, according to the second lemma given at the beginning of the paragraph, always effect the arrangement, so that $\zeta = \infty$ is not an elliptic fixpoint.

[4] This figure illustrates the case, that the domain covered by the net N is simply connected. However, the arrangement holds without further ado also in the case, that the net N has infinitely high connectivity.

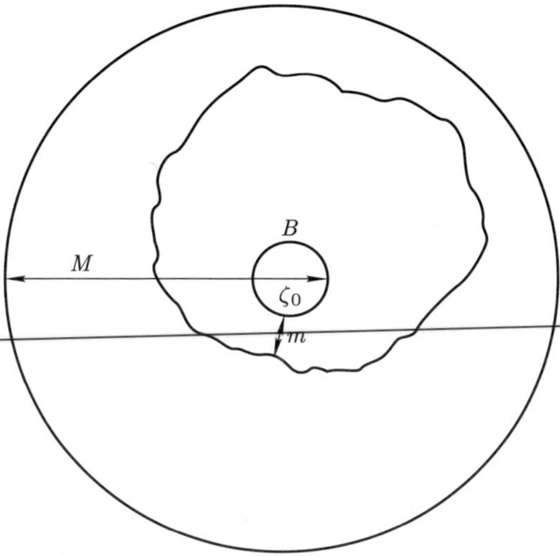

Fig. 24.

For the investigated series (4) we have:

$$\sum_k{}' \left| \frac{dV_k(\zeta)}{d\zeta} \right|^{\frac{d'}{2}} < C^{\frac{d'}{2}} J_0^{\frac{d'}{4}} \cdot \sum_k{}' J_k^{\frac{d'}{4}}. \tag{8}$$

This inequality is valid at the place ζ considered. If now, first of all, $d' = 4$, then one has the series, $\Sigma'_k J_k$, and this is certainly convergent; for the domains B_k do not collide with one another, and the entire net N runs in the finite part of the plane. With this, since $\lim_{k=\infty} J_k = 0$, the convergence is proved, and it follows, for all values $d' > 4$.

Since furthermore, as emphasized, the place ζ considered was arbitrarily chosen in the domain B, the uniformity of the convergence in the domain B is also proved by the previous consideration.

According to the arguments given at the beginning of the paragraph, we are consequently able to state the following general convergence theorem: *for each polygon group* Γ, *all Poincaré series of the dimension* $d \leq -4$ *converge, in the above sense, unconditionally and uniformly in the entire interior the polygon net* N, *i.e., in the neighborhood of each interior point finitely distant from the boundary of the net.*

The main point of view of the first Poincaré convergence study, hereby brought to an end, consists in the reduction of the investigation to the convergence of the series $\sum J_k$, which is geometrically evident. Thereby, the J_k signify the "Euclidean" surface area of the equivalent domains B_k. In the second Poincaré convergence study, a similarly basic use is made of the "hyperbolic" surface area of the domains B (cf. §7).

3.3 Behavior of the Poincaré series at parabolic cusps

In order to be able to bring into general application the theorem on the automorphic character of a Poincaré series, we must also *count as part of the polygon net* N, in the usual manner, *the possible parabolic cusps of the polygons.* For such kinds of parabolic points, however, the behavior of the series is not yet determined by the consideration just given.

[148] Since the convergence of the individual Poincaré series is independent of the exercise of an arbitrary linear transformation of ζ, we place the proposed parabolic point at $\zeta=\infty$ for closer investigation and put the associated generating substitutions, according to (5) on p. 65, into the form:

$$\zeta_1' = \zeta_1 + 2i\pi\zeta_2, \qquad \zeta_2' = \zeta_2. \tag{1}$$

Let the corresponding multiplicator be $e^{2i\pi\frac{\lambda}{l}}$, *understanding by λ one of the numbers* $0,1,2,\ldots,(l-1)$. As development quantities of the automorphic function and forms for the "neighborhood" of the parabolic point, we have, according to (10) on p. 70, to bring in:

$$t = e^{\zeta}.$$

We now assume a Poincaré series to be proposed, whose convergence is already proved in the "interior" of the net N. This is indeed the case in general for $d' = -d \geq 4$. *But by only requiring $d' \geq 2$,* our next arguments will also hold without further ado for the series with $d' = 3$ and 2, still convergent for certain special groups.

Since, upon exercise of the parabolic substitution, $\zeta_2' = \zeta_2$, then the Poincaré series multiplied by $\zeta_2^{d'}$, which indeed converges uniformly in the "interior" of the net, represents a *function* of ζ there, which assumes the factor $e^{2i\pi\frac{\lambda}{l}}$ when applying the substitution $\zeta' = \zeta + 2i\pi$. For the "neighborhood" of the parabolic cusp, we argue to the existence of a uniquely determined representation:

$$\sum_k \mu_k^{-1} H\left(\frac{\alpha_k\zeta+\beta_k}{\gamma_k\zeta+\delta_k}\right)(\gamma_k\zeta+\delta_k)^{-d'} = t^{\frac{\lambda}{l}} \sum_{h=-\infty}^{+\infty} a_h t^h, \tag{2}$$

where the *Laurent series* standing on the right is convergent inside a certain finite circle about $t=0$, apart from the point $t=0$ itself, if need be.

It is now most important, *that we are able to transform the Poincaré series into the power series thus intended in a direct manner.* The development required for this divides itself in the following manner:

Since the Poincaré series is unconditionally convergent in the interior of the net N, we are able, without attention to the value of the sum, to put the terms of the series into an arbitrary arrangement. Notice, that inside Γ, the substitutions $\zeta' = \zeta + 2i\pi\nu$ (with $\nu = -\infty,\ldots,+\infty$) furnish a subgroup of index ∞. For this subgroup, suppose we have
[149] chosen a representative system in the substitutions V_κ (with $\kappa = 0,1,\ldots$); and indeed, in such a way, that, in:

$$V_\kappa(\zeta + 2i\pi\nu); \qquad \kappa = 0,1,2,\ldots; \nu = -\infty,\ldots,+\infty,$$

we obtain all substitutions of the group, and each only once. Then the following arrangement of the terms of the series is most expedient for us:

$$\sum_\kappa \left[\mu_\kappa^{-1} \sum_{\nu=-\infty}^{+\infty} e^{2\nu i\pi\frac{\lambda}{l}} H\left(\frac{\alpha_\kappa\zeta+\beta_\kappa-2\nu i\pi\alpha_\kappa}{\gamma_\kappa\zeta+\delta_\kappa-2\nu i\pi\gamma_\kappa}\right)(\gamma_\kappa\zeta+\delta_\kappa-2\nu i\pi\gamma_\kappa)^{-d'}\right].$$

If, for abbreviation, we put:

$$\sum_{v=-\infty}^{+\infty} H\left(\frac{\alpha_\kappa\zeta+\beta_\kappa-2vli\pi\alpha_\kappa}{\gamma_n\zeta+\delta_\kappa-2vli\pi\gamma_\kappa}\right)(\gamma_\kappa\zeta+\delta_\kappa-2vli\pi\gamma_n)^{-d'}=\Phi_\kappa(\zeta), \tag{3}$$

then, with this, our series to be investigated has assumed the sample form:

$$\sum_\kappa\left[\mu_\kappa^{-1}\sum_{\sigma=0}^{+\infty}e^{2\sigma i\pi\frac{\lambda}{l}}\Phi_\kappa(\zeta-2i\pi\sigma)\right]. \tag{4}$$

Since the rational function $H(\zeta)$ is pole-free everywhere on the boundary, and since, further d' was assumed integral and ≥ 2, then:

$$H\left(\frac{\alpha_\kappa\zeta+\beta_\kappa}{\gamma_\kappa\zeta+\delta_\kappa}\right)(\gamma_\kappa\zeta+\delta_\kappa)^{-d'}=R(\zeta). \tag{5}$$

For the individual index κ, we have a rational function of ζ, which at $\zeta=\infty$ has *a zero of at least the second order*. One wants to convince oneself, that this rule suffers no exception for $\kappa=0$, i.e., for the substitution $V_0=1$. Here:

$$R(\zeta)=H(\zeta)=\zeta_2^{d'}H_d(\zeta_1,\zeta_2). \tag{6}$$

Now since $H_d(\zeta_1,\zeta_2)$ was to remain finite for $\zeta_2=0$, i.e., at the parabolic boundary point, and $d'\geq 2$ holds, then, as asserted, $H(\zeta)$, and therefore $R(\zeta)$, has a zero at ∞ of second or higher order.

Now suppose $R(\zeta)$ to have s poles all together,which lie at the finite places a_1, a_2, $\cdots$, a_s, and let the hth of them be of order p_h. Then, with consideration of the behavior of $R(\zeta)$ at $\zeta=\infty$, the partial fraction decomposition holds at $\zeta=\infty$:

$$R(\zeta)=\sum_{h=1}^{s}\left[\frac{A_h^{(1)}}{\zeta-a_h}+\frac{A_h^{(2)}}{(\zeta-a_h)^2}+\cdots+\frac{A_h^{(p_h)}}{(\zeta-a_h)^{p_h}}\right] \tag{7}$$

with vanishing sum of all numerators $A_h^{(1)}$: [150]

$$\sum_{h=1}^{s}A_h^{(1)}=0. \tag{8}$$

Now write formula (3) thus:

$$\Phi_\kappa(\zeta)=R(\zeta)+\sum_{v=1}^{\infty}[R(\zeta-2i\pi lv)+R(\zeta+2i\pi lv)]. \tag{9}$$

The summation indicated here on the right is to be carried through for the individual terms of the developed expression (7) of $R(\zeta)$. For this purpose utilize the formula:

$$\frac{1}{\zeta-a}+\sum_{v=1}^{\infty}\left[\frac{1}{\zeta-a-2i\pi lv}+\frac{1}{\zeta-a+2i\pi lv}\right]=\frac{1}{2il}\operatorname{cotg}\left(\frac{\zeta-a}{2il}\right).$$

If we express the function cotg by the exponential function and write:

$$e^{\frac{\zeta}{l}} = t^{\frac{1}{l}} = t', \tag{10}$$

then this equation assumes the form:

$$\frac{1}{\zeta - a} + \sum_{\nu=1}^{\infty}\left[\frac{1}{\zeta - a - 2i\pi l\nu} + \frac{1}{\zeta - a + 2i\pi l\nu}\right] = -\frac{1}{2l}\frac{1 + t'\cdot e^{-\frac{a}{l}}}{1 - t'\cdot e^{-\frac{a}{l}}}. \tag{11}$$

The sum standing on the left therefore furnishes a rational function of t', which assumes the value $-\frac{1}{2l}$, *independent of the position of the pole a,* at the parabolic cusp, i.e., for $t' = 0$.

The series, standing on the left in the last equation may be termwise differentiated according to ζ, without harming its uniform convergence. On the right, a rational function of t' hereby enters, which vanishes for $t' = 0$. Exactly the same state of affairs also remains subsisting for multiple repetition of this differentiation process; each of the series

$$\sum_{\nu=-\infty}^{+\infty}\left(\frac{1}{\zeta - a - 2i\pi\lambda\nu}\right)^2, \sum_{\nu=-\infty}^{+\infty}\left(\frac{1}{\zeta - a - 2i\pi\lambda\nu}\right)^3, \dots$$

furnishes a rational function of t', *which vanishes for* $t' = 0$.

With utilization of the relation (8), and with consideration of the behavior of the expression standing on the right in (11) at $\zeta = \infty$, we accordingly find, for the right side of the equation (9), a rational function of t', which vanishes for $t' = 0$. With this, $\Phi_\kappa(\zeta)$ permits a development according to powers of t':

$$\Phi_\kappa(\zeta) = b_1^{(\kappa)} t' + b_2^{(\kappa)} t'^2 + b_3^{(\kappa)} t'^3 + \cdots,$$

convergent inside a finitely extended circle about $t' = 0$.

[151] By means of this expression for $\Phi_\kappa(\zeta)$, now transform the sum (4); one finds, first of all:

$$\sum_{\sigma=0}^{l-1} e^{2\sigma i\pi\frac{\lambda}{l}}\Phi_\kappa(\zeta - 2i\pi\sigma) = \sum_{\sigma=0}^{l-1} e^{2\sigma i\pi\frac{\lambda}{l}}\left[b_1^{(\kappa)} t' e^{-\frac{2i\pi\sigma}{l}} + b_2^{(\kappa)} t'^2 e^{-\frac{4i\pi\sigma}{l}} + \cdots\right].$$

In the summation with respect to σ, all those terms, on the right side simply fall out, for which the exponent of t' is not congruent with λ modulo l. It thus follows:

$$\mu_\kappa^{-1}\sum_{\sigma=0}^{l-1} e^{2\sigma i\pi\frac{\lambda}{l}}\Phi_\kappa(\zeta - 2i\pi\sigma) = \mu_\kappa^{-1} l t'^\lambda\left[b_\lambda^{(\kappa)} + b_{\lambda+l}^{(\kappa)} t'^l + b_{\lambda+2l}^{(\kappa)} t'^{2l} + \cdots\right],$$

whereby, in the case $\lambda = 0$, the initial coefficient $b_\lambda^{(\kappa)}$ vanishes.

The sum of all expressions standing here on the right for $\kappa = 0, 1, 2, \dots$ furnishes the value of the convergent series (4). This may, which has been proved by Weierstrass[5], be rearranged according to ascending powers of t, and with this, provides the desired representation of our Poincaré series in the form:

$$\sum_k \mu_k^{-1} H\left(\frac{\alpha_k\zeta + \beta_k}{\gamma_k\zeta + \delta_k}\right)(\gamma_k\zeta + \delta_k)^{-d'} = t^{\frac{\lambda}{l}}(a_0 + a_1 t + a_2 t^2 + \cdots), \tag{12}$$

5 Monatsberichte der Berliner Akademie 1880 or Weierstrass' Mathematische Werke, Vol. 2 on p. 205.

where λ is a number from the series $0,1,2,\ldots,l-1$ and a_0 *vanishes in the case* $\lambda=0$.

Hereby, we have obtained the important theorem: *a Poincaré series of dimension* $d\leq -2$, *unconditionally convergent in the "interior" of the net N, also never exhibits an essential singular point at the individual parabolic cusp possibly coming up. According to the basic theorem set up on the p. 140, the series will thus represent an automorphic form; and indeed, under all circumstances, even if the multiplicator* $\mu=1$ *occurs, this form has a zero at the parabolic point.* In particular, this circumstance will be very rich in consequences later on.

The calculations of the present paragraph also give rise to the following preliminary consideration. If we allow, that the rational form $H_d(\zeta_1,\zeta_2)$ *has a pole at the parabolic point* $\zeta=\infty$ *at most of the order* $d'-2$, then all all individual steps of our analytic development remain formally unaltered. The rational function $R(\zeta)$ represented in (6) and (in case we allow still further poles at boundary points) the function (5) for arbitrary κ, [152]
preserve their zeros of at least the second order at $\zeta=\infty$; and, assuming the absolute and uniform convergence, in the interior the net, of the series coming into question, one finally arrives again at the result (12) in its full extent and to the theorems attracted thereto.

Now, in fact, the requisite convergence proof has been presented by H. v. Mangoldt[6] for the Poincaréé series for $d'\geq 4$ thus generalized. The consideration bases itself in simple fashion on calculations, as we presented them in the formula (3) ff. One must only fix the representative system V_κ of the parabolic subgroup (contrary to the measure followed above) in such a manner, that, in:

$$V_\kappa(\zeta)+2i\pi\nu;\qquad \kappa=0,1,2,\ldots;\qquad \nu=-\infty,\ldots,+\infty.$$

One now gets all substitutions of the group and each only once.

In the result thus described, there initially appears, to be sure, a contradiction, insofar as, *in spite of the pole of* $H_d(\zeta_1,\zeta_2)$ *at the parabolic point*, the automorphic form represented by the series:

$$\sum_k \mu_k^{-1} H_d(\alpha_k\zeta_1+\beta_k\zeta_2,\gamma_k\zeta_1+\delta_k\zeta_2)$$

always has a zero right there. However, one wants to remind oneself here of our determinations above concerning zeros of the forms of parabolic cusps(cf. p. 70). There holds, as power series development of the individual form, at the place in question:

$$\zeta_2^{-d'}\, t^{\frac{\lambda}{l}}(a_0+a_1t+a_2t^2+\cdots),$$

and here again,one finds the pole of the form H_d in the pole of the first factor $\zeta_2^{-d'}$, which, on the basis of the agreement mentioned, we do not count in giving the behavior of the automorphic form.

Since, furthermore, according to this, the series considered by v. Mangoldt are also *not* free of the restricting properly, of furnishing zeros at parabolic cusps, then these

[6] Compare the paper, "Über ein Verfahren zur Darstellung elliptischer Modulfunctionen durch unendliche Producte nebst einer Ausdehnung dieses Verfashions auf allgemeinere Functionen" Göttinger Nachr. of 1886, p. 1.

[153] series, as we shall later represent in even more detail[7], are not able to accomplish more, than the series considered by us so far. On this account, it appears justified, if from now on, we again assume the *rational form* $H_d(\zeta_1,\zeta_2)$ *as pole-free at all boundary points of the group.*

3.4 The Poincaré series of $(-2)^{\text{nd}}$ dimension for groups Γ with boundary curves

In the studies of the previous paragraph, we had already considered the series of $(-2)^{\text{nd}}$ and $(-3)^{\text{nd}}$ dimensions. There lie numerous investigations before us, principally concerning the series with $d = -2$, concerning which we will now report.

First, we give the negative theorem, *that the series of* $(-2)^{\text{nd}}$ *dimension are certainly no longer absolutely convergent, if the polygon net has one or infinitely many boundary curves*[8].

The proof yields itself easily from the approaches developed in the paragraph before last, whose approaches we again take up, on this account, in their full extent. However, instead of the poles, one now wants to mark the zeros of the rational function $H(\zeta)$. One wants to surround each individual one of these with a small circle, and remove from the net N all those polygons which have a finitely extended portion of surface in common with at least one of these circles. Should one zero or several zeros of $H(\zeta)$ find themselves on a boundary curve of N, then infinitely many polygons are removed in this way. We are then able to choose the circles to be situated about the zeros so small, *that finitely extended stretches of the boundary of the net remain uncovered.*

We now again choose ζ anywhere in the domains denoted above (p. 144) by B_0 or B, and retain only those terms of the proposed series of the $(-2)^{\text{nd}}$ dimension, whose domains B_k belong to the rest of the net. The shortening of the series may again be indicated by an upper prime on the summation sign. The smallest value K, which $|H(\zeta)|$ assumes in the rest of the net retained, is > 0. There we have:

$$\sum_k{}' |\mu_k^{-1} H[V_k(\zeta)]| \cdot \left| \frac{dV_k(\zeta)}{d\zeta} \right| \geq K \sum_k{}' \left| \frac{dV_k(\zeta)}{d\zeta} \right|,$$

[154] so that our series will not be absolutely convergent, in the event the series:

$$\sum_k{}' \left| \frac{dV_k(\zeta)}{d\zeta} \right| \tag{1}$$

is divergent.

The domain B could not come infinitely close to any boundary point of the group F, and was to be so fixed, that the domains B_k, equivalent to it, never reached over one another. The choice of B, however, is subjected to no further restriction. We suffice with the requirements named, by taking B *straightaway identical with the polygon of*

[7] See the theorems to be developed at the close of the chapter, as well as in the next chapter, concerning representation of arbitrary automorphic forms by Poincaré series.

[8] Ritter has set up this theorem in §12 of his oft-cited work in Vol. 41 of the Mathem. Annalen.

the net, to which the chosen point ζ belongs[9]. Except, at the possible parabolic tips, one wants to cut off small segments along associated boundary curves, in order that B not come infinitely near any boundary points.

Let the circumference of B be L_0, and L_k have the corresponding meaning for B_k. Then, in each case:

$$\left|\frac{dV_k(\zeta)}{d\zeta}\right|_m L_0 < L_k \leq \left|\frac{dV_k(\zeta)}{d\zeta}\right|_M L_0,$$

with utilization of (5) on p. 146, we have, for our point ζ belonging to the domain B:

$$\left|\frac{dV_k(\zeta)}{d\zeta}\right| \geq \left|\frac{dV_k(\zeta)}{d\zeta}\right|_m > C^{-1}\left|\frac{dV_k(\zeta)}{d\zeta}\right|_M \geq C^{-1}\frac{L_k}{L_0}.$$

For the series (1) to be investigated, we thus have the inequality:

$$\sum_k{}'\left|\frac{dV_k(\zeta)}{d\zeta}\right| > \frac{1}{CL_0}\sum_k{}' L_k. \tag{2}$$

Now for the sake of greater clarity, first of all, we return for the moment, once more, to the unshortened polygon net N, and convince ourselves without trouble, *that the sum of the circumferences of all polygons $P_0, P_1, P_2, \ldots$ of the net N is infinitely great.* Consider the polygons in the following arrangement. Lay a closed wreath of polygons, neighboring with P_0, about P_0. If P_0 is multiple connected, then it suffices to lay this polygon wreath about the outer boundary of P_0. The polygon P_0 has such a wreath, since the point $\zeta = \infty$ is situated outside the net N, and therefore outside the polygon P_0. Around the polygon complex constructed, lay anew a closed wreath of further polygons, and continue in the same manner. First after we have arranged next to one another infinitely many polygon wreaths of this kind surrounding one another, do we [155]
come infinitely close to the boundary of the net N.

Now, first of all, for the individual such polygon wreath, sum the circumstances of all the polygons taking part. However this partial sum might come out for the individual wreath, in any case, it will *not* approach the limit zero, if we approach the boundary of the net in the series of the polygon wreaths.

The sum $\sum_k L_k$, coming into consideration for us, to be sure, does, not refer itself to all polygons just drawn into consideration and (in the case of parabolic tips) not to be complete polygons. But it clearly suffices, that the net N, even after the shortening, still takes part in a *finitely* extended portion of the original outer boundary of the net, in order to also recognize, from the representations given, the divergence of the series $\sum_k{}' L_k$. The theorem given at the beginning of the paragraph is thus proved.

We attach here, further, a more parenthetical remark. *In the case of the modular functions, the Eisenstein series of the* $(-2)^{\text{nd}}$ *dimension, closely related to the Poincaré series, are "conditionally" convergent.* The definition of the Eisenstein series[10] is given by:

9 If ζ lies on a side or at a vertex of the polygon division, then choose of the polygons taking part as the domain B.

10 Compare the paper "*Genaue Untersuchung der unendlichen Doppelproducte, aus welchen die elliptischen Function als. Quotienten zusammengesetzt sind*", Journ. Math. Vol. 35 (1847).

$$\sum_{m_1,m_2}\left(\frac{1}{m_1\zeta_1+m_2\zeta_2}\right)^{d'}, \tag{3}$$

where the sum refers itself to all pairs of whole numbers m_1, m_2, with the exclusion of the combination $m_1 = 0$, $m_2 = 0$. Corresponding to this series (3), we have the following series:

$$\sum_{k}\left(\frac{1}{\gamma_k\zeta_1+\delta_k\zeta_2}\right)^{d'}, \tag{4}$$

which emerges from (3), if one leaves out all those terms of the first series, in which m_1 and m_2 have a common divisor different from 1. Notice, that in (4), we do not directly have a Poincaré series of the kind considered so far. Namely, the rational form $H_d = \zeta_2^d$ underlying it has the property, of remaining unaltered by a parabolic subgroup of the
[156] modular group. On this account, in order to obtain an automorphic form, one will restrict oneself to a representative system belonging to that subgroup. Thereby, however, one is led right to the series (4).

The relation of both series (3) and (4) is easy to give in the case of unconditional convergence[11]. By putting, in place of the individual number pair γ_k, δ_k, all pairs $t\gamma_k, t\delta_k$ with $t = 1,2,3,\ldots$, one obviously obtains all pairs m_1, m_2. Thus, if the series (3) is unconditionally convergent, then arrange all the terms, which refer themselves to the same pair of relatively prime numbers γ_k, δ_k, together and find in this way:

$$\sum_{m_1,m_2}\left(\frac{1}{m_1\zeta_1+m_2\zeta_2}\right)^{d'} = \sum_{t=1}^{\infty}\left(\frac{1}{t}\right)^{d'}\cdot\sum_{k}\left(\frac{1}{\gamma_k\zeta_1+\delta_k\zeta_2}\right)^{d'},$$

so that, the two series differ only by a constant factor.

However, as is known, this unconditional convergence enters only for $d' > 2$; on the other hand, the series (3) is conditionally convergent for $d' = 2$. If, e.g., one favors that term arrangement, which we had utilized on p. 149 for the purpose of the power series development in the neighborhood of a parabolic cusp, then we have [12]:

$$\sum_{m_1,m_2}\left(\frac{1}{m_1\zeta_1+m_2\zeta_2}\right)^{2} = \zeta_2^{-2}\left[2\sum_{m=1}^{\infty}\left(\frac{1}{m}\right)^2 - 8\pi^2\sum_{n=1}^{\infty}\frac{nq^{2n}}{1-q^{2n}}\right],$$

understanding by q the development quantity $e^{\pi i\zeta}$, current in the theory of elliptic functions, for the neighborhood of the parabolic cusp $\zeta = i\infty$. Here on the right, in fact, stands a convergent series. But if one chooses another arrangement of the series terms, then one arrives at more and more different values of the sum, as, e.g., this is investigated in detail in the work of A. Hurwitz (Kap. 2. §4) just cited.

The series of type (4) arise for each group Γ with parabolic substitutions. Thereby, it is natural to expect, that the series of $(-2)^{\text{nd}}$ dimension might be conditionally convergent in this manner in each case. However, this question has not yet been subjected to any detailed investigation.

---

[11] Compare, for this, the paper of O. Rausenberger "*Notiz zur theorie der Modulfunctionen*", Math. Annalen, Vol. 20 (1882).

[12] Cf. A. Herwitz "*Grundlagen einer independenten Theorie der elliptischen Modulfunctionen u.s.w.*", Math. Annalen, Vol. 18 (1881).

3.5 The Poincaré series of $(-2)^{\mathrm{nd}}$ dimension for principal-circle groups with isolatedly situated boundary points

[157]

A first positive result on the series of $(-2)^{\mathrm{nd}}$ dimension is contained in the following theorem: *for those principal-circle groups which have one net N spread out over the entire plane, for which, therefore, the boundary points do not cover the principal circle everywhere densely, but lie isolatedly on it, the Poincaré series of $(-2)^{\mathrm{nd}}$ dimension are always unconditionally and uniformly convergent in the entire net*[13].

The theorem set up is obviously significant in principle for all the categories of automorphic models cited in 2. on p. 31. The proof succeeds easily, on the basis of the viewpoint developed in the first Poincaré convergence proof (cf. §2 on p. 142 ff.).

One wants to make the principal circle into the real ζ-axis, which is allowed by the second theorem set up in §2 (p. 143), and to take care, that the point $\zeta = \infty$ is not a boundary point of Γ. Thereby, we could easily yet attain, that this point $\zeta = \infty$ is situated in the interior of a polygon, i.e., not on the boundary between two.

Let the series proposed for investigation be given by:

$$\sum_k \mu_k^{-1} H\left(\frac{\alpha_k\zeta+\beta_k}{\gamma_k\zeta+\delta_k}\right)(\gamma_k\zeta+\delta_k)^{-2}. \tag{1}$$

This is to be considered in the neighborhood of any finite places ζ_0 in the interior of the net N. As above, let the neighborhood of ζ_0 be represented by a domain B_0, which we will immediately fix even more closely. Beforehand, mark those polygons of the net, in which poles of $H(\zeta)$ are situated. The corresponding terms of the series (1) are to be left out. We proceed similarly with that term, whose associated polygon carries $\zeta = \infty$. Let the shortening of the series be indicated by a prime on the summation sign. There are, all together, only *finitely many* terms left out, which have no influence on the convergence of the series (1).

The next thing to do, is to choose the domains $B_0, B_1, \ldots$, in suitable fashion. [158]

To this end, we introduce a net of *normal polygons*, by, in particular, making the chosen place ζ_0 the center of a first polygon P (cf. I. pp. 106 and 243). This polygon is now intersected by at least one segment s_0 of the real ζ-axis. i.e., the principal circle, whereby the two endpoints of s_0 correspond by means of a hyperbolic generator V of Γ. The two level circles of V through the endpoints of s_0 which are marked heavily in figure 25, take part in the boundary of the polygon P, at least in the vicinity of s_0. In figure 25, the level curve of V running through ζ_0 is also indicated dottedly, which intersects the principal circle at the place A of the segment s_0 and at the point A' outside the polygon P. Through A' and the two endpoints of s_0 lay the two circles perpendicular to the real axis, denoted in the figure by K_1 and K_2.

After these preparations, we determine, that B_0 is any circle to be laid about the point ζ_0, which is not to reach out over the crescent formed domain bounded by K_1 and K_2, and naturally, also, not over the polygon of the net belonging to ζ_0. The immediate

[13] W. Burnside and E. Ritter have found this theorem independently of one another, and published it almost simultaneously. The Burnside development is contained in the work, to be cited often, "*On a class of automorphic functions*" (proceedings of the London Math. Soc., Nov. 1891); for Ritter, see, §12 of the work in Vol. 41 of Mathem. Annal. (New York, 1892).

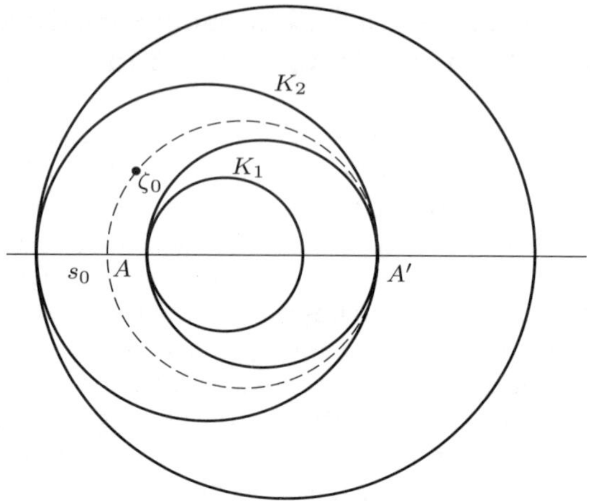

Fig. 25.

perception of the figure informs us, that one can choose a *finitely extended circle of this kind*, and *that this can never be greater than the circle with the diameter* s_0. For the circumference L_0 of B_0, we have:

$$L_0 \leq \pi s_0. \tag{2}$$

The determinations hereby made have the character of invariance with respect to the circle-relation and therefore, for the transition to the domains B_K equivalent to B_0.
[159] With this, we obtain, without further ado, for the length L_k of the circumference of the domain B_k, the inequality:

$$L_k \leq \pi s_k, \tag{3}$$

if s_k signifies the segment of the kth polygon equivalent with s_0, resp, the length of this segment.

The uniform convergence of the series of absolute values:

$$\sum_k{}' |H[V_k(\zeta)]| \cdot \left| \frac{dV_k(\zeta)}{d\zeta} \right| \tag{4}$$

for any place ζ inside B_0 is now easily proved. Next, the study carried out on p. 145 ff., concerning the oscillation of the mapping modulus upon transition from B_0 to B_k, is also valid here. The coefficient γ_k is different from zero in each substitution V_k of Γ, apart from the identity. The points equivalent with ∞ are scattered on the real axis, and $-\frac{\delta_k}{\gamma_k}$ will furnish a point situated in the interior of the domain B_0, at most for one single k. Were this to occur, then simply leave out the concerned series term. Then, here too, there holds an inequality of the form (5) on p. 146 with finite C, independent of k. We are able, further, in consequence of the exclusion of the polygons containing the poles of $H(\zeta)$, to give a finite number G, beyond which $|H[V_k\zeta]|$ cannot grow in any of the domains B_k. But then we find, analogously, as above (p. 146 ff.), the subsistence of the inequality:

$$\sum_k{}' |H[V_k(\zeta)]| \cdot \left| \frac{dV_k(\zeta)}{d\zeta} \right| < GC \cdot L_0^{-1} \sum_k{}' L_k,$$

i.e., by means of (3):

$$\sum_k{}' |H[V_k(\zeta)]| \cdot \left|\frac{dV_k(\zeta)}{d\zeta}\right| < \pi G C L_0^{-1} \sum_k{}' s_k,$$

The s_k, coming into consideration in the last sum, cover, all together, only a *finite* portion of the real axis, since the one segment running into the infinite is left out. With this, $\sum_k{}' s_k$ is finite, and therefore, since ζ was arbitrarily choosable in the domain B_0, our series is, in fact, absolutely and uniformly convergent there[14].

The behavior of our series of $(-2)^{\text{nd}}$ dimension, recognized as convergent, at pos- [160]
sible cusps is already made clear by the developments of §3 (p. 147 ff.). We may immediately state the theorem: *for each of the principal-circle groups considered in this paragraph, one of our Poincaré series of $(-2)^{\text{nd}}$ dimension represents an automorphic form, which has a zero at the possible parabolic cusps.*

3.6 Convergence of the Poincaré series of $(-2)^{\text{nd}}$ dimension for certain groups without boundary curves and without principal circle

The series of $(-2)^{\text{nd}}$ dimension are yet to be investigated from various aspects for a further important class of automorphic models. We attach to an *arbitrary fundamental domain, which is bounded by $2p$ full circles running separately and pairwise coordinated to one another by loxodromic or hyperbolic substitutions* (cf. figure 159 in I). Here again, as for the principal-circle groups just considered, the net N will cover the entire ζ-plane up to isolatedly situated boundary points. We need not exclude here principal-circle polygons of the kind mentioned; however, those cases are already disposed of by the previous paragraph.

The question of the absolute and uniform convergence of the series of $(-2)^{\text{nd}}$ dimension has been decided at least for a portion of the groups in question by an interesting consideration of W. Burnside's in the work cited on p. 157. On the other hand, the work of Schottky's "Uber eine Function, welche bei einer bestimunten linearen transformation inres arguments unverändert bleibt"[15]. The following representation attaches itself most closely to the developments of Schottky's.

Let a proposed series of $(-2)^{\text{nd}}$ dimension:

$$\sum_k \mu_k^{-1} H\left(\frac{\alpha_k\zeta+\beta_k}{\gamma_k\zeta+\delta_k}\right)(\gamma_k\zeta+\delta_k)^{-2} \tag{1}$$

be investigated in the neighborhood of the point ζ_0, which naturally is not to be a boundary point of the group. We so effect the arrangement on the basis of the theorem cited on p. 143, that ζ_0 lies in the finite part of the plane, and that $\zeta=\infty$ is not a

[14] We yet draw attention to the work of H. Weber "*Ein Beitrag zu Poincaré's Theorie der Fuchs'schen Functionen*" (Göttinger Nachrichten of 1886 on p. 359), to which we shall return later on in even more detail. In this work, for special principal circle groups, with isolatedly situated boundary points, infinite products are brought into account, whose convergence proof is based on the same geometric viewpoint, which also furnishes the foundation for the investigations of the text.

[15] Jounal für Mathematik, Vol. 101, p. 22 (1887).

boundary point, but rather, is situated in the "interior" of a polygon. Besides this, we assume for the sake of simplicity, that ζ_0 is not situated right on the boundary between two polygons of the net N, which lets itself be attained, if necessary, by an allowed alteration of the polygons of the net.

[161] Let the domains $B_0, B_1, \ldots$, which we situated about ζ_0 and the points equivalent to ζ_0, be circles, each individual one of which is to be situated entirely inside its polygon. Next, we then have to again make the same preparations, as in the foregoing cases. By leaving out, if necessary, a finite number of series terms, which are able to play no role in the question of convergence, we obtain the theorem: *the proposed series (3) will be unconditionally and uniformly convergent in the neighborhood B_0 of ζ_0, in the case the sum $\sum_k{}' L_k$ of the domain circumferences, corresponding to the shortened series, has a finite value.*

We begin the summation with the circumferences L_k of that polygon, which contains the point $\zeta = \infty$. If we only first position this polygon into the ζ-plane, then $2p$ open circular discs still remain, in which $2p$ further polygons attach themselves to the first one. The circles B_k situated in these are obviously smaller then the $2p$ circles just mentioned. If we replace these $2p$ circumferences L_k, to the extent that they come up in the sum $\sum_k{}' L_k$, by the periphery of the $2p$ circles just cited, then $\sum_k{}' L_k$ will thereby be increased. There now still remain $2p(2p-1)$ open circular discs, and in each of these a further polygon attaches itself directly to the net so far. We again put the peripheries of those $2p(p-1)$ circular in place of $2p(2p-1)$ next following circumferences L_k, through which our sum of the circumferences becomes increased anew. We continue in the same fashion.

The convergence of the proposed series will accordingly be proved, *if the sum of all circular circumstances, or (what amounts to the same thing) the sum of all circular radii of our net N has a finite value.*

Now Schottky l.c. under a certain assumption concerning the shape of the individual polygon, has actually been able to prove the finiteness of the sum in question of the circular radii. The "Schottky Condition" is the following: *in the individual polygon of the net, there are to let themselves be added additional circles running separately from the $2p$ boundary circles as well as from one another, in such manner, that the polygon appears decomposed into exclusively "threefold" connected partial domains.*

For the principal-circle groups, which subsume themselves under our kind of
[162] groups coming into question here the "Schottky condition" is always fulfillable. Figure 26, e.g., shows, for a polygon with eight boundary circles, the decomposition by five further circles, which are marked heavily in the figure, into six domains, connected threefold in this specific case. By analogous arrangement of the supplementary circles, but also in even a multitude of other ways, the desired decomposition lets itself be effected for each principal-circle group of our kind.

Schottky presents l.c. yet a second special example. First of all, in place of $2p$ boundary circles let there be $2p$ points, which are entirely arbitrarily situated in the ζ-plane, and which let be called $A_1, A_2, \ldots, A_{2p}$ in order to be immediately prescribed more closely. Further, mark a point C, which does not have the same distances from any two of the points A; we imagine the $2p$ points denoted in such an order by $A_1, A_2, \ldots$, that A_1 lies closest to the point C, then A_2 follows, etc.. Now lay a system of $(2p-3)$ concentric circles about the common center C in such a manner, that the

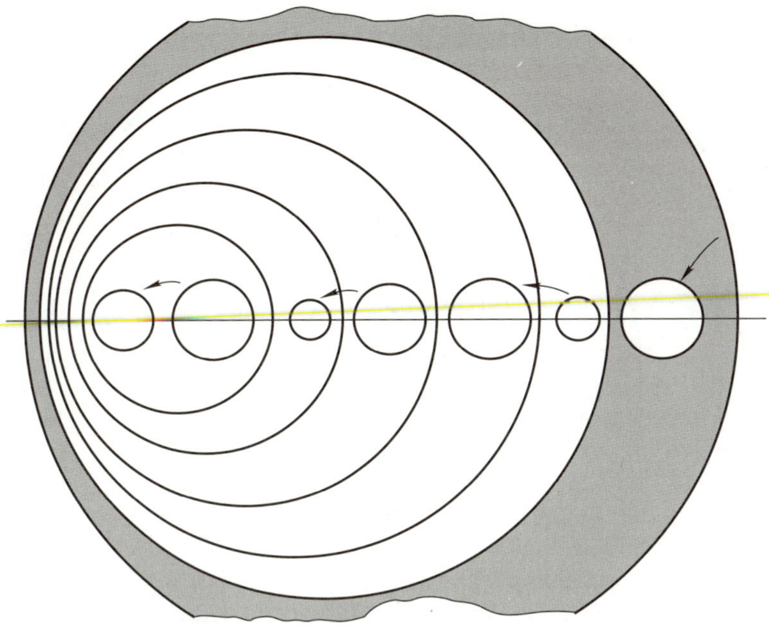

Fig. 26.

periphery of the smallest circle runs between A_2 and A_3 and therefore separates the points A_1 and A_2 from the remaining ones, that the following then runs between A_3 and A_4 etc., while the last circle runs between A_{2p-2} and A_{2p-1}. Now we extend the $2p$ [163]
points $A_1, A_2, ..., A_{2p}$ into just as many circles $K_1, K_2, ..., K_{2p}$; however, these are never to reach the $(2p-3)$ circles already drawn, and K_1, and K_2 are also to run separately from one another, just as K_{2p-1} and K_{2p}, as all this is brought into realization in figure 27. Finally, if we coordinate the $2p$ circles pairwise and loxodromically in any manner to one another, then, with this, we have obtained a fundamental domain, which fulfills the Schottky condition.

Furthermore, naturally, in place of the concentric circles about C utilized here, we may also have any other circles, which decompose the ζ-plane, become $2p$-fold disconnected by removal of the points A, into a series of threefold connected domains.

Now let there be an arbitrary net N of polygons $P_0, P_1, P_2, ...$, which fulfill the Schottky condition. We decompose P_0 in the given manner into $(2p-2)$ threefold connected partial domains $T_0', T_0'', T_0''', ..., T_0^{(2p-2)}$ and carry this division over, in equivalent fashion, to all polygons of N. The ζ-plane appears covered in such a way with a net of threefold connected domains, each individual one of which is equivalent with one of the $(2p-2)$ domains $T_0^{(i)}$ and, as this one, appears bounded by three full circles running [164]
separately from one another. Naturally, we shall so effect the arrangement, that $\zeta = \infty$ is situated in the "interior" of one of the threefold connected domains. *If we have, that the sum of the radii of all circles of the net now obtained is finite, then, all the more, the sum of the radii of all boundaries of the polygon net $P_0, P_1, P_2, \dots$ is finite and, with this, our series converges.*

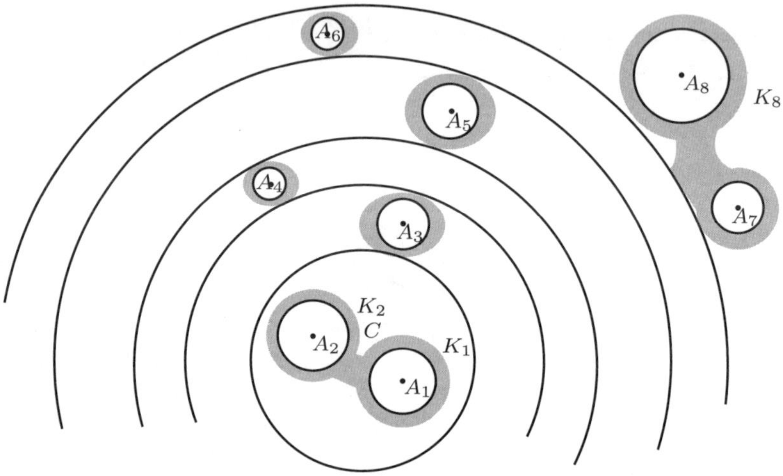

Fig. 27.

Let an arbitrary circle of our complete net have the radius R. A threefold connected domain is included by this circle, which is bounded in the interior by two circles of the radii r and r'. Then, obviously, we always have $r + r' < R$. However, before we are able to bring this inequality into application, it must first be developed even more precisely.

For this purpose, we denote the distance between the centers of the two circles of the radii R and r under discussion here by e, while e' has the same meaning for the including circle of the radius R and the inner circle of the radius r'.

One now wants to remind himself, *that two circles of the ζ-plane have an invariant with respect to arbitrary linear transformations of ζ*. One can bring in as such the invariant $j = \alpha + \delta$, to be defined according to I on p. 337, of that substitution, which arises by combination of the reflections in those two circles. In our case, this substitution is always hyperbolic and therefore j is real and absolutely > 2.

In order, in particular, to calculate the invariant; for the two circles of the radii R and r, we temporarily translate the circle system in such a way, that the center of the smaller circle comes to lie at the origin, that of the large circle on the positive real ζ-axis. Then the two reflections are represented by:

$$\zeta' = \frac{r}{r^{-1}\bar{\zeta}}, \qquad \zeta' = \frac{eR^{-1}\bar{\zeta} - (e^2R^{-1} - R)}{R^{-1}\bar{\zeta} - eR^{-1}}, \tag{2}$$

and, for the invariant, we find the expression:

$$j = Rr^{-1} + rR^{-1} - e^2r^{-1}R^{-1} = \frac{R^2 + r^2 - e^2}{rR}. \tag{3}$$

In this term, j is positive, and therefore the inequality $j > 2$ holds.

Each threefold connected domain of our net is now equivalent with one of the domains $T_0', T_0'', \ldots, T_0^{(2p-2)}$. Therefore, only a *bounded number of invariants* comes into consideration here; and we call j_0 the smallest value represented by them. Thus,for the
[165] pair of circles considered above:

$$R^2 + r^2 - e^2 \geq Rr\,j_0, \qquad j_0 > 2. \tag{4}$$

Since, on the other hand, $R - e > r$ occurs, then it follows:

$$R^2 - r^2 + e^2 > 2Re. \tag{5}$$

By combination of the inequalities (4) and (5) arises:

$$e < R - \frac{1}{2} j_0 r.$$

Next to this, there arranges itself analogously:

$$e' < R - \frac{1}{2} j_0 r',$$

So that, by addition of both inequalities, we obtain:

$$e + e' < 2R - \frac{1}{2} j_0 (r + r').$$

But now, it yields itself, from the shape of our threefold connected domain (insofar, namely, as the two inner boundary circles run separately), that $r + r' < e + e'$. We have, with this:

$$r + r' < e + e' < 2R - \frac{1}{2} j_0 (r + r')$$
$$r + r' < \frac{4}{2 + j_0} \cdot R.$$

One calls the quotient, multiplied on the right here by R, ε, and then obviously has $0 < \varepsilon < 1$, on account of the second inequality (4). *In place of the preliminary inequality* $r + r' < R$, *we accordingly get now*:

$$r + r' < \varepsilon R, \tag{6}$$

where ε *is a number belonging to the interval* $0 < \varepsilon < 1$, *uniquely determined for our domain net.*

Now start with the domain, which contains $\zeta = \infty$ and is bounded by three circles excluding one another. Let an arbitrary one of these three circles have the radius R, and conceive the totality of the further circles of the net included b this circle. The two circles initially attaching themselves furnish a radius sum of $r + r' < \varepsilon R$. Inside each of the two latter circles, two further circles attach themselves, whose radii sum is $< \varepsilon r$ resp. $< \varepsilon r'$. The sum of the radii of all these four circles is thus $< \varepsilon(r + r')$ and therefore, in consequence of (6), also $< \varepsilon^2 R$. If we again go a step further, then eight circles follow with a radii sum $< \varepsilon^3 R$; and the law coming to light there also remains subsisting further on. The first named circle and *all* the circles included by it accordingly furnish a radii sum which is [166]

$$< R + \varepsilon R + \varepsilon^2 R + \varepsilon^3 R + \cdots \text{and therefore} < \frac{R}{1 - \varepsilon}.$$

with this is actually proved the unconditional and uniform convergence of the Poincaré series of $(-2)^{\mathrm{nd}}$ *dimension for all groups of our kind fulfilling the "Schottky condition".*

As was already emphasized at the beginning of the paragraph, W.Burnside has also carried out convergence studies belonging here, later then Schottky, but completely independently of the latter; it is a question of the "second proof" in Art. 2 of Burnside's work cited on p. 157. The author shows, which is very simple, that it is sufficient, to prove the convergence of the series:

$$\sum_k \left| \frac{1}{\gamma_k^2} \right|.$$

This series is then investigated directly by an interesting consideration, concerning which one wants to consult the details l.c..

Unfortunately, in this way, the convergence is only first proved for the case, that the "multipliers" of the p generating substitutions are sufficiently large, which we are able to formulate, tying in with the geometric perceptions utilized above, by saying, that the $2p$ boundary circles of the polygon P must be sufficiently close or sufficiently far from one another.

One is immediately reminded of the example of Schottky's illustrated by figure 27 on p. 163. However the precise relation of the "Burnside condition" to the "Schottky condition" formulated above (p. 161) has not yet been determined. In two directions, Burnside may have gone further than Schottky: *first, Burnside announces, that he had been able to carry through his convergence study also for examples of such non-principal-circle groups, which have "parabolic" generators besides possible loxodromic ones.* In Schottky, this limiting case is not comprehended, at least not directly. On the other hand, there emerges from Burnside's considerations the remarkable result, *That for each p, there are also groups of our kind with absolutely convergent Poincaré series of* $(-1)^{\mathrm{st}}$ *dimension.*

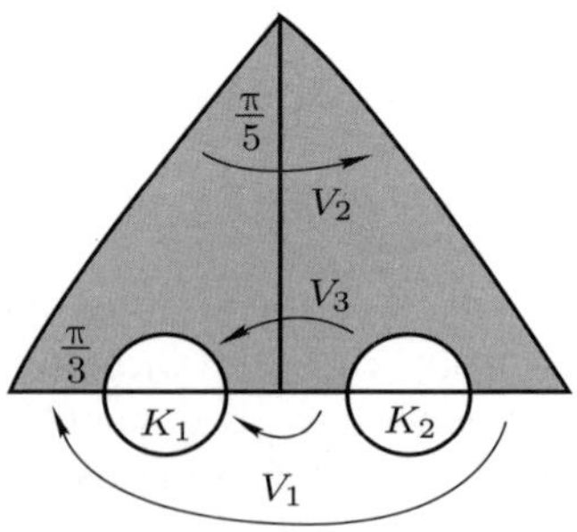

Fig. 28.

That, furthermore, groups with *elliptic* generators also occur, for which the series of $(-2)^{\mathrm{nd}}$ dimension are absolutely convergent, is shown by the example of figure 28.
[167] Here, V_1 and V_2 alone generate an icosahedral group. The latter is composed with the hyperbolic substitution V_3, which is transformed by V_1 into V_3^{-1}. One understands

these groups best, in case one starts from the 30 hyperbolic substitutions, which proceed from V_3 by transformation with the substitutions of the icosahedral group. These 30 hyperbolic substitutions generate a group, which, for circles K_1, K_2 (cf. figure 28) chosen sufficiently small directly subsumes itself in the Schottky examples illustrated in figure 27 on p. 163, and therefore has unconditionally convergent series of $(-2)^{\mathrm{nd}}$ dimension. The fundamental domain of the group in question now shows the icosahedral regularity; and the addition of the icosahedral group furnishes the group with the fundamental domain of figure 28. *Here, therefore, we have a group of genus zero and signature* $(0,5;2,2,2,3,5)$ *for which the series of* $(-2)^{\mathrm{nd}}$ *dimension certainly converges absolutely; for there is a subgroup of "finite" index* 60 *for which the convergence in question subsists.*

All these indications make it probable, that the Poincaré series of $(-2)^{\mathrm{nd}}$ dimension are still convergent for far more groups, then was recognized so far. The conjecture has even been occasionally uttered, that the series are convergent for *all* groups without boundary curves; and, for proving this conjecture, one has tried to evaluate other viewpoints, e.g., certain developments of potential theory. However, these attempts, to date, have not yet led to palpable results.

3.7 Second convergence study in the principal-circle case. Continuous dependence of the Poincaré series on the group moduli

Poincaré, in his work in the Acta mathematica Vol.1 on p. 198 ff., has carried out yet a second convergence study of his series for the principal-circle groups; this second investigation obtains, a result of fundamental significance, since for it, not only the argument ζ, but also the *moduli of the group* Γ count *as variable.*[16]. Accordingly, let, at the same time, a *"family" of groups or automorphic models with principal circle* be [168]
proposed, whereby it is immaterial, whether the principal circle is a boundary circle for this model or not. It is thereby a question of a *continuum* of groups resp. models, whose dimension number was determined in I, Part II, Chapter 2. The individual model of this continuum we could uniquely represent by *a system of real moduli* $j_1, j_2, \ldots, j_\nu$, between which, insofar as this system, in general, contains superfluous moduli[17], a number of equations subsists. Besides this, in order that to a proposed system of moduli there also actually corresponds a principal-circle model, these moduli must always fulfill certain *inequalities* which are given exhaustively in I l.c.. if one or more of these

[16] Moreover, one could also have fashioned Poincaré's first convergence proof, referring to arbitrary polygon groups, as well as the Schottky by developments, which were reported on in §6, in such a manner, that variability of the group moduli were allowed. If, in the text, we refer this development to principal-circle groups, then we do this because we have a complete overview of the manifold of the individual species only for these latter groups, while for the groups without principal circle, we must leave aside as too difficult a development of the modular theory in I.

[17] In particular, we also imagine the irrationalities (moreover uniquely determined) initially coming up, as e.g., $\sqrt{D_{123}}$ or j_{123} in I on p. 377, simply absorbed into the modular system $j_1, j_2, \ldots, j_\nu$, in order afterwards to be able to utilize the theorem, that generators, and with this, all substitutions of Γ generally, are built up *rationally* in the $j_1, j_2, \ldots, j_\nu$.

inequalities go over into equations, then we have to do with *limiting cases of the family. These kinds of limiting cases are, for the present, to remain excluded from consideration.*

Moreover, we yet remind ourselves, that to the individual admissible modular system there always belongs an entire *class* of groups, which are equivalent with one another, insofar as they may be carried over into one another by linear transformation of ζ. *We may represent this class by one group, picked out arbitrarily.* The coefficients in the substitutions $V_k(\zeta)$ of the latter, as already remarked above, are built up *rationally* in the moduli, $j_1, j_2, \ldots, j_\nu$, according to I l.c.

Let an arbitrary particular modular system $j_1^{(0)}, j_2^{(0)}, \ldots, j_\nu^{(0)}$ furnish the group $\Gamma^{(0)}$ of the proposed family. Around the values of this system we now delimit, with numbers $d_1, d_2, \ldots, d_\nu$, non-vanishing but chosen sufficiently small, intervals, inside which the j, to the extent that they are not fixed[18], are to be considered as continuously variable real quantities:

$$j_1^{(0)} - d_1 < j_1 < j_1^{(0)} + d_1, \ldots, j_\nu^{(0)} - d_\nu < j_\nu < j_\nu^{(0)} + d_\nu. \tag{1}$$

Naturally, the equations between the moduli are always hereby to remain subsisting; also, we shall imagine the d chosen so small, that the characteristic inequalities, always hold. If, in the substitutions of the representing group $\Gamma^{(0)}$, *without otherwise altering the former,* we let an arbitrary system, $j_1, \ldots, j_\nu$ satisfying the inequalities (1) take the
[169] place of $j_1^{(0)}, j_2^{(0)}, \ldots, j_\nu^{(0)}$, then there arises a new group Γ of the family; and all modular systems thus considered will make up *the "full neighborhood" of the group* $\Gamma^{(0)}$ *inside the family.*

Thereby, we may in particular assume, *that all those groups of the neighborhood of* $\Gamma^{(0)}$ *have one and the same principal circle.* Namely, if we initially prescribe, that the principal circle of $\Gamma^{(0)}$ is to be the real axis, then all the substitution coefficients are real expressions in the $j_1^{(0)}, j_2^{(0)}, \ldots, j_\nu^{(0)}$. But since the $j_1, \ldots, j_\nu$ always assume only real values, then the remaining groups under discussion will have real substitution coefficients throughout and therefore again have the real axis as principal circle. Our theorem on the commonality of the principal circle for all groups Γ of the neighborhood $\Gamma^{(0)}$ will then also immediately hold, in case, in place of the real ζ-axis, we choose any other principal circle by means of an arbitrary linear transformation. Except one must naturally assume the coefficients of this transformation itself as not variable with the moduli j.

We finally make note, that for continuous alteration of the moduli, the group generators likewise alter themselves continuously. The theory of the discontinuity domain in I, upon this, furnishes the theorem: *we are always able to follow each continuous alteration of the moduli by a similarly continuous alteration of the fundamental domain.*

For the sequel, it is now expedient, to choose the *unit circle* of the ζ-plane as the principal circle. If the point ζ_0, to whose neighborhood we will refer the convergence study, does not lie right on the principal circle itself, then we may assume it (p. 143) situated in the interior of this circle. Points of the principal circle itself, which to be sure, in case the latter is not a boundary circle, are able to belong to the "interior" of the polygon net N, are consequently to remain excluded for the time being.We assume, first of all, besides this, that ζ_0 is not an elliptic fixed point of $\Gamma^{(0)}$. Moreover, should the

18 Thus, e.g., such moduli, which are invariants of elliptic generators, remain constant.

group also contain substitutions, which interchange the two sides of the principal circle, then refer the investigation, which is completely sufficient, instead of to the given group, to that subgroup of index 2, whose substitutions only carry the principal circle over into itself again.

We so choose the polygon net N of $\Gamma^{(0)}$, that ζ_0 does not lie right on the boundary between two polygons[19]. As neighborhood ζ_0, we consider the points of a circular disc [170]
B_0 to be laid about ζ_0 as center, which never comes infinitely close to the boundary the polygon belonging to ζ_0 and also to the principal circle.

By now also holding the point ζ_0 and the domain B_0 fixed for the remaining groups Γ of the full neighborhood of $\Gamma^{(0)}$, we are able, according to the arguments presented before, to assume the $d_1, d_2, \ldots, d_\nu$ as positive quantities different from 0 and so small, *that upon running through all neighboring groups, only such alternations of the polygon surrounding B_0 enter, for which the polygon boundary consistently remains at a finite distance from the circle B_0.*

Now let the series proposed for investigation be:

$$\sum_k \mu_k^{-1} H[V_k(\zeta)] \left(\frac{dV_k(\zeta)}{d\zeta} \right)^{\frac{d'}{2}}. \tag{2}$$

We ask, whether this is unconditionally and uniformly convergent for all groups Γ of the neighborhood of $\Gamma^{(0)}$ and all points ζ of the circular disc B_0. By also considering the prepatory developments on p. 144 ff., as carried out here, we shall have to investigate the series of the absolute values:

$$\sum_k \left| \frac{dV_k(\zeta)}{d\zeta} \right|^{\frac{d'}{2}} \tag{3}$$

for an *arbitrary* group Γ of this neighborhood and for an *arbitrary* point ζ of B_0.

Here now, with Poincaré, we make a most interesting use of the *hyperbolic determination of measure* for the interior of the principal circle, for which the points of the principal circle itself play the role of the infinitely distant elements. Statements concerning length, which refer themselves to this determination of measure, are always to be made known by the additional word "hyperbolic".

The basic idea in the application of the determination of measure in question to the investigation of our series is the following. We describe about $\zeta = 0$ a system of concentric circles $K_1, K_2, K_3, \ldots$ with hyperbolic radii, $r, 2r, 3r, \ldots$, understanding by r a finite positive value. All terms of the series (3), whose $\zeta^{(k)} = V_k(\zeta)$ lies inside K_1, furnish the sum S_1; and corresponding, in general, let all terms, whose $\zeta^{(k)} = V_k(\zeta)$ lies between K_{n-1} and K_n or on K_{n-1}, yield the sum S_n. Hereby, we mean by ζ the point chosen before in B_0. Since, in the individual circular ring, there always find themselves only finitely many points $\zeta^{(k)}$ equivalent with the chosen point ζ, then all numbers $S_1, S_2, \ldots$, are finite. But since, further, the series (3) only contains positive terms, then its term may be arbitrarily rearranged; and therefore, we have: [171]

$$\sum_k \left(\frac{dV_k(\zeta)}{d\zeta} \right)^{\frac{d'}{2}} = S_1 + S_2 + S_3 + \cdots. \tag{4}$$

19 We could, e.g., work with the normal polygon of the center ζ_0.

We shall be able to prove, *that the series standing on the right here is more rapidly convergent than a determinate convergent geometric series, which is independent of the particular group* Γ *chosen from the neighborhood of* $\Gamma^{(0)}$ *as well, too, as of the particular point* ζ *picked out from* B_0.

The hyperbolic arc element $d\sigma$ and surface element $d\tau$ were given in I on p. 28 under (7) for the determination of measure established in the ζ-halfplane. But we immediately map the ζ-halfplane onto the surface of the unit circle of the ζ'-plane by means of the transformation:

$$\zeta = i\frac{i+\zeta'}{1-\zeta'}.$$

From this, if we write $\zeta = \xi + i\eta$ and $\zeta' = \xi' + i\eta'$, it follows directly, first:

$$\eta = \frac{1-\xi'^2-\eta'^2}{(\xi'-1)^2+\eta'^2},$$

then by differentiation:

$$|d\zeta| = \frac{|d\zeta'|}{|\xi'-1+i\eta'|^2} = \frac{|d\zeta'|}{(\xi'-1)^2+\eta'^2}.$$

For the arc element $d\sigma$ of the hyperbolic determination of measure corresponding to the unit circle, we thus obtain the expression:

$$d\sigma = \frac{|d\zeta|}{\eta} = \frac{|d\zeta'|}{1-\xi'^2-\eta'^2}.$$

For the abbreviation of the formulas, we introduce polar coordinates ρ, ω in the ζ'-plane, by setting $\zeta' = \rho e^{i\omega}$. Moreover, we again leave out the upper index in ζ', since, further on, we shall only be concerned with the determination of measure corresponding to the unit circle. By also calculating the surface element $d\tau$ in known fashion, with utilization of the formula for $d\sigma$, it follows: *the hyperbolic arc element* $d\sigma$ *and surface element* $d\tau$ *are represented by*:

$$d\sigma = \frac{|d\zeta|}{1-\rho^2}, \qquad d\tau = \frac{\rho\, d\rho\, d\omega}{(1-\rho^2)^2}. \tag{5}$$

A circle with ordinary radius ρ about $\zeta = 0$ with $\rho < 1$, consequently receives the hyperbolic radius:

$$R = \int_0^\rho \frac{d\rho}{1-\rho^2} = \frac{1}{2}\log\frac{1+\rho}{1-\rho}. \tag{6}$$

[172] Conversely ρ is calculated from R by mean, of the formula:

$$\rho = \frac{e^{2R}-1}{e^{2R}+1}. \tag{7}$$

The hyperbolic area J of the circle under consideration is:

$$J = \int_0^{2\pi}\left(d\omega\int_0^\rho \frac{\rho\, d\rho}{(1-\rho^2)^2}\right) = \frac{\pi\rho^2}{1-\rho^2}. \tag{8}$$

It follows: the hyperbolic area J of the circle of the hyperbolic radius R is:

$$J = \frac{\pi}{4}(e^{2R} + e^{-2R} - 2). \tag{9}$$

For the individual term of the series (3) to be investigated, we have;

$$\left|\frac{dV_k(\zeta)}{d\zeta}\right| = \left|\frac{d\xi^{(k)}}{d\xi}\right| = \frac{|d\zeta^{(k)}|}{|d\zeta|}.$$

Now arc elements, equivalent with respect to the group Γ, are directly congruent in the case of the hyperbolic determinations of measure. Thus, from the first formula (5), it follows;

$$|d\zeta| = (1 - |\zeta|^2)d\sigma, \qquad |d\zeta^{(k)}| = (1 - |\zeta^{(k)}|^2)d\sigma.$$

We notice that equation

$$\left|\frac{dV_k(\zeta)}{d\zeta}\right| = \frac{1 - |\zeta^{(k)}|^2}{1 - |\zeta|^2}. \tag{10}$$

If we denote by R and R_k true hyperbolic distances of the points ζ and $\zeta^{(k)}$ from the origin $\zeta = 0$, then from (10), with use of formula (7), here is calculated, *for the individual term of the series* (3), *the representation*:

$$\left|\frac{dV_k(\zeta)}{d\zeta}\right|^{\frac{d'}{2}} = \left(\frac{e^{2R} + e^{-2R} + 2}{e^{2R_k} + e^{-2R_k} + 2}\right)^{\frac{d'}{2}} = \left(\frac{e^{R} + e^{-R}}{e^{R_k} + e^{-R_k}}\right)^{d'}. \tag{11}$$

Now we return to the domain B_0, which will have a determinate hyperbolic area τ, finite and different from 0. All domains B_k equivalent to B_0 with respect to Γ are congruent to B_0 in the sense of the hyperbolic determination of measure, and they consequently all have the same area τ. Further, we again bring in the circles, already introduced above, $K_1, K_2, \ldots$, described about $\zeta = 0$, of the hyperbolic radii $r, 2r, \ldots$, and will call the number of all points $\zeta^{(k)}$, equivalent with ζ, which lie inside K_n, A_n; by ζ we naturally understand here again the point chosen in B_0, for which we study the series [173]
(3). The domains B_k belonging to these $\zeta^{(k)}$ either lie entirely inside K_n or are also able to reach out partially over K_n. If D (considered in the sense of the hyperbolic determination of measure) is the greatest distance between two boundary points of the domain B_0, then, in any sense, all the domains B_k belonging to the $\zeta^{(k)}$ sought will be entirely inside a circle about $\zeta = 0$ with the hyperbolic radius $(nr + D)$. Thereby, whatever group Γ we may have picked out of the neighborhood of $\Gamma^{(0)}$, these domains will not collide with one another. Accordingly, the total surface $A_n\tau$ of these B_k will be smaller than the surface of the circle with the hyperbolic radius $(nr + D)$, to be calculated according to (9). *From this, we have, for the number A_n of all points $\zeta^{(k)}$ situated inside the circle K_n, the inequality*:

$$A_n < \frac{\pi}{4\tau}(e^{2nr+2D} + e^{-2nr-2D} - 2). \tag{12}$$

The number A'_n of all $\zeta^{(k)}$ in the ring included by K_{n-1} and K_n or on K_{n-1} is $\le A_n$; it follows, for A'_n:

$$A'_n < \frac{\pi}{4\tau}(e^{2nr+2D} + e^{-2nr-2D} - 2) < \frac{\pi}{4\tau}e^{e^{2nr+2D}}. \tag{13}$$

The individual amounts of these A'_n terms of the series (3), however, since the associated $\zeta^{(k)}$ do not lie inside the circle K_{n-1} in consequence of (11) are

$$\leq \left(\frac{e^R + e^{-R}}{e^{(n-1)r} + e^{-(n-1)r}}\right)^{d'} < (e^R + e^{-R})^{d'} \cdot e^{-d'(n-1)r}.$$

For the partial sum, denoted above by S_n, of the series terms belonging between K_{n-1} and K_n, we have the inequality:

$$S_n < \frac{\pi}{4\tau}(e^R + e^{-R})^{d'} e^{d'r+2D} \cdot e^{-(d'-2)rn}.$$

With this and the consideration of formula(4), we come to the result, that the following inequality subsists:

$$\sum_k \left|\frac{dV_k(\zeta)}{d\zeta}\right|^{\frac{d'}{2}} < \frac{\pi}{4\tau}(e^R + e^{-R})^{d'} e^{d'r+2D} \sum_{n=1}^{\infty} e^{-(d'-2)rn}. \tag{14}$$

If $d' > 2$, then here on the right, in the last place, stands a *convergent* geometrical series. All the quantities coming up in the right of the inequality (14) are independent of the particular choice of the group Γ in the neighborhood of $\Gamma^{(0)}$. Only R is dependent on the choice of the point ζ in B_0; however, R remains under an obtainable number. We are consequently able to formulate the following result: *the Poincaré series of the*
[174] *dimension $d < -2$, for all principal-circle groups Γ in the neighborhood of $\Gamma^{(0)}$ and for all points ζ in the neighborhood of ζ_0, are unconditionally and uniformly convergent.*

In the initially excluded case, that ζ_0 is *an elliptic vertex*, one has the present a series of modifications of the previous considerations, without the basic ideas of the proof or the result undergoing an alteration. Above all one has to bound the domain B_0, as a polygon segment lying at ζ_0, by an associated orbit curve of the elliptic substitution, say, and must, for the continuous alteration of the moduli, alter the representing group Γ along with this, in such a way, that the elliptic fixed point ζ_0 remains fixed. Concerning the possible *parabolic* vertices, we shall simply refer ourselves to the considerations on p. 148 ff., whose validity was also explicitly stressed above *for the series of* $(-3)^{\text{rd}}$ *dimension.*

Furthermore, the arguments developed break down, because of their nature, for the groups without boundary circle, for the points ζ on the principal circle. Here, now as before, we are directed to the investigation on p. 157 ff., through which the unconditional convergence of the series was proved for $d \leq -2$.

The most important result of the development given, however, consists in the knowledge of the *uniformity of the convergence of our series for alternation of the moduli* $j_1, j_2, \ldots, j_\nu$. With this, these series furnish automorphic forms, which, in their dependency on the moduli $j_i, j_2, \ldots, j_\nu$, are *continuous functions*[20].

[20] For points ζ of the principal circle, if such belong to the "interior" of the net, the uniformity of the series convergence, to be sure, has not yet been shown with respect to the moduli. However, it is easy, to also draw into consideration the studies carried out in the text with respect to the alteration of moduli, e.g., for the convergence investigations developed in §5 on p. 157. For the theorem just stated (in the text), therefore, the points ζ of the principal circle will by no means play an exceptional role.

We state this result, at the same time, in a form especially usable for later. The *algebraic model* belonging to the individual group, possesses, in the sense of the theory of algebraic functions, a series of "*moduli*" namely, first the Riemann "*class moduli*", whose number depends, in known fashion, on the genus p of the model, then the "*absolute invariants*" of those places of the model, which correspond to the *fixed polygon vertices.* Combining we will designate these class moduli and invariants as the "*algebraic moduli*" of the models, which might be called the "*group moduli*" in constrast to the $j_1, j_2, \ldots, j_\nu$. Then we are able to state the fundamental theorem: *if the representing* [175]
group runs continuously through the proposed family of principal-circle groups, without hereby coming infinitely close to the limiting cases (cf. p. 168), then corresponding to this, the algebraic moduli of the associated automorphic model also experience only "continuous" alterations.

E. Ritter, in his very extensive investigation[21], without calling in the Poincaré series, has considered the alteration of the automorphic functions for continuous alteration of the fundamental domain. The Ritter investigation works with Riemann methods of perception, as we analogously applied them above (p. 8ff.) to the direct existence proof of the automorphic functions. The consequence is, that the continuity theorems proved by Ritter hold, not only for discontinuity domains of groups, but also for more general fundamental domains (cf. p. 52). The Poincaré continuity study, which we reproduced above, in contrast to this, refers itself solely to single-valued automorphic functions with principal circle. In this restricted domain, Poincaré's study has the advantage of greater brevity.

3.8 Poles of the Poincaré series and the possibility of its vanishing identically. Discussion for the case $p = 0$

An individual convergent Poincaré series of dimension d and the multiplicator system M, as we have seen, furnishes for the proposed group Γ, a *single-valued automorphic form of dimension d and the multiplicator system* M:

$$\sum_k \mu_k^{-1} H_d(\alpha_k\zeta_1 + \beta_k\zeta_2, \gamma_k\zeta_1 + \delta_k\zeta_2) = \varphi_d(\zeta_1, \zeta_2). \tag{1}$$

However, this approach, to which we shall now have to immediately turn our attention, can become unusable, by the series standing on the left in (1) *identically vanishing.* This possibility of the identical vanishing of Poincaré series has even been the principal difficulty in the further development of the theory of these series. For the present, we are only able to state the theorem, *that a convergent Poincaré series, in any case, does not then identically vanish, if it has one or more poles in the fundamental domain.* A pair of further remarks, concerning these poles, are to follow here next.

A pole of the rational form $H_d(\zeta_1, \zeta_2)$ in the interior of the net N is, in general, a [176]
pole of the same order for the automorphic form $\varphi_d(\zeta_2, \zeta_2)$ represented in (1). An exception to this rule can only then enter, if two or more poles of H_d lie at equivalent

21 *"Die Stetigkeit der automorphen Functionen bei stetiger Abänderung des Fundamentalbereichs"*, Parts I and II, Mathemati. Annalen, Vol. 45, p. 473 and Vol. 46, p. 200 (1894).

places ζ. Then, namely, correspondingly many terms of the series become infinite at the individual one of these places, and it can be, that in summing them, a reduction of the order of the poles, indeed even a dropping out of the latter enter. *If, however, $H_d(\zeta_1,\zeta_2)$ becomes infinite at least at one place of the net N, but is furthermore pole-free at its equivalent places, then $\varphi_d(\zeta_1,\zeta_2)$ will likewise have a pole there; it follows, that in this sense, certainly, one has to do with a non-identically vanishing in series.*

The position of the pole of $H_d(\zeta_1,\zeta_2)$, and with this, of $\varphi_d(\zeta_1,\zeta_2)$, in the fundamental domain is, moreover, subjected to no further restriction, than this: *one of these can never be situated at a possible parabolic cusp.* In fact, we have indeed already in §3.3 (p. 147 ff.) discussed in detail the important fact, *that a Poincaré series always exhibits a zero at a parabolic cusp.*

This state of affairs makes an insignificant modification in the way of writing the formulas on p. 97 ff. desirable, which refers itself to the representation of the single-valued forms of a *model of genus zero* by the associated prime- and groundforms. This representation was furnished us by:

$$\varphi_d(\zeta_1,\zeta_2) = R_m(z_1,z_2)\prod_{k=1}^{n}{}' (z,e_k)^{\frac{m}{l_k}}, \tag{2}$$

whereby, in particular, we now return to the prescription $0 \le m'_k < l'_k$, given for the parabolic vertices. If the form $\varphi_d(\zeta_1,\zeta_2)$ is furnished by a Poincaré series, then a zero would always lie at the individual parabolic cusp possibly occurring. We bring this to expression outwardly most simply, *by letting the following inequality take the place of the one just mentioned*:

$$0 < m'_k \le l'_k, \tag{3}$$

which obviously signifies a modification only in the case of the parabolic multiplicator
[177] 1[22]. Moreover, naturally, the poles of the rational form $R_m(z_1,z_2)$ will now be restricted, to the extent, that none fall at a parabolic cusp.

The new determination (3) will effect a modification of the dimension m of the rational form $R_m(z_1,z_2)$. *Obviously, for the forms $\varphi(\zeta_1,\zeta_2)$ of the dimension $d \le -2$ furnished by Poincaré series, m will become smaller, then earlier (p. 99 ff.) by the number ε of the parabolic multiplicators* 1*, which come up in the underlying system* M.

The altered definition of m has, further, a pair of alterations in the formulas on p. 102 ff. as consequence. First of all, we return once more to the pairs of mutually inverse multiplicator systems M, $\overline{\mathrm{M}}$, and, as on p. 102, hold fast, concerning the corresponding dimensions, to the relation:

$$d + \overline{d} + 2 = 0. \tag{4}$$

Let $d \le -2$ hold for the dimension d, and let the associated number m be calculated by utilization of the determination (3). Since $\overline{d} \ge 0$, then there is no cause, to calculate the number $\overline{m}$, belonging to $\overline{d}$ and $\overline{\mathrm{M}}$, according to any other rule than that followed on p. 101. *Under these circumstances, the formulas (8) and (9) on p. 104 are to be replaced by*:

[22] Ritter already follows the determination (3) in the general theory of automorphic forms. However, this measure is avoided above, since there lies therein the consideration of a matter, which is foreign to the automorphic forms by themselves.

$$\frac{m_k}{l_k}+\frac{\overline{m_k}}{l_k}=1-\frac{1}{l_k} \qquad \text{resp.} \qquad \frac{m'_k}{l'_k}+\frac{\overline{m'_k}}{l'_k}=1, \tag{5}$$

$$m+\overline{m}+2=0. \tag{6}$$

A simplification takes place for the theorems on p. 105. We take note of what one immediately verifies by means of the calculations at that time: *for* $d<-2$, $m\geq -1$ *holds; this same inequality* $m\geq -1$ *subsists for* $d=-2$, *in case we do not have the multiplicator system* $\mathrm{M}_0=1$. *On the other hand, all "properly" automorphic Poincaré series of* $(-2)^{\text{nd}}$ *dimension have* $m=-2$.

We now return in more detail to the possibility of the identical vanishing of Poincaré series. Thereby, we restrict ourselves to the case $p=0$, but entirely analogous considerations hold for models of arbitrary genus p. Let there be proposed, say, a group Γ *with one boundary curve.* Then, for a given multiplicator system M, form a series of $(-4)^{\text{th}}$ dimension, all of which are to have poles of the first order at $\sigma, \eta', \eta'', \ldots, \eta^{(\sigma)}$ prescribed places of the fundamental domain. To this end, one will so choose the rational form $H_{-4}(\zeta_1,\zeta_2)$, that:

$$(\zeta,\eta')(\zeta,\eta'')\cdots(\zeta,\eta^{(\sigma)})\cdot H_{-4}(\zeta_1,\zeta_2)$$

is pole-free in the interior of the net. However, outside the latter, arbitrarily many, let us say ν, poles of the form H_{-4} may yet occur. Then the numerator of H_{-4} will have the dimension $(\sigma+\nu-4)$, so that the most general rational form H_{-4} admissible here lets itself be built up linearly and homogeneously form $(\sigma+\nu-3)$ linearly independent forms. [178]

Now, according to p. 107, there are only $(m+\sigma+1)$ linearly independent automorphic forms $\varphi_{-4}(\zeta_1,\zeta_2)$ with the given poles $\eta', \eta'', \ldots, \eta^{(\sigma)}$, where m calculates itself uniquely from the dimension $d=-4$ and the multiplicator system M. By sufficiently large choice of ν, however, one is able to exceed the value $(m+\sigma+1)$ with the number $(\sigma+\nu-3)$ arbitrarily. Between the automorphic forms φ_{-4} furnished by the $(\sigma+\nu-3)$, linear relations must thus subsist. *Each such identity leads us, by corresponding combination of the participating* H_{-4}, *to an identically vanishing Poincaré series; the associated rational form* H_{-4} *will indeed be pole free in the interior of the net, but certainly not vanish identically*[23].

The possibility that series, which are formed with non-identically vanishing $H_d(\zeta_1,\zeta_2)$ pole-free in the interior of the net, furnish, for their part, identically vanishing sums, is therefore inevitable. From this, however, arises a difficulty in the treatment of the question, most important by now, *to what extent arbitrarily chosen automorphic forms* $\varphi_d(\zeta_1,\zeta_2)$ *of admissible dimensions with zeros at the parabolic tips are individually representable by Poincaré series.* The investigation of this question will consequently make necessary yet a few special preparations.

---

[23] The part of the ζ-plane running outside the boundary curve carries a second net belonging to the group in question. Inside this, net, the individual one of the identically vanishing series just produced is not pole-free, and will therefore, by no means, also identically vanish.

3.9 Construction of one-pole Poincaré series

The relations just discussed cause us, for the present, not to work with polefree series, but to place the *one-poled Poincaré series* at the center of the investigation.

The following investigations, where nothing further is remarked, hold, at the same time, for *models of an arbitrary genus p*. Whether, for each dimension d, where Poincaré series are convergent, we are always able to form one-poled series, must first be shown by the further development. Should we immediately bring in the example of the models with $p = 0$, then, according to a theorem of the previous paragraph, the properly
[179] automorphic series, of $(-2)^{\text{nd}}$ dimension would always have $m = -2$, and therefore be at least two-poled in the fundamental domain. Here, therefore, there can be no one-poled series. It will reveal itself, *that for arbitrary p, apart from this one exceptional case* $d = -2, \mathrm{M} = 1$, *we are always, completely generally, in the position of forming one-poled series.*

First of all, this is completely simple, in the case that the polygon net N does *not* cover the entire ζ-plane, rather, is bounded by one or infinitely many boundary curves. Let a first pole be prescribed at an *arbitrary* place ξ in the "interior" of the net N, which, however, for the moment, is not to be a fixed polygon vertex. We split ξ into the quotient $\xi_1 : \xi_2$ and utilize, as earlier, the abbreviated way of writing (ζ, ξ) for the expression $(\zeta_1\xi_2 - \zeta_2\xi_1)$. Let $H_{d+1}(\zeta_1, \zeta_2)$ be a rational form of dimension $d+1$, all of whose poles are to lie *outside* the net N, of which is arbitrarily choosable except for this. *Then:*

$$\sum_k \mu_k^{-1} \frac{H_{d+1}(\zeta_1^{(k)}, \zeta_2^{(k)})}{(\zeta^{(k)}, \xi)}, \tag{1}$$

for $d \leq -4$, *is a given a convergent Poincaré series with a prescribed multiplicator system* M, *which only exhibits a pole of the first order at one place of the fundamental domain equivalent with* ξ, *and is therefore, in fact, "one-poled".* If, moreover, we have a group with one boundary circle, then, as is known, we also need not exclude the case $d = -3$.

The simplest choice of the rational form $H_{d+1}(\zeta_1, \zeta_2)$ is obviously this, *that we assume it as the reciprocal value of an integral rational form* $G_{-d-1}(\zeta_1, \zeta_2)$ *of the dimension* $(-d-1)$, *whose* $(-d-1)$ *zeros lie outside the net* N. We consequently note for ourselves, immediately, the one-poled series in the form:

$$\sum \mu_k^{-1} \frac{1}{(\zeta^{(k)}, \xi) G_{-d-1}(\zeta_1^{(k)}, \zeta_2^{(k)})}. \tag{2}$$

More circumstantial considerations are necessary in the other cases, where the net N covers the *entire* ζ-plane up to isolatedly situated boundary points. We again imagine ξ chosen arbitrarily, but different from the fixed polygon vertices, and hold fast to the scheme (1). The rational form $H_{d+1}(\zeta_1, \zeta_2)$, which is nevertheless of negative dimension, then necessarily has a pole in the *interior* of the net. In order that none of the latter be a pole of the automorphic form represented in (1), then one has to choose H in such a manner, *that the pole under discussion upon summing the terms of the series (1), exactly removes itself.* This task lets itself be solved, e.g., in the following manner.
[180] In the "interior" of the net N, we choose a *place* η, *arbitrarily, but not equivalent with* ξ *and also not coinciding with a fixed vertex,* and understanding, by $V_1, V_2, \ldots,$

V_{-d-1}, any $(-d-1)$ substitutions of Γ, in whose choice, however, a few conditions, to be named immediately, are to be observed. Let the $-d$ places:

$$\eta, \eta' = V_1(\eta), \eta'' = V_2(\eta), \ldots, \eta^{(-d-1)} = V_{-d-1}(\eta) \tag{3}$$

be poles of the first order of H. Corresponding to this, we give this form in the form:

$$H_{d+1}(\zeta_1, \zeta_2) = \frac{G_1(\zeta_1, \zeta_2)}{(\zeta, \eta)(\zeta, \eta') \cdots (\zeta, \eta^{(-d-1)})} \tag{4}$$

understanding by $G_1(\zeta_1, \zeta_2)$ a linear integral form. *The $(-d-1)$ substitutions (3) are to be different form one another and from the identity, the third coefficient γ is not to vanish in all of them, and finally, the associated multiplicator $\mu_1, \mu_2, \ldots, \mu_{-d-1}$, as long as the system* $\mathrm{M}_0 = 1$ *is not present, are not all to be equal to* 1. These conditions, to be brought into application further on, are able to be easily observed in the choice of the substitutions in question.

Our task now amounts *to determining the linear form*:

$$G_1(\zeta_1, \zeta_2) = a_1\zeta_1 + a_2\zeta_2 \tag{5}$$

in such a way, that the automorphic form represented by:

$$\varphi_d(\zeta_1, \zeta_2) = \sum \mu_k^{-1} \frac{G_1(\zeta_1^{(k)}, \zeta_2^{(k)})}{(\zeta^{(k)}, \eta)(\zeta^{(k)}, \eta') \cdots (\zeta^{(k)}, \eta^{(-d-1)})} (\zeta^{(k)}, \xi)^{-1} \tag{6}$$

is polefree at the place η, while the pole remains maintained at ξ. If the first condition is satisfied, then because of the automorphic behavior $\varphi_d(\zeta_1, \zeta_2)$, also at

$$\eta', \eta'', \ldots, \eta^{(-d-1)},$$

no poles come up. But if the last condition were not satisfied, then $\varphi_d(\zeta_1, \zeta_2)$ would be pole-free, and the identical vanishing of the series would not be excluded.

Since, on account of the unimodular form of the homogeneous substitutions, $(\zeta', \eta') = (\zeta, \eta)$, $(\zeta'', \eta'') = (\zeta, \eta), \ldots$ takes place, then, at the place η, only the terms of the series (6) belonging to the substitutions $1, V_1, V_2, \ldots, V_{-d-1}$ will exhibit poles. If, in the neighborhood of η, we develop the form $\varphi_d(\zeta_1, \zeta_2)$ according to powers of (ζ, η), then the coefficient of the power $(\zeta, \eta)^{-1}$, to be denoted from now on by $-A$, will be given by the equation:

$$A = \sum_{k=0}^{-d-1} \mu_k^{-1} \frac{G_1(\eta_1^{(k)}, \eta_2^{(k)})}{(\eta^{(k)}, \eta) \cdots (\eta^{(k)}, \eta^{(k-1)})(\eta^{(k)}, \eta^{(k+1)}) \cdots (\eta^{(k)}, \eta^{(-d-1)})} (\xi, \eta^{(k)})^{-1}. \tag{7}$$

If, on the other hand, one develops $\varphi_d(\zeta_1, \zeta_2)$ in the neighborhood of ξ according to powers of (ζ, ξ), then the coefficient of $(\zeta, \xi)^{-1}$ has the form: [181]

$$\frac{G_1(\xi_1, \xi_2)}{(\xi, \eta)(\xi, \eta') \cdots (\xi, \eta^{(-d-1)})}, \tag{8}$$

we write:

$$\frac{\xi_2^{d+1}}{\eta_2 \cdot \eta_2' \cdots \eta_2^{(-d-1)}} \cdot \frac{G_1(\xi, 1)}{(\xi - \eta)(\xi - \eta') \cdots (\xi - \eta^{(-d-1)})} \tag{9}$$

for this and develop the rational function of ξ, standing on the right here as the second factor, whose denominator has at least the second degree, into the partial fraction:

$$\frac{G_1(\eta, 1)}{(\eta - \eta') \cdots (\eta - \eta^{(-d-1)})} (\xi - \eta)^{-1} + \cdots$$
$$+ \frac{G_1(\eta^{(-d-1)}, 1)}{(\eta^{(-d-1)} - \eta) \cdots (\eta^{(-d-1)} - \eta^{(-d-2)})} (\xi - \eta^{(-d-1)})^{-1}.$$

Now again add the factor standing at the first place in (9), and again introduce the homogeneous terminology everywhere. If, for short, one calls, the product of the expression (8) with $(\frac{\eta_2}{\xi_2})^{d+2}$, B, then it follows:

$$B = \sum_{k=0}^{-d-1} \Big[(\gamma_k \eta + \delta_k)^{-d-2} \tag{10}$$
$$\frac{G_1(\eta_1^{(k)}, \eta_2^{(k)})}{(\eta^{(k)}, \eta) \cdots (\eta^{(k)}, \eta^{(k-1)})(\eta^{(k)}, \eta^{(k+1)}) \cdots (\eta^{(k)}, \eta^{(-d-1)})} (\xi, \eta^{(k)})^{-1} \Big].$$

Here, as in (7), $\eta_1^{(0)}, \eta_2^{(0)}$ naturally signify η_1, η_2. Our task is now, *to determine the coefficients a_1, a_2 of G_1, in such a manner, that $A = 0$ and $B \neq 0$.*

This task is unsolvable, in case we have $d = -2$ and $M_0 = 1$; for in this case $A = B$. With this, we return to the case, already discussed above, of the properly automorphic series of $(-2)^{\text{nd}}$ dimension.

By *excluding this exceptional case $d = -2, M_0 = 1$ from now on,* we arrange A and B according to the coefficients a_1, a_2 of G_1:

$$\begin{cases} A = a_1 \Phi_1(\eta_1, \eta_2) + a_2 \Phi_2(\eta_1, \eta_2), \\ B = a_1 \psi_1(\eta_1, \eta_2) + a_2 \psi_2(\eta_1, \eta_2). \end{cases} \tag{11}$$

The expressions multiplied here by a_1 and a_2, *for fixed ξ_1, ξ_2 and place η considered variable, resp, variable η_1, η_2 we shall consider as forms of these η_1, η_2.* The meaning of these forms emerges immediately from (7) and (10)[24]; in particular, they are all of the dimension $(2d+2)$.

[182] The condition $A = 0$ now requires that we take a_1, a_2 proportional with $\Phi_2 - \Phi_1$:

$$a_1 : a_2 = \Phi_2(\eta_1, \eta_2) : -\Phi_1(\eta_1, \eta_2). \tag{12}$$

Two mutually exclusive cases are then able to occur. *Either the equation*:

$$\frac{\Phi_1(\eta_1, \eta_2)}{\psi_1(\eta_1, \eta_2)} = \frac{\Phi_2(\eta_1, \eta_2)}{\psi_2(\eta_1, \eta_2)} \tag{13}$$

subsists identically in η, and then, with A, B is also always equal to zero; or the equation (13) subsists for only finitely many η, and then η lets itself be so chosen, that $B \neq 0$.

[24] One hereby wants only to consider, that $\eta_1^{(k)}, \eta_2^{(k)}$ are linear and homogeneous in η_1, η_2.

If we proceed from the occurrence of the first possibility, then let us denote the expressions, identical with one another, on the right and left sides of (13), as rational in $\eta_1 : \eta_2 = \eta$, by $R(\eta)$. From (11), the conclusion is then to be drawn, that whatever may be the meaning of the a_1, a_2, the quotient $A : B$ is always identical with this rational function $R(\eta)$. The expression $R(\eta)$ can, on this account, no longer contain ξ_1, ξ_2. If, namely, we choose $a_1 = \xi_2$, $a_2 = -\xi_1$, then the ξ_1, ξ_2 drop entirely out of A as well as B, in consequence of (7) and (10).

In this situation, the quotient $A : B$ of these expressions A and B, rational in the ξ and η, then also maintains its value $R(\eta)$, free of ξ, if we now subsequently alter ξ arbitrarily. But if one lets ξ move to the place $\eta^{(k)}$, then from (7) and (10) follows:

$$\frac{A}{B} = R(\eta) = \frac{(\gamma_k \eta + \delta_k)^{d+2}}{\mu_k}, \tag{14}$$

valid for all $-d$ indices $k = 0, 1, 2, \ldots, -d-1$. If we now choose $k = 0$ here, then $R(\eta)$ shows itself to be identical with the constant 1. If we then take a k with $\gamma_k \neq 0$ (and a substitution V_k satisfying this requirement among the $V_1, V_2, \ldots, V_{-d-1}$), then $d = -2$ follows; for in all remaining cases, the right side of (14) is variable with η. But if we substitute $d = -2$ and $R(\eta) = 1$ into (14), then it follows:

$$\mu_1 = 1, \mu_2 = 1, \ldots, \mu_{-d-1} = 1,$$

and therefore the multiplicator system $\mathrm{M}_0 = 1$ is present (cf. p. 180).

Since we are hereby led back to the exceptional case already excluded above, then it follows, *that by avoiding finitely many places η with a coefficients pair a_1, a_2 corresponding to condition (12), we are always able to fulfill the requirement $B \neq 0$ in the desired fashion.*

It is now most expedient, to comply with the proportion (12) in the following manner: [183]

$$\begin{cases} a_1 = \displaystyle\sum_{k=0}^{-d-1} \mu_k^{-1} \eta_2^{(k)} \frac{(\xi,\eta)\cdots(\xi,\eta^{(k-1)})(\xi,\eta^{(k+1)})\cdots(\xi,\eta^{(-d-1)})}{(\eta^{(k)},\eta)\cdots(\eta^{(k)},\eta^{(k-1)})(\eta^{(k)},\eta^{(k+1)})\cdots(\eta^{(k)},\eta^{(-d-1)})}, \\ a_2 = -\displaystyle\sum_{k=0}^{-d-1} \mu_k^{-1} \eta_1^{(k)} \frac{(\xi,\eta)\cdots(\xi,\eta^{(k-1)})(\xi,\eta^{(k+1)})\cdots(\xi,\eta^{(-d-1)})}{(\eta^{(k)},\eta)\cdots(\eta^{(k)},\eta^{(k-1)})(\eta^{(k)},\eta^{(k+1)})\cdots(\eta^{(k)},\eta^{(-d-1)})} \end{cases} \tag{15}$$

through this, G_1, in the ξ_1, ξ_2, changes into a rational integral form of the dimension $-d-1$; On this account, we denote this, in more detail, by $\mathrm{G}_1^{(-d-1)}(\zeta_1, \zeta_2; \xi_1, \xi_2)$. As the principal result, however, we state: also in the case of a group without boundary curves, for an arbitrarily prescribed pole ξ, outside the fixed vertices, apart from the repeatedly mentioned exceptional case $d = -2, \mathrm{M}_0 = 1$, we are able, for each dimension d corresponding to the general convergence theorem and for each multiplicator system M, to form one-poled series in the form:

$$\sum_k \mu_k^{-1} \frac{\mathrm{G}_1^{(-d-1)}(\zeta_1^{(k)}, \zeta_2^{(k)}; \xi_1, \xi_2)}{(\zeta^{(k)},\xi)(\zeta^{(k)},\eta)(\zeta^{(k)},\eta')\cdots(\zeta^{(k)},\eta^{(-d-1)})}. \tag{16}$$

3.10 One-poled series with poles at elliptic vertices

In the developments of the previous paragraph, it was explicitly stipulated, that ξ was to be different from each fixed vertex of our polygon net. In consideration of the *parabolic* points, we must also initially hold fast from now on to this determination; for we have stipulated above, for the convergence studies of the Poincaré series, that the rational form $H_d(\zeta_1,\zeta_2)$, utilized in the formation of the individual series, was always to be pole-free at parabolic boundary points. It is otherwise with *elliptic* vertices. If ξ moves into such a vertex, then the one-poled series produced here remain convergent; but relations enter here, which require a closer investigation.

Accordingly, let ε be an arbitrary elliptic vertex, for which we use the symbols $(\zeta,\varepsilon),(\zeta,\varepsilon')$ in the earlier sense (cf. (3) on p. 64). Then let the one-poled series (2) on p. 64 be formed under the assumption $\xi=\varepsilon$; if we imagine the integral rational form G_{-d-1} transformed to $(\zeta,\varepsilon),(\zeta,\varepsilon')$ then it obtains the form:

$$\sum_k \mu_k^{-1}\frac{1}{(\zeta^{(k)},\varepsilon)G_{-d-1}[(\zeta^{(k)},\varepsilon),(\zeta^{(k)},\varepsilon')]}. \tag{1}$$

Here, at the place ε, l terms of the series all together have a pole of the first order, un-
[184] derstanding by l the period of the substitution belonging to the vertex in question; and
indeed, the initial term is given approximately at $\zeta=\varepsilon$, up to a constant factor, by:

$$(\zeta,\varepsilon)^{-1}(\zeta,\varepsilon')^{d+1}. \tag{2}$$

From this arise the remaining $(l-1)$ terms polar at ε, by repeatedly applying the elliptic generator (cf.(4) on p. 64) and each time dividing by its multiplicator $\mu=e^{\pi i\frac{\lambda}{l}}$. The effect of these operations is the addition of the multiplicative root of unity:

$$e^{\frac{-\pi i}{l}-\frac{(d+1)\pi i}{l}-\frac{\pi i\lambda}{l}}=e^{-\frac{\pi i}{l}(\lambda+d+2)} \tag{3}$$

and its powers up to the $(l-1)^{\text{st}}$ to the expression (2). We recognize immediately: *only if λ fulfills the congruence*:

$$\lambda\equiv -d-2 \pmod{2l}, \tag{4}$$

does our series actually have a pole at the vertex ε, and indeed, measured in the polygon, one such of order $\frac{1}{l}$, if, on the contrary, the congruence (4) is not fulfilled, then the series (1) becomes pole-free at ε and consequently everywhere and then its identical vanishing is not precluded.

One now convinces oneself easily, *that this theorem remains directly subsisting for the series (16) on p. 183.* In fact, apart from a factor independent of the ζ_1,ζ_2, the approximate value (2) also holds here again for the initial term of the series. One has only to consider, that the product $(\zeta,\eta)(\zeta,\eta')\cdots(\zeta,\eta^{(-d-1)})$ represents an integral form of dimension $-d$ not vanishing at the place ε, and can therefore be replaced approximately by $C(\zeta,\varepsilon')^{-d}$ with C different from 0, and that, on the other hand, the linear form $G^{(-d-1)}$ in the numerator of the initial term (16) on p. 183 does not vanish for $\zeta=\xi=\varepsilon$.

We have now advocated above (p. 100ff.) the standpoint, that a zero of the order determined by (5) on p. 100 is always to lie at the individual elliptic vertex, whose order is equal to $\frac{l-1}{l}$ in case of the congruence (4). On the strength of this, we have to form the conception, *that upon the validity of the congruence (4), a pole of the "first" order is added to this zero.* Moreover, if we hold fast to this conception already stated on p. 106, then it is attained, *that our series is, in fact, a "one-poled one" in the case of the congruence (4).*

On the basis of this method of perception, whether the boundary curve case might occur or not, we are finally, easily able, also for the non-validity of the congruence (4), to produce one-poled series for elliptic vertices $\zeta = \varepsilon$. We calculate for ourselves, ac- [185]
cording to (5) on p. 100, for the vertex ε considered, the associated number m[25] from:

$$\frac{m}{l} = \frac{d+\lambda}{2l} - E\Big[\frac{d+\lambda}{2l}\Big].$$

Then we form the series:

$$\sum_k \frac{\mu_k^{-1}}{(\zeta^{(k)},\varepsilon)^{l-m} G_{-d+m-l}(\zeta_1^{(k)},\zeta_2^{(k)})} \tag{5}$$

in whose place, however, for the case that $-d+m-l<0$ was to hold, we must get the series:

$$\sum_k \mu_k^{-1} \frac{G_{-d+m-l}(\zeta_1^{(k)},\zeta_2^{(k)})}{(\zeta^{(k)},\varepsilon)^{l-m}}. \tag{6}$$

The zeros of the integral rational form G_{-d+m-l} utilized in (5), are to be neither at fixed vertices, not at boundary points of the group; also, in the boundary curve case, let all of them be situated outside the net N. In the case (6), it is only to be prescribed, that $G_{d-m+l}(\zeta_1,\zeta_2)$ is not to exhibit a zero right at the vertex ε.

One next calculates easily, that the series (5) resp (6), in the sense of the conception above, has a pole of the first order at ε. This is a consequence of the congruence, now valid in place of (4):

$$\lambda \equiv -d+2m \pmod{2l}. \tag{7}$$

Only the case of a group without boundary curve now requires the arguments to be carried further, if, besides this, $-d+m-l \geq 1$ and therefore the scheme (5) comes into question. Then we yet prescribe, that $G_{-d+m-l}(\zeta_1,\zeta_2)$ is only to have simple zeros. These will correspondingly furnish new poles of the first order for the series (5). But it is an easy matter, for each of those additional poles, to form a one-poled series of the form (16) on p. 183, therefore, by substitution of these series, multiplied by suitable coefficients, from the series (5), one then obtains a series pole-tree outside the vertices equivalent with ε. *Accordingly, for each arbitrary position of the pole ζ in the interior of the net N, with the exclusion only of the parabolic vertices, we have succeeded in the production of an associated one-pole series.* In the paragraph after next, we return in detail to the case that ζ moves into a parabolic vertex.

3.11 Introduction of the elementary forms $\Omega(\zeta_1,\zeta_2;\xi_1,\xi_2)$

[

We now return to the one-pole series of the paragraph before last,and indeed, in particular, to the series (1) on p. 179. In order to invest the form represented by this series with the simplest possible behavior at the place ξ, we will yet normalize it with a factor independent of ζ_1,ζ_2; and indeed, we choose for this the reciprocal value of the rational form H_{d+1} standing in the numerator of the initial term, formed for $\zeta_1=\xi_1,\zeta_2=\xi_2$. *The addition of this factor, to be immediately considered even more closely, furnishes us the expression, designated from now as an "elementary form"*:

$$\Omega(\zeta_1,\zeta_2;\xi_1,\xi_2)=\sum_k \mu_k^{-1}\frac{H_{d+1}(\zeta_1^{(k)},\zeta_2^{(k)})}{(\zeta_1^{(k)},\xi)H_{d+1}(\zeta_1,\zeta_2)}. \tag{1}$$

The automorphic form of the ζ_1,ζ_2 and of the dimension d and the multiplicator system M hereby represented, obviously has the property, *that upon approach to the pole ξ, and indeed for* $\lim\zeta_1=\xi_1,\lim\zeta_2=\xi_2$, *the limit equation occurs*:

$$\lim[(\zeta,\xi)\cdot(\zeta_1,\zeta_2;\xi_1,\xi_2)]=1. \tag{2}$$

We say for this, by imagining the development of Ω carried out according to powers of $(\zeta-\xi)$, that the form Ω, at the place ξ, has a pole of the first order *"of the coefficient* 1*"*. Correspondingly, a pole of first order of the coefficient μ_k will lie at the place ξ_k.

In more detail, in the *case of a group with one or infinitely many boundary curves*, the form of the factor to be added emerges from the following formula for the elementary form:

$$\Omega(\zeta_1,\zeta_2;\xi_1,\xi_2)=\sum_k \mu_k^{-1}\frac{G_{-d-1}(\xi_1,\xi_2)}{(\zeta^{(k)},\xi)G_{-d-1}(\zeta_1^{(k)},\zeta_2^{(k)})}, \tag{3}$$

which corresponds to the one-poled series (2) on p. 179. Hereby, one yet wants to notice, that in consequence of the agreement made on p. 179, the rational integral form $G_{-d-1}(\xi_1,\xi_2)$ is different from 0 for *each* place ξ in the interior of the net.

For a *group without boundary curve* we shall correspondingly attach ourselves to the form (16) on p. 183 of the one-poled series. Here matters lie especially simply, insofar as $H_{d+1}(\xi_1,\xi_2)$ becomes *independent* not only of ζ_1,ζ_2, *but also of* ξ_1,ξ_2. In fact we
[187] have from (15) on p. 183

$$\begin{aligned}H_{d+1}(\xi_1,\xi_2)&=\frac{G_1^{(-d-1)}(\xi_1,\xi_2;\xi_1,\xi_2)}{\prod\limits_{k=0}^{-d-1}(\xi,\eta^{(k)})}\\&=\sum_{k=0}^{-d-1}\frac{\mu_k^{-1}}{(\eta^{(k)},\eta)\cdots(\eta^{(k)},\eta^{(k-1)})(\eta^{(k)}\eta^{(k+1)})\cdots(\eta^{(k)}\eta^{(-d-1)})}\end{aligned} \tag{4}$$

where the sum standing on the right, according to the developments of the paragraph before last, apart from the case $d=-2,\mathrm{M}_0=1$, assumes a value different from zero

[25] This number, otherwise denoted by m_k, one does not want to contuse with the number in the sense of (7) on p. 101.

by suitable choice of η. *We will now hold fast from now on to the η chosen already*, and then simply absorb the value (4), different form 0, as denominator, into the coefficients a_1, a_2 of the linear from G. It will scarcely give rise to confusions, if, after this alteration, we keep to the notation $G_1^{(-d-1)}$ for our rational integral form. With this, next to formula (3) there arranges itself as the *representation of the elementary form in the case of a group without boundary curves*:

$$\Omega(\zeta_1,\zeta_2;\xi_1,\xi_2) = \sum_k \mu_k^{-1} \frac{G_1^{(-d-1)}(\zeta_1^{(k)},\zeta_2^{(k)};\xi_1,\xi_2)}{(\zeta^{(k)},\xi)(\zeta^{(k)},\eta)(\zeta^{(k)},\eta')\cdots(\zeta_l^{(k)},\eta^{(-d-1)})}. \tag{5}$$

It is now a very essential step, *that from now on, besides the place ζ, we will also consider the place ζ movable in the net N and, according to the fundamental theorems developed for the ζ_1,ζ_2 on p. 64, the binary quantities ξ_1,ξ_2 as variable.* In the description of the behavior of the elementary form $\Omega(\zeta_1,\zeta_2;\xi_1,\xi_2)$ as a form of two variable pairs, we are able to partially rely upon calculations already carried through; besides this, to be sure, a few more further investigations will be requisite.

If, first of all, ξ moves *into an elliptic vertex*, then Ω, as a form of the ζ_1,ζ_2, will only then preserve its pole, if the corresponding multiplicator μ satisfies the congruence (4) on p. 184. *If, however, this congruence is not fulfilled, then Ω becomes pole-free, and then identical,i.e., taking place for all ζ, vanishing of the form Ω is not excluded.*

If ξ moves *into a parabolic* cusp, then our prescription, completely observed elsewhere, that the forms $H(\zeta_1,\zeta_2)$, utilized for the production of Poincaré series, are to be pole-free at all boundary points, is broken. Now, to be sure, also in this case, at least for all dimensions $d \leq -3$, the convergence of the series is clear from the theorem of v. Mangoldt cited on p. 152. But it will be necessary, that we know explicitly the precise transformation law of the form Ω in dependency on the ξ_1,ξ_2 for approach of the place ξ to the parabolic cusp, in order to sufficiently establish the function theoretic argu- [188]
ments following later. We come back to the investigation of this formation law, which is bound up with a few complications, in the next paragraph.

Finally, in the sense of a group without boundary curves, *the place η and the places equivalent to η* were initially unapproachable for ξ. But we are able to let this restriction drop, as the following argument shows:

If, first of all, we let ξ move into one of the $-d$ places $\eta,\eta',\ldots,\eta^{(-d-1)}$, which we call $\eta^{(i)}$, and specialize formula (16) on p. 183 for this case, then it follows, according to (15) on p. 183:

$$a_1\zeta_1 + a_2\zeta_2 = \mu_i^{-1}(\zeta,\eta^{(i)}).$$

The formula (16), mentioned on p. 183, with this, assumes the form:

$$\mu_i^{-1}\sum_k \frac{\mu_k^{-1}}{(\zeta^{(k)},\eta)(\zeta^{(k)},\eta')\cdots(\zeta^{(k)},\eta^{(-d-1)})}.$$

Here we have a convergent and easily approachable Poincaré series before us, which preserves its pole of first order at the place η. If we divide, as we must, by the expression (4), then one calculates immediately, for the place η, a pole of the coefficient μ_i^{-1}, *so that the elementary form actually retains its pole of the first order of the coefficient* 1 *at the place $\xi = \eta^{(i)}$.*

If, on the other hand, ξ moves into a place $\eta^{(l)}$, which is equivalent with η, but different from $\eta, \eta^1, \ldots, \eta^{(-d-1)}$, then equation (16) on p. 183 furnishes the convergent Poincaré series:

$$\sum \mu_k^{-1} \frac{G_1^{(-d-1)}(\zeta_1^{(k)}, \zeta_2^{(k)}; \eta_1^{(l)}, \eta_2^{(l)})}{(\zeta^{(k)}, \eta)(\zeta^{(k)}, \eta^1) \cdots (\zeta^{(k)}, \eta^{(-d-1)})(\zeta^{(k)}, \eta^{(l)})}.$$

If we develop the form represented by this at η according to powers of (ζ, η), then the coefficient of $(\zeta, \eta)^{-1}$ becomes:

$$\mu_l^{-1} \frac{G(\eta_1^{(l)}, \eta_2^{(l)}; \eta_1^{(l)}, \eta_2^{(l)})}{(\eta^{(l)}, \eta) \cdots (\eta^{(l)}, \eta^{(-d-1)})}$$
$$+ \sum_{k=0}^{-d-1} \mu_k^{-1} \frac{G(\eta_1^{(k)}, \eta_2^{(k)}; \eta_1^{(l)}, \eta_2^{(l)})}{(\eta^{(k)}, \eta^{(l)})(\eta^{(k)}, \eta) \cdots (\eta^{(k)}, \eta^{(k-1)})(\eta^{(k)}, \eta^{(k+1)}) \cdots (\eta^{(k)}, \eta^{(-d-1)})}.$$

The sum coming up here in the second place is equal to the expression A represented in (7) on p. 180, if one replaces ξ_1, ξ_2 by $\eta_1^{(l)}, \eta_2^{(l)}$ in the latter. But since the expression A vanishes identically in ξ_1, ξ_2, the value of the sum also remains subsisting after this
[189] replacement. With this, the coefficient of $(\zeta, \eta)^{-1}$, according to (15) on p. 183, reduces itself to:

$$\mu_l^{-1} \frac{G(\eta_1^{(l)}, \eta_2^{(l)}; \eta_1^{(l)}, \eta_2^{(l)})}{(\eta^{(l)}, \eta) \cdots (\eta^{(l)}, \eta^{(-d-1)})}$$
$$= \mu_l^{-1} \sum_{k=0}^{-d-1} \frac{\mu_k^{-1}}{(\eta^{(k)}, \eta) \cdots (\eta^{(k)}, \eta^{(k-1)})(\eta^{(k)}, \eta^{(k+1)}) \cdots (\eta^{(k)}, \eta^{(-d-1)})},$$

i.e., to the product of μ_l^{-1} and the value different from zero represented in (4) on p. 187. If, again, by means of division by this latter value, we go from the series (16) on p. 183 to the elementary form, then the latter obtains a pole of the coefficient μ_l^{-1} at η, and therefore, at $\xi = \eta^{(l)}$, one such of the coefficient 1.

In each case, therefore, the theorem holds: *the fundamental property of the elementary form $\Omega(\zeta_1, \zeta_2; \xi_1, \xi_2)$, of having, as a form of ζ_1, ζ_2, a pole of the first order of the coefficient 1 at the place ξ, remains maintained unaltered, even if ξ moves into any place equivalent with η.*

Further, we describe the properties, which the *elementary form has as a form of the ξ_1, ξ_2 for fixed ζ_1, ζ_2*. If we thereby restrict ξ to the "interior" of the net N, then we will thereby bring to expression, that parabolic cusps are to be initially avoided. If ζ lies *neither at a fixed vertex, nor at a place equivalent with η*, then the following basic theorem holds: *the elementary form Ω, apart from the poles to be described immediately even more closely, is a continuous analytic form in the ξ_1, ξ_2 everywhere in the "interior" of the net N, of the non-negative dimension*:

$$\overline{d} = -d - 2. \tag{6}$$

According to the general convergence studies of the Poincaré series (cf. p. 142 ff.), namely, we recognize very easily the uniform convergence of our series (3) and (5), defining the elementary forms, in case ξ remains restricted to the "interior" of the net. But concerning the pole just mentioned, the following theorem holds: *the form Ω of the ξ_1,ξ_2 has a pole of first order in each polygon of the net, and indeed, these poles are obviously the places $\zeta,\zeta^{(1)},\zeta^{(2)},\ldots$, equivalent with ζ*. Thereby, as one immediately calculates from formula (3) as well as in the case (5), $-\Omega$ has, in more detail, a pole of first order of the coefficient $\overline{\mu_k}=\mu_k^{-1}$ at the place $\zeta^{(k)}$. One can therefore say, *that $-\Omega$, in its manner of becoming infinite, exhibits the behavior of a one-poled automorphic form of dimension $\overline{d}$ and of the multiplicator system $\overline{\mathrm{M}}$ inverse to* M. Of the particular positions of ζ, we have first to consider the *fixed polygon vertices.*

In this regard, first of all, the theorem holds: *if ζ lies at a parabolic cusp or at such an elliptic vertex ε_k, for which (cf. (5) on p. 100) the number $(d+\lambda_k)$ is not divisible by $2l_k$, then Ω vanishes identically as a form of ξ_1,ξ_2*. In both cases, namely, as an automorphic form of ζ_1,ζ_2,, the form Ω has a zero at the vertex in question, independently of ξ; in the elliptic case, the zero has the order $\frac{m_k}{l_k}$ determined in (5) on p. 100 which is only then equal to zero, if $2l_k$ divides $(d+\lambda_k)$. [190]

If follows further: *if ξ lies at an elliptic vertex ε_k, for which $2l_k$ divides $(d+\lambda_k)$, then Ω represents a non-identically vanishing form of ξ_1,ξ_2. Now, namely, a pole of the form Ω of ξ_1,ξ_2 remains maintained at ε_k and each of the equivalent pints.* One will verify this by analogous calculations, as we carried them out on p. 184. In order, say, tying in with formula (3) on p. 186, to give a few brief indications, call the elliptic vertex ε and understand ε', in the usual way, the second fixed point of the associated substitution. In consequence of the identical equation:

$$(\varepsilon,\varepsilon')(\zeta,\xi)=(\zeta,\varepsilon)(\xi,\varepsilon')-(\zeta,\varepsilon')(\xi,\varepsilon),$$

and since the rational integral form G_{-d-1} does not vanish at the place ε, we are able to write the initial term of the series (3), for a ζ situated in the immediate vicinity of ε, up to a numerical factor, thus:

$$(\zeta,\varepsilon')^d\frac{G_{-d-1}(\xi_1,\xi_2)}{(\xi,\varepsilon)}$$

upon exercise of the elliptic substitution $(\zeta,\varepsilon')^d$ takes on the factor $e^{-\pi i\frac{d}{l}}$, and besides this, the multiplicator $\mu^{-1}=e^{-\pi i\frac{\lambda}{l}}$ is to be added. Both factors compensate each other in the present case:

$$e^{-\pi i\frac{\lambda}{l}}\cdot e^{-\pi i\frac{d}{l}}=e^{-2i\pi\frac{(d+\lambda)}{2l}}=1,$$

so that, on account of $G_{-d-1}(\varepsilon_1,\varepsilon_2)\neq 0$, the pole remains subsisting in fact, at $\xi=\varepsilon$.

Besides the fixed vertices, in the case of a group without boundary curves, *the place η and the places equivalent with it* also yield special positions of ζ. In this regard, first of all, it is known from the developments on p. 180 ff., that in case ζ moves into a place equivalent with η, the polarly discontinuous terms of the series exactly cancel one another for this transition. *Accordingly, even if ζ moves into any place equivalent with η, we have in Ω a form of the ξ_1,ξ_2, which is pole-free in the "interior" of the net N, assuming that ξ itself does not become equivalent with η.* There only remains to investigate here beforehand, how the form of the ζ_1,ζ_2 thus arising behaves itself at the places equivalent with η. [191]

Since Ω is automorphic in the ζ_1,ζ_2, then it is sufficient, if we let ζ move into the place η. If ξ then wanders into the place $\eta^{(l)}$ equivalent with η, which is different from the $-d$ place $\eta,\eta',\dots,\eta^{(-d-1)}$, then hereby, the behavior of the series (16) on p. 183 is exceedingly easy to look over. The first $-d$ terms become polarly discontinuous individually for $\lim\zeta=\eta$, but their sum remains continuous. Among the following terms, only that one for $k=l$ is discontinuous, on account of the vanishing factor $(\eta^{(l)},\xi)$ in the denominator. We find here, for the series (16), a pole of the first order of the coefficient(cf. (15) on p. 183):

$$-\mu_l^{-1}\frac{G(\eta_1^{(l)},\eta_2^{(l)};\eta_1^{(l)},\eta_2^{(l)})}{(\eta^{(l)},\eta)\cdots(\eta^{(l)},\eta^{(-d-1)})}$$
$$=-\mu_l^{-1}\sum_{k=0}^{-d-1}\frac{\mu_k^{-1}}{(\eta^{(k)},\eta)\cdots(\eta^{(k)},\eta^{(k-1)})(\eta^{(k)},\eta^{(k+1)})\cdots(\eta^{(k)},\eta^{(-d-1)})},$$

which has as consequence, for the elementary form itself, a pole of the first order of the coefficient $-\mu_k^{-1}$ situated at $\xi=\eta^{(l)}$.

In a somewhat more circumstantial manner, we arrive at the analogous result, in case ξ wanders into an arbitrary one of the places $\eta_1,\dots,\eta^{(-d-1)}$, which we again call $\eta^{(i)}$. If we again tie into the series (16) on p. 183, then, in the present case, the sum of the first $-d$ terms is to be investigated; for the remaining part of the series has a finite summation value. Thus, in the sum:

$$\sum_{k=0}^{-d-1}\mu_k^{-1}\frac{a_1\zeta_1^{(k)}+a_2\zeta_2^{(k)}}{(\zeta^{(k)},\xi)(\zeta^{(k)},\eta)(\zeta^{(k)},\eta')\cdots(\zeta^{(k)},\eta^{(-d-1)})},\tag{7}$$

we first have to let ζ move into the place η, and then to investigate the expression thus arising for $\lim\xi=\eta^{(i)}$. If one now develops the expression (7) according to powers of $(\zeta,\eta)=(\zeta^{(k)},\eta^{(k)})$, then as we know, the coefficient of $(\zeta,\eta)^{-1}$ will be identical with zero. As the absolute term of this power series, and therefore, as the expression of the sum (7) for $\zeta=\eta$, we obtain after elementary intermediate calculation:

$$\sum_{k=0}^{-d-1}\frac{\mu_k^{-1}\eta_2^{(k-1)}(\eta^{(k)},\xi)^{-1}}{(\eta^{(k)},\eta)\cdots(\eta^{(k)},\eta^{(k-1)})(\eta^{(k)},\eta^{(k+1)})\cdots(\eta^{(k)},\eta^{(-d-1)})}$$
$$\left[a_1-(a_1\eta^{(k)}+a_2)\left(\frac{1}{\eta^{(k)}-\xi}+\frac{1}{\eta^{(k)}-\eta}+\cdots+\frac{1}{\eta^{(k)}-\eta^{(k-1)}}\right.\right.$$
$$\left.\left.+\frac{1}{\eta^{(k)}-\eta^{(k+1)}}+\cdots+\frac{1}{\eta^{(k)}-\eta^{(-d-1)}}\right)\right].\tag{8}$$

[192] If we now investigate this expression in its dependency on ξ, then for $\xi=\eta^{(i)}$, only the one term with $k=i$ becomes polarly discontinuous. It thereby follows from (15) on p. 183, that $a_1\eta_1^{(i)}+a_2\eta_2^{(i)}$ receives the factor $(\xi,\eta^{(i)})$. The expression (8), therefore, at $\xi=\eta^{(i)}$, has a pole of the *first* order of the coefficient:

$$\frac{\mu_i^{-1}}{(\eta^{(i)},\eta)\cdots(\eta^{(i)},\eta^{(i-1)})(\eta^{(i)},\eta^{(i+1)})\cdots(\eta^{(i)},\eta^{(-d-1)})}\left[\frac{a_1}{\eta_2^{(i)}}+\lim_{\xi=\eta^{(i)}}\left(\frac{a_1\eta_1^{(i)}+a_2\eta_2^{(i)}}{(\xi,\eta^{(i)})}\right)\right].$$

From (15) on p. 183, however, $a_1 = \mu_i^{-1}\eta_2^{(i)}$ immediately finds itself for $\xi = \eta^{(i)}$, and

$$\lim_{\xi=\eta^{(i)}} \left(\frac{a_1\eta_1^{(i)} + a_2\eta_2^{(i)}}{(\xi,\eta^{(i)})} \right)$$
$$= \sum_{k=0}^{-d-1}{}' \mu_k^{-1} \frac{(\eta^{(i)},\eta)\cdots(\eta^{(i)},\eta^{(i-1)})(\eta^{(i)},\eta^{(i+1)})\cdots(\eta^{(i)},\eta^{(-d-1)})}{(\eta^{(k)},\eta)\cdots(\eta^{(k)},\eta^{(k-1)})(\eta^{(k)},\eta^{(k+1)})\cdots(\eta^{(k)},\eta^{(-d-1)})}$$

where, by the prime on the summation sign, it is to be indicated, that the value $k = i$ is to be left out. But for $k = i$, $\mu_i^{-1} = \frac{a_1}{\eta_2^{(i)}}$ would enter as the value of the corresponding term of the sum. Thus, the coefficient of the pole (8) situated at $\xi = \eta^{(i)}$ is simply:

$$\mu_i^{-1} \sum_{k=0}^{-d-1} \frac{\mu_k^{-1}}{(\eta^{(k)},\eta)\cdots(\eta^{(k)},\eta^{(k-1)})(\eta^{(k)},\eta^{(k+1)})\cdots(\eta^{(k)},\eta^{(-d-1)})}. \tag{9}$$

The division of the series (16) on p. 183 by the sum (whose value is $\neq 0$) multiplied here in (9) by μ_i^{-1} now yields the elementary form.

Recapitulating, we have arrived at the result, *that the theorems stated above concerning the poles of the form Ω of ξ_1,ξ_2 also then remain subsisting without any alteration, if the point ζ moves into any place of the polygon net equivalent with η.*

3.12 Behavior of the elementary form $\Omega(\zeta_1,\zeta_2;\xi_1,\xi_2)$ at a parabolic cusp ξ

It now yet remains for us, to determine the behavior of the elementary form $\Omega(\zeta_1,\zeta_2;\xi_1,\xi_2)$ upon *approach of ξ to a parabolic cusp.* According to a principle set up earlier (p. 143), we may place a cusp of this kind, proposed for investigation, at $\zeta = \infty$ and, as on p. 148, utilize as the generating parabolic substitution: [193]

$$\zeta_1' = \zeta_1 + 2i\pi\zeta_2, \qquad \zeta_2' = \zeta_2. \tag{1}$$

Let the associated multiplicator be $e^{2i\pi\frac{\lambda}{l}}$, whereby λ is chosen in agreement with $0 < \lambda \leq l$, say.

We will now obtain, *for $\xi_2^{-\overline{d}}\Omega$, an analytic representation in ξ, valid in the neighborhood of the parabolic point in question,* which permits the behavior of Ω to be given for $\xi = \infty$, i.e., at the parabolic cusp itself. Hereby, we process in such a fashion, that we first of all represent the $(-d-1)^{\text{st}}$ derivative of $\xi_2^{-\overline{d}}\Omega$ taken with respect to ξ, and arrive from here, backwards, by integration, to an expression for $\xi_2^{-\overline{d}}\Omega$ itself. The reason for this procedure will become clear immediately from the following development.

For the purpose mentioned, we write the initial form of the series (3) on p. 186 in the form:

$$-\zeta_2^d\xi_2^{\overline{d}} \cdot \frac{G_{-d-1}(\xi,1)}{(\xi-\zeta)G_{-d-1}(\zeta,1)}, \tag{2}$$

resp. that of the series (5) on p. 187 in the form:

$$-\zeta_2^d \xi_2^{\overline{d}} \cdot \frac{G_1^{(-d-1)}(\zeta,1;\xi,1)}{(\zeta-\xi)(\zeta\eta_2-\eta_1)(\zeta\eta_2'-\eta_1')\cdots(\zeta\eta_2^{(-d-1)}-\eta_1^{(-d-1)})} \tag{3}$$

and notice, that the quotient, standing here in the second place, conceived of as a rational function of ζ, exhibits a zero of order $-d$ for $\zeta=\infty$. On account of the factor ζ_2^d, this is requisite for the convergence proof of the series (3) and (5) on p. 186 ff., insofar as the form (2) resp (3) is to be pole-free at the parabolic boundary point. In fact, indeed, since $\zeta=\infty$ represent, a boundary point of the group, $G_{-d-1}(\zeta_1,\zeta_2)$ different from zero for $\zeta_2=0$; and for the same reason, in the expression (3), the $\eta_2,\eta_2',\dots,\eta_2^{(-d-1)}$ as well, too, as ξ_2, will not vanish.

But now notice, that the derivatives of first and higher order taken with respect to ξ of the quotients multiplied by $-\xi_2^d\xi_2^{\overline{d}}$ in (2) and (3) furnish rational expressions in ζ, which exhibit zeros of order at least $(-d)$ at $\zeta=\infty$. Thereby, new poles, different from the ones so far, do not come up, so that these derivatives, multiplied by ζ_2^d, represent forms of dimension d in ζ_1,ζ_2, which are suited for the formation of the convergent Poincaré series. From this, however, according to well-known theorems concerning uniformly convergent series, follows: *the series for $\xi_2^{-\overline{d}}\Omega$ arising from (3) and (5) on p. 186 ff. are able, without destroying their uniform convergence in the interior of the net, to be repeatedly differentiated according to ξ termwise, and thereby furnish the derivatives of $\xi_2^{-\overline{d}}\Omega$ taken according to ξ.*

[194] Before we now carry out this differentiation, we decompose the rational function of ξ multiplicator by $-\xi_2^d\xi_2^{\overline{d}}$ in (2) resp. (3) into the sum of a partial fraction with respect to the pole ζ and an integral function ξ ascending to the degree $-d-2$. Thereby, we imagine ζ fixed anywhere in the "interior" of the net, i.e., ζ, for the moment, is to remain distant from the parabolic points in question and the equivalent places. in both cases, i.e., for (2) as well as (3), we arrive at an expression of the form:

$$\frac{1}{\xi-\zeta}+R_0(\zeta)+\xi R_1(\zeta)+\cdots+\xi^{-d-2}R_{-d-2}(\zeta), \tag{4}$$

where the $R(\zeta)$ are rational functions of ζ, whose method of construction one will follow easily in the first as well as the second case. For $-\xi_2^{-\overline{d}}\Omega$ there arises the representation:

$$-\xi_2^{-d}\Omega=\sum_k \mu_k^{-1}(\zeta_2^{(k)})^d\left[\frac{1}{\xi-\zeta^{(k)}}+R_0(\zeta^{(k)})+\xi R_1(\zeta^{(k)})+\cdots+\xi^{-d-2}R_{-d-2}(\zeta^{(k)})\right], \tag{5}$$

and, from now on, we are able to treat the two cases, separated until now, together.

In passing, let it be remarked, that the separate summing of the terms standing in (5) is inadmissible. Thus, e.g., for:

$$\sum_k \mu_k^{-1}(\zeta_2^{(k)})^d\cdot\frac{1}{\xi-\zeta^{(k)}},$$

there would be a form $H_d(\zeta_1,\zeta_2)$, which exhibits a pole of order $-d-1$ at the parabolic boundary point $\zeta_2=0$. This series would thus elude the convergence studies on p. 140 ff., as well, too, as the v. Mangoldt theorem (see p. 152). Rather, one has to consider the $R_0, \xi R_1, \ldots$, as supplementary terms, which are inextricably bound up with $\frac{1}{\xi-\zeta^{(k)}}$, and by this means, make possible the applicability of the convergence theorems cited. However, the complicated method of construction hereby described of the series (5) vanishes, if we differentiate $(-d-1)$ times according to ξ; for this reason, we next turn ourselves to the series arising from this differentiation.

Let the derivative of the νth order taken according to ξ now be denoted by the symbol D_ν. Equation (5) then furnishes, on the basis of the argument presented above:

$$D_{-d-1}[\xi_2^{-d}\Omega] = (-1)^d(-d-1)!\sum_k \mu_k^{-1}(\zeta_2^{(k)})^d \cdot \frac{1}{(\xi-\zeta^{(k)})^{-d}}. \tag{6}$$

As the representation of the derivative of the $(-d-1)^{\text{st}}$ order. But the series standing here on the right permits a rearrangement into a form more suitable for our further purposes.

Notice first of all, that the series (6) is absolutely convergent. We may therefore arrange this, without alteration of the value of the sum, by means of the procedure followed on p. 149, according to the parabolic subgroup at hand. With utilization of the notations defined l.c., it follows: [195]

$$D_{-d-1}[\xi_2^{-d}\Omega] = (-1)^d(-d-1)!\sum_\kappa \left[\mu_\kappa^{-1}(\zeta_2^{(\kappa)})^d \cdot \sum_{\nu=-\infty}^{+\infty} \frac{e^{2\nu i\pi\frac{\lambda}{l}}}{(\xi-\zeta^{(\kappa)}-2\nu i\pi)^{-d}}\right]. \tag{7}$$

The inner sum on the right side we develop further thus:

$$\sum_{\nu=-\infty}^{+\infty} \frac{e^{2\nu i\pi\frac{\lambda}{l}}}{(\xi-\zeta^{(\kappa)}-2\nu i\pi)^{-d}} = \sum_{\sigma=0}^{l-1}\left[e^{2\sigma i\pi\frac{\lambda}{l}} \sum_{\nu=-\infty}^{+\infty} \frac{1}{(\xi-\zeta^{(\kappa)}-2\sigma i\pi-2\nu i\pi l)^{-d}}\right]. \tag{8}$$

Now we already had occasion above (p. 150), to make use of the following equation:

$$\sum_{\nu=-\infty}^{+\infty} \frac{1}{(\xi-a-2\nu i\pi l)^{-d}} = -\frac{1}{2l}D_{-d-1}\left[\frac{1+e^{\frac{\xi}{l}}\cdot e^{-\frac{a}{l}}}{1-e^{\frac{\xi}{l}}\cdot e^{-\frac{a}{l}}}\right], \tag{9}$$

where on the right side, after the $(-d-1)$ differentiation according to ξ, there is a rational function of $\zeta^{\frac{\zeta}{l}}$, which vanishes with $e^{\frac{\xi}{l}}$. For abbreviation, we set:

$$e^{\frac{\xi}{l}} = t,$$

and develop the quotient on the right in (9), to be differentiated, according to powers of t, before carrying out this differentiation. In this way, one obtains easily:

$$\sum_{\nu=-\infty}^{+\infty} \frac{1}{(\xi-a-2\nu i\pi l)^{-d}} = -l^d\left[e^{-\frac{a}{l}}t + \frac{e^{-\frac{2a}{l}}}{2^{d+1}}t^2 + \frac{e^{-\frac{3a}{l}}}{3^{d+1}}t^3 + \cdots\right], \tag{10}$$

where the series, standing here on the right, in consequence of the stipulation made concerning the position of ζ, is always uniformly convergent in the neighborhood of the point $t=0$, which corresponds to the parabolic cusp.

Now put in $a=\zeta^{(\kappa)}+2\sigma i\pi$, and further, substitute $\sigma=0,1,2,\dots,l-1$ into (10) one after the other, and add, after appending the factors $e^{2\sigma i\pi\frac{\lambda}{l}}$ prescribed on the right in
[196] (8) to all l equations. There thus yields itself:

$$\sum_{\nu=-\infty}^{+\infty}\frac{e^{2\nu i\pi\frac{\lambda}{l}}}{(\xi-\zeta^{(\kappa)}-2\nu i\pi)^{-d}}$$
$$=-l^{d+1}e^{\frac{-\lambda\zeta^{(\kappa)}}{l}}t^{\lambda}\left[\frac{1}{\lambda^{d+1}}+\frac{e^{-\zeta^{(\kappa)}}}{(\lambda+l)^{d+1}}t^{l}+\frac{e^{-2\zeta^{(\kappa)}}}{(\lambda+2l)^{d+1}}t^{2l}+\cdots\right].$$

Notice, that the stipulation $0<\lambda\leq l$ hereby comes into force.

Equation (7) now furnishes, further:

$$D_{-d-1}\left[\xi_2^{-d}\Omega\right]=(-1)^{d+1}(-d-1)!\,l^{d+1}\sum_{\kappa}\mu_{\kappa}^{-1}(\zeta_2^{(\kappa)})^{d}e^{\frac{-\lambda\zeta^{(\kappa)}}{l}}t^{\lambda}\cdot\left[\frac{1}{\lambda^{d+1}}+\frac{e^{-\zeta^{(\kappa)}}}{(\lambda+l)^{d+1}}t^{l}+\frac{e^{-2\zeta^{(\kappa)}}}{(\lambda+2l)^{d+1}}t^{2l}+\cdots\right].\tag{11}$$

Here on the right stands a sum of infinitely many power series, uniformly convergent in the neighborhood of $t=0$. *This sum is always uniformly convergent in the interior of a finitely extended ring-shaped domain about* $t=0$; for it represents, in the form of the right side of equation (7), a convergent Poincaré series. In this situation, on the basis of the theorem of Weierstrass already utilized above (p. 151), we are able to undertake the rearrangement of the right side of equation (11) according to ascending powers of $t=e^{\frac{\xi}{l}}$. The coefficients become forms of dimension d in ζ_1,ζ_2, i.e., we arrive at a representation:

$$D_{-d-1}\left[\xi_2^{-d}\Omega\right]=e^{\frac{\lambda\xi}{l}}\left[\chi_d^{(0)}(\zeta_1,\zeta_2)+\chi_d^{(1)}(\zeta_1,\zeta_2)e^{\xi}+\chi_d^{(2)}(\zeta_1,\zeta_2)e^{2\xi}+\cdots\right]\tag{12}$$

whereby the series standing on the right is uniformly convergent to the same extent, *and obviously, also with inclusion of the parabolic cusp itself,* as the series (11).

Inside the domain of uniform convergence, we may now repeatedly integrate the series (12) termwise. After carrying out this operation $(-d-1)$ times, we finally arrive at the desired goal: *the elementary form* $\Omega(\zeta_1,\zeta_2;\xi_1,\xi_2)$ *as a form of the* ξ_1,ξ_2, *permits the development in the neighborhood of the parabolic cusp investigated*:

$$\Omega(\zeta_1,\zeta_2;\xi_1,\xi_2)=\xi_2^{\overline{d}}\Big\{e^{\frac{\lambda\xi}{l}}\left[\varphi_d^{(0)}(\zeta_1,\zeta_2)+\varphi_d^{(1)}(\zeta_1,\zeta_2)e^{\xi}+\varphi_d^{(2)}(\zeta_1,\zeta_2)e^{2\xi}+\cdots\right]+\psi_d^{(0)}(\zeta_1,\zeta_2)+\psi_d^{(1)}(\zeta_1,\zeta_2)\xi+\cdots+\psi_d^{(-d-2)}(\zeta_1,\zeta_2)\xi^{-d-2}\Big\},\tag{13}$$

whereby the series progressing according to powers of e^{ξ} *is uniformly convergent in the neighborhood mentioned, and where the coefficients* φ *and* ψ *are forms of the* ζ_1,ζ_2 *of*
[197] *dimension* d. With this, the behavior of the elementary form Ω for approach of ξ to the parabolic cusp is completely clarified.

The nature of the terms $\varphi_d(\zeta_1,\zeta_2)$ and $\psi_d(\zeta_1,\zeta_2)$ coming up here we need not exhaustively exhibit. Thus, e.g., it would be meaningless, to ask for the behavior of the

$\varphi_d(\zeta_1,\zeta_2)$ themselves at the parabolic cusp; for the foregoing developments assume as a convergence condition, that the real component of each $\zeta^{(\kappa)}$ is greater than that of ξ, so that ζ must remain distant from the parabolic cusp. All the same, the following theorem is remarkable, which emerges from later developments, namely from the behavior of $\Omega(\zeta_1,\zeta_2;\xi_1,\xi_2)$ upon exercise of the parabolic generator (1) on ξ_1,ξ_2: *for $\overline{d}>0$, the last $\overline{d}$ of the forms ψ, i.e., the coefficients $\psi_d^{(1)}(\zeta_1,\zeta_2),\dots,\psi_d^{(-d-2)}(\zeta_1,\zeta_2)$ of $\xi,\xi^2,\dots,\xi^{-d-2}$ in (13), are "pole-free" Poincaré.* Whether this holds for $\psi_d^{(0)}(\zeta_1,\zeta_2)$, must remain to be seen.

To this theorem, we attach yet the following interesting consequence . If ξ reaches the parabolic cusp, i.e., if we take $\xi_2=0$, then from (13) follows:

$$\Omega(\zeta_1,\zeta_2;\xi_1,0)=\xi_1^{-d-2}\cdot\psi_d^{(-d-2)}(\zeta_1,\zeta_2). \tag{14}$$

If now $d<-2$, then here on the right, in any case, there stands a pole-free form, representable as a Poincaré series[26]. This vanishes, in case ζ itself moves into the parabolic cusp. *Here, therefore, the theorem, otherwise valid (i.e., in the "interior" of the polygon net), that $\Omega(\zeta_1,\zeta_2;\xi_1,\xi_2)$, as a form of the ζ_1,ζ_2, has a pole at the place ξ, suffers an exception.*

3.13 Behavior of the elementary forms upon exercise of substitutions of the group Γ on ξ_1,ξ_2. Discussions for the models of genus $p=0$

The elementary form $\Omega(\zeta_1,\zeta_2;\xi_1,\xi_2)$,in case it represents a non-identically vanishing form of the ξ_1,ξ_2, has the property, of having a pole of the first order in each polygon of the net. These poles lie at the equivalent places $\zeta,\zeta',\zeta'',\dots$; and indeed, the coefficient of the infinity at the individual one of these places is one such, as if Ω were an automorphic form in the ξ_1,ξ_2 of dimension $\overline{d}$ and the multiplicator system in M inverse to $\overline{\text{M}}$ (cf. p. 102). In this state of affairs, it is natural to ask, *how does $\Omega(\zeta_1,\zeta_2;\xi_1,\xi_2)$ behave itself under exercise of the operation U_i of the group Γ on ξ_1,ξ_2.* We shall find, that Ω, in certain cases, actually represents an automorphic form in ξ_1,ξ_2 as well. Moreover, exactly those cases, for which Ω is not automorphic in the ξ_1,ξ_2, by means of an ingenious turn of thought originating form Poincaré, will afterwards become the most important ones. [198]

By U_i let an arbitrarily chosen substitution of the homogenous group Γ be understood. The result of the exercise of U_i on any integral rational form $G(\zeta_1,\zeta_2)$ is $G(\zeta_1^{(i)},\zeta_2^{(i)})$, but is nevertheless also to be denoted symbolically form now on also by $G[U_i(\zeta_1,\zeta_2)]$. Make use of the movability of the place of ξ in this fashion, by exercising the substitution U_i, initially in the boundary circle case, upon the argument ξ_1,ξ_2 of the form $\Omega(\zeta_1,\zeta_2;\xi_1,\xi_2)$ given by (3) on p. 186. There yields itself:

$$\Omega(\zeta_1,\zeta_2;\xi_1^{(i)},\xi_2^{(i)})=\sum_k\mu_k^{-1}\frac{G_{-d-1}[U_i(\xi_1,\xi_2)]}{(\zeta^{(k)},\xi^{(i)})G_{-d-1}(\zeta_1^{(k)},\zeta_2^{(k)})}. \tag{1}$$

[26] It would also follow from the v. Mangoldt theorem (p. 152), that the series (3) resp (5) on p. 186 ff. remains a *convergent* Poincaré series for $\xi_2=0$. Furthermore, naturally, the possibility remains substituting throughout, that the series denoted in the text by $\psi_d^{(-d-2)}$ identically vanishes.

For the further transformation of the right side we call U_l the substitution arising by combination of U_i^{-1} and U_k; one then has:

$$U_i^{-1}U_k = U_l, \qquad \mu_i^{-1}\mu_k = \overline{\mu_i}\mu_k = \mu_l, \tag{2}$$

and it holds, further:

$$(\zeta^{(k)}, \xi^{(i)}) = (\zeta^{(l)}, \xi), G(\zeta_1^{(k)}, \zeta_2^{(k)}) = G[U_i(\zeta_1^{(l)}, \zeta_2^{(l)})], \tag{3}$$

whereby, on account of the first equation (3), one wants only to consider, that we are concerned with unimodular substitutions.

If we utilize (2) and (3) and observe, that with U_k, obviously, the substitution U_l also runs through the entire homogeneous group Γ, then equation (1) takes on the form:

$$\Omega(\zeta_1, \zeta_2; \xi_1^{(i)}, \xi_2^{(i)}) = \overline{\mu_i} \sum_l \mu_l^{-1} \frac{G_{-d-1}[U_i(\xi_1, \xi_2)]}{(\zeta^{(l)}, \xi) G_{-d-1}[U_i(\zeta_1^{(l)}, \zeta_2^{(l)})]}. \tag{4}$$

The sum standing here on the right obviously again furnishes an elementary form referred to the place ξ; only in the continuation of this form, in place of $G_{-d-1}(\zeta_1, \zeta_2)$, the rational integral form:

$$G_{-d-1}(\zeta_1^{(i)}, \zeta_2^{(i)}) = G_{-d-1}[U_i(\zeta_1, \zeta_2)]$$

has been applied, whose zeros, as those of the form:

$$G_{-d-1}(\zeta_1, \zeta_2)$$

lie outside the net N. If we call the form thus arising $\Omega_i(\zeta_1, \zeta_2; \xi_1, \xi_2)$, then these yield, as
[199] the *behavior of the elementary form* $\Omega(\zeta_1, \zeta_2; \xi_1, \xi_2)$ *under the substitution* U_i *on* ξ_1, ξ_2, *the following general formula*:

$$\Omega(\zeta_1, \zeta_2; \xi_1^{(i)}, \xi_2^{(i)}) = \overline{\mu_i} \Omega_i(\zeta_1, \zeta_2; \xi_1, \xi_2). \tag{5}$$

This rule now also remains maintained in the case of a group without boundary curves. It is hereby a question of the transformation of the elementary form given by (5) on p. 187, with utilization of the equations (2) and (3), we find easily:

$$\Omega(\zeta_1, \zeta_2; \xi_1^{(i)}, \xi_2^{(i)}) = \overline{\mu_i} \sum_l \mu_l^{-1} \frac{G_1^{(-d-1)}[U_i(\zeta_1^{(l)}, \zeta_2^{(l)}); U_i(\zeta_1, \zeta_2)]}{(\zeta^{(l)}, \xi)(\zeta^{(l)}, \overline{\eta})(\zeta^{(l)}, \overline{\eta}') \cdots (\zeta^{(l)}, \overline{\eta})^{(-d-1)}}, \tag{6}$$

if we are to denote, by $\overline{\eta_1}^{(m)}, \overline{\eta_2}^{(m)}$, the quantities arising from $\overline{\eta_1}^{(m)}, \overline{\eta_2}^{(m)}$ by U_i^{-1}. Here, we have to understand by $\Omega_i(\zeta_1, \zeta_2; \xi_1, \xi_2)$ that elementary form, which arises from (5) on p. 187, in case, in place of the form:

$$\frac{G_1^{(-d-1)}(\zeta_1, \zeta_2; \xi_1, \xi_2)}{(\zeta, \eta)(\zeta, \eta') \cdots (\zeta, \eta^{(-d-1)})}, \tag{7}$$

we put the form arising from it by application the substitution U_i to ζ_1, ζ_2 as ξ_1, ξ_2. That this new rational form is suited just as well as the original form for the construction of an elementary form, emerges from formula (6) itself, insofar as an elementary form stands on the left side of this. Obviously, formula (6) shows, *that the rule (5) in fact, also remains the groups without boundary curves.*

If now, $\Omega_i(\zeta_1,\zeta_2;\xi_1,\xi_2)$ is identical with $\Omega(\zeta_1,\zeta_2;\xi_1,\xi_2)$ for all substitutions U_i of the group and indeed, for arbitrary ζ,ξ, then our elementary form *with respect to the exercise of the substitutions of* Γ *on* ξ_1,ξ_2, shows *the behavior of an automorphic form of* ξ_1,ξ_2 *with the multiplicator system* $\overline{\text{M}}$. The elementary form will then be, without further ado, an *automorphic form* of the ξ_1,ξ_2. In fact, we convince ourselves, that Ω presents, at all places of the fundamental domain, the behavior of an automorphic form $\xi \to 0$ discussed on p. 66 ff., i.e., is free of essential singularties. After the foregoing developments, we only need to devote a brief consideration to possible parabolic cusps. But if we develop Ω at such a cusp in form (13) on p. 196, then one immediately argues from the automorphic behavior of Ω with respect to the parabolic generators to the identical vanishing of the forms $\psi_d^{(1)},\psi_d^{(2)},\ldots,\psi_d^{(-d-2)}$; and obviously $\psi_d^{(0)}$ will also vanish identically in case the parabolic multiplicator $\mu = 1$ is not present. We therefore come back to the required formation law (12) on p. 70, and yet make note, that a possible automorphic form Ω of ξ_1,ξ_2 can never have a pole at a parabolic cusp. [200]

But the assumption, that:

$$\Omega_i(\zeta_1,\zeta_2;\xi_1,\xi_2) = \Omega(\zeta_1,\zeta_2;\xi_1,\xi_2),$$

only takes place in individual cases. We will carry this through here initially only for the case $p = 0$, and thereby evaluate a few theorems developed earlier for the automorphic forms of this genus. First of all, the rules developed in the formula (4) ff. on p. 177 fit immediately to the relations present here. Ω is an automorphic form in ζ_1,ζ_2 of dimension d and the multiplicator system M. Should Ω also be an automorphic form in ξ_1,ξ_2, then it has the dimension $\overline{d} = -d-2$ and the multiplicator system $\overline{\text{M}}$ inverse to M. The whole numbers in $m,\overline{m},m_1,\overline{m}_1,\ldots$ coming up in the representation of these automorphic forms by the prime- and groundforms of the model consequently fulfill the relations:

$$\frac{m_k}{l_k}+\frac{\overline{m_k}}{l_k}=1-\frac{1}{l_k},\quad \text{resp.} \qquad \frac{m'_k}{l'_k}+\frac{\overline{m}'_k}{l_k}=1, \tag{8}$$

$$m+\overline{m}+2=0. \tag{9}$$

Now Ω is a one-pole automorphic form of ζ_1,ζ_2; therefore $m \geq -1$. The consequence is, that $\overline{m} \leq -1$ takes place. But, in consequence of the represented behavior in the fundamental domain, Ω would also be one-pole in ξ_1,ξ_2, that we argue to $\overline{m} \geq -1$. Therefore it follows: *should* Ω *be an automorphic form of* ξ_1,ξ_2*, then, necessarily,* $\overline{m} = -1$ *and, with this, also* $m = -1$. Remember thereby, that the number m is determined from the dimension d and the multiplicator system M by means of equation (7) on p. 101. However, on account of the altered meaning of the number m on p. 177, an insignificant modification of this equation is to be undertaken; we return to this immediately. Beforehand, we show that Ω, at the place in question is also actually automorphic in the ξ_1,ξ_2.

If, namely, we have the condition $m = -1$, then this means, that an automorphic form of dimension d and the multiplicator system M has at least one pole. Now the two automorphic forms $\Omega_i(\zeta_1,\zeta_2;\xi_1,\xi_2)$ and $\Omega(\zeta_1,\zeta_2;\xi_1,\xi_2)$ of ζ_1,ζ_2 both have the place

ξ as a pole of first of order of the coefficient 1. From this, there arises the important consequence, *that the difference*:

$$\Omega_i(\zeta_1,\zeta_2;\xi_1,\xi_2)-\Omega(\zeta_1,\zeta_2;\xi_1,\xi_2) \tag{10}$$

[201] *becomes pole-free at the place* ξ, *and accordingly either represents an integral automorphic form of* ζ_1,ζ_2 *or vanishes identically.*[27] But we just noticed, that an integral form of dimension d and the multiplicator system M does not exist in the present case $m=-1$. *The difference* (10) *consequently vanishes identically, in fact, so that* $\Omega_i=\Omega$ *takes place.* Recapitulating, we are led to the theorem: *For a model of genus zero, the elementary form* $\Omega(\zeta_1,\zeta_2;\xi_1,\xi_2)$ *is always then and only then also an automorphic form in* ξ_1,ξ_2, *and indeed one-poled of dimension* $\overline{d}$, *and has multiplicator system* $\overline{\mathrm{M}}$, *if* $m=-1$.

The *representation of the elementary form* $\Omega(\zeta_1,\zeta_2;\xi_1,\xi_2)$ *by means of the prime and groundforms in the case* $m=-1$ *in question* is easily carried out according to the general rules on p. 98 ff. We calculate for ourselves, according to the prescription on p. 100 resp p. 176, the exponents $m_1,m_2,\ldots,m_n$, which refer themselves to the fixed polygon vertices, and with this, to the groundforms. Let the value x_1,x_2 of the primeforms introduced on p. 73 belong to the arguments ξ_1,ξ_2. Then, first of all, from (6) on p. 99, it follows, as a representation of Ω:

$$\Omega(\zeta_1,\zeta_2;\xi_1,\xi_2)=\frac{A}{(z,x)}\prod_{k=1}^{n}{}'(z,e_k)^{\frac{m_k}{l_k}}, \tag{11}$$

insofar as Ω is an automorphic form of the ζ_1,ζ_2. Hereby, A is independent of ζ_1,ζ_2, but naturally, by no means of the ξ_1,ξ_2. The dependency of the factor A on ξ_1,ξ_2, since Ω is, at the same time, an automorphic form of the ξ_1,ξ_2, recognized according to the same rule, it follows: *the representation of the elementary form* $\Omega(\zeta_1,\zeta_2;\xi_1,\xi_2)$ *by the prime- and groundforms, is, in the present case* $m=-1$, *accomplished by the equation*:

$$\Omega(\zeta_1,\zeta_2;\xi_1,\xi_2)=C\frac{\prod_{k=1}^{n}{}'(x,e_k)^{\frac{\overline{m_k}}{l_k}}\cdot\prod_{k=1}^{n}{}'(z,e_k)^{\frac{m_k}{l_k}}}{(z,x)}, \tag{12}$$

hereby, C *denotes a constant independent of* ζ *and* ξ, *and the numbers* $\overline{m_k}$ *are to be determined from the* m_k *by means of formulas (8).*

The calculation of m from d and M is thereby effected by means of the equation (7) on p. 101 mentioned before. In the latter, for the individual parabolic vertex, we must replace $\frac{\lambda_k}{2l_k}$ by $\frac{\lambda_k'}{l_k}$; from now on, we will again remember this replacement, say, by an
[202] upper prime on the summation sign. For parabolic vertices of multiplicator $\mu=1$, the determinations developed on p. 176 hold, in consequence of which, the number now denoted by m is smaller than the number m meant in (7) on p. 101 by the number ε of the parabolic multiplicators of the system $\mathrm{M}=1$. We thus have:

$$2m=d(2-n)-\sum_{k=1}^{n}{}'\frac{\lambda_k}{l_k}+2\sum_{k=1}^{n}{}'E\left[\frac{d+\lambda_k}{2l_k}\right]-2\varepsilon. \tag{13}$$

---

[27] This theorem obviously subsists for models of each arbitrary genus p.

Now should the case $m=-1$ just considered be present, then one hasbeen a Diophantine equation for d, n, l_k and λ_k.

We check this, say, for the multiplicator system $\mathrm{M}_0 = 1$, always coming up for even dimension. All λ_k equal to zero. We restrict ourselves further, say, to those dimensions d, for which $-d \leq 2l_k$ for $k = 1, 2, \ldots, n$; for $d = -2$ and -4, this is always the case. Under those circumstances:

$$2\sum_{k=1}^{n}{}' E\left[\frac{d+\lambda_k}{2l_k}\right] - 2\varepsilon = -2n.$$

And this formula (13) furnishes:

$$-2 = d(2-n) - 2n. \tag{14}$$

The only possibility of solving equation (14) in a whole number m and an even number d, is $n = 3$, $d = -4$. The associated numbers m_k, in consequence of (5) on p. 100, are equal to $(l_k - 2)$, whereupon, by means of (8) on p. 200, the value 1 yields itself for $\overline{m_k}$. There thus arises the remarkable theorem: *in the case of the groups of the circular-arc triangles, the properly automorphic elementary forms of dimension $d = -4$ are always also properly automorphic in the ξ_1, ξ_2, and indeed, of the dimension $\overline{d} = 2$; their representation in the prime- and groundforms is*:

$$\sum_k \frac{G_3(\xi_1,\xi_2)}{(\zeta^{(k)},\xi)\, G_3(\zeta_1^{(k)},\zeta_2^{(k)})} = C\frac{\prod_{k=1}^{3}(x,e_k)^{\frac{1}{l_k}}\prod_{k=1}^{3}(z,e_k)^{1-\frac{2}{l_k}}}{(z,x)}, \tag{15}$$

whatever cubic form G_3 we might have chosen according to the prescription on p. 179. Moreover, one has here directly before one's eyes, that Ω, if ξ or ζ moves into an elliptic vertex, vanishes, identically as a rule, as a form of the other variable pair in this specific case.

Apart form this, formula (15), referred to the "properly automorphic" elementary forms of all groups of the character (0, 3), we always have only one isolated case with $m = -1$. If, e.g., we consider the group of the signature (0, 3; 6, 10, 15), then (cf. p. 91) this has, in all, 60 multiplicator systems, of which 30 belong to even dimension d. Among [203]
the latter, beside the system $\mathrm{M}_0 = 1$ already disposed of, only *three* are obtained, for which $m = -1$, and indeed for $d = -4$; these are the multiplicator systems:

$$\mu_1 = e^{\frac{\pi i}{3}}, \quad \mu_2 = e^{\frac{\pi i}{5}}, \quad \mu_3 = e^{\frac{22\pi i}{15}},$$
$$\mu_1 = e^{\frac{5\pi i}{3}}, \quad \mu_2 = e^{\frac{\pi i}{5}}, \quad \mu_3 = e^{\frac{2\pi i}{15}},$$
$$\mu_1 = e^{\frac{\pi i}{3}}, \quad \mu_2 = e^{\frac{9\pi i}{5}}, \quad \mu_3 = e^{\frac{28\pi i}{15}}.$$

Further, among the thirty systems of odd dimension, only *one* finds itself, for which $m = -1$, and indeed, $d = -5$: this system is:

$$\mu_1 = e^{\frac{\pi i}{2}}, \quad \mu_2 = e^{\frac{3\pi i}{10}}, \quad \mu_3 = e^{\frac{\pi i}{5}}.$$

We now ignore the exceptional case $m = -1$ considered, and with this, turn ourselves to the case, far more important for the continuation of our investigation, *that*

$\Omega(\zeta_1,\zeta_2;\xi_1,\xi_2)$ *is not automorphic in* ξ_1,ξ_2. Here, therefore, one can choose a substitution U_i from the group Γ, and indeed, already among its generators, for which the difference:

$$\Omega(\zeta_1,\zeta_2;\xi_1^{(i)},\xi_2^{(i)}) - \overline{\mu_i}\Omega(\zeta_1,\zeta_2;\xi_1,\xi_2) \tag{16}$$

does not identically vanish. In dependency on ζ, this difference represents an automorphic form $\varphi_d(\zeta_1,\zeta_2)$, which we are also able to write thus:

$$\varphi_d(\zeta_1,\zeta_2) = \overline{\mu_i}[\Omega_i(\zeta_1,\zeta_2;\xi_1,\xi_2) - \Omega(\zeta_1,\zeta_2;\xi_1,\xi_2)], \tag{17}$$

since the elementary forms standing here in the parentheses both exhibit poles of first order at ξ of the coefficient 1, then their difference, and therefore, also the form φ_d, becomes pole-free at ξ, and consequently represents an *integral* automorphic form.

These representations lead us to a very important step forward with respect to the possibility, discussed on p. 175, of the identical vanishing of the pole-free Poincaré series: *In fact, in the right side of equation (17), we have before us a first example, of a pole-free Poincaré series, which certainly does not vanish identically.* If we yet add, that in the exceptional case $m = -1$, integral automorphic forms with zeros at the possible parabolic cusps do not exist at all, then the important theorem yields itself: *in the case of a model of the genus zero, we are always able, for all the dimensions d satisfying the convergence theorem and for each associated multiplicator system* M, *in case any associ-*
[204] *ated integral automorphic forms with zeros at the possible parabolic cusps exist at all, to also form non-identically vanishing pole-free Poincaré series.* We return, in the course of the next paragraph, to the question, whether perhaps, in the individual case, *all* existing integral forms, which vanish at the possible parabolic cusps, are represented by series of this kind[28].

3.14 Concerning the representability of arbitrary automorphic forms of genus zero by the elementary forms and the Poincaré series

The important question just raised concerning the representation of the *integral* automorphic forms by pole-free Poincaré series we shall now be able to answer at the opportunity of an investigation, which has, as its subject, the yet somewhat more general problem of the representation of *arbitrary* automorphic forms by the elementary forms, and with this, by Poincaré series. The treatment of these questions we refer here, for the present, to a model of the genus zero; nevertheless, these partially hold directly for arbitrary p as well, are able to be easily generalized.

First of all, let a *negative* dimension d be given, which is only subjected to the one restriction, that the series of dimension d belonging to our group Γ are convergent. As well, let a form $\varphi_d(\zeta_1,\zeta_2)$ belonging to this dimension, of the multiplicator system M, only first have simple poles; in the fundamental domain. Let there be found σ such poles at the places ${}^{(1)}\xi, {}^{(2)}\xi, \ldots, {}^{(\sigma)}\xi$, of which, however, none may represent a parabolic

[28] We already draw attention here to the fact, that in case $\Omega(\zeta_1,\zeta_2;\xi_1,\xi_2)$ in the ξ_1,ξ_2 is *not* automorphic, then we are also obviously able to produce pole-free Poincaré series according to the principle utilized in the text for models of an arbitrary genus p.

cusp, and also, for the moment, none are to lie at an elliptic vertex. From the elementary forms belonging to these σ places:

$$\Omega(\zeta_1,\zeta_2;{}^{(1)}\xi_1,{}^{(1)}\xi_2),\ldots,\Omega(\zeta_1,\zeta_2;{}^{(\sigma)}\xi_1,{}^{(\sigma)}\xi_2)$$

of the given dimension d and the multiplicator system M then σ constants $A_1, A_2,\ldots,$ A_σ independent of the ζ_1,ζ_2 let themselves be chosen in such a way, that the difference of the proposed form $\varphi_d(\zeta_1,\zeta_2)$ and the sum: [205]

$$A_1\Omega(\zeta_1,\zeta_2;{}^{(1)}\xi_1,{}^{(1)}\xi_2)+A_2\Omega(\zeta_1,\zeta_2;{}^{(2)}\xi_1,{}^{(2)}\xi_2)+\cdots+A_\sigma\Omega(\zeta_1,\zeta_2;{}^{(\sigma)}\xi_1,{}^{(\sigma)}\xi_2)$$

becomes pole-free. We conclude from this, that $\varphi_d(\zeta_1,\zeta_2)$ lets itself be represented in the form:

$$\varphi_d(\zeta_1,\zeta_2)=\sum_{\nu=1}^{\sigma}A_\nu\Omega(\zeta_1,\zeta_2;{}^{(\nu)}\xi_1,{}^{(\nu)}\xi_2)+\text{ an integral automorphic form}(\zeta_1,\zeta_2). \quad (1)$$

We now also allow poles of higher order for $\varphi_d(\zeta_1,\zeta_2)$, and the poles, from now on, may also be situated at elliptic vertices, except that, in such a case, we must estimate the order of the pole according to the prescription given on p. 106. At possible parabolic cusps, however, $\varphi_d(\zeta_1,\zeta_2)$, now as before, is to be pole-free. It offers no difficulty, to form a series of dimension d and the multiplicator system M, which has a pole of given order at the prescribed place, but except for this, exhibits only simple poles in the fundamental domain. If the prescribed place is an elliptic vertex of the fundamental domain, then the calculations on p. 183 ff. come into force. By combination of such series, with addition of elementary forms themselves, we are then able to obtain an expression, which, subtracted from $\varphi_d(\zeta_1,\zeta_2)$, furnishes a pole-free difference. It follows as a preliminary theorem: *each automorphic form $\varphi_d(\zeta_1,\zeta_2)$ of a dimension d satisfying the convergence theorems, with arbitrary poles, but not lying at parabolic cusps, is representable by a Poincaré series, up to an additive integral automorphic form.*

Having just represented the forms of negative dimension by the Ω as forms of ζ_1,ζ_2, we shall, on the other hand, be able to express the forms of non-negative dimensions d by the Ω as forms of the ξ_1,ξ_2. Hereby, first of all, we must turn our attention to a very interesting *analogy between these latter forms Ω of our automorphic model and the integrals of the second kind of an algebraic model.* We attach to the normal integrals of this kind (cf, "M" I on p. 531), each individual one of which, on the Riemann surface belonging to the model, has a pole of the first order "of the coefficient 1" at only one place, and only exhibits periods different from zero for half the canonical crosscuts. Only in the case $p=0$ is this integral a *single-valued* function on the surface. For $p>0$, this alters itself by associated periods for closed circuits; hereby, the details of the relations represented l.c. come into play.

One will immediately recognize the full analogy of these relations to those for our elementary forms $\Omega(\zeta_1,\zeta_2;\xi_1,\xi_2)$. The individual Ω, as a form of the ξ_1,ξ_2, has *a pole of the first order of the coefficient L at only one place* of the fundamental domain. But [206]
only for $m=-1$ have we to do with an *automorphic* form of the ξ_1,ξ_2. For $m>-1$, Ω alters itself, apart from the multiplicative constant $\overline{\mu_i}$, with respect to an arbitrary substitution U_i of the group, by an additive constant, which represents, in ζ_1,ζ_2, a *pole-free series*: in fact, from the considerations on p. 203, it follows immediately:

$$\Omega(\zeta_1,\zeta_2;\xi_1^{(i)},\xi_2^{(i)}) = \overline{\mu_i}\Omega(\zeta_1,\zeta_2;\xi_1,\xi_2) + \text{a pole-free series } (\zeta_1,\zeta_2)_i. \qquad (2)$$

The pole-free series hereby entering are therefore those, which correspond to the periods of the elementary integrals.

Now, as is known, one builds up, from linear combinations of the normal integrals mentioned, algebraic functions of the surface(cf. "M" I on p. 540); one only has to take care, that the combination of integrals exhibits consistently vanishing periods. *Exactly thus, here for* $m > -1$, *we shall produce automorphic forms of the* ξ_1, ξ_2 *of the dimension* $\overline{d}$ *by linear combination of elementary forms.* Thereby, correspondingly, for each such combination, the additive expressions occurring with respect to U_i must obviously all have the value zero.

We now enter in detail into the representation in question of forms of the dimension $\overline{d}$ in the case $m > -1$.

According to p. 107, for the dimension d, there are, all together, $(m+1)$ linearly independent integral automorphic "forms". If, at the same time, we again restrict ourselves here to forms, which consistently have zeros at parabolic cusps, then in order that this theorem remain subsisting, we must calculate the number m according to the prescription on p. 177. It has not yet been proved of the number, to be denoted by τ, of the linearly independent pole-free "series" of the dimension d, that it actually attains to value $(m+1)$. In consideration of the theorem set up at the close of §13 on p. 203, we are only first able to note:

$$0 < \tau \leq m+1; \qquad (3)$$

But it will be our goal, to show, that $\tau = m+1$ always takes place here, in fact.

We imagine any particular system of τ linearly independent pole-free series proposed in:

$$\omega_1(\zeta_1,\zeta_2), \omega_2(\zeta_1,\zeta_2), \ldots, \omega_\tau(\zeta_1,\zeta_2), \qquad (4)$$

those are to be free of ξ_1, ξ_2. On the strength of this, we are able to explicitly give the
[207] individual equation (2) the form:

$$\Omega(\zeta_1,\zeta_2;\xi_1^{(i)},\xi_2^{(i)}) = \overline{\mu_i}\Omega(\zeta_1,\zeta_2;\xi_1,\xi_2) + \sum_{\nu=1}^{\tau} f_\nu^{(i)}(\xi_1,\xi_2)\omega_\nu(\zeta_1,\zeta_2), \qquad (5)$$

where the $f_\nu^{(i)}(\xi_1,\xi_2)$ are forms of the ξ_1, ξ_2 alone[29].

If we are now to build up a σ-poled automorphic form $\psi_{\overline{d}}(\xi_1,\xi_2)$ of the multiplicator system $\overline{\mathrm{M}}$ from the elementary forms, then, in consequence of p. 107, the condition $\overline{m} + \sigma \geq 0$ must be satisfied; for the number σ we therefore have:

$$\sigma \geq -\overline{m} = m+2. \qquad (6)$$

[29] Specialize this formula (5) for a parabolic substitution U_i, by returning to the formation-law (13) on p. 196 of the elementary form in the vicinity of a parabolic vertex ξ. The calculation shows, that the forms of the ζ_1, ζ_2, denoted at that time by $\psi_d^{(1)}, \ldots, \psi_d^{(\overline{d})}$, represent themselves hereby as linear combinations of the $\omega_1, \ldots, \omega_\tau$, and are therefore, in fact, pole-free series. The forms $f(\xi_1,\xi_2)$, for their part, are integral rational homogeneous forms in the case in question of a parabolic substitution U_i.

All σ poles are to be of first order and situated in the fundamental domain at the σ places ${}^{(1)}\zeta, {}^{(2)}\zeta, \ldots, {}^{(\sigma)}\zeta$, which, naturally, again let be different from parabolic vertices and, for the sake of simplicity, also from elliptic fixpoints, in addition to this, however, are expressly to be assumed as movable independently of one another. Now, with the aid of σ quantities B, which are to be independent of ξ_1, ξ_2, we form the linear combinations of elementary forms, which exhibit poles of the first order at the prescribed places:

$$B_1\Omega({}^{(1)}\zeta_1, {}^{(1)}\zeta_2; \xi_1, \xi_2) + B_2\Omega({}^{(2)}\zeta_1, {}^{(2)}\zeta_2; \xi_1, \xi_2) + \cdots + B_\sigma\Omega({}^{(\sigma)}\zeta_1, {}^{(\sigma)}\zeta_2; \xi_1, \xi_2) \qquad (7)$$

and try to determine line B in such a manner, that this expression (7) is automorphic in ξ_1, ξ_2. For this, it is necessary and sufficient, that the additive amounts, by which the aggregate (7) alters itself with respect to the U_i, all vanish. That the expression (7), under these circumstances, also has the character of an automorphic form at possible parabolic cusps, follows from the formation law (13) on p. 196 by means of the argument already utilized on p. 199 for an individual elementary form. Namely, for the expression (7), put down the representation, following from (13) on p. 196, the vicinity of the parabolic cusp. Then if the sum (7) is to show automorphic behavior with respect to the parabolic substitution U_i, then the terms with the powers $\xi, \xi^2, \ldots, \xi^{\bar{d}}$ must vanish identically, and therefore one comes back to the formation law of the automorphic forms.

Now, in consequence of (5), the additive quantity, which enters upon exercise of the substitution U_i on (7), is the following: [208]

$$\sum_{\nu=1}^{\tau} f_\nu^{(i)}(\xi_1, \xi_2)[B_1\omega_\nu({}^{(1)}\xi_1, {}^{(1)}\xi_2) + B_2\omega_\nu({}^{(2)}\xi_1, {}^{(2)}\xi_2) + \cdots + B_\sigma\omega_\nu({}^{(\sigma)}\xi_1. {}^{(\sigma)}\xi_2)]. \qquad (8)$$

If we denote the parenthetical expression multiplied here by $f_\nu^{(i)}$ by C_ν:

$$C_\nu = B_1\omega_\nu({}^{(1)}\xi_1, {}^{(1)}\xi_2) + B_2\omega_\nu({}^{(2)}\xi_1, {}^{(2)}\xi_2) + \cdots + B_\sigma\omega_\nu({}^{(\sigma)}\xi_1, {}^{(\sigma)}\xi_2) \qquad (9)$$

then we have to require the equations:

$$\sum_{\nu=1}^{\tau} C_\nu f_\nu^{(i)}(\xi_1, \xi_2) = 0.$$

As substitutions U_i we obviously need only admit a system of *generators* of the group Γ. Among them, there is certainly at least one substitution U_i, for which the $f_\nu^{(i)}(\xi_1, \xi_2)$, standing on the right in (5), do not all vanish identically; let this take place, in all, for κ of the generators, which we call $U_1, \ldots, U_\kappa$, say. Then obviously our task is, to determine the B is such a manner, that the κ equations:

$$\begin{cases} C_1 f_1^{(1)}(\xi_1, \xi_2) + C_2 f_2^{(1)}(\xi_1, \xi_2) + \cdots + C_\tau f_\tau^{(1)}(\xi_1, \xi_2) = 0, \\ \cdots\cdots\cdots\cdots\cdots\cdots \\ C_1 f_1^{(\kappa)}(\xi_1, \xi_2) + C_2 f_2^{(\kappa)}(\xi_1, \xi_2) + \cdots + C_\tau f_\tau^{(\kappa)}(\xi_1, \xi_2) = 0, \end{cases} \qquad (10)$$

are valid for variable ξ_1, ξ_2.

The equations (10) are fulfilled, in case all $C_1, C_2, \ldots, C_\tau$ vanish; but, for their part, they do not always necessarily require the vanishing of all the C. Namely, it is first of all not excluded, that the forms $f(\xi_1,\xi_2)$ of the individual equation are linearly dependent of one another, i.e., that one or more linear identities with constant coefficients subsist between them. On this account, we may only first draw the conclusion from the equations (10). That ρ particular equations subsist between the $C_1, C_2, \ldots, C_\tau$:

$$\begin{cases} \alpha_{11}C_1 + \alpha_{12}C_2 + \cdots + \alpha_{1\tau}C_\tau = 0, \\ \alpha_{21}C_1 + \alpha_{22}C_2 + \cdots + \alpha_{2\tau}C_\tau = 0, \\ \cdots\cdots\cdots\cdots \\ \alpha_{\rho 1}C_1 + \alpha_{\rho 2}C_2 + \cdots + \alpha_{\rho\tau}C_\tau = 0, \end{cases} \tag{11}$$

with numerical, i.e., independent of the ξ_1,ξ_2 and ζ_1,ζ_2, coefficients α, whereby $\rho \geq 1$ and $\rho \leq \tau$:

$$1 \leq \rho \leq \tau \leq m+1. \tag{12}$$

[209] The equations (11) we may consider as linearly independent of one another, since in the contrary case, nothing stands in the way of a reduction of the system (11). We utilize this circumstance, in order to produce, from the τ linearly independent pole-free series (4), in:

$$\begin{cases} \omega_1'(\zeta_1,\zeta_2) = \alpha_{11}\omega_1(\zeta_1,\zeta_2) + \alpha_{12}\omega_2(\zeta_1,\zeta_2) + \cdots + \alpha_{1\tau}\omega_\tau(\zeta_1,\zeta_2), \\ \cdots\cdots\cdots\cdots\cdots \\ \omega_\rho'(\zeta_1,\zeta_2) = \alpha_{\rho 1}\omega_1(\zeta_1,\zeta_2) + \alpha_{\rho 2}\omega_2(\zeta_1,\zeta_2) + \cdots + \alpha_{\rho\tau}\omega_\tau(\zeta_1,\zeta_2), \end{cases} \tag{13}$$

ρ new linearly independent pole-free series $\omega_1'(\zeta_1,\zeta_2),\ldots,\omega_\rho'(\zeta_1,\zeta_2)$. With the aid of these series, the equations (11) to be satisfied by the B write themselves explicitly thus:

$$\begin{cases} B_1\omega_1'({}^{(1)}\zeta_1, {}^{(1)}\zeta_2) + B_2\omega_1'({}^{(2)}\zeta_1, {}^{(2)}\zeta_2) + \cdots + B_\sigma\omega_1'({}^{(\sigma)}\zeta_1, {}^{(\sigma)}\zeta_2), \\ \cdots\cdots\cdots\cdots\cdots \\ B_1\omega_\rho'({}^{(1)}\zeta_1, {}^{(1)}\zeta_2) + B_2\omega_\rho'({}^{(2)}\zeta_1, {}^{(2)}\zeta_2) + \cdots + B_\sigma\omega_\rho'({}^{(\sigma)}\zeta_1, {}^{(\sigma)}\zeta_2). \end{cases} \tag{14}$$

These developments now place us in the position, of making the decisive arguments for the solution of our question of the representability of the integral automorphic forms by pole-free series. Namely, as soon as $\sigma \geq \rho+1$, the equations (14) are always fulfilled by non-identically vanishing values B, which naturally depend on ${}^{(1)}\zeta_1,\ldots,{}^{(\sigma)}\zeta_2$. We are also able, from the linear independence of the pole-free series (13) and the movability of the places ${}^{(1)}\zeta,\ldots,{}^{(\sigma)}\zeta$, which were assumed above, to prove the *necessity of the condition* $\sigma \geq \rho+1$. However, we first of all also arrive at our goal without using this latter circumstance by utilizing the fact, that for $\sigma = \rho+1$, a system of usable quantities B always exists, for which we have a non-identically vanishing automorphic form in (7). The number of the poles of this form is at most equal to $\rho+1$ (namely, if all B are different from zero); and since this number of poles, according to (6), must be at least equal to $m+2$, then we obtain the condition for ρ:

$$\rho \geq m+1. \tag{15}$$

Now compare this condition with the one given in (12) and immediately draw the highly important conclusion, that $\rho = \tau = m+1$.[30] We have, in fact, arrived in this man-
[210] ner at the fundamental theorem already announced above: *the number τ of the linearly*

[30] Notice the analogy of the procedure of reasoning followed in the text to the proof of the Brill. Noether reciprocity theorem in the theory of special functions on a Riemann surface; see the details in "M" I on p. 552 ff.

independent pole-free Poincaré series is equal to the number $(m+1)$ *of the linearly independent integral automorphic forms, which vanish at the possible parabolic cusps.*

By means of a simple theorem in determinants, there arise from this the fundamental theorem: *each integral automorphic form* $\varphi_d(\zeta_1,\zeta_2)$ *of the dimension* d *of our group* Γ, *vanishing at the possible parabolic cusps, is represented by a Poincaré series*:

$$\varphi_d(\zeta_1,\zeta_2) = \sum_k \mu_k^{-1} H_d(\zeta_1,\zeta_2), \tag{16}$$

Assuming naturally, that the dimension d *satisfies the convergence theorems for the group* Γ.[31] The theorem already stated above concerning representability of the forms of negative dimension by Poincaré series can accordingly be completed thus: *Each automorphic form* $\varphi_d(\zeta_1,\zeta_2)$, *vanishing at the possible parabolic cusps, of a negative dimension satisfying the convergence theorems, can be represented in the form (16) by a Poincaré series.* This obviously basic theorem gives us the conclusion of the theory of the Poincaré series of the groups Γ of the genus zero; this will be proved in the next chapter for automorphic models of arbitrary genus p.

We are now also able to carry to a conclusion the *representation of the* σ*-poled automorphic forms* $\varphi_{\overline{d}}(\xi_1,\xi_2)$ *of positive dimension* $\overline{d}$ *by the elementary forms,* just interrupted. For this purpose, we now have to put $\rho = m+1$ in the equation system (14) and now utilize the linear independence of the $(m+1)$ pole-free series ω' and the independence of the σ places ${}^{(1)}\zeta,\dots,{}^{(\sigma)}\zeta$ from one another. From this, one can draw the conclusion, that of the $(m+1)$-termed determinants of the matrix:

$$\begin{vmatrix} \omega'_1({}^{(1)}\zeta_1,{}^{(1)}\zeta_2) & \omega'_1({}^{(2)}\zeta_1,{}^{(2)}\zeta_2) & \cdots & \omega'_1({}^{(\sigma)}\zeta_1,{}^{(\sigma)}\zeta_2) \\ \vdots & \vdots & & \vdots \\ \omega'_{m+1}({}^{(1)}\zeta_1,{}^{(1)}\zeta_2) & \omega'_{m+1}({}^{(2)}\zeta_1,{}^{(2)}\zeta_2) & \cdots & \omega'_{m+1}({}^{(\sigma)}\zeta_1,{}^{(\sigma)}\zeta_2) \end{vmatrix}$$

None can vanish identically in the ${}^{(1)}\zeta_1,\dots,{}^{(\sigma)}\zeta_2$. The consequence is, that of the σ quantities B,

$$\sigma - m - 1 = \sigma + \overline{m} + 1 \tag{17}$$

may be arbitrarily chosen, while the remaining ones are then uniquely, and indeed, linearly and homogeneously determined in those. *There are* $(\sigma+\overline{m}+1)$ *linearly independent automorphic expressions (7), which have poles of the first order at* σ *arbitrarily choosable places* ${}^{(1)}\zeta,\dots,{}^{(\sigma)}\zeta$ *outside the fixed vertices of the polygon net.* [211]

According to p. 107, $(\overline{m}+\sigma+1)$ is now exactly also the number of linearly independent σ-place forms of the dimension $\overline{d}$. From this, follows, by a sequence of reasoning already utilized: *Each automorphic form* $\varphi_{\overline{d}}(\xi_1,\xi_2)$ *of the non-negative dimension* $\overline{d}$, *which only has simple poles not falling at the fixed vertices of the net* N, *can be represented as an aggregate of the form (7) of elementary forms, in case the convergence theorems permit the construction of elementary forms* Ω *for this dimension* $\overline{d}$ *at all.*

It is natural to conjecture here, that the previous results concerning the representation of the forms of positive dimensions only hold "in general", that therefore, similarly, as for the Riemann-Roch theorem, increases in the number $(\sigma+\overline{m}+1)$ of linearly

[31] This latter rider is necessary. That there are also *integral* forms for $d=-1$, is shown by the example of the number form $\sqrt[12]{\Delta}$.

independent automorphic expression (7) might also enter for special point systems ${}^{(1)}\zeta, {}^{(2)}\zeta, \cdots, {}^{(\sigma)}\zeta$. In the individual case, there would obviously have to be added the number of linearly independent integral forms of dimension d, each individual one of which vanished simultaneously at the σ points ${}^{(1)}\zeta, \ldots, {}^{(\sigma)}\zeta$. But according to p. 98, an *integral* form of dimension d has only m zeros in the fundamental domain, while $\sigma \geq m+2$ takes place. *The stated conjecture is consequently groundless.*

With the preceding developments, we have brought the theory of the Poincaré series for $p=0$ to a conclusion. And with this, our form-theoretic investigations generally of the automorphic models of genus zero. Let a pair of historical remarks on the elementary forms find place here at the close. These forms first came up in Poincaré[32] in non-homogeneous form, and were utilized by him for the proof of the fundamental theorems concerning representability of the automorphic functions by his series. In fact, that which Poincaré designates, l.c. as "élement simple", corresponds, directly to our elementary form, the form-theoretic development has also been given here by Ritter[33].

[212] Ritter thereby takes the path exactly opposite to Poincaré's, original development. The latter (homogeneously speaking) attached to automorphic forms of a non-negative dimension $\overline{d}$, and represented the individual such form by an infinite partial fraction, whose terms correspond to the poles of the form. By gathering together those partial fractions, which refer themselves to the individual system of equivalent poles $\zeta, \zeta^{(1)}, \zeta^{(2)}, \ldots$, Poincaré then arrives at an expression, which shows itself to be automorphic in dependency on this place of discontinuity ζ, and even represented an "élément simple". In contradistinction to this, we have, with Ritter, placed the elementary form Ω at the forefront.

Moreover, in his publication mentioned, Ritter has not yet broken through to the simplest form of the theory of the elementary forms; rather, this theory has yet, in many respects to be established more firmly, to be simplified. and to be developed further[34].

---

32 Compare the note, cited on p. 46, in the Comptes rendus of 1881 and 1882, as well as the papers "*Ménioire sur le, function fuchsienne;*" p. 5 ff., Acta mathematica Vol.1 p. 242 ff. (1882), "*Mémeine Sur les fonction zétafuchsienne*" § 6, Acta Mathem., Vol. 5. p. 252.ff.(1884).

33 "*Die eindeutigen automorphen Formen des Gechlechtes nuli*" §17, Mathem, Annalen, Vol. 41, p. 69ff. (1891).

34 Compare, for this, the note of the author "*Die autmorphen Elementarformen*" in the Göttinger Nachrichlen of the year 1900, Heft. 3.

4

The automorphic forms and their analytic representations for models of arbitrary genus

In the generalization of the foregoing developments to the automorphic models of arbitrary genus p, to the extent that these do not already hold for this case as well, we shall by and large again find the viewpoints attained for $p = 0$. To be sure, in the establishment of the theory of the automorphic forms for arbitrary p, we shall have to slightly alter the train of thought vis-a-vis the developments given on p. 66ff. This lies in the difficulty of producing a primeform for an arbitrary algebraic model with $p > 0$. For the genus $p = 0$, we simply had to split a *principal function* z into the quotient $z_1 : z_2$ of two binary variables, in order to possess, in $(z_1e_2 - z_2e_1)$, a primeform with zero e to be prescribed arbitrarily. For $p > 0$, we have to attach to the *transcendent primeform*, which was defined in "M" II on p. 502ff., and we have yet to develop the theory of this primeform somewhat further first. [213]

The ζ_1, ζ_2, for their part, we shall then conceive, tying into the developments on pp. 43 and 109ff. as *"polymorphic forms" on the Riemann surface* and bring them to representation in the primeform. Through this, then, at the same time, the possibility is offered, of considering the automorphic forms of ζ_1, ζ_2, as forms on the Riemann surface. They receive there the character of the so-called *"multiplicative forms"* whose theory E. Ritter, in a work[1] to be cited yet more often further on, based on primeform. Hereby, *Representations of the automorphic forms in the primeform* will arise, which furnish the precise generalization of the equations (4) and (6) on p. 98ff, valid for $p = 0$. These representations allow us to carry through the generalization of the studies on p. [214]
99. referred to $p = 0$, which treat the relations between the multiplicator system, the dimension, the number of the poles and zeros etc. of an automorphic form.

Our developments on the *convergence of the Poincaré series* (p. 142ff), which was repeatedly referred to above are valid for models of each genus p, and too, the investigations on one-poled series and elementary forms (p. 178 ff.) are, for the most part, independent of the genus number p. By utilization of the elementary forms, we are able, at the close of the chapter, to turn ourselves to the important question, to what extent the forms of a proposed model with $p > 0$ are representable by Poincaré series.

[1] *Die multiplicativen Formen auf algebraischem Gebilde beliebigen Geschlechtes mit Anwendung auf die Theorie der automorphen Formen"*, Mathem. Annalen, Vol. 44 p. 261 (1893).

We devoted a separate study to the series of $(-2)^{\text{nd}}$ dimension in the cases, where these converge. Those series are most closely connected with the *integrals of the automorphic model*, and in direct tie-in with this, we shall yet become acquainted with very remarkable *product developments for the primeforms*. These matters have an exceedingly richly developed literature, on which we report in more detail below.

4.1 Recapitulation concerning the groups of arbitrary genus p and their generation

Let this be the place, analogously, as at the beginning of the second chapter, so also here, to present beforehand a brief recapitulation concerning the groups, which underlay the automorphic models of arbitrary genus p.

According to the developments in I on p. 182ff., we shall in each case be able to attach to a *canonical fundamental domain*, whose "character" we denote as usual by (p, n). The generators of the non-homogenous group Γ we write:

$$V_1, V_2, \ldots, V_n, V_{a_1}, V_{b_1}, \ldots, V_{a_p}, V_{b_p}.$$

To these correspond the homogeneous substitutions:

$$U_1, U_2, \ldots, U_n, U_{a_1}, U_{b_1}, \ldots, U_{a_p}, U_{b_p},$$

which we imagine as written unimodularly. Thereby, exactly as above for $p = 0$, we shall fix the individual one of the substitutions $U_1, U_2, \ldots, U_n$ according to the prescription in I on p. 198. How we will imagine the individual one of the substitutions $U_{a_1}, U_{b_1}, \ldots,$ chosen between each of the two possibilities, which there are, is only first to be given below.

[215] As concerns the *relations for the homogeneous generators*, naturally, the same thing holds for the elliptic U_k as for $p = 0$. The "last primary relation", however, according to the developments in I on p. 200 ff. will assume the form:

$$\prod_{k=1}^{n} U_k \cdot \prod_{g=1}^{p} U_{a_g}^{-1} U_{b_g} U_{a_g} U_{b_g}^{-1} = (-1)^n. \tag{1}$$

We best illustrate this relation, as well as the circumstance of possible *secondary relations*, immediately for a few of the most important types of groups.

If the *polygon net N is simply connected* and therefore bounded by a single boundary curve, then the relations lie especially simply. Then the individual fundamental domain is also simply connected, and no secondary relations at all come up.

If we have a *fundamental domain with $2p$ separately running boundary curves,* which are pairwise mapped onto one another hyperbolically or loxodromically, then one has $n = 0$, and the p generators $V_{a_1}, V_{a_2}, \ldots, V_{a_p}$ are all equal to the identical substitution 1. Since U_{a_g} is equal to $+1$ or -1, then:

$$U_{a_g}^{-1} U_{b_g} U_{a_g} = U_{b_g}$$

will always occur. The relation (1) will thus be fulfilled, by:

$$U_{a_g}^{-1}U_{b_g}U_{a_g}U_{b_g}^{-1} = U_{b_g}U_{b_g}^{-1} = 1$$

holding for the individual index g and additional relations for the p remaining generators $U_{b_1}, U_{b_2}, \ldots, U_{b_p}$ will not come up.

We further consider the case of a *group with principal circle,* but which is *not a boundary circle,* whereby we come to the generalization of the considerations carried through above (p. 57 ff.) for $p = 0$. We hereby pursue, say, only the case, that no substitutions come up in Γ, which interchange the two sides of the principal circle. In the projective plane (to which we attach on p. 26), we have a polygon of the character (p', n') with $\mu' > 0$ hyperbolic vertices. The elliptic or parabolic generators, which are furnished from the projective polygon, let be $V_1, V_2, \ldots, V_{n'-\mu'}$. The $V_{n'-\mu'+1}, \ldots, V_n'$ are then hyperbolic, naturally too the

$$V_{a_1}, V_{b_1}, \ldots, V_{a_{p'}}, V_{b_{p'}},$$

one calculates the character of the fundamental domain in the ζ-plane from: [216]

$$n = 2(n' - \mu'), \qquad p = 2p' + \mu' - 1. \tag{2}$$

The character of the group is recognized most clearly, if one generalizes the form of the fundamental domain given in figure 16 for arbitrary p. One will easily blaze a path for this generalization starting out from the projective polygon, by pushing a certain cycle of accidental vertices over the ellipse, whereby, then, one of the hyperbolic generators, say, $V_{n'}$, drops out[2]. We find: *The group under discussion lets itself be generated by composition from:*

1. $n' - \mu'$ *cyclic elliptic or parabolic groups,*
2. $\mu' - 1$ *cyclic hyperbolic groups,*
3. p' *symmetric groups of the character* $(2, 0)$.

Figure 29 illustrates the character of this composition of groups for the example:

$$n' = 2, \quad \mu' = 1, \quad p' = 1.$$

We easily subsume these relations under the canonical fundamental domain, taken as a starting point at the beginning of the paragraph. The n fixed vertices are furnished by the $(n' - \mu')$ fixpoint pairs of the $V_1, V_2, \ldots, V_{n'-\mu'}$; thereby, the individual parabolic point is to be counted doubly, insofar as the polygon reaches up to it with two cusps. The n generators V_k will therefore here, in particular, be the following: [217]

$$V_1, V_2, \ldots, V_{n'-\mu'}, V_{n'-\mu'+1} = V_1^{-1}, \ldots, V_n' = V_{n'-\mu'}^{-1}.$$

We have the $(n' - \mu')$ *secondary relations:*

$$V_1 \cdot V_{n'-\mu'+1}' = 1, \quad V_2 \cdot V_{n'-\mu'+2}' = 1, \ldots, V_{n'-\mu'}' \cdot V_n' = 1. \tag{3}$$

[2] Compare the representation, given in detail above (p. 59), of the present relations in the case $p = 0$.

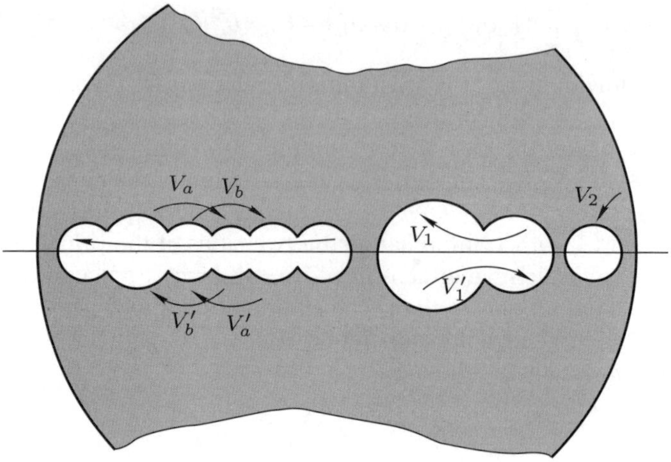

Fig. 29.

The p pairs V_a, V_b are furnished first by $(\mu' - 1)$ pairs:

$$V''_{a_h} = 1, \quad V''_{b_h} = V_{n'-\mu'+h} \qquad (h = 1, 2, \ldots, \mu' - 1),$$

corresponding to the $\mu' - 1$ hyperbolic cyclic groups mentioned in 2. Next to these, furnished by the p' symmetric groups $(2, 0)$, there arrange themselves p' quadruples:

$$V_{a_g}, V_{b_g}, V'_{a_g} = V_{b_g}^{-1}, V'_{b_g} = V_{a_g}^{-1} \qquad (g = 1, 2, \ldots, p').$$

In the equations:

$$V_{a_g} V'_{b_g} = 1, \qquad V_{b_g} V'_{a_g} = 1 \qquad (g = 1, 2, \ldots, p'). \tag{4}$$

We recognize $2p'$ *additional secondary relations*, so that, all together, we obtain $(2p' + n' - \mu')$ relations of this kind. From the equations (4), there immediately yields itself the relation, belonging to the individual group of the character (2, 0):

$$V_{a_g}^{-1} V_{b_g} V_{a_g} V_{b_g}^{-1} \cdot V'^{-1}_{a_g} V'_{b_g} V'_{a_g} V'^{-1}_{b_g} = 1,$$

as then, moreover, the last primary relation (1) is a simple consequence of the equations (3) and (4).

For the homogeneous way of writing determined above, the relations (3) will assume the form:

$$U_1 U'_{n'-\mu'+1} = 1, \quad U_2 U'_{n'-\mu'+2} = 1, \ldots, U_{n'-\mu'} U'_n = 1; \tag{5}$$

corresponding to the equations (4), we are only first able to put down:

$$U_{a_g} U'_{b_g} = \pm 1, \quad U_{b_g} U'_{a_g} = \pm 1 \qquad (g = 1, 2, \ldots, p'), \tag{6}$$

since the homogeneous $U_{a_g}, U_{b_g}, \ldots,$ are not yet finally fixed.

The kind of groups just considered furthermore represents only a special case of the *most general kind of group* Γ *of the character* (p,n), *which we are able to produce by the process of composition or shuffling into one another form finitely many groups* $(p_1,n_1),(p_2,n_2),\ldots$. We mean thereby, that simple case of composition, for which the [218]
boundaries of the remaining polygons lie in the "interior" of the polygon of the individual group to be composed, while each of these individual polygons, taken by themselves, represent a simply connected domain. The individual group to be composed may naturally also be a cyclic or an elliptic or parabolic rotation group. The "last primary relation" of the individual group to be composed will hereby, from the standpoint of the composed group (p,n), furthermore represents a *secondary relation*. The "last primary relation" of the total group (p,n) then appears as a simple consequence of those secondary relations.

To this study of the "most general" groups, let there yet be attached a remark concerning such automorphic models with $p>0$, which let themselves be treated without difficulty by models of genus zero. We mention, in this regard, that certain *elliptic models of the character* $(1,1)$ *are hyperelliptic ones of the character* $(p,0)$.

Automorphic models of the character $(1,0)$ are found only for the parabolic rotation groups. In the boundary circle case, in order, say, to go into this, the groups of the character $(1,1)$ were the simplest belonging to the genus $p=1$. To these groups we have devoted a detailed investigation in I on p. 254 ff. We saw there, *that each group of the signature* $(1,1;l)$ *is a distinguished subgroup of index 2 inside the group of the signature* $(0,4;2,2,2,2l)$. Consequently, one will be able to reduce the function-theoretic treatment of the group $(1,1,l)$ to this latter group of genus $p=0$, resp. to the associated automorphic model.

Secondly, let a hyperbolic model of the character $(p,0)$ be proposed, whereby, therefore, the group Γ will not exhibit elliptic or parabolic generators. Each such model permits *a fixed determinate transformation of period* 2 *into itself.* In the simplest cases, this transformation represents itself as a linear substitution in ζ, which, to be sure, we are not yet able to investigate decisively here. The consequence will then be, *that the group* Γ *is a distinguished subgroup of index* 2 *in a group of genus 0 and the signature*:

$$(0,2p+2;2,2,2,\ldots,2), \tag{7}$$

which lets itself be generated by $(2p+2)$ *substitutions of period two.* The two most important special cases belonging here are the following:

First, we start out, in the category of the *boundary-circle groups*, from a polygon of [219]
the signature given in (7). One can draw this as a $(2p+2)$-angle, whereby the $(2p+2)$ "side centers" are the fixpoints of the generators . If one exercises one of these generators, say V_1, and adds the polygon arising to the first, then one has obtained a fundamental domain of the corresponding hyperelliptic model. Thereby of the $2\cdot(2p+1)$ sides still free, each two which are diametral with respect to the fixpoint of V_1 are mapped onto one another by means of the substitutions $V_1V_2, V_1V_3,\ldots,V_1V_{2p+2}$.[3]

The second example belongs to the *principal-circle groups with isolatedly situated boundary points*. We mean here that hyperelliptic model, which we already discovered

[3] Automorphic models of this kind are considered in detail by E.T.Whittaker in the work "*On the connection of algebraic functions with automorphic functions*", Philosophical transactions, Vol, 192(1898).

above (p. 62 ff.). We start out from $(p+1)$ full circles running separately and outside one another, which all cut a principal circle orthogonally. The $(p+1)$ associated reflections generate a group of the second kind, whose maximal subgroup of the first kind furnishes a hyperelliptic model. The addition of the reflection in the principal circle itself to the $(p+1)$ reflections already cited yields (as we have represented in detail on p. 62 ff.) a group of the second kind, whose maximal subgroup of the first kind has the genus $p=0$[4].

4.2 Recapitulation and extension of the theory of the primeform for an arbitrary algebraic model

In additions to the theory of the primeform, we attach to an entirely arbitrary algebraic model of the genus p. We choose an associated algebraic function z, which let be w-valued, and map the model onto the corresponding w-sheeted Riemann surface F_w positioned over the z-plane. Let the function z be so chosen, that the surface F_w is free of branch-points at $z=\infty$.

We provide the surface F_w with a canonical system of crosscuts $a_1, b_1, \ldots, a_p, b_p$.
[220] We are able to represent a schematic picture of the cut up surface by a fundamental domain of our kind of the character $(p,0)$, as appended figure 30 gives this, e.g., for $p=3$. Naturally, with this, it is not yet to be asserted, that the surface F_w always itself be mapped conformally onto such a domain. The V_{a_k}, V_{b_k} corresponding to the arrows are consequently not yet to signify ζ-substitutions here, but are only symbols for the running-through of the elementary periodpaths on the closed surface. Thereby, V_{a_k} (see the figure in I on p. 184 ff.) will join the banks of the cut a_k and therefore correspond to the path b_k.

In "M" II on p. 476 ff., we have studied the *integral of the third kind*, and, in particular, the *normal integrals* $\prod_{\xi,\eta}^{x,y}$ on the Riemann surface. Here, the upper indices x, y represent the integral limits, the lower indices ξ, η, however, the "parameter", i.e., the two logarithmic discontinuity points of the integral. These two places of the surface we imagine joined by a line directed from η to ξ, which, for the sake of simplicity, let meet none of the cuts a_k, b_k (cf. figure 30).

The *period properties* of the integral $\prod_{\xi,\eta}^{x,y}$, which we consider, say, as a function of the upper limit x, are treated in "M" II on p. 479. If we add, to the crosscuts a_k, b_k so far, one more cut carried along Λ, then our integral is single-valued on the surface thus cut up, and exhibits the constant difference $-2i\pi$ along the cut Λ, whereby the excess of the value of the integral at the right bank over the value lying opposite is meant. With respect to V_{a_k}, the integral experiences no alteration at all, while V_{b_k} leads from $\prod_{\xi,\eta}^{x,y}$ to:

$$\overline{\prod}_{\xi,\eta}^{x,y} = \prod_{\xi,\eta}^{x,y} + 2i\pi j_k^{\xi,\eta}. \tag{1}$$

[4] H. Weber investigated the hyperbolic models of the last category in the essay, already cited on p. 159 "*Ein Beitrag Pioncaré's Theorieder Fuchs'schen Functionen*", Göttinger Nachrichten of 1886, p. 359. We return once more below, as already placed in prospect, to the product developments of automorphic functions considered in this work.

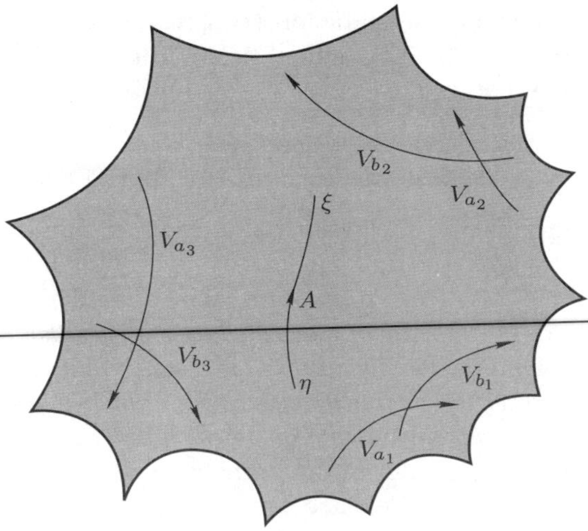

Fig. 30.

Hereby, by $j_1, j_2, \ldots, j_\rho$ are understood the normal integrals of the first kind belonging [221]
to the canonical cut system a_k, b_k, and, in particular, $j_k^{\xi,\eta}$ signifies the definite integral j_k extended along Λ from η to ξ.

We remember, finally, the *theorem on the permutability of parameter and argument.* If the integrator curve L, running from y to x, meets neither Λ nor the remaining cuts a_k, b_k, then the equation holds:

$$\prod_{\xi,\eta}^{x,y} = \prod_{x,y}^{\xi,\eta}, \tag{2}$$

where, for the integral standing on the right, Λ is the integration curve, while a cut to be carried along L has taken the place of Λ.

The formation law of the integral $\prod_{\xi,\eta}^{x,y}$ first becomes entirely clear upon taking up the *form-theoretic method of study,* which was established in "M" II on p. 484 ff. We split z into the quotient $\frac{z_1}{z_2}$ of two homogeneous quantities z_1, z_2 with the usual stipulation, that z_1, z_2 are to be always finite and may never vanish simultaneously. The differential $z_1 dz_2 - z_2 dz_1$ denoted in "M" II on p. 493 by $d\zeta$, we shall better denote here, in detail, by (z, dz), in order to avoid confusion with the argument ζ of the automorphic functions. Our integral of the third kind then permits (cf. "M" II on p. 496) the representation in the form of the following double integral:

$$\prod_{\xi,\eta}^{x,y} = \int_y^x \int_\eta^\xi \frac{\psi(z,z')}{(z_1 z_2' - z_2 z_1')^2} (z_1, dz)(z', dz'), \tag{3}$$

carried out along the curves just denoted by Λ and L. Hereby, $\psi(z,z')$ is a determinate algebraic function of our model in dependency on z' as well as z.

Let the *primeform* be denoted, in the form originally introduced by Klein, by $\Omega(x_1, x_2; y_1, y_2)$ or, of short, $\Omega(x, y)$.[5] It is defined in "M" II on p. 505 by formula(8) or (9) (where, moreover, the notation P was used in stead of Ω), of which we reproduce the
[222] first, say:

$$\Omega(x,y) = \sqrt{-(x,dx)(y,dy)e^{-\Pi_{x,y}^{x+dx,y+dy}}} \quad (\lim dx = 0, dy = 0). \tag{4}$$

We may hereby assume, that the integration paths from y to x and from $y+dy$ to $x+dx$, without meeting one another, run right up next to one another on the Riemann surface.

Of the properties of this "Klein primeform", which are derived l.c., we first of all reproduce the following. *In dependency on the place x of the Riemann surface, resp. as a form of the* x_1, x_2, *it remains everywhere finite and continuous; it has a zero of the first order at the place y; further zeros come up only at the branch-points of* F_w, *and indeed, at the individual one, a zero of the order* $\frac{1}{2}(\kappa-1)$, *in case* κ *sheets hang together there*[6].

In order to fashion the behavior of the primeform into one as simple as possible with respect to closed circuits of the place x resp. the x_1, x_2 on the surface, we go over, by further refashioning of $\Omega(x, y)$, to a new form of the primeform, which E.Ritter has first investigated in detail and utilized in his paper cited on p. 213. The following consideration effects the transition to this "Ritter primeform", which, at the same time, will
[223] represent the periodicity properties of the primeforms in an exhaustive fashion, as was done in "M" II on p. 506.

We mark, on the surface F_w, four arbitrary places y, x, ξ, η and, on the surface cut up by the a_k, b_k (cf. figure 31), draw a curve, consisting of three portions L, C, Λ, from y over x and ξ to η.[7] The adjacent dotted line, consisting of L', C', Λ', is to indicated a path from $y+dy$ over $x+dx$ and $\xi+d\xi$ to $\eta+d\eta$. If one will, one can consider the $dx, dy, d\xi, d\eta$ here as finite numbers of sufficiently small values.

Now form the following aggregate of integrals, to be denoted by $\prod$:

$$\prod = \prod_{x,\eta}^{x+dx,\eta+d\eta} + \prod_{y,\xi}^{y+dy,\xi+d\xi} - \prod_{x,\xi}^{x+dx,\xi+d\xi} - \prod_{y,\eta}^{y+dy,\eta+d\eta},$$

whereby, for the determination of each individual one of these four terms, which, according to (3), represent double integrals, the lines just described are to be utilized as integration curves. Therefore, in the first term of our aggregate, the integration is to take place from η to x along Λ and C, from $\eta+d\eta$ to $x+dx$, correspondingly, over Λ' and C' etc. Now separate the integration path from η to x just mentioned into its two components Λ and C, and proceed correspondingly in the last term of $\prod$; in this manner, it follows:

---

[5] The notation Ω for the primeform is only utilized here temporarily, so that a confusion with the elementary forms, likewise denoted by Ω, is not to be feared.

[6] In the original representation of Klein (see his paper "*Zur Theorie der Abel'schen Functionen*", Mathem. Ann. Vol. 36 (1889) p. 1ff.) an invariant construction of the primeform on the foundation of the normal curve of the forms φ is given, whereby the zeros situated at the branch-points of our surface, just mentioned in the text, naturally do not come up. These zeros, in fact, first originated, by our given preference, in "M" II l.c., in the establishment of the form theory, to a special surface F_w of the algebraic model.

[7] Naturally, one can again employ the line L, Λ already utilized before.

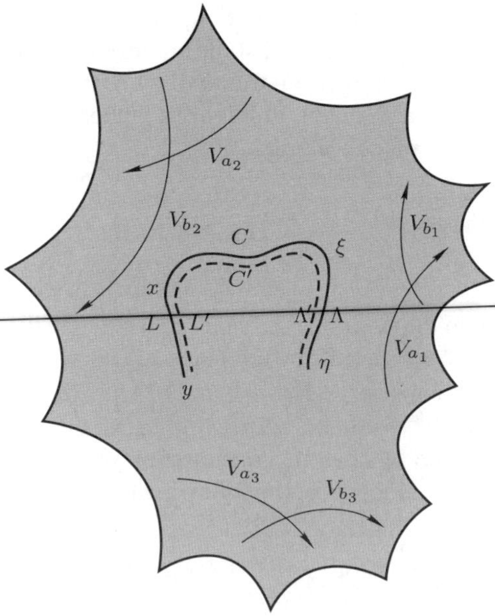

Fig. 31.

$$\prod = \prod_{x,\xi}^{x+dx,\eta+d\eta} + \prod_{\xi,\eta}^{x+dx,\eta+d\eta} + \prod_{y,\xi}^{y+dy,\xi+d\xi} - \prod_{x,\xi}^{x+dx,\xi+d\xi} - \prod_{\xi,\eta}^{y+dy,\eta+d\eta} - \prod_{y,\xi}^{y+dy,\eta+d\eta} .$$

Here, in each pair of double integrals standing under one another, the limits of one integration (and naturally, also the associated integration curve) agree. By gathering together each pair of these integrals, there thus follows:

$$\prod = \prod_{x,\xi}^{\xi+d\xi,\eta+d\eta} + \prod_{\xi,\eta}^{x+dx,y+dy} + \prod_{\xi,y}^{\xi+d\xi,\eta+d\eta} ,$$

whereby, furthermore, in the last term, the two directions of integration we both simultaneously reversed. By drawing the first and third terms together, it follows, finally, for $\lim dx = 0, \ldots$:

$$\lim \prod = \prod_{x,y}^{\xi,\eta} + \prod_{\xi,\eta}^{x,y} = 2\prod_{\xi,\eta}^{x,y} .$$

We are now able to put the original equation given for the integral aggregate $\prod$ into the form: [224]

$$\prod = \log \frac{e^{-\Pi_{x,\xi}^{x+dx,\xi+d\xi}} \cdot e^{-\Pi_{y,\eta}^{y+dy,\eta+d\eta}}}{e^{-\Pi_{x,\eta}^{x+dx,\eta+d\eta}} \cdot e^{-\Pi_{y,\xi}^{y+dy,\xi+d\xi}}} .$$

If one now extends the fraction standing here under the logarithmic sign by $(x,dx)(\xi,d\xi)(y,dy)(\eta,d\eta)$ and goes over to the limit $dx = 0, \ldots$, then:

$$\prod_{\xi,\eta}^{x,y} = \log \frac{\Omega(x,\xi)\Omega(y,\eta)}{\Omega(x,\eta)\Omega(y,\xi)} . \tag{5}$$

In this, *representation of the integral of the third kind by the Klein primeform* (cf. formula (3) in "M" II on p. 511), the integral standing on the left is *uniquely* determined by the paths Λ and L, the primeforms on the right by $\Lambda, C, \ldots, L'$. Finally, the logarithm standing on the right is to be so taken, that it approaches the limit zero, in case x, for fixed ξ, y, η, approaches the place y along L.

In consequence of (5), the function:

$$\log \frac{\Omega(x,\xi)}{\Omega(x,\eta)}, \tag{6}$$

which we will conceive in dependency on x for fixed y, ξ, η, will have the same periods as $\prod_{\xi,\eta}^{x,y}$ with respect to closed circuits by the place x. Therefore, if we have an arbitrary closed circuit before us, which passes over the cut Λ μ times more often from left to right than from right to left[8], while it cuts the line b_k, ν_k times more often from left to right than conversely[9], then by analytic continuation over this circuit, we arrive from the initial branch (6) of our function at the following branch:

$$\overline{\log} \frac{\Omega(x,\xi)}{\Omega(x,\eta)} = \log \frac{\Omega(x,\xi)}{\Omega(x,\eta)} + 2i\pi\mu - 2i\pi \sum_{k=1}^{p} \nu_k j_k^{\xi,\eta}. \tag{7}$$

We now turn our attention to the circumstance, that for the primeform quotient $\Omega(x,\xi) : \Omega(x,\eta)$, the zeros of $\Omega(x,\xi)$ and $\Omega(x,\eta)$ situated at the branch points of the surface F_w exactly cancel. For this, to be sure, first of all, a pole of the first order situated at the place η comes up. In order to again eliminate the latter, we proceed in the following manner:

[225] We will initially give the place η a special position on the surface F_w, and for this purpose, choose one of the w places ∞. Let the w places lying over one another at $z = \infty$ (the surface is not to exhibit a branch-point here) be called $\infty_1, \infty_2, \ldots, \infty_w$, an arbitrary one of them, however, ∞_h. Also, from now on, let the place ξ be characterized as fixed by the notation; on this account, we write e instead of ξ. In addition to this, at the same time, we will draw w lines $\Lambda_1, \Lambda_2, \ldots, \Lambda_w$ from the w places ∞_h to w (cf. figure 32); and for an arbitrary prescribed closed circuit by x, let the number denoted above by μ be called, in particular, μ_h for Λ_h. Finally, we will again replace the notation of the variable place x by the original terminology z. For the circuit considered; there then arise from (7) the w equations:

$$\overline{\log} \frac{\Omega(z,e)}{\Omega(z,\infty_h)} = \log \frac{\Omega(z,e)}{\Omega(z,\infty_h)} + 2i\pi\mu_h - 2i\pi \sum_{k=1}^{p} \nu_k j_k^{e,\infty_h}. \tag{8}$$

Now the product of all w primeform quotients considered here will represent a function of z, which has a zero of order w at e and exhibits a pole of the first order at each of the w places ∞_h. If we also add the factor z_2, which has the places just named as zeros of the first order, and extract the wth root of the product in one of the w ways, then there arises, in:

[8] We thereby imagine, again, the cut Λ directed from η to ξ.

[9] According to the figures in I on p. 183ff., V_{b_k} points from the left to right bank of b_k.

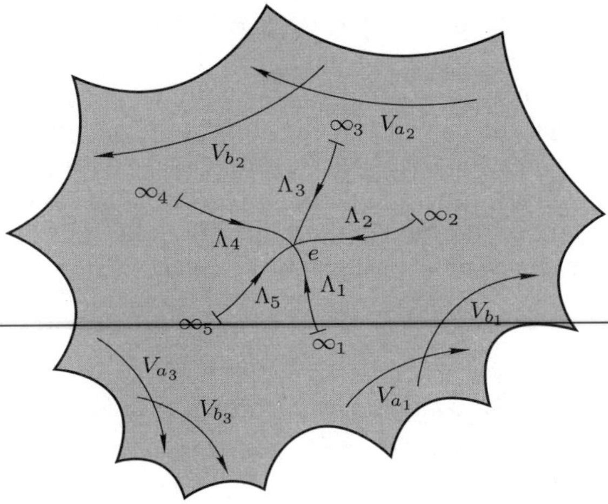

Fig. 32.

$$\Omega(z,e)\sqrt[w]{\frac{z_2}{\Omega(z,\infty_1)\Omega(z,\infty_2)\cdots\Omega(z,\infty_w)}} = P(z,e), \tag{9}$$

a *form of dimension* $\frac{1}{w}$, *which is continuous everywhere and exhibits only a single zero, of the first order, situated at the arbitrary place e of the surface.*[10] In the form $P(z,e)$, or, written in more detail, $P(z_1,z_2;e_1,e_2)$, we have now obtained the "Ritter primeform", taken into prospect before. This has the advantage over the primeform $\Omega(z,e)$, *that it together, as* Ω, *exhibits fixed zeros at the branch-points of the surface* F_w[11]. The principal advantage of $P(z,e)$ over $\Omega(z,e)$, however, we shall recognize upon consideration of the periodicity. [226]

In the latter respect, formula (8) allows us almost immediately to bring into discovery the behavior of the logarithm of the Ritter primeform, with respect to circuits by the place z. We add the w equations entering for $h = 1,2,\ldots,w$, and divide the result by w. Thereby, for abbreviation, we set:

$$\mu_1+\mu_2+\cdots+\mu_w=\sigma \tag{10}$$

$$\frac{1}{w}(j_k^{z,\infty_1}+j_k^{z,\infty_2}+\cdots+j_k^{z,\infty_w}) = W_k^z, \tag{11}$$

so that we have before us, in $W_1^z, W_2^z, \ldots, W_p^z$, a particular system of normal integrals of the first kind for our canonical cut up surface. The equation (8) then furnishes, in the manner described:

$$\overline{\log}\left(\frac{P(z,e)}{\sqrt[w]{z_2}}\right) = \log\left(\frac{P(z,e)}{\sqrt[w]{z_2}}\right) + 2i\pi\frac{\sigma}{w} - 2i\pi\sum_{k=1}^{p}\nu_k W_k^e,$$

---

[10] Naturally, e may also be situated, without further ado, at a branch-point of the surface F_w or at the place ∞. The estimation of the order 1 of the zero then refers itself, self-evidently, to the Riemann surface and not, say, to the z-plane.

[11] In this respect, therefore, the Ritter primeform shares the simplicity of the original primeform utilized by Klein (see the note on p. 222).

whereby the value of the integral W_k^e is to be calculated by utilizing the integration curves $\Lambda_1, \Lambda_2, \ldots, \Lambda_w$ in (11) on the left side.

For a circuit by the place z, we shall assume both homogeneous variables z_1, z_2 as simultaneously variable and indeed, not only their quotient, but z_1 and z_2 individually are finally to return to their initial values. We must avoid a place $z = \infty$ with our circuit, so that the point z_2 will not run through the zero in its plane. let the path of this point, without being pushed over the zero in the z_2-plane, let itself be drawn together to σ_1 simple circuits about $z_2 = 0$ in the positive sense. Then:

$$\overline{\log}\ \sqrt[w]{z_2} = \log\ \sqrt[w]{z_2} + 2i\pi\frac{\sigma_1}{w},$$

[227] and we obtain the result: *the logarithm of the Ritter primeform* $\log P(z,e)$, *upon continuation along a closed circuit, goes over into*:

$$\overline{\log} P(z,e) = \log P(z,e) + 2i\pi\frac{\sigma+\sigma_1}{w} - 2i\pi\sum_{k=1}^{P} \nu_k W_k^e, \tag{12}$$

if this circuit passes over the cuts $\Lambda_1, \Lambda_2, \ldots, \Lambda_w$, σ *times, all together, more often from the left to the right than conversely, and similarly, the cut* b_k, ν_k *times more often the left to the right than conversely, and if, finally, the closed path of* z_2 *in the* z_2*-plane lets itself be drawn together into* σ_1, *circuits in the positive sense about the zero* $z_2 = 0$.

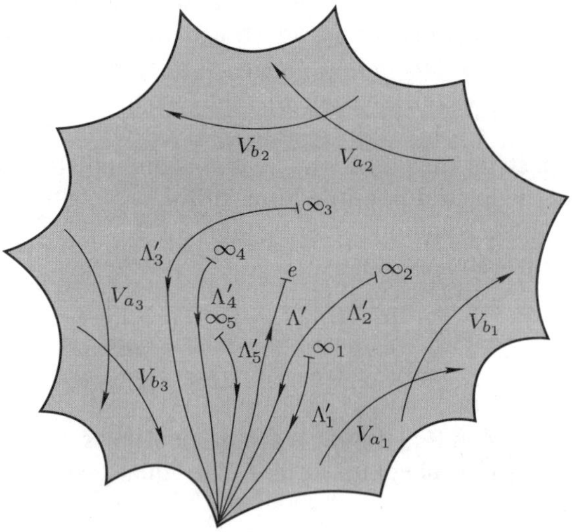

Fig. 33.

If an investigation refers itself, besides to $P(z,e)$, to yet further primeforms, then it will be expedient, to alter the lines $\Lambda_1, \ldots, \Lambda_w$ a bit in their sense. First of all, say, in the manner represented by figure 33, we carry w lines $\Lambda_1', \Lambda_2', \ldots, \Lambda_w'$ from $\infty_1, \infty_2, \ldots, \infty_w$ to a vertex of the cut system of the a_k, b_k, this vertex chosen fixed once and for all, and draw from here a single line Λ' to e, which now furnishes, together with Λ'_h, the cut Λ_h in the new form. Then there obviously holds:

$$\overline{\log} P(z,e) = \log P(z,e) + 2i\pi\frac{\sigma' + \sigma_1}{w} + 2i\pi\sigma'' - 2i\pi\sum_{k=1}^{p} \nu_k W_k^e, \tag{13}$$

if the circuit passes over the cuts $\Lambda'_1, \Lambda'_2, \ldots, \Lambda'_w$ σ' times, all together, more often from the left to right than conversely, and similarly, the cut Λ' σ'' times more often from left to right than conversely.

Notice, that with the behavior of $\log P(z,e)$ with respect to circuits by z, we not only precisely know the corresponding behavior of the primeform $P(z,e)$, but that of *each arbitrary roots of $P(z,e)$*. In this respect, therefore, the result of our present investiga- [228]
tion goes considerably beyond the representations in "M" II on p. 506. The decisive step forward by the Ritter primeform, however, is grounded in the circumstance, *that the multiplicator, which this primeform assumes for any circuit by z, is independent of z itself,* while the multiplicator of the Klein primeform represents a function of z (cf. "M" II on p. 507). The advantage of the Ritter primeform given by this will, in fact, obtain a fundamental significance further on.

4.3 The polymorphic forms ζ_1, ζ_2 for a model of arbitrary genus p

Let an arbitrarily proposed automorphic model be carried over by a function $z = \varphi(\zeta)$ contained in the model onto the associated Riemann surface F_w, on which $\zeta = f(z)$ is then a linearly-polymorphic function. In order to obtain the *transition to the polymorphic forms,* we first form, as in (1) on p. 114, the two "functions":

$$Z_1 = \frac{\zeta}{\sqrt{\frac{d\zeta}{dz}}}, \qquad Z_2 = \frac{1}{\sqrt{\frac{d\zeta}{dz}}}, \tag{1}$$

whereby we imagine the square root in the denominator chosen in one of the two ways for an initial value z. As was already emphasized on p. 114, the Z_1, Z_2, whose quotient is ζ, substitute themselves binarily and unimodularly for circuits by z on F_w; *in this fashion, they yield the homogeneous group Γ of our automorphic model in the unimodular form*:

Naturally, the Z_1, Z_2 are not yet suited to serve as arguments of the automorphic forms. As we know here from the genus $p = 0$, they do not yet satisfy the requirement, of always representing only "allowed value pairs" for arbitrary paths of z. However, self-evidently, we shall hold fast here to this requirement, *that the arguments ζ_1, ζ_2 of the automorphic forms are to remain steadily finite and never vanish simultaneously.*

In order to be able to easily describe the behavior of Z_1, Z_2 on the Riemann surface, it is expedient to take up again the language of "M" II on p. 493 concerning *the differentials on the surface F_w*. If the neighborhood of a plane of the surface is mapped by the associated function t onto the simply and completely covered neighborhood of a point t_0 of the t-plane situated in the finite part of the plane, then the differential dt of [229]
this function at the place in question is called finite and non-vanishing. A second differential dt' is similarly finite and different from zero at this place or at a zero or pole

of the order m, in case $\frac{dt'}{dt}$ is finite and non-vanishing resp. exhibits a zero or pole of this order.

The differential dz, according to "M" II on p. 494, has a pole of the second order at each place ∞ of F_w and a zero of the order $(\kappa - 1)$ at the individual branch-point, in case κ sheets hang together there. From this, common poles and zeros of the Z_1, Z_2 originate, which by no means adhere to the automorphic models as such, but are only introduced by the particular function $z = \varphi(\zeta)$ chosen here. In order, first of all, to remove these poles and zeros, we return to the form-theoretic preparations of the previous paragraph, and are able, by means of these, to easily fashion the differential dz into an "*everywhere finite and non-vanishing differential*"[12] (cf. "M" II on p. 494). We get:

$$d\omega_z = \frac{(z, dz)}{P(z, v_1)^{\kappa_1 - 1} P(z, v_2)^{\kappa_2 - 1} \ldots}, \tag{2}$$

where the symbol (z, dz) is used in the usual manner as abbreviation for $z_1 dz_2 - z_2 dz_1 = -z_2^2 dz$, and where the factors of the denominator refer themselves to be various branch points of F_w.

We now determine, at the same time, the dimension of this homogeneous, everywhere finite differential $d\omega_z$. *Hereby, from now on, we will relate all statements concerning dimension on the surface F_w, not to z_1, z_2, but to the primeform.* If the latter is of dimension w^{-1} in the z_1, z_2, then the z_1, z_2 become forms of dimension w from now on, since one immediately reads off from (9) on p. 225 the representation:

$$z_2 = p(z, \infty_1) p(z, \infty_2) \cdots p(z, \infty_w).$$

If we take heed of the known relation between, p, w and the numbers $\kappa_1, \kappa_2, \ldots$ of the sheets hanging together at the branch-points (cf. "M" I on p. 494), it follows: *the everywhere finite differential $d\omega_z$ is of the dimension*:

$$2w - \sum(\kappa - 1) = 2 - 2p. \tag{3}$$

[230] For $p = 1$, we are led back to the differential of the first kind, and in case $p = 0$, which is also included here, one simply has $d\omega_z = (z, dz)$, as long as one understands by z a principal function.

We now turn ourselves back to the splitting of ζ into the quotient of two polymorphic forms ζ_1, ζ_2. The differential $d\zeta$, at the individual place e_k, which originates from a fixed polygon vertex, has a pole of the order $(1 - \frac{1}{l_k})$, as follows from the calculations on p. 44. Should the point ∞ of the ζ-plane find an image on F_w, the $d\zeta$ has a pole of second order at this place. Except for this, however, $d\zeta$ is everywhere finite and different from zero.

These considerations now lead us to the desired splitting of ζ. *In fact, in*:

$$\begin{cases} \zeta_1 = \dfrac{\zeta}{\sqrt{\frac{d\zeta}{d\omega_z}}} \displaystyle\prod_{k=1}^{n} P(z, e_k)^{-\frac{1}{2}(1 - \frac{1}{l_k})}, \\ \zeta_2 = \dfrac{1}{\sqrt{\frac{d\zeta}{d\omega_z}}} \displaystyle\prod_{k=1}^{n} P(z, e_k)^{-\frac{1}{2}(1 - \frac{1}{l_k})}, \end{cases} \tag{4}$$

[12] Klein, in his investigation cited on p. 222, places the everywhere finite differential at the forefront, by basing its construction on the normal curve of the φ.

we obtain two forms of dimension:

$$1 - p - \frac{1}{2}\sum_{k=1}^{n}\left(1 - \frac{1}{l_k}\right) = \nu^{-1}, \tag{5}$$

whose quotient is equal to ζ, *and which, for allowed value pairs of the* z_1, z_2, *invariably again represent allowed value pairs themselves.* With this, we have produced a pair of polymorphic forms on our surface F_w, which, in the lowest spacial case $p = 0$, go over directly into forms (2) on p. 110, if one takes the constant factor C in these latter formulas identical with the imaginary unit $-i$.

We also make note, at the same time, of the relation of these polymorphic forms ζ_1, ζ_2 to the Z_1, Z_2 defined in (1):

$$\frac{\zeta_1}{Z_1} = \frac{\zeta_2}{Z_2} = i z_2 \prod_h P(z, v_h)^{-\frac{1}{2}(\kappa_h - 1)} \cdot \prod_k P(z, e_k)^{-\frac{1}{2}\left(1 - \frac{1}{l_k}\right)}. \tag{6}$$

The expression standing here on the right, with respect to circuits of z resp. z_1, z_2, reproduces itself up to a *constant* factor, i.e., independent of z_1, z_2; we say, with Ritter, this expression represents a "*multiplicative form*" or it "behaves itself". "multiplicatively" with respect to circuits of z. The consequence is, *that the polymorphic forms* [231]
ζ_1, ζ_2, *for these circuits, substitute themselves binarily, but no longer necessarily unimodularly.* The group of the substitutions of ζ_1, ζ_2 arising will accordingly be mapped onto our homogeneous Γ *homomorphically, but in general no longer isomorphically.* As we shall yet see, this will not be a hindrance in the utilization of the ζ_1, ζ_2 as arguments of the automorphic forms.

We will, at the same time, also produce the *generators of this group of the* ζ_1, ζ_2-*substitutions*; we call these generators, corresponding to the U_k, U_{a_g}, U_{b_g} of the unimodular group Γ, $\boldsymbol{U}_k, \boldsymbol{U}_{a_g}, \boldsymbol{U}_{b_g}$. First, the substitution $\boldsymbol{U}_k$ corresponds to simple circuit of the place e_k, which corresponds to the polygon vertex ε_k. We come back exactly to the relations discussed on p. 112 ff. for $p = 0$. Namely, if one develops (Z, e_k) and (Z, e'_k), on the basis of (1) and upon this, (ζ, ε_k) and (ζ, ε'_k) by means of (6), in the neighborhood of the place e_k, then one returns, in the case of an elliptic vertex, to the representation (7) on p. 112, but for a parabolic vertex to (9) on p. 113. Consequently, for $\boldsymbol{U}_k$, we have, in the elliptic case:

$$(\zeta', \varepsilon_k) = e^{\frac{2i\pi}{l_k}}(\zeta, \varepsilon_k), \qquad (\zeta', \varepsilon'_k) = (\zeta, \varepsilon'_k), \tag{7}$$

in the parabolic one, however:

$$(\zeta', \varepsilon_k) = (\zeta, \varepsilon_k), \qquad (\zeta', \varepsilon'_k) = 2i\pi(\zeta, \varepsilon_k) + (\zeta, \varepsilon'_k). \tag{8}$$

Further, let the U_{a_g}, U_{b_g} be *those unimodular substitutions, which the* Z_1, Z_2 *experience with respect to the period denoted in the figures on p. 220ff. by* V_{a_g}, V_{b_g}[13]. The $\boldsymbol{U}_{a_g}, \boldsymbol{U}_{b_g}$ then simply deviate from the $\boldsymbol{U}_{a_g}, \boldsymbol{U}_{b_g}$ by those constant factors, which the multiplicative form on the right side of (6) assumes for the circuits in question. With the help of the rules of the previous paragraph, we find, on the strength of this, *as the form of the last* $2p$ *generators* $\boldsymbol{U}_{a_g}, \boldsymbol{U}_{b_g}$:

[13] Hereby, the final determination of the unimodular U_{a_g}, U_{b_g}, still left undone above (p. 214), is disposed of.

$$\begin{cases} U_{a_g} = U_{a_g} \\ U_{b_g} = U_{b_g} \cdot e^{-i\pi[\sum_h (\kappa_h - 1) W_g^{v_h} + \sum_k (1 - \frac{1}{l_k}) W_g^{e_k}]}. \end{cases} \tag{9}$$

If z describes a closed single circuit on the surface F_w about a branch-point, then
[232] ζ_1, ζ_2 will suffer a sign change or remain unaltered, according as the associated number of sheets κ is even or odd. With respect to a circuit about an individual place ∞, the ζ_1, ζ_2 assume the factor $e^{\frac{2i\pi}{wv}}$. Notice hereby, that $(wv)^{-1}$ is the dimension of the ζ_1, ζ_2 in the z_1, z_2.

For all these circuits, we imagine z_2 constant or permit the point z_2 only such circuits in the z_2-plane, which let themselves, without pushing then over the origin, be drawn together into a point different from this origin. If there are circuits about $z_2 = 0$, then one will easily determine the multiplicators of the ζ_1, ζ_2 originating hereby from the form of the right side of (6) on the basis of the rule (13) on p. 227.

We now treat here, finally, the question, whether the polymorphic forms ζ_1, ζ_2 can be considered as uniquely determined from their essential properties or not. As essential properties, we are hereby to consider, *that* $\zeta = \zeta_1 : \zeta_2$ *for allowed value pairs of* z_1, z_2, *always again correspond only allowed value pairs of the forms* ζ_1, ζ_2, *and that the* ζ_1, ζ_2, *with respect to a closed circuit, substitute themselves in the form*:

$$\zeta_1' = \tau(\alpha\zeta_1 + \beta\zeta_2), \qquad \zeta_2' = \tau(\gamma\zeta_1 + \delta\zeta_2), \tag{10}$$

where τ *is a constant factor (to be determined separately for each circuit), and* $\alpha, \beta, \gamma, \delta$ *are the coefficients of the corresponding unimodular substitution furnished by the* Z_1, Z_2.

Suppose the forms $\overline{\zeta_1}, \overline{\zeta_2}$, have the properties of the ζ_1, ζ_2, in that, in particular, with respect to the circuit considered, experience the substitution:

$$\overline{\zeta_1'} = \overline{\tau}(\alpha\overline{\zeta_1} + \beta\overline{\zeta_2}), \qquad \overline{\zeta_2'} = \overline{\tau}(\gamma\overline{\zeta_1} + \delta\overline{\zeta_2}).$$

Then denote the quotient $\frac{\overline{\zeta_1}}{\zeta_1}$, which is identical with $\frac{\overline{\zeta_2}}{\zeta_2}$, as a form on the Riemann surface F_w, by $f(z_1, z_2)$. This form $f(z_1, z_2)$ is everywhere finite and different from zero on the surface, and it behaves itself multiplicatively with respect to closed circuits. To the circuit just considered corresponds, in particular, obviously, the multiplicator $\frac{\overline{\tau}}{\tau}$ of the form $f(z_1, z_2)$.

The following important theorem, yet to be applied often further on, now holds: *A multiplicative form* $F(z_1, z_2)$, *which is everywhere free of poles and zeros on the surface* F_w, *and which, as is always assumed here, never has essential singular points on* F_w, *is*
[233] *the exponential of an integral of the first kind.* Namely, if we first assume, that $F(z_1, z_2)$ has a dimension different from 0, then:

$$[F(z_1, z_2)]^{-\frac{1}{d}} \cdot P(z, e),$$

with arbitrarily chosen place e, represents a multiplicative "function" on F_w, which is pole-free and exhibits only one zero of first order, situated at e. The logarithm of this function, with respect to circuits by z, behaves itself "additively", i.e. it represents *an integral of the surface* F_w, and indeed, this would have *one* logarithmic discontinuity point situated at e. But there is no such integral; and therefore, the assumption, that the dimension d of $F(z_1, z_2)$ is different from 0, is inadmissible. But if this form is of the dimension 0, often one directly recognizes in its logarithm an integral of the first kind, with which our assertion in proved.

As a system of everywhere finite integrals we now suitably utilize the normal integrals $W_1^z, W_2^z, \ldots, W_p^z$, which otherwise came into application for the primeform. As a result, then it follows: *the most general manner of splitting of the polymorphic function ζ into two polymorphic forms ζ_1, ζ_2, which furnishes a substitution group homomorphic with the unimodular Γ, and which, for allowed value pairs z_1, z_2, itself invariably assumes only allowed value pairs, is furnished by:*

$$\begin{cases} \zeta_1 = e^{c_0+c_1W_1^z+\cdots c_pW_p^z} \cdot \dfrac{\zeta}{\sqrt{\frac{d\zeta}{d\omega_z}}} \prod\limits_{k=1}^{n} P(z,e_k)^{\frac{1}{2}(1-\frac{1}{l_k})}, \\ \zeta_2 = e^{c_0+c_1W_1^z+\cdots c_pW_p^z} \cdot \dfrac{1}{\sqrt{\frac{d\zeta}{d\omega_z}}} \prod\limits_{k=1}^{n} P(z,e_k)^{-\frac{1}{2}(1-\frac{1}{l_k})}, \end{cases} \tag{11}$$

whereby the $c_0, c_1, \ldots, c_p$ are arbitrarily choosable constants.

In the case $p=0$, where the exponential factor coming up in (11) on the right reduces itself to a constant we refer to our earlier formulas (2) on p. 110.

4.4 Differential equations of the polymorphic functions and forms for models with $p>0$

For the individual one of our polymorphic functions $\zeta(z)$ we obtained, in the first chapter (p. 48ff.), a *differential equation of the third order*. Namely, it revealed itself, that the Schwarzian differential expression $[\zeta]_z$, defined in (1) on p. 47, becomes an associated algebraic function:

$$[\zeta]_z = 2R(s,z).^{14} \tag{1}$$

on the Riemann surface, for which we again take up the notation F_w utilized on p. 219 [234]
ff.

We have already enumerated on p. 49 all the poles of this function $R(s,z)$, as well as a few zeros. To be sure, we have only first pursued the construction of this function more precisely l.c. for $p=0$. We arrived here at the differential equation(3) on p. 50, on whose right side an integral function $G_{n-4}(z)$ of degree $(n-4)$ still remained undetermined. The $(n-3)$ temporarily unknown coefficients of this function we designated as the "*accessory parameters*" of the differential equation.

Next, it will not be hard at all, to also enumerate, for arbitrary p, the *number of the accessory parameters*. For this purpose, we must only return somewhat more precisely to the statements on p. 49 ff. We choose the terminology suitably, in agreement with our investigation just carried through in §3.

The poles of our algebraic function $[\zeta]_z$ lie exclusively at the n places $e_1, e_2, \ldots, e_n$ and at the branch-points v of the surface F_w. There holds, for the individual place e, according to (6) on p. 49:

---

[14] In order to avoid confusions with the number w of sheets of the surface F_w, we have here called the algebraic function, denoted on p. 49 by w, s.

$$[\zeta]_z = \frac{1}{2}\left(1 - \frac{1}{l^2}\right)(z-e)^{-2}[1 + a_1(z-e) + a_2(z-e)^2 + \cdots]; \tag{2}$$

the developments arising from (5) on p. 48:

$$[\zeta]_z = \frac{\kappa^2 - 1}{2\kappa^2}(\sqrt[\kappa]{z-v})^{-2\kappa}[1 + * + b_2(\sqrt[\kappa]{z-v})^2 + b_3(\sqrt[\kappa]{z-v})^3 + \cdots] \tag{3}$$

refers to the neighborhood of the individual branch-point v, at which κ sheets hang together, in which series, it is to be emphasized by the symbol $*$, that the term with the first power of the development quantity $\sqrt[\kappa]{z-v}$ drops out. We take note: the number of the poles, and, with this, the "volume" of our algebraic function $[\zeta]_z = 2R(s,z)$ is :

$$2n + 2\sum_h \kappa_h, \tag{4}$$

whereby the sum refers to the branch-points of the w-sheeted surface F_w.

[235] Between the number w of sheets, the genus p and the multiplicities κ of the branch-points, there now (cf. "M" I on p. 494) subsists the relation:

$$-2w + \sum_h (\kappa_h - 1) = 2p - 2. \tag{5}$$

the valence of $[\zeta]_z$ given in (4) is accordingly always $> 2p-2$; and therefore, the Riemann-Roch theorem (cf. "M" I on p. 549) shows that the most general algebraic function of F_w, which has the same poles as $[\zeta]_z$, still contains:

$$2n - p + 1 + 2\sum_h \kappa_h \tag{6}$$

arbitrary constants linearly and homogeneously.[15]

For our particular function $[\zeta]_z$, however, this number reduces itself first on the basis of the developments (2) and (3), and on the other hand, since a zero of the fourth order lies at the infinitely distant place of each of w sheets (cf. p. 49). In (2), the initial coefficient is determined, in (3) even the first two coefficients, so that for each place e, one unit, for each branchpoint, two units are to be subtracted from the number (6). If, at the same time, corresponding to the w zeros of the fourth order at $z = \infty$, we subtract $4w$ from the number (6), then there remains:

$$n - p + 1 - 4w + 2\sum_h (\kappa_h - 1),$$

a number which, with the help of relations (5), one immediately clothes in the form $(n + 3p - 3)$. Therefore, the result arises: *the number of the accessory parameters in the differential equation (1), i.e. those parameters, which are not yet determined by the poles and the zeros of $R(s,z)$ situated at $z = \infty$ is, for arbitrary p*:

$$n + 3p - 3. \tag{7}$$

---

[15] For the case $p = 1$, which was initially excluded in the formulation of the Riemann-Roch in "M" I on p. 549, this statements remains valid directly.

For $p = 0$, we come back to the number $(n-3)$ of the coefficients of the function $G_{n-4}(z)$.

Since we now go more precisely into the form of the function $R(s,z)$, we also put ourselves in connection at the same time, with the developments on p. 117 ff. Concerning the *linear differential equations of second order for polymorphic forms*. In particular, for the forms $f(z_1,z_2)$ of the linear family: [236]

$$f(z_1,z_2) = A\zeta_1 + B\zeta_2 \tag{8}$$

we have obtained, in(12) on p. 125, an *invariant* form of the differential equation, which takes into account the circumstance, that any linearly independent combinations az_1+bz_2, cz_1+dz_2 are able to enter in place of z_1,z_2, as arguments of our polymorphic forms for $p=0$.

The investigation[16] carried out for $p>0$, in corresponding fashion, brings into force the invariant character, which is present in the individual case, as well for the differential equations (1) of the polymorphic functions, as for the linear differential equations of the polymorphic forms. The following cases come into consideration first and foremost:

1. elliptic models of the character $(1,n)$,
2. hyperelliptic models of the character $(p,0)$
3. non-hyperelliptic models of the character $(3,0)$.

In the first two cases, one will choose z as a *two-valued* function, so that we obtain a Riemann surface F_2 with $(2p+2)$ branchpoints. The method of construction of the differential equations must then again be invariant with respect to the transformation:

$$z_1' = az_1 + bz_2, \qquad z_2' = cz_1 + dz_2; \tag{9}$$

for, besides $z = z_1 : z_2$, $z' = z_1' : z_2'$ is also always a two-valued function on the same footing as z.

In the two cases 1 and 2, we accordingly find ourselves in the domain of *binary* invariant theory. In this regard, we make note, at the same time, *that* $z_2^{-4}[\zeta]_z$ *is an invariant form of* $(-4)^{\text{th}}$ *dimension of the* z_1,z_2; in fact, one very easily obtains, from (3) on p. 48:

$$z_2^{-4}[\zeta]_z = (ad-bc)^2 z_2'^{-4}[\zeta]_{z'}. \tag{10}$$

Further, as on p. 125, we shall again build up the differential equations of second order from *transvectants*.

In the case 3, one will attach to the "*normal curve of the* φ" (cf. "M" I on p. 569), which, for $p=3$, becomes a singularity-free plane curve of the fourth order. Here, therefore, one lands in the domain of *ternary* invariant theory.

For an elliptic model of the character $(1,n)$, we denote the form of fourth dimen- [237]

---

[16] Besides the papers cited in detail on p. 126, there yet comes into consideration the lectures of Klein "*Über lineare Differentialgleichungen zweiter Ordnung*" of the summer semester of 1894, as well as the paper of G. Pick "*Zur Theorie der zu einem algebraischen Gebilde gehörenden Formen*", Mathem. Ann., Vol. 50 (1897). Also see the note of Klein "*Autographierte Vorlesungshefte II*", Mathem. Ann. Vol. 46 (p. 77 ff.), 1894.

sion, whose zeros are the four branch-points[17], by $\varphi_4(z_1, z_2)$:

$$\varphi_4(z_1, z_2) = \prod_{h=1}^{4} (z_1 v_h^{(2)} - z_2 v_h^{(1)}). \tag{11}$$

Then we have, in the product:

$$z_2^{-4}[\zeta]_z \cdot \varphi_4(z_1, z_2). \tag{12}$$

An algebraic *function* of the model, which, at $z = \infty$, is finite and different from zero in both sheets, and which, at the individual branch-points, exhibits only one more pole, of second order. We now understand, by:

$$H_4(z_1, z_2) = \frac{\partial^2 \varphi}{\partial z_1^2} \cdot \frac{\partial^2 \varphi}{\partial z_2^2} - \left(\frac{\partial^2 \varphi}{\partial z_1 \partial z_2}\right)^2. \tag{13}$$

The Hessian form belonging to the form φ, which is similarly of the fourth dimension in the z_1, z_2. The calculation, which one most expediently attaches to one of the normal forms of φ_4 (cf. "M" I on p. 13ff.), then shows, that the "function" $-\frac{H}{24\varphi}$ has poles exclusively at the branch-points and indeed, poles of the same coefficients as the function represented in (12) [see formula (3) above]. The consequence is, that:

$$z_2^{-4}[\zeta]_z \varphi + \frac{1}{24} \frac{H}{\varphi} = Q(s, z)$$

is an algebraic function of our model, which has poles only at the n places e, of the second order in each case. If one applies to the representation of this function the known theorems concerning the expression of the algebraic functions of an elliptic model in the integral of the first kind, and hereby brings in the initial coefficients of the developments (2), then it reveals itself, that exactly n coefficients remain unknown in the expression of $Q(s, z)$.[18] *With this, we have, in fact, obtained the n accessory parameters; and the differential equation of third order obtains the invariant form*:

$$z_2^{-4}[\zeta]_z = -\frac{1}{24} \frac{H}{\varphi^2} + \frac{Q(s, z)}{\varphi}. \tag{14}$$

[238] We arrive at entirely analogous relations in the case of a hyperelliptic model of the character $(p, 0)$. To the $(2p+2)$ branch-points, there here corresponds the form:

$$\varphi_{2p+2}(z_1, z_2) = \prod_{h=1}^{2p+2} (z_1 v_h^{(2)} - z_2 v_h^{(1)}), \tag{15}$$

of the dimension $(2p+2)$ in the z_1, z_2. Now we have, in $z_2^{-4}[\zeta]_z \varphi_{2p+2}$, a form of the dimension $(2p-2)$ which is only still polarly discontinuous at the branch-points and

[17] For the sake of simplicity, we hold fast to the argument made on p. 219, that one of the branch-points is situated at $z = \infty$.

[18] The closer discussion of this development one finds in the lectures of Klein on p. 67, cited on p. 236.

indeed, of the order 2 at each. The Hessian form H_{4p}, belonging to φ_{2p+2}, has the dimension $4p$ in the z_1, z_2, so that the quotient $\frac{H}{\varphi}$, which similarly belongs to the dimension $(2p-2)$ in the z_1, z_2, again becomes polarly discontinuous at the same places and in the same order as $z_2^{-4}[\zeta]_z\varphi_{2p+2}$. The more detailed calculation thereby shows, that

$$z_2^{-4}[\zeta]_z\varphi + \frac{3}{8(2p+1)^2}\frac{H}{\varphi}$$

remains finite at the branch-points and consequently represents an *integral* algebraic form of our model. The latter is representable in the form:

$$\psi_{2p-2}(z_1,z_2) + \chi_{p-3}(z_1,z_2)\sqrt{\varphi_{2p+2}(z_1,z_2)}$$

where ψ and χ are integral rational forms of the dimensions $(2p-2)$ and $(p-3)$ in the z_1, z_2. *The* $2p-1+p-2=3p-3$ *coefficients of these forms are the accessory parameters coming into consideration here; the differential equation then obtains the form:*

$$z_2^{-4}[\zeta]_z = -\frac{3}{8(2p+1)^2}\frac{H}{\varphi^2} + \frac{\psi+\chi\sqrt{\varphi}}{\varphi}. \tag{16}$$

The transition to the *differential equation of the second order of the polymorphic forms* we will sketch more closely, say, for the case of the hyperelliptic model $(p,0)$ just considered. If, first of all:

$$F(z_1,z_2) = AZ_1 + BZ_2$$

is an arbitrary form the linear family to be formed by means of the two "forms of zero dimension" (1) on p. 228, then according to p. 118ff., we have:

$$\frac{d^2F}{dz^2} + R(s,z)F = 0, \tag{17}$$

where $R(s,z)$ is the function standing on the right in (1), i.e., in the present case, the right side of equation (16) multiplied by $\frac{1}{2}z_2^4$. The polymorphic forms preferred by us above, however, are not the Z_1, Z_2, but the ζ_1, ζ_2, which in the case in question, according to (4) on p. 230, have the form:

$$\zeta_1 = \zeta\sqrt{\frac{d\omega_z}{d\zeta}}, \qquad \zeta_2 = \sqrt{\frac{d\omega_z}{d\zeta}}.$$

Now, in the hyperelliptic case, we are also able to give the everywhere finite and nowhere vanishing differential $d\omega_z$ the algebraic form of expression: [239]

$$d\omega_z = \frac{(z,dz)}{\sqrt{\varphi_{2p+2}(z_1,z_2)}}, \tag{18}$$

so that the polymorphic forms ζ_1, ζ_2 present the dimension $\frac{1-p}{2}$ in the z_1, z_2. The forms $f(z_1,z_2)$ of family (8) are accordingly bound together, with the forms F of zero dimension just mentioned, simply by the relation:

$$F = z_2^{-1} \sqrt[4]{\varphi_{2p+2}} \cdot f. \tag{19}$$

But if we transform the differential equation (17) to f by means of this relation, the intermediate calculation, which we suppress here, leads to the following result: *the invariant form of the differential equation of second order of the polymorphic form* $f(z_1, z_2)$ *of the dimension* $\frac{1-p}{2}$ *(in the* z_1, z_2*), in the case of a hyperelliptic model* $(p, 0)$*, is the following*:

$$(f, \varphi)_2 = (\psi_{2p-2} + \chi_{p-3}\sqrt{\varphi}) \cdot f, \tag{20}$$

where, on the left, the second transvectant of the forms f *and* φ *is meant.*

To the equation (20), we yet attach the following parenthetical remark. The hyperelliptic model permits a single-valued transformation of period 2 into itself, which effects a sign change of the function $\sqrt{\varphi}$, for unaltered z. In many cases, this transformation represent itself as a *linear substitution* of period 2 in the polymorphic function ζ, whose addition to be underlying group Γ of the character $(p, 0)$ generates a group Γ' of the signature.

$$(0, 2p+2; 2, 2, \ldots, 2).$$ [19]

If this is the case, then in (20), the form χ_{p-3} vanishes identically, and the differential equation (20) reduces itself to:

$$(f, \varphi)_2 = f \cdot \psi_{2p-2}.$$

This is none other, than the differential equation, set up in general in (12) on p. 125, which belongs to the present group of genus zero and the signature $(0, 2p+2; 2, 2, \ldots, 2)$. In fact, one also easily shows directly from p. 123, that the form denoted l.c. by $v(z_1, z_2)$ vanishes identically for the signature under discussion here.

[240] A non-hyperelliptic model of character (3, 0) was named above (p. 236) in the third place; for the algebraic treatment, we would take the normal curve of fourth order as a point of departure, which let be given by $g_4(z_1, z_2, z_3) = 0$. In order, next to obtain a tie-in with the binary invariant theory, one can project the curve just named onto the axis $z_3 = 0$, where z_1, z_2 are then the binary variables. Klein takes up this standpoint in his lectures cited on p. 236, where one finds the differential equation of third order set up on p. 97, that of the second order for the forms denoted here by $f(z_1, z_2)$ on p. 100, which, moreover, have the dimension $-\frac{1}{2}$ in the z_1, z_2, z_3[20]. But Gordan[21] has then subjected this latter differential equation to a transformation, which treats the ternary variables z_1, z_2, z_3 uniformly. If we denote the underlying normal curve of fourth order symbolically by:

---

[19] Here it must remain to be seen, for which groups Γ the relations described in the text occur.

[20] Just notice, that the everywhere finite and never vanishing differential has the dimension -1 in the z_1, z_2, z_3 here, insofar as it can be put into the form:

$$d\omega_z = \frac{(z, dz)}{\left(\frac{\partial g_4}{\partial z_3}\right)},$$

understanding by $g_4(z_1, z_2, z_3) = 0$, as int the text, the equation of the normal curve of the fourth order.

[21] "*Über unverzweigte lineare Differentialgleichungen zweiter Ordnung auf ebenen Curven vierter Ordnung*" Mathem. Annalen, Vol. 46, p. 606 (1895).

$$g_4(z_1, z_2, z_3) = a_z^4 = b_z^4 = 0,$$

Then, in the symbols of ternary invariant theory, the Gordan form of the differential equation of second order for the polymorphic forms of for a model (3,0) *is*:

$$a_z^2 b_z^2 (abf)^2 + \psi_2(z_1, z_2, z_3) \cdot f = 0; \tag{21}$$

hereby, ψ_2 *signifies a ternary integral form of second degree, remaining unknown, whose six coefficients are the six accessory parameters entering in this case.*[22]

4.5 Representation of all unbranched automorphic forms of a group Γ of arbitrary genus by the prime-and groundforms

[241]

In §3 of the second chapter (p. 66 ff.), the concept of the *unbranched automorphic forms* $\varphi_d(\zeta_1,\zeta_2)$ of arbitrary integral or fractional, positive or negative dimensions d was determined for models of arbitrary genus and developed in more detail. Such a form, upon exercise of a substitution of the unimodular homogeneous group, assumed a constant factor μ. If it is a question of a *single-valued* form, the multiplicator μ is unambiguously determined by the substitution exercised; otherwise, one still has to state, along which path the original arguments ζ_1,ζ_2 are to be carried over into the equivalent ones ζ_1',ζ_2'.

The arguments ζ_1,ζ_2 of the automorphic forms $\varphi_d(\zeta_1,\zeta_2)$, now as before, will be able to represent only "allowed value pairs" (cf. p. 64). But, in addition to this, we will proceed a bit more generally here and not restrict ourselves only to the use of unimodular substitutions. *If the substitution being exercised deviates from the unimodular one by factor* τ, *as in (10) on p. 232, then the associated multiplicator of the form will be given by*:

$$\mu' = \mu \cdot \tau^d, \tag{1}$$

understanding by μ *the multiplicator of the form* $\varphi_d(\zeta_1,\zeta_2)$ *belonging to the unimodular substitution.* If d is integral (and this is always the case for the *single-valued* forms, solely coming into consideration later), then the multiplicators μ and μ' are directly and uniquely calculable from one another.

This consideration makes possible for us the *use of the forms* ζ_1,ζ_2, *produced in the previous paragraph, as arguments of the automorphic forms.* And by then going back to the representations (4) on p. 230 of the ζ_1,ζ_2, *the individual unbranched automorphic forms* $\varphi_d(\zeta_1,\zeta_2)$, *by mediation of these* ζ_1,ζ_2, *represents a form on the Riemann surface* F_w, *which is free of essential singular places, and which behaves itself multiplicatively with respect to closed circuits on the surface.*

This conception is in agreement with our method followed from the beginning, of estimating the orders of the zeros and poles of an automorphic form on the fundamental domain, or, as we are also able to say, on the Riemann surface F_w. We remember, that at the n places e_k, which originate from the fixed polygon vertices, poles or zeros [242]

[22] Pick, in the paper cited on p. 236, has developed a series of schemes concerning the general form of the differential equations of second order for any algebraic models, which schemes are then brought to a more detailed discussion for binaries given models.

of *fractional* orders could come up (cf. p. 69). If the place e_k is furnished by an *elliptic* vertex, then, for the individual form $\varphi_d(\zeta_1,\zeta_2)$, we have to assume a zero of order $\frac{m_k}{l_k}$ here, where m_k is a whole number from the internal $0 \le m_k < l_k$, and l_k signifies the period of the associated elliptic generator. But if e_k corresponds to a parabolic vertex, then we imagine l'_k chosen arbitrarily, and assume a zero for $\varphi_d(\zeta_1,\zeta_2)$ of the order $\frac{m'_k}{l'_k}$ at e_k, where again, $0 \le m'_k < l'_k$ holds. In both cases, in addition to the zeros determined in such a manner, any pole or zero of *integral* order can then occur too.

We thus discover: *the unbranched automorphic forms* $\varphi_d(\zeta_1,\zeta_2)$ *furnish multiplicative forms on the Riemann surface* F_w*, which are branched at the n places* $e_1, e_2, \ldots, e_n$ *in the manner just described, but except for this, remain unbranched on the entire surface.* Thereby, the dimension (measured in the primeform) of the multiplicative form furnished by $\varphi_d(\zeta_1,\zeta_2)$ becomes $d\cdot\nu^{-1}$, where ν is defined by (5) on p. 230.

It is now all important, that the converse of the theorem thus stated is also obviously valid without further ado: *any multiplicative form, of the surface* F_w*, branched as described, furnishes, interpreted in dependency on* ζ_1,ζ_2*, an unbranched automorphic form of our group* Γ*, whose dimension d in the* ζ_1,ζ_2 *arises from the dimension of that multiplicative form measured in the primeform by attaching the factor* ν.

On the basis of those theorems, we are in the position, of *expressing* the unbranched automorphic forms of the proposed group Γ *in the most general fashion by means of the primeform* and with this, to survey, at the same time, the totality of these automorphic forms. Suppose, for a form $\varphi_d(\zeta_1,\zeta_2)$, the numbers m_k resp. m'_k, l'_k to be fixed, and besides this, let the form have t simple zeros at the places $x_1, x_2, \ldots, x_t$ of the surface and s simple poles at $y_1, y_2, \ldots, y_s$. Naturally, the simple zeros and similarly, the poles too, may unite themselves by coincidence into zeros resp. poles of higher order. We are then able to produce a multiplicative form from the primeform, which agrees with $\varphi_d(\zeta_1,\zeta_2)$ in respect of the poles and zeros in a such a manner, that the quotient of the two forms is an everywhere finite and non-vanishing multiplicative form on F_w. Such a form, however, according to the fundamental theorem proved on p. 233, represents the
[243] exponential function of an everywhere finite integral. There thus arises the theorem: *Each unbranched automorphic form* $\varphi_d(\zeta_1,\zeta_2)$ *permits a representation of the following kind:*

$$\varphi_d(\zeta_1,\zeta_2) = e^{c_0 + c_1 W_1^z + \cdots + c_p W_p^z} \frac{P(z,x_1)\cdots P(z,x_t)}{P(z,y_1)\cdots P(z,y_s)} \prod_{k=1}^{n}{}' P(z,e_k)^{\frac{m_k}{l_k}}, \tag{2}$$

where, by the upper prime on the product sign is to be remembered, that in the case of a parabolic place e_k*, the exponent* $\frac{m_k}{l_k}$ *is to be replaced by* $\frac{m'_k}{l'_k}$. The constants $c_0, c_1, \ldots, c_p$ remain undermined here, i.e. They are initially arbitrary for a given position of the poles and zeros of $\varphi_d(\zeta_1,\zeta_2)$.

By means of the theorem before last, it now also follows, conversely: *Each expression of the form of the right side of equation (2), in dependency on* ζ_1,ζ_2*, represents an unbranched automorphic form of our model.* Thereby, the dimension d of this form in ζ_1,ζ_2 will be given by the equation:

$$d = \nu\left(m + \sum_{k=1}^{n}{}' \frac{m_k}{l_k}\right), \tag{3}$$

where $m = t - s$ represents the excess of the number of free zeros of $\varphi_d(\zeta_1, \zeta_2)$ over that of the free poles, and ν is the number furnished by (5) on p. 230.

All these results, as one will have perceived, correspond directly to the relations met for $p = 0$ (cf. p. 97 ff.). As on p. 74, as the simplest unbranched forms $\varphi_d(\zeta_1, \zeta_2)$ of our model, we have to consider the *primeforms* and the *groundforms*. *We designate an unbranched pole-free form of the automorphic model as a primeform, or, it is desirable for distinction, as an "automorphic" primeform, if it exhibits only a zero of the first order in the fundamental domain.* These primeforms, in consequence of (2), are furnished directly by:

$$\varphi_\nu(\zeta_1, \zeta_2) = e^{c_0 + c_1 W_1^z + \cdots + c_p W_p^z} P(z, x). \tag{4}$$

One will immediately set up a corresponding definition for the *groundforms of elliptic points e_k*, and obtain the representation:

$$\varphi_{\frac{\nu}{l_k}}(\zeta_1, \zeta_2) = e^{c_0 + c_1 W_1^z + \cdots + c_p W_p^z} P(z, e_k)^{\frac{1}{l_k}} \tag{5}$$

In consistent fashion, one would have to consider $P(z, e_k)$ as a *ground form of a parabolic vertex*; however, this expression does not behave itself multiplicatively with respect to closed circuits on F_w, but additively, and is accordingly not to be counted [244]
among the automorphic forms resp. functions in our sense (p. 66). Actually, also for the representation (2) of the form $\varphi_d(\zeta_1, \zeta_2)$, not $\log P(z, e_k)$, but $P(z, e_k)^{1/l'_k}$ comes into utilization; accordingly, we shall designate this last expression, which we also multiple, for the sake of generality, by the exponential function of an everywhere finite integral, as a groundform of the parabolic vertex.

After these determinations, formula (2) effects the *representation of an arbitrary unbranched automorphic form of the model in the "automorphic" prime - and groundforms.*

4.6 The single-valued automorphic forms and their multiplicator systems for a group of arbitrary genus

Among all the unbranched automorphic forms of Γ we will now, in particular, pursue the "*single-valued ones*" further. As for $p = 0$ (cf. p. 83), we place the following definition at the head: *An unbranched automorphic form $\varphi_d(\zeta_1, \zeta_2)$ of integral dimension d is called single-valued, in case it always assumes one and the same factor μ with respect to a homogeneous substitution U of Γ, along whatever path, lying inside the domain of admissible values ζ_1, ζ_2, the original arguments ζ_1, ζ_2 may be carried over into the equivalent ζ'_1, ζ'_2.*

For a proposed single-valued form $\varphi_d(\zeta_1, \zeta_2)$, let the multiplicators $\mu_1, \mu_2, \ldots, \mu_n$ correspond to the n elliptic resp. parabolic generators $U_1, U_2, \ldots, U_n$; we designate these, with Ritter, as "*branch multiplicators*", since they are furnished on the Riemann surface F_w by circuits about the branch-points $e_1, e_2, \ldots, e_k$ of the form $\varphi_d(\zeta_1, \zeta_2)$. The "*period multiplicators*" $\mu_{a_1}, \mu_{b_1}, \ldots, \mu_{a_p}, \mu_{b_p}$ correspondingly belong to the U_{a_1}, U_{b_1}, $\ldots$, U_{a_p}, U_{b_p}.

Now, on the basis of the argument carried through on p. 85, *equations for the multiplicators* run parallel to the relations between the generators. First of all, the branch multiplicators μ_k belonging to the elliptic resp. parabolic vertices determine themselves, exactly as for $p = 0$, in the form (6) and (7) and (8) on p. 85. Above and beyond these, we have to bring in, as the "last primary relation":

$$\prod_{k=1}^{n} U_k \cdot \prod_{g=1}^{p} U_{a_g}^{-1} U_{b_g} U_{a_g} U_{b_g}^{-1} = (-1)^n. \tag{1}$$

Now too, however one might have chosen the two period multiplicators μ_{a_g}, μ_{b_g}, the
[245] multiplicator belonging to:

$$U_{a_g}^{-1} U_{b_g} U_{a_g} U_{b_g}^{-1}$$

is always equal to 1; for U_{a_g} furnishes the multiplicator $\mu_{a_g}^{-1}$ etc. The consequence is, that the relation for the multiplicators, corresponding to equation (1), reduces itself to the same form, which we have already discussed above (p. 85 ff.) for $p = 0$: *For the form of the branch multiplicators and the number of theoretically possible multiplicator systems $\mu_1, \mu_2, \ldots, \mu_n$, all discussions given and results obtained for p = 0 hold without restriction; the period multiplicators $\mu_{a_1}, \mu_{b_1}, \ldots, \mu_{a_p}, \mu_{b_p}$ remain, in each case, as fully arbitrarily choosable complex constants.* The number of all theoretically possible multiplicator systems M is consequently, even in the case that we only have finitely many systems $\mu_1, \mu_2, \ldots, \mu_n$, always infinitely great for $p > 0$.

As concerns possible *secondary relations*, the same remarks hold for those again, as for the models of genus zero. *Inside the total Abelian group of all theoretically possible multiplicator systems* M *for the signature*:

$$(p, n; l_1, l_2, \ldots, l_n),$$

A subgroup is separated out by the supervention of secondary relations, which includes all multiplicator systems still now admissible.

We immediately pursue these relations, say, for the *example of the principal-circle groups with isolatedly situated boundary points.* This category of groups was last discussed in detail on p. 216 ff.; we again take up here the notations used there.

One has, in all $(n' - \mu' + 2p')$ secondary relations, of which the first $(n' - \mu')$ concern the branch multiplicators, and dispose of themselves especially easily. These relations are represented in (5) on p. 217, and yield, for the *branch multiplicators,* the relations:

$$\mu_1 \cdot \mu_{n'-\mu'+1} = 1,\ \mu_2 \cdot \mu_{n'-\mu'+2} = 1, \ldots, \mu_{n'-\mu'} \cdot \mu_n = 1. \tag{2}$$

The $2p'$ secondary relations still remaining concern the p' symmetric groups of the character (2, 0), which, according to p. 216, take part in the production of our present group by composition. For the quadruple of generators V_a, V_b, V_a', V_b' of the individual one of these groups, according to (4) on p. 217, the two relations hold:

$$V_a \cdot V_b' = 1, \qquad V_b V_a' = 1. \tag{3}$$

Do not fail to make clear to oneself the closed circuits in the polygon net, as they cor-
[246] respond to these relations, say with the help of figure 29 on p. 216. The principal circle, which in figure 29 is the real ζ-axis, intersects the fundamental domain in four segments. On the second of these (from the left), fix an arbitrary point ζ_0. From here, describe a path to the equivalent point $V_b'(\zeta_0)$ through the lower semi-polygon and the

respective neighboring semi-polygon, whose endpoint $V_b'(\zeta_0)$ is again situated on the real axis. In the upper halfplane, let the path situated symmetrically to the first be described from $V_b(\zeta_0)$ to ζ_0. *In this way there obviously arises the closed circuit, which leads to the first relation (3).*

The homogeneous substitutions $U_{a_1}, U_{b_1}, \ldots, U_{a_p}, U_{b_p}$ were now to be furnished by the binary "forms" (1) on p. 228. If we choose the function z so that it comes out real on the principal-circle segments of the fundamental domain, which we are free to do, then for conjugate complex ζ, the value Z_1 itself becomes conjugate complex and similarly, that of Z_2. According to this, the coefficients, e.g., of the substitutions U_b' and U_a^{-1} will come out pairwise conjugate complex. Since the latter coefficients, however, are now real, insofar as the real ζ-axis is to furnish the principal circle, then $U_b' = U_a^{-1}$ follows. Generally, one finds in this fashion, as the *homogeneous form of the* $2p'$ *secondary relations in question*:

$$U_{a_g} U'_{b_g} = 1, \quad U_{b_g} U'_{a_g} = 1 \qquad (g = 1, 2, \ldots, p'). \tag{4}$$

There follow, *for the associated period multiplicators, the* $2p'$ *equations*:

$$\mu_{a_g} \mu'_{b_g} = 1, \quad \mu_{b_g} \mu'_{a_g} = 1 \qquad (g = 1, 2, \ldots, p'). \tag{5}$$

Further relations for the multiplicators are not present for the groups under discussion.

4.7 Existence of the single-valued forms for a given multiplicator system in the case of an arbitrary genus

Now again, let an arbitrary group Γ be proposed, for which a canonical fundamental domain may be chosen. This furnishes, for the unimodular homogeneous group Γ, the generators:

$$U_1, U_2, \ldots, U_n, U_{a_1}, U_{b_1}, \ldots, U_{a_p}, U_{b_p},$$

between which, besides the possible secondary relations, there subsist, in any case, the primary relations in the form known by us. Let an arbitrary associated system M , i.e., satisfying *all* relations coming into consideration, of even or odd dimension d consist of the multiplicators:

$$\mu_1, \mu_2, \ldots, \mu_n, \mu_{a_1}, \mu_{b_1}, \ldots, \mu_{a_p}, \mu_{b_p}.$$

According to p. 100, for the individual elliptic generator, we have to set, in more detail: [247]

$$\mu_k = e^{2i\pi \frac{\lambda_k}{2l_k}}, \quad \lambda_k \equiv d \pmod 2), \tag{1}$$

but for the individual parabolic substitution U_k:

$$\mu_k = e^{2i\pi \frac{\lambda'_k}{l'_k}}. \tag{2}$$

We shall usually subsume formula (2) under (1), by imagining, for $l_k = \infty$, λ_k also chosen infinitely great in such a manner, that $\frac{\lambda_k}{2l_k}$ becomes equal to the rational fraction $\frac{\lambda'_k}{l'_k}$.

From the “last primary relation” it yields itself, that the sum $\sum_{k=1}^{n}{}' \frac{\lambda_k}{l_k}$ is a whole number satisfying the congruence:

$$\sum_{k=1}^{n}{}' \frac{\lambda_k}{l_k} \equiv dn \pmod 2. \tag{3}$$

The secondary relations occurring, say, will naturally have as consequence even further restrictions for the multiplicators of the system M .

If there is now a single-valued form $\varphi_d(\zeta_1,\zeta_2)$ of this multiplicator system M , then according to p. 243, as an unbranched automorphic form, this will let itself be represented in the form:

$$\varphi_d(\zeta_1,\zeta_2) = e^{c_0+c_1W_1^z+\cdot+c_pW_p^z} \frac{P(z,x_1)\cdots P(z,x_t)}{P(z,y_1)\cdots P(z,y_s)} \prod_{k=1}^{n}{}' P(z,e_k)^{\frac{m_k}{l_k}}. \tag{4}$$

Hereby, the difference $m = t - s$ of the numbers t and s, the numbers m_k and the dimension d are connected by the relation:

$$d = \nu\left(m + \sum_{k=1}^{n}{}' \frac{m_k}{l_k}\right), \tag{5}$$

whereby, the meaning of ν emerges from:

$$\frac{1}{\nu} = 1 - p - \frac{1}{2}\sum_{k=1}^{n}\left(1 - \frac{1}{l_k}\right). \tag{6}$$

We now turn ourselves to the investigation of the question, whether there actually exists a single-valued form (4) for the proposed multiplicator system M . For this purpose, we determine the multiplicators of the expression standing on the right in equation (4) for circuits by the place z on the closed surface. Thereby, we naturally have to
[248] observe, that the substitution of the ζ_1,ζ_2 corresponding to one such circuit, deviates from the unimodular substitution in each case by a factor τ (cf. p. 232), and that, on this account, the multiplicator μ' of the form represented in (4) is connected with the corresponding multiplicator μ of the system M by means of relation, considered on p. 241:

$$\mu' = \mu \cdot \tau^d. \tag{7}$$

We now investigate, first of all, a *circuit about the place* e_k in the positive sense. Hereby, only the factor $P(z,e_k)^{\frac{m_k}{l_k}}$ will experience an alteration j and indeed, the circuit in question amounts to this, that we pass over the cut Λ leading to e_k once from left to right. In the formula (13) on p. 227, referred to $\log P(z,e_k)$, we thus have to take $\sigma' = \sigma_1 = \nu_k = 0$ and set $\sigma'' = 1$. For $P(z,e_k)^{\frac{m_k}{l_k}}$, and with this, for the right sides of (4), there arises from this the multiplicator $\mu' = e^{2i\pi\frac{m_k}{l_k}}$. Since the associated multiplicator $\tau = e^{\frac{\pi i}{l_k}}$, then follows: *For the “branch multiplicator”* μ_k, *the new representation yields itself from the development (4) of our form* φ_d:

$$\mu_k = e^{2i\pi\left(\frac{m_k}{l_k} - \frac{d}{2l_k}\right)}. \tag{8}$$

The parabolic case is also included here. For it, in fact, according to p. 231, $\tau_k = 1$, and therefore formula (8), in the form:

$$\mu_k = e^{2i\pi \frac{m'_k}{l'_k}}, \tag{9}$$

furnishes the connect multiplicator.

Among the *period multiplicators,* the μ_{a_g}, first of all, determine themselves especially easily. The operation U_{a_g} signifies, for the use of formula (13) on p. 227, a passage over the cut a_g from the right to left on the Riemann surface. Since the $W_1^z, \ldots, W_p^z$ only deviate from the normal integrals $j_1^z, \ldots, j_p^z$ by an additive constant, then these share the period properties of the j^z (cf. "M" I on p. 530). Accordingly, with respect to U_{a_g}, only W_g^z experiences an alteration, and indeed, grows by one unit. Since all the prime - and groundforms contained on the right in (4) remain unaltered with respect to U_{a_g} and the associated multiplicator τ is equal to 1 according to (9) on p. 231, then follows: the "period multiplicator" μ_{a_g} is equal to the exponential function of the coefficient c_g, coming up on the right in (4):

$$\mu_{a_g} = e^{c_g}. \tag{10}$$

The determination of the period multiplicator μ_{b_g} is somewhat more circumstantial. The operation U_{b_g} amounts to a single passage over the crosscut b_g from the right to the left. let the $W_1^z, \ldots, W_p^z$ hereby obtain the increases $\omega_{g1}, \ldots, \omega_{gp}$. For the individual primeform, for the use of the formula (13) on p. 227, a ν_g is different from 0 and is to be taken equal to -1; All remaining ν, as well as the numbers σ, are equal to 0. The multiplicator τ is given in (9) on p. 231. We find, in this manner: *the period multiplicator μ_{b_g} is the exponential function of the following expression:* [249]

$$\sum_{\alpha=1}^{p} c_\alpha \omega_{\gamma\alpha} + 2i\pi\left[\sum_{\beta=1}^{t} W_g^{x_\beta} - \sum_{\gamma=1}^{s} W_g^{y_\gamma}\right] + \pi i d \sum_h (\kappa_h - 1) W_g^{v_h} + 2i\pi \sum_{k=1}^{n}{}' \left[\frac{m_k}{l_k} + \frac{d}{2}\left(1 - \frac{1}{l_k}\right)\right] W_g^{e_k}. \tag{11}$$

For the integrals W are meant everywhere here those values, which are obtained by restriction of the integration paths to the canonically cut up surface.

The formulas obtained now permit us, conversely, for given M and d, to first of all determine the quantities $m_k, c_g, \cdots$ of formula (4) the development serving this purpose divides itself in the following manner:

1. If we first just put the expressions (1) and (2) for μ_k into (8) resp. (9) and take account of the determinations $0 \le m_k < l_k$ and $0 \le m'_k < l'_k$, then it follows: *The numbers m_k determine themselves uniquely from the multiplicator system* M *and the dimension d by means of the formula:*

$$\frac{m_k}{l_k} = \frac{d + \lambda_k}{2l_k} - E\left[\frac{d + \lambda_k}{2l_k}\right], \tag{12}$$

which, for the parabolic case, assumes the form;

$$\frac{m'_k}{l'_k} = \frac{\lambda'_k}{l'_k} - E\left[\frac{\lambda'_k}{l'_k}\right]. \tag{13}$$

2. On the strength of this, the relation (5) gives, further, *a uniquely determined value* m. One convinces oneself easily, that this, as it must be, is *integral.* In fact, in consequence of (5), (12) and (13), one has

$$d=\nu\left\{m+\sum_{k=1}^{n}{}'\left(\frac{d+\lambda_k}{2l_k}\right)-\sum_{k=1}^{n}{}' E\left[\frac{d+\lambda_k}{2l_k}\right]\right\},$$

$$d\left(\frac{2}{\nu}-\sum_{k=1}^{n}\frac{1}{l_k}\right)=2m+\sum_{k=1}^{n}{}'\frac{\lambda_k}{l_k}-2\sum_{k=1}^{n}{}' E\left[\frac{d+\lambda_k}{2l_k}\right].$$

[250] From (6) yields itself:

$$\frac{2}{\nu}-\sum_{k=1}^{n}\frac{1}{l_k}=2-2p-n,$$

so that one obtains for $2m$:

$$2m=2d(1-p)-dn-\sum_{k=1}^{n}{}'\frac{\lambda_k}{l_k}-2\sum_{k=1}^{n}{}' E\left[\frac{d+\lambda_k}{2l_k}\right]. \tag{14}$$

But here on the right, in fact, in consequence of (3), there stands an *even* whole number.

3. By inversion of the equation (10) one immediately obtains, further: *the coefficients $c_1, c_2, \ldots, c_p$ in the exponential factor of the right side of (4) are given by:*

$$c_g=\log\mu_{a_g}+2\nu_g i\pi, \tag{15}$$

where $\nu_1, \nu_2, \ldots, \nu_p$ are indeterminate whole numbers.[23] The coefficient c_0 is naturally independent of M and d.

Further, however, from the values of the multiplicators μ of the proposed system M and the dimension d, there yield themselves highly remarkable relations for the *position of the zeros x and the poles y*. We have, up until now, only first determined the difference $m=t-s$ of their numbers t and s. Bringing in the period multiplicators μ_{b_g} now furnishes us the p equations:

$$\sum_{\beta=1}^{t} W_g^{x_\beta}-\sum_{\gamma=1}^{s} W_g^{y_\gamma}=A_g+\nu'_g-\sum_{\alpha=1}^{p}\nu_g\omega_{g\alpha}, \tag{16}$$

where, besides the ν_g, p further indeterminate whole numbers ν'_g are introduced, and where A_g is to be uniquely calculated from:

$$2i\pi A_g=\log\mu_{b_g}-\sum_{\alpha=1}^{p}\omega_{g\alpha}\log\mu_{a_g}-\pi i d\sum_{h}(\kappa_h-1)W_g^{v_h}-2i\pi\sum_{k=1}^{n}{}'\left[\frac{m_k}{l_k}+\frac{d}{2}\left(1-\frac{1}{l_k}\right)\right]W_g^{e_k}. \tag{17}$$

These formulas now lead us into the *domain of the Abelian functions, in particular, of the Jacobi inversion problem*[24]. We need treat here only of the solvability of the
[251] equations (16). We will, first of all, choose the number $t \geq p$ in such a manner, that

[23] In equation (15) and the next following formulas, by symbol log is always to be understood the principal value of the logarithm. Naturally, the numbers $\nu_1, \nu_2, \ldots, \nu_p$ have nothing to do with the numbers so denoted in (13) on p. 227.

[24] The solution of the inversion problem with the aid of the theta functions one finds treated in an especially accessible and detailed form in H. Stahl's "*Theorie der Abel'schen Function*" (Leipzig, 1896). cf. also "M" II on p. 512.

$s = t - m \geq 0$ always holds. We further let the s places $y_1, y_2, \ldots, y_s$ be arbitrarily fixed. There exists, on the basis of the theory named, a family of solution systems $x_1, x_2, \ldots, x_t$, whose multiplicity one has to determine according to the Riemann-Roch theorem (cf. "M" II on p. 513).

The individual solution system, substituted into (4), now furnishes us a form φ_d, which we now recognize without trouble as a *single-valued* automorphic form. We have only yet to prove, that neither the circling of a place x or y, nor that of one of the w places $\infty_1, \infty_2, \ldots, \infty_w$ effects an alteration of φ_d. The places x and y dispose of themselves immediately; for upon a circuit about x_β, $P(z, x_\beta)$ does not alter itself, according to (13) on p. 227, and self-evidently, no further factor of the right side of (4) alters itself. For a circuit about a place ∞, however, the primeform product in (4) alters itself by the factor:

$$e^{\frac{2i\pi}{w}\left(m+\sum_{k=1}^{n}{}'\frac{m_k}{l_k}\right)}.$$

We determined the associated multiplicator τ on p. 232 to be $e^{\frac{2i\pi}{w\nu}}$, so that we find here, in particular, for the equation (7):

$$e^{\frac{2i\pi}{w}\left(m+\sum_{k=1}^{n}{}'\frac{m_k}{l_k}\right)} = \mu \cdot e^{\frac{2i\pi}{w\nu}\cdot d}.$$

The relation (5) immediately furnishes $\mu = 1$; i.e., φ_d, as a form of the ζ_1, ζ_2, substituting themselves "unimodularly", remains unaltered upon a circuit of the place ∞ in question.

The developments of the preceding paragraph lead to the following fundamental existence theorem: *For each theoretically possible multiplicator system* M *of our group* Γ, *and for each even resp. odd*[25] *dimension d, there are associated single-valued automorphic forms.*

4.8 More on single-valued automorphic forms for arbitrary p. The p forms $\Phi_{-2}(\zeta_1,\zeta_2)$

For the numbers m_k, m'_k, in the case here, the same inequalities:

$$0 \leq \frac{m_k}{l_k} \leq 1 - \frac{1}{l_k}, \qquad 0 \leq \frac{m'_k}{l'_k} < 1, \tag{1}$$

hold as for $p = 0$. Since, in addition to this, the relation (5) on p. 247 between $d, m, m_k, \ldots$ subsists unaltered for $p > 0$, in the same form as for $p = 0$, then follows, as above (p. 105): for the number $m = t - s$, for the individual single-valued form $\varphi_d(\zeta_1, \zeta_2)$, the inequality holds: [252]

$$\frac{d}{\nu} \geq m \geq \frac{d}{\nu} - \sum_{k=1}^{n}\left(1 - \frac{1}{l_k}\right), \tag{2}$$

where the second equality sign can only hold, if all generators are elliptic. If, however, in place of the second inequality (1), one utilizes the following:

[25] According as M is a multiplicator system of each or odd dimension.

$$0 < \frac{m'_k}{l'_k} \leq 1, \tag{3}$$

which is useful in the investigation of the forms representable by Poincaré series (cf. p. 176), *then the second equality sign can hold in (2), but only first then, if no parabolic generations are present.*

For all groups Γ coming into consideration here, v has a *negative finite* value. The inequalities (2), for a given dimension d, consequently permit a few immediately arising consequences to be stated concerning the difference m of the number t of the zeros and the number s of poles of the associated forms.

First, if $d > 0$, then the negative number $d \cdot v^{-1}$ is greater than m, so that $m \leq -1$ necessarily takes place. *A single-valued automorphic form of "positive" dimension is never pole-free.* Further statements concerning the forms with $d \geq 0$ follow at the close of the paragraph.

For $d = 0$, one has $m \leq 0$, whereby $m = 0$ only belongs to the multiplicator system $\mathrm{M} = 1$. We conclude easily: *The "forms of zero dimension" are also never pole-free, if one will then count a constant as a pole-free property of automorphic functions.*

Among the negative dimensions d, on account of the relation to the Poincaré series to be considered soon, those satisfying the condition $d \leq -2$ will mainly interest us. Accordingly, we leave $d = -1$ entirely out of consideration here.

If $d = -2$, then we are able to make the inequality (2), in consideration of (6) on p. 247, precise thus:

$$-\frac{2}{v} \geq m \geq -\frac{2}{v} - \sum_{k=1}^{n}\left(1 - \frac{1}{l_k}\right) = 2p - 2.$$

Should the equation $m = 2p - 2$ hold, then for all indices k, the equations:

$$\frac{m_k}{l_k} = 1 - \frac{1}{l_k} \qquad \text{resp.} \qquad \frac{m'_k}{l'_k} = 1$$

subsists. As above (p. 105), we conclude from this, that we have the multiplicator sys-
[253] tem $\mathrm{M} = 1$: *For the properly automorphic forms of the dimension -2, which have zeros at the parabolic vertices, $m = 2p - 2$,*[26] *all remaining forms of the* $(-2)^{\text{nd}}$ *dimension have* $m \geq 2p - 1$.

Among the properly automorphic $\varphi_{-2}(\zeta_1, \zeta_2)$ just considered, in particular, the *integral or pole-free forms* are important. An individual one of them we denote by $\Phi_{-2}(\zeta_1, \zeta_2)$. This has $2p - 2$ movable zeros and exhibits a zero at the place e_k of the order $1 - \frac{1}{l_k}$. Now we determined on p. 72, that the differential $(\zeta, d\zeta) = \zeta_1 d\zeta_2 - \zeta_2 d\zeta_1$, has a pole of the order $1 - \frac{1}{l_k}$ at the place e_k, but except for this, is everywhere finite and different from zero. The consequence is, that the differential $\Phi_{-2}(\zeta_1, \zeta_2)(\zeta, d\zeta)$ of zero dimension, apart from $2p - 2$ movable zeros, is everywhere finite and different from zero. With this, we are led back to well-known relations. In:

$$\int \Phi_{-2}(\zeta_1, \zeta_2)(\zeta, d\zeta) = W(\zeta), \tag{4}$$

[26] As long as one considers the stipulation (3) as valid.

one obtains *an integral of the first kind belonging to the present automorphic model* (cf. p. 36); and one also perceives immediately, that conversely, from an arbitrary integral $W(\zeta)$ of the first kind, by means of the "differentiation precess" there arises a form of our kind:

$$\frac{dW(\zeta)}{(\zeta,d\zeta)}=\Phi_{-2}(\zeta_1,\zeta_2). \tag{5}$$

There are thus p linearly independent properly automorphic pole-free forms of the $(-2)^{\text{nd}}$ *dimension, which exhibit zeros at possible parabolic vertices.* We shall designate these forms, for short, from now on, as the "*forms* $\Phi_{-2}(\zeta_1,\zeta_2)$".

We remember in passing, that $p-1$ of the $2p-2$ zeros, $x_1,x_2,\dots,x_{2p-2}$ of a form Φ_{-2} are uniquely determined by the remaining $p-1$ freely choosable ones (cf. "M" I on p. 569). In transcendental form, the relations (16) on p. 250, specialized for this case:

$$W_g^{x_1}+W_g^{x_2}+\cdots+W_g^{x_{2p-2}}=\sum_h(\kappa_h-1)W_g^{v_h}+\nu'_g-\sum_{\alpha=1}^{p}\nu_g\omega_{g\alpha},$$

gives the relations subsisting between the zeros of x. Here, the ν_g,ν'_g are uniquely determined whole numbers. In the lowest case $p=1$, with the use of a two-sheeted Riemann surface, these let themselves be easily calculated directly from the meaning of the sum $\sum_h$.

If we ignore the properly automorphic forms of $(-2)^{\text{nd}}$ dimension, then in the inequalities: [254]

$$m\geq\frac{d}{\nu}-\sum_{k=1}^{n}\left(1-\frac{1}{l_k}\right)\geq 2p-2,$$

the equality signs are not able to hold simultaneously at both places. *For all single-valued automorphic forms of the dimensions* $d\leq-2$, *excepting only the properly automorphic forms of* $(-2)^{\text{nd}}$ *dimension,* $m\geq 2p-1$.

We also consider here, next, the *integral forms* again, in order to determine their manifold for given d and M. The individual integral form has $m\geq 2p-1$ zeros x_1, x_2, …, x_m. For these, there subsist the p equations (16) on p. 250, which, by leaving out the system of simultaneous periods $\nu'_g-\sum_{\alpha=1}^{p}\nu_g\omega_{g\alpha}$, we write in the usual fashion as congruences:

$$W_g^{x_1}+W_g^{x_2}+\cdots+W_g^{x_m}\equiv A_g \qquad (g=1,2,\dots,p).$$

For given d, M the A_g are uniquely determined. Since $m\geq p$, and since, at the same time, $m>2p-2$ occurs, then according to the inversion theorem, there are, all together, ∞^{m-p} point systems $x_1,x_2,\dots,x_m$ satisfying the conditions (6), which, in the sense of "M" I on p. 562, form a "*group of equivalent point systems*". If we hold fast to the integration paths prescribed for the W (cf. p. 249), then, with this, the numbers ν,ν', coming up in the equations:

$$W_g^{x_1}+W_g^{x_2}+\cdots+W_g^{x_m}\equiv A_g+\nu'_g-\sum_{\alpha=1}^{p}\nu_g\omega_{g\alpha} \qquad (g=1,2,\dots,p). \tag{6}$$

are uniquely determined. In this manner, we have obtained the following result: *For a given dimension* $d\leq-2$, *and given multiplicator system* M, *there are, in all,* $(m-p+1)$

linearly independent integral automorphic forms s. Naturally, as determined before, the case of the form Φ_{-2} already discussed is excluded here.

In the same fashion, we obtains the somewhat more general theorem: *For a given dimension $d \leq -2$ and given* M , *there are, in all,* $(m+s-p+1)$ *linearly independent forms, which may have poles of the first order at s prescribed places.* By the terminology of this theorem, it is indicated, that the forms, which remain finite at a few or even at all of the s places, are also counted here.

We are also able to push the study of the *forms of non-negative dimensions* even
[255] somewhat further. We start out from an individual $\varphi_d(\zeta_1,\zeta_2)$ with $d \geq 0$, which exhibits the s poles $y_1, y_2, \ldots, y_s$ and the t zeros $x_1, x_2, \ldots, x_t$ (besides fixed zeros situated at the vertices). The x then satisfy the p conditions:

$$\sum_{\beta=1}^{t} W_g^{x_\beta} \equiv \sum_{\gamma=1}^{s} W_g^{y_\gamma} + A_g \qquad (g = 1, 2, \ldots, p).$$

If we hold the $y_1, \ldots, y_s$ fixed, then the $\chi_1, \chi_2, \ldots, \chi_t$ represent a system contained a group of $\infty^{t-p+\sigma}$ equivalent point-systems, where σ, in consequence of the Riemann-Roch theorem, is the number of the linearly independent forms Φ_{-2}, which vanish simultaneously at the t points $\chi_1, \chi_2, \ldots, \chi_t$ of an individual one of these systems. For $t > 2p-2$, naturally, σ is always $= 0$. There arises the theorem: *For each form $\varphi_d(\zeta_1,\zeta_2)$ of non-negative dimension with the s poles $y_1, y_2, \ldots, y_s$, these are*

$$m+s-p+\sigma+1$$

linearly independent forms of the same dimension d and the same multiplicator system M , *which may have poles of the first order at those s places $y_1, y_2, \ldots, y_s$; σ hereby signifies the number of linearly independent forms Φ_{-2}, which vanish simultaneously at the t zeros of these forms.* In the next paragraph we shall be able to give this theorem in yet another, far more usable version.

4.9 Concept of conjugate forms. Extended Riemann-Roch theorem and applications of it

One convinces oneself easily, that the considerations carried out on p. 102 ff. concerning the mutual *relation between multiplicator systems* M *and* $\overline{\mathrm{M}} = \mathrm{M}^{-1}$, *inverse to one another*, remain subsisting here for arbitrary p, with an insignificant modification of the final formula. The multiplicators of $\overline{\mathrm{M}}$ are all $\overline{\mu}_k, \overline{\mu}_{a_g}, \overline{\mu}_{b_g}$ for these we have:

$$\mu_k \cdot \overline{\mu}_k = 1, \quad \mu_{a_g} \cdot \overline{\mu}_{a_g} = 1, \quad \mu_{b_g} \cdot \overline{\mu}_{b_g} = 1, \tag{1}$$

if $\mu_k, \mu_{a_g}, \mu_{b_g}$ are the multiplicators of M. For the associated dimensions $d, \overline{d}$, we hold fast to the stipulation:

$$d + \overline{d} + 2 = 0. \tag{2}$$

At l.c., we then find, that for the corresponding numbers $m_k, m_k', \ldots$, the relations:

$$\frac{m_k}{l_k} + \frac{\overline{m}_k}{l_k} = 1 - \frac{1}{l_k}, \qquad \frac{m'_k}{l'_k} + \frac{\overline{m}'_k}{l_k} = \varepsilon_k \tag{3}$$

hold, while the numbers $m, \overline{m}$ are connected by: [256]

$$m + \overline{m} = 2p - 2 + \varepsilon. \tag{4}$$

Here ε denotes the number of the parabolic multiplicators 1; and ε_k is, in general, equal to 1, and is only then equal to 0, if the parabolic multiplicator $\mu_k = 1$. The formula (9) on p. 104 subsumes itself directly under (4).

Also, the modification of the formula under discussion, which we spoke about on p. 177, carries itself over here without further ado. if we assume $d \leq -2$, and for forms of this dimension, stipulate the prescription:

$$0 < \frac{m'_k}{l'_k} \leqslant 1, \tag{5}$$

Then, in place of (3), we shall have *the relations between the numbers* $m_k, \overline{m}_k$:

$$\frac{m_k}{l_k} + \frac{\overline{m}_k}{l_k} = 1 - \frac{1}{l_k}, \qquad \frac{m'_k}{l'_k} + \frac{\overline{m}'_k}{l'_k} = 1, \tag{6}$$

while *the equation between* m *and* $\overline{m}$ obtains the form:

$$m + \overline{m} = 2p - 2. \tag{7}$$

We now agree upon the terminology, *that two single-valued automorphic forms* $\varphi_d(\zeta_1,\zeta_2)$ *and* $\psi_{\overline{d}}(\zeta_1,\zeta_2)$ *for whose dimensions* $d, \overline{d}$ *and multiplicator systems* $\mathrm{M}, \overline{\mathrm{M}}$ *the relations (2) and (1) subsist, are to be called "conjugate" to one another.* It is thereby expedient, for one of the two forms, say φ_d, whose dimension d we might assume ≤ -2, *to make use of the determination (5)* for parabolic vertices, from now on; The rules (6) and (7) hold invariably. *The product of two conjugate forms is obviously a properly automorphic form of dimension* -2, *which has a zero of order* $(1 - \frac{1}{l_k})$ *at the individual vertex* e_k *(the parabolic vertices included), and for which the difference of the number of free zeros and that of free poles is equal to* $2p - 2$.

If we now assume $d \leq -2$, then from this follows $\overline{d} \geq 0$. The form $\varphi_{\overline{d}}$ is then certainly not pole-free; its poles, whose number let be called $\overline{s}$, let lie at $y_1, y_2, \ldots, y_{\overline{s}}$. If φ_d is an integral form conjugate to $\psi_{\overline{d}}$, of whose m free zeros $\overline{s}$ are situated at $y_1, y_2, \ldots, y_{\overline{s}}$, then $\varphi_d \cdot \psi_{\overline{d}}$ will be a pole-free form of $(-2)^{\text{nd}}$ dimension, and indeed, obviously, *a form* Φ_{-2}, *which vanishes at the* $\overline{t}$ *zeros of* $\psi_{\overline{d}}$ [27]. Conversely, if, besides, $\psi_{\overline{d}}$, any form Φ_{-2} of this kind is proposed, then by: [257]

$$\varphi_d(\zeta_1,\zeta_2) = \frac{\Phi_{-2}(\zeta_1,\zeta_2)}{\psi_{\overline{d}}(\zeta_1,\zeta_2)} \tag{8}$$

will always be furnished an integral form conjugate to $\psi_{\overline{d}}$, which has zeros at the $\overline{s}$ poles of $\psi_{\overline{d}}$. From this yields itself the possibility, of also defining the number denoted

[27] In the arguments of the text, the zeros and poles are always to be thought of as simple. Nevertheless, these may unite themselves into zeros or poles of higher order by coincidence.

by σ in the closing theorem of the previous paragraph by means of the forms φ_d just mentioned. Obviously, the theorem yields itself: *For* $\overline{d} \geq 0$, *the individual single-valued automorphic form* $\psi_{\overline{d}}$, *which has the s places* $y_1, y_2, \ldots, y_{\overline{s}}$ *as poles, is contained in a family producible from* $(\overline{m} + \overline{s} - p + \sigma + 1)$ *linearly independent forms with the same poles*[28], *the same dimension* $\overline{d}$ *and the same multiplicator system* $\overline{\mathrm{M}}$, if σ *is the number of the linearly independent integral forms* φ_d *conjugate to* $\psi_{\overline{d}}$, *which vanish simultaneously at the* $\overline{s}$ *places* $y_1, y_2, \ldots, y_s$.[29]

For $\overline{d} = 0$ and $\overline{\mathrm{M}} = 1$, $\overline{m}$ becomes $= 0$, and the integral forms conjugate with $\psi_{\overline{d}}$ now become our forms Φ_{-2} obtained above. For this special case, out theorem simply represents the Riemann-Roch theorem. This state of affairs caused Ritter[30] to call the theorem in question the "*extended Riemann-Roch theorem*", a terminology, which we adopt from this point on.

To this extended Riemann-Roch theorem we attach the question, fundamental for the further development, of the *least number of "freely movable" poles*, which an automorphic form with $\overline{d} \geq 0$ is able to exhibit. If we initially exclude $\overline{d} = 0, \overline{\mathrm{M}} = 1$, then, according to p. 254, there are, in all, $(m-p+1)$ integral forms φ_d conjugate with $\psi_{\overline{d}}$, which are linearly independent. We imagine such a system chosen in $\varphi_d^{(1)}, \varphi_d^{(2)}, \ldots, \varphi_d^{(m-p+1)}$; an arbitrary one of these forms φ_d then has the form:

$$\varphi_d(\zeta_1, \zeta_2) = c_1 \varphi_d^{(1)}(\zeta_1, \zeta_2) + c_2 \varphi_d^{(2)}(\zeta_1, \zeta_2) + \cdots + c_{m-p+1} \varphi_d^{(m-p+1)}(\zeta_1, \zeta_2), \tag{9}$$

where the c signify constants, the $\overline{s}$ poles of $\psi_{\overline{d}}$, whose arguments pairs we call ${}^{(1)}\zeta_1$;
[258] ${}^{(1)}\zeta_2; \ldots, {}^{(\overline{s})}\zeta_1, {}^{(\overline{s})}\zeta_2$, are to be *movable independently of one another*. If then, the form (9) is zero at these $\overline{s}$ places simultaneously, then the c satisfy the $\overline{s}$ equations:

$$c_1 \varphi_d^{(1)}({}^{(k)}\zeta_1, {}^{(k)}\zeta_2) + \cdots + c_{m-p+1} \varphi_d^{(m-p+1)}({}^{(k)}\zeta_1, {}^{(k)}\zeta_2) = 0 \quad (k = 1, 2, \ldots, \overline{s}). \tag{10}$$

First now, let $\overline{s} \geq m - p + 1$. Under this condition, we utilize the fact, that in consequence of the linear independence of the $\varphi_d^{(1)}, \ldots, \varphi_d^{(m-p+1)}$, the determinant:

$$|\varphi_d^{(i)}({}^{(k)}\zeta_1, {}^{(k)}\zeta_2)| \quad (i, k = 1, 2, \ldots, m - p + 1)$$

does not vanish identically for freely movable ${}^{(1)}\zeta_1, {}^{(1)}\zeta_2; {}^{(2)}\zeta_1, {}^{(2)}\zeta_2; \cdots$. With this, the coefficients c for $\overline{s} \geq m - p + 1$ determine themselves from (10) as simultaneously identical to 0, so that the number σ, in the present case, is equal to zero. Since, on the other hand:

---

28 In the family, forms are also able to occur, which remain finite at some of the places $y_1, y_2, \ldots, y_{\overline{s}}$, the terminology of the text is justified, insofar as one can obviously always choose $(\overline{m} + \overline{s} - p + \sigma + 1)$ linearly independent ones among those forms of the family, which actually have poles at all the places $y_1, y_2, \ldots, y_{\overline{s}}$.

29 In the present form, the theorem would also remain subsisting directly for the forms $\psi_{\overline{d}}$ with $\overline{d} < -2$. Here, namely, $\sigma = 0$ in general, so that one arrives back at the number $(\overline{m} + \overline{s} - p + 1)$ already determined on p. 254. Only for $\overline{d} = -2$, $\overline{\mathrm{M}} = 1$ and $\overline{s} = 0$ does one have to take $\sigma = 1$ in order to obtain the p linearly independent forms Φ_{-2}, whereby (corresponding to just the value $\sigma = 1$) a constant counts as a pole-free properly automorphic form of zero dimension.

30 Cf. §11 in the paper cited on p. 213.

$$\overline{t} = \overline{m} + \overline{s} \geq m + \overline{m} - p + 1 = p - 1$$

holds, then in the case of the inequality sign, the solvability of the inversion problem in the above sense (cf, p. 251) is always guaranteed. The extended Riemann-Roch theorem consequently yields, as the number of linearly independent forms $\psi_{\overline{d}}$:

$$\overline{m} + \overline{s} - p + 1 > m + \overline{m} - 2p + 2 = 0,$$

while this number become equal to zero for $\overline{s} = m - p + 1$. *In the latter case, there are accordingly no forms $\psi_{\overline{d}}$, while for $\overline{s} > m - p + 1$, such always exist.*

If now $\overline{s} < m - p + 1$, then it follows, from the movability of the poles and the linear independence of the $\varphi_d^{(1)}, \varphi_d^{(2)}, \ldots$, for the number σ, the value $m - p + 1 - \overline{s}$.

By considering both cases together, we find the following theorem: *For forms $\psi_{\overline{d}}$ of non-negative dimension $\overline{d}$ with freely movable poles, the number $\overline{s}$ of the latter is bound by the condition:*

$$\overline{s} \geq m - p + 2 = p - \overline{m}, \tag{11}$$

whereby, however, the combination $\overline{d} = 0, \overline{\mathrm{M}} = 1$ is excluded.

But for $\overline{d} = 0, \overline{\mathrm{M}} = 1$, and we are led back to the automorphic functions in the original sense, and with this to the known theorems concerning the algebraic functions on a Riemann surface. *Here, for free movability of the poles, there holds, for their number:*

$$\overline{s} \geq p + 1. \tag{12}$$

With this, one would, in general, be able to state the theorem, *that the least number of the poles, for free movability, in each case $d \geq 0$, is greater by one unit than the number of linearly independent conjugate integral forms.*[31] [259]

A further remarkable *application of the concept of conjugate forms* we discuss here only in passing. If an arbitrary form φ_d, for which we hold fast to the prescription (5) on p. 256 for possible parabolic vertices, has a pole of first order at the place given by ${}^{(1)}\zeta_1, {}^{(2)}\zeta_2$, which for the sake of brevity may be assumed different from a polygon vertex, then in the neighborhood of the place ${}^{(1)}\zeta = {}^{(1)}\zeta_1 : {}^{(1)}\zeta_2$, the development holds:

$$\zeta_2^{-d} \cdot \varphi_d(\zeta_1, \zeta_2) = \frac{C}{\zeta - {}^{(1)}\zeta} + C_0 + C_1(\zeta - {}^{(1)}\zeta) + \cdots. \tag{13}$$

Here, let C be called *the "initial coefficient belonging to the Pole ${}^{(1)}\zeta$".*

If, in particular, φ_d is a properly automorphic form of the dimension -2, then:

$$\int \varphi_{-2}(\zeta_1, \zeta_2)(\zeta, d\zeta) = -C \log(\zeta - {}^{(1)}\zeta) + \cdots \tag{14}$$

[31] It is very interesting, that the theorem in its present form also holds for $\overline{d} \leq -2$. Now the conjugate forms have $d \geq 0$; but here the number of the integral forms is equal to 0 in general, and only then equal to 1, if $d = 0, \mathrm{M} = 1$ is present. In fact, we shall soon form one-poled forms for the dimensions ≤ -2 with a freely movable pole, apart from the case of the properly automorphic forms of $(-2)^{\mathrm{nd}}$ dimension, where the least number of the poles will be 2.

Moreover, let it yet be mentioned, that Ritter, in §22 of his work cited on p. 213, also makes a few remarks concerning the least number of the poles for restricted movability of the latter.

becomes an *integral of the third kind* belonging to the automorphic model, which exhibits a logarithmic discontinuity point at ${}^{(1)}\zeta$ "of the residue $-C$". If now, φ_{-2} has s poles all together in the fundamental domain, which are all of the first order and have the initial coefficients $C^{(1)}, C^{(2)}, \dots, C^{(s)}$, then the values of the coefficients, taken negatively, will furnish the residues of all logarithmic discontinuity points of the integral (14). According to a well-known theorem, the sum of those initial coefficients consequently vanishes for our properly automorphic form $\varphi_{-2}(\zeta_1, \zeta_2)$ of the dimension -2:

$$C^{(1)} + C^{(2)} + \dots + C^{(s)} = 0. \tag{15}$$

If now, $\psi_{\overline{d}}$ is again an $\overline{s}$-poled form of non-negative dimension, and if ${}^{(1)}\zeta_1, {}^{(1)}\zeta_2; \cdots; {}^{(\overline{s})}\zeta_1, {}^{(\overline{s})}\zeta_2$ signify the poles, which are naturally all to be of the first order, then, understanding by $\varphi_d^{(i)}$ one of the linearly independent integral forms conjugate to $\psi_{\overline{d}}$, apply
[260] the preceding consideration to the properly automorphic form of $(-2)^{\text{nd}}$ dimension:

$$\varphi_d^{(i)}(\zeta_1, \zeta_2) \cdot \psi_{\overline{d}}(\zeta_1, \zeta_2) = \varphi_{-2}(\zeta_1, \zeta_2).$$

If the initial coefficient $C^{(k)}$ belongs to the pole ${}^{(k)}\zeta$ of $\psi_{\overline{d}}$, then to the same Pole of φ_{-2}:

$${}^{(k)}\zeta_2^{-d} \varphi_d^{(i)} \left({}^{(k)}\zeta_1, {}^{(k)}\zeta_2 \right) \cdot C^{(k)}$$

belongs as initial coefficient. But since φ_{-2} obviously does not exhibit further poles besides the $\overline{s}$ poles of $\psi_{\overline{d}}$, then on the basis of equation (15) arises the theorem: *For the $\overline{s}$ initial coefficients $C^{(k)}$, which belong to the poles of the forms $\psi_{\overline{d}}$, there subsist the linear homogeneous relations*:

$${}^{(1)}\zeta_2^{-d} \varphi_d^{(i)} \left({}^{(1)}\zeta_1, {}^{(1)}\zeta_2 \right) \cdot C^{(1)} + \dots + {}^{(\overline{s})}\zeta_2^{-d} \varphi_d^{(i)} \left({}^{(\overline{s})}\zeta_1, {}^{(\overline{s})}\zeta_2 \right) \cdot C^{(\overline{s})} = 0, \tag{16}$$

where for $\varphi_d^{(i)}$ are to be substituted, one after the other, all the linearly independent integral forms conjugate to $\psi_{\overline{d}}$ from any fixed system of such forms. If, furthermore, there are σ linearly independent integral forms conjugate $\varphi_{\overline{d}}$, which vanish simultaneously at the $\overline{s}$ poles of $\varphi_{\overline{d}}$, then among the $(m-p+1)$ resp. p equations (16), σ are obviously dependent on the remaining ones.

From equation (15) arises yet a further consequence, to which we shall return once more immediately. *One-poled properly automorphic forms of $(-2)^{\text{nd}}$ dimension do not exist.* If fact, indeed, for $s = 1$, from (15) would emerge the result $c^{(1)} = 0$. This theorem is also provable in the following manner. Since we consider the prescription (5) on p. 256 as fulfilled for possible parabolic vertices, then for a properly automorphic form φ_{-2}, which is one-poled, $t = m + 1 = 2p - 1$. Consequently, according to p. 254, the number of linearly independent φ_{-2}, which may have a pole at the prescribed place, is $2p - 1 - p + 1 = p$. But this is already the number of the *integral* forms Φ_{-2}, which are also to be counted here; therefore, there are no φ_{-2}, which have one pole at the place in question.

The investigations hereby brought to an end, as one will have noticed, represent the generalization of the developments of the second chapter to arbitrary p. We are informed in detail concerning the concept, existence and manifold of the forms

$\varphi_d(\zeta_1,\zeta_2)$ of an automorphic model of arbitrary p, as well as concerning the polymorphic forms inverse to them. This is approximately the place, up to which Ritter has developed the theory of automorphic models in his work, cited on p. 213, on multiplicative forms. Our discussions following further are now to tie themselves into the developments of the third chapter (p. 138 ff.).

4.10 The Poincaré series and the elementary forms for p. Unimultiplicative forms

[261]

We now turn ourselves to the study of the *Poincaré series* for the groups Γ of arbitrary genus p. All convergence investigations carried out above were independent of the number p of the genus (cf. p. 139ff.). Our result obtained at that time, *That the series of the dimension $d \leq -4$ are always absolutely convergent, that for groups with limit circles the series with $d=-3$ also absolutely converge, and for principal-circle groups with isolatedly situated boundary points, as well as certain groups with isolatedly situated boundary points satisfying the "Schottky conditions" (cf. p. 161), also the series with* $d=-2$, consequently converge in genera! for each p.

We must now turn our attention immediately to a very important circumstance. We have carried out the convergence proof of the Poincaré series above only under the assumption, *that all the multiplicator μ_k are numbers of the absolute value* 1. This indeed always holds, now, in the foregoing investigations on the forms belonging to a group of arbitrary p, for the branch multiplicators $\mu_1,\mu_2,\ldots,\mu_n$; on the other hand, the period multiplicators were any complex numbers. A single-valued automorphic form $\varphi_d(\zeta_1,\zeta_2)$, whose period multiplicators, in particular, all have the absolute value 1, we will designate from now on as "*unimultiplicative*". *Through Poincaré series, then, we are only able, in any case, to represent unimultiplicative forms.*

In order to measure the degree of the restriction lying herein, we set up the theorem: *Each single-valued automorphic form φ_d, by multiplication by the exponential function of an everywhere finite integral $c_1W_1^z+c_2W_2^z+\cdots+c_pW_p^z$, can be changed into a unimultiplicative form, and indeed the coefficients $c_1,\ldots,c_p$ are hereby uniquely determined.*

Namely, upon addition of this exponential factor, the branch multiplicators experience no alteration. The period multiplicators of the form:

$$\varphi_d' = e^{c_1W_1^z+\cdots+c_pW_p^z}\varphi_d \tag{1}$$

are then obviously:

$$\mu_{a_g}' = e^{c_g}\mu_{a_g}, \qquad \mu_{b_g}' = e^{c_1\omega_{g1}+\cdots+c_p\omega_{gp}}\mu_{b_g} \tag{2}$$

separating real and imaginary components, we write:

$$c_g = c_g' + ic_g'', \qquad \omega_{gh} = \omega_{gh}' + i\omega_{gh}''.$$

If $|\mu_{a_g}'|=1$, then follows: [262]

$$c_g' = -\log|\mu_{a_g}|, \tag{3}$$

whereby the real components of the c_g are uniquely determined. The requirement, that all p values $|\mu'_{b_g}|$ are also equal to 1, with utilization of the values of the c'_g just obtained, clothes itself in the form of the p equations:

$$c''_1\omega''_{g1} + c''_2\omega''_{g2} + \cdots + c''_p\omega''_{gp} = \log|\mu_{b_g}| - \sum_{h=1}^{p} \omega'_{gh}\log|\mu_{a_h}| \tag{4}$$

for $g = 1, 2, \ldots, p$. The c'' determine themselves from these p equations uniquely, since, according to a known theorem from the theory of Abelian integrals, the p-termed determinant $|\omega''_{gh}|$ is different from zero.

Another restriction for the forms representable by Poincaré series is known to us from earlier; it is this, *that the forms $\varphi_d(\zeta_1, \zeta_2)$ furnished by Poincaré series always have zeros at the parabolic cusps.* We deal with this matter, for the forms with $d \leq -2$ (as already in the preceding paragraph), by setting up the representation (2) on p. 243, to be accomplished by means of the primeform, according to the prescription (5) on p. 256 for possible cusps.

The deeper lying theorems concerning representation of forms by Poincaré series we had obtained above for $p = 0$ on the basis of the theory of the *one-poled series* and the *elementary forms* $\Omega(\zeta_1, \zeta_2; \xi_1, \xi_2)$ to be produced from them. An elementary form was a one-poled series, for which "the coefficient belonging to the pole ξ" has the value 1:

$$\zeta_2^{-d}\xi_2^{-\bar{d}}\Omega(\zeta_1, \zeta_2; \xi_1, \xi_2) = \frac{1}{\zeta - \xi} + C_0 + C_1(\zeta - \xi) + \cdots. \tag{5}$$

If we have a group with one or infinitely many boundary curves, then we define the elementary $\Omega(\zeta_1, \zeta_2; \xi_1, \xi_2)$ by the series (3) on p. 186. If the polygon spans the entire ζ-plane up to the boundary points, then $\Omega(\zeta_1, \zeta_2; \xi_1, \xi_2)$ has the form (5) on p. 187. All development on p. 179 ff. referring to the construction of the elementary forms, as well, too, as the investigations concerning production of series with poles at elliptic vertices, those concerning the behavior of the elementary forms upon approach of ξ to a parabolic vertex, concerning the behavior upon application of the substitutions U_i on ξ_1, ξ_2, etc., as we already repeatedly emphasized above, hold directly for groups of an arbitrary genus p.

[263] The special discussions on p. 200ff. referring to the genus $p = 0$ we shall now have to generalize to arbitrary p.

First of all, let us remember, *that for $d = -2, \mathrm{M} = 1$, elementary forms are not producable* (cf. p. 181). This agrees, now, with theorem set up at the close of the previous paragraph, that one-poled properly automorphic forms do not exist. We shall return even more often to this matter.

For the exercise of the substitution U_i on ξ_1, ξ_2, the question is, in particular, treated above (p. 199), whether $\Omega(\zeta_1, \zeta_2; \xi_1, \xi_2)$ can also be an automorphic form in the ξ_1, ξ_2. We shall be easily able to show here: *For no group Γ with $p > 0$ can there be an elementary form Ω, which is directly an automorphic form in the ξ_1, ξ_2.* Namely, since the case of the properly automorphic forms of $(-2)^{\text{nd}}$ dimension remain excluded, then an individual $\Omega(\zeta_1, \zeta_2; \xi_1, \xi_2)$, as a form of dimension $d \leq -2$ in the ζ_1, ζ_2, has an $m \geq 2p - 1$. From the relation:

$$m + \overline{m} = 2p - 2$$

consequently follows $m \leq -1$. But since Ω, as a form of the ξ_1, ξ_2, would also be one-poled, then we would have to have $\overline{m} \geq -1$, so that $\overline{m} = -1$ would follow. This one pole would possess free movability. We nevertheless found, that the least number of freely movable poles is

$$p - \overline{m} = p + 1.$$

As soon as $p > 0$, this latter fact contradicts the assumption, that Ω be automorphic in the ξ_1, ξ_2; this assumption, therefore, connot take place.

Through exactly the same argument, as on p. 203 for $p = 0$, by initially *excluding the combination* $d = -2, \mathrm{M} = 1$, the fundamental result yields itself: *For all dimensions $d \leq -2$ finding themselves in agreement with the convergence theorems, and for each associated system "M", there are pole-free, non-identically vanishing Poincaré series.*

For the behavior of the elementary form $\Omega(\zeta_1, \zeta_2; \xi_1, \xi_2)$, belonging to given d and M , upon exercise of the substitutions U_i on ξ_1, ξ_2, which regulate themselves according to the developments on p. 198 ff., we recapitulate the following: if, for this dimension d and this multiplicator system M , there are, all together, τ linearly independent pole-free Poincaré series, which we will again (as on p. 206) denote symbolically by:

$$\omega_1(\zeta_1, \zeta_2), \omega_2(\zeta_1, \zeta_2), \ldots, \omega_\tau(\zeta_1, \zeta_2), \tag{6}$$

then $\Omega(\zeta_1, \zeta_2; \xi_1, \xi_2)$, *upon exercise of* U_i *on* ξ_1, ξ_2, *goes over into:* [264]

$$\Omega(\zeta_1, \zeta_2; \xi_1^{(i)}, \xi_2^{(i)}) = \overline{\mu_i}\Omega(\zeta_1, \zeta_2; \xi_1, \xi_2) + \sum_{\nu=1}^{\tau} f_\nu^{(i)}(\xi_1, \xi_2)\omega_\nu(\zeta_1, \zeta_2), \tag{7}$$

where the $f_\nu^{(i)}(\xi_1, \xi_2)$ *are forms in the* ξ_1, ξ_2 *alone of the dimension* $\bar{d} = -d-2$, *which in no case* d, M, *all vanish identically.* In fact, as we just saw, Ω indeed cannot possess the behavior of an automorphic form ξ_1, ξ_2 with respect to all group generators U_i.

Starting from this point, we shall be able to decide the question, *whether all forms resp, which forms of a dimension d finding itself in agrement with the convergence theorems, are representable by a Poincaré series.* To be sure, the case $d = -2, \mathrm{M} = 1$, excluded for the elementary forms, which nevertheless satisfies the convergence theorems under certain assumptions(cf. p. 157 ff.), still does not let itself be disposed of hereby. First of all, we must go one more step further and draw more closely into consideration *the properly automorphic "two-poled" series of* $(-2)^{\text{nd}}$ *dimension.*

4.11 Two-poled series of $(-2)^{\text{nd}}$ dimension and integrals of the 2nd kind for automorphic models of arbitrary genus p

The groups Γ, whose Poincaré series of $(-2)^{\text{nd}}$ dimension are convergent, necessarily possess nets N, as we know, which cover the entire ζ-plane up to isolatedly situated boundary points. For these groups, we are led to automorphic models, which have been repeatedly investigated. In order to present here at the beginning a few historical notes, first let the bequeathed note of Riemann's be remembered "*Gleichgewicht der Elektricität auf Cylindern mit kreisfömigem Querschnitt und parallelen Axen*"[32].

[32] Riemann's gesammelte Werke, p. 413 (the first ed.), p. 440 (the second ed.).

This refers to a domain bounded by $(p+1)$ symmetry circles running separately, which is first completed to a fundamental domain (of the "first" kind) by reproduction in one of those circles. The great significance of this work has been repeatedly referred to, thus e.g., on p. 114, as well as in I on p. 438. Further let us remember the two great works of Schottky's: "*Über conforme Abbildmgen mehrfach zusammenhängender ebener Flächen*" and "*Über eine specielle Function, welche bei einer bestimmten linearer*
[265] *transformation ihres Argumentes unverändert bleibt*"[33], which have already been cited often, and to the second of which we shall tie into in more detail. Shortly before the last paper there appeared the likewise repeatedly cited note of Weber's "*Ein Beitrag zu Poincaré's Theorie der Funchs'schen Function*"[34], in which, for a special class of hyperbolic models, the Schottky results were anticipated. On the other hand, four years after the second Schottky work, W. Burnside let his work: "*On a class of Automorphic Functions*" appear, cited on p. 157, which, besides the cases considered by Riemann and Schottky, also drew into consideration a the principal-circle groups admissible here, and in which the treatment of the associated automorphic models was based on the convergent Poincaré series of $(-2)^{\text{nd}}$ dimension, which we shall immediately study. Further references to the literature, namely such, which principally concern certain product developments yet to be considered in more detail, better follow below.

The remarkable thing about the *"properly automorphic" Poincaré series of* $(-2)^{\text{nd}}$ *dimension* to be studied here is, that, starting out from them, one arrives with great ease at *series developments for the integrals of the three kinds*, which belong to the automorphic model in question, as well as to *product representations of their respective primeforms*. We only first investigate, in the present paragraph, the series of $(-2)^{\text{nd}}$ dimension in their relation to the integrals of the second kind.

In this regard, the most suitable point of departure is offered by the form $\psi(\zeta_1,\zeta_2;\xi_1,\xi_2)$ of two variable pairs ζ_1,ζ_2 and ξ_1,ξ_2 defined by the following series:

$$\psi(\zeta_1,\zeta_2;\xi_1,\xi_2)=\sum_k \frac{1}{(\zeta^{(k)},\xi)^2}. \tag{1}$$

The principal properties of this form are the following: *In each of the two variable pairs, ψ represents a properly automorphic form of $(-2)^{\text{nd}}$ dimension, which is obviously two-poled (insofar as it exhibits a pole of 2nd order at ξ resp. ζ), and which remains unaltered upon interchange of the two variable pairs*:

$$\psi(\xi_1,\xi_2;\zeta_1,\zeta_2)=\psi(\zeta_1,\zeta_2;\xi_1,\xi_2). \tag{2}$$

If ξ is held fixed at a place different from a polygon vertex, then ψ, as form of the ζ_1,ζ_2,
[266] *has(besides $2p$ free zeros) a zero of order $(1-\frac{1}{l_k})$ at the individual elliptic or parabolic polygon vertex. The corresponding thing, naturally, also holds for ψ as a form of the ξ_1,ξ_2 for fixed ζ.*

Now the invariant differential $(\xi,d\xi)$ has a pole of the order $(1-\frac{1}{l_k})$ at the individual vertex, so that the product of ψ and $(\xi,d\xi)$, up to a pole of second order lying at ζ, is pole-free in the entire fundamental domain. This product (except for ζ_1,ζ_2) is only

[33] Journal für Math. Vol. 33 (1877) and Vol. 101 (1887).

[34] Göttinger Nachrichten of 1886 on p. 359.

dependent on the quotient $\xi = \xi_1 : \xi_2$. Now integrate, for fixed ζ_1, ζ_2, with respect to ξ, between the limits ξ_0 and ξ, say; let the integral arising the called $Y(\zeta_1, \zeta_2; \xi, \xi_0)$:

$$Y(\zeta_1,\zeta_2;\xi,\xi_0) = \int_{\xi_0}^{\xi} \psi(\zeta_1,\zeta_2;\xi_1,\xi_2)(\xi, d\xi). \tag{3}$$

Since the series (1) may be integrated termwise in the entire domain of its uniform convergence, i.e., in the interior of the polygon net (in the entire ξ-plane apart from the boundary points of the group), there follows, as a *representation of Y, the Poincaré series*:

$$Y(\zeta_1,\zeta_2;\xi,\xi_0) = -\sum_k \frac{(\xi,\xi_0)}{(\zeta^{(k)},\xi)(\zeta^{(k)},\xi_0)}. \tag{4}$$

The absolute convergence of this series, in case ξ and ξ_0 do not lie at possible parabolic cusps, is also immediately clear according to the earlier convergence theorems. But even the parabolic points offer no difficulty here. Namely, in case ξ moves into the vicinity of a parabolic cusp to be positioned at ∞, say, if we develop the "automorphic" from ψ of ξ_1, ξ_2 according to power of e^{ξ}, as on p. 151, then the integration leads to a convergent representation for Y:

$$Y(\zeta_1,\zeta_2;\xi,\xi_0) = \varphi_{-2}^{(0)}(\zeta_1,\zeta_2) + \varphi_{-2}^{(1)}(\zeta_1,\zeta_2)e^{\xi} + \varphi_{-2}^{(2)}(\zeta_1,\zeta_2)e^{2\xi} + \cdots, \tag{5}$$

which clarifies the behavior of Y upon approach of ξ to the cusp in question, and which subsumes itself under the development of the elementary forms at parabolic cusps(cf. p. 196).

As a function of ξ, Y has *one pole of the first order* in the fundamental domain, situated at ζ, and indeed, $\zeta_2^2 \cdot Y$ behaves itself at ζ as $(\xi - \zeta)^{-1}$. Upon exercise of substitutions of the group on ξ, Y, as an integral (3), increases itself by *periods, which are independent of ξ*. We return immediately to the form of these periods. *One consequently recognizes, in $Y(\zeta_1,\zeta_2;\xi,\xi_0)$, considered as a function of ξ, a one-poled integral of the second kind.* If we have a fundamental domain, which is bounded by $2p$ full circles running separately, then these circle pairs furnish p reentrant cuts for the closed surface, which we are able to consider as the cuts $b_1, b_2, \ldots, b_p$ of a canonical cut system. From the [267] single-valuedness of Y in ξ follows, that Y remains unaltered, in case ξ runs through one of the cuts b. But an integral of the second kind, which (considered on the cut up surface) exhibits no value differences different from zero anywhere along the p cuts a, is called a "normal integral" belonging to the canonical cut system (cf. "M" I on p. 531). Therefore follows: *In the cases of a fundamental domain with $2p$ separately running full circles as boundary curves, $Y(\zeta_1,\zeta_2;\xi,\xi_0)$, as a function of ξ, is a one-pole "normal integral" of the second kind.*

The integral of the second kind, moreover, is not provided to us directly in that form, which arises from the theory of algebraic functions, but deviates from the latter by a certain factor, dependent on ζ. Notice, namely, *that Y, in dependence on the discontinuity place ζ, represents a properly automorphic form of $(-2)^{\text{nd}}$ dimension, which is two-poled.* Such a form has a zero of order $(1 - \frac{1}{l_k})$ at the individual fixed polygon vertex. If, therefore, ζ moves into a fixed vertex, then Y will identically vanish as a function of ξ. Wc immediately return once more to this deviant behavior of Y vis-a-vis the integrals of the second kind in the algebraic theory, for the integrals of the third kind.

Furthermore, one wants yet to notice, *that the "periods" of the integral Y are indeed independent of ξ, but furnish, in ζ_1,ζ_2, pole-free properly automorphic forms of* $(-2)^{\text{nd}}$ *dimension with zeros at the possible parabolic cusps.* We are therefore led back here to the forms $\Phi_{-2}(\zeta_1,\zeta_2)$, which is in agreement with the behavior of the integrals of the second kind, well-known in the theory of the algebraic functions (cf. "M" I on p. 532 and II p. 497). If, in particular, we exercise the substitution U_i on ξ and write:

$$Y(\zeta_1,\zeta_2;\xi^{(i)},\xi_0) = Y(\zeta_1,\zeta_2;\xi,\xi_0) + \Phi_{-2}(\zeta_1,\zeta_2), \tag{6}$$

Then there arises, for the form Φ entering here, from (4) after a short intermediate calculation, the series:

$$\Phi_{-2}(\zeta_1,\zeta_2) = \sum_k \frac{(\xi_0,\xi_0^{(i)})}{(\zeta^{(k)},\xi_0)(\zeta^{(k)},\xi_0^{(i)})} \tag{7}$$

That we have to do here with a pole-free series, also emerges easily and directly from the method of construction of the series, in consideration of the equivalence of ξ_0, and $\xi_0^{(i)}$ (comp. the arguments on p. 180 ff.). Since, because of $p>0$, the integral Y cannot remain unaltered with respect to all substitutions U_i exercised on ξ, the important result follows, *that not all pole-free series (7) are* able to vanish identically. If there are, in
[268] all, τ linearly independent pole-free Poincaré series of the $(-2)^{\text{nd}}$ dimension, then let us denote a particular such system by:

$$\Phi_{-2}^{(1)}(\zeta_1,\zeta_2), \Phi_{-2}^{(2)}(\zeta_1,\zeta_2), \ldots, \Phi_{-2}^{(\tau)}(\zeta_1,\zeta_2).$$

In place of (6), we then get, *with respect to the behavior of Y upon exercise of the substitution U_i on ξ, the more detailed equation*:

$$Y(\zeta_1,\zeta_2;\xi^{(i)},\xi_0) = Y(\zeta_1,\zeta_2;\xi,\xi_0) + \sum_{\nu=1}^{\tau} c_{i\nu}\Phi_{-2}^{(\nu)}(\zeta_1,\zeta_2), \tag{8}$$

where the c are constants independent of ξ and ζ. For the number τ here holds, initially, $1 \le \tau \le p$; in the paragraph after next, we shall be able to show, without difficulty, the occurrence of the equation $\tau = p$.

4.12 The integrals of the first and third kinds. Product representation for the primeform

With reservation of a short addition concerning the integrals of the second kind, we now turn ourselves first of all, to the study of the *integrals of the first and third kinds* for those automorphic models, whose series of $(-2)^{\text{nd}}$ dimension are convergent.

The differential $\Phi_{-2}(\zeta_1,\zeta_2)(\zeta,d\zeta)$, which we are able to produce from the form $\Phi_{-2}(\zeta_1,\zeta_2)$ furnished in equation (7) of the previous paragraph, as one easily determines, is also pole-free at the polygon vertices. This will directly represent, conceived in dependency on ζ, a differential of the first kind belonging to the algebraic model. By integration between the limits ζ_0 and ζ, let the *integral $J(\zeta,\zeta_0)$ of the first kind* arise:

$$J(\zeta,\zeta_0)=\int_{\zeta_0}^{\zeta}\Phi_{-2}(\zeta_1,\zeta_2)(\zeta,d\zeta). \tag{1}$$

If we utilize the invariance of the differential $(\zeta,d\zeta)$ with respect to U_k, then by termwise integration of the series for $\Phi_{-2}(\zeta_1,\zeta_2)(\zeta,d\zeta)$ furnished by (7) on p. 267, as *a representation of the integral* $J(\zeta,\zeta_0)$ *of the first kind, we get the following series*:

$$J(\zeta,\zeta_0)=-\sum_k \log\frac{(\zeta^{(k)},\xi_0)(\zeta_0^{(k)},\xi_0^{(i)})}{(\zeta_0^{(k)},\xi_0)(\zeta^{(k)},\xi_0^{(i)})}, \tag{2}$$

which can also be written thus:

$$J(\zeta,\zeta_0)=-\sum_k \log\left(\frac{\zeta^{(k)}-\xi_0}{\zeta^{(k)}-\xi_0^{(i)}}\cdot\frac{\zeta_0^{(k)}-\xi_0^{(i)}}{\zeta_0^{(k)}-\xi_0}\right). \tag{3}$$

If we already assume as proved here, that the number, denoted at the close of the previous paragraph by τ, is equal to p, then we find sufficiently many substitutions among the generators U_i of the group, in order to produce, in the manner described, p linearly [269]
independent integrals of the first kind.

Here too, *a fundamental domain bounded by* $2p$ *separately running full circles* presents the opportunity for a separate study. If we understand thereby, by U_i, one of the p generators, then $\xi_0^{(i)}$ is the point in the neighboring domain equivalent with the place ξ_0. Let the latter domain have the circle K_i in the common with the initial domain. Corresponding to the multiple-valuedness of $\log\left(\frac{\zeta-\xi_0}{\zeta-\xi_0^{(i)}}\right)$, we carry a crosscut from ξ_0 to $\xi_0^{(i)}$, which passes over the circle K_i, without meeting any other circle of the net. If ζ now runs through the circle K_i in the correct direction, then the initial term of the series (3), corresponding to the value $k=0$, increases by the amount $2\pi\sqrt{-1}$. On the other hand, the points $\zeta^{(1)},\zeta^{(2)},\ldots$, equivalent with ζ, describe circles of the net, which do not meet the crosscut just drawn; all further following terms of the series (3) remain unaltered with respect to the circuit by ζ in question. If ζ describes the circle K_i, then according to this, $J(\zeta,\zeta_0)$ alters itself by the period $2\pi\sqrt{-1}$. One finds, by easy continuation of the argument, that $J(\zeta,\zeta_0)$ remains unaltered, in case ζ describes one of those boundary circles of the fundamental domain, which belong to the $(p-1)$ remaining generators. We perceive immediately *the p group generators, in the case in question, furnish a system of p normal integrals of the first kind* [cf. "M" I on p. 530[35]]; *and since, as such, these are linearly independent, then the same thing holds for the p pole-free series of dimension* -2 *corresponding to them.* In this case, therefore, the proof of our assertion, that the number $\tau=p$, is already furnished.

In order, further, to arrive at the *integral of the third kind*, we form the differential $Y(\zeta_1,\zeta_2;\xi,\xi_0)\cdot(\zeta,d\zeta)$, which is homogeneous of dimension zero in ζ_1,ζ_2, and therefore only depends on the quotient ζ. This has two poles of the first order in the fundamental domain, at ξ and ξ_0, while it is pole-free at the vertices of the latter. The integration now furnishes us directly, in:

35 Moreover, at this place, as well, too, as otherwise noted in the text, the normal integrals of the first kind used deviate from the integrals J under discussion here by the common factor $2\pi\sqrt{-1}$.

$$Q(\zeta,\zeta_0;\xi,\xi_0) = \int_{\zeta_0}^{\zeta} Y(\zeta_1,\zeta_2;\xi,\xi_0)(\zeta,d\zeta), \tag{4}$$

[270] *An integral of the third kind with two logarithmic discontinuity points ξ and ξ_0.*

As series development of this integral, it follows from (4) on p. 266:

$$Q(\zeta,\zeta_0;\xi,\xi_0) = \sum_k \left[\log\left(\frac{\zeta^{(k)}-\xi}{\zeta^{(k)}-\xi_0}\right) - \log\left(\frac{\zeta_0^{(k)}-\xi}{\zeta_0^{(k)}-\xi_0}\right)\right]. \tag{5}$$

One can also write for this:

$$Q(\zeta,\zeta_0;\xi,\xi_0) = \sum_k \log\left(\frac{(\zeta^{(k)},\xi)(\zeta_0^{(k)},\xi_0)}{(\zeta^{(k)},\xi_0)(\zeta_0^{(k)},\xi)}\right),$$

where a double-ratio of four points stands under the logarithm sign of the individual term. If one takes into consideration the invariance of this double-ratio upon simultaneous application of one and the same substitution U_i on the two variable pairs contained in it, then, from the last representation of Q, one can easily argue to the following one:

$$Q(\zeta,\zeta_0;\xi,\xi_0) = \sum_k \left[\log\left(\frac{\zeta-\xi^{(k)}}{\zeta-\xi_0^{(k)}}\right) - \log\left(\frac{\zeta_0-\xi^{(k)}}{\zeta_0-\xi_0^{(k)}}\right)\right]. \tag{6}$$

The comparison with (5) shows that we have to do with an integral of the third kind, which permits *interchange of "parameter" and "argument"*:

$$Q = (\xi,\xi_0;\zeta,\zeta_0) = Q(\zeta,\zeta_0;\xi,\xi_0). \tag{7}$$

The equation arising from (3) on p. 266:

$$Q(\zeta,\zeta_0;\xi,\xi_0) = \int_{\zeta_0}^{\zeta}\int_{\xi_0}^{\xi} \psi(\zeta_1,\zeta_2;\xi_1,\xi_2)(\xi,d\xi)(\zeta,d\zeta) \tag{8}$$

verifies this directly, in consideration of (2) on p. 265.

In the case of a fundamental domain bounded by $2p$ isolatedly running full circles, as is to be seen immediately, $Q(\zeta,\zeta_0;\xi,\xi_0)$ is a *"normal integral" of the third kind* belonging to the canonical cut system, whereby, corresponding to the measure followed above(p. 220), one will assume the two places ξ and ξ_0 in one and the same polygon of the net.

In addition, we yet notice, *that Q (in contrast to the integral Y), conceived in dependence on ζ and ξ, is to be considered directly as an integral of the third kind arising from the algebraic theory.* But, if now, on the Riemann surface, one will obtain by differentiation the transition from Q to the integral of the second kind, then one will avail oneself of the differential $d\omega_z$ (cf. p. 229), everywhere finite and different from zero. By

[271] utilizing here, instead of this, the equation leading to our Y:

$$\frac{dQ(\zeta,\zeta_0;\xi,\xi_0)}{(\zeta,d\zeta)} = Y(\zeta_1,\zeta_2;\xi,\xi_0),$$

in contrast to the integral $\frac{dQ}{d\omega_z}$ the *integral Y of the second kind,* conceived in dependency on its discontinuity place ζ, receives *the zeros at the fixed polygon vertices already noticed above,* as these correspond to the poles of the differential $(\zeta, d\zeta)$ to be found there. With this, the deviation of Y from the integral $\frac{dQ}{d\omega_z}$ of the second kind of the algebraic'theory, only first discussed in passing before (p. 267), is fully cleared up.

Starting out from the integral $Q(\zeta,\zeta_0;\xi,\xi_0)$ of the third kind, we finally arrive, very easily, at that *primefunciton,* which Schottky utilized in the second of the works cited on p. 264 for obtaining *product developments of automorphic functions.*

Suppose we have the *case of a fundamental domain with* $2p$ *full circles* as boundary curves, to which Schottky restricts himself l.c. Under these circumstances, in the nonhomogeneously written group Γ, apart form the identity $V_0 = 1$, no substitution V_k is identical with its inverse V_k^{-1}. Accordingly, if of each two mutually inverse substitutions, we initially allow only one, and call the substitutions thus chosen V_κ in general, then in $V_0 = 1$, then V_κ and then V_κ^{-1}, we shall have exhaustively named the entire group and each operation only once.

If, instead of V_κ^{-1}, we write the symbol $V_{-\kappa}$, then the series representations of the integral Q, upon ongoing over to the exponential function of Q, furnish, first of all, the following product developments for e^Q:

$$e^{Q(\zeta,\zeta_0;\xi,\xi_0)} = \frac{(\zeta,\xi)(\zeta_0,\xi_0)}{(\zeta,\xi_0)(\zeta_0,\xi)} \prod_\kappa \frac{(\zeta^{(-\kappa)},\xi)(\zeta_0^{(-\kappa)},\xi_0)}{(\zeta^{(-\kappa)},\xi_0)(\zeta_0^{(-\kappa)},\xi)} \cdot \prod_\kappa \frac{(\xi,\zeta^{(\kappa)})(\xi_0,\zeta_0^{(\kappa)})}{(\xi,\zeta_0^{(\kappa)})(\xi_0,\zeta^{(\kappa)})}.$$

By refashioning the parentheses in the first product on the right side on the basis of the invariant property of those symbols, there further follows for e^Q the expression:

$$\frac{(\zeta,\xi)(\zeta_0,\xi_0)}{(\zeta,\xi_0)(\zeta_0,\xi)} \prod_\kappa \frac{(\zeta,\xi^{(\kappa)})(\zeta_0,\xi_0^{(\kappa)})}{(\zeta,\xi_0^{(\kappa)})(\zeta_0,\xi^{(\kappa)})} \cdot \prod_\kappa \frac{(\xi,\zeta^{(\kappa)})(\xi_0,\zeta_0^{(\kappa)})}{(\xi,\zeta_0^{(\kappa)})(\xi_0,\zeta^{(\kappa)})}.$$

Now, *the primefunction* $E(\zeta,\zeta_0)$ *utilized by Schottky is to be defined* thus:

$$E(\zeta,\zeta_0) = \lim_{\substack{\xi=\zeta \\ \xi_0=\zeta_0}} \sqrt{-(\zeta-\xi)(\zeta_0-\xi_0)e^{-Q(\zeta,\zeta_0;\xi,\xi_0)}}. \tag{9}$$

Obviously, for the passage to the limit prescribed here, the two products in the expression last given for e^Q become equal to one another, and we obtain, as a *a convergent product representation of the primefunction* $E(\zeta,\zeta_0)$. [272]

$$E(\zeta,\zeta_0) = (\zeta-\zeta_0) \prod_\kappa \frac{(\zeta-\zeta_0^{(\kappa)})(\zeta_0-\zeta^{(\kappa)})}{(\zeta-\zeta^{(\kappa)})(\zeta_0-\zeta_0^{(\kappa)})}. \tag{10}$$

The relation of the primeform $\Omega(z,z_0)$ defined in (4) on p. 222 to the primefunction $E(\zeta,\zeta_0)$ is easily shown. First of all, in the construction of that primeform, in order to eliminate the zeros lying at the branch-points of the Riemann surface, we will let *the everywhere finite differential* $d\omega_z$ (cf. p. 229) take the place of the differential (z, dz). Let the primeform thus arising be called $\Omega'(z,z_0)$, whereby the places z, z_0 are to correspond to the arguments ζ,ζ_0; this $\Omega'(z,z_0)$ is identical with that form of the primeform, mentioned often on p. 222 ff., which Klein constructed by means of the homogeneous viewpoint based on the normal curve of the φ.[36] For the present group, the relations now lie very simply, insofar *as* $(\zeta, d\zeta)$ *directly represents an everywhere finite differential.* It follows: *The relation between the primeform* $\Omega'(z, z_0)$ *and the Schottky primefunction* $E(\zeta,\zeta_0)$ *is given by*:

$$\Omega'(z, z_0) = \zeta_2 \cdot \zeta_2^{(0)} \cdot E(\zeta,\zeta_0), \tag{11}$$

whereby we imagine ζ_0 split into $\zeta_1^{(0)} : \zeta_2^{(0)}$. One sees, that the Schottky primefunction will itself have poles at ∞ in dependence on ζ_0 as well as ζ, while $\Omega'(z, z_0)$ has the advantage of being everywhere finite. Apart from this (i.e. from the factor $\zeta_2\zeta_2^{(0)}$), however, Schottky, in his $E(\zeta,\zeta_0)$, for the particular model under discussion here, has obtained exactly that expression $\Omega'(z, z_0)$, which Klein has produced l.c., for arbitrary algebraic models[37].

For the automorphic functions of our model under discussion, by mediation of the primeform and on the basis of the formula (10), there arise from this, *representations by infinite products*, which directly bring into expression the zeros and poles of the individual function. However, it will scarcely be necessary, to also give these product developments here explicitly[38].

4.13 On the representability of the automorphic forms of arbitrary genus *p* by the elementary forms and the Poincaré series

[273]

It finally yet remains for us to answer the question exhaustively, *which automorphic forms* $\varphi_d(\zeta_1,\zeta_2)$ *of a proposed model with* $p > 0$ *are representable as convergent Poincaré series.* According to p. 261, it is self-evident, that *only unimultiplicative forms* are able to come into consideration here. In addition, naturally, the dimension d must be chosen in agreement with the convergence theorems; we shall assume $d \leq -2$, since we have also not pursued further above the possibly converging series of $(-1)^{\text{st}}$ dimension (cf. p. 166). In the case $d = -2$, the multiplicators system $\mathrm{M} = 1$, and with this, the properly automorphic forms, are to be excluded for the present.

---

[36] See the paper "*Zur Theorie der Abel'schen Functionen*", Mathem. Ann. bd. 36., p. 1 ff. (1889).

[37] Compare, moreover, the even more wide-ranging remarks of Klein l.c. on p. 13.

[38] Other attempts, of arriving at product developments from the Poincaré series, which are provided with transcendental, convergence-generating, supplementary factors, for the automorphic functions have been made by H. Mangoldt in the work "*Über ein Verfahren zur Darstellung elliptischer Modulfunctionen durch unendliche Producte etz.*" (Göttinger Nachrichten of 1886 Nr.1) and H.Stahl in the paper "*Über die Darstellung der eindevtigen Functionen, die sich durch lineare Substitutionen reproducieren, durch underdlicre Producte*" (Mathen. Annalen Vol. 33, 1888).

Recently, J.C.Kluyver has begun a continuation of the Schottky investigations (cf. p. 264); compare Kluyver's paper "*A special case of Dirichlet's problem for two dimensions*" (Acta mathematica, Vol. 21, 1897). In particular, the symmetric models are treated here, for which the halved fundamental domains are bounded by $p + 1$ symmetry circles. The potentials of these half-domains are drawn into investigation, whereby an analytic representation of these potentials is developed on the basis of the general Schottky products.

The question raised is treated above (p. 204 ff.) for the models of genus zero; we shall now show, that the argument there, with entirely insignificant alterations, also leads to our goal for $p > 0$.

We claim, first of all, that the studies concerning the production of a Poincaré series, which becomes infinite at the same places and in the same fashion as a proposed form $\varphi_d(\zeta_1,\zeta_2)$, remain valid without further ado. The assumption here is, as there, that none of the poles is situated at a possible parabolic vertex of the fundamental domain: *Each automorphic form $\varphi_d(\zeta_1,\zeta_2)$ of a dimension d satisfying the convergence theorems with poles which are arbitrary but not lying at the parabolic cusps, is representable by a Poincaré series, up to an additive integral automorphic form.*

In this manner, we have reduced our question to the special case, that $\varphi_d(\zeta_1,\zeta_2)$ is an *integral* automorphic form. Here then, an indirect procedure of reasoning comes into play, as was established above on p. 207 ff. for $p = 0$ on the consideration of the forms of *positive* dimension $\overline{d}$. Thereby, as above, let $\overline{d} = -d-2$ and let the multiplicator system $\overline{\text{M}}$ be inverse to that of $\varphi_d(\zeta_1,\zeta_2)$. For the models coming into consideration for the dimension $d = -2$, the case $\overline{d} = 0$ is naturally also admissible. Now, let $\overline{s}$ places of the fundamental domain be denoted by ${}^{(1)}\zeta, {}^{(2)}\zeta, \ldots, {}^{(\overline{s})}\zeta$. We assume, *that these $\overline{s}$ places are independent of one another and freely movable*; however, they are not to be able to move into possible parabolic vertices, and for the sake of simplicity, let them also remain distant from elliptic vertices of the fundamental domain. If we succeed in constructing *an automorphic form $\psi_{\overline{d}}$* of the non-negative dimension $\overline{d}$ and the multiplicator system $\overline{\text{M}}$, *which exhibits poles at these $\overline{s}$ places*, then according to a theorem proved on p. 258: [274]

$$\overline{s} \geq m - p + 2 = p - \overline{m}. \tag{1}$$

With the aid of the elementary form $\Omega(\zeta_1,\zeta_2;\xi_1,\xi_2)$ belonging to the present d,M, we now build up, for the production of a desired form $\psi_{\overline{d}}$, the expression:

$$B_1\Omega({}^{(1)}\zeta_1,{}^{(1)}\zeta_2;\xi_1,\xi_2) + B_2\Omega({}^{(2)}\zeta_1,{}^{(2)}\zeta_2;\xi_1,\xi_2) + \cdots + B_{(\overline{s})}\Omega({}^{(\overline{s})}\zeta_1,{}^{(\overline{s})}\zeta_2;\xi_1,\xi_2) \tag{2}$$

in which the B are to be quantities independent of the ξ_1,ξ_2. In the further discussion of this expression, the argument on p. 207 ff. remains subsisting in all essential points. The B are to be determined there, in such a manner, that the expression (2) represents an automorphic form of ξ_1,ξ_2 with respect to the operations of the group. The behavior of the individual elementary form upon exercise of U_i on ξ_1,ξ_2 is given in (7) on p. 264, if τ signifies the number of *linearly independent polefree Poincaré series* for d and M , and a system of such series is given by $\omega_1(\zeta_1,\zeta_2),\omega_2(\zeta_1,\zeta_2),\ldots,\omega_\tau(\zeta_1,\zeta_2)$.

The corresponding number of the *integral automorphic forms* existing, which vanish at the possible parabolic cusps, is $m-p+1$, according to a theorem stated on p. 254, where m is to be calculated according to the prescription given on p. 256. With this, we make note of the inequality for the number τ:

$$\tau \leq m - p + 1. \tag{3}$$

Now, for the discussion of the expression (2), one wants to reproduce exactly the argument, which is attached on p. 207 to the similar expression (7). One only has to put $\overline{s}$ instead of σ everywhere here for the number of poles, and, following the condition given in (3), must let the following inequality take the place of the one in (12) on p. 208:

$$1 \le \rho \le \tau \le m - p + 1. \tag{4}$$

[275] The free movability of the $\overline{s}$ places ${}^{(1)}\zeta, {}^{(2)}\zeta, \ldots$ and the linear independence of the $\omega'_1, \omega'_2, \ldots, \omega'_\rho$ (cf. p. 209) lead, as l.c., to the inequality $\overline{s} \ge \rho + 1$. But already for the lowest case $\overline{s} = \rho + 1$, we are actually successful in constructing a form $\psi_{\overline{d}}$. Therefore, according to the condition (1) just set up, it will also follow:

$$\overline{s} = \rho + 1 \ge m - p + 2, \qquad \rho \ge m - p + 1;$$

but then the comparison with (4) furnishes:

$$\rho = \tau = m - p + 1. \tag{5}$$

We therefore have the central theorem: *the number of linearly independent pole-free series for given d, M is equal to the corresponding number of the integral forms, which vanish at the parabolic cusps, which occur, say.*

On the strength of this, by means of an elementary argument, there arises, as the completion of the theorem set up preliminarily at the beginning of the paragraph, the proposition: *Each unimultiplicative automorphic form of our model of arbitrary p, vanishing at the possible parabolic cusps, as long as its dimensions satisfy the convergence theorems, can be represented in the form of a Poincaré series.*

In order to carry the study of the *forms $\psi_{\overline{d}}$ of non-negative dimension* to a conclusion, in the equation system (14) on p. 209 we have to take the number $\rho = m - p + 1$ for the $B_1, B_2, \ldots, B_{\overline{s}}$. We conclude, as at that time (p. 210), that of the $\overline{s}$ quantities B:

$$\overline{s} - m + p - 1 = \overline{s} + \overline{m} - p + 1 \tag{6}$$

are arbitrarily choosable, while the remaining ones then represent themselves linearly and homogeneously in those: *these are $(\overline{s} + \overline{m} - p + 1)$ linearly independent automorphic expressions (2), which exhibit poles of the first order at $\overline{s}$ freely movable places outside the fixed polygon vertices.*

According to the extended Riemann-Roch theorem (cf. p. 257), $(\overline{s} + \overline{m} - p + \sigma + 1)$ is now the number of linearly independent forms $\psi_{\overline{d}}$, which have poles of the first order at the places ${}^{(1)}\zeta, {}^{(2)}\zeta, \ldots, {}^{(\overline{s})}\zeta$. Thereby, σ denotes the number of linearly independent integral form φ_d, conjugate to $\psi_{\overline{d}}$, which vanish simultaneously at the $\overline{s}$ poles of $\psi_{\overline{d}}$. Now according to the investigations above, the most general integral form φ_d contains $(m - p + 1)$ constants linearly and homogeneously. The $\overline{s}$ required zeros furnish just as many linear homogeneous equations for the latter. If we now again lay stress on the free movability of the places ${}^{(1)}\zeta, \ldots, {}^{(\overline{s})}\zeta$, then those $\overline{s}$ equations are to be considered as independent of one another. In order then, that there be any desired integral forms φ_d
[276] at all, $m - p + 1 > \overline{s}$ is to be required. But this contradicts the condition (1), so that $\sigma = 0$ must take place.

With this, the forms of non-negative dimension are also disposed of: *the number of the linearly independent $\psi_{\overline{d}}$ is equal to $(\overline{s} + \overline{m} - p + 1)$ and therefore identical to the number of the linearly independent automorphic expressions (2) to be construced from elementary forms.*

The result obtained we are now also easily able to clothe in the form of the following theorem: *By ignoring for the moment, the properly automorphic "functions", we are* *able to represent each family of automorphic forms $\psi_{\overline{d}}(\xi_1,\xi_2)$ of non-negative dimension $\overline{d}$, with $\overline{s}$ "freely movable" poles of the first order outside the fixed vertices, in the form (2), as an aggregate of elementary forms, assuming that the convergence theorems of the Poincaré series, for the dimension $\overline{d}=-d-2$ in the ξ_1,ξ_2, permit the production of elementary forms at all.*

Last of all, it is also no longer difficult, to carry our question to a conclusion in the initially excluded case $d=-2, \mathrm{M}=1$.

Here, as we know, a non-integral form $\varphi_{-2}(\zeta_1,\zeta_2)$ has at least two poles in the fundamental domain. If in general , the number of the poles is $s(\geq 2)$, and these lie at:

$$\eta^{(1)},\eta^{(2)},\ldots,\eta^{(s)},$$

then as long as the poles are all of the first order and not lying at fixed polygon vertices, we form, with help of the associated coefficients $C_1,C_2,\ldots,C_s$ (cf. p. 259), the rational function:

$$\frac{C_1}{\zeta-\eta^{(1)}}+\frac{C_2}{\zeta-\eta^{(2)}}+\cdots+\frac{C_s}{\zeta-\eta^{(s)}}. \tag{7}$$

According to the theorem on the vanishing of the sum of all C (cf. (15) on p. 259), the sum of the s fractions (7) furnishes an expression, whose numerator is an integral function $G_{s-2}(\zeta)$ of the $(s-2)^{\mathrm{nd}}$ degree. If one then writes:

$$\zeta_2^{s-2}G_{s-2}(\zeta)=G_{s-2}(\zeta_1,\zeta_2),$$

then one has, in:

$$\sum_k \frac{G_{s-2}(\zeta_1^{(k)},\zeta_2^{(k)})}{(\zeta^{(k)},\eta^{(1)})\ldots(\zeta^{(k)},\eta^{(s)})}$$

A convergent Poincaré series, which becomes infinite at the same places and in the same manner as $\varphi_{-2}(\zeta_1,\zeta_2)$.

If poles of higher order are present outside lie fixed vertices, then, in the scheme (7), one will simply let a corresponding partial fraction of higher degree be added. Finally, it also poses no kind of difficulty, to even let poles supervene at the elliptic vertices, whose orders are then naturally to be read off from the rule (cf. p. 106) relevant in each case. For the construction of the series in these cases, the measures come into force, which were discussed for the elementary forms on p. 134 ff. According to this, here again arises the preliminary theorem, *that the individual automorphic form $\varphi_{-2}(\zeta_1,\zeta_2)$, with any poles situated outside possible parabolic cusps, is representable, up to an additive integral form, by a Poincaré series.* [277]

With this, everything is reduced to the question of the representability of the p *integral* forms $\Phi_{-2}(\zeta_1,\zeta_2)$ by Poincaré series. This question, however, disposes of itself correspondingly, as above, on the occasion of the representation of the properly automorphic "functions" $\psi(\xi)$ by the integrals of the second kind. We will write the latter, for short, by leaving out the lower limit ξ_0, $Y(\zeta_1,\zeta_2;\xi)$ (cf. p. 266). Let the function $\psi(\xi)$ which we want to consider, have only simple poles, at ${}^{(1)}\zeta,{}^{(2)}\zeta,\ldots,{}^{(\overline{s})}\zeta$. These $\overline{s}$ places are to remain distant from possible fixed polygon vertices, *but otherwise are to be independent of one another and freely movable.*

The function $\psi(\xi)$ we shall attempt to produce by means of an expression:

$$B_0 + B_1 Y({}^{(1)}\zeta_1, {}^{(1)}\zeta_2; \xi) + \cdots + B_{\overline{s}} Y({}^{(\overline{s})}\zeta_1, {}^{(\overline{s})}\zeta_2; \xi), \tag{8}$$

for whose closer investigation, the arguments followed on p. 207 ff. are again to come into force throughout. The behavior of the individual integral of the second kind, upon exercise of a substitution V of the group on ξ, is determined by equation (8) on p. 268. Here, τ signifies the number of linearly independent, properly automorphic, pole-free series of $(-2)^{\text{nd}}$ dimension, so that $1 \le \tau \le p$ occurs. Corresponding to the relations (14) on p. 209, we arrive at ρ independent linear homogeneous equations for the $B_1, B_2, \ldots, B_{\overline{s}}$, whereby:

$$1 \le \rho \le \tau \le p \tag{9}$$

occurs. The solvability of these equations require, as a necessary and sufficient condition, $\overline{s} \ge \rho + 1$. For $\overline{s} = \rho + 1$, we are also able to actually produce functions with freely movable poles. Now, however, as is known (cf. formula (12) on p. 258), just on account of the free movability of the poles, $\overline{s} = \rho + 1 \ge p + 1$, so that $\rho \ge p$ follows, and therefore, in consideration of (9), also, obviously $\rho = \tau = p$. With this, we find the theorem, already proved in another way above (p. 269) for a special species of groups belonging here: *the number of linearly independent pole-free Poincaré series of* $(-2)^{\text{nd}}$ *dimension*
[278] *with* $\mathrm{M} = 1$ *is (as long as these series converse at all) equal to* p *and therefore, just as great as the number of the linearly independent forms* $\Phi_{-2}(\zeta_1, \zeta_2)$, *i.e., those integral forms of* $(-2)^{\text{nd}}$ *dimension which belong to* $\mathrm{M} = 1$ *and vanish at possible parabolic cusps.*

Now, again, we easily complete the theorem stated preliminarily before: *Each properly automorphic form of* $(-2)^{\text{nd}}$ *dimension, vanishing at possible parabolic cusps, of an arbitrary model of genus* p, *is representable in the form of a Poincaré series.*

By considering both cases, namely $d < -2$ and $d = -2$, together, we arrive at the following conclusive theorem: *For the dimensions d finding themselves in agreement with the convergence theorems, all unimultiplicative forms, which vanish at the possible parabolic vertices, and only those forms, are representable as Poincaré series.*

Finally, we could continue our discussion of the expression (8), and with this, the representability of arbitrary functions $\psi(\xi)$ in the $Y(\zeta_1, \zeta_2; \xi)$. However, we return here simply to the well-known theorems on the *representation of algebraic functions of a Riemann surface in the associated integrals of the second kind.*

4.14 Closing remarks

The section of our lectures hereby brought to a close has furnished us the concepts and existence theorems of the automorphic functions and forms and of the polymorphic functions and forms inverse to them. In addition, we have subjected the properties of these quantities to a multifaceted study, and we have, in particular, also obtained a series of important results concerning the analytic ways of representing a series.

For our later purposes, of the applications of our functions, be it in the domain of theoretical development or be it for practical problems, the question of the *usability of those analytic representations for the calculation of the function values* will naturally move quite prominently into the foreground. We will yet add here a few remarks pertinent to this.

In the theory of the modular functions, and indeed, in particular, for the transformation theory, we have made the most extensive use of *power series developments* according to the development quantity $e^{2\pi i\omega}$, denoted in "M" by r. The usability of these series in the solution of the problems of the transformation theory was grounded in the circumstance, that the coefficients, in many cases, were whole numbers of easily [279]
surveyable arithmetic formation laws (see, e.g., "M" II. on p. 576 ff.).

We have now produced above (p. 148ff.), from the Poincaré series, power series of the automorphic forms as special cases. But also, in the neighborhood of any other place of the polygon net, it would be possible, to obtain power series developments from the Poincaré series, as we postulated them on pp. 68 and 69 for the automorphic forms. The individual term of the Poincaré series, as a rational function of ζ, admits a power series, referred to the neighborhood of the place in question, according to the development quantity t, defined l.c.. The sum of all these power series, corresponding to the Poincaré series, then permits (cf. p. 151) rearrangement according to powers of t, through which the desired power series is obtained.

Unfortunately, apart from the special case of the modular functions mentioned, the power series thus obtained have up until now, not yet been able to be practically evaluated. Namely, the development coefficients again clothe themselves in the form of infinite series, which, in the individual case, can initially only ba calculated approximately. Under these circumstances, there is no talk of that easy manipulation of the series of modular forms above, as it was obtained from the integrality of the development coefficients.

In this situation, one will do well, *to always reach back to the Poincaré series themselves for the representation and calculation of the automorphic forms.* These series not only have their important theoretical significance, as the two preceding chapters sufficiently prove, but it is not to be doubted, that they are of prominent value for the *numerical calculation* of the automorphic forms as well.

For possible applications, one will naturally not be able to start out, in complete mastery of the arithmetic formation law of a proposed group Γ (cf. I on p. 446 ff.). Rather, it is assumed, that one only knows the generators, from which one can calculate, by combination, additional substitutions of Γ, as they belong, say, to a certain number of polygon wreaths about the initial domain. By then summing the terms of a proposed Poincaré series, corresponding to these substitutions, one obtains an approximate value of the represented automorphic form; and here it is the convergence [280]
studies on p. 142 ff., 153 ff., 161 ff., 168 ff., which place at our disposal all important vantage points for *deciding the degree of the approximation reached.*

In this latter regard, i.e., concerning the rapidity of the convergence of the Poincaré series, one notices a highly heterogeneous behavior for different automorphic models. In particular, for the principal-circle groups, from the considerations on p. 168 ff., there arises the general principle, *that the series converge all the more rapidly, the greater is the hyperbolic area of the fundamental domain.* Also, the convergence investigation on p. 142 ff. leads to the same result. Let the principal-circle group of the signature (0, 3; 2, 3, 7) serve as an example, whose net is indicated adjacently in figure 34. The conver- [281]
gence of the series of the $(-4)^{\text{th}}$ dimension, in order to consider only these, are to be estimated by the sum of the areas of the triangles. Here there is only very slow convergence; for one has to collect a considerable number of triangles, in order to obtain a

fraction of the surface of the principal circle coming somewhat close to unity. The net actually drawn in the figure already includes 350 double triangles and yet only covers approximately $\frac{4}{5}$ of the entire surface of the circle.

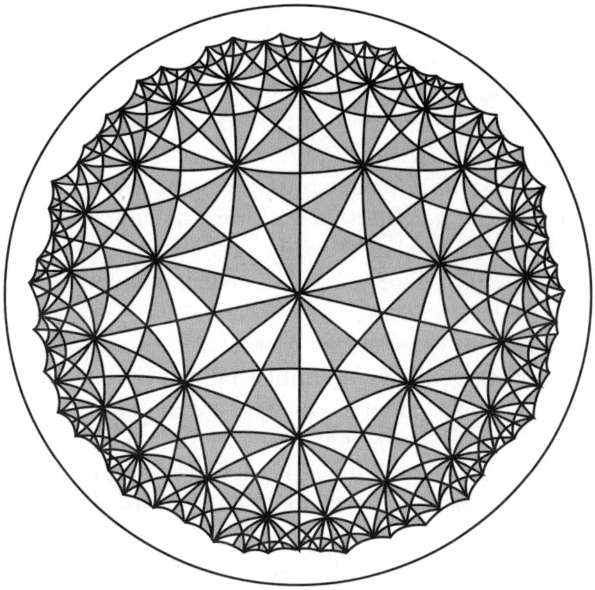

Fig. 34.

If, moreover, in the case of the example just considered; we go to a subgroup, say to that distinguished subgroup of index 168, studied most carefully in "M" I, then (in agreement with our general principle stated before) the rapidity of the convergence is most extraordinarily heightened. A single wreath of polygons about the initial polygon here already furnishes an approximation, which far exceeds the one just estimated.

This circumstance establishes, in general, an important significance, which the *subgroups* receive for the numerical calculation of the automorphic forms. We also met with the fact in the theory of the modular functions, that the power series of the modular functions belonging to subgroups of higher indices, such as $\sqrt[4]{k}$, $\sqrt[8]{1-k^2}$, converge far more rapidly, then the power series of the modular function $J(\omega)$ of the first level, provided with integral numerical coefficients. For the calculation of the functions of the total group from those of the subgroup, to be sure, rational calculations always still have to be exercised. These calculations, however, in comparison to the complications, which are attached to the evaluation of poorly convergent series, scarcely come into consideration.

Furthermore, for numerical calculations, just as in the theoretical developments of the foregoing chapter, one will prepare a preferential place for the integral forms, and with this, the pole-free series. In this regard, the *possibility of the identical vanishing of a pole-free series* furnished a very considerable difficulty for our developments above on p. 175 ff., the overcoming of which succeeded only by our far extended investiga-

tion of elementary forms. In contrast to this, it is very remarkable, *that in the praxis of numerical calculations that difficulty almost completely loses its significance.* While we possessed no direct means in the general theory, of determining that a proposed pole-free series does not identically vanish, this investigation usually offers no difficulty at all for a numerically given group with rapidly converging series. If one has summed the first m turns of the series to be investigated, then the convergence investigations on p. [282]
142 ff . . . furnish an upper limit for the corresponding remainder of the series. As soon as that sum value has greater value than the remainder of the series, then we have a non-identically vanishing series.

After these investigations, one may hope, that the automorphic functions, just by means of their representation by Poincaré series, are called upon to play a role in the domain of the applications, similar to the doubly periodic functions.

Part II Fundamental theorems concerning the existence of polymorphic functions on Riemann surfaces

[283] The concept of a linearly-polymorphic or, briefly, a "*polymorphic*" function $\zeta = f(z)$ on a Riemann surface F, as it yields itself upon inversion of the single-valued automorphic functions $z = \varphi(\zeta)$, has been discussed above (p. 43 ff). The function $\zeta = f(z)$ maps the suitably cut up surface F backwards single-valuedly and conformally onto the polygon P of the ζ-plane, which was originally given to us as the discontinuity domain of a group Γ. The substitutions:

$$\zeta' = \frac{\alpha\zeta + \beta}{\gamma\zeta + \delta}$$

of this group Γ make their appearance on the Riemann surface F in this fashion, by furnishing the infinitely many "branches" of the function $\zeta = f(z)$, which, in fact, substitute themselves linearly for closed circuits on the surface F. To there infinitely many branches correspond the infinitely many polygons P, which compose our polygon net N.

It is hereby to be considered as the most important fact, that not only the individual polygon, but also, the entire net of the polygons represents only a *simply* covered domain of the ζ-plane. Upon this base themselves the general "single-valuedness theorems" on p. 40 ff., as well as the theorems immediately preceding these on the single-valuedness of the integrals of an automorphic model in dependence on ζ.

Because of the great significance, thereby established, which the polymorphic function $\zeta = f(z)$ possesses for the investigation of the remaining functions of the surface F, Klein and Poincaré had early on turned themselves to the question, whether perhaps, *on each arbitrarily chosen Riemann surface F* (which, as is known, can serve
[284] as the definition of an associated algebraic model), there might always exist polymorphic functions $\zeta = f(z)$ of the kind above.[39] This question we have specialized above (p. 45 ff) in the form of three special "*fundamental problems*", whose wording in given there under I, II and III. The problems I and II referred themselves to automorphic models with principal circle and among them, problem I referred, in particular, to the case, that the principal circle is, at the same time, a limit circle of the net. Problem III concerned polygons P with $2p$ closed boundary curves, pairwise mapped onto one another loxodromically.

The solution to these problems is contained in three "*fundamental theorems*" on the existence of polymorphic functions on Riemann surfaces, which we correspondingly provided with the numerals I, II and III, and of which we also denote the first as the "*boundary-circle theorem*", the second as the "*principal-circle theorem*".

Furthermore, these theorems are contained as special cases in a general fundamental theorem, which Klein has set up in the fourth section of this paper "*Neue Beiträge zur Riemann'schen Functionentheorie*"[40]. This "general fundamental theorem" answers that problem, which we have sketched above, on p. 47 at the close of §11.

[39] The works of Klein and Poincaré coming into consideration are cited on p. 46 below the text.
[40] Cf. the footnote on p. 46.

1

Continuity studies in the domain of the principal-circle groups

The first proof of the fundamental theorem, conceived more or less simultaneously by Klein and Poincaré, rests upon continuity perceptions. The basic idea of this proof is, to place the individual continuum of conjugate polygons resp. groups opposite the coordinated continuum of Riemann surfaces, in order, from those properties, which one knows about the relation of the two continua to one another from the beginning, to argue to the mutual single-valuedness of this relation. [285]

Klein, in the year 1882, in the paper just cited "*Neve Beiträge zur Riemann'schen Functionentheorie*", has developed his ideas concerning that "continuity proof". Poincaré, in 1883, in the paper "*Sur les groupes des équations linéaires*" has gone into the continuity proof in application to the boundary-circle theorem, and, in a manner remarkable for a first penetration, has discovered the depth and difficulties of the continuity proof and devised ideas for its conquest.

An essential difficulty of the procedure of proof in question consists in this, that the continua, of which something is to be proved, are hardly surveyable. Concerning the continua of the groups, to be sure, in the *principal-circle* case, the *polygon theory of the principal-circle groups*, developed in I on pp. 210 to 398 permits us an extremely deep going insight. In opposition to this, the methods of representing the continua of Riemann surfaces, in spite of the far greater age of the approaches coming into consideration, are much less developed. If, corresponding to this, the continuity method, in consideration of conclusive and general proofs of the fundamental theorems, lags behind the methods to be developed in the second chapter, then these become very valuable for us all the same. By offering us a fairly extensively developed knowledge of the continua of automorphic models, and with this (just on account of the fundamental theorems), at the same time, of the continua of algebraic models. [286]

We must ignore a corresponding treatment of the continua of models without principal circle, since, in this domain, the results attained in Vol. I are still inadequate. We have to restrict ourselves here, accordingly, first of all, to the problems I and II, and therefore to the boundary-circle and the principal-circle theorems.

The following representation stems essentially from two papers of the author, of which one concerns the polygon continua of the principal-circle groups, while the other furnishes the actual effectuation of the continuity proof in a few lower cases.[1]

1.1 Recapitulation of the polygon theory of the principal-circle groups

In the polygon theory of the principal-circle groups developed in I on p. 210 ff., the operations of our group Γ are not assumed in their original form as ξ-substitutions, but in projective form, in which form they represent collineations of the ellipse, given by:

$$z_1 z_3 - z_2^2 = 0,$$

into itself (cf. I on p. 12 ff.). The advantage of this measure consists, first, in this, that in the place of the circle-geometry of the ζ-halfplane resp. the ζ-plane, we get the *rectilinear* projective geometry of the interior of the ellipse, more accessible to the eye. Secondly, for the projective treatment, those two cases of principal-circle groups, which we have to distinguish, according as the principal circle is, at the same time, a boundary circle or not (cf. our problems I and II, on p. 45 ff.), are only so unessentially different, that a common treatment of both cases appears possible, which was then also fundamentally carried through in I l.c.. The individual polygon P runs either entirely in the interior of the ellipse, by reaching the ellipse at most at a few parabolic cusps (the principal circle is also a boundary circle), or it reaches, with one or more hyperbolic cusps, over beyond the ellipse, while, except for this, it runs in the interior of the ellipse (the principal circle is not a boundary circle).

Let the genus attaching to the projective polygon P, according to I on p. 179, be
[287] called p; if, moreover, P has n fixed vertices(elliptic, parabolic or hyperbolic), then we called (p, n) the "*character*" of polygon (cf. I on p. 262). If, of the n fixed vertices, ν in all are elliptic and therefore situated in the interior of the ellipse, then the polygon angles taking place here are aliquot ports $\frac{2\pi}{l_1}, \frac{2\pi}{l_2}, \ldots, \frac{2\pi}{l_\nu}$ of 2π, where therefore, the $l_1, l_2, \ldots, l_\nu$ are finite positive whole numbers ≥ 2. The collection $(p, n; l_1, l_2, \ldots, l_\nu)$ we designate as the "*signature*" of the polygon P (cf., e.g., I on p. 353).[2]

Of the various shapes of the polygons P, we have, first of all, to utilize the "*normal polygons*", which were introduced in I on p. 106, and whose detailed theory finds itself developed in I on p. 210 ff. The individual normal polygon has a center C, and its inner points are able to be defined as the totality of all those places of the polygon net N,

[1] "*Über die in der Theorie der automorphen Functionen auftretenden Polygon-continua*", Göttinger Nachrichten 1903, Heft. 5. "*Beiträge zum Continuitätsbeweise der Existenz linear-polymorpher Functionen auf Riemann'schen Flächen*" Math. Ann., Vol. 59 (1904).

[2] In the text, we assume the numbers l as *finite* and consequently coordinate the parabolic vertices to the hyperbolic ones as limit cases. Naturally, one can also allow $l = \infty$, and with this, consider the parabolic cusps as limit cases of the elliptic vertices. However, we shall usually hold fast to the former conception.

which are situated closer to the center C than to any point equivalent to C.[3] A normal polygon of the character (p, n) has, in general, $(12p+4n-6)$ sides, n fixed vertices, and $(4p+n-2)$ three-termed cycles of movable vertices; the sides are all straight lines, the angles consistently concave.[4] For particular positions of the center C, decreases of the stated number of sides are able to enter, with simultaneous increases of the numbers of the polygons surrounding the vertices (so-called "special types" or "transitional types", cf. I on p. 263 ff.).

Secondly, we have to work with "*canonical polygons*" P, so called, since the boundary of P furnishes a "canonical" cut system on the associated closed surface F (cf. figure 40 in I on p. 183). One can attain a decrease of the number of sides of the canonical polygon $(2n+4p)$, by drawing all cuts c, in the figure just cited, together into a point, and pushing the crossing-place of each two cuts a and b to the point E. The detailed theory of these canonical polygons is developed in I on p. 284ff. for the principal-circle groups; it peaks in the theorem obtained in I on p. 319, that in each case (i.e., however one might choose the cut system on the closed surface), one can fashion the canonical [288]
polygon as a *rectilinear* polygon with $(2n+4p)$ sides and with all *concave* angles.

First on the foundation of the theory of the canonical polygons could we, in I on p. 335 ff., obtain an overview of the *continua of all principal-circle polygons of given signatures* and prepare the way for an *invariant representation of these continua.* In the former connection, we arrived at a general theorem in I on p. 389, according to which, to the individual signature $(p, n; l_1, l_2, \dots, l_\nu)$, there always belongs a single $(3n-\nu+6p-6)$-fold infinite continuum of polygons.

The transition from the polygons P to the groups Γ corresponding to them, is to be represented in detail below (which will occur in §8). Already here, we remember, that two signatures with equal p, n, ν, which distinguish themselves only in arrangement of the whole numbers l, indeed furnish distinct polygon continua, but one and the same group continuum; for we could, for the canonical cut system on the closed surface (cf. figure 40 in I on p. 183), give the cuts d to the points e a completely arbitrary arrangement. Further, remember the effect of the transformation of the canonical polygons discussed in I on p. 320 ff. The infinitely many canonical polygons, which proceed from one of them by transformation, furnish infinitely many distinct polygons in the polygon continuum; but all these polygons always corresponded to one and the same group Γ.

Consequently, by keeping the concepts of the polygon continua and the group continua sharply separated from one another at the very beginning, we return, first of all, to the *theory of the polygon continua,* which we have to essentially expand vis-a-vis the representation adhered to in Vol. I. The fundamental viewpoints for the transition to the group continua we shall then develop (as already remarked) in §8.

For the sake of completeness, we also remember, that for $p=0$, the number n must be ≥ 3, and that, correspondingly, for $p=1$, the number n is to be taken ≥ 1, in order that we have to do with "principal circle groups". Also, for the character (0, 3), we have to exclude the signature $(0,3; l_1, l_2, l_3)$ as elementary; for here, the continuum of the

[3] Self-evidently, these statements concerning measure always refer themselves, as in I on p. 107 ff., to the projective determination of measure, which starts out from the ellipse $z_1 z_3 - z_2^2 = 0$ as the "absolute conic section".

[4] With the inclusion of the limit case of a recumbent angle.

polygons reduces to the single double triangle, composed of two circular-arc triangles of the angles $\frac{\pi}{l_1}, \frac{\pi}{l_2}, \frac{\pi}{l_3}$, symmetric to one another.

1.2 The polygon continua of the character (0, 3)

[289] The invariant theory of the polygon P of the character (0, 3) is treated in I on p. 341 ff. To the individual polygon P, as figure 35 indicates, we are able to give the shape of a hexangle. The coordination of the angle sides furnish the three group generators V_1, V_2, V_3, which are bound by the relation.

$$V_1 V_2 V_3 = 1.$$

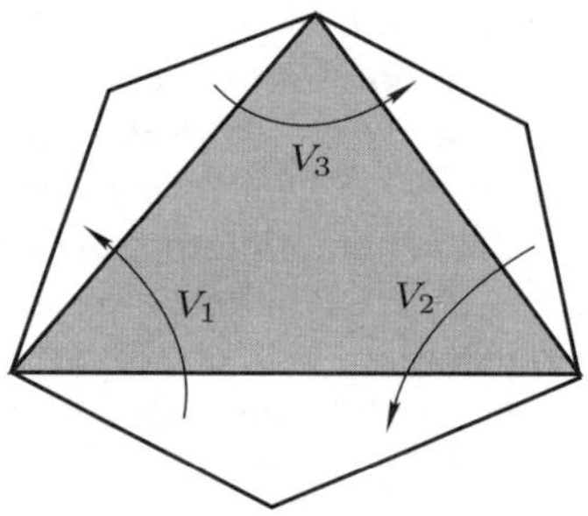

Fig. 35.

The three invariants j_1, j_2, j_3 of these substitutions are, at the same time, all the invariants of the polygon P. For the three invariants j_{12}, j_{23}, j_{31} of the substitution pairs, there hold, according to I on p. 347:

$$j_{12} = -j_3, \qquad j_{23} = -j_1, \qquad j_{31} = -j_2.$$

For these j_1, j_2, j_3, we found, in I on p. 348, the conditions:

$$\begin{cases} j_1 \geq 0, \quad j_2 \geq 0, \quad j_3 \geq 0, \\ j_1^2 + j_2^2 + j_3^2 + j_1 j_2 j_3 > 4, \end{cases} \tag{1}$$

with the addition, that for $j_\kappa < 2$, the equation:

$$j_\kappa = 2\cos\frac{\pi}{l_\kappa} \tag{2}$$

must hold. Also, conversely, each system of real numbers j_1, j_2, j_3, which satisfy the conditions just named, furnished a determinate polygon. On this account, we designated the relations in question as "*the characteristic conditions*" for the invariants of the polygons of the character (0, 3).

Since, in addition, according to the closing remark in §1, the case that all three $j_\kappa < 2$ is to be excluded here as elementary, then at least one of the invariants j_κ, say j_3, is ≥ 2. Accordingly, if the fourth inequality (1) is always fulfilled spontaneously, then we are dealing with the lowest case still admissible, $j_1 = j_2 = 0, j_3 = 2$. This belongs as a limiting case (parabolic dihedral group) to the groups $j_1 = j_2 = 0, j_3 > 2$ (hyperbolic dihedral groups), for which the fourth relation (1) is fulfilled. *The system of the characteristic conditions consequently reduces itself, for us (apart form the equations (2) possibly present), to the first three inequalities (1).*

For the sake of easier visualization, in the case $j_\kappa \geq 2$, we will introduce, in place of j_κ, the quantity:

$$t_\kappa = \frac{j_\kappa - 2}{j_\kappa} \tag{3}$$

and utilize it as an invariant of the polygon. The set of values characterized by $j_\kappa \geq 2$ [290] thereby carries over single-valuedly and continuously onto the set given by $0 \leq t_\kappa < 1$.

For the purpose of a geometrical language, we interpret the t as rectangular coordinates, and obtain thus the surveyable theorem: *the one-dimensional continuum of the polygons of each individual signature* $(0,3;l_1,l_2)$ *appears single-valuedly and continuously mapped onto the points of the rectilinear segment* $0 \leq t_3 < 1$*; the two-dimensional continuum of the polygons of the individual signature* $(0,3;l_1)$ *is single-valuedly and continuously mapped onto the points of a square* $0 \leq t_2 < 1, 0 \leq t_3 < 1$*; the three-dimensional continuum of the polygons of the signature* $(0,3)$ *lets itself be mapped single-valuedly and continuously onto the points of the cube* $0 \leq t_1 < 1, 0 \leq t_2 < 1, 0 \leq t_3 < 1$.

If, in the individual case, one fixes a variable invariant, then with this, from the model in question, one such of the next lower number of dimensions is cut out, therefore, from the cube a square, etc.

In consideration of the investigations immediately following, we must yet procure ourselves the form of the generators V_κ as ζ-substitutions, whereby the real ζ-axis is counted as the principal circle. We may thereby assume the substitution V_3 as hyperbolic (in consideration of the future application of the formulas to be derived), ie., we exclude the parabolic limiting case $j_3 = 2$. Corresponding to the equation $V_3^{-1} = V_1 V_2$, we would rather write $-j_{12}$ for the invariant j_3, which will show itself to be expedient. Introduce ζ, so that the fixpoints of V_3 fall at $\zeta = 0$ and $\zeta = \infty$, and so that $|\alpha_3| < |\delta_3|$, understanding by α_3 and δ_3 the first and last coefficients of V_3. V_3 is then uniquely determined from the invariant $j_{12}(= -j_3)$, and indeed, one finds:

$$V_1 V_2 = V_3^{-1} = \begin{pmatrix} \frac{1}{2}(j_{12} - \sqrt{j_{12}^2 - 4}) & 0 \\ 0 & \frac{1}{2}(j_{12} + \sqrt{j_{12}^2 - 4}) \end{pmatrix} \tag{4}$$

whereby the last $\sqrt{j_{12}^2 - 4}$ is to be taken positively.

Without altering V_3, we are now still able to transform ζ by means of any hyperbolic substitution with the fixpoints 0 and ∞. The latter transformation, in particular, lets itself be so chosen, that the equation $\beta_2 + \gamma_2 = 0$ holds for the coefficients of V_2 (cf. the discussions on the invariants in the case (0, 3) in I on pp. 335 ff and 341 ff.). With consideration of the arrangement of the V_1, V_2, V_3 determined by figure 35, the calculation [291] leads to the following form of V_2:

$$V_2 = \begin{pmatrix} \frac{1}{2}(j_2 - \frac{j_2 j_{12} - 2j_1}{\sqrt{j_{12}^2-4}}) & \frac{J_{12}}{\sqrt{j_{12}^2-4}} \\ -\frac{J_{12}}{\sqrt{j_{12}^2-4}} & \frac{1}{2}(j_2 + \frac{j_2 j_{12} - 2j_1}{\sqrt{j_{12}^2-4}}) \end{pmatrix} \tag{5}$$

where J_{12} is an abbreviation for the following square root, to be taken positively:

$$J_{12} = \sqrt{j_1^2 + j_2^2 + j_{12}^2 - j_1 j_2 j_{12} - 4}. \tag{6}$$

The geometric meaning of this form of V_2 is this, that in the elliptic case ($j_2 < 2$), the fixpoints of this substitution lie on the unit circle of the ζ-plane to the left of the imaginary axis; in the parabolic case ($j_2 = 2$), the fixpoints move together at $\zeta = -1$, while, finally, in the hyperbolic case ($j_2 > 2$), they more away from one another again onto the real ζ-axis, and are mirror images with respect to the unit circle.

The substitution $V_1 (= V_3^{-1} V_2^{-1})$ writes itself most complicatedly for the choice of ζ made:

$$V_1 = \begin{pmatrix} \frac{1}{2}(j_1 - \frac{j_1 j_{12} - 2j_2}{\sqrt{j_{12}^2-4}}) & \frac{1}{2}(-j_{12} + \sqrt{j_{12}^2 - 4}) \cdot \frac{J_{12}}{\sqrt{j_{12}^2-4}} \\ \frac{1}{2}(j_{12} + \sqrt{j_{12}^2 - 4}) \cdot \frac{J_{12}}{\sqrt{j_{12}^2-4}} & \frac{1}{2}(j_1 + \frac{j_1 j_{12} - 2j_2}{\sqrt{j_{12}^2-4}}) \end{pmatrix}. \tag{7}$$

1.3 The polygon continua of the character (0, 4)

One obtains a suitable overview of the continua of the polygons of the character (0, 4) on the basis of the *principle of composition*, already brought into application many times in I (cf. I on pp. 190 ff., 363 ff.).

Let an individual polygon P of the character $(0,4)$ furnish the four generating substitutions V_1, V_2, V_3, V_4. Then the three substitutions $V_1, V_2, (V_1 V_2)^{-1}$, taken by themselves, generate a group of the character (0, 3). Thereby, in each case, $j_{12} < -2$, i.e., the third generator $(V_1 V_2)^{-1}$ of this group of the character (0, 3) is hyperbolic. But the substitutions $V_3, V_4, (V_3 V_4)^{-1}$ also generate a corresponding group; and indeed, on account of $(V_1 V_2) \cdot (V_3 V_4) = 1$, the third (hyperbolic) generators $(V_1 V_2)^{-1}$ and $(V_3 V_4)^{-1}$ are hereby inverse to one another, so that $j_{12} = j_{34}$ holds.

One can now produce the proposed polygon P of the character $(0,4)$ *by composition from the two polygons of the groups of the character* $(0,3)$ *just named* (cf. I on p.
[292] 363). The geometric relations hereby coming into consideration one surveys best in the projective plane. The polar to be denoted by Π, say (cf. figure 36) of the fixpoint of the common hyperbolic generator of the two groups of the character (0, 3) hereby serves suitably for orientation with respect to the ellipse. The individual one of the polygons to be composed lies with five of its vertices on one side of Π and sends over Π only one hyperbolic cusp to the fixpoint, lying outside the ellipse, of the common hyperbolic generator (to the pole of Π, cf. figure 36). Thereby, both polygons to be composed lie opposite one another with respect to Π, so that one of the two hyperbolic cusps (in figure 36, that belonging to the generator $(V_3 V_4)^{-1}$) runs through the point at infinity[5].

[5] The reader wants to consult the closer details of the composition process in I, where, in particular, the argument on p. 313 ff. comes into consideration.

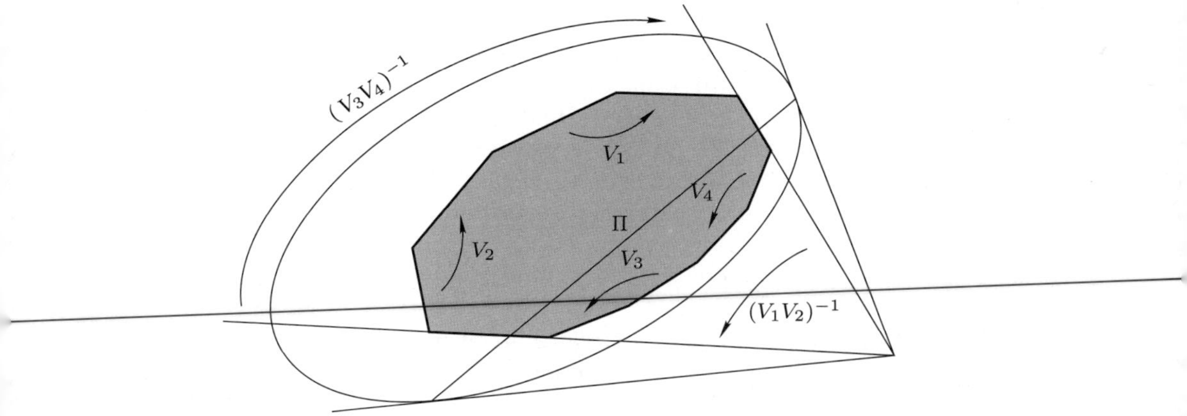

Fig. 36.

The preceding argument is now invertible. The theorem holds, *that only two polygons of the character* (0,3) *of our kind with equal third (hyperbolic) invariant are always composable, in infinitely many ways into a polygon of the character* (0,4). Namely, choose an arbitrary chord as the polar Π and make its pole into the fixpoint of the common hyperbolic generator of the two groups to be composed. Next, arrange the two polygons of the character (0, 3) on opposite sides of Π, but except for this, arbitrarily. Their two hyperbolic cusps in question have equal breadth (naturally, measure in the sense of the projective determination of measure), but do not need to coincide. Nevertheless, for one of the two polygons, by means of allowed modification, one can push the hyperbolic cusp continuously along Π, a measure, which was described most [293]
carefully in I on p. 298 (to be sure, in application to the polygons of the character (1, 1)). Carry this pushing so far, until the hyperbolic cusp of the allowed polygon coincides with that of the other polygon, remaining fixed, whereupon the composition can be smoothly effected. The composed polygon hereby comes out of this, in general, in a distorted form; nevertheless, according to the theorems mentioned in §1 on p. 288, concerning the shape of the canonical polygons, this can be changed around into a rectilinear octangle with all concave angles.

We now establish the *invariant theory of the polygons of the character* (0,4), with utilization of the preceding composition considerations, thus: We first give j_1, j_2, j_{12}, and calculate the associated substitutions V_1, V_2 in the forms (7) and (5) on p. 291. the polar Π then furnishes the imaginary axis in the ζ-halfplane.

It is expedient for understanding the composition, to remember, that a group of the character (0, 3) is always capable of extension by reflections. In the present case, the unit circle of the ζ-plane is a first symmetry circle of the group to be generated from V_1 and V_2; for this circle, carried over onto the projective plane, furnishes the joining-line of the two fixpoints[6] of V_2 and $(V_1V_2)^{-1}$.

[6] If we speak of "the fixpoint" of a substitution in the projective plane, then in the case of a hyperbolic substitution, the fixpoint situated outside the ellipse is always meant.

If we now arrange the second polygon to be composed corresponding to the invariants $j_3, j_4, j_{34} = j_{12}$, on the other side of the imaginary ζ-axis (polar Π) in such a manner, that in the projective plane, the fixpoint of V_3 likewise lies on the line corresponding to the ζ-unit circle, then the second group of the character (0, 3), to be generated from V_3, V_4, also has the unit circle as symmetry circle, and here arises from the composition a group of the character (0, 4), capable of extension by reflections. Visualize for oneself, that thereby, in particular, the position of the two elementary triangles belonging to $V_1, V_2, (V_1 V_2)^{-1}$ resp $V_3, V_4, (V_3 V_4)^{-1}$ (symmetrically halved polygons) is such a one, that, for the composition, they directly furnish the elementary quadrangle of the composed group.

But now, by holding V_1, V_2 fast, one can still transform the substitutions V_3, V_4 just chosen by an arbitrary substitution $V = \begin{pmatrix} \tau & 0 \\ 0 & \tau^{-1} \end{pmatrix}$ with $\tau > 0$, without the intermediately
[294] transformed polygon of the second group (the one to be generated from V_3, V_4) ceasing to have the third (hyperbolic) generator in common with the polygon belonging to V_1, V_2. Each such transformed polygon, as we saw, is then still composable with the first polygon, on which the transformation V was not exercised. The previously mentioned symmetric shape of the polygon of the character (0, 4) thereby appears with $\tau = 1$ as the middle position.

Hereupon, first of all, we have invariants j_1, j_2, j_3, j_4, corresponding individually to the generators V_1, V_2, V_3, V_4, which are subjected to the known conditions, that either:

$$j_\kappa \geq 2 \qquad \text{or} \qquad j_\kappa = 2\cos\frac{\pi}{l_\kappa}.$$

holds. Let these four invariants be designated as "*vertex invariants*". There follows the invariant $j_{12} (= j_{34})$ with the condition $j_{12} < -2$; let j_{12} be called a "*transversal invariant*", insofar as the orbit curve Π belonging to j_{12} resp. $V_1 V_2 = (V_3 V_4)^{-1}$ is a transversal of the polygon of the character (0, 4). Finally, we have in $(\tau + \tau^{-1})$, or, which is more to the purpose, in τ itself, a sixth invariant, which we want to call the "*composition invariant*".

For the purpose of a surveyable representation in geometric form, we set (as in §2):

$$\frac{j_1 - 2}{j_1} = t_1, \ldots, \frac{j_4 - 2}{j_4} = t_4, \quad \frac{j_{12} + 2}{j_{12}} = t_5, \quad \frac{\tau}{\tau + 1} = t_6. \tag{1}$$

And again interpret the t as rectangular coordinates. On the strength of this, the following theorem yields itself concerning the continua of the polygons of the character (0, 4): *Each two dimensional continuum of the polygons of signature* $(0, 4; l_1, l_2, l_3, l_4)$ *is single-valuedly and continuously mappable onto the square* $0 < t_5 < 1$, $0 < t_6 < 1$; *each three dimensional continuum of the polygons of a signature* $(0, 4; l_1, l_2, l_3)$ *is single-valuedly and continuously mappable onto the cube* $0 \leq t_4 < 1, 0 < t_5 < 1, 0 < t_6 < 1$ *etc.; finally: the six dimensional continuum of the polygons of the signature* $(0, 4)$, *corresponding to the case* $\nu = 0$ *with only hyperbolic vertices, is single-valuedly and continuously mappable onto the "six dimensional cube"* $0 \leq t_1 < 1, \ldots, 0 \leq t_4 < 1, 0 < t_5 < 1, 0 < t_6 < 1$. We have hereby designated the regular solid of a number of dimensions > 3, corresponding to the ordinary (three dimensional) cube, as a "cube" again.

The occurrence of the equality signs stipulated for the vertex invariants furnish polygons with parabolic cusps. If one would, moreover, hold one of the vertex invariants, considered variable, fixed, then a cube of the text lower number of dimensions is thereby cut of the particular model at hand.

1.4 The polygon continua of the character (0, *n*)

The argument carried through is capable of generalization without further ado. [295]

Let a polygon of the character (0, 4) and the generators V_1, V_2, V_3, V_4' be proposed, of which, in any case, the last, V_4', is to be hyperbolic. Let the invariants be $j_1, j_2, j_3, j_4', j_{34}'$, τ_1, whose meaning proceeds from §3. Secondly, let a polygon of the character (0, 3) with the generators $V_4, V_5, V_4'^{-1}$ be chosen, of which the third, $V_4'^{-1}$, as the notation already indicates, is to be inverse to the fourth generator V_4' of the first group. The two polygons are then immediately composable into a polygon of the character (0, 5), and conversely, we are able to imagine each polygon of the character (0, 5) as arising in this manner[7].

If we permit one of the polygons to be composed, say that of the character (0, 3), movability along the polar Π of the fixed point of V_4', then it is clear, at the same time, that the composition of the two polygons can again be effected in simply infinitely many ways.

As invariants of the composed polygon we bring in, first of all, the six invariants above of the polygon of the character (0, 4). Only hereby, with consideration of the second polygon to be composed, the invariant j_4' is expediently replaced by $j_{45} (= -j_4')$, and in place of j_{34}', we better utilize the notation j_{345}, since it is a question here of the invariant of the substitution $V_3 V_4 V_5$. From the second of the polygons to be composed, there then also originate the invariants j_4, j_5, and finally, the composition itself furnishes one more invariant τ_2.[8] All together, therefore, we have five vertex invariants $j_1, j_2, \ldots, j_5$, two transversal invariants j_{345}, j_{45} and two composition invariants τ_1, τ_2. The inequalities valid for these invariants are precisely the same as those given in §3 for the corresponding invariants of the polygons of the character (0, 4).

It holds not the slightest difficulty, to carry the composition process and the production of the invariants further in the same fashion. Obviously, one arrives at the following general theorem: *To the polygons of the character* $(0, n)$ *belong* $(3n-6)$ *invariants, namely n "vertex invariants"* $j_1, j_2, \ldots, j_n$, *subjected to the conditions*: [296]

$$j_\kappa \geq 2 \quad resp. \quad j_\kappa = 2\cos\frac{\pi}{l_\kappa}, \tag{1}$$

[7] One finds the details of this composition process in I on p. 361 ff.

[8] The first composed polygon, corresponding to the value $\tau_2 = 1$, we imagine chosen arbitrarily in the text among all simply infinitely many polygons of the character (0, 5) arising by composition of the two polygons of the characters (0, 4) and (0, 3). If one will remove the indeterminacy, which thereby adheres to the introduction of the composition invariant τ_2, then one can stipulate, say, that $\tau_2 = 1$ for the generators $(V_1 V_2)$, V_3, V_4, V_5 furnishes the symmetric case of the associated polygon of the character (0, 4). This prescription would immediately attach itself to the introduction of the invariant τ in §3.

$(n-3)$ *further "transversal invariants", which we logically denoted by* $j_{3,4,\ldots,n}, j_{4,\ldots,n}, \ldots, j_{n-1,n}$, *and which satisfy the inequalities*:

$$j_{3,4,\ldots,n} < -2,\ j_{4,\ldots,n} < -2, \ldots, j_{n-1,n} < -2, \tag{2}$$

finally, $(n-3)$ *"composition invariants"* $\tau_1, \tau_2, \ldots, \tau_{n-3}$, *with the conditions*:

$$\tau_1 > 0, \tau_2 > 0, \ldots, \tau_{n-3} > 0. \tag{3}$$

All $(3n-6)$ *invariants are independent of one another and are subjected to no other conditions than those given.*

If, in the case of an arbitrary n, one introduces $(3n-6)$ quantities t exactly as in (1) on p. 294 for $n=4$ and interprets the t as rectangular coordinates, then one arrives at the following conclusive theorem for the manifold of the polygons of the character $(0,n)$: *the polygons belonging to the individual signature* $(0,n;l_1,l_2,\ldots,l_\nu)$ *for a* $(3n-\nu-6)$ *dimensional continuum, which is single-valuededly and continuously mappable onto the* $(3n-\nu-6)$ *dimensional cube represented by the inequalities*:

$$0 \le t_{\nu+1} < 1, \ldots, 0 \le t_n < 1, 0 < t_{n+1} < 1, \ldots, 0 < t_{3n-6} < 1.$$

Such boundary points of the cube, which correspond to the equality signs stipulated for the vertex invariants, furnish polygons with one or more parabolic cusps. Also here, the fixing of a vertex invariant initially considered variable leads to a cube of the next lower number of dimensions. This is of importance for the later continuations of the composition procedure.

1.5 Another representation of the polygon continua of the character (0,4)

The special simplicity of the results for the preceding representation of the polygon continua rests on the *unsymmetric* introduction of the invariants. In I on p. 361 ff., on the other hand, the invariants were introduced in a fashion corresponding more to the requirements of symmetry. At least in the case of the character $(0,4)$. The relation of the two invariant representations (the present unsymmetric ones and the one preferred in
[297] I l.c.) is to be pursued, and indeed, mainly since we arrive at relations here, which we shall immediately evaluate in the discussion of the polygons of the character $(1,1)$.

Before (§3 on p. 291 ff.), for the polygons of the character (0, 4), we had started out form the six invariants:

$$j_1, j_2, j_3, j_4, j_{12} = j_{34}, \tau, \tag{1}$$

which were independent of one another. In I (see e.g., on p. 394), in contrast to this, the system of the eight invariants

$$j_1, j_2, j_3, j_4, j_{12} = j_{34}, j_{23} = j_{41}, j_{13}, j_{24} \tag{2}$$

was chosen, which is fashioned symmetrically, insofar as the cyclic interchange of the indices 1, 2, 3, 4 carries the system over into itself. The two relations subsisting between the invariants (2):

$$\begin{cases} j_{12}^2+j_{23}^2-j_{13}j_{24}-(j_1j_2+j_3j_4)j_{12}-(j_1j_4+j_2j_3)j_{23} \\ \quad +(j_1j_2j_3j_4+j_1^2+j_2^2+j_3^2+j_4^2-4)=0, \\ j_{12}j_{23}+j_{13}+j_{24}-(j_1j_3+j_2j_4)=0, \end{cases} \tag{3}$$

naturally go over into themselves, similarly, for cyclic interchange of indices 1, 2, 3, 4.

The first five invariants (2) are also contained among the invariants (1). The last three invariants (2), namely j_{23}, j_{12}, j_{24}, must be representable uniquely in the individual invariants (1).

In order to accomplish these representations, we have to calculate the generators V_1, V_2, V_3, V_4, whereby we are able to take the substitutions V_1, V_2 over into the form given by (7) and (5) on p. 291. First calculate this substitutions V_3, V_4 for the middle position of a polygon capable of extension by reflections (compare, for this, §3 on p. 293) and, with strict observance of the representation given on p. 293 ff., transform the two substitutions arising by $\begin{pmatrix}\tau & 0\\ 0 & \tau^{-1}\end{pmatrix}$. In place of j_{34}, write j_{12} in the coefficients of V_3 and V_4; in addition, corresponding to the abbreviated notation J_{12} introduced on p. 291, use the abbreviation of J_{34} for the root, to be taken positive:

$$J_{34}=\sqrt{j_3^2+j_4^2+j_{34}^2-j_3j_4j_{34}-4}. \tag{4}$$

One will find the following forms of the two generators still lacking:

$$V_3=\begin{pmatrix} \frac{1}{2}\left(j_3+\frac{j_3j_{12}-2j_4}{\sqrt{j_{12}^2-4}}\right) & \frac{J_{34}\cdot\tau^{-2}}{\sqrt{j_{12}^2-4}} \\ -\frac{J_{34}\cdot\tau^{2}}{\sqrt{j_{12}^2-4}} & \frac{1}{2}\left(j_3-\frac{j_3j_{12}-2j_4}{\sqrt{j_{12}^2-4}}\right) \end{pmatrix}, \tag{5}$$

$$V_4=\begin{pmatrix} \frac{1}{2}\left(j_4+\frac{j_4j_{12}-2j_3}{\sqrt{j_{12}^2-4}}\right) & \frac{1}{2}\left(-j_{12}+\sqrt{j_{12}^2-4}\right)\frac{J_{34}\cdot\tau^{-2}}{\sqrt{j_{12}^2-4}} \\ \frac{1}{2}\left(j_{12}+\sqrt{j_{12}^2-4}\right)\frac{J_{34}\cdot\tau^{2}}{\sqrt{j_{12}^2-4}} & \frac{1}{2}\left(j_4-\frac{j_4j_{12}-2j_3}{\sqrt{j_{12}^2-4}}\right) \end{pmatrix}, \tag{6}$$

whereby, as also above (p. 290), the root $\sqrt{j_{12}^2-4}$ is to be taken positive. [298]

The calculation of invariants j_{13}, j_{23}, j_{24} from the generators is now effected simply, according to the relevant prescriptions in I on p. 337 ff.; one finds:

$$\begin{cases} j_{13}(j_{12}^2-4)= j_{12}(j_1j_4+j_2j_3)-2(j_1j_3+j_2j_4) \\ \qquad +\frac{1}{2}\left\{(\tau^2+\tau^{-2})j_{12}-(\tau^2-\tau^{-2})\sqrt{j_{12}^2-4}\right\}J_{12}J_{34}, \\ j_{23}(j_{12}^2-4)= j_{12}(j_1j_3+j_2j_4)-2(j_1j_4+j_2j_3)-(\tau^2+\tau^{-2})J_{12}J_{34}, \\ j_{24}(j_{12}^2-4)= j_{12}(j_1j_4+j_2j_3)-2(j_1j_3+j_2j_4) \\ \qquad +\frac{1}{2}\left\{(\tau^2+\tau^{-2})j_{12}+(\tau^2-\tau^{-2})\sqrt{j_{12}^2-4}\right\}J_{12}J_{34}. \end{cases} \tag{7}$$

In consideration of the determination, that the square roots $\sqrt{j_{12}^2-4}, J_{12}, J_{34}$ were to be taken positive, we have therefore, in fact, obtained unique representations of the invariants (2) in the independent invariants (1).

If one were to substitute the expressions of j_{13}, j_{23}, j_{24} arising from (7) into the relations (3), then these would be identically fulfilled. In fact, one verifies this without difficulty.

In consideration of the further developments following concerning the polygons of the character (1, 1), we attach here the following arguments, for which we imagine the *vertex invariants* j_1, j_2, j_3, j_4 *chosen fixed* in any fashion corresponding to their conditions. We eliminate j_{24} from the two relations (3), by substituting into the first relation the expression to be calculated for j_{24} from the second. Further, we set:

$$j_{12} = -x, \quad j_{13} = -y, \quad j_{23} = -z \tag{8}$$

and interpret these x, y, z as rectangular space coordinates. The result of the elimination of j_{24} writes itself thus

$$x^2 + y^2 + z^2 - xyz + (j_1 j_2 + j_3 j_4)x + (j_1 j_3 + j_2 j_4)y + (j_1 j_4 + j_2 j_3)z \\ + (j_1 j_2 j_3 j_4 + j_1^2 + j_2^2 + j_3^2 + j_4^2 - 4) = 0 \tag{9}$$

and furnishes, for the prescribed geometric interpretation of the x, y, z, a *surface of third degree.*

[299] The course of this surface we determine below in detail. It will be found, that in the space octants coming into consideration for us, which is represented by the inequalities valid here:

$$-j_{12} = x > 2, -j_{13} = y > 2, -j_{23} = z > 2,$$

there runs a single simply-connected sheet of the surface (9). Let this part of the surface of third degree be called Φ_3; it is, in a new form, the image of the polygon continuum. By reserving the detailed study of our surface of third degree, we make note of the theorem: *All polygons of the character* $(0, 4)$, *with given vertex invariants* j_1, j_2, j_3, j_4, *form a two dimensional simply connected continuum, which one can map single-valuedly and continuously onto the sheet of the surface of third degree represented by (9) situated in the space octant* $x > 2, y > 2, z > 2$ *and denoted by* Φ_3.

In the treatment of the polygons of the character (0, 4) in §3, the fixing of the vertex invariants amount to holding the t_1, t_2, t_3, t_4 fixed. All polygons with the chosen vertex invariants are consequently furnished by the points of the square $0 < t_5 < 1, 0 < t_6 < 1$: *this square must therefore be reversibly single-valuedly mapped onto the sheet* Φ_3 *of the surface of third degree.*

For the representation of this mapping between the t_5, t_6 and the x, y, z, all means are already furnished by the foregoing developments. In order, e.g., to represent x, y, z by t_5 and t_6, first calculate, from (1) on p. 294, for $x(= -j_{12})$ and τ in t_5 and t_6:

$$x = \frac{2}{1 - t_5}, \qquad \tau = \frac{t_6}{1 - t_6}, \tag{10}$$

determine, corresponding to the previous agreement, that the square roots:

$$\sqrt{x^2 - 4}, \quad J_{12} = \sqrt{j_1^2 + j_2^2 + x^2 + j_1 j_2 x - 4}, \quad J_{34} = \sqrt{j_3^2 + j_4^2 + x^2 + j_3 j_4 x - 4},$$

whose radicands are always positive for admissible values j_1, j_2, j_3, j_4, x, are themselves to be taken positively and finally, calculate y and z, according to (7), in the following form:

$$\begin{cases} y(x^2-4) = (j_1 j_4 + j_2 j_3)x + 2(j_1 j_3 + j_2 j_4) \\ \qquad\qquad + \frac{1}{2}\left\{(\tau^2+\tau^{-2})x + (\tau^2-\tau^{-2})\sqrt{x^2-4}\right\} J_{12}J_{34}, \\ z(x^2-4) = (j_1 j_3 + j_2 j_4)x + 2(j_1 j_4 + j_2 j_3) + (\tau^2+\tau^{-2})J_{12}J_{34}. \end{cases} \tag{11}$$

We shall immediately have to apply these last developments in another connection.

1.6 The polygon continua of the character (1, 1)

[300]

The invariant theory of the polygons of the character (1, 1) is laid out in I on p. 354 ff. The individual polygon had, first of all, three invariants j_a, j_b, j_{ab}, for which the inequalities held:

$$j_a > 2, \qquad j_b > 2, \qquad j_{ab} > 2. \tag{1}$$

From them, we produced yet a fourth invariant j_c, the "vertex invariant" of the other fixed vertex of the polygon, on the basis of the relation:

$$j_a^2 + j_b^2 + j_{ab}^2 - j_a j_b j_{ab} - 2 = j_c. \tag{2}$$

For the individual signature $(1,1;l)$, the fixed vertex of the polygon is *elliptic* and there holds:

$$j_c = -2\cos\frac{\pi}{l}, \tag{3}$$

so that here, the equation (2) represents a relation between the j_a, j_b, j_{ab}. For the signature (1, 1), i.e., for polygons whose fixed vertex is *hyperbolic* resp. in the limiting case, *parabolic*, there holds:

$$j_c \leq -2, \tag{4}$$

so that, from (2), the inequality for the j_a, j_b, j_{ab} arises:

$$j_a^2 + j_b^2 + j_{ab}^2 - j_a j_b j_{ab} \leq 0. \tag{5}$$

These relations now let themselves be immediately put into connection with the developments concerning the polygons of the character (0, 4), in following manner, *that we are able to consider the invariant theory of the polygons of the character* $(1,1)$ *as a special case of the invariant theory of the polygons of the character* $(0,4)$.

This rests on the fact, already mentioned in I on p. 293 and pursued more closely on p. 355, that each group of the character (1, 1), by addition of a suitable elliptic substitution of period 2, is extended to a determinate group of the character (0, 4). Thereby, the latter has the signature (0, 4; 2, 2, 2), if $j_c \leq -2$;[9] on the other hand, we arrive at the signature $(0,4;2l,2,2,2)$, in case we have the signature $(1,1;l)$ for the given group[10]. [301]

[9] The individual polygon of the character (1, 1) with *hyperbolic* vertex, as we present below in more detail, is carried over, by an associated "two-valued" automorphic function, onto a two-sheeted orthosymmetric Riemann surface of the genus $p = 2$ with 6 branchpoints situated pairwise symmetrically. The half of this surface cut off by the symmetry line corresponds to the portion of the polygon lying in the interior of ellipse of the projective plane. The three branchpoints belonging to that half of the surface then furnish exactly the three newly supervening vertices for the extended group of the signature (0, 4; 2, 2, 2), while the individual sheet of the initially doubly sheeted half of the surface corresponds exactly to the polygon of the signature (0, 4; 2, 2, 2).

In order to actually produce the connection of the formula (1) ff. with those of the previous paragraph, set:

$$j_a = x, j_b = y, j_{ab} = z, j_c = 2 - j_1^2, \tag{6}$$

whereby the last equation, in the case of the signature $(1,1;l)$, amounts to $l_1 = 2l$. From (2), there now yields itself:

$$x^2 + y^2 + z^2 - xyz + (j_1^2 - 4) = 0, \tag{7}$$

an equation, in fact, which one can subsume as a special case of equation (9) on p. 298, namely for $j_2 = j_3 = j_4 = 0$.

Conversely, however, each group of the signature (0, 4; 2, 2, 2) resp. $(0,4;2l,2,2,2)$ contains a determinate distinguished subgroup of index 2 of the character (1, 1), so that full single-valuedness of the relation subsists. Namely, from a double triangle belonging to the first group, chosen, say, according to I on p. 299ff., by arrangement of a further double triangle next to the first, one must easily produces the discontinuity of the character (1, 1).

On the basis of the preceding results, it is admissible, for the representation of the polygon continua of the character (1, 1), to bring in our above results concerning the case of the character (0, 4) in their full extent. Thus, we may also bring in the invariants t_5, t_6 resp. t_1, t_5, t_6[11] of the polygons of the character (0, 4) for the polygons of the character (1, 1) in place of the x, y, z, j_c. The relations between the two kinds of invariants emerge immediately from the equations (10) and (11) on p. 299. One finds, for the
[302] x, y, z:

$$\begin{cases} x = j_a = \frac{2}{1-t_5}, \tau = \frac{t_6}{1-t_6}, \sqrt{2-j_c} = j_1 = \frac{2}{1-t_1}, \\ y\sqrt{x^2-4} = \frac{1}{2}\left\{(\tau^2+\tau^{-2})x + (\tau^2-\tau^{-2})\sqrt{x^2-4}\right\}\sqrt{x^2-j_c-2}, \\ z\sqrt{x^2-4} = (\tau^2+\tau^{-2})\sqrt{x^2-j_c-2}. \end{cases} \tag{8}$$

By inversion of these equations, one obtains, as representations of the t_1, t_5, t_6 in the x, y, z, j_c:

$$t_1 = \frac{-2+\sqrt{2-j_c}}{\sqrt{2-j_c}}, t_5 = \frac{x-2}{x}, t_6 = \frac{\tau}{1+\tau}, \tag{9}$$

whereby τ is given by the root, to be taken positively:

$$\tau = \sqrt{\frac{2y - xz + z\sqrt{x^2-4}}{2\sqrt{x^2-j_c-2}}}. \tag{10}$$

[10] For the polygons of the character (1, 1) with *elliptic* or *parabolic* vertices, we return here to known theorems from the theory of elliptic functions. An automorphic function belonging to such a polygon maps this onto a Riemann surface of genus $p = 1$. If one denotes the integral of the first kind of this surface by u, then as is known, the surface permits the simply infinitely many transforms $u' = -u + C$ of period 2 into itself (cf. "M" II on p. 241). In order to obtain the transformation intended in the text, we have to observe, that the Riemann surface with *one* point originating from the fixed polygon vertex is "signed". One has to so choose the constant C, that one of the four fixed points of the transformation $u' = -u + C$ falls at the point of the Riemann surface just named.

[11] The vertex invariants j_2, j_3, j_4 are equal to 0 in both cases.

Furthermore, the radicand of this root is certainly positive, since the expression standing in the numerator is > 0. From the relation (7), namely, with consideration of $x > 2$ and $z > 2$, here follows:

$$y^2 + z^2 - xyz = -j_1^2 - (x^2 - 4) < 0,$$

$$4y^2 - 4xyz + x^2z^2 < z^2(x^2 - 4),$$

$$|2y - xz| < z\sqrt{x^2 - 4}.$$

Accordingly, should $2y - xz < 0$ also hold, then there nevertheless follows from the last inequality, as was asserted:

$$2y - xz + z\sqrt{x^2 - 4} > 0.$$

By reserving bringing int the invariants x, y, z until below, during the investigation of the group continua of the character (1, 1), we will here utilize the invariants t_5, t_6 resp. t_1, t_5, t_6 in the representation of the polygon continua. We arrive at the following result: *All the polygons of a given signature* $(1,1;l)$ *form a two dimensional continuum, which is mappable single-valuedly and continuously onto the points of the square* $0 < t_5 < 1, 0 < t_6 < 1$; *similarly all polygons of the signature* $(1,1)$ *(with* $j_c \leq -2$*) form a three dimensional continuum, which is mappable single-valuedly and continuously onto the points of the cube* $0 \leq t_1 < 1, 0 < t_5 < 1, 0 < t_6 < 1$.

Moreover, the important remark again holds for the composition, that by fixing the value $j_1 \geq 2$ or $j_c \leq -2$, a square is cut out of the last named cube.

1.7 The polygon continua of the character (p, n)

Through the developments so far, the way is now prepared, in order, by means of the principle of composition, to arrive at the most general continuum of the principal-circle polygons. Let an arbitrary signature be proposed, whereby the number v of the elliptic vertices is any one of the numbers $0, 1, 2, \ldots, n$. According to I on p. 313, one can imagine any associated polygon P as arising by composition of a polygon of the signature $(p-1, n+1; l_1, l_2, \ldots, l_v)$ and a polygon of the signature $(1, 1)$, which have a hyperbolic generator in common. The details of the composition process are described precisely l.c. [303]

Conversely, one can always compose two polygons of the kind described, with common hyperbolic generator, in simply infinitely many ways into a polygon of the character (p, n). Hereby, if need be, the distortion already mentioned (above) on p. 293 is to be undertaken for one of the polygons, say that of the character (1, 1), which distortion was moreover described in detail in I on p. 298 precisely for the polygon of the character (1, 1). According to the theorem in I on p. 319, one can naturally always alter the composed polygon afterwards into a rectilinear $(2n+4p)$-angle with only concave angles again.

The invariants of the polygon of the character (1, 1) are discussed in the previous paragraph. Of these invariants, j_c is already given by the polygon of the character $(p-1, n+1)$, namely as a "vertex invariant" of this. With the polygon of the character (1, 1), there next supervene three more invariants j_a, j_b, j_{ab}, bound to one another by the relation (2) on p. 300; and further, one has to observe, that the two polygons are compassable in simply infinitely many ways, so that one more invariant τ, of the kind discussed on p. 293 ff., supervenes.

In order to adopt and extend our different notations for the various invariants, then j_c, after the composition, has lost its character as a vertex invariant and has changed into a "*transversal invariant*", since period-points on the Riemann surface correspond to it (cf. I on p. 183). Finally, τ retains its character as "*composition invariant*".

Now nothing stands in the way, of calculating the invariants j_a, j_b, j_{ab}, newly su-
[304] pervening with the polygon of the character (1, 1), which are bound to one another by *one* relation[12], in two invariant t_5 and t_6, independent of one another. For this purpose, we have to substitute $x = j_a, y = j_b, z = j_{ab}$ in the formulas (9) and (10) on p. 302, and calculate from these two formulas the two independent "period invariants" t_5 and t_6 [13]. Finally, as earlier [14], we are able to replace the invariant τ, coming up in the composition of the two oft-named polygons, by an invariant to be denoted for the moment, say, by t_7, which is variable in the interval $0 < t_7 < 1$. As an intermediate result, we take note, *that the addition of the polygon of the character* $(1,1)$ *to that of the character* $(p-1, n+1)$ *makes the introduction of three further invariants t necessary, which are independent of one another and variable in the internal* $0 < t < 1$.

By repeated application of this theorem, we are now able, starting out from §4 the polygons of the character $(0, n)$ to arrive at each arbitrary character (p, n) resp. to each arbitrary signature $(p, n; l_1, l_2, \ldots, l_\nu)$. We tie into the signature $(0, n+p; l_1, l_2, \ldots, l_\nu)$, and with this, into a continuum of the polygons of the genus 0 with $(n+p)$ fixed vertices, of which $\nu (\leq n)$ are elliptic and have the angles $\frac{2\pi}{l_1}, \frac{2\pi}{l_2}, \ldots, \frac{2\pi}{l_\nu}$, while $(n-\nu)$ further fixed vertices are hyperbolic or parabolic and the last p are hyperbolic. We further add p polygons, one after the other, of the character (1, 1), in the manner described. The first polygon, of the character $(0, n+p)$ has $(3(n+p)-\nu-6)$ independently variable invariants t; each polygon of the character (1, 1) furnishes three further invariants t, ζ so that we finally arrive at the correct number $(3n-\nu+6p-6)$.

By now suitably denoting the *variable* invariants of our polygon by $t_{\nu+1}, t_{\nu+2}, \ldots,$ $t_{3n+6p-6}$, we determine once more, that these are arbitrary and variable independently of one another in the intervals:

$$\begin{aligned} &0 \leq t_{\nu+1} < 1, \ldots, 0 \leq t_n < 1, \\ &0 < t_{n+1} < 1, \ldots, 0 < t_{3n+6p-6} < 1, \end{aligned} \tag{1}$$

whereby, the equality signs stipulated for the first $n-\nu$ invariants indicate the occur-
[305] rence of parabolic cusps. We have hereby arrived at the conclusive theorem: *All the principal-circle polygons, belonging to any proposed signature* $(p, n; l_1, l_2, \ldots, l_\nu)$, *form a* $(3n-\nu+6p-6)$ *dimensional continuum, which is mappable continuously and single-valuedly onto the* $(3n-\nu+6p-6)$ *dimensional cube represented by the inequalities (1).*

---

[12] The invariant occurring in this relation is already guen by the polygon of the character $(p-1, n+1)$.

[13] We do not fail to refer to the fact, that our present composition invariant τ, to be mentioned in the text again immediately, is not identical with the invariant thus denoted in the formulas on p. 302.

[14] See e.g. that last formula (1) on p. 294.

1.8 Transition from the polygon continua to the group continua

Each of our canonical polygons P furnishes *one and only one* principal-circle group Γ. One and the same group, however, is furnished, conversely, by *infinitely many different* canonical polygons. In obtaining the *polygon continua,* therefore, by no means do we yet obtain a final insight into *group continua*; rather, the mapping of the two continua onto one another must be penetratingly discussed.

It is a question here of a point of principle in the historical development of the fundamental theorem.

Klein, in his work "*Neue Beiträge zur Riemmannschen Functionentheorie*", Part III §5, has touched upon the relation between polygons and groups, and for the reasons developed in more detail there, derived the plan, of "always proceeding from the classification of domains" in the continuation of this investigation. Accordingly, the continua, with which Klein works in the proof of the fundamental theorem, are the continua of the *polygons* and the continua of the *cut up* Riemann surfaces belonging to them[15].

In contradistinction to this, in his work "Sur les groupes des équations linéaires" (§8ff), Poincaré took the *group continua* as a point of departure in the proof theorem. Namely, I.e., Poincaré recognizes the polygon continua as an *unclosed* manifolds[16]; On the other hand, the intends showing, and indeed, what is characteristic, "*par une discussion spéciale in à chaque cas particulier*", that the group continua considered by him are *closed* manifolds, as we shall find this verified below in examples of the boundary circle cases. Herein, with justification, Poincaré sees essential simplification of the continuity proof in the use of the group continua.

The transition from a first polygon P to all remaining canonical polygons, which [306]
represent the same group, is developed in detail in I on p. 320 ff. in the "transformation theory of canonical polygons". It is a question of those transformations, which correspond to the essential alterations of the canonical cut system on the closed surface belonging to P.

The invariants themselves, with respect to those transformations of the polygon, experience *binational transformations*[17]. All the transformations of this kind entering for the character (p,n) furnished us, in I on p. 391, *the "modular group" belonging to the character* (p,n). The name of this group originates from the fact, that in I, we also denoted the invariants of the polygon P as its "moduli"[18]. Moreover, notice too, that the modular group of the character (p,n) stands in exactly the same relation to the principal-circle groups of this character, as, in the theory of the elliptic functions (theory of the "parabolic rotation groups" (cf. p. 24 and I on p. 165)), the "modular group" coming up there to the group of the substitutions:

$$u' = u + m_1\omega_1 + m_2\omega_2 \quad (m_1, m_2 = 0, \pm1, \pm2, \ldots),$$

[15] See e.g., also Part IV §2 of the work of Klein just cited, where it is explicitly stated: "Wir haben einerseits die Riemannschen Flächen (p, n, l_κ), die wir uns nach einem bestimmten Typus zerschnitten denken. Sie bilden eine erste Mannigfaltigkeit M_1".

[16] The cube obtained in §7 has a $(3n - \nu + 6p - 7)$ dimensional boundary.

[17] See the additional discussions concerning these transformations in I on p. 389 ff.

[18] For the sake of better distinction, we will reserve the name "modulus" for the "moduli of the Riemann surface" in the present chapter.

ie. the parabolic rotation groups.

In order, moreover, to put the developments of the foregoing paragraphs concerning polygon continua into the correct relation to the cited investigations from Vol. I, it is to be yet noticed, that the vertex invariants, which belong to the elliptic generators, also take part in the birational transformations of the invariants, insofar as any prescribed permutation of the ν polygon vertices coming into consideration was attainable by a suitable transformation (cf. I on p. 301). In the study of the group continua, correspondingly, the arrangement of the numbers l in the signature $(p, n; l_1, l_2, \ldots, l_\nu)$ was immaterial and capable of alteration in arbitrary fashion.

In contradistinction to this, in the production of our polygon continua above, we have had to hold fast to the arrangement of the elliptic vertices and, with this, to the signature, once chosen. *In this situation, for obtaining the individual group continuum, we may attach at our pleasure to one of several polygon continua, and indeed, in the*
[307] *highest case, of* $\nu!$ *continua, if, namely, all* l_κ *are different and* $p > 0$*, or if for* $p = 0$*, the number* $\nu < n$[19] In the modular group, we consequently have *a distinguished subgroup of the corresponding finite index, which leaves the signature unaltered and therefore always carries a polygon over again into one such of the same continuum.* The subgroup thus obtained we are able to designate as "*the modular group belonging to the signature* $(p, n; l_1, l_2, \ldots, l_\nu)$". To be sure, the reason for an unsymmetry of our representation is jumbled up with this unavoidable step, which even remains subsisting, if we effect the introduction of the invariants (as in I) symmetrically with respect to the individual polygon.

1.9 The discontinuity of the modular group

The modular group belonging to a proposed signature represents all birational transformations of the invariants which, calculated in the invariants t, carry our $(3n - \nu + 6p - 6)$ dimensional cube, obtained in §7, over into itself. Thereby the question of the *discontinuity of the modular group in the interior the cube* steps into the foreground. In fact, *a discontinuity domain of the modular group* will, indeed, be *a single-valued and continuous image of the manifold of the groups belonging to the proposed signature.*

Poincaré, on p. 263 l.c., designates the modular group as "*évidemment discontinu*". This is to be so understood, that *in the interior of the cube* or in a domain equivalent to the cube, entering for another choice of the invariants, the modular group is *"properly" discontinuous* (cf. I on p. 62), i.e., *has there a finitely extended* $(3n - \nu + 6p - 6)$ *dimensional discontinuity domain, whose reproductions fill up the entire interior of the cube.*

Without holding this assertion to be immediately evident, we will ask beforehand, what means we have for the actual production of a discontinuity domain of the modular group.

For this purpose we shall have to pick out, among all the canonical polygons P equivalent with respect to the modular group (which polygons thus furnish the same

---

[19] For $p = 0, \nu = n$, we have, in the highest case, $(n-1)!$ continua, insofar as now the $\nu = n$ vertices form a closed cycle.

principal-circle group Γ), one, as a so-called "*reduced*" polygon. Poincaré, on p. 265 l.c., gives a prescription for the definition of a "*polygon redvit*", which, with the utilization of the notations and conceptions of the paragraph before last (§7, the general case (p,n)), lets itself be described thus: understand, by $\Phi(t_{\nu+1}, t_{\nu+2},\ldots)$, a real continuous [308]
and singlevalues function of the invariants $t_{\nu+1},\ldots,t_{3n+6p-6}$; and indeed, this function Φ is to be equal to 1 everywhere on the boundary of the cube, vanish at the center of the cube, and except for this, lie between 0 and 1 in the interior of the cube. Among all the polygons P furnishing the same principal circle group Γ, according to Poincaré's definition, that one is called reduced, whose invariants $t_{\nu+1},\ldots,t_{3n+6p-6}$ furnish the smallest value of the function Φ[20].

Now (as was already said in I on p. 393) it is more natural for us to introduce a theory of reduced polygons in another fashion, and indeed, by returning to the theory of the "*normal*" and the "*natural polygons P*" represented in I on p. 241 ff. and 275 ff.. The individual normal polygon of a proposed group Γ is uniquely determined by a center C, and centers, which are equivalent with respect to Γ, furnish (for the invariants standpoint) equal polygons. If, on the strength of this, one lets the center run through a first arbitrarily chosen polygon, then one already obtains all normal polygons occurring for this group Γ.

As far as the shape of the normal polygons goes, we had to distinguish a series of different *types* in I on p. 262 ff., between which, upon transition from one type to the other, the *special* or *transitional types* with a lowered number of sides interposed themselves. In the case of each individual character (p,n), the number of different types is *finite*; to actually determine the number, is a combinational problem, which we carried through in I on p. 266 ff. for a few lower cases.

Now let a normal polygon P_0 belong to a first center C_0, which does not have a transitional type. Upon continuous alteration of the center C from the initial position C_0, the polygon P will also alter itself continuously. We shall consequently be able to bound about C_0 a *finitely extended* domain[21], so that all points of this domain furnish [309]
polygons P of the same type and the same generators as P_0. If we now ask, to what extent C may wander inside P_0, until we come to a first transitional type, then the answer to this is already given in I on p. 275ff. We arrive at a domain surrounding C_0, denoted l.c. by T_0, for which we will use here lie name "*natural elementary domain*", and which (see also I on p. 245ff.) we found bounded by finitely many pieces, to be obtained from the generators of P_0, of lines, conic sections, or curves of third order. To the boundary

---

20 If one sets, which is most natural:

$$\Phi = 1-4^{3n-\nu+6p-6}\cdot\prod_\kappa t_\kappa(1-t_\kappa) = 1-\prod_\kappa(1-(2t_\kappa-1)^2),$$

Then the production of the discontinuity domain of the modular group amounts to the principle utilized in the formation of the "normal domains" in I on p. 106ff. The individual point in the interior of the cube will always and only then be assigned to be discontinuity domain, if it lies closer to the center of the cube, or at least not further, than from its equivalent points.

21 For the moment, as the determination of the measure in the projective plane, the ordinary, not the projective one is to be utilized, in order that domains, which reach up to the ellipse or over beyond this, retain finite surface area.

curves of T_0 corresponded the various transitional types, which attached themselves to the type of P_0.

The finiteness of the surface-extent of such a natural elementary domain T is essential. Since P_0 is also finitely extended (in the sense of the elementary determination of measure utilized temporarily), *then in P_0, only a finite number of such domains T find place*[22]*: the number of the natural elementary domains, inequivalent with respect to Γ, is finite.* We denoted this number (I on p. 278) by μ, the domains themselves by $T_0, T_0', T_0'', \ldots, T_0^{(\mu-1)}$. Such a system of inequivalent domains T furnished us l.c. A *"natural" discontinuity domain* for Γ.

Now since all centers C of an individual natural elementary domain T furnish polygons of the same type and the same generating system, then we have, for our group Γ, only μ essentially different normal polygons. Consequently, if we choose, first of all, as *reduced* polygons, the *normal polygons of the generating types*, then the theorem arises without further ado, *that for our group Γ, the number of reduced polygons is finite, namely, equal to μ.*

To be sure, these reduced polygons do not in general yet have the canonical form. But nothing stands in the way, of prescribing, for each type, a uniquely determined measure for transformation into a canonical polygon, which we will then use, from now on, as a reduced polygon, instead of the normal polygon. In this manner, for each principal-circle group Γ, we arrive, even if not to a single one, *nevertheless to a finite*
[310] *number of μ reduced polygons.* How we are further able, among these μ reduced polygons, to seek out an individual one, better remains reserved for the immediately following detailed investigations.

However, if we imagine this unique choice of the reduced polygon made, then with this, to each point in the interior of the $(3n-v+6p-6)$ dimensional cube, and with this, to each non-degenerate polygon, there will be coordinated one and only one point in the domain of the reduced polygons, and with this, *the discontinuity domain of the modular group in the cube* determined.

It is now also easy to see, that this discontinuity domain is likewise $(3n-v+6p-6)$ dimensional, whereby *the proper discontinuity of the modular group* would be recognized. One wants only to observe, that with respect to continuous alteration of the invariants, our polygons P also continuously alter themselves. If we now choose such a point in the cube, which belongs to a reduced normal polygon, which is finitely distant from each transitional type, and we bound a $(3n-v+6p-6)$ dimensional domain about the chosen point, then one can choose the latter so small, that for no point of this do we obtain a normal polygon with transitional type. The polygons belonging to the points of that domain are consequently all reduced; but, at the same time, no two of them are equivalent with respect to the modular group, since they do not have exactly the same invariants. The discontinuity of the modular group in the cube is accordingly to be recognized as a proper one.

The preceding investigations now require, throughout, a closer explanatory effectuation in the individual cases (p, n). The lowest case (0, 3) is thereby to be excluded

[22] In order to meet an obvious objection to the procedure of proof in the text, assume the number of the different domains T inside P_0 as infinite. These domains T must then necessarily have at least one accumulation point in P_0. However, one easily recognizes one such as impossible, by choosing it itself as the center C and considering the normal polygon.

as elementary. Here we have to do exclusively with. "triangle groups" capable of extension by reflections (comp. e.g. I on p. 342), whose canonical polygons are uniquely determined. In this case the modular group reduces itself to the identical transformation, so that its discontinuity is furnished by the entire cube (for $\nu = 0$) resp. then by the square ($\nu = 1$) or the entire straight line segment ($\upsilon = 2$).[23]

1.10 The reduced polygons of the character (1, 1)

According to I on p. 265, for the normal polygons of the character (1, 1) there is only *one* general type, namely a decangle with the coordination of sides represented in figure 37. Besides the fixed vertex e, we have three cycles of movable vertices; the vertices of the first cycle are called $\varepsilon_1, \varepsilon_1', \varepsilon_1''$ in the figure, those of the second $\varepsilon_2, \varepsilon_2'$, etc. There is a distinction, coming into play below, between the first and the other two cycles, which one easily perceives, if, in this case, one lays the three polygons about the vertices $\varepsilon_1, \varepsilon_2, \varepsilon_3$: *of the three polygons surrounding a point* ε_1*, two have the fixed vertex e in common; the three polygons surrounding a point* ε_2 *and* ε_3 *have three different fixed vertices e.* [311]

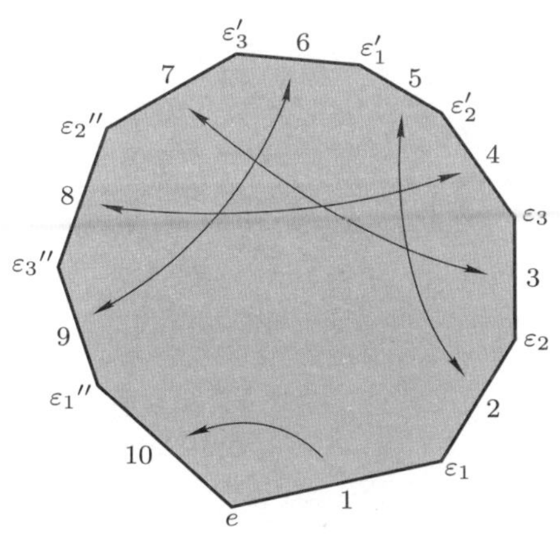

Fig. 37.

We best base the choice of a reduced polygon for the individual group Γ here on the *special types* of the normal polygons, and indeed, we favor polygons with a three-termed cycle of fixed vertices. The origination of such a special type from the general type we have to picture to ourselves thus: if we are to arrive continuously at a polygon with a two-termed cycle of fixed vertices, then we have to move the center C in such a manner, that the side 1 and 10 of our decangle represented in figure 37 draw themselves together into points. The vertices ε_1 and ε_1'' come into coincidence with the vertex e, and simultaneously, ε_1' has reached a second fixed vertex e'; the polygon

[23] Compare the investigation carried through above in §2, as well as the relevant discussions following below (§§27, 28 and 30) concerning the associated fundamental theorems.

has changed into the octangle of figure 38. The continuous transformation of the octangle into a polygon, which, besides e and e', obtained yet a third fixed vertex, is now possible in two ways, either yet a third fixed vertex, is possible in two ways, either by drawing the sides 2 and 5 together into points, or by the sides 6 and 9 vanishing. *In both cases we obtain a hexangle, whose opposite sides are coordinated to one another, and whose vertices form a three-termed cycle of fixed vertices and one such of movable vertices.* figure 39 represents this normal hexangle; the fixed vertices are denoted here by e_1, e_2, e_3, the accidental ones by $\varepsilon', \varepsilon'', \varepsilon'''$.

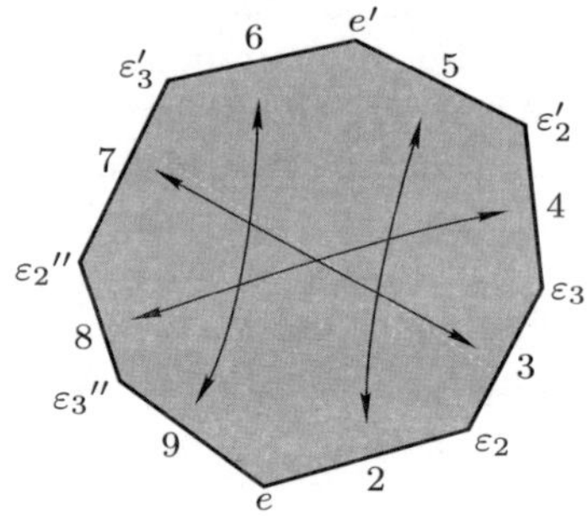

Fig. 38.

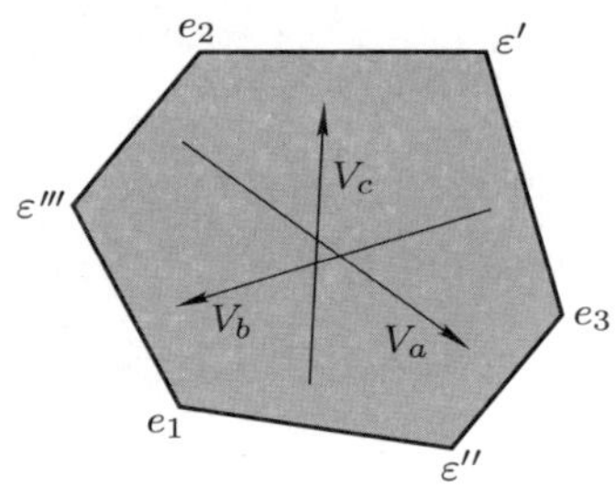

Fig. 39.

[312] The question now makes its appearance: *Are there such hexangles, and now many essentially different ones exist for our group* Γ?

The answer to this question is contained in the theory of those normal domains Q, investigated in I on p. 258 ff., which arise, in case, the center C is chosen at the fixed vertex e. The individual domain Q appears put together from a system of normal polygons surrounding the point e, therefore, in the elliptic case, by l in the parabolic and hyperbolic case by infinitely many normal polygons. The domain Q shows the shape of the ordinary normal polygons, insofar as it is bounded rectilinearly, as well as having only concave angles and three-termed cycles of accidental vertices.

From Q, we cut out an ordinary normal polygon by any two level lines running out from e of the generator V belonging to e, of which the one is carried over into the other by V (cf. the lightly drawn sides 1 and 10 of the polygon in figure 40). We arrive at this normal polygon, if we carry C along the (indicated in the figure by an arrow) angle-bisector[24] of e. Since the points ε_1 and ε_1'', in general, are not vertices in the net of the domains Q, the angle sum $\sphericalangle\varepsilon_1 + \sphericalangle\varepsilon_1''$ is equal to π, and therefore the angle at ε_1' has changed into a straight one. The cycle of vertices $\varepsilon_1, \varepsilon_1', \varepsilon_1''$, which we already distinguished above from the other two, does not, as such, belong to the domain net of the Q, but first supervenes upon cutting out the special polygon of figure 40.[25]

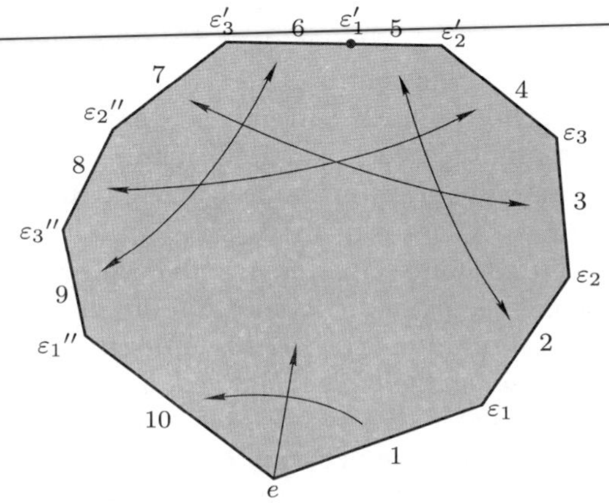

Fig. 40.

To the normal polygon of figure 40, we now apply the reciprocity theorem set up in I on p. 260, it follows immediately: if one chooses C in the interior of that polygon or on the sides 1 or 10, excepting $\varepsilon_1, e, \varepsilon_1''$, then the normal polygon associated to this C has only one fixed vertex e. If one chooses C on one of the sides $2, 3, \ldots, 9$, but different from a vertex, excepting $\varepsilon_1, \varepsilon_1', \varepsilon_1''$, then the normal polygon corresponding to this C has a two-termed cycle of the fixed vertices (cf. figure 38). Finally, if one chooses one of the six vertices $\varepsilon_2, \varepsilon_2', \ldots, \varepsilon_3''$ as the center C, then one arrives at a normal hexangle of the kind sketched in figure 39. But since $\varepsilon_2, \varepsilon_2', \varepsilon_2''$ are equivalent points, and the same holds for $\varepsilon_3, \varepsilon_3', \varepsilon_3''$, then follows: *For our group* Γ, *there are two and only two normal hexangles with a three-termed cycle of fixed vertices.* [313]

We now remember the fact, determined in I on p. 293 and 355, that our group Γ is a distinguished subgroup of index 2 in a determinate group Γ_0 of the character $(0,4)$ (see also p. 300ff.). This group Γ_0 has the signature $(0,4;2,2,2,2l)$ or $(0, 4; 2, 2, 2)$, according as the elliptic case is present for Γ or not. Since, as already remarked, Γ is contained distinguishably in Γ_0, then each substitution of Γ_0 not contained in Γ will also transform the net of the domains Q into itself; for this net is uniquely determined with Γ. The

[24] In the parabolic case, the angle-bisector is that line, with respect to which the sides 1 and 10 are symmetric to one another.

[25] If we choose the line $e\varepsilon_2$ as the side 1, then the normal polygon becomes an octangle all of whose vertices, apart from the vertex e itself, belong to the net of the domains Q.

individual such substitution will therefore either interchange the two normal hexangle nets of Γ or carry each over into itself.

If now, in particular, V is the generator of Γ belonging to the fixpoint e (in figure 40), which carries the side 1 into 10, then $V_0 = V^{\frac{1}{2}}$ is a generator of Γ_0. This substitution V_0 transforms the boundary of the domain Q surrounding e into itself in such a manner, that e.g., the vertex ε_2 goes over into ε_3' and vertex ε_3 into ε_2''; the two vertex cycles $\varepsilon_2, \varepsilon_2', \varepsilon_2''$ and $\varepsilon_3, \varepsilon_3', \varepsilon_3''$ are consequently interchanged. It follows: by the substitutions of Γ_0 not already contained in Γ, the two normal hexangle nets of the group Γ are interchanged. The two hexangle nets are accordingly congruent and have the same invariants: *from the invariant standpoint, there is only one normal hexangle with a three-termed cycle of fixed vertices for the individual group* Γ *of the character* $(1,1)$*; this hexangle we choose as the "reduced" polygon of the group* Γ.[26]

If U is that substitution of Γ, which carries the side 7 (in figure 40) over into the side 3, then, since conversely, V_0 transforms the side 3 into the side 7, the substitution UV_0, composed from V_0 and U, transforms the side 3 into itself, and indeed, in such
[314] manner, that the endpoints ε_2 and ε_3 interchange themselves. According to this, UV_0 is elliptic of period 2, and the fixpoint of this substitution is midpoint of the side 3. *The fixpoints of the elliptic substitutions of period* 2 *contained in* Γ_0 *are the midpoints of the sides in the net of the domain Q; the three sides, inequivalent with respect with respect to* Γ*, in the net of the Q furnish the three classes of those substitutions.*

If, for the moment, we denote the midpoint of the side 3 by e' and write the associated substitution $UV_0 = V'$, then the domain Q with the center e is transformed by this substitution V' into the domain Q_1 neighboring along the side 3, whose center let be called e_1. Since V' is of period 2, then the three points e, e' and e_1 lie on one line. Now the side 3 is the common side of the two normal domains Q and Q_1, of the centers e and e_1; it therefore follows from the general properties of the normal domains, *that the lines* ee_1 *and* $\varepsilon_2\varepsilon_3$ *intersect one another perpendicularly at the fixpoint* e' *of* V'. This consideration carries itself over immediately to all remaining sides in the domain net of the Q.

One now wants to move the center C, say, into the that point, which is denoted in figure 40 by ε_2, and imagine the associated hexangle as actually produced (cf. figure 41). In this figure, let e_1 be the vertex just furnished with this notation, so that the present vertex e_3 is the point previously denoted by e, and e_2 is the center of the domain Q, neighboring along the side 2 in figure 40 with the domain Q here. By joining the three fixed vertices e_1, e_2, e_3 rectilinearly, we produce, as in figure 41, the triangle $e_1e_2e_3$ belonging to the hexangle.

Further, the normal domain of the group Γ_0 belonging to the same center $C = \varepsilon_2$ is to be produced, which is a component of the hexangle. The net, belonging to the center $e(= e_3)$, of the domains Q of the group Γ_0 is none other than the previous net Q; with this, the normal domain of the center $C = \varepsilon_2$ for Γ_0 will likewise reach to the points e_1, e_2, e_3. The triangle e_1, e_2, e_3 already named is consequently also a component of the

[26] This is, indeed, not yet directly a canonical polygon. However, the transition to one such e, as e.g., in figure 43 (at the close of the present paragraph), is easily effected. Moreover, this transition is effectable in three way, of which we have chosen one below (at the end of §12).

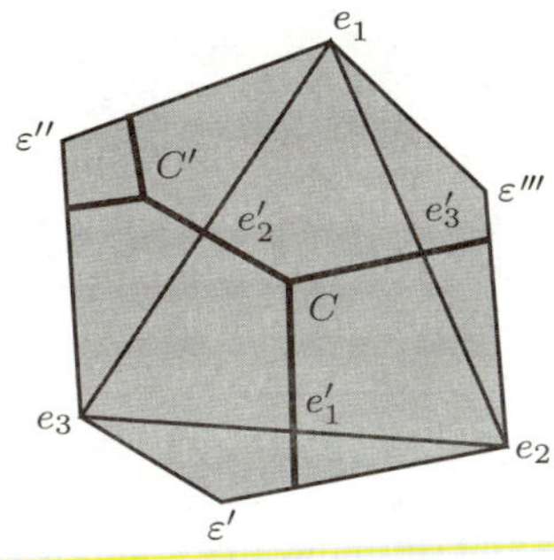

Fig. 41.

normal domain of Γ_0. *But this triangle is now straight away identical with the normal domain of* Γ_0 *belonging to* $C = \varepsilon_2$.

Namely, if one chooses the point denoted in figure 40 by ε_3, equivalent with ε_2 with respect to Γ_0, as the center C' of a normal domain of Γ_0, then the latter will necessarily also reach to the point $e(= e_3)$ and e_1. Now, however, the self-evident principle subsists [315]
from the nature of the normal domains, that the line joining two points of a normal domain belongs to this domain entirely, even if only as a boundary line. The joining line $e(= e_3)e_1$ therefore belongs to the two domains belonging to C and C' and is thus the boundary line of these. This consideration also carries itself over spontaneously to the triangle sides e_1e_2 and e_2e_3, so that the triangle $e_1e_2e_3$ actually is the normal domain of the center ε_2 for Γ_0.

We yet remember, now, from the point C, one can obtain the linear skeleton of the domain net of the Q, to the extent that this belongs to the hexangle. One has simply to let three perpendiculars from C fall on the sides of the triangle $e_1e_2e_3$ (which cut out, on these sides, the fixpoints e_1', e_2', e_3' (cf. figure 41) of the elliptic generators of period 2), and to extend two of these perpendiculars beyond the boundary of the hexangle according to the coordination of the opposite sides, until they meet at the point C' (cf. figure 41).

The developments so far give us the means, at the same time, for an *independent definition of the reduced polygons of all our groups* Γ *and* Γ_0. Only one must hereby separate the elliptic and hyperbolic cases form the parabolic ones, which latter require a separate consideration.

Accordingly, first of all, let any triangle $e_1e_2e_3$ be chosen, whose vertices lie simultaneously inside or outside the ellipse. However, in the first case, the sum of the angles of the triangle is to equal to π/l, and in the second case (although the vertices lie outside the ellipse), the sides are nevertheless to represent secants of the ellipse. The midpoints e_1', e_2', e_3' of the sides of the triangle are determinate in each case and lie in the interior of the ellipse. The three perpendiculars erected on the sides at these midpoints intersect one another at a point C, *then center of the circle circumscribed by the triangle.* We set the additional essential requirement, *that this center of the circumscribed circle is to belong to the interior or even on the boundary of the triangle.* This indeed occurs, as we know, for our normal triangles; but now, conversely, the theorem also holds: *Each such triangle is the normal domain, belonging to* C, *of a group* Γ *of our kind.*

Namely, denote the elliptic substitutions of period 2 belonging to the fixpoints e_1', e_2', e_3' by V_1', V_2', V_3'. Our triangle, with the arrangement of the boundary points given in figure 42 on p. 316, then represents the discontinuity domain of a group of signature
[316] $(0,4;2,2,2,2l)$ resp. (0, 4; 2, 2, 2), which has V_1', V_2', V_3' and $V_4' = (V_1'V_2'V_3')^{-1}$ as generators. In this domain, we now easily recognize, in fact, the normal domain of the group in question belonging C as center. Namely, if one constructs the three (also given in figure 42) points $C_1 = V_1'(C), C_2 = V_2'(C), C_3 = V_3'(C)$, then it is clear, that our triangle is the common component of the three normal domains, belonging to C, of the cyclic elliptic groups arising from V_1', V_2', V_3', taken individually. According to the principle set up in I on p. 245, that the normal domain, belonging to a center C, of any group Γ is the common component of all normal domains, belonging to the same C, of all cyclic subgroups contained in Γ, therefore, for our present group of the character (0, 4), the normal polygon belonging to C must be a component of the triangle $e_1e_2e_3$. But it cannot be smaller than this triangle, since otherwise, it would not furnish a discontinuity domain of the group. Therefore, as was to be proved, the triangle itself is the normal domain.

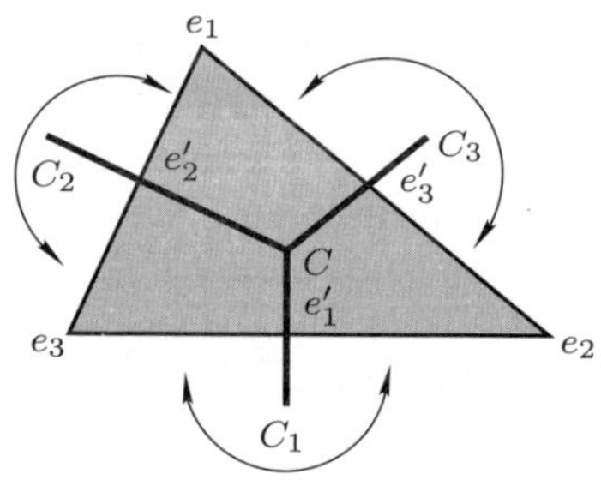

Fig. 42.

In the parabolic case, one has to attach to a triangle whose vertices e_1, e_2, e_3 lie *on* the ellipse. A triangle of this kind does not yet have an invariant, as such; each one is transformable into each other one. Also, there are no determinate midpoints of the sides, since the sides have infinite lengths. The previous construction accordingly breaks down.

Now in order here, all the same, to arrive at a discontinuity domain for a group of the character (0, 4) from an arbitrarily chosen triangle of the kind described, we mark, on the three sides of the triangles, three fixpoints e_1', e_2', e_3' for three elliptic substitutions V_1', V_2', V_3' of period 2 given by these points. The choice of the points e_1', e_2', e_3' is not fully arbitrary; rather, the condition is to be satisfied, that the substitution $V_1'V_2'V_3' = V_4'^{-1}$, which has the point e_2 (cf. figure 42) situated on the ellipse as a fixpoint, must not be hyperbolic but parabolic.

There now subsists the following simple theorem: *the requirement, that* V_4' *is parabolic, i.e., that the invariant*:

$$\begin{aligned} j_{123} = &\alpha_1\alpha_2\alpha_3 + \delta_1\delta_2\delta_3 + \alpha_1\beta_2\gamma_3 + \delta_1\beta_3\gamma_2 + \alpha_2\beta_3\gamma_1 \\ &+ \delta_2\beta_1\gamma_3 + \alpha_3\beta_1\gamma_2 + \delta_3\beta_2\gamma_1 \end{aligned}$$

of the triple V_1', V_2', V_3' (cf. I on p. 365 equation (2)) has the value ± 2, clothes itself in the geometrical form, that the three perpendiculars erected upon the sides of the triangles at the points e_1', e_2', e_3' intersect at ont point. [317]

The proof of this theorem we carry through by calculation, in which we temporarily also utilize the ζ-halfplane. We introduce ζ so that our triangle is bounded in the positive ζ-halfplane by the unit-circle and the two lines erected perpendicularly to the real ζ-axis at the $\zeta = \pm 1$. On these three sides, let the points e' be situated at the following places:

$$e_1' = \cos\alpha + i\sin\alpha, \quad e_2' = -1 + ib, \quad e_3' = 1 + ic.$$

For the three substitutions V', one then finds:

$$V_1' = \begin{pmatrix} \operatorname{ctg}\alpha & -\frac{1}{\sin\alpha} \\ \frac{1}{\sin\alpha} & -\operatorname{ctg}\alpha \end{pmatrix}, \quad V_2' = \begin{pmatrix} b^{-1} & b + b^{-1} \\ -b^{-1} & -b^{-1} \end{pmatrix}, \quad V_3' = \begin{pmatrix} c^{-1} & -c - c^{-1} \\ c^{-1} & -c^{-1} \end{pmatrix}$$

so that the requirement $j_{123} = \pm 2$ clothes itself in the form:

$$\operatorname{ctg}\alpha(bc^{-1} - b^{-1}c) - \frac{1}{\sin\alpha}(bc^{-1} + b^{-1}c) = \pm 2,$$

or (refashioned a bit):

$$\cos\alpha(b^2 - c^2) - (b^2 + c^2) = \pm 2bc\sin\alpha.$$

this, relation lets itself be written still somewhat more simply, if one substitutes $\sin\alpha = \sqrt{1 - \cos^2\alpha}$, squares etc.; one finds:

$$\cos\alpha(b^2 + c^2) = b^2 - c^2. \tag{1}$$

The perpendiculars on the triangle side at the points e_1', e_2', e_3' are, in the ζ-halfplane, the circles:

$$\begin{aligned} \left(\xi - \frac{1}{\cos\alpha}\right)^2 + \eta^2 &= \operatorname{tg}^2\alpha, \\ (\xi + 1)^2 + \eta^2 &= b^2, \\ (\xi - 1)^2 + \eta^2 &= c^2, \end{aligned}$$

whereby, as earlier, we have written $\zeta = \xi + i\eta$. According to formula (3) in I on p. 21, we have to effect the transition to the homogeneous coordinates z_1, z_2, z_3 of the projective plane by the proportion:

$$(\xi^2 + \eta^2) : \xi : 1 = z_1 : z_2 : z_3.$$

In the projective plane, therefore, the three perpendiculars in question are given by the equations:

$$\begin{aligned} \cos\alpha \cdot z_1 - 2z_2 + \cos\alpha \cdot z_3 &= 0, \\ z_1 + 2z_2 + (1 - b^2)z_3 &= 0, \\ z_1 - 2z_2 + (1 - c^2)z_3 &= 0. \end{aligned}$$

the relation (1) now says, in fact, that these three lines run through *one* point. We de- [318]

note this point from now on, at the same time, by C again.

We now put the further restriction on the choice of the three points e_1', e_2', e_3', *that the point C corresponding to them is to belong to the interior or even on the boundary of the triangle.* This indeed takes places for our normal triangle but again, conversely, the theorem holds, *that each such triangle is the normal domain belonging to C of a group* Γ_0 *of the signature* $(0,4;2,2,2)$. The proof of this assertion one can carry through by verbatim representation of the argument utilized above in the non-parabolic cases.

Now there only still remains to notice, that with the normal triangle $e_1 e_2 e_3$, that the normal hexangles surrounding this triangle of the subgroup Γ of index 2 of the character (1, 1), contained distinguishably in Γ_0, is uniquely given in each case. Namely, the normal polygon of Γ belonging to C will reach up to the three vertices and therefore furnish a normal hexangle. If, as in figure 39 on p. 311, we again call the generators of the latter group Γ, V_a, V_b, V_c, then these represent themselves by means of the generators V_1', V_2', V_3' of Γ_0 as follows:

$$V_a = V_2' V_3', \quad V_b = V_3' V_1', \quad V_c = V_1' V_2'. \tag{2}$$

Between them subsists the relation:

$$V_a V_b V_c = 1, \tag{3}$$

and the three substitutions, on an equal footing, inside Γ, belonging to the fixpoints e_1, e_2, e_3, are, as we take note at the same time —

$$V_1 = V_b V_a V_c, \qquad V_2 = V_c V_b V_a, \qquad V_3 = V_a V_c V_b. \tag{4}$$

In order to understand the preceding developments concerning the independent presentation of our reduced domains in their full significance, just remember the general theory of the canonical domains of the character (1, 1) developed in I on p. 294 ff. and 354 ff. Each of the infinitely many canonical hexangles of our group Γ we are able to change around into an associated hexangle with a three-termed cycle of fixed vertices. In figure 43, where this transformation is carried out, the canonical hexangle was heavily marked and provided with a coordination of sides, the hexangle with
[319] the three-termed cycle of fixed vertices e_1, e_2, e_3 shaded. The triangle $e_1 e_2 e_3$ belonging to the hexangle is also, here again, a discontinuity domain of the extended group Γ_0; for it is half of the quadrangle[27], produced in figure 92 in I on p. 295 from the canonical hexangle, with opposite sides coordinated to one another. *Among all the infinitely many triangles, which we are able to choose in this manner as discontinuity domains of our group* Γ_0, *then just our reduced triangle alone is thereby distinguished, that for it, the point C, defined independently as above, belongs to the interior or the boundary of the triangle.*

1.11 The surface Φ_3 of third degree coming up for the character (1, 1)

In §6 (p. 300ff.), the polygon continua of the character (1, 1) are represented by squares resp. by a cube which made a transformation of the initial invariants j_a, j_b, j_{ab}, j_c

[27] In I l.c., we have formed these quadrangle nets of the group Γ, which corresponded precisely to the parallelogram nets in the theory of the doubly periodic unctions (theory of the parabolic rotation groups), to the hexangles now utilized in the text.

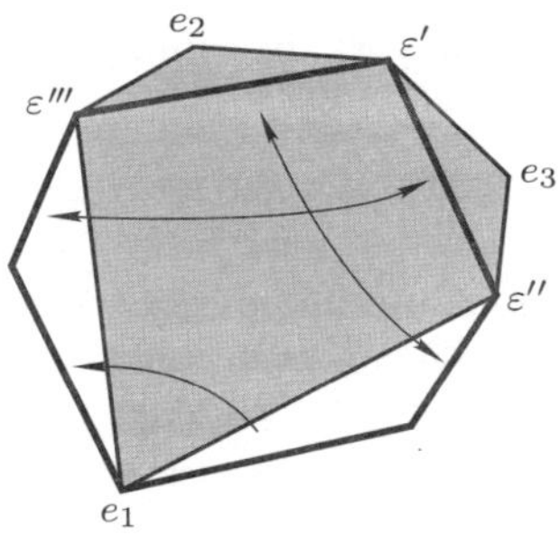

Fig. 43.

necessary (cf. p. 300). Here, for the sake of symmetry, we will retain the invariants j_a, j_b, j_{ab}, j_c, but will, in addition, rather write j_e instead of j_c, in order that we are able to use the notation j_c for j_{ab}. Then the j_a, j_b, j_c are the invariants of our three group generators V_a, V_b, V_c and j_e is an invariant of the substitutions just represented in §10 under (4) with the fixpoints e_1, e_2, e_3.

The relations, which characterize the polygon continua, are (cf. p. 301) the following: *First, the equation*

$$j_a^2 + j_b^2 + j_c^2 - j_a j_b j_c = j_e + 2, \tag{1}$$

secondly, for the j_a, j_b, j_c, the inequalities:

$$j_a > 2, \qquad j_b > 2, \qquad j_c > 2. \tag{2}$$

thirdly, for j_e, the condition:

$$j_e = -2\cos\frac{\pi}{l} \quad \text{resp.} \quad j_e \leq -2. \tag{3}$$

For the introduction of a geometrical method of conception, we interpret (as earlier p. 301) the j_a, j_b, j_c as *space coordinates*, add indeed, we utilize, as a rule, homogeneous coordinates x, y, z, t by setting:

$$j_a : j_b : j_c : 1 = x : y : z : t. \tag{4}$$

If, instead of these, ordinary rectangular coordinates are to be used, then we set $t = 1$. [320]
The fourth invariant j_e, however, we conceive of as a *parameter*.

The relation (1) now furnishes, in the form:

$$t(x^2 + y^2 + z^2) - xyz - (j_e + 2)t^3 = 0, \tag{5}$$

for fixed value of the parameter j_e, a surface of third degree (cf. I on p. 397), which however only comes into consideration for us, to the extent that the inequalities:

$$x > 2t, \qquad y > 2t, \qquad z > 2t,$$

following from (2), hold. Upon the use of rectangular coordinates, we accordingly find ourselves in the space octant represented by $x > 2, y > 2, z > 2$, which, in the projective space, let be given by the tetrahedron, lying entirely in the finite part of the space and bounded by the four planes:

$$x-2t=0, \quad y-2t=0, \quad z-2t=0, \quad t=0. \tag{6}$$

We will call this tetrahedron T for short and denote the tetrahedron T belonging to the part of the surface of the third degree (5) by Φ_3 (cf. p. 299).

The plane $t=0$ cuts our surface of third degree in the three lines $x=0, y=0, z=0$ (edges of T); this plane is thus a tritangential plane of the surface, whose three tangency points lie at the tetrahedral vertices belonging to the surface $t=0$.

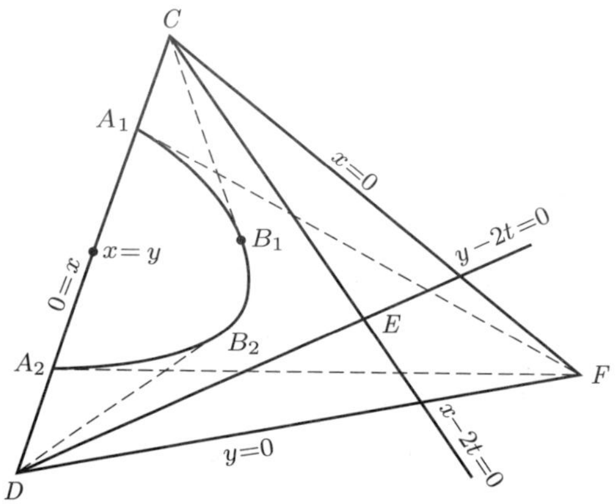

Fig. 44.

In order to further determine the course of the surface Φ_3 in the interior T, we cut Φ_3 with the planes of the pencil $z-2\lambda t=0$, whereby, in order to describe the interior of the tetrahedron T from the side $z-2t=0$ to the side $t=0$, we have to let the parameter λ grow from 1 to ∞. The individual plane $z-2\lambda t=0$ cuts out, besides the tetrahedral edges $z=0, t=0$. yet a curve of the second degree, which is represented in that plane by:

$$x^2-2\lambda xy+y^2+(4\lambda^2-j_e-2)t^2=0.$$

As one easily determines, the course of this curve is given in figure 44. The triangle CDE represents here the cut through the interior of the tetrahedron T. The line $t=0$ (tetrahedral edges $t=0, z=0$) cuts the curve in the two points, situated symmetrically
[321] to the midpoint of the edge:

$$y=(\lambda\pm\sqrt{\lambda^2-1})x;$$

The tangents of the curve at these points, denoted in the figure by A_1 and A_2, intersect one another at the point F of the coordinates $x=0, y=0$[28]. These two points A_1 and A_2, in case one lets λ decrease from ∞ to 1, move together from their initial positions C

[28] For rectangular coordinates ($t=1$), these two tangents, given by

$$y=(\lambda\pm\sqrt{\lambda^2-1})x,$$

and D, and coincide, for $\lambda = 1$, at the midpoint $x = y$ of the edge. For $\lambda = \infty$, our curve, as known, degenerates into the pair of the tetrahedral edges $x = 0, t = 0$ and $y = 0, t = 0$; in the other limiting case, $\lambda = 1$, the curve draws itself together into the edge midpoint $x = y, z = 0, t = 0$. Determine, above all, that our curve, for all intermediate values λ does not reach, with its arc lying in the triangle CDE, the sides CE and DE of this triangle (cf. figure 44). For the sake of an additional conclusion to be drawn, visualize for oneself, too, the tangency points B_1, B_2 of the tangents from the points C and D to our curve, running inside T, as these are given in the figure.

If we yet add, therefore, that our surface of third degree is invariant with respect to the six permutations of the x, y, z, and that, therefore, in particular, all results just stated for the pencil $z-2\lambda t = 0$ will also hold for the pencils of the plane $x-2\lambda t = 0$ and $y-2\lambda t = 0$, there arises a very visualizable image of the course of the Φ_3: *the portion* Φ_3 *of our surface of third degree belonging to the tetrahedron* T *is a surface stretched tautly at the boundary of the tetrahedral side* $t = 0$*, and hanging slackly in the interior of* T*, which, likewise, has the planes of symmetry* $y-z = 0, z-x = 0, x-y = 0$ *of* T *as symmetry planes, and therefore goes over into itself by a rotation about the axis* $x = y = z$ *by the angle* $\frac{2\pi}{3}$. As was to be expected from the theorems of §6 on p. 302, the sheet Φ_3 of the surface thus obtained represents a *simply connected domain.*

The point denoted in figure 44 by D represents the tetrahedral vertex $y = 0, z = 0, t = 0$. Since this vertex D belongs to the surface of third degree, then a ray running out from this vertex will meet the surface in at most two more points. If, in particular, we join the vertex D with a point P of Φ_3 with coordinates x, y, z, t, then the ray DP will [322]
cut the surface in a further point P' with coordinates x', y', z', t', *which necessarily also lies on the sheet* Φ_3. Namely, if we map figure 44 onto the plane $z-2\lambda t = 0$ determined by the tetrahedral edge CD and the point P, then clearly that additional intersection point P' always belongs, with P, to the arc $A_1B_1B_2A_2$ of our surface of second degree considered above.

In order to calculate the point P' from P, we may, insofar as the chosen ray runs out from the vertex $y = 0, z = 0, t = 0$, set $y' = y, z' = z, t' = t$. There accordingly holds:

$$\begin{aligned} t(x^2+y^2+z^2)-xyz-(j_e+2)t^3 &= 0,\\ t(x'^2+y^2+z^2)-x'yz-(j_e+2)t^3 &= 0, \end{aligned}$$

and therefore follows:

$$t(x'^2-x^2)-yz(x'-x) = 0,$$

whereupon (besides the solution $x' = x$ leading back to the point P) it follows:

$$x' = -x+\frac{yz}{t}.$$

By avoiding denominators, we better set, for the coordinates of P':

$$x' : y' : z' : t' = (yz-xt) : ty : tz : t^2. \qquad (7)$$

---

become the asymtotes of the hyperbola:

$$x^2-2\lambda xy+y^2+(4\lambda^2-j_e-2) = 0,$$

which is cut out by the "horizontal plane" $z = 2\lambda$ on the surface of third degree.

We make note: *By the quadratic transformation (7), our surface of third degree is transformed into itself in such a fashion, that, in particular, the sheet Φ_3 of the surface goes over into itself.*

The transformation (7) has the character of a *symmetric transformation* or *reflection* of the Φ_3 into itself and on this account is to be denoted by S. The associated symmetry line, which let ba called K, is furnished by the tangency points of the tangents, which run from the tetrahedral vertex $y = 0, z = 0, t = 0$ to the Φ_3. By substitution $x' = x, y' = y, z' = z, t' = t$ in (7), there follows immediately: *the symmetry line K of the reflection S is cut out on the surface Φ_3 by the surface of the second degree:*

$$yz - 2xt = 0, \tag{8}$$

which, as a hyperboloid[29], *we will call H for short.*

In the plane $z - 2\lambda t = 0$ of figure 44, B_2 is a point of the curve K. We read off from this figure, that for all finite $\lambda > 1$, the associated points of the curve K satisfy the condition
[323] $x > y$.[30] However, for $\lambda = 1$, with the point B_2 and therefore the curve K, we reach the side midpoint $x = y, z = 0, t = 0$. By correspondingly investigating the intersection of the planes of the pencil $y - 2\mu t = 0$ with the curve K, we determine, for the points of this curve, the inequality $x > z$, with the exception of the endpoint of the curve K on the sheet Φ_3[31] lying at the midpoint $x = z$ of the edge $y = 0, t = 0$. We make note: *The symmetry line K ends, on the sheet Φ_3 of the surface, at the two edge midpoints $x = y, z = 0, t = 0$ and $x = z, y = 0, t = 0$ of the tetrahedron* T *and except for this, runs entirely in the third of our tetrahedron* T *defined by $x > y, x > z$.*

Remember now, finally, that our surface Φ_3 is carried over into itself by means of the two rotations about the axis $x = y = z$ through the angles $\frac{2\pi}{3}$ and $\frac{4\pi}{3}$. It follows immediately: *the surface Φ_3 permits the three reflections into itself, to the denoted now by S_1, S_2, S_3:*

$$\begin{cases} (S_1) & x' : y' : z' : t' = (yz - xt) : yt : zt : t^2, \\ (S_2) & x' : y' : z' : t' = xt : (zx - yt) : zt : t^2, \\ (S_3) & x' : y' : z' : t' = xt : yt : (xy - zt) : t^2, \end{cases} \tag{9}$$

which, in space, have the character of central projections of the Φ_3 into itself from the three tetrahedral vertices lying in the plane $t = 0$. The three associated symmetry lines K_1, K_2, K_3 are set out on the Φ_3 by the three hyperboloids H_1, H_2, H_3:

$$\begin{cases} yz - 2xt = 0, \quad zx - 2yt = 0, \\ xy - 2zt = 0. \end{cases} \tag{10}$$

These three curves bound, on Φ_3, a triangular domain, which reach to the boundary of
[324] *the sheet Φ_3 with three cusps at three edge midpoints of the tetrahedral* T, *and which*

29 In the rectangular system ($t = 1$), it becomes a hyperbolic paraboloid.

30 Notice e.g., that the symmetry plane $x = y$ of figure 44 runs through the intersection point of the tangents CB_1, DB_2 and through the points E and F.

31 We remark in passing: the intersection of the hyperboloid H with our surface of third order decomposes into the two tetrahedral edges $y = 0, t = 0$ and $z = 0, t = 0$ and into another curve of fourth degree, which is purely representable as the intersection of the two surfaces of the second degree:

$$x^2 - y^2 - z^2 + (j_e + 2)t^2 = 0, \ \ yz - 2xt = 0.$$

this latter curve furnishes our symmetry line K.

is decomposed by the three symmetry planes $y = z, z = x, x = y$ into six smaller alternatively symmetric and congruent triangles. We shall immediately be led back to this domain, represented in figure 45, in another connection.

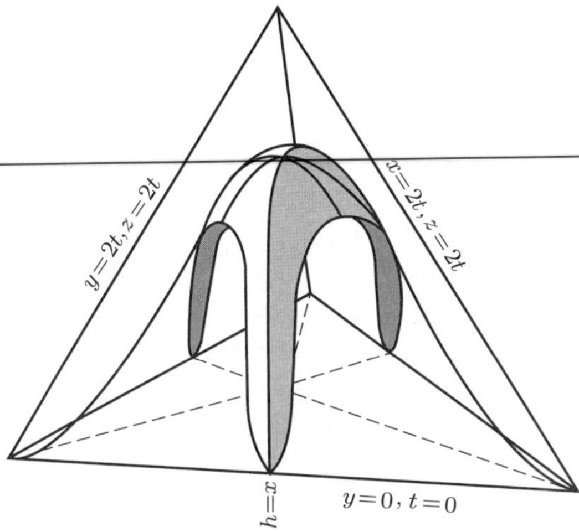

Fig. 45.

1.12 The discontinuity domain of the modular group and the character (1, 1)

For the transformation of a canonical polygon of the character (1, 1), the angle at the fixed vertex of the polygon, and with this, the invariant j_e, remains unaltered. All infinitely many canonical polygons, which furnish one and same group, therefore correspond, in our space of the invariants $j_\alpha = \frac{x}{t}, \ldots$, to points on the same sheet Φ_3 of the surface. In order to obtain the group continua of the character (1, 1), we shall thus, first of all, solve the problem, *of determining, on the individual sheet Φ_3 of the surface, the discontinuity domain of the modular group.*

We begin with determining, on the Φ_3, the domain, consisting of one, or possibly also of several, pieces of those points, which furnish reduced normal hexangles of the kind discussed in §10. But if, to a point (x, y, z, t) of Φ_3, corresponds such a hexangle, then the point C of this hexangle, defined independently above (p. 315), will belong to the triangle e_1, e_2, e_3. If, in particular, C lies in the "interior" (not on a side) of this triangle, then the immediate neighborhood of the point (x, y, z, t) of Φ_3 also furnishes reduced hexangles. Namely, for continuous alteration of the point (x, y, z, t), our hexangle and its point C also alter themselves continuously; we are consequently not able to immediately arrive at a hexangle, for which C moves onto a side or even out of the triangle. Conversely, we perceive, *that a boundary point of the domain to be constructed on the Φ_3 must always furnish a hexangle, for which C belongs to a side of the triangle*

e_1, e_2, e_3. By thus determining *all* points of the Φ_3, which give hexangles of this special kind, we shall, in any case, also obtain the entire boundary of the domain sought on the Φ_3.

We have three cases to distinguish, possibly to be treated fully the same, according as C moves onto the triangles e_2e_3 or e_3e_1 or e_1e_2. If we have the first case, say, then C coincides with the point denoted above by e_1'. The consequence is (visualize, in figure 41 on p. 314, the coincidence of the points C and C' with e_2'), that in the net of the
[325] domains Q, one of the three inequivalent sides draws itself together into a point, and that the two initial vertices of this net fuse into a vertex surrounded by *four* domains Q. The normal polygon belonging to the center $C = e_1'$ consequently reaches up to four fixed vertices e_1, e_2, e_3, e_4; and indeed, if we use the notations of figure 41 on p. 314, ε' has changed into the fourth fixed vertex e_4, while the sides $e_2\varepsilon'''$ and $e_3\varepsilon''$ have drawn themselves together into points[32].

In figure 46, our normal quadrangle of the center $C = e_1'$ is represented. The opposite sides are coordinated to one another by the generators V_b and V_c. Since it is a question here of a normal polygon then the sides e_1e_2 and e_3e_4 furnish level lines of the hyperbolic substitution V_b, which lie symmetrically with respect to the level line of V_b going through the center $C = e_1'$. If $\overline{V}$ is the reflection in this latter level line, then V_b is transformed by $\overline{V}$ into $V_b^{-1} = \overline{V}V_b\overline{V}$. By $\overline{V}$, however, not only will the two level lines of V_b furnished by the sides e_1e_2 and e_3e_4 of our quadrangle be interchanged, but, in particular, the points e_2 and e_4 are interchanged and, similarly, the points e_1 and e_3; for e.g., e_2 and e_4 lie on one and the same orbit curve of V_b, which, however, is transformed into itself by $\overline{V}$. We perceive: *the reflection $\overline{V}$ transformations our quadrangle $e_1e_2e_3e_4$ into itself; the associated group Γ is consequently capable of extension by $\overline{V}$ and thus furnishes the group $\overline{\Gamma}$ of the second kind* (cf. I on p. 137).

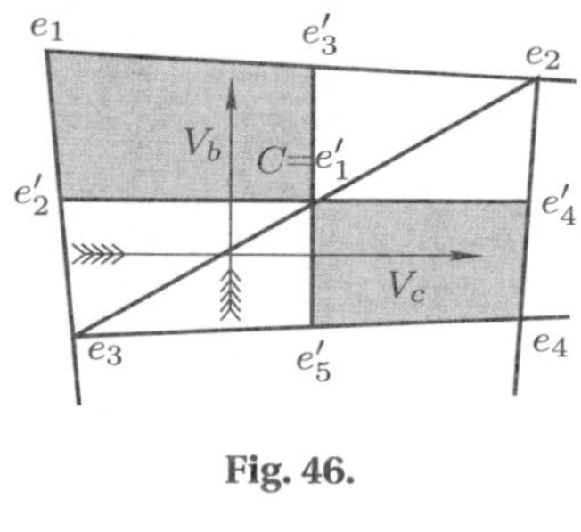

Fig. 46.

Since the side e_1e_3 of our quadrangle is transformed into itself by $\overline{V}$, then this side stands perpendicular to the symmetry line of $\overline{V}$. But the perpendicular from $C(= e_1')$ to e_1e_3 cuts out the fixpoint e_2' of the substitution V_2' on e_1e_3. If we call the intersection point of e_2e_4 with the symmetry line of $\overline{V}$, e_4' say, *then the quadrangle $e_1e_2'e_4'e_2$ furnishes a discontinuity domain for the group $\overline{\Gamma}$ of the second kind.* The side e_1e_2 is hereby the symmetry line of the generating reflection $V_b\overline{V}$ (cf. figure 46).

[32] If we write $C' = V_1'(C)$ (deviating from figure 41), the two hexangles belonging to C and C', equivalent in general, become identical in the case at hand, and assume the form of the quadrangle e_1, e_2, e_3, e_4.

Further by $\overline{V}$, the two level lines of V_c furnished by e_1e_3 and e_2e_4 and therefore also the fixpoint of this hyperbolic substitution lying outside the ellipse, are transformed into themselves. We find: the reflection $\overline{V}$ transforms V_c into itself, $V_c = \overline{V}V_c\overline{V}$, the [326] symmetry of $\overline{V}$ is the line among the orbit curves of V_c. The level line of this substitution running through $C = e_1'$ and the fixpoint of V_c cuts the symmetry line $e_2'e_4'$ of $\overline{V}$ in e_1' at a right angle. The reflection $\overline{V}'$ in this level line of V_c not only transforms the quadrangle $e_1e_2e_3e_4$ into itself, but also our discontinuity domain $e_1e_2'e_4'e_2$. The addition of $\overline{V}'$ to $\overline{\Gamma}$ leads to *a group* $\overline{\Gamma_0}$ *of the second kind, as whose discontinuity domain we are able to choose the quadrangle with the vertices* e_1, e_2', e_1', e_3', *bounded by the symmetry lines of the four generating reflections* $\overline{V}, \overline{V}', V_b\overline{V}, \overline{V}'V_c$.

The subgroup Γ_0 of index 2 contained in $\overline{\Gamma_0}$ contains our group Γ and besides this, in any case, the substitution $V_1' = \overline{V}\overline{V}'$. This group Γ_0 is therefore identical with our group of the character (0, 4) thus denoted above (p. 313 ff.). As result, we make note: *If the point C of the reduced hexangle moves onto the triangle side* e_2e_3, *then the group* Γ_0 *of the character* (0,4) *is capable of extension to a group* $\overline{\Gamma_0}$, *whose discontinuity domain is a quadrangle* $e_1e_2'e_1'e_3'$ *bounded by four symmetry lines; and indeed, hereby, which is characteristic, the sequence of vertices is one such, that* e_1' *lies on the boundary of the quadrangle opposite the vertex* e_1, *also characteristic of the group* Γ.

The result obtained is reversible: *If the group* Γ_0 *is capable of extensions by reflections to the group* Γ_0, *whose discontinuity domain is a quadrangle* $e_1e_2'e_1'e_3'$ *bounded by four symmetry lines, and if the arrangement of the vertices of this quadrangle is so that* e_1' *lies opposite* e_1, *the elliptic vertex of the angle* $\frac{\pi}{2l}$ *resp. the hyperbolic or parabolic vertex, then for the reduced normal hexangle of* Γ, *the point C lies on the side* e_2e_3 *of the triangle* e_1, e_2, e_3. Namely, if we lay the four quadrangles surrounding the point e_1' together into a larger quadrangle $e_1e_2e_4e_3$, then this will give a discontinuity domain of the subgroup Γ contained in $\overline{\Gamma_0}$, whereby the opposite sides of the larger quadrangle are mapped onto one another by the generators V_b and V_c of Γ. This quadrangle now immediately shows, itself to be the normal domain of Γ belonging to the center $C = e_1'$. Namely, the lines furnished by the quadrangle sides e_1e_2 and e_3e_4 bound a normal domain of the center $C = e_1'$ for the cyclic group to be generated from V_b, and the lines e_1e_3 and e_2e_4 accomplish the same thing for V_c. The quadrangle $e_1e_2e_4e_3$ is therefore the common component of these two cyclic normal domains, so that, according to I on p. 245, the normal domain of Γ belonging e_1', for its part, will be a component of the [327] quadrangle $e_1e_2e_4e_3$. It must consequently be straightaway identical with this quadrangle, since both domains are also complete discontinuity domains of Γ. Therefore, we actually have the special case before us, that the normal hexangle of the group Γ degenerates into a quadrangle reaching up to four fixed vertices, and indeed with a C on the triangle side e_2e_3.

In order to calculate the invariants in the present symmetric case, we must explicitly give the generators V_b and V_c. We recognized above, in the symmetry line $e_2'e_4'$ (cf. figure 46), the line among the orbit curves of V_c, and similarly, $e_3'e_5'$ is the "orbit line" of V_b. If we so choose ζ, that this latter line represents the imaginary ζ-axis, and the "orbit line" of V_c the unit circle of the ζ-halfplane, then we have the schema:

$$V_b = \begin{pmatrix} \alpha & 0 \\ 0 & \delta \end{pmatrix}, \; V_c = \begin{pmatrix} \alpha' & \beta' \\ \beta' & \alpha' \end{pmatrix}, \; V_a = V_c^{-1}V_b^{-1} = \begin{pmatrix} \delta\alpha' & -\alpha\beta' \\ -\delta\beta' & \alpha\alpha' \end{pmatrix},$$

from which we deduce:

$$j_a = \frac{x}{t} = \alpha'(\alpha+\delta),\ j_b = \frac{y}{t} = \alpha+\delta,\ j_c = \frac{z}{t} = 2\alpha',$$

Accordingly, the relation subsists:

$$yz - 2xt = 0, \tag{1}$$

i.e., we find ourselves on the curve of Φ_3, denoted above (p. 323) by K_1.

Also, this result is again capable of reversal. If there be proposed, for the Γ_0, a discontinuity domain in the shape of a triangle e_1, e_2, e_3 (cf. figure 42 on p. 316), be this a normal triangle or not, then join the fixpoints e_1', e_3', situated on the triangle sides, rectilinearly and, in consequence of $V_3' V_1' = V_b$, thus obtain, in each case, the orbit line of the associated substitution V_b. Similarly, the line $e_2' e_1'$ furnishes the orbit line of V_c. This latter we again make into the unit circle in the ζ-halfplane, while we may only generally require of the former, say, that it furnish a semicircle of the ζ-halfplane running out from $\zeta = 0$ (one of the two of fixpoints of V_b lies at $\zeta = 0$). To this choice of ζ then corresponds the schema:

$$V_b = \begin{pmatrix} \alpha & 0 \\ \gamma & \delta \end{pmatrix}, V_c = \begin{pmatrix} \alpha' & \beta' \\ \beta' & \alpha' \end{pmatrix}, V_a = V_c^{-1} V_b^{-1} = \begin{pmatrix} \alpha'\delta + \beta'\gamma & -\alpha\beta' \\ -\beta'\delta - \alpha'\gamma & \alpha\alpha' \end{pmatrix},$$

The invariants are given by:

$$j_a = \frac{x}{t} = \alpha'(\alpha+\delta) + \beta'\gamma,\ j_b = \frac{y}{t} = \alpha+\delta, j_c = \frac{z}{t} = 2\alpha',$$

[328] from which one calculates:

$$\frac{yz - 2xt}{t^2} = -2\beta'\gamma.$$

Consequently, should the equation (1) subsist, then it follows, since β' is certainly different from 0, that γ is necessarily equal to zero. One immediately recognizes the imaginary ζ-axis and unit circle as symmetry circles; for, by the reflections in them, the two generators V_b, V_c go over into themselves resp. their inverse substitutions. Therefore our symmetric case is present: *To each reduced normal polygon, for which C moves onto the side $e_2 e_3$ of the triangle $e_1 e_2 e_3$, corresponds, on the sheet Φ_3 of the surface, a point of the curve K_1, and conversely, each point of this curve K_1 furnishes a symmetric normal polygon of our kind.*

The other cases, that C reaches the point e_2' on the line $e_3 e_1$ or the point e_3' on the line $e_1 e_2$, furnish corresponding results: *the points on the curve K_2 give exactly and exhaustively those symmetric cases, for which C moves onto the side $e_3 e_1$, for which, therefore, on the boundary of the elementary quadrangle of the Γ_0, the vertex e_2' lies opposite the vertex e_1, also coming up for Γ; finally, the curve K_3 similarly furnishes the third case, that C lies on $e_1 e_2$ and e_3' lies opposite the point e_1.*

According to the argument at the beginning of the paragraph, the total boundary of the domain of the reduced normal hexangles to be bounded on Φ_3 will now be furnished by the three curves K_1, K_2, K_3. But according to p. 323, the only domain bounded by these curves is the triangle represented in figure 45, in whose boundary the K_1, K_2, K_3 take part in their entire extent: *the triangle, bounded on the sheet Φ_3 of the surface by three curves K_1, K_2, K_3, represents the domain of the reduced normal hexangles.*

The discontinuity domain of the modular group on the sheet Φ_3 of the surface is now obtained immediately. Notice, that our normal hexangle is indeed uniquely determined for the invariant standpoint for the individual group. For the substitution V_a, however, we still have the free choice among the three generators; and first after we have made this choice, are V_b and V_c also then uniquely determined[33]. If, for this purpose, we determine, say, that j_a is to be neither greater than j_b, nor greater then j_c:

$$j_a \le j_b, \qquad j_a \le j_c, \tag{2}$$

then we are restricted to the double triangle with the symmetry line $y = z$, outlined darkly and characterized by $x \le y, x \le z$, in the triangular domain schematically represented by figure 47: *This double triangle is a discontinuity domain of the modular group on the sheet Φ_3 of the surface.* [329]

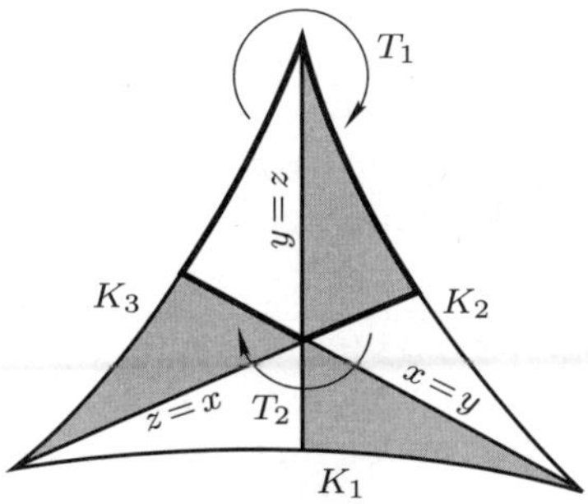

Fig. 47.

If we would also then consider two polygons as equivalent, if the one goes over into the other by a substitution of the second kind (a reversal)[34], then to this would correspond the *extension of our modular group* by the reflection in the symmetry line $y = z$ (interchange of the invariants j_b and j_c), to be denoted, say, by S_1'. The discontinuity domain of the modular group thus extended would then be, say, the elementary triangle shaded in figure 47, which is bounded by the symmetry lines $y = z, z = x$ and the curve K_2.

To the generators of this extended modular group, in any case, there then belong the reflections, to be denoted by S_1' and S_2', in the two symmetry lines $y = z$ and $z = x$:

$$\begin{cases} (S_1') & x' : y' : z' : t' = x : z : y : t, \\ (S_2') & x' : y' : z' : t' = z : y : x : t. \end{cases} \tag{3}$$

33 Compare, for this, the remark on p. 313 concerning the choice of one reduced canonical polygon among three possible ones.

34 However, this corresponds to the approaches in Vol. I and to the conceptions of the theory of algebraic functions, to only then consider polygons as equivalent, if they go over into one another by transformation of the *first* kind.

Further, however, the conjecture is natural, that the third generator of the extended modular group is the reflection S_2 of the surface Φ_3 into itself, given in (7) on p. 322. *Then the generators of the modular group (of the first kind), denoted in figure 47 by T_1 and T_2, would represent themselves in the form $T_1 = S_2 S_1'$, $T_2 = S_1' S_2'$, or, explicitly:*

$$\begin{cases} (T_1) & x' : y' : z' : t' = tz : (xy - zt) : yt : t^2, \\ (T_2) & x' : y' : z' : t' = z : x : y : t. \end{cases} \tag{4}$$

But, in fact, these are the generators of the modular group, as we had derived them in I on p. 397 from the transformations of the canonical polygons (modifications of the canonical crosscut systems on the closed surface)[35].

To the production of all transformations of the modular group from T_1 and T_2 corresponds a simple and gapless covering of the entire surface Φ_3 with a net of equivalent [330] double triangles. The analogy to the triangle net of the modular group coming up in the theory of the elliptic functions lies close at hand. That it is even a question here of a single-valued correspondence of the two kinds of triangle nets (with a reversibly single-valued mapping between the domains covered by them), will emerge below from the fundamental theorem for the signature $(1, 1; l)$.

Yet notice in passing, that the symmetry lines $x = y$, $y = z$ and $z = x$ furnish symmetric models. If, e.g., $y = z$, ie., if the two substitutions $V_b = V_3' V_1'$ and $V_c = V_1' V_2'$ have the same invariants, then in the normal hexangle of figure 41 on p. 314, the segments $e_1' e_2'$ and $e_1' e_3'$ are equally long. The reflection in the bisector of the angle $\sphericalangle e_2' e_1' e_3'$ thus transforms V_1', V_2', V_3' into V_1', V_3', V_2' and therefore $V_1^{\frac{1}{2}} = V_3' V_1' V_2'$ into $V_1^{-\frac{1}{2}}$, so that it runs through the vertex e_1 (cf. figure 41) If one completes the triangles $e_1 e_2 e_3$ of this figure to the quadrangle $e_1 e_2 e_4 e_3$, by positioning the triangle $e_2 e_3 e_4 = V_1'(e_3 e_2 e_1)$ along the side $e_2 e_3$, *then this quadrangle is equilateral (rhombus) and has the diagonals $e_1 e_4$ and $e_2 e_3$ as symmetry lines.*

Finally, if we now ask about the representation of our *group continua of the character* $(1, 1)$, then this representation is accomplished immediately in the *elliptic* case. Here $j_e = -2\cos\frac{\pi}{l}$ is given fixed: *On the associated surface Φ_3, our double triangle just bounded immediately furnishes the antitype of the two dimensional group continuum in this case.* In the *hyperbolic* case (including the *parabolic* limiting one), j_e is to be considered as variable, corresponding to the inequality $j_e \leq -2$. Here we have to arrange the respective surfaces Φ_3 next to one another continuously, and on each one, bound off our double triangle (discontinuity domain of the modular group). For the $j_e = -\infty$, the surface Φ_3 reduces itself to the side $t = 0$ of our tetrahedron T, often mentioned above (p. 320 ff). We find: *The three dimensional group continuum of the character* $(1, 1)$ *with $j_e \leq -2$ is represented by a hexahedron, which is bounded by the three planes:*

$$x - y = 0, \quad x - z = 0, \quad t = 0, \tag{5}$$

the two surfaces of the second degree (hyperboloids H_3 and H_2):

$$xy - 2zt = 0, \qquad xz - 2yt = 0 \tag{6}$$

[35] The inessential deviations, in the manner of writing the transformations, from I on p. 397, rest upon the fact, that, at that time, the discontinuity domain of the modular group was fixed somewhat differently.

and the surface of the third degree (belonging to the parabolic limiting case):

$$t(x^2 + y^2 + z^2) - xyz = 0. \tag{7}$$

1.13 Connectivity and boundary of the individual group continuum of the character (1, 1)

[331]

The antitype of an individual one of our group continua (double triangle on a Φ_3 resp. the hexahedron just obtained) is, taken by itself, *a simply connected* domain, which, however, only first represents the associated group continuum *in cut up form*, insofar as the four boundary curves of the double triangle, as well as the four boundary surfaces of the hexahedron corresponding to them, are coordinated to one another in pairs by the generating transformations T_1 and T_2 of the modular group. The cutting up of our continua present here takes place along "*symmetric*" models; for, as we know, the boundary curves K_2, K_3 resp. the boundary surfaces H_2, H_3, furnish quadrangles $e_1e_2e_4e_3$ (cf. figure 46 on p. 325) with "*midline symmetry*". As we are able to say for short, while the boundary curves resp. surface coordinates onto one another by T_2, give quadrangles with "diagonal symmetry".

In order to represent somewhat more closely the realization of this closing together of the boundary models, say for the example of the transformations T_1, we have, in figure 48 (cf. p. 314ff.), first completed the off-mentioned triangle $e_1e_2e_3$ along the side e_1e_3 to the quadrangle $e_1e_2e_3e_4$ with the two generators V_a, V_b (darkly outlined in the figure), then secondly, along the side e_2e_3 to the quadrangle $e_1e_2e_5e_3$ with the generators V_a, V_c (shaded in the figure). The transformation T_1, in its original meaning (modification of the cut system on the closed surface) is the transformation of the second of these quadrangles to the first.

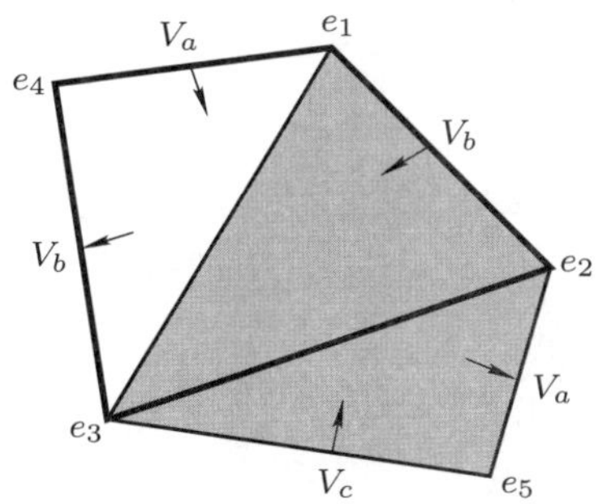

Fig. 48.

A point on the boundary H_2 furnishes now a quadrangle $e_1e_2e_5e_3$ with midline symmetry. If we go from here, say, along the associated surface Φ_3 to the corresponding boundary point on H_3, then to this there always runs parallel a continuous transforation of the discontinuity domain, for which the quadrangle $e_1e_2e_3e_4$ finally goes over into the shape $e_1e_2e_5e_3$ of figure 48. We thus actually come back to the same group Γ at the end, from which we started out, only that now the quadrangle with the generators V_a, V_b is that one, which directly exhibits the midline symmetry.

Now, as was already imparted in §8 on p. 305, the assertion was set up by *Poincaré,*
[332] *that the individual continuum of "boundary circle groups" was a closed one.* In our present case there subsists the yet somewhat further reaching theorem, *that for given invariant j_e, the associated two dimensional group continuum is a closed one and has the connectivity of the surface of the sphere.* But we are not able to conclude this theorem without further ado from the closing together of the boundary curves of the double triangle on the surface Φ_3; for this double triangle reaches with a cusp up to *boundary of the polygon continuum.* So long as we move in the "interior" of the polygon continuum, to each point uniquely corresponds its associated group Γ, which, for a continuously wandering point, itself also continuously alters along with this. but as soon as we arrive at the boundary of the polygon continuum, we obtain degenerate models, which we will designate as the "boundary models" of the group ocntinuum, and which, just on this account, always require a special consideration, since our investigations, valid so far only for non-degenerate models, are not binding for the boundary models.

Now, first of all, let the value of this invariant j_e also remain constant for $j_e \leq -2$. In the associated double triangle, we then move to the cusp in question, by effecting the passage to the limit:

$$\lim j_a = 2, \qquad \lim j_b = \infty, \qquad \lim j_c = \infty. \tag{1}$$

In order to determine the degeneration entering here, we tie into a normal polygon of figure 40 on p. 312, but, at the same time, put the vertex ε_1 at a vertex peculiar to the domain net Q, so that the normal polygon obtains the shape sketched in figure 49. The generators called V_a, V_b, V_c need not be exactly identical with the substitutions thus denoted so far, but are conjugate to them, and therefore have the same invariants. By suitable choice of the two polygon sides coordinated to one another by V, we are always able to obtain the arrangement of the V_a, V_b, V_c given in figure 49.

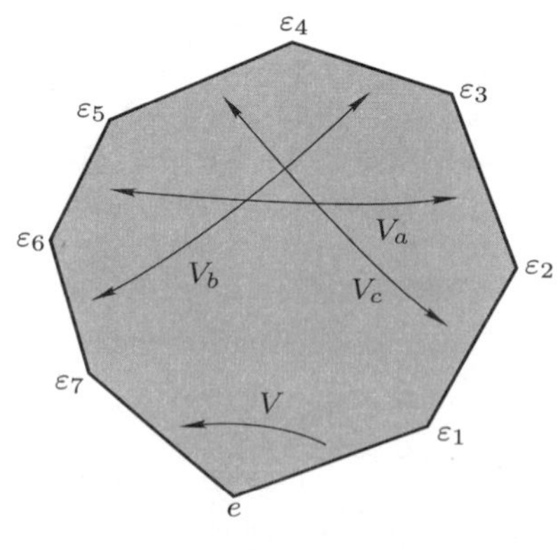

Fig. 49.

As we know, the six sides coordinated to one another by V_a, V_b, V_c always lie in the interior of the ellipse[36]. Thereby the sides $\varepsilon_1\varepsilon_2$ and $\varepsilon_4\varepsilon_5$ are portions of two level lines of [333] V_c, which bound a cyclic *normal domain* belonging to the center e. If, in the passage to the limit (1), we hold the point e, say, fixed in its place, then the two level lines of V_c just mentioned, for $\lim j_c = \infty$, become ellipse tangents, and therefore the two sides $\varepsilon_1\varepsilon_2$ and $\varepsilon_4\varepsilon_5$ draw themselves together into points on the periphery of the ellipse. Since, on account of $\lim j_b = \infty$, the same thing holds for the sides mapped onto one another by V_b, then, in the case of the passage to the limit (1), the vertices $\varepsilon_1, \varepsilon_2$ will fuse into one point ε'_1 of the periphery of the ellipse, the points $\varepsilon_3, \varepsilon_4, \varepsilon_5$ into a similar point ε'_2, finally $\varepsilon_6, \varepsilon_7$, to a third point ε'_3, which similarly lies on the ellipse (cf. figure 50).

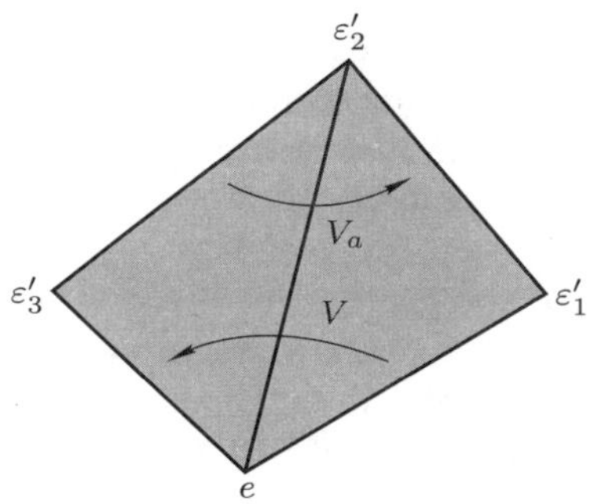

Fig. 50.

The question now makes its appearance, whether these vertices lying on the ellipse have *parabolic* (and not hyperbolic, say) character, which is necessary for the usability of the boundary polygon according to I on p. 143. For the vertex ε'_2 this is immediately clear; for it is the fixpoint of the substitution V_a, for which indeed, $\lim j_\alpha = 2$ holds. For the two vertices $\varepsilon'_1, \varepsilon'_3$, belonging to a cycle, notice, that the diagonal $\varepsilon'_2 e$ (cf. figure 50) joins the fixpoint ε'_2 of V_a with the polygon center e. The reflection $\overline{V}$ in this diagonal consequently interchanges the level lines $\varepsilon'_2\varepsilon'_1$ and $\varepsilon'_2\varepsilon'_3$ of V_a and on then, in particular, the ellipse points ε'_1 and ε'_3. The reflection $\overline{V}$ will accordingly also interchange the two lines $e\varepsilon'_1$ and $e\varepsilon'_3$. We conclude from this, that the substitution V_aV with the fixpoint ε'_1 is actually parabolic; for one can represent V_aV in the form:

$$V_aV = (V_a\overline{V})(\overline{V}V),$$

as the product of two reflections $V_a\overline{V}$ and $\overline{V}V$, whose symmetry lines $\varepsilon'_1\varepsilon'_2$ and $\varepsilon'_1 e$ intersect at the point ε'_1.

In this manner, we have arrived at the result: *If, from the interior of the double triangle, by passage to the limit (1), one goes in any manner to the cusp reaching up to the boundary of the polygon continuum, then to the groups of the character* (1,1), *there attaches itself here single-valued and continuously, as a degeneration, a group of the*

[36] See e.g. figure 41 on p. 314, where the e'_1, e'_2, e'_3 are elliptic fixpoints and the three sides Ce'_1, Ce'_2, Ce'_3 of the net of the domains Q, running together at C, form concave angles at C.

character (0,3), *which, besides the previous generator V, has two parabolic generators.*
[334] By considering this "boundary group" as belonging to our continuum, it follows the theorem set up above concerning the connectivity of the continuum of all groups belonging to a given j_e.

The elliptic cases, and also the parabolic limiting case, if we want to consider it by itself, are already completely disposed of by the result just obtained. Now, however, for $j_e \leq -2$, we still have the three dimensional continuum of all hyperbolic cases, which we had to visualize by the hexadron represented at the close of §12 on p. 330. This hexahedron, corresponding to all values $j_e \leq -2$ of the "parameter" j_e, represents a linear continuum of double triangles, for each individual one of which the result just obtained holds. Correspondingly, we shall consider the hexadron, as the image of the group manifold, closed at least to the extent, *that the two boundary surfaces H_2 and H_3, coordinated to one another by the transformation T_1, are not to count as different, and similarly, for the two boundary surfaces assigned to one another by T_2.*

But the hexahedron, besides this, is also bounded by the surface $t = 0$ of our tetrahedron T and the surface Φ_3 represented by (7) on p. 330. The latter surface, with the double triangles furnishing the parabolic case ($j_e = -2$), takes part in the boundary of our hexahedron. We arrive here at a two dimensional group continuum, which, according to the nature of the matter, remains unconverted at the boundary of the hexahedron: *the three-dimensional continuum of principal circle groups "without boundary circle", corresponding to the values $j_e \leq -2$, is accordingly to no longer a "closed" continuum.*

The boundary surface $t = 0$ of the hexadron again requires a special consideration, since it also belongs to the boundary of the polygon continuum. In order to determine the boundary models present here, visualize, first of all, the position of the two hyperboloids H_2 and H_3 and determine, how these cut the surface Φ_3 for j_e decreasing further and further to $-\infty$.[37] On the Φ_3, the triangular domain of the net of the mod-
[335] ular group, apart from the three double triangles positioned about the point $x = y = z$, will always press more and more to the boundary of the sheet Φ_3. *For $j_e = -\infty$, Φ_3 has changed into the side $t = 0$ of* T, *and of the entire triangle net, only those three double triangles are still remaining, which already completely fill up the boundary surface Φ_3 in question; the modular group, apart form the cyclic permutations of the j_a, j_b, j_c, has reduced itself to the identity alone.*

---

[37] The curve denoted on p. 322 by K, according to p. 323 (footnote), was purely represented by:

$$x^2 - y^2 - z^2 + (j_e + 2)t^2 = 0, \ yz - 2xt = 0.$$

If, for the sake of easier visualization, we first take $x = 0$ as the infinitely distant plane, set $x = 1$ and interpret t, y, z as rectangular coordinates, then the first of these equations, for $j_e < -2$, represents a family of ellipsoids of rotation, which all cut the plane $t = 0$ in the circle $y^2 + z^2 = 1$, and which, for

$$\lim j_e = -\infty,$$

reduce themselves to the doubly counted surface of this circle. By making this clear for the tetrahedron T of figure 45 on p. 323 and comparing with the course of the hyperboloid H, one perceives, that for the passage to the limit $\lim j_e = \infty$, the curve K continuously approaches the pairs of lines $yz = 0$, from which the statements in the text arise.

The degeneration present here must accordingly be very simple. To clarify matters further, we next approach an "inner" point of the tetrahedral side $t = 0$. As discontinuity domain of Γ, we utilize the quadrangle $e_1e_2e_3e_4$, to be produced from the normal hexangle of figure 41 on p. 314, with the generators V_a, V_b (cf. figure 48 on p. 331) and for the sake of easier visualization, carry this quadrangle over onto the ζ-plane, where it furnishes the discontinuity domain of figure 51. The passage to the limit to be undertaken now requires $\lim j_a = \infty$, $\lim j_b = \infty$. Hereby, the four boundary circles of the domain described draw themselves together into points on the real ζ-axis, and the discontinuity domain finally covers the entire ζ-plane. *The boundary model attained, which attached itself continuously to the foregoing groups of the character* (1,1), *and is to naturally again to be counted as part of the continuum, is the group* Γ *consisting of the identity 1 alone, irrespective of which interior point of the tetrahedral surface* $t = 0$ *we have reached.*

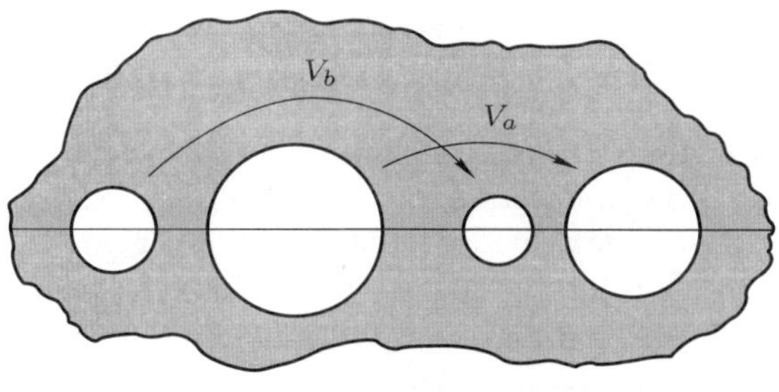

Fig. 51.

The tetrahedral surface $t = 0$ takes part in the bounding of our hexahedron of the reduced polygons only with that third, which is characterized by $x \le y, x \le z$. But to this third belongs the tetrahedral edge $x = 0, t = 0$ at which the hexahedral surface $t = 0$ meets the boundary surface H_2 and H_3, in its entire extent. What boundary models do we reach upon approaching a point of this edge?

The two halves of the edge (which are coordinated to one another by T_1), separated by the midpoint $y = z$, behave themselves symmetrically, so that we restrict ourselves, say, to be half described by $z \le y$. If, first of all, (y_0, z_0) is any interior point of this half, [336]
then set:

$$\frac{2y_0}{z_0} = j_a^{(0)}, \tag{2}$$

and observe, that the line of the surface H_2 represented by:

$$x = j_a^{(0)}t, \qquad 2y = j_a^{(0)}z$$

runs out from this point (y_0, z_0). Consequently, if we so approach the point (y_0, z_0), that the path is tangent to a determinate plane:

$$x - j_a t = 0, \tag{3}$$

running through the edge $x = 0, t = 0$, then, since the approach is to take place in the hexadron (including the boundary), the "parameter" j_a of equation (3) is restricted to the interval $j_a \ge j_a^{(0)}$, the upper limit $j_a = \infty$ included.

But which boundary models we reach upon tangent approach to the plane (3), is shown by a glance at figure 51 on p. 335, in which $\lim j_b = \infty$ is to be taken, while the invariant of V_a finally becomes equal to the parameter in Equation (3): *To the edge point* (y_0, z_0) *corresponds continuously and multiple valuedly the linear continuum of the cyclic hyperbolic groups* Γ *of all parameter values* $j_a \geq j_a^{(0)}$. Continuous modification of the plane (3) stipulates continuous modification of the boundary model reached. The continuity and single-valuedness of the boundary models obtained remain fully preserved, if we only discriminate from one another the infinitely many approaches to (y_0, z_0), corresponding to the places (3).

To the endpoint $z_0 = 0$ of our half-edge corresponds $j_a^{(0)} = \infty$, and therefore only the one group Γ, consisting of the identity 1. The hexahedral vertex coming into consideration here we are only able to reach tangentially to the plane $t = 0$. At the edge midpoint $y_0 = z_0$, we even come upon the parabolic limiting case $j_a^{(0)} = 2$ of the cyclic hyperbolic groups. Do not forget, that to this point $(y = z)$, all double triangles of the surface Φ_3 reach with boundary groups of the character $(0,3)$ of one hyperbolic and two parabolic generators. *But even here the continuity and single-valuedness of the boundary mod-*
[337] *els reached remains presented.* In the former respect, one need only cast a glance at figure 52, in which the double triangle of figure 50 on p. 333 is carried over onto the ζ-halfplane. For $\lim j_e = -\infty$, after wandering through all surfaces Φ_3, the two semicircles $a\varepsilon_3'$ and $b\varepsilon_1'$ have drawn themselves together into points, and the parabolic limiting case of our cyclic hyperbolic groups already obtained makes its appearance. For preserving the uniqueness, however, it is only requisite, that upon approach to the point $y = z$ tangentially to the tetrahedral side $x - 2t = 0$, we still distinguish which of the surfaces Φ_3 is finally touched by the path.

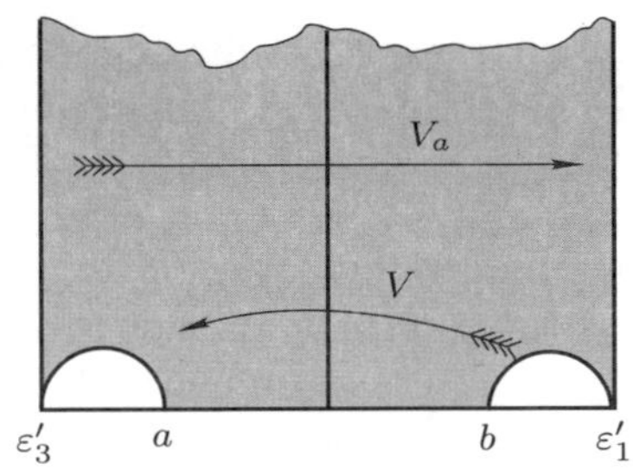

Fig. 52.

We have thus arrived at the result, fundamental for the continuity proof, *that we have also assigned continuously, and in this sense described, single-valuedly, a closed continuum of boundary models to the entire boundary of the hexahedron.* If we will, we also need not exclude the boundary surfaces assigned to one another by T_1 and T_2.

1.14 The reduced polygons of the character (0, 4)

According to I on p. 263, a normal polygon of the character $(0,4)$ of ordinary type is a decangle with four fixed vertices and two three-termed cycles of movable vertices. The single type hereby coming up is represented in figure 53, which, at the same time, let define the notations of the generators, the fixed vertices and a few movable vertices.

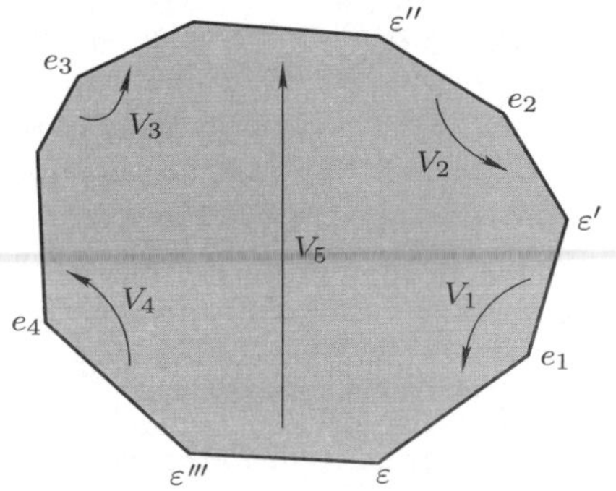

Fig. 53.

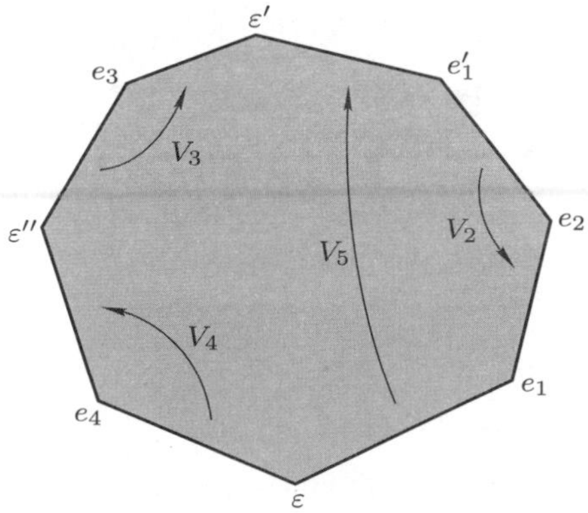

Fig. 54.

If, upon movement of the center C, a type change enters, with alteration of the generators V_1, V_2, V_3, V_4, say the first one V_1, then at this moment, the sides of the decangle coordinated to one another by V_1 vanish, and obtain, as a special type, the octangle represented in figure 54 with three fixed vertices e_2, e_3, e_4, one two-termed cycle [338]
of movable vertices e_1, e'_1 and one three-termed cycle of movable vertices. the center C, in the present situation, lies on the boundary line between the two domains Q belonging to e_1 and e'_1, which we call Q_1, Q'_1, corresponding to the vertices e_1, e'_1.

Upon carrying C further along the boundary line of Q_1 and Q'_1, the cycle e_1, e'_1, and with this, the pair of sides e_1e_2, e'_1e_2, remains maintained. A further type change can accordingly only enter in such a way, that either the sides of the octange coordinated to one another by V_3 or by V_4 or by V_5 vanish. The first two cases permit the same treatment; let us accordingly assume, say, that the sides mapped onto one another by

V_3 have just drawn themselves together into points. The normal polygon then has the shape of the hexangle represented in figure 55. We will prove, *that this is transformed into itself by the reflection* $\overline{V}$ *in the diagonal* e_2e_4. Namely, join the center C with the vertex e_2 and call $\overline{V}$ the reflection in the line Ce_2[38]; this reflection obviously transforms the points e_1 and e_1' into one another. Now, at the same time, C is the center of that cyclic normal domain belonging to V_5, which is bounded by the level lines e_1e_3 and $e_1'e_3'$ of V_5. We obtain the middle symmetry of this cyclic normal domain running through C and the fixpoint of V_5, e.g. by letting a perpendicular drop from C onto the line joining the points e_1 and $e_1' = V_5^{-1}(e_1)$; this perpendicular, however, as we already know, is our line Ce_2, of which we thus perceive, that it runs through the fixpoint of V_5. In the same way, one finds, that the line Ce_4 runs through the fixpoint just named. Therefore C lies on the diagonal e_2e_4, which is a symmetry line of the hexangle. Γ, *in the present case, is capable of extension by reflections to a group* $\overline{\Gamma}$ *of the second kind, as whose discontinuity domain, the quadrangle* $e_1e_2e_4e_3$ *with the four generating reflections* $\overline{V}$, $\overline{V}V_2$, $V_5\overline{V}$, $V_4\overline{V}$ *can serve.*

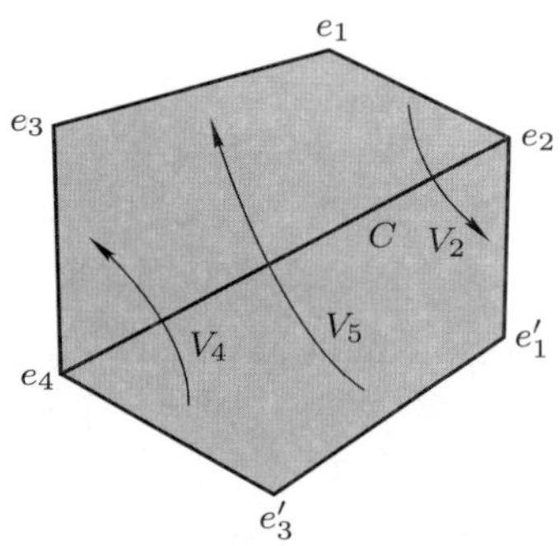

Fig. 55.

The symmetric special case thus obtained will now naturally subsume itself under
[339] the general (unsymmetric) case. Accordingly, by initially excluding it, there remains, for the movement of C along the boundary of Q_1 and Q_1', for the entrance of a further special type, only the other possibility, that the sides coordinated to one another in figure 54 by V_5 draw themselves together into points. *We arrive at the hexangle of figure 56 with three fixed vertices* e_2, e_3, e_4 *and one three-termed cycle of movable vertices* e_1, e_1', e_1''; *we will designate this hexangle as an "unsymmetrical normal hexangle".* The center C now finds itself at the three-termed vertex of the net of the domains Q belonging to $e_1, e_1', e_1'', \ldots$.

The question, how often, after preferral of the vertex e_1, we are able to choose such an unsymmetrical normal hexanlge for the individual group Γ, we answer again by going into the associated net of the domains $Q_1, Q_1', Q_1'', \ldots$. Accordingly, if we imagine the polygon of figure 53 as a normal polygon with center C at the vertex e_1, then the

[38] Should C coincide with e_2, then let Ce_2 be the bisector of the angle $\sphericalangle e_1e_2e_1'$, which is uniquely determined, also for parabolic V_2, as a line, with respect to which e_1e_2 and $e_1'e_2$ are symmetric to one another.

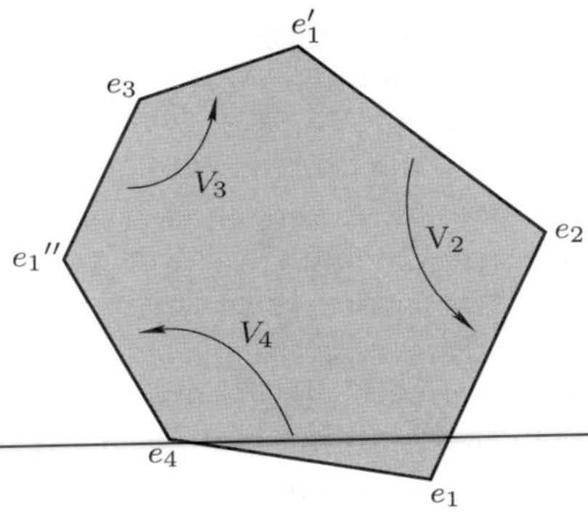

Fig. 56.

sides of the polygon, apart form $e_1\varepsilon$ and $e_1\varepsilon'$, will belong to the net of the Q; the two sides $e_1\varepsilon'$ and $e_1\varepsilon = V_1(e_1\varepsilon')$, however, as level lines of V_1, are added arbitrarily (cf. above on p. 312 ff.). We conclude (as l.c.), that the two angles at ε and ε' give the sum π, so that the third angle at ε'' belonging to the cycle is equal to π by itself. If we then, further, push the vertex ε' to e_2, say, then our normal polygon has assumed the shape of the octangle in figure 57, in which the heavily emphasized sides belong to the net of the Q.

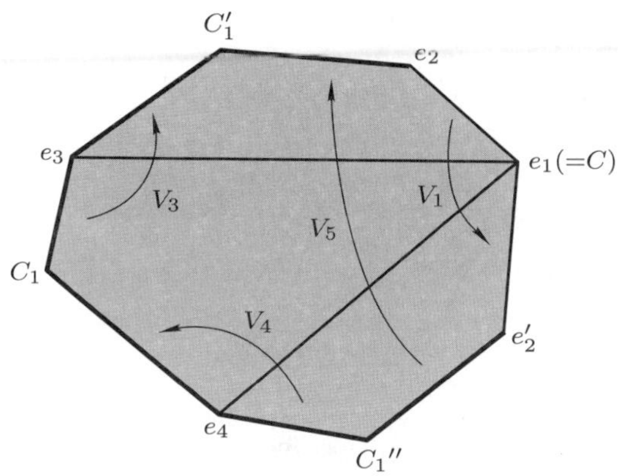

Fig. 57.

In this polygon, are the two sides coordinated to one another by V_3 or V_4 or V_5 able to vanish? It is no restriction of generality, if we only investigate the vanishing of the two sides joined by V_3. Namely, V_4 and V_3 directly represent themselves in figure 57 as coordinated. But if the vanishing pair of sides belongs to V_5, then in going from figure [340]
53 to figure 57, one wants to choose the line e_1e_4 as bounding level line of V_1 in place of $e_1\varepsilon$ whereupon the vanishing pair of sides obtains the position, which the pairs of sides belonging to V_3 has in figure 57. But now, as soon as this pair of sides belonging to V_3 vanishes in figure 57, there arises a hexangle of the type of figure 55 with the center

C at the vertex e_1. All arguments attached to figure 55 remain in force, we obtain a hexangle symmetric with respect to its diagonal $e_1 e_4$ and thus arrive at the symmetric case, counting as excluded:*In the general (unsymmetric* case, the sides of the normal octangle of figure 57 belonging to the center e_1 are never able to the drawn together into points[39].

A glance at figure 57 shows now, that there is *one* cycle of accidental vertices: *For our "unsymmetric" group* Γ, *there is always one and only one normal hexangle with a three-termed cycle of fixed vertices* e_1, e_1', e_1''.

The production of the entire net of these normal hexangles from the net of the $Q_1, Q_1', Q_1'', \dots$ lets itself be effected in this fashion, *by one drawing, in each domain Q, rectilinearly rays from the centrum* e_1 *to the fixed vertices* $\cdots, e_2, e_3, e_4, e_2', \dots$, *(cf. figure 57) and then then imagining all remaining lines taken away.* Thereby, e.g. for the center denoted in figure 57 by C_1, we arrive at the hexangle of figure 58, which is built up out of the three quadrangles:

$$C_1 e_4 e_1 e_3,\ C_1 e_3 e_1' e_2'' = V_3^{-1}(C_1' e_3 e_1 e_2),\ C_1 e_2'' e_1'' e_4 = V_4(C_1'' e_2' e_1 e_4).$$

Notice, in fact, that in figure 57, the line $e_1 e_3$ halves the angle $\sphericalangle C_1 e_3 C_1'$; this line is accordingly the boundary of the two cyclic normal domains of V_3 belonging to C_1 and C_1'. Exactly thus, one recognizes the remaining five sides of the hexangle of figure 58 as the boundary line of cyclic normal domains inside Γ, from which, according to a method of argument already effected often, the hexangle of figure 58 yields itself as the normal domain of the center C_1.

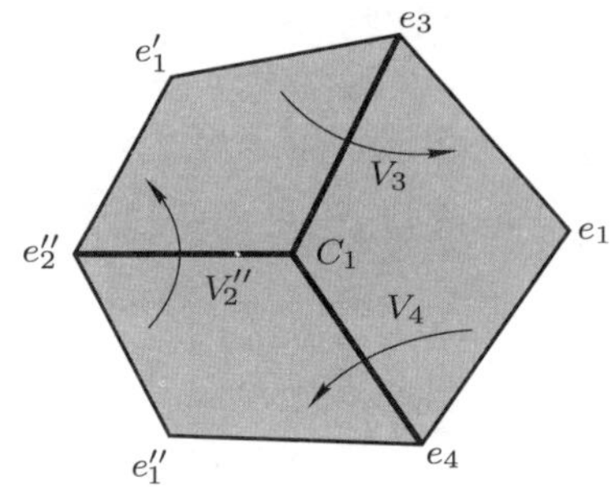

Fig. 58.

[341] As one sees, we have here (comp. the last footnote) arrived at another arrangement of the vertices e_2, e_3, e_4, then in figure 56. Upon continuous transition to figure 57, accordingly, a modified vertex arrangement would have some out. By returning to the original arrangement, our hexangle has the shape sketched in figure 59. We yet determine, at the same time: *the center* C_1 *lies in the interior of the triangle* $e_2 e_3 e_4$, *can only*

[39] For the sake of simplicity, we have assumed the arrangement of the fixed vertices on the boundary of the polygon in figure 57 exactly as in figure 53. Actually, we shall immediately see and later explain, that upon continuously carrying the polygons considered in figure 53 through 56 over into the octangle of figure 57, a type change with arrangement of vertices enters in between.

then move into a vertex of it in case the symmetric case is present, and is furthermore never situated on a side of the triangle. In fact, the three lines C_1e_2, C_1e_3, C_1e_4 are indeed the three inequivalent sides in the domain net of the $Q_1, Q_1', Q_1'', \ldots$, which form concave angles at the point C_1. The limiting case, that one of these angles, e.g., $\sphericalangle e_2C_1e_4$, becomes equal to π, whereby C_1, would move into the side e_2e_4, is excluded. Namely, the line C_1e_1 goes by reflection in C_1e_2 into C_1e_1', by reflection in C_1e_4 over into C_1e_1'', so that in the limiting case $\sphericalangle e_2C_1e_4 = \pi$, the lines C_1e_1' and C_1e_1'' coincide. Accordingly, a degenerating by coincidence of fixed vertices (e_1' and e_3) would enter, which initially remains excluded, but to be sure, is to be considered separately below. The other part of our assertion. That upon the vanishing of one of the sides C_1e_2, C_1e_3, C_1e_4, the symmetric case is present, is already proved above.

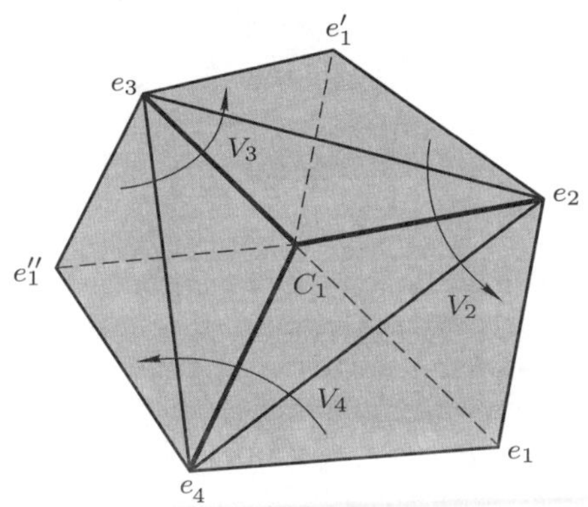

Fig. 59.

By the three diagonals e_2e_3, e_3e_4, e_4e_2, the hexangle is decomposed into four triangles. If we imagine the same division effected in all hexangles of the net, then a triangle net arises, which we will denote by N_3, and which, after preferral of the vertex e_1, is always uniquely determined. There now subsists the important fact: *the triangle net N_3 is generally uniquely determined for our unsymmetric group* Γ. Namely, one can let the net N_3 arise from the domains $Q_1, Q_1', \ldots$, by joining the center e_1 of the individual domain Q rectilinearly with the fixed vertices, and besides this, joining each two fixed vertices following one another by a diagonal (also see figure 57). Now let the center C wander rectilinearly from C_1 to e_2. The normal polygon will lose vertex e_1'', but retain e_1 and e_1', also the vertex e_2, which indeed now lies even closer to the center. We consequently obtain, as a special type, an octangle of figure 54, whereby, even the notations of the vertices and generators agree precisely. In the wandering of the center [342]
C to e_2, the two sides coordinated to one another by V_5 are not able to vanish, since the polygon cannot reach up to a third point equivalent to e_1, e_1'. But also the sides coordinated to one another by V_3 and V_4 are not able to vanish, since the symmetric case is excluded (cf. figure 55). At the conclusion we obtains the octangle of figure 60, whose vertices e_1, e_2, e_1', e_3, e_4 are exactly the same points, which carry these notations in figure 59. but the octangle of figure 60 is a section of the domain Q_2 belonging to the center $C = e_2$. If, in this, from the center e_2, we draw the rays $e_2e_1', e_2e_3, e_2e_4, e_2e_1$,

as well as the diagonals $e_1' e_3, e_3 e_4, e_4 e_1$ of fixed vertices following one another, then we arrive at the associated triangle net. But the lines drawn, in fact, are just those, which our first triangle net N_3 furnished in figure 59.

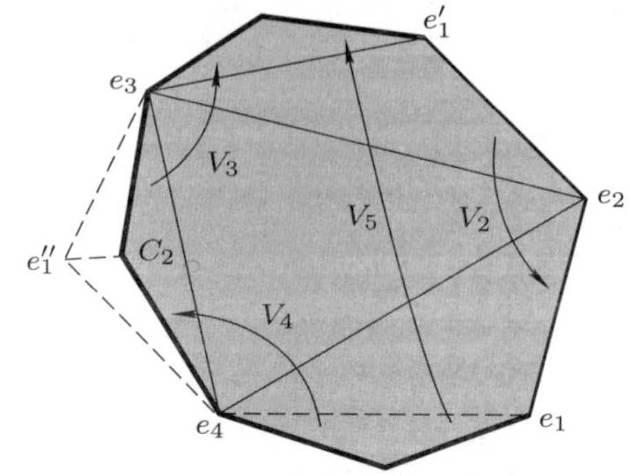

Fig. 60.

On the basis of unique determinateness of the triangle net N_3, we conquer the unsymmetry, which adheres to the consideration so far, by favoring the vertex e_1. In the net N_3 there are *four* inequivalent triangles; and we always obtain four such, if we surround the individual triangle with the four neighboring triangles. This is possible in four inequivalent ways: *We obtain the four different normal hexangles existing for our "unsymmetric" group* Γ, *as they correspond to the preferrals of the vertices* e_1, e_2, e_3, e_4.

Let the four associated centers be called C_1, C_2, C_3, C_4; the individual center C_i belongs to the inner triangle of its hexangle. In figure 61, the individual center is bound by heavily marked lines to the vertices of its triangle; imagine the same construction carried out in all remaining triangles of the net N_3: *One obtains a quadrangle net* N_4 *uniquely associated to our "unsymmetric" group* Γ, *which represents the result of the superimposition of all four nets of the domains* $Q_i, Q_i', Q_i'', \ldots$, *belonging to the vertices* e_i. According to p. 309, we designate these quadrangles as "*natural elementary do-*
[343] *mains*"[40]. *Six inequivalent ones among them (cf. figure 61) comprise a "natural" discontinuity domain for* Γ. Upon bending together the sides mapped onto one another, the latter obtains the shape of a *hexahedron,* whose six surfaces are the six natural elementary domains, while the eight vertices, which, by means of the two inscribed tetrahedra, fall into two quadruples, furnishing the four vertices e_1, e_2, e_3, e_4 and the four centers C_1, C_2, C_3, C_4, corresponding to this partition.

If C moves into the interior of one of the six quadrangles, then the polygon either has the general type in figure 53 on p. 337, or a type change, by means of the vanishing of the two sides coordinated to one another by fifth generator, comes up, without the

[40] Hereby, it is to be remarked, that in the text, we have to leave aside such transitional types, for which the two sides bound by V_5 (cf. figure 53 on p. 337) vanish, and therefore a four-termed cycle of accidental vertices enters, as inessential with respect to the arrangements of the fixed vertices. Upon addition of those transitional types, one more subdivision of the quadrangles of the text by certain curves of third order would enter (cf. I on p. 251 ff.).

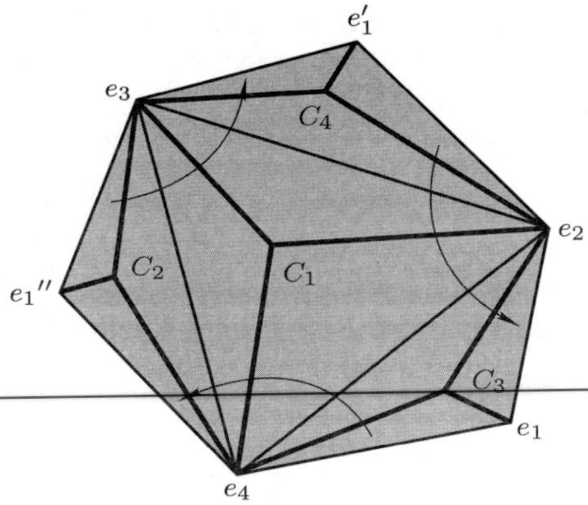

Fig. 61.

four generators V_1, V_2, V_3, V_4 being altered. Now, however, from the normal polygon of figure 53, we produced a *canonical octangle*, by arbitrarily fixing an accidental vertex ε_4, say, on the side $\varepsilon\varepsilon'''$ participating in V_5, next to which are the three further vertices $\varepsilon_3 = V_4(\varepsilon_4), \varepsilon_2 = V_3(\varepsilon_3), \varepsilon_1 = V_2(\varepsilon_2)$. The canonical octangle then has only four generators denoted above by V_1, V_2, V_3, V_4. The canonical octangles to be obtained thus are now to be chosen as *"reduced" polygons: corresponding to the six natural elementary domains (surfaces of the hexahedron above), we obtain, for the proposed unsymmetric group* Γ*, all together, "six" reduced polygons.*

Now remember, that in the theory of the canonical polygons (cf. I on p. 299ff.), none of the different arrangements of the fixed vertices $e_1, e_2, \ldots$ on the polygon boundary was distinguished above the remaining ones. Since, in our case, there are *four* vertices forming a cycle, then exactly 3! = 6 arrangements are combinatorially possible: *Our six surfaces of the hexahedron of the natural elementary domains furnish precisely the six different combinatorially possible vertex arrangements, so that, for each arrangement,* [344]
we obtain one reduced octangle. In fact, figure 61 informs us of the following correspondence of the quadrangles and the vertex arrangements:

quadranlge	vertex arrangement
$e_2C_1e_4C_3$	e_1, e_2, e_3, e_4
$e_3C_2(e_1'', e_1')C_4$	e_4, e_3, e_2, e_1
$e_3C_1e_2C_4$	e_1, e_3, e_4, e_2
$e_4C_2(e_1'', e_1)C_3$	e_2, e_4, e_3, e_1
$e_4C_1e_3C_2$	e_1, e_4, e_2, e_3
$e_2C_3(e_1, e_1')C_4$	e_3, e_2, e_4, e_1

One determines this most simply in this fashion, by fixing, in the individual quadrangle, the one diagonal, which is a side in the triangle net N_3, and which arranges the two adjacent triangles together into one quadrangle. This is then, in each case, the quadrangle of the fixed vertices inscribed in the associated canonical octangle.

We also draw the conclusion: *Opposite faces of the hexahedron of the natural elementary domains always furnish two reduced octangles with opposite vertex arrangements.* Two such canonical octangles we call "*conjugate*" to one another according to I on p. 306. Their production from one another, resp. from an associated "*double quadrangle*" (cf. figure 102 in I on p. 305) is discussed more closely there.

In order to subsume the *symmetric cases*, assume that the center C_1 moves into the vertex e_3 of the hexangle of figure 61. Since $C_1 e_2$ halves the hexangle angle at e_2, then in the case in question:

$$\sphericalangle e_1 e_2 e_3 = \sphericalangle e_1' e_2 e_3.$$

Since now, on the other hand, $\sphericalangle C_3 e_2 e_3$ is equal to the sum of the angles $\sphericalangle C_3 e_2 e_1$ and $\sphericalangle e_1' e_2 e_3$ (insofar as $C_3 e_2$ also halves the angle at e_2 in the associated hexangle), then it follows in our case, that $e_2 C_3$ falls on $e_2 e_1$ and therefore C_3 at e_1[41]. There holds, further (see also figure 59 on p. 341):

$$\sphericalangle e_2 C_1 e_4 = \sphericalangle e_1' C_1 e_2 + \sphericalangle e_1'' C_1 e_4,$$

in each case. Therefore, it follows for $C_1 = e_3$, in particular:

$$\sphericalangle e_2 e_3 e_4 = \sphericalangle e_1' e_3 e_2 + \sphericalangle e_1'' e_3 e_4.$$

[345] Since on the other hand:

$$\sphericalangle C_4 e_3 e_4 = \sphericalangle C_4 e_3 e_1' + \sphericalangle e_1'' e_3 e_4$$

holds, then $e_3 C_4$ necessarily falls on $e_3 e_2$, and one finds, by continuation of the argument, that C_4 moves to e_2, and C_2 to e_4.

The hexahedron of the natural elementary domains degenerates in such a fashion, that only *two* opposite surfaces remain maintained, while on the other hand four draw themselves together into lines (common sides of those two opposite surfaces): *the net N_4 contains only two inequivalent, and indeed symmetric, quadrangles; it directly represents the regular-symmetric quadrangle net of the group $\overline{\Gamma}$ of the second kind, which arises form the present group Γ, capable of extension.*

Corresponding to the three pairs of opposite faces of the hexahedron, we have *three kinds of such symmetric groups.* We put together tabularly at the same time, for these three cases, the vertex coincidences on the hexadron and the vertex arrangements of each of the two symmetric elementary quadrangles:

Coincidences	vertex arrangements
$C_1 = e_3, C_2 = e_4, C_3 = e_1, C_4 = e_2$	$(e_1, e_2, e_3, e_4)(e_4, e_3, e_2, e_1)$
$C_1 = e_4, C_2 = e_3, C_3 = e_2, C_4 = e_1'$	$(e_1, e_3, e_4, e_2)(e_2, e_4, e_3, e_1)$
$C_1 = e_2, C_2 = e_1'', C_3 = e_4, C_4 = e_3$	$(e_1, e_4, e_2, e_3)(e_3, e_2, e_4, e_1)$

Furthermore, hereby, the symmetric cases are also all exhausted. Namely, if one starts from a regular-symmetric net and chooses C on the boundary of two neighboring, and therefore symmetric, quadrangles, then the associated normal polygon has the shape of figure 55 on p. 338, and therefore furnishes the double quadrangle consisting of those two quadrangles. From this one concludes without trouble, that the associated

[41] The objection, whether perhaps C_3 might fall at the other vertex e_2 of the side $e_2 e_1$, one meets by consideration of the vertex e_4, where one finds $e_4 C_3$ situated on the side $e_4 e_1$.

net N_4 is just the proposed regular-symmetric quadrangle net, whereby our assertion is shown.

In the *symmetric special case,* the position of the two reduced polygons among all infinitely many canonical octangles of the group Γ is easily characterized. *The reduced octangles are just those, for which the inscribed quadrangle of the fixed vertices form the two symmetric elementary quadrangles of the regular-symmetric net.*

If this special case is not present, then before we distinguish the reduced octangle of the individual vertex arrangement, for the present group Γ, above the infinitely many further canonical octangles of the same vertex arrangement, we must first, beforehand, yet draw attention to a remarkable unsymmetry of our investigation. As an example, we choose the arrangement e_1, e_2, e_3, e_4 and attach the investigation to figure 61 on p. 342. The reduced octangle has the fixed vertices, denoted in this figure by e_1, e_2, e_3, e_4. If we [346]
draw the inscribed quadrangle of these vertices, repeat the same construction in all octangles of the associated net and imagine all remaining lines obliterated, then according to I on p. 299 ff., a quadrangle net remains, in which any two quadrangles lying next to one another furnish a discontinuity domain of Γ in the shape of an unsymmetric double quadrangle. We choose, say, the quadrangle $e_1e_2e_3e_4$ taken from figure 61, which we will call P for short, and the quadrangle neighboring along e_2e_3, to be called $\overline{P}$ say, which has vertices at the points $e_3e_2e_1'$ and a point e_4' equivalent with e_4.

If we would now arrive from this quadrangle net to the triangle net denoted above by N_3 and uniquely associated to our group Γ, then we must draw the diagonal e_2e_4 in P and therefore not the other diagonal, in $\overline{P}$ the diagonal $e_1'e_3$ and not e_2e_4'. Now reflect the entire net in the line running through e_2 and e_3, whereby, let P go over into a quadrangle $\overline{P}'$, and $\overline{P}$ into a quadrangle P' of the vertex arrangement e_1, e_2, e_3, e_4, while Γ is transformed into a group Γ', which, according to our earlier fundamental theorem (cf. the footnote on p. 329), counts as essentially distinct from Γ. In the new quadrangle net, for the production of the triangle net N_3, we now have to draw the diagonal e_1e_3 and no longer e_2e_4 in the quadrangle P' of the arrangement e_1, e_2, e_3, e_4, and naturally, in $\overline{P}'$, similarly, the other diagonal e_2e_4. *After exclusion of the symmetric case, we consequently have two kinds of reduced octangles (and with this, of groups Γ), according as, in the inscribed quadrangle of the fixed vertices, the one or the other diagonal leads to the net N_3.*

Each of the infinitely many canonical octangle nets of prescribed vertex arrangement, which exist for the group Γ, furnishes an associated net of double quadrangles. Each such net furnishes two triangle nets, according to the choice of the diagonal pairs in the individual double quadrangle. Finally, in each such triangle net, exactly as above in the net N_3, we are able to produce a hexangle in four ways with a three-termed cycle of fixed vertices e_i, e_i', e_i'' and the three additional fixed vertices e_k, e_l, e_m. *If the three bisectors of the hexangle angles at the vertices e_k, e_l, e_m then meet one another at an "interior" point C_i of the middle triangle, then it was the reduced octangle, from which we started*; for one recognizes in the hexangle, according to a method of reasoning repeatedly effected[42], the normal polygon of the center C_i. *In all remaining cases, those three* [347]
angle bisectors can thus not meet at the interior point of the middle triangle.

[42] On the basis of the principle, that a normal polygon of the center C is the common component of all normal domains of the center C of the cyclic subgroups contained in the group.

1.15 The surfaces Φ_3 of the third degree coming up for the character (0,4)

The invariant representation of the polygon continua of the character (0,4) was given in §§3 and 5 (p. 291 ff.). Thereby, one of the six vertex arrangements was given preference at the very beginning. We attach ourselves here to the representation of §5, so that, besides the vertex invariants j_1, j_2, j_3, j_4, we first of all use the four transversal invariants next to one another:

$$j_{12} = j_{34}, \quad j_{23} = j_{41}, \quad j_{13}, \quad j_{24}. \tag{1}$$

For abbreviation, we write:

$$J = j_1 j_2 j_3 j_4 + j_1^2 + j_2^2 + j_3^2 + j_4^2 - 4,$$

$$J_1 = j_1 j_2 + j_3 j_4, J_2 = j_1 j_3 + j_2 j_4, J_3 = j_1 j_4 + j_2 j_3,$$

so that the two relations (3) on p. 297 transcribe themselves into:

$$\begin{cases} j_{12}^2 + j_{23}^2 - j_{13} j_{24} - J_1 j_{12} - J_3 j_{23} + J = 0, \\ j_{12} j_{23} + j_{13} + j_{24} - J_2 = 0. \end{cases} \tag{2}$$

Since the values of the vertex invariants are subjected to the conditions (1) on p. 296, then one deduces, for the values J, J_1, J_2, J_3, the conditions:

$$J \geq -3, \ J_1 \geq 0, \ J_2 \geq 0, \ J_3 \geq 0. \tag{3}$$

the values $J = -3, J_1 = J_2 = J_3 = 0$ are, in fact, attained for the signature (0, 4; 2, 2, 2, 3)[43]. The values of the transversal invariants are all < -2.

We are able to eliminate either j_{13} or j_{24} from the two relations (2). In the first case, of transversal invariants, we still have the three $j_{34} (= j_{12}), j_{42}, j_{23}$, and write, for the introduction of a geometrical language, already prepared on p. 298:

$$j_{34} : j_{42} : j_{23} : -1 = x : y : z : t. \tag{4}$$

In the second case, we have the three transversal invariants $j_{12} (= j_{34}), j_{13}, j_{23}$, and set:

$$j_{12} : j_{13} : j_{23} : -1 = x' : y' : z' : t'. \tag{5}$$

These two possibilities correspond to the two ways, of decomposing the quadrangle,
[348] inscribed in the canonical octangle, with the fixed vertices e_1, e_2, e_3, e_4, into two triangles by either the one or the other diagonal we return to this later[44].

If we interpret the x, y, z, t as homogeneous space coordinates, then on account of $j_{ik} < -2$, we immediately find ourselves again in the *tetrahedron* T bounded by the four planes (6) on p. 320 and already introduced there. The individual polygon (apart from

[43] The signature (0, 4; 2, 2, 2, 2) remains excluded, since it leads to the parabolic rotation groups (groups of the doubly periods functions).

[44] Notice, that to the triangle e_2, e_3, e_4 corresponds the choice j_{34}, j_{42}, j_{23}, similarly, to the triangle e_1, e_2, e_3, the choice j_{12}, j_{13}, j_{23}.

the vertex invariants, which have to be given separately) is then represented either by the point (x, y, z, t) or the point (x', y', z', t'), of which the second proceeds from the first, according to (4), (5) and the second relation (2), by the transformation:

$$x' : y' : z' : t' = xt : (xz - yt - J_2 t^2) : zt : t^2. \tag{6}$$

This transformation, which is to be called S_2, proved itself to be one such of period 2, as one easily shows by calculation; it therefore, conversely, transforms the point (x', y', z', t') into (x, y, z, t).

If we now go from the individual polygon to the associated polygon continuum, then in case the latter has more than two dimensions, i.e., in case at least one vertex invariant $j_\kappa \geq 2$ is therefore variable, we will *initially hold the vertex invariants* j_1, j_2, j_3, j_4 *fast*, whereby, from the multiply dimensional continuum, a *two dimensional* one is cut out. This procedure (exactly as above on p. 324) is to be justified by the fact, that the vertex invariants experience only permutations with respect to the transformations of the modular group, and that consequently, those transformations of the modular group, which leave the vertex arrangement invariant and therefore transform the polygon continuum into itself, simultaneously leave the values of the vertex invariants unaltered.

After these determinations, only the transversal invariants, and therefore the coordinates x, y, z, t resp. x', y', z', t', are still variable. *Thereby, as one finds by elimination of* j_{13} *from the relations (2), the* x, y, z, t *are bound to one another by the relation*:

$$(x^2 + y^2 + z^2)t - xyz + (J_1 x + J_2 y + J_3 z)t^2 + J t^3 = 0, \tag{7}$$

and one shows, by elimination of j_{24} *from the relations (2), that exactly the same relation is also valid for the coordinates* x', y', z', t'.

By the relation (7), a *surface of third degree* is represented; as above (p. 320) we will denote that part of this surface , which comes into consideration us corresponding to the values of the transversal invariants, by Φ_3. Since the points (x, y, z, t) and (x', y', z', t'), as we saw, both lie on the surface Φ_3, we find immediately: *The surface* [349]
Φ_3 *goes over into itself by the transformation* S_2 *given in (6).* In order to determine the points of Φ_3, which correspond to themselves for the transformation S_2, one substitutes x, y, z, t for x', y', z', t' in the left side of (6) and finds: *On the surface* Φ_3, *the points of that curve remain unaltered with respect to S, which are cut out by the surface of the second degree.*

$$xz - 2yt - J_2 t^2 = 0. \tag{8}$$

We will again call this surface a *hyperboloid* and denote it by H_2; let the curve cut out, with whose shape we shall soon become acquainted more closely, be called K_2.

The analogy of these results to those in §11 on p. 319 ff. lies close at hand. This is grounded in the circumstance that the investigation at that time, as concerning the character (1, 1), or (what amounts to the same thing), the signature $(0,4;2,2,2,2l)$ resp $(0,4;2,2,2)$, subsumes itself under the present developments as the simplest special case. We shall accordingly also have to expect here, from now on, results similar to the ones at that time.

One again wants to make clear to oneself the course of the surface of third degree (7) in the tetrahedron T in the following manner, by cutting this with the planes of the pencil:

$$x - 2\lambda t = 0, \quad \lambda = 1, \ldots, +\infty.$$

The individual one of these planes furnishes as intersection, apart form the tetrahedral edge $t = 0, x = 0$, the curve of second degree:

$$y^2 - 2\lambda yz + z^2 + (J_2 y + J_3 z)t + (4\lambda^2 + 2\lambda J_1 + J)t^2 = 0. \tag{9}$$

In order to disclose the course of this curve, we are again able to bring in figure 44 on p. 320, with inessential alteration[45]. We find: the curve (9) cuts the line CD (of figure 44) with the equation $t = 0$ in the two points:

$$z = (\lambda \pm \sqrt{\lambda^2 - 1})y,$$

symmetric with respect to the midpoint $y = z$, which are denoted in the figure by A_1 and A_2. The two tangents to our curve at these points no longer meet one another, to
[350] be sure, at the point F with coordinates $y = 0, z = 0$, but at the point:

$$y : z : t = (J_2 + \lambda J_3) : (\lambda J_2 + J_3) : 2(\lambda^2 - 1).^{46}$$

The intersection of the tetrahedron T with the plane $x - 2\lambda t = 0$ is the triangle CDE of figure 44, included by the lines $t = 0, y - 2t = 0, z - 2t = 0$. Inside this triangle, the curve (9) runs as figure 44 reveals, i.e. above all: *The curve does not reach the sides CE and DE.* Namely, if, corresponding to the side CE say, we put $y = 2t$ into (9), then follows:

$$(z - 2\lambda t)^2 + J_3 zt + (2\lambda J_1 + 2J_2 + J + 4)t^2 = 0,$$

an equation which, in consideration of the positive value λ and the inequalities (3), is fulfillable by no positive quotient $z : t$. Exactly as above (p. 321), for the one boundary value $\lambda = 1$, our curve (9) also draws itself together into the midpoint $y = z$ of the line $t = 0$, while for the other boundary value $\lambda = \infty$, it degenerates into the pair of lines $yz = 0$ in the plane $t = 0$.

We find exactly analogous results, if we cut the surface of third degree with the other two pencils of planes, which have as axes the edges $y = 0, t = 0$ and $z = 0, t = 0$ of the tetrahedron T axen. We again arrive at the same visualizable image of the course of the surface as on p. 321: *The portion of our surface of third degree belonging to the tetrahedron* T *is a surface tautly stretched in the boundary of the facial surface* $t = 0$ *of the tetrahedron, and hanging loosely in the interior of* T*, which as such, represents a simply connected domain.* Compare the indications of figure 45 on p. 323, whereby however, it is to be observed, that the symmetries in the planes $x = y, y = z, z = x$ now drop out. We shall immediately return in detail to the course of the curves K_1, K_2, K_3.

It has now so far not yet been explicitly proved, *that this entire domain (and not, say, a part of this): is the image of the polygon continuum.* In order to now present this proof, we avail ourselves, in the introduction of the surface of the third degree, of the

[45] However, in the figure in question, the coordinates x, y, z are to be permuted cyclically in this manner, that y takes the place of x etc; in the text, we have formed the pencil $x - 2\lambda t = 0$, and not as l.c., the pencil of planes $z - 2\lambda t = 0$, and indeed, for a reason emerging below.

[46] Midpoint of the hyperbola (9), which is present in rectiangular coordinates y, z after the substitution $t = 1$.

proportion (5), since this assumption underlay the equations in §5 on p. 296 ff. The polygon continuum of given vertex invariants was represented there, initially, by the square $0 < t_5 < 1, 0 < t_6 < 1$; the mapping of this square surface onto the surface of third degree was accomplished by formulas (10) and (11) on p. 299. The character of this mapping is now easily described. If we lay the transversal $t_5 = \mu$ through the square, [351] where μ is any value between 0 and 1, then to this corresponds, in case we set $(1-\mu)^{-1} = \lambda$, to the prescription $x - 2\lambda t = 0$ for our present homogeneous coordinates, i.e., the transversal finds its image on the curve (9). We now describe the transversal completely, if we let t_6 wander from 0 to 1, therefore τ (cf. (10) on p. 299) from 0 to ∞. Hereby, however, the point $y:z:t$, to be determined from the recalculated equations (11) on p. 299:

$$\frac{y}{t}(4\lambda^2-4) = 2J_2 + 2\lambda J_3 + J_{12}J_{34}\left\{(\tau^2+\tau^{-2})\lambda + (\tau^2-\tau^{-2})\sqrt{\lambda^2-1}\right\},$$
$$\frac{z}{t}(4\lambda^2-4) = 2\lambda J_2 + 2J_3 + (\tau^2+\tau^{-2})J_{12}J_{34},$$

as one easily determines, describes simply and completely the portion of our curve (9) lying in the tetrahedron T. One has now only has yet to notice, that the covering of the square with transversalis $\mu = 0,\ldots,1$ stands in correspondence to the covering of the portion of the surface of third degree running in T with curves (9) for $\lambda = 1,\ldots,\infty$, so that, in fact, the entire surface Φ_3 running in T is the image of the square.

All discussions on p. 321 ff. concerning the geometrical meaning of in (6) the transformation S_2 of our surface Φ_3 into itself hold without alteration. In particular, $x':z':t' = x:z:t$ holds for S_2. In space, therefore, this transformation again has the character of a "central projection" of the surface Φ_3 into itself from the tetrahedral vertex $x=0, z=0, t=0$. On the surface itself, our transformation has the character of a "reflection", whose symmetry line K_2 is cut out by the "hyperboloid" H_2 given by (8). K_2 is a space curve of fourth degree[47], which is formed by the tangency points of all tangents running from the tetrahedral vertex $x=0, z=0, t=0$ to the surface Φ_3. These are the tangency points B_2 of the tangents in figure 44 on p. 320 laid from D to the plane intersections of the Φ_3; on the strength of this, the course of the curve K_2 on the surface Φ_3 again yields itself exactly in the manner described above (p. 322 ff.).

Also the conclusion of the development fashions itself exactly as above: the transformation S_2 is one of three coordinated operations. We arrive at the following result: *the surface Φ_3 admits the three "reflections" into itself to be denoted by S_1, S_2, S_3*: [352]

$$\begin{cases} (S_1) \quad x':y':z':t' = (yz - xt - J_1t^2):yt:zt:t^2, \\ (S_2) \quad x':y':z':t' = xt:(zx - yt - J_2t^2):zt:t^2, \\ (S_3) \quad x':y':z':t' = xt:yt:(xy - zt - J_3t^2):t^2, \end{cases} \tag{10}$$

which, in space, possess the character of "central projections" of the Φ_3 into itself from the three tetrahedral vertices situated in the plane $t=0$. The three associated symmetry lines K_1, K_2, K_3 are cut out on the Φ_3 by the "hyperboloid" H_1, H_2, H_3:

[47] Namely, it will be purely represented as the intersection of the hyperboloid H_2 with the surface of second degree:

$$x^2 - y^2 + z^2 + J_1xt + J_3zt + Jt^2 = 0.$$

$$\begin{cases} yz - 2xt - J_1 t^2 = 0, \\ zx - 2yt - J_2 t^2 = 0, \\ xy - 2zt - J_3 t^2 = 0. \end{cases} \tag{11}$$

these three curves bound a simply connected three-sided domain on the Φ_3, which reaches, with three cusps, to the boundary of the surface, and indeed to the midpoint of the three edges of the tetrahedron T *situated in the plane* $t = 0$.

1.16 The discontinuity domain of the modular group and the group continua of the character (0, 4)

In order to obtain the group continua of the character (0, 4), we first retain the assumption of constant vertex invariants made in the previous paragraph. By holding fast to the vertex arrangement, we will then fix a discontinuity domain of modular group on the associated surface Φ_3. As one such discontinuity domain we are able to choose the totality of those points (x, y, z, t), which furnish reduced polygons; for, according to p. 344, the reduced polygon is *uniquely* determined for each group Γ for prescribed vertex arrangement. However, it is to be noticed, that this approach only holds, then, *if we first of all assume the four vertex invariants consistently different from one another.* However, as soon as the vertex invariants are no longer all different from one another, the number of different vertex arrangements reduces itself, and therefore, it is to be expected, that more than one reduced polygon comes up for the individual arrangement. These special cases we shall have to consider separately below.

Now, according to p. 346, if we initially ignore the symmetric cases, there are two kinds of reduced polygons for the vertex arrangement e_1, e_2, e_3, e_4. For one kind (cf. figure 61 on p. 342), the diagonal $e_2 e_4$ is to be drawn in the quadrangle of the fixed vertices e_1, e_2, e_3, e_4 if the triangles net N_3, uniquely belonging to the group Γ according
[353] to p. 341, is to be obtained. For the other kind, the diagonal $e_1 e_3$ plays this distinguished role. If we first of all consider the former kind of reduced polygons, then for the purpose of introducing the surface Φ_3, we shall avail ourselves of the proportion (4) on p. 347, since the sides of the triangle $e_2 e_3 e_4$ surrounding the center C_1 (cf. figure 61 on p. 343), for the sake of symmetry, recommend the utilization of the invariants j_{34}, j_{42}, j_{23}.

As a reduction condition, we determined on p. 346 for the hexangle of figure 61, that the bisectors of the hexangle at e_2, e_3, e_4 had to meet in point C_1 of the triangle $e_2 e_3 e_4$. For a first polygon, for which C_1 lies neither at a side nor at a vertex of this triangle $e_2 e_3 e_4$, and therefore in the interior of this, we construct the associated point (x, y, z, t) of the surface Φ_3. Since upon continuous alteration of this point on the Φ_3, the polygon is likewise only able to alter itself continuously, then C_1 will not immediately be able to move into the boundary of the triangle $e_2 e_3 e_4$, i.e. the immediate neighborhood of that first point (x, y, z, t) on the Φ_3 also furnishes reduced polygons. Conversely, the entire boundary of that domain, which corresponds to reduced polygons of the kind in question, will consequently only be able to be furnished by such points, whose polygons C_1 lie on the boundary of the triangle $e_2 e_3 e_4$.

Now we already determined on pp. 341 and 345: *the three cases, that* C_1 *moves into a vertex of the triangle* $e_2 e_3 e_4$, *furnish the three linear continua of symmetric models,* *which are already collected tabularly l.c.; however, if C_1 moves onto a side of the triangle (but not at a vertex), then degeneration enters by coincidence of fixed polygon vertices, and therefore we find ourselves at the boundary of the polygon continuum.* By pushing aside this investigation, we consider first of all the symmetric cases.

We have described above (p. 346) the transition from our group Γ to a Γ', in whose reduced polygons the diagonal e_1e_3 was distinguished, and indeed, we effected this transition by the reflection, to be called $\overline{V}$ say, in the line e_2e_3 of figure 61. If we arrange the discontinuity domain for Γ' about the quadrangle of the fixed vertices, denoted l.c. by P', then one has, as is easy to determine, as generators of Γ':

$$V_1' = \overline{V}V_2^{-1}V_1^{-1}V_2\overline{V}, V_2' = V_2, V_3' = V_3, V_4' = \overline{V}V_3V_4^{-1}V_3^{-1}\overline{V}.$$

From this follows, in consideration of $\overline{V}V_2 = V_2^{-1}\overline{V}$ and $V_1V_2V_3V_4 = 1$:

$$V_1'V_2' = \overline{V}V_2^{-1}V_1^{-1}V_2\overline{V}V_2 = \overline{V}(V_1V_2)^{-1}\overline{V},$$

$$V_2'V_3' = V_2V_3,$$

$$V_4'V_2' = \overline{V}V_3V_4^{-1}V_3^{-1}\overline{V}V_2 = \overline{V}V_3V_4^{-1}V_3^{-1}V_2^{-1}\overline{V} = \overline{V}(V_3V_1)\overline{V},$$

there thus holds $j_{12}' = j_{12}$ and therefore $j_{34}' = j_{34}$, further $j_{23}' = j_{23}$ and $j_{24}' = j_{13}$. With consideration of the two relations (2) on p. 347, we therefore find invariants of the group Γ' from those of the group Γ by means of: [354]

$$j_{34}' = j_{34}, j_{24}' = -j_{24} - j_{12}j_{23} + J_2, j_{23}' = j_{23}. \tag{1}$$

Now if, in the table on p. 345, we have the first named symmetric case (C_1 at the vertex e_3), then Γ is transformed into itself by $\overline{V}$, and in particular, the quadrangles denoted on p. 346 by P and P' coincide, so that, besides the first and third relations (1), $j_{24}' = j_{24}$ also holds, But in the unsymmetric case, as we know from he theory of the reduced polygons, the group Γ and Γ' are essentially distinct, and therefore the equation $j_{24}' = j_{24}$ does not hold then. But if one substitutes $j_{24}' = j_{24}$ in the second relation (1), then $j_{12}j_{23} + 2j_{24} - J_2 = 0$ follows, or, in homogeneous terminology:

$$xz - 2yt - J_2t^2 = 0,$$

ie. we find ourselves on the curve denoted above by K_2: *The linear continuum of symmetric models named at the first place in the table on p. 345 is completely represented precisely by the points of the curve K_2 lying on the Φ_3.*

If C_1 moves into the vertex e_4, then we have the second symmetric case of the table on p. 345, for which the quadrangle $e_1'e_3e_4e_2$ of figure 61 on p. 343 becomes the elementary quadrangle. In order to be able to directly evaluate the foregoing calculations, we use, for the moment, the notations e_1', e_2', e_3', e_4' for e_1', e_3, e_4, e_2. The associated invariants, to be correspondingly denoted by $j_1', j_2', \ldots$, are:

$$j_1' = j_1, j_2' = j_3, j_3' = j_4, j_4' = j_2, j_{34}' = j_{24}, j_{24}' = j_{23}, j_{23}' = j_{34},$$

from which $J_2' = J_3$ also calculates itself. For the symmetric case, now, according to the calculation carried out above, the relation:

$$j'_{34}j'_{23}+2j'_{24}-J'_2=0$$

is characteristic. This calculates itself around into $j_{24}j_{34}+2j_{23}-J_3=0$, or, written homogeneously:

$$xy-2zt-J_3t^2=0,$$

ie. we arrive at the curve K_3.

By disposing of the third symmetric case exactly thus, we find: *the three linear continua of symmetric models, for the use of the reduced polygon, are precisely exhaustively represented by the three curves K_1, K_2, K_3, which bound the simply connected domain*
[355] *obtained on the surface Φ_3 of the close of the previous paragraph* (p. 352). To the three cusps, with which this domain reaches to the boundary of the polygon continuum, correspond the three degenerations, for which C_1 moves onto a side of the off-named triangle $e_2e_3e_4$. We come back to this below.

Now on the Φ_3, the entire boundary of the domain of reduced polygons with distinguished diagonal e_2e_4 is furnished by these curves K_1, K_2, K_3. But these curves bound only our off-named triangular domain and hereby also come into play in their entire (situated on the surface Φ_3) extent: *reduced polygons of the vertex arrangement e_1, e_2, e_3, e_4, for which the diagonal e_2e_4 leads to the net N_3.*

For the reduced polygon of the other kind (with distinguished diagonal e_1e_3), we avail ourselves the proportion (5) on p. 347 for introduction of the surface Φ_3, which, for the moment, tying into the notation x', y', z', t' of the coordinates, we call Φ'_3. The previous considerations then carry themselves over immediately, and again give, as the domain of the reduced polygons, the triangle bounded by the curves K_1, K_2, K_3. But the relation between the surface Φ_3 and Φ'_3 is in (6) p. 348 given the transformation S_2 is presented. So we stay in the surface Φ_3, i.e. if we hold fast to the proportion (4) for the introduction of this, then the domain of the reduced polygons of the kind now under discussion is *the triangle neighboring the triangle of the sides K_1, K_2, K_3 along K_2, proceeding from it by S_2, which, besides K_2, is bounded by the sides $K'_1=S_2(K_1)$ and $K'_3=S_2(K_3)$.* Two of the vertices of this new triangle are the edge midpoints of the tetrahedron T furnished by the endpoints of the curves K_2; the third vertex, at which the two curves K'_1 and K'_3 run together, is the tetrahedral vertex $x=0, z=0, t=0$.[48]

With the surfaces of the two triangles named, all the reduced polygons are exhausted: *On the surface Φ_3, a discontinuity domain of the modular group is formed from the double triangle bounded by the curves K_1, K_3, K'_1, K'_3, which is schematically represented adjacently in figure 62. The boundary curves of the double triangle, as the*
[356] *figure indicates, are coordinated to one another by the generators $T_1=S_2S_3$, $T_3=S_1S_2$ of the modular group.* It is even somewhat more symmetrical, to utilize as generators of the modular group, the three transformations, bound to one another by the relation $T_1T_2T_3=1$:

$$T_1=S_2S_3,\ T_2=S_3S_1,\ T_3=S_1S_2.$$

Explicitly, from (10) on p. 352, one finds for T_1:

$$x':y':z':t'=xt^2:(x^2y-xzt-yt^2-J_3xt^2-J_2t^3):(xyt-zt^2-J_3t^3):t^3, \qquad (2)$$

48 That we obtained just this triangle neighboring with the first and neither of the other two, lies in the special manner, in which we produced the second reduced polygon (reflection $\overline{V}$ in the line e_2e_3) from the first (the one with the distinguished diagonal e_2e_4).

while T_2 and T_3 arise from this by cyclic permutation of the coordinates x, y, z and the indices $1, 2, 3$.

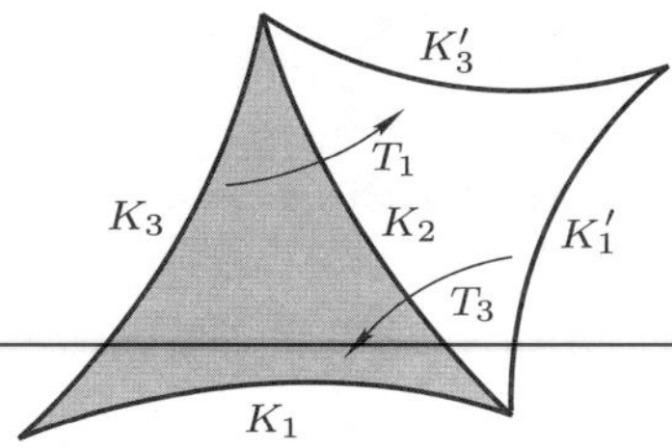

Fig. 62.

Hereby, in fact, we have arrived back at the original meaning of the transformations of the modular group, which we indeed derived in vol. I from the alterations of the canonical cut systems. The transformation denoted here by T_1, namely, is none other than the square of the transformation denoted by T_1 in I on p. 395. The latter, which we better call $T_1^{1/2}$ from now on, is given by:

$$\begin{cases} x' : y' : z' : t' = xt : (xy - zt - J_3 t^2) : yt : t^2, \\ J_1' = J_1,\ J_2' = J_3,\ J_3' = J_2,\ J' = J, \end{cases} \tag{3}$$

as whose square, in fact, one immediately obtains the substitution (2) again. Moreover, the transformation (3) carries the surface Φ_3 over into the surface given by:

$$(x^2 + y^2 + z^2)t - xyz + (J_1 x + J_3 y + J_2 z)t^2 + Jt^3 = 0 \tag{4}$$

to be denoted say by Φ_3', which one can generate by reflection of the Φ_3 in the plane $y = z$, and which corresponds to the vertex arrangement e_2, e_1, e_3, e_4. Since we favored the vertex arrangement e_1, e_2, e_3, e_4 here, then the transformation $T_1^{1/2}$ could not directly come up in the present development.

The last considerations lead us back directly to the special cases already mentioned above (p. 352), that two or even more vertex invariants become equal to one another. If, two, but not three, vertex invariants are equal, then among the three quantities J_1, J_2, J_3, two become equal to one another, while the third is different from them, if at least three vertex invariants are equal to one another, then $J_1 = J_2 = J_3$ holds. Again, understanding by S_1', S_2' the reflections in the symmetry planes $y = z$ and $z = x$ of the tetrahedron T, introduced in (3) on p. 329, and define S_3' correspondingly for the plane $x = y$. If now $J_2 = J_3$ say, then not only is the surface Φ_3 transformed into itself by S_1', *but the extended modular group to be generated from* S_1, S_2, S_3 *is capable of extension* [357]
anew by the addition of S_1', since S_1 is transformed into itself by S_1', but S_2 and S_3, on account of $J_2 = J_3$, are carried over into one another by S_1'. The discontinuity domain of this extended group is a half of the triangle of the sides K_1, K_2, K_3 cut off by the "altitude" $y = z$. If we even have $J_1 = J_2 = J_3$, *then extension by* S_1', S_2' *and* S_3' *is possible.* The triangle just named is then to be decomposed into six partial triangles by the three

"altitudes" $y = z$, $z = x$, $x = y$, of which one makes up the discontinuity domain of the group last obtained.[49]

Now, first of all, let $j_1 = j_2$, $j_3 \gtrless j_1$, $j_4 \gtrless j_1$, hold, and therefore $J_2 = J_3$, $J_1 \gtrless J_2$. Of the six vertex arrangements, (e_1, e_2, e_3, e_4) and (e_2, e_1, e_3, e_4) then fuse,[50] while the remaining four are still different from the those two. The reduction theory accordingly now furnishes *two* reduced polygons, which, in figure 61 on p. 342, we have to position around the quadrangles of the fixed vertices e_1, e_2, e_3, e_4 and e_2, e_1', e_3, e_4. If the generators of the first polygon are, as before, V_1, V_2, V_3, V_4, then we have for the second:

$$V_1' = V_2,\ V_2' = V_2^{-1} V_1 V_2,\ V_3' = V_3,\ V_4' = V_4,$$

which furnishes the following transformation of the invariants:

$$j_1' = j_2,\ j_2' = j_1,\ j_3' = j_3, j_4' = j_4, j_{12}' = j_{12},\ j_{23}' = j_{24},\ j_{13}' = j_{23},\ j_{14}' = j_{24},$$

which, in consideration of the relations subsisting between the invariants, immediately yields itself as the transformation $T_1^{1/2}$ given in (3). On the other hand, one shows, in the present special case $J_2 = J_3$, without further ado, the substituting of the relation:

$$T_1^{1/2} = S_1' S_3 = S_2 S_1'.$$

The discontinuity domain of the modular group, according to this, in the special case $j_1 = j_2$ here, is the double triangle represented in figure 63 with the generators:

$$T_1^{1/2} = S_1' S_3 = S_2 S_1',\ T_3 T_1^{1/2} = S_1 S_1'.$$

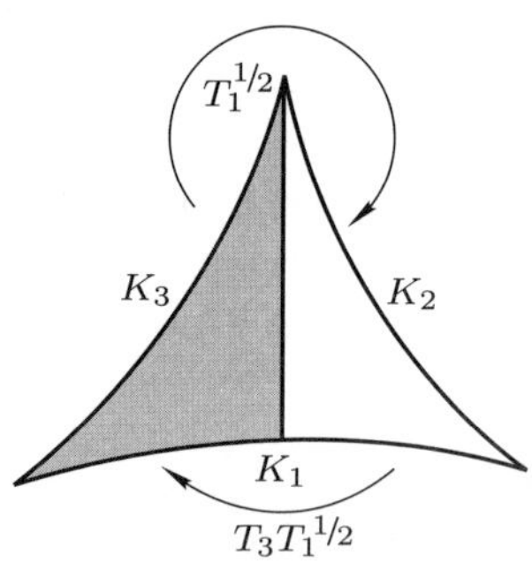

Fig. 63.

[358] This result remains subsisting without alteration, in case, besides $j_1 = j_2$, the equa-

[49] It is a question here of the same transformation, which in the theory of the elliptic modular functions led back from the principal congruence group of the second level(cf. "M" I on p. 270 ff) to the total group. Furthermore(which, however, we do not further evaluate in the text), the modular group of the text to be generated from S_1, S_2, S_3 is isomorphic with that principal congruence group(in the form extended by reflection), while the group extended by addition of S_1', S_2', S_3' is isomorphic with the total modular group(of be second kind) of the elliptic functions.

[50] I.e., the two associated polygon continua become identical.

tion $j_3 = j_4$ also holds, if only $j_1 \gtrless j_3$ holds. Indeed, along with the two vertex arrangements considered before, the additional ones (e_2, e_1, e_4, e_3) and (e_1, e_2, e_4, e_3) now also fuse. However, the transition from the originally chosen reduced polygon to that of the arrangement e_2, e_1, e_4, e_3, as one easily determines with the help of figure 61 on p. 343, is represented by:

$$V_1' = V_2,\ V_2' = V_2^{-1} V_1 V_2,\ V_3' = V_3 V_4 V_3^{-1},\ V_4' = V_3,$$

from which follows, for the invariants, the transformation:

$$j_1' = j_2,\ j_2' = j_1,\ j_3' = j_4,\ j_4' = j_3,\ j_{34}' = j_{34},\ j_{42}' = j_{42},\ j_{23}' = j_{23},$$

thus, in the space of the x, y, z, t, we arrive at the identical transformation. But the transformation to the reduced polygon of the arrangement (e_1, e_2, e_4, e_3) again amounts to the transformation $T_1^{1/2}$.

If, finally, $j_1 = j_2 = j_3$ holds, then irrespectively of whether j_4 is also equal to those invariants or not, we have only *one* vertex arrangement, and therefore all six reduced polygons belong to the same arrangement. The modular group also contains, besides $T_1^{1/2}$, the coordinated transformations:

$$T_2^{1/2} = S_2' S_1 = S_3 S_2',\ T_3^{1/2} = S_3' S_2 = S_1 S_3'.$$

The discontinuity domain of the modular group can be chosen in the shape of the double triangle of figure 64, corresponding to the two generators:

$$(T_1^{1/2})\quad x' : y' : z' : t' = xt : (xy - zt - J_1 t^2) : yt : t^2$$

$$(S_1' S_2')\quad x' : y' : z' : t' = z : x : y : t.$$

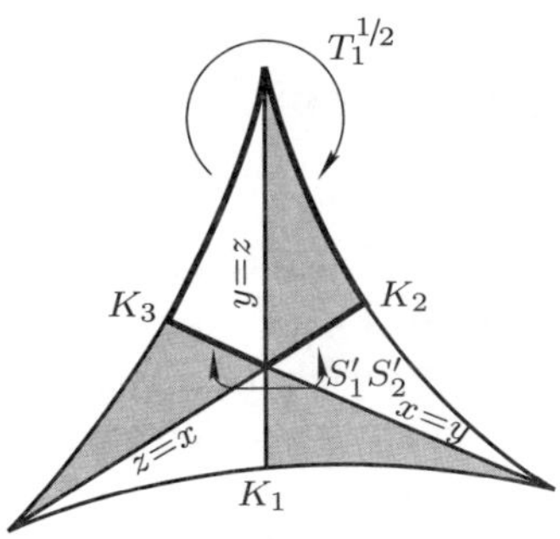

Fig. 64.

In the lowest case $J_1 = 0$, we shall be led back to the two generators (4) on p. 329 of the modular group of the character $(1, 1)$.

If now, all vertex invariants $j_\kappa \le 2$, then[51] as we know, these are not variable. *The discontinuity domain of the modular group bounded on the associated surface Φ_3 then*

[51] We are able here to include the parabolic case as well.

directly and single-valuedly represents the continuum of the groups of the signature $(0,4;l_1,l_2,l_3,l_4)$*, two dimensional in this case.*

If, on the contrary, one or more invariants $j_\kappa \geq 2$, then these are freely variable in
[359] the interval $j_\kappa \geq 2$, and we arrive at three or more dimensional group continua. We then proceed most simply, by again bringing in the invariants:

$$t_\kappa = \frac{j_\kappa - 2}{j_\kappa},$$

already introduced for the polygon continua, and in addition, in order not to obtain the group Γ multiple, always set one vertex fixed, for which the condition:

$$j_{\kappa-1} \leq j_\kappa \tag{5}$$

is fulfilled. If we again interpret the t as rectangular coordinates, then as soon as at least two hyperbolic vertices are present, corresponding to the condition(5) and therefore to the inequality $t_{\kappa-1} \leq t_\kappa$, we do not have to allow, as above (p. 294), the entire cube. $0 \leq t_\kappa \leq 1$, but only the half represented by $t_3 \leq t_4$, the sixth furnished by $t_2 \leq t_3 \leq t_4$ or the twenty-fourth of the cube determined by $t_1 \leq t_2 \leq t_3 \leq t_4$.

In each case, however, we arrive at the single-valued and complete representation of the group continuum, if we now consider each point of the allowed portion of the cube as a representative of the discontinuity domain of the modular group bounded on the associated surface Φ_3.

The boundary of the discontinuity domain of the modular group, besides boundary surfaces of the cube, then also consists of one resp. two or three inner planes $t_3 = t_4, \ldots$ of the the cube. In the generation of the modular group, these planes play the role of symmetry planes. In fact, those transformations, for which the vertex invariants are permuted, now also come either partially, or in their totality, into application.

1.17 Boundary and connectivity of the individual group continuum of the character (0, 4)

To the "boundary planes" $t_3 = t_4, \ldots$ of the discontinuity domain of the modular group, just obtained and situated in the interior of the cube, encloses the group continuum itself. We are only able to reach an actual boundary point of the individual group continuum, in fact, where we simultaneously find ourselves at the boundary of the polygon continuum.

In order, correspondingly, to determine the genuine limiting cases in the group continuum, we first of all again choose the four vertex invariants fixed, and indeed, the hyperbolic invariants j_κ, in any case, not at their upper limit ∞. The low dimensional group continuum thereby fixed we represented by the double triangle of figure
[360] 62 resp. the double triangles of figures 63 and 64 (p. 355 ff.). These domains, however, reach with three resp. two or one vertex to the boundary of the polygon continuum. It is consequently to be determined, what becomes of our group at such a vertex.

In order to answer, we first consider a symmetric polygon, which is indeed always reduced. In figure 65, corresponding to the curve K_2 on the surface Φ_3, an elementary quadrangle of the vertices e_1, e_2, e_3, e_4 is drawn.[52] The two endpoints of the curve K_2 must then furnish two invariant-theoretically different limiting cases; and indeed, in the first case let[53]:

$$\lim j_{12} = \lim j_{34} = -2,$$

and in the second:

$$\lim j_{23} = \lim j_{14} = -2.$$

Each of these two passages to the limit is realizable geometrically in three ways; and indeed, in two cases, we obtain, as the group Γ, *one such of the character* (0,3) *with at least one parabolic generator; in the third case, however, we arrive at a degeneration, i.e., in the limiting case, we no longer have a group at all.*

Namely, j_{12} is the invariant of that substitution, which arises by combination of the reflections in the circles e_1e_4 and e_2e_3 of figure 65. Accordingly, should $j_{12} = -2$, then this substitution must be either parabolic or = 1. The former alternative requires, that either the circle e_3e_4 or the circle e_1e_2 (but not both simultaneously) draws itself together into a point of the real ζ-axis, whereby the two symmetry circles of the reflections just mentioned come into at either the point $e_3 = e_4$ or the point $e_1 = e_2$. In the first case, the passage to the limit leads to a group of the character (0,3) and the generators $V_1, V_2, (V_1V_2)^{-1}$, the last one of which is parabolic; in the second case, correspondingly, we get a group of the character (0,3) and the generators V_3, V_4 and the parabolic $(V_3V_4)^{-1}$. There still remains, as the third case, that of the degeneration, for which the two circles e_1e_2 and e_3e_4 simultaneously become infinitely small and the two circles e_1e_4 and e_2e_3 come into coincidence; the polygon ceases to be the discontinuity [361]
domain of a group.[54]

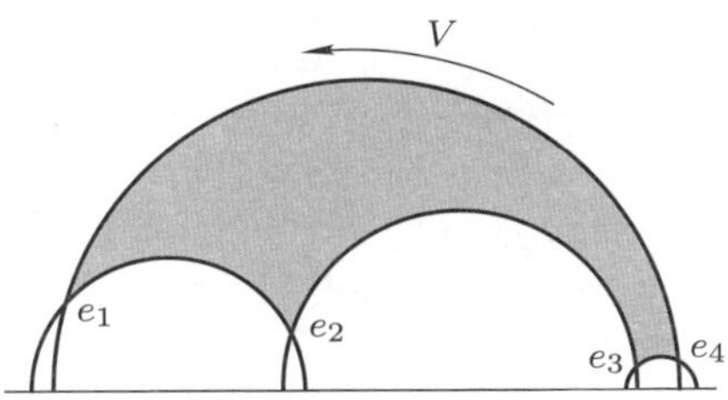

Fig. 65.

We return once more to the first two limiting cases with the groups of the character (0,3), and assume that we already find ourselves, on the curve K_2 of the symmetric

[52] In order that the vertices emerge as such in the ζ-sphere, we have chosen them elliptically; the statements of the text are independent of this assumption.

[53] The surface Φ_3 as emerges from the developments on p. 349 ff., touches the plane $x = 2t$ of the tetrahedron T at the edge midpoint $y = z$.

[54] I.e., a group with principal circle. On the other hand, if one transforms ζ so that the negative ζ-half plane initially becomes a finite circle, and if one then draws this circle together into a point in the passage to the limit $\lim j_{12} = -2$, then as the limiting case there emerges the discontinuity domain of a cyclic parabolic group.

models of our kind, very close to the endpoint $j_{12} = -2$, without, however, already having reached it. Two quadrangles $e_1e_2e_3e_4$ only slightly different from the triangles finally coming out will then always be equivalent too. Thus, in figure 65, e.g., the side e_3e_4 is already assumed very small. We are then able here, however, by means of a hyperbolic substitution V, which has the circle e_4e_1 as orbit curve and moves in the direction of the arrow in figure 65, i.e., from e_4 to e_1, as long as we only choose the invariant of V sufficiently large, to transform in such a manner, that the side e_1e_2 becomes very small, while e_3e_4 correspondingly increases. Only in the limiting case itself is this transformation no longer effectable for the invariant; of the hyperbolic substitution just mentioned, for $\lim j_{12} = -2$, would itself have to be infinite.

In this consideration, the third limiting case (the degeneration) also lets itself easily be included. For a quadrangle close to degeneration, the vertices e_1, e_2 have come very close to one another, as have the vertices e_3, e_4. If one now effects a motion by means of the substitution just denoted by V, then e_3 and e_4 remove themselves from one another, while e_1 and e_2 approach one another. We therefore come close to the second limiting case of the character $(0,3)$. Therefore, as long as we approach the three limiting cases, the equivalence of the polygons remains preserved; however, this is lost in the limit itself.

Also in the general(unsymmetric) case, we come to the same result.[55] We avail our-
[362] selves here with advantage of the normal polygon of figure 54 on p. 337, which corresponds to the assumption of a center C on the tetrahedron edge C_1e_2 of figure 61 on p. 343. Here, the sides coordinated to one another by V_5 are level curves of this hyperbolic substitution. If we extend these sides over ε and ε'(into the infinite) up to the fix point Φ of V_5 lying outside the ellipse, and leave out the sides coordinated to one another by V_3 and V_4, then we have the polygon P' of the group Γ' of the character $(0,3)$, to be generated from $V_1, V_2, V_5 = (V_1V_2)^{-1}$, and indeed, this polygon P' is bounded exclusively by symmetry lines of this group, always capable of extension by reflections(C lies on the line joining the vertices e_2 and Φ, which similarly is a symmetry line of Γ'.) If, on the other hand, we extend the sides coordinated by V_5 over e_1 and e_1' to Φ and leave out the two polygon sides belonging to V_2, there arises the polygon P'' of the group Γ'' of the character $(0,3)$, to be generated from $V_3, V_4, V_5 = V_3V_4$ which composed with Γ' gives our group Γ again.

The polar Π of the fix point Φ (linear orbit curve of V_5), in each of the polygons P' and P'', separates the hyperbolic vertex Φ from the remaining fixed vertices. According to the first theorem in I on p. 308, we are also able(if need be, by inessential alteration

[55] We already determined on p. 341, what happens, in case the point C, of figure 59 on p. 340 moves onto the side e_2e_4 (but not into one of the vertices e_2, e_4). There yielded itself $\sphericalangle e_1'C_1e_3 = \sphericalangle e_1''C_1e_3 = 0$, so that the edge e_1e_3 of the tetrahedron, arising by bending together the sides of the four triangles of figure 59, draws itself together into a point. The two triangles neighboring this edge vanish; the triangle net denoted above (p. 341) by N_3 no longer exhibits four but only two inequivalent triangles and furnishes, directly, the regular symmetric triangle net of the group of the character $(0,3)$ coming out in the limiting case. The six edges of the tetrahedron just named furnish six limiting cases, of which each two belong together and are not distinguishable by our invariants. The three pairs of limiting cases correspond to the three cusps of the discontinuity domain of the modular group. The considerations of the text will verify these preliminary statements.

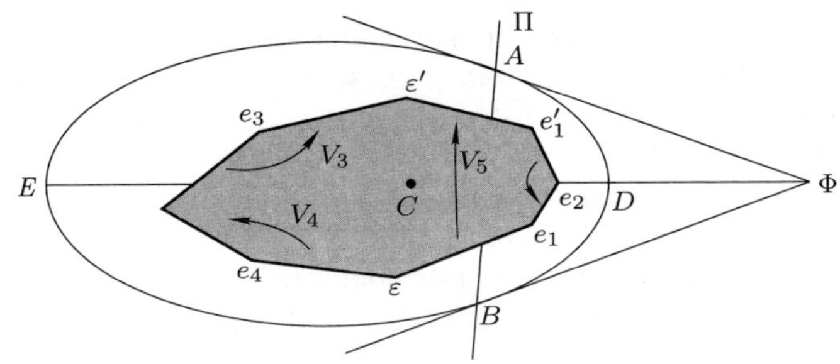

Fig. 66.

of P'') to so assume the accidental vertices of P'', that they are separated by the polar Π from the vertex Φ. The polygon P of Γ then has, in the projective plane, the shape schematically indicated in figure 66; here, A and B are the intersection points of the polar Π with the ellipse and therefore the fix points of V_5 situated on the latter; by D and E are denoted the intersection points of the ellipse and that level line of V_5, which runs through the polygon center C and the vertex e_2. [363]

The passage to the limit $\lim j_{12} = -2$ now requires, that V_5 become either a parabolic or the identical substitution. The first case can initially enter, by the triangle $AB\Phi$ drawing itself together into a point. The fix points $e_1 e_2 e'_1$ come into coincidence, and indeed, it is a question hereby, since the sides $\varepsilon e_1, e_1 e_2, e_2 e'_1, e'_1 \varepsilon', e_2\Phi$ are symmetry lines of Γ', of exactly the same geometrically perspicuous transition to the parabolic cusp (by vanishing of the sides $e_1 e_2$ and $e'_1 e_2$ as well as the "tangency" of $\varepsilon e_1, Ce_2, \varepsilon' e'_1$) as we become acquainted with in the symmetric case, in connection with figure 65 on p. 360. For the polygon P'', the limiting position is far simpler, insofar as an essential alteration of this does not enter, rather only the hyperbolic cusp D of this goes into a parabolic one. *As one sees, we obtain, as limiting case, a group* Γ'' *of the character* $(0,3)$ *and the generators* V_3, V_4, V_5, *of which, in any case,* V_5 *is parabolic.*

Secondly, the prescribed passage to the limit to a parabolic substitution V_5 is yet effectable in the manner, that that triangle $AB\Phi$, which attaches itself to the left on AB and reaches through the infinite to the vertex Φ, becomes drawn together into a point.[56] *Here, we then obtain, correspondingly, as a limiting case of our group* Γ, *a group* Γ' *of the character* $(0,3)$ *and the generators* V_1, V_2, V_5, *whose last one, in any case, is parabolic.* Upon the use of a normal polygon, whose center C is situated on one of the hexahedron edges e_3C_4 or e_4C_3 (figure 61 on p. 343), this transition will again fashion itself in a geometrically perspicuous manner, as in the previous case.

If, finally, the substitution V_5, upon the passage to the limit, change into the identity, then the polygon degenerates into an ellipse chord. Thereby, as we determined this above in the symmetric case of figure 65 on p. 360, and just on this account, see directly for the vertices $e_1 e_2$ in figure 66 and indirectly for the remaining vertices (namely, by the other shape just mentioned of the normal polygon with the center C on e_3C_4 or e_4C_3), all vertices move onto the ellipse; and indeed, the vertices e_1, e_2, e'_1 furnish one

56 Transform the polygon of figure 66, first of all, by a hyperbolic substitution V, which has DE as orbit line and moves D to E, so that the triangle $AB\Phi$ intended in the text no longer goes through the infinite, and therefore Φ lies on the extension of DE over E.

endpoint of the ellipse chord finally subsisting, the remaining vertices the other end-
[364] point of this. *The polygon has therefore degenerated in this case, i.e., no longer represents the discontinuity domain of a principal-circle group.* [57]

Also, the *equivalence class of three polygons, which are very close to the first two limiting cases and the degeneration,* is again clear, as in the symmetric case. Namely, if the triangle $AB\Phi$, drawn finite in figure 66 , is already very small, then by transformation by means of the substitution denoted in the footnote before last by V, for sufficiently large invariant of this, one can always go over to a polygon, for which the triangle $AB\Phi$, going through the infinite in figure 66, becomes very small. Similarly, in case we have a polygon very near the degeneration, one can carry this over, by a substitution V with sufficiently large invariant, into a shape close to a boundary polygon of the character $(0,3)$. Only for the limiting cases themselves do the equivalences cease, since the invariants of the substitutions V coming into application become infinitely large.

To the individual vertex of the discontinuity domain of the modular group we find coordinated in this manner two groups Γ' and Γ'' of the character $(0,3)$ and one degeneration (cyclic parabolic group). We will count the system of these limiting cases as one single group of the character $(0,4)$ representing the continuum of these groups. Under this assumption, the theorem holds: *All groups of the character* $(0,4)$ *with given vertex invariants form a two dimensional "closed" "continuum" of the connectivity of the surface of the sphere.*

By now returning to the question developed at the introduction of the paragraph (p. 359), we have to determine, that in consequence of the result just obtained, we are never able, also for the multi-dimensional group continua of the character $(0,4)$, to arrive at an actual boundary point of the continuum, as long as the variable vertex invariants t_κ find themselves in the "interior" of their intervals $0 < t_\kappa < 1$. Accordingly, in order to actually arrive at the boundary of the individual continuum, we now have to investigate those groups Γ, which correspond to the extreme value, $t_\kappa = 0$ and $t_\kappa = 1$ of the vertex invariants.

The individual boundary surfaces $t_\kappa = 0$ offer no difficulty. Here, the previous hyperbolic vertex e_κ becomes parabolic, and we arrive at a boundary manifold with a
[365] number of dimensions decreased by one unit. *At this boundary manifold, the group continuum just considered is unclosed.*[58]

The investigation at an individual boundary surface $t_\kappa = 1$ fashions itself more circumstantially; for here, j_κ becomes ∞, and therefore the generator V_κ degenerates.

We ask, first of all, in the case of group of the character $(0,3)$, how a vertex invariant, e.g., j_3, can become infinite. Such a group is always capable of extension by reflection, and furnishes in the ζ-half plane, as discontinuity domain of the extended group, the domain shaded in figure 67, whereby, as an example, $j_1 > 2$ and $j_2 < 2$ was assumed. If the passage to the limit, $\lim j_3 = \infty$, is now to be effected, then either the semicircle $e_2 a$ or the semicircle $e_1 b$ must draw itself together into a point of the real ζ-axis, or both must happen at the same time. Therefore, we also obtain, here again, three limiting cases: *First, the proposed group* Γ *of the character* $(0,3)$ *reduces itself to the cyclic group to be generated from* V_1*, secondly, to the cyclic group to be generated from* V_2*; in the third*

[57] See, however, the note on p. 361 .

[58] See the corresponding result above on p. 334 for the continuum of the signature $(1,1)$.

case, however, the group reduces itself to the identity 1, *since the discontinuity domain extends itself over the entire* ζ*-half plane.*

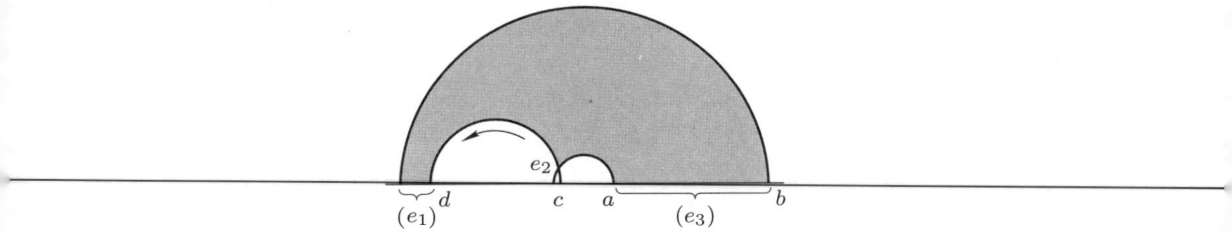

Fig. 67.

Also, concerning the approach to the limiting cases, we again find the same relations, as in the passages to the limit discussed above. If the semicircle $e_2 a$ is already very small, then whether the semicircle $e_1 b$ also already be very small or not, one can always, by means of a hyperbolic substitution with the fix points c and d (cf. figure 67), which moves c toward d, for invariant; of V chosen sufficiently large, transform, so that circle $e_2 a$ increases arbitrarily, while the circle $e_1 b$ correspondingly draws itself together. Only the limiting cases themselves no longer permit such transitions, since $\lim j$ would become ∞.

We now turn back to the individual group continuum of the character (0,4), for which $j_4 > 2$ is always to hold, remember, first of all, the relation:

$$j_{12}^2 + j_{13}^2 + j_{23}^2 + j_{12} j_{13} j_{23} - J_1 j_{12} - J_2 j_{13} - J_3 j_{23} + J = 0,$$

as well as the inequalities:

$$J_k \geq 0, \; j_\kappa \geq 0, j_{\iota\kappa} \leq -2.$$

since, for $\lim j_4 = \infty$, $\lim J = \infty$ also holds and moreover: [366]

$$-J_1 j_{12} - J_2 j_{13} - J_3 j_{23} \geq 0$$

certainly takes place, then for the passage to the limit, $\lim j_4 = \infty$, at the same time, we must have:

$$\lim(j_{12}^2 + j_{13}^2 + j_{23}^2 + j_{12} j_{13} j_{23}) = -\infty.$$

Two of the three transversal invariants taking part here must consequently become infinite of the same order, so that only one of them can remain finite, and with this, variable; however it is not excluded, that the third invariant also becomes infinite.

In order to make the nature of these limiting cases understandable, we choose the center C of the normal polygon on the hexahedron edge $C_4 e_3$ of figure 61, p. 343, and receive the displayed polygon in figure 68, by means of which, as above(p. 362), we effect the decomposition of Γ into two groups Γ' and Γ'' of the character (0,3). Let Γ' be the group of the generators V_1, V_2, V_5 and Γ'', that of the generators V_3, V_4, V_5; for the latter group Γ'', the symmetry lines are drawn heavily.

By means of this figure, we are able to pursue the passage to the limit, $\lim j_4 = \infty$, for the case that the invariant j_{12} of V_5 receives an arbitrary value; for those considerations

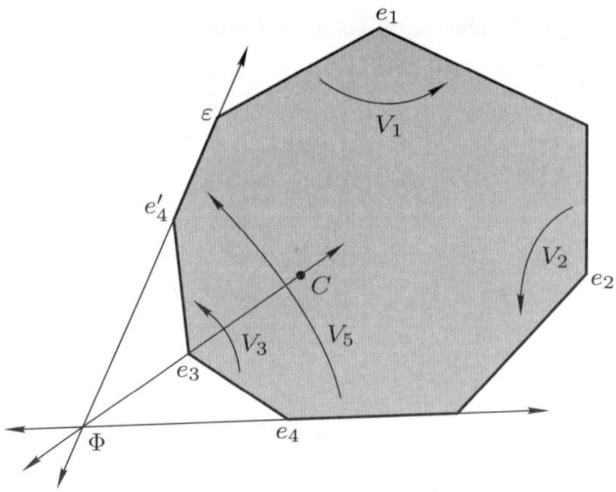

Fig. 68.

directly find application to the heavily drawn double triangle of figure 68, which we attached to figure 67. Let A and B be the intersection points of the line $\Phi e_3 C$ with the ellipse, and indeed, A the one situated between Φ and C. Further, let V be a hyperbolic substitution, which has the ellipse chord AB as a linear orbit curve and moves from A to B.

The first possibility of carrying out the passage to the limit, $\lim j_4 = \infty$, is then this, that the vertex e_3 moves into the point A while the lines $e_3 e_4$ and $e_3 e_4'$ come to lie on the tangent to the ellipse at the point A. As limiting model of the proposed continuum, we obtain a group of the character $(0,3)$ and the generators V_1, V_2, V_5, of which, in any case, the last is hyperbolic, and indeed, of the invariant j_{12}. The second kind of passage to the limit we are able to effect with exactly the same invariants, by only taking care, in the exercise of a hyperbolic substitution V with sufficiently large invariant j, that the
[367] point e_3 remain fixed right there. The consequence is as we know here from figure 67, that now the point Φ becomes pressed up to the ellipse point B, and that the lines Φe_4 and $\Phi e_4'$ come to lie on the ellipse tangent at the point B. Of the vertices of our polygon in figure 68, e_3 has remained in its place, e_4 and e_4' line outside the ellipse, while all remaining vertices coincide at B. As a limiting case there appears the cyclic group of the generator V_3. The third possibility still remaining appears as a combination of the first two: e_3 moves toward A, but Φ toward B; the group Γ reduces itself, in this case, to the identical substitution. *Should the bound $t_4 = 1$ of our continuum be reached for transversal invariant j_{12} remaining arbitrary, then we arrive at the system of the following three groups. First, the group of the character $(0,3)$ and the invariants j_1, j_2, j_{12} secondly, the cyclic group of the invariant j_3, thirdly, the group consisting of the identity alone.*

In the other two cases, that either j_{13} or j_{23} remains arbitrary, one attaches completely analogous arguments to two normal polygons, whose centers belong to the hexahedron edges $C_4 e_2$ and $C_4 e_1'$ in figure 61 (p. 343). *The results are exactly corresponding to the ones just stated.*

Finally, we are able to effect the passage to the limit $t_4 = 1$ in yet the following manner, that all three invariants j_{12}, j_{13}, j_{23} simultaneously become infinite. *The system of* *the attainable limiting cases then composes itself of the three cyclic groups of the invariants j_1, j_2, j_3 and of the group containing only the identity.*

If we hold the first three vertex invariants fixed, while t_4 runs through its interval $0 \le t_4 \le 1$ or(what amounts to the same thing), j_4 runs through its interval $2 \le j_4 \le \infty$, then we are able to lend the preceding consideration in the tetrahedron T introduced on p. 320 a visual interpretation in the same way, as we did this on p. 334 ff. in the case of the signature $(1, 1)$. To the different values of the "parameter" j_4 corresponds a family of surfaces (7) on p. 348, of which, as one easily shows, no two run through one and the same point in the interior of T. If, on each of these Φ_3, one arrives at ta hexahedron, of whose six facial surfaces, two remain uncovered, namely those belonging to $j_4 = 2$ and $j_4 = \infty$, while the remaining ones are mapped onto one another by the generators of the modular group.

The preceding limit study then concerns the boundary plane $t = 0$ of the hexahedron, belonging to $j_4 = \infty$. Upon approaching an interior point of this boundary plane, [368]
we have the passage to the limit just considered, with the three invariants j_{12}, j_{13}, j_{23} simultaneously becoming infinite. The approaches to the three edges of the tetrahedral surface $t = 0$ correspond to the three kinds of passage to the limit $j_4 = \infty$ with one invariant, j_{12} or j_{13} or j_{23}, remaining finite, the relations fashion themselves in detail entirely analogously to those of the signature $(1, 1)$, which were discussed meticulously on p. 335 ff.[59]

Finally, we remember our determination $t_1 \le t_2 \le t_3 \le t_4$. Should, for example, $t_3 = 1$, then we also find ourselves in the limiting case $t_4 = 1$. For the boundary groups obtained for this latter boundary case, the reader will further easily effect the passage to the limit indicated by $\lim t_3 = 1$ on the basis of those statements, which we developed on p. 365 concerning the vertex invariants becoming infinite for groups of the character $(0, 3)$.

1.18 The normal and the reduced polygons of the character $(0, n)$

In the generalization of the preceding developments, we next turn ourselves to the general case of the genus $p = 0$ with n fixed vertices and here again, also take the concept of a *normal polygon of the character* $(0, n)$ *as a point of departure* for the reduction theory of the polygons.[60]

A normal polygon of the character $(0, n)$ of the general type has $(4n - 6)$ sides and therefore, besides n fixed vertices, $(n-2)$ more three-termed cycles of movable vertices.

---

[59] The theory of the signature $(1, 1)$, in fact, is only a special case of the development of the text. In addition, the present investigation is laid out more generally, insofar here, upon transition to a boundary point $t = 0$ of the hexahedron, we have always set up the "system of all boundary groups hereby attainable", while there only *one* boundary group was always favored, and indeed, with the intention of coordinating a continuum of uniquely determinate boundary groups to the boundary of the hexahedron.

[60] Poincaré's reduction condition (cf, above on p. 307) is not geometrically accessible to the same degree. In the symmetric case, Poincaré utilizes the symmetric double n-angle as reduced, which as normal polygons, are also reduced in the sense of the text. Here, Poincaré was, after all, successful in easily actually carrying through the continuity study.

In the case $p = 0$, there are *three* different kinds of cycles of movable vertices, according
[369] as one, two or three hyperbolic generators take part in the cycle, which do not belong to fixed vertices of the polygon[61] These are presented in the figures 69 through 71, where e_1, e_2 are fixed, $\varepsilon, \varepsilon', \varepsilon''$ movable vertices.

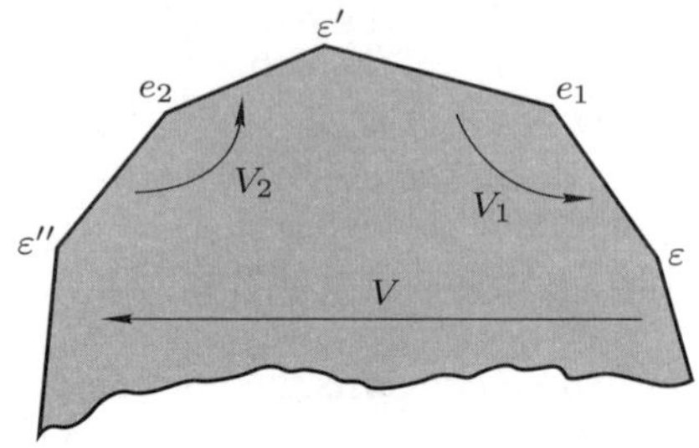

Fig. 69.

If one imagines, in an interruption of one of these polygon parcels, the chain of the polygon sides entering here hung on, then these sides are only able to belong to one another. Namely, e.g., in the case of a parcel of the second kind(cf. figure 70), if a side finding itself on the other side of the right interruption were mapped onto a side on the other side of the left interruption, then the genus p could ont be $= 0$, for the substitution belonging to those two sides and the one denoted in figure 70 by V would furnish a pair of conjugate re-entrant cuts on the closed surface belonging to the polygon. The total polygon therefore arises by smooth arrangement next to one another(composition) of polygon parcels of these three kinds, whereby every two neighbouring parcels have a hyperbolic generator in common. If, hereby, λ_1 parcels of the first kind, λ_2 of the second and λ_3 of the third come into application, then the polygon has $(2\lambda_1 + \lambda_2)$ fixed vertices and $1 + 2(\lambda_1 + \lambda_2 + \lambda_3)$ generators, from which one concludes:

$$2\lambda_1 + \lambda_2 = n, \ 1 + 2(\lambda_1 + \lambda_2 + \lambda_3) = 2n - 3.$$

Accordingly, for the whole numbers λ, the conditions hold:

$$2 \leq \lambda_1 \leq \frac{n}{2}, \ \lambda_2 = n - 2\lambda_1, \ \lambda_3 = \lambda_1 - 2, \tag{1}$$

so that, in particular, λ_2 and λ_3 are determined by λ_1.

On the basis of this enumeration, one can easily put together the various general types of normal polygons for the lower cases $n = 5, 6, 7, \ldots$, whereby, in addition, one
[370] will find that to the individual number triple $\lambda_1, \lambda_2, \lambda_3$, there belong several types in general. In order to disclose here the incorporation of the symmetric polygons[62] we have to take the number triple, always entering, $\lambda_1 = 2, \lambda_2 = n-4, \lambda_3 = 0$ and to position

[61] Except for $n = 3$, all three generators of the one cycle coming up here belong to fixed vertices of the hexangle.

[62] As in the note before last, the extreme case is always meant here, that the extended group is generatable from n reflections.

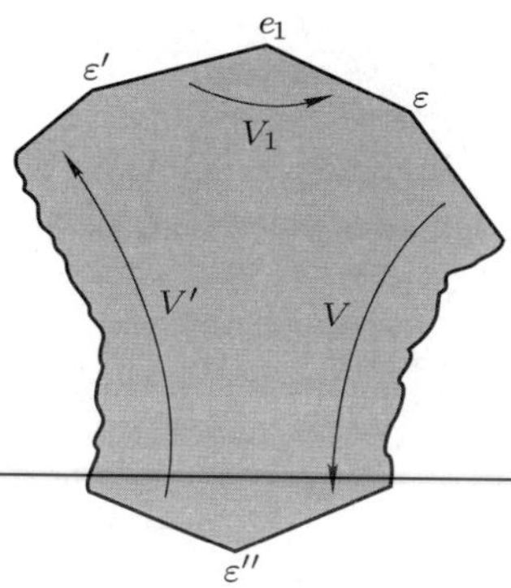

Fig. 70.

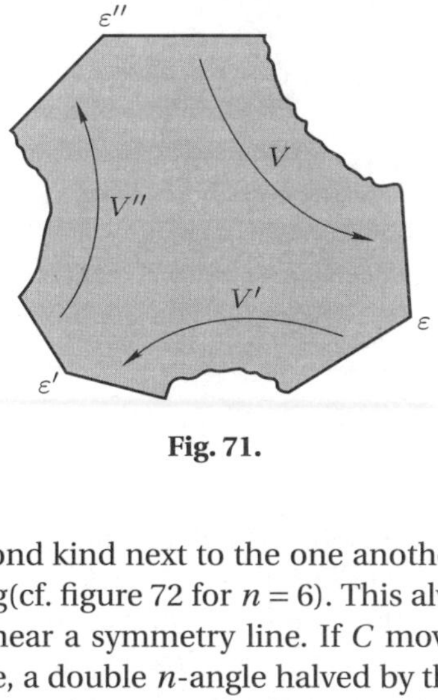

Fig. 71.

the $n-4$ parcels of the second kind next to the one another, so that the fixed vertices always become neighboring(cf. figure 72 for $n=6$). This always makes its appearance, if the polygon center C is near a symmetry line. If C moves onto the symmetry line itself, then, as a special type, a double n-angle halved by this line comes up. If, e.g., in figure 72, center C moves onto the symmetry line $e_1 e_2$, then the sides coordinated to one another by V_3, by V_4, by V_5 and finally by V_6 simultaneously vanish, and it follows the symmetric double hexangle of figure 73.

In contradistinction to this extreme case, for each group Γ, transitional types come up by a single pair of sides vanishing. If it is thereby a question of two non-neighboring sides, then two cycles of accidental vertices fuse into one four-termed cycle of such vertices. If, however, two neighbouring sides situated at the vertex e_i have vanished, then we have a two-termed cycle of fixed vertices e_i and e_i', while one of the $(n-2)$ cycles of movable vertices has vanished. These latter transitional types become especially important further on.

For our proposed group Γ, to each normal polygon belongs a *unique canonical polygon*, which has the same fixed vertices as the normal polygon and, in particular, in the same arrangement, which is present on the boundary of the normal polygon. Accordingly, if we alter the center C, then the canonical polygon remains essentially unaltered, as long as no type change of the second kind just named(with a two-termed

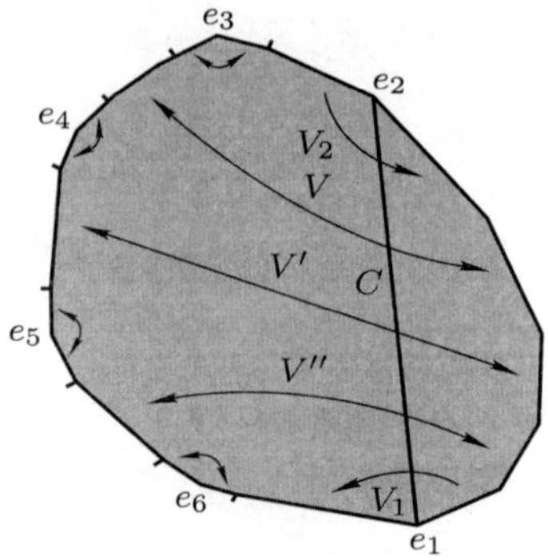

Fig. 72.

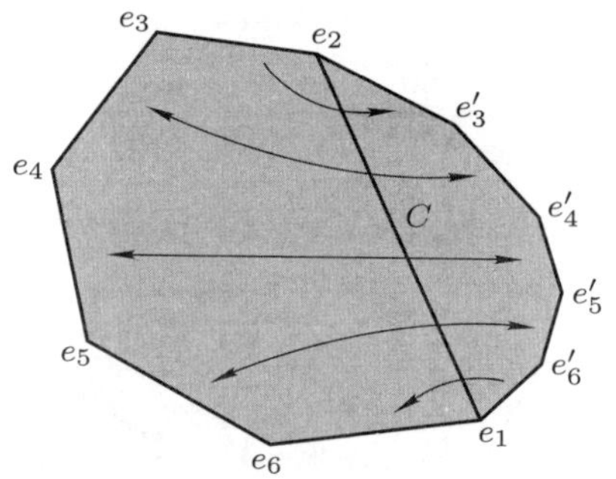

Fig. 73.

[371] cycle of fixed vertices e_i, e'_i enters for the normal polygon. For this latter type change, with alteration of the arrangement of the fixed vertices on the polygon boundary, the nets of the domains Q, whose theory is developed in I on p. 258 ff., are fundamental, as in the case $(0,4)$, so too in the general case of the character $(0,n)$).

If the center C moves into the vertex e_1, then a domain Q takes the place of the normal polygon, which appears put together from a chain of l_1 (in the elliptic case) resp. from infinitely many (in the parabolic and hyperbolic cases) normal polygons surrounding the point e_1. The boundary of Q is transformed into itself by the generator V_1 belonging to e_1. If ε is any accidental vertex of this boundary, then a portion of the boundary is delimited by the two points ε and $\varepsilon' = V_1(\varepsilon)$, which consists exclusively of points inequivalent with respect to V_1. Along this portion there find themselves the $(n-1)$ vertices $e_2, e_3, \ldots, e_n$ in a determinate arrangement stipulated by the present group Γ, intermixed with $3(n-3)$ accidental vertices. Therefore, on the boundary of Q, there are, in all, $(4n-10)$ sides inequivalent with respect to V_1, which belong together in pairs with respect to Γ. If, in particular, we call the domain Q surrounding e_1, Q_1 and correspondingly arrange, next to the points $e'_1, e''_2, \ldots$, equivalent to e_1, the equivalent domains $Q'_1, Q''_1, \ldots$, *then in the net* $Q_1, Q'_1, Q''_1, \ldots$, *there find themselves, in all,* $(2n-5)$ *sides inequivalent with respect to the total group.* However, these remarks only hold "in general", namely, if the general type is present for Q_1. If, however, Q_1 shows a special type, then the number of inequivalent sides in the net $Q_1, Q'_1, Q''_1, \ldots$, is *smaller than*

$(2n-5)$. These special types will serve below for bounding the discontinuity domain of the modular group.

Now, form, similarly, for each of the remaining vertices e_i, the net $Q_i, Q_i', Q_i'', \ldots$ and put the n nets over one another. Thus arises *the net of the natural elementary domains peculiar to our group* Γ.[63] If for no domain Q_i is there a special type, then the net of the natural elementary domains exhibits $n(2n-5)$ inequivalent lines, which run together in sets of $(n-1)$ at the individual vertex e_i, in threes at each of the $n(n-3)$ accidental vertices and besides this, may cross over in a manner not generally surveyable. For $n = 4$, such crossings did not yet come up; here we come back to the perception of the hexahedron often cited on p. 343 ff. But in each case, as one deduces from the finiteness of the number $n(2n-5)$ of inequivalent rectilinear segments, *the number* μ *of inequivalent natural elementary domains is recognized as a finite one.* [372]

The preceding numerical statements modify themselves, if there are special types for one or more domains $Q_1, Q_2, \ldots, Q_n$; in particular, in this case, the number of inequivalent rectilinear segments becomes $< n(2n-5)$. As extreme case, the *symmetric one* is again to be mentioned, *where the net of the natural elementary domains reduces itself simply to the regular-symmetric net of the double n-angles, and therefore* $\mu = 2$.

If we choose the center C in an individual natural elementary domain, then the associated normal polygon has n fixed vertices, which retain their position and their arrangement on the polygon boundary, as long as C moves itself in that elementary domain. Consequently, to each of the μ natural elementary domains belongs a uniquely determined canonical polygon. Therefore, if we only first determine, in general, that *the canonical polygons, belonging to the normal polygons(without special type) of our proposed group* Γ, are to be used as *reduced polygons, then in each case we have a finite number* μ *of reduced polygons.*

But this number is always > 1. Even in the most favorable special case of a symmetric group, we found $\mu = 2$ and, correspondingly, two reduced polygons with vertex arrangements opposite to one another. But in general, μ is even far greater; for we already found $\mu = 6$ for $n = 4$.

Accordingly, if, for bounding a discontinuity domain of the modular group, we have to seek out one of the μ reduced polygons, then as also in the case $n = 4$, an unsymmetric mode of the procedure is no longer to be avoided. For this purpose, we will first of all assume, for the simplification of the developments following next, *that the n vertex invariants of the proposed group* Γ *are to be different from one another.* The addenda in the cases of equal vertex invariants will cause no particular difficulty later. The smallest vertex invariant we call j_1, the second smallest j_2; the associated vertices themselves are called e_1 and e_2. Form the domain Q_1, belonging to the vertex e_1. Concerning the sequence of the vertices on the boundary of this domain Q_1, we have two cases to distinguish. In general, along a boundary portion of Q_1, which we delimit as above by two points ε and $\varepsilon' = V_1(\varepsilon)$, there find themselves *only* $(n-1)$ *fixed vertices.* We then cut out of Q_1, by means of two lines running out from e_1, of which the one is transformed into the other by V_1, a normal polygon in such a way, *that* e_1 *becomes the fixed vertex following* e_1 *in the usual sense(see e.g., figure 102 in I on p. 305.) the canonical polygon,* [373]

[63] The concept "natural elementary domain" is thought of here in the same sense as on p. 343 for $n = 4$; comp. the footnote there.

given directly with the normal polygon, which is uniquely determined for our group Γ, *we choose as the reduced polygon* P. In particular cases, it can occur, that on the boundary of Q_1, between ε and $\varepsilon' = V_1(\varepsilon)$, more than $(n-1)$ fixed vertices are situated. The measure just followed would then even yield several canonical polygons, as one will easily survey. We shall immediately meet up again with the relations present here and need not be pursued here further.

1.19 The continua of the reduced polygons of the character $(0, n)$ for given vertex invariants and fixed vertex arrangement

The consideration up until now concerned the choice of a reduced polygon for the individual group Γ. Now going back to the group continua, we nevertheless hold fast, until further notice, to the assumption, *that the vertex invariants* $j_1, j_2, \ldots, j_n$ *be constant.* Thus, only the $(2n-6)$ independent composition and transversal invariants will be variable. We assume, that a first group Γ is proposed, for which the special type mentioned at the close of the previous paragraph does not occur. By further holding fast, for the present, to the *assumption of consistently different vertex invariants,* the reduced polygon P of Γ is uniquely determined. Let the vertices following e_2 on the boundary of P be called $e_3, e_4, \ldots, e_n$ in their present sequence. P belongs to a polygon continuum, which, corresponding to the solely variable composition and transversal invariants, is $(2n-6)$ dimensional and was represented above in the shape of a cube of this number of dimensions; we will call this cube W_{2n-6} for short.

Next, we now have to consider the $(n-3)$ composition invariants more closely. The construction of an arbitrary canonical polygon of the character $(0, n)$, effected on p.295 ff., consisted in the repetition of $(n-3)$ similar compositions. In the individual one of these $(n-3)$ steps, we had to compose a polygon to be called P_ν say of the character $(0, \nu)$, and the generators:

$$V_1, V_2, \ldots, V_{\nu-1}, V = (V_\nu V_{\nu+1} \cdots V_n)$$

with a polygon P_3 of the character $(0, 3)$ and the generators:

$$V_\nu,\ V' = (V_{\nu+1} V_{\nu+2} \cdots V_n), V^{-1},$$

which has the hyperbolic generators V in common with P_ν, into a polygon $P_{\nu+1}$. In
[374] place of P_3, however, we could also hereby take any other polygon $P_3' = V^\sigma(P_3)$, where σ represents an entirely arbitrary exponent between $-\infty$ and $+\infty$. For the individual polygon of the linear continuum entering thus, the associated exponent σ is the composition invariant in a first form. Notice, that the definition of σ presupposes the giving of a first polygon $P_{\nu+1}$, which then receives the value $\sigma = 0$, which, however, besides this, can be any one of the linear continuum of polygons of the character $(0, \nu+1)$ coming up here.

Two σ, differing by a whole number now indeed furnish different polygons $P_{\nu+1}$, but one immediately recognizes the groups Γ associated to them as identical. Namely, if for fixed P_3, we move the polygon P_3 through a discontinuity domain of the common hyperbolic generator V, then the group belonging to the moved polygon $V(P_3)$ has obviously again become identical with the group of P_3. If, as at the introduction to the paragraph, P is a normal polygon of the center C, then the vertices $e_1, e_2, \ldots, e_\nu$, and, as emerges from the shape of the normal hexangle P_3 of the center C, also the vertex belonging to V', lie in the interior of the cyclic normal domain of the center C of the hyperbolic substitution V; both polygons $P_{\nu-1}$ and P_3 are consequently able to be arranged inside this cyclic domain. If we therefore introduce σ in such a fashion, that $\sigma = 0$ belongs to any polygon $P_{\nu+1}$, for which those two polygons to be composed lie in the cyclic normal domain in question, then the composition invariant σ under discussion of our normal polygon P can never be $> +1$ and never < -1. We arrive at the result, *that the composition invariants, also in their form utilized on p. 296 ff. as quantities t for our reduced polygons W_{2n-6}, are never able to come near the limits $t = 0$ and $t = 1$, present for the polygon continuum.*

The $(n-3)$ transversal invariants $j_{3,4,\ldots,n}, j_{4,\ldots,n}, \ldots, j_{n-1,n}$ behave themselves otherwise. If we have a canonical polygon with $2n$ non-vanishing sides, for which, therefore, no two of the fixed vertices have come infinitely close to one another, *then $j_{\nu,\nu+1,\ldots,n}$, as invariant of the hyperbolic substitution $V_\nu V_{\nu+1} \cdots V_n$, has a finite value, which lies between -2 and $-\infty$, excluding the limits.* In contrast to this, we shall soon see, *that already for the symmetric polygons with coincidence of fixed vertices, the transversal invariants are able to reach their limiting values -2 resp. $-\infty$.*

We now go back to the normal polygon of the fixed vertices $e_1, e_2, \ldots, e_n$, introduced at the beginning of the paragraph, which we had cut out of the domain Q_1 of the center e_1 by two level lines of V_1, and remember, that the $2n-5$ inequivalent sides of Q_1 are [375]
initially consistently non-vanishing. Accordingly, to the reduced polygon P in question will correspond a point in the "interior" of the cube W_{2n-6}. If, starting out from here, we alter the invariants continuously, then P likewise alters itself continuously. Thus, none of the $(2n-5)$ sides can vanish suddenly, and therefore the points of the cube W_{2n-6} in the immediate neighborhood of the place fixed similarly furnish reduced polygons. Accordingly, then P likewise alters itself continuously, then W_{2n-6} in the immediate neighborhood of the place fixed similarly furnish reduced polygons. Accordingly, if we gather into a domain B all those points of W_{2n-6}, which furnish reduced polygons of our kind with the vertex arrangement $e_1, e_2, \ldots, e_n$, then this domain B will consist of one or even of several parts, *which are similarly* $(2n-6)$ *dimensional,* and to whose boundaries, we are only able to come over, *by either a type change with rearrangement of the fixed vertices entering for the domain Q_1, or by at least one of the transversal invariants attaining one or the other of its boundary values.*

We turn ourselves to the first of those two possibilities and assume, that the two sides of Q_1 meeting at e_i have just vanished. Previously, i.e., before the vanishing has occurred, we first modify P in this manner, that, in place of the previous level lines of V_1, we cut out the polygon P from Q_1, by means of the lines $e_1 e_i$ and $e_1 e_i' = V_1(e_1 e_i)$. P then has the shape sketched in figure 74, in which $\varepsilon, \varepsilon'$ and ε'' form a cycle of movable vertices. If the two sides $e_i \varepsilon'$ and $e_i' \varepsilon''$ now vanish, then at this moment ε changes into a vertex e_i'' equivalent to e_i and e_i', and P assumes the shape given in figure 75.

The sides assigned to one another by V_2' are level lines of this substitution and upon extension, intersect one another at the fixpoint ε_2 of this. The corresponding thing [376]
holds for the sides belonging to V_4'. It is not precluded that ε_2 is a vertex, and indeed then a fixed vertex of P, in case, namely, the vertices e_i and ε_2 $(= e_{i+1})$ form a polygon

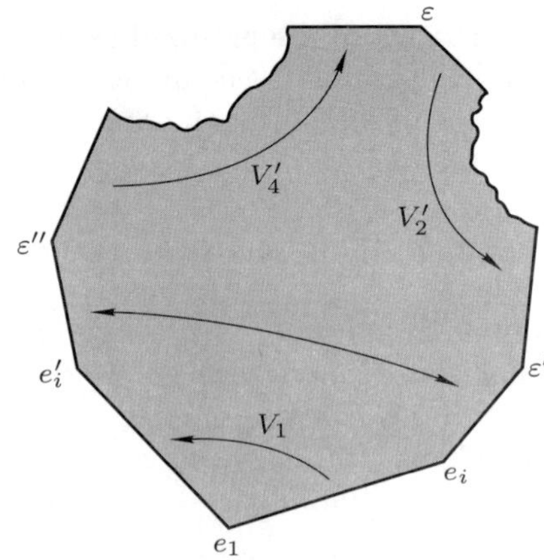

Fig. 74.

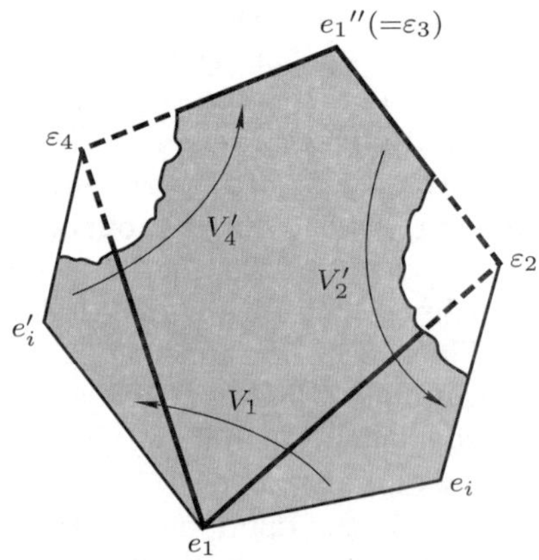

Fig. 75.

parcel of the first kind (cf. figure 69 on p. 368). But in any case, since n is assumed > 4, P cannot reach up to both vertices ε_2 and ε_4.

One now perceives, that the four substitutions:

$$V_1, V_2', V_3' = V_2'^{-1} V_1^{-1} V_4'^{-1}, V_4'$$

with the fixpoints $e_1, \varepsilon_2, \varepsilon_3 (= e_i''), \varepsilon_4$, generate a subgroup of the character $(0,4)$ contained in Γ for which, in figure 75, there is a normal hexangle with a three-termed cycle of fixed vertices. The center C of this normal hexangle lies at the fixed vertex e_1; in consequence of this, the theory of the normal polygons of the character $(0,4)$ (cf. p. 339) yields the result: *for the group in question of the character* $(0,4)$, *we have the symmetric case, whereby the quadrangle* $e_1\varepsilon_2\varepsilon_3\varepsilon_4$, *heavily bounded in figure 75 on p. 375, furnishes an elementary quadrangle.* We will describe this state of affairs by saying, that we have a "*partial symmetry*" of our reduced polygon P with the sequence of vertices $e_1, e_2, \ldots, e_n$.

Let the invariants of the polygon of the character $(0,4)$ be called $j_1'(=j_1), j_2', j_3'(=j_i), j_4', j_{12}',\ldots$. According to p. 354 there then subsists the symmetry condition:

$$j_{12}'j_{23}'+2j_{24}'-J_2'=0. \tag{1}$$

These invariants either belong directly to those of our reduced polygon P of the character $(0,n)$, or they are rationally representable in the invariants of the latter. The equation (1), by means of these representations, calculates itself around into *an algebraic relation*:

$$F(j_1,j_2,\ldots,j_{12},\ldots,j_{n-1,n})=0 \tag{2}$$

between the invariants of P, which is fulfilled in the case of the partial symmetry found. This relation (2) *represents, in the case* W_{2n-6}, *a* $(2n-7)$ *dimensional boundary* R_{2n-7} *of the domain B of our reduced polygons P with the vertex arrangement* $e_1,e_2,e_3,\ldots,e_n$. After passing over this boundary, a displacement of the vertex e_i has occurred in the normal domain, so that the canonical polygon with the vertex arrangement $e_1,e_2,\ldots,e_n$, as long as we will retain this, has ceased to belong to the reduced polygons with the vertex arrangement.

The place to which e_i is displaced, regulates itself according to the polygon parcels [377]
in which e_i had part immediately before the occurrence of the partial symmetry (see the accidental vertex ε in the polygon of figure 74 on p. 375). One now wants to observe, that in the interior of B, such type changes of the normal domain Q_1, in which only sides bounded by accidental vertices take part, are by no means excluded. That kind of type change leaves the vertex arrangement $e_2,e_3,\ldots,e_n$ on the boundary of Q_1 unaltered, but modifies the polygon parcels. One can accordingly only say in general, that e_i is displaced into any one of the $(n-3)$ intervals remaining free between the other vertices $e_{i+1},e_{i+2},\ldots,e_n,e_2,\ldots,e_{i-1}$, which is the highest case, furnishes $(n-3)$ partial symmetries. Since, in addition, e_i is one of the vertices $e_2,e_3,\ldots,e_n$, then all together, the number of the partial symmetries $\leq (n-1)(n-3)$.

Now remember, that besides this, one can only yet arrive at the boundary of the domain B, by at least one of the transversal invariants reaching one or the other of their limiting values, whereby we find ourselves, at the same time, at the boundary of the polygon continuum W_{2n-6}. We arrive at the important knowledge: *In the interior of* W_{2n-6}, *ie. apart from such boundary of the polygon continuum* W_{2n-6}, *the* $(2n-6)$ *dimensional domain B of our reduced polygons of the vertex arrangement* $e_1,e_2,\ldots,e_n$ *is completely bounded by certain* $(2n-7)$ *dimensional "algebraic boundaries"* R_{2n-7}, *whose number is* $\leq (n-1)(n-3)$. In the case $n=4$, the upper bound $(n-1)(n-3)=3$ is actually reached; here the three models R_1 are the curves K_1,K_2,K_3, which bound the domain B as *one* continuum, namely, as the oft-named triangle of p. 355 ff. on the surface Φ_3 (cf. figure 45 on p. 323)[64].

[64] Consider the position of the symmetric polygons of the vertex arrangement $e_1,e_2,\ldots,e_n$ by means of the figure 73 on p. 370, drawn for $n=6$. On the boundary of Q_1, between two points e_n and $e_n'=V_1^{-1}(e_n)$ lie two points e_3,e_3', two points $e_4,e_4',\ldots$, finally, two points e_{n-1},e_{n-1}', but only one point e_2. With this, $(n-3)$ partial symmetries come up at the same time, so that we arrive at the $(2n-6)-(n-3)=(n-3)$ dimensional intersection of $(n-3)$ particular boundaries R_{2n-7}. As one easily determines directly for the symmetric polygon, this intersection forms *one* continuum.

We now investigate those reduced boundary polygons, which enter, in case a trans-
[378] versal invariant reaches one or the other of its boundary values. The investigations entering here are, throughout, analogous and not essentially more complicated than the corresponding ones for $n = 4$.

To our reduced polygons P, in particular, always belong two *symmetric polygons* of the vertex arrangement $e_1, e_2, \ldots, e_n$. Let an elementary n-angle with this vertex arrangement be called $\overline{P}$, the reflection in the side e_{i-1}, e_i be called $\overline{V_i}$, that:

$$V_i = \overline{V}_i \overline{V}_{i+1}.$$

holds. The transversal invariant $j_{1,2,\ldots,i-1} = j_{i,i+1,\ldots,n}$ is the invariant of the hyperbolic substitution:

$$V_1 V_3 \cdots V_{i-1} = (\overline{V}_1 \overline{V}_2)(\overline{V}_2 \overline{V}_3) \cdots (\overline{V}_{i-1} \overline{V}_i) = (\overline{V}_1 \overline{V}_i),$$

which arises by combination of the reflections in the first and i^{th} sides. For abbreviation, we call transversal invariant picked out j for short.

For the boundary value $j = -2$ exactly the same thing holds now as for $n = 4$ (cf. p. 360): *one can carry out the passage the limit* $\lim j = -2$ *in three ways, not to be distinguished by the invariants, either in this manner, that with coincidence of* $e_i, e_{i+1}, \ldots, e_n$ *at a point of the real* ζ*-axis, the symmetry circles of* $\overline{V}_1$ *and* $\overline{V}_i$ *come into tangency at the coincidence place, or that with coincidence of* $e_1, e_2, \ldots, e_{i-1}$*, the two circles in question finally touch one another at this latter place of the real* ζ*-axis, or finally, by these two vertex coincidences simultaneously take place, while the two symmetry semicircles come into juxtaposition. In the last case,* $\overline{V}_1 \overline{V}_i$ *changes into the identical substitution, and the polygon degenerates*[65]; the first two cases furnish the two possibilities, that $\overline{V}_1 \overline{V}_i$ becomes parabolic: *we obtain two symmetric polygons of the characters* $(0, i)$ *and* $(0, n - i + 2)$*, where both times, a parabolic cusp has entered at the coincidence place of the previously separated vertices.* Also, it again becomes clear, that the three limit passages are not distinguishable by the invariants. Thus, e.g., for $n = 7$, $i = 4$ in figure 76, a
[379] polygon near to the parabolic limiting case is drawn. By transformation by means of a hyperbolic substitution, which has the symmetry circle of $\overline{V}_4$ as an orbit curve, moves from e_4 toward e_3 and has a sufficiently large invariant, we are always able to carry this polygon over into a polygon near the other parabolic limiting case.

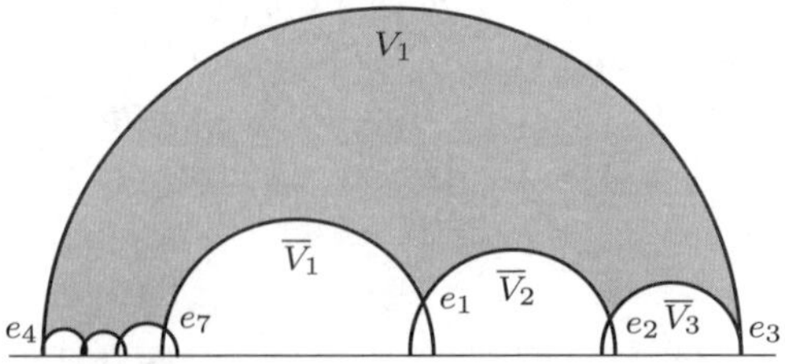

Fig. 76.

For the passage to the limit just effected, the transversal invariants:

[65] I.e. as principal-circle polygon. But see the footnote on p. 361.

$$j_{i+1,\dots,n}, \dots, j_{n-1,n}$$

following:

$$j = j_{i,i+1,\dots,n},$$

will all lose their meaning for the first boundary polygon, that with the generators $V_1, V_2, \dots, V_{i-1}, (V_i V_{i+1} \cdots V_n)$; rather, it is the $(n-i-1)$ transversal invariants of the other boundary polygon, for which the first $(i-3)$ transversal invariants no longer come into consideration. If, further, for any tow indices k, l satisfying the inequalities:

$$1 < k < i, \ i < l,$$

we form the transversal invariant $j_{k,k+1,\dots,l-1}$, i.e., the invariant of the hyperbolic substitution $\overline{V}_k \, \overline{V}_l$, which, to be sure, we did not assume among the $(n-3)$ independent transversal invariants, then this invariant, as a glance at figure 76 informs us, will itself become infinite upon passage to the limit line $j = -2$.

Also conversely, in case a transversal invariant should become infinite, then necessarily, at least one often will attain the limiting value -2. Should e.g. our invariant j selected above become $j = -\infty$, then at least one of the two symmetry circles, which belong to $\overline{V}_1$ and $\overline{V}_i$, must draw itself together into a point on the real ζ-axis. According to this, either the points e_n and e_1 or e_{i-1} and e_i fuse, as long as both do not happen at the same time. One will easily pursue the relations entering here further and arrive at the result: *boundary polygons, other than the symmetric polygons of the lower characters produced above by fixed vertices moving together, do not occur at all in the symmetric case of the character* $(0, n)$.

In the unsymmetric case, we arrive at the corresponding result. We again investigate, say, the transversal invariant:

$$j = j_{1,2,\dots,i-1} = j_{i,i+1,\dots,n},$$

which belongs to the hyperbolic substitution:

$$V = V_1 V_2 \cdots V_{i-1} = (V_i V_{i+1} \cdots V_n)^{-1}.$$

To our reduced polygons P, with retention of the fixed vertices, we give the shape sketched in figure 77, which, according to the first theorem in I on p. 308, is always [380] attainable. Here Φ is the fixpoint of V and Π the polar of Φ, which cuts the ellipse at the points A and B. There are only two cycles of movable vertices, one in each individual one of the two boundary parcel's of P cut off by Π. The one cycle includes the vertices ε', $V_{i-1}(\varepsilon')$, $V_{i-2}V_{i-1}(\varepsilon'), \dots, \varepsilon = V_1 V_2 \cdots V_{i-1}(\varepsilon')$, the other the vertices $\varepsilon''', V_n(\varepsilon'''), \dots, \varepsilon'' = V_i V_{i+1} \cdots V_n(\varepsilon''')$.

The polygon P let itself be decomposed here very easily into two partial polygons P' and P'' of the characters $(0, i)$ and $(0, n-i+2)$ and the generators $V_1, V_2, \dots, V_{i-1}, V^{-1}$ and $V, V_i, V_{i+1}, \dots, V_n$. E.g., P'' arises by leaving out the sides of P between ε and ε' on the right side (in figure 77) and adding the sides $\varepsilon\Phi$ and $\varepsilon'\Phi$, corresponding by V; correspondingly, one obtains the polygon P' (cf. p. 362).

For the passage to the limit $\lim j = -2$, the argument carried through above (p. 363) in the case $n = 4$ now holds word for word. *We arrive at two boundary polygons* P', P''

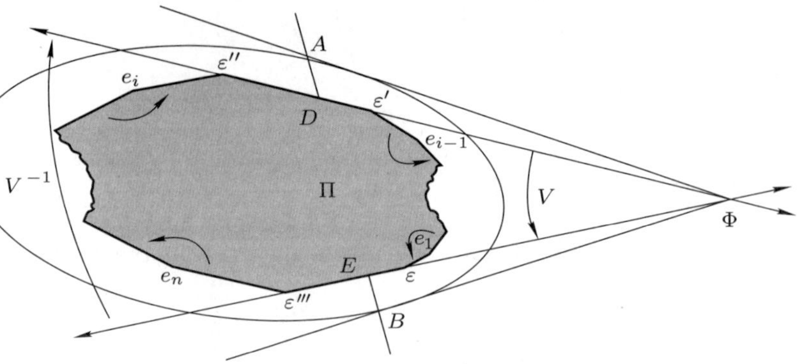

Fig. 77.

of the characters $(0,i)$ *and* $(0,n-i+2)$ *and one degeneration*[66]*; of the* $(n-3)$ *transversal invariants, the first* $(i-3)$*'s* $j_{1,2}, j_{1,2,3}, \ldots, j_{1,2,\ldots,i-2}$ *have become the transversal invariants of* P'*, the* $(i-2)^{\mathrm{nd}}$ *invariant has become the vertex invariant for* P' *and* P'' *having the value* -2*, and the result of the* $(n-i-1)$ *invariants* $j_{i+1,\ldots,n}, \ldots, j_{n-1,n}$ *furnish the transversal invariants for* P''.

For the investigation of the passage to the limit $\lim j = -\infty$, it is expedient to draw the sides $\varepsilon'\varepsilon''$ and $\varepsilon\varepsilon'''$, coordinated to one another by V, together into the points D
[381] and E of Π, so that P becomes a canonical $2n$-angle. In the passage to the limit, we are able to attain, if need be, by displacement of P along Π, that the vertex E say, initially remains in its place. Then consider the group of the character $(0,3)$, which has, as generators:

$$V_1 V_2 \cdots V_{i-2}, V_{i-1}, V_i V_{i+1} \cdots V_n.$$

Of the three associated invariants, the first is the transversal invariant $j_{1,2,\ldots,i-2}$, not to be further restricted, the second, as the vertex invariant j_{i-1} of P, is fixed, while the third of our invariants is according to the results at that time, the vertex e_{i-1} is pushed toward the ellipse point A (cf. figure 77), since on account of the vertex E being held fixed, the fixpoint of $V_1 V_2 \cdots V_{i-2}$ always remains distant from B.

One can carry through exactly the corresponding argument for the group of the character $(0,3)$ to be generated from:

$$V_1 V_2 \cdots V_{i-1}, V_i, V_{i+1} V_{i+2} \cdots V_n.$$

We come to the result, that e_{i-1} and e_i come into coincidence at the ellipse point A. From the theory of the character $(0,3)$ follows, that for this coincidence, V_{i-1} and V_i lead to a *parabolic* substitution $V_{i-1}V_i$.[67] Accordingly, we arrive here at a limiting case of exactly the same kind, as in the case $\lim j = -2$ considered before: *the boundary cases*

[66] Cf. the footnote on p. 361.

[67] Notice say, that $V_{i-1}, V_i, (V_{i-1}V_i)^{-1}$ furnish a subgroup of the character $(0,3)$ contained in our group.

obtained above in the case $j=-2$ are the only ones, which are able to come up for our reduced polygons of the character $(0,n)$.[68]

In order to determine the number of the invariants of the individual boundary manifold coming up for our domain B of the reduced polygons P, notice first of all, that the degeneration furnishes no invariants[69]. Further, the polygon P' of the character $(0,i)$, taken by itself, has $(2i-6)$ independent invariants and the polygon P'' of the character $(0,n-i+2)$, $(2(n-i+2)-6)$ such. *The total number of the invariants of the boundary manifold is consequently*:

$$(2i-6)+(2(n-i+2)-6)=2n-8;$$

it is therefore smaller by two units than the number of dimensions of the domain B itself. This lowering by two units is also understandable in this way, that in the limiting case, [382]
not only one of the transversal invariants assumes a fixed value, but that, in addition to this, the composition invariant, entering in general for the composition of P' and P'' into the polygon P, drops out for the limiting case.

Since the assertion of Poincaré's, mentioned on p. 305, that the group continuum is a closed one, is connected with the relations considered here, we go into these yet somewhat more closely.

Denote, as above, (p. 374), the composition invariant, which enters by uniting P' and P'', by σ, and remember that σ, since P is a reduced polygon, was restricted to the interval $\sigma_1 \leq \sigma \leq \sigma_2$, which came close to the boundary values of the interval prescribed for σ for the polygon continuum, neither with the one endpoint σ_1 nor with the other one σ_2.

If we now find ourselves in a boundary, considered before, of the domain B, which belongs to a $(2n-7)$ dimensional boundary of the cube W_{2n-6}, then we are able, attaching to any boundary model (a combination of special polygons P', P'' and the degeneration), by holding the $(2n-8)$ invariants of P' and P'' fixed, to move the invariant σ from σ_1 to σ_2, without coming to another boundary model. We have thus described a line in the boundary of B, whose points all represent the same model. *The domain B, at its boundary in question, if no longer in the position, of furnishing each individual one of our reduced boundary polygons only once by means of its points.*

In the case $n=4$, we had, on this account, constricted the domain B, not in the square W_2, but on the surface Φ_3. Here the boundary under consideration of the domain B was directly only a single point (cusp of the triangle in figure 45 on p. 323). Without disposing of the question, whether also for arbitrary n, we are able, in this regard, to make a more suitable choice of the invariants, we may, in any case, conceive the system of our groups (reduced polygons) in abstracto, i.e., independently of that representation by invariants, as a continuous manifold. *This manifold, corresponds to the passage to the limit in question, then certainly has a* $(2n-8)$ *dimensional boundary.* We immediately return to the meaning of this result for the Poincaré assertion in question.

[68] Naturally, it is not to be precluded, that not only one, but several series of vertices simultaneously fuse into just as many parabolic points. Then the individual boundary polygon, in its manifold of lower character, itself represents a limiting case.

[69] The cyclic parabolic group possesses no invariants; each parabolic substitution is transformable into each other one.

1.20 The discontinuity domain of the modular group and the group continua of the character $(0, n)$

The results just obtained now lead us easily to the discontinuity of the modular group
[383] for the character $(0, n)$. The domain B either itself represents a continuum or fell into finitely many, let us say ν, partial domains $B_1, B_2, \ldots, B_\nu$, each individual one of which is a $(2n-6)$ dimensional continuum. If now, in the cube W_{2n-6}, we start out from any one of these partial domains, say B_i, by passing over a boundary R_{2n-7}, then at this moment there enters a displacement of a fixed vertex on the boundary of the normal domain Q_1.

If the displaced vertex is not equal to e_2, then we retain the two previous level lines of V_1, which cut the polygon P out of the domain Q_1. If, on the contrary, e_2 is displaced, then in order that the displaced vertex e_2 also follows e_1 on the new polygon P, we correspondingly bring in other level lines of V_1. In each case, one sees, that another arrangement of the $(n-2)$ vertices $e_3, e_4, \ldots, e_n$ is present for the new polygon.

But now all polygons of the new vertex arrangement furnish a second cube W'_{2n-6}; in this is situated the representing point for the new reduced polygon P. But if we carry over the theory of the previous paragraph to this cube, correspondingly, let $B'_1, B'_2, \ldots, B'_{\nu'}$ be the ν' continua of the reduced polygons of this new vertex arrangement. We have arrived in one of these continua, as we had passed over the boundary R_{2n-7} of B_i; let the continuum thus reached be B'_k say.

Now there are, in all $(n-2)!$ arrangements of the vertices $e_3, \ldots, e_n$, and for each arrangement, reduced polygons come up; for symmetric polygons are found for each one. In the individual one of the $(n-2)!$ cubes $W^{(\lambda)}_{2n-6}$, imagine the $\nu^{(\lambda)}$ continuous partial domains $B^{(\lambda)}$ of the reduced polygons bounded: *there arises thus a finite number of continua $B^{(\lambda)}_k$, each individual one of which is bounded by finitely many boundaries R_{2n-7}*. These boundaries are coordinated to one another in pairs, since two corresponding boundary points furnish one and the same polygon with transitional type. By considering two such boundaries as not different, the $\nu + \nu' + \nu'' + \cdots$ domains $B^{(\lambda)}_k$ arrange themselves together one larger continuum or into several such.

We assert *that it can hereby only be a question of the single continuum*. Namely, if P and P' are any two reduced polygons, which belong to two of our partial domains $B^{(\lambda)}_k$, then, for the two associated groups Γ and Γ', choose two polygons P_0 and P'_0 in an individual one of our cubes W_{2n-6},[70] and describe in W_{2n-6}, from P_0 to P'_0, any
[384] path (say the rectilinear one). Carry this domain over, by returning point for point to the reduced polygon, to the total domain of all reduced polygons. Here, on account of the coordination of the boundaries R_{2n-7}, it changes into a similarly connected path, which leads from P to P'. Our assertion is therefore correct.

In the case $n = 4$, $(n-2)! = 2$. For this case (cf. p. 352 ff.), we had two squares W_2, W'_2, or, in their place, the two surfaces Φ_3 and Φ'_3, and on each of these, a domain consisting of only one continuum (the triangle of the reduced polygons). These two domains B and B', taken together, give *a closed continuum*. We actually produced this closedness above (p. 355) in this fashion, by carrying B' (in one of three ways) over onto the surface Φ_3 and uniting it with B into a double triangle. The four sides still open then belonged

[70] For each of our groups Γ, these are infinitely many polygons in W_{2n-6}.

together in pairs and furnished in such a manner the generators of the modular group, whose discontinuity domain was just that double triangle.

We are able to proceed precisely analogously in the general case $(0, n)$. If, in the cube W_{2n-6} utilized in the previous paragraph say, we start out from a first partial domain B_1, then let the passage over a boundary R_{2n-7}, corresponding to the previous perception, lead to a partial domain $B_k^{(\lambda)}$ in the cube $W_{2n-6}^{(\lambda)}$. This domain $B_k^{(\lambda)}$ is equivalent to a domain positioned directly next to B_1 in W_{2n-6}, which furnishes exactly the same groups as $B_k^{(\lambda)}$, but no longer, to be sure, with reduced polygons.

One continues this consideration without any trouble and finds the cube W_{2n-6} filled out simply and gaplessly by a "net" of infinitely many partial domains. But among these partial domains, there are only a finite number $m = \nu + \nu' + \nu'' + \cdots$, inequivalent with respect to the modular group. We are able, in a certain finite number of ways, beginning with that first partial domain B_1, to arrange together m inequivalent partial domains into a continuum B: *this continuum* B *of* $(2n-6)$ *dimensions forms a discontinuity domain of our present modular group; the boundaries* R_{2n-7} *of this domain* B *still remaining free belong together in pairs and furnish the generators of the modular group, whose number, as one sees, is finite.*

The boundary models, which correspond to the points of B lying in the boundary of the cube, we imagine, as above for $n = 4$, as belonging to our group continuum. For these boundary surfaces remaining uncovered, the discussions hold which were given at the close of the previous paragraph. The individual boundary surface is $(2n-7)$ dimensional, but in the sense described here, represents only a $(2n-8)$ dimensional continuum of boundary models. In the case $n = 4$, it is a question of the three vertices of [385]
a discontinuity domain of the modular group. Upon bending the boundary curves of this domain together, those three vertices cease to be on boundary curves, and the group continuum obtained the connectivity of the surface of a sphere, and is therefore a closed continuum. If we conceive the group continuum, as at the close of §19, in abstracto, then for this continuum there are at most those finitely many $(2n-8)$ dimensional boundaries, which correspond to the boundary models situated on the boundary of W_{2n-6}. Now[71] an individual point can never be the boundary of a two dimensional continuum, in case it belongs to the continuum, and an individual line can never be the boundary of a three dimensional continuum, in case it belongs to this continuum. If, correspondingly, an m-dimensional continuum is never boundable by an $(m-2)$ dimensional continuum belonging to it, then the Poincaré assertion, that our group continuum is a thoroughly closed one, is proved.

Up until now, we had assumed the vertex invariants as consistently different. There still remains to refer to the particularities which enter, if this is no longer the case. The representations on p. 356 ff. for the case $n = 4$ are typical here. *There enter regular subdivisions of the domain* B *and corresponding extensions of the regular group.* If e.g. $j_{n-1} = j_n$, then a two-part division enters; in place of each two cubes W_{2n-6}, whose vertex arrangements go over into one another by interchange of e_{n-1} and e_n (cf. I on p.

[71] The following argument raises no objections regarding rigor. If one conceives the group continuum (in order that the boundary models furnish $(2n-8)$ dimensional continua), independently of the representation by invariants, abstractly as a continuous set, then one is directed to the theorems of the general doctrine of sets, which has not yet set up theorems concerning the nature of the boundary of n-dimensional continua.

301). Except for this, all essential results just obtained concerning the modular group and its discontinuity domain remain maintained.

For the sake of fixed vertex invariants, and therefore, in particular, then, if hyperbolic vertices e_i are not present, then the theory is hereby isolated. We now assume, only the first ν invariants $j_1, j_2, \ldots, j_\nu$ to be fixed[72], while the rest of the vertex invariants, which are then certainly ≥ 2, to be freely variable. For the treatment of the extended group continuum hereby entering, the corresponding theory for the case $n = 4$
[386] (cf. p. 358 ff.) is to be a prototype. As there, we utilize the variable vertex invariants in the form:

$$t_\kappa = \frac{j_\kappa - 2}{j_\kappa},$$

and interpret the t_κ as rectangular coordinates. In order not to obtain the individual group multiple, we fix a vertex arrangement, for which $j_{\kappa-1} \leq j_\kappa$ and therefore $t_{\kappa-1} \leq t_\kappa$ holds. Of the $(n-\nu)$ dimensional cube $W_{n-\nu}$ of the t_κ, only the portion defined by:

$$t_{\nu+1} \leq t_{\nu+2} \leq \cdots \leq t_n, \tag{1}$$

then comes into consideration for us. To each point of this portion now belongs a $(2n-6)$ dimensional group continuum B, which alters itself continuously for continuous alteration of that point in $W_{n-\nu}$. *We arrive at a single-valued and complete representation of the* $(3n-\nu-6)$*-dimensional group continuum entering here, if we conceive each point of the admissible portion of the cube* $W_{n-\nu}$ *as a representative of the associated* $(2n-6)$*-dimensional continuum* B.

The boundary of the admissible portion of the cube $W_{n-\nu}$ then consists, besides boundary planes, also of inner planes $t_{n-1} = t_n, \ldots$ of this. *But these latter planes are, in any case, not boundary surfaces for our group continuum.* They are rather, as we already perceived in the case $n = 4$ (cf. p. 359), symmetry planes for the generators of the modular group, and indeed, transformations hereby come into consideration, which make their appearance for the regular divisions of the domain B in the case of equal vertex invariants.

Further, on account of the inequalities (1), we have only the boundary surfaces $t_{\nu+1} = 0$ and $t_n = 1$. The former disposes of itself, as in the case $n = 4$ (cf. p. 364), immediately: *for* $t_{\nu+1} = 0$*, the vertex* $e_{\nu+1}$*, hyperbolic in general, becomes parabolic, and we arrive at a proper boundary manifold of our group continuum, consisting of a* $(3n-\nu-7)$*-dimensional group continuum of our kind.* Our group continuum therefore, in any case, is no longer a closed one.

The passage to the limit $\lim t_n = 1$ must be investigated somewhat more meticulously for the individual one of our reduced normal polygons P. The vertex e_n belongs to a polygon parcel of P, which we imagine picked out in figure 78. We call the point e_n and e'_3 here for the present, and the substitution V_n, V'_3. The two other substitutions
[387] taking part in the parcel are V'_1 and V'_2. Let n_1 fixed vertices on the boundary of polygon P interpose themselves in the interruption belonging to V'_1, of which ν_1 have fixed invariants, and correspondingly let n_2 fixed vertices follow in the other interruption with ν_2 fixed invariants; then there always holds:

$$n_1 + n_2 + 1 = n, \quad \nu_1 + \nu_2 = \nu. \tag{2}$$

[72] The case $\nu = 0$ counts as included here.

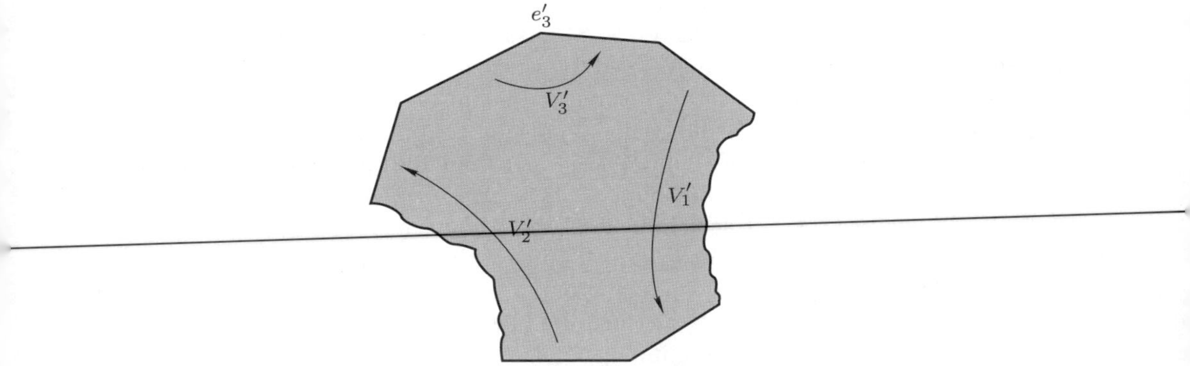

Fig. 78.

The substitutions V'_1, V'_2, V'_3 now form a group of the character $(0,3)$, whose third invariant $j'_3 = j_n$. The passage to the limit $\lim j'_3 = \infty$ was investigated above (p. 365) and three invariant-theoretically indistinguishable limiting cases of the polygon, and with this the group: the cyclic group of the generator V'_1 that of the generator V'_2, and thirdly, the group consisting of the identity alone. Apply this to figure 78, and imagine, at the same time, the chains of sides of the polygon P hung into the interruptions. It follows: *the passage to the limit* $\lim j_n = \infty$ *leads to the invariant-theoretically indistinguishable boundary forms of our polygon: a polygon of the character* $(0, n_1 + 1)$ *with* ν_1 *fixed vertex invariants, a polygon of the character* $(0, n_2 + 1)$ *with* ν_2 *fixed vertex invariants, thirdly and finally, corresponding to the entire surface of the ellipse, the group consisting of the identity alone.* If one of the two numbers n_1, n_2, say the first, equal to 1, then the first of the polygons just mentioned is simply the discontinuity domain of V'_1.

As the number of the invariants of the boundary manifold obtained, in case $n_1 > 1$ and $n_2 > 1$ hold, we yet:

$$(3(n_1 + 1) - \nu_1 - 6) + (3(n_2 + 1) - \nu_2 - 6),$$

and therefore, in consideration of the relations (2), the value $(3n - \nu - 9)$. If, on the contrary, one of the numbers n_1, n_2, say n_1 again, is equal to 1, then we find, as the number in question:

$$(1 - \nu_1) + (3(n - 1) - \nu_2 - 6) = 3n - \nu - 8.$$

1.21 The group continua of the character (p, n)

For a detailed study of the group continua of the character (p, n), the theory of the normal polygons is to be taken as a point of departure, exactly as in the cases considered so far. Fora a proposed group Γ, according to p. 309, there is always only a finite number μ of essentially different normal polygons, which corresponded to the μ inequivalent

natural elementary domains. One succeeds in choosing one of these μ normal polygons, for $n > 0$ say, again in this manner, by choosing a first fixed vertex e_1 as the center
[388] C. But since this approach breaks down for $n = 0$, we better retain all μ normal polygons next to one another and choose as "reduced" any canonical polygon P, which can then be produced by μ well-determined transformations from those normal polygons.

If, for the moment, the vertex invariants are constant, then this reduced polygon P belongs to a $(2n+6p-6)$-dimensional polygon continuum, which we represented above (p. 304) in the shape of a cube $W_{2n+6p-6}$. In this cube the discontinuity domain of the modular group is then to be bounded. As in case, $(0, n)$, there exists, in $W_{2n+6p-6}$, a "net" of domains $B_1, B_2, \ldots$, each individual one of which has a finite number of boundaries $R_{2n+6p-7}$, and among those, only a finite number B_1, B_2, can be picked out inequivalent with respect to the modular group.

These boundaries $R_{2n+6p-6}$ are furnished by those special groups Γ, for which a decrease of the number μ, present in general, by at least one of the μ natural elementary domains receiving vanishing content (cf. the case $(0,4)$ above on p. 345). In this case, at least one of the μ normal polygons will assume a special type, which cannot be removed by alteration of the center C.

For the more precise investigation of the relations, however, furthergoing preparations are still necessary. If one introduces, as above (p. 368), the polygon parcels belonging to the three-termed cycles of accidental vertices, then it reveals itself, that two kinds of parcels are to be distinguished, which one has to designate as the parcels of the genus $p = 0$ and such of the genus $p = 1$. For the discussion of those special types of normal polygons, which furnish the boundaries $R_{2n+6p-7}$, one has also to investigate more closely the pairs of neighboring polygon parcels; and hereby, it reveals itself, that the effectability of the consideration in the general case (p, n) first presupposes the more precise knowledge of the special theory of the character $(1,2)$.[73]

In this state of affairs, we ignore carrying through the theory presently indicated, rather we make an attempt to give an account of the group continua of the character
[389] (p, n) in a less fundamental way, to the extent that it appears desirable, in consideration of our further studies.

In order to immediately put before our eyes the most general case, we assume, that of the n vertex invariants v are given fixed, and will use the symbol (p, n, v) as an abbreviated notation for the signature. For an associated group let a first canonical polygon P be proposed, which furnishes, as generators of the group, $(n+2p)$ substitutions $V_1, \ldots, V_n, V_{a_1}, V_{b_1}, \ldots, V_{a_p}, V_{b_p}$. From $V_{a_\kappa}, V_{b_\kappa}$, according to I on p. 311, we produce the substitution:

$$V_{c_\kappa} = V_{b_\kappa} V_{a_\kappa}^{-1} V_{b_\kappa}^{-1} V_{a_\kappa}.$$

On the basis of the composition theory of I on p. 314 ff., we now subject this polygon P to a reduction, which, to be sure, we shall be able to designate only as an "incomplete one". We produce the polygon P l.c. by composition from a polygon of the character $(0, n+p)$ and the generators:

$$V_1, \ldots, V_n, V_{n+1} = V_{c_1}^{-1}, \ldots, V_{n+p} = V_{c_p}^{-1}.$$

---

73 Two neighboring parcels, of which the one belongs to the genus 0 and the other to the genus 1, taken by themselves, form a normal polygon of the character $(1,1)$; on the other hand, two neighboring parcels of genus 1 form a normal polygon of the character $(1,2)$.

and from p polygons of the character $(1,1)$, each individual one of which gave a triple V_{a_κ}, V_{b_κ}, V_{c_κ} as generators. The first of these polygons, whose signature we would have to denote by $(0, n+p, \nu)$, furnishes a group, which belongs to a $(3(n+p)-\nu-6)$-dimensional group continuum. The latter we are able to represent, in the cube $W_{3(n+p)-\nu-6}$ of the polygon of the signature $(0, n+p, \nu)$ under discussion by a continuous domain B_0, which appears bounded in the interior of the cube, but on the boundary of this has $(3(n+p)-\nu-7)$-dimensional uncovered boundary manifolds.

As a first step of the reduction to be undertaken, we replace the polygon of the signature $(0, n+p, \nu)$ initially present, by that equivalent polygon, which belongs to the domain B_0. The generators $V'_1, \ldots, V'_n, V'_{n+1} = V'^{-1}_{c_1}, \ldots, V'_{n+p} = V'^{-1}_{c_p}$ of the new polygon are conjugate with $V_1, \ldots, V_n, V^{-1}_{c_1}, \ldots, V^{-1}_{c_p}$ in the group of the signature $(0, n+p, \nu)$ and therefore, also in our original total group Γ. Thereby, let the substitution, which carries V_{c_κ} over into V'_{c_κ}, at the same time, transform V_{a_κ} into V'_{a_κ} and V_{b_κ} into V'_{b_κ}.

The second step of the reduction refers itself to each individual polygon of the character $(1,1)$, which we have to hang on. According to the reduction theory of the character $(1,1)$, we may replace the polygon belonging to V'_{a_κ}, V'_{b_κ}, V'_{c_κ} by a uniquely determined reduced polygon, which belongs to a closed two-dimensional domain, which we will call B_κ. Hereby, let the generators V''_{a_κ} and V''_{b_κ} take the place of V'_{a_κ} and V'_{b_κ}, [390]
while V'_{c_κ} remain, maintained.

The third step of the reduction concerns the composition invariants, which enter for the p compositions in question, and with this, the transformation of the individual polygon of the character $(1,1)$ by the associated substitution V_{c_κ}. If we fix the composition invariant as above (p. 374) in the form of an exponent σ_κ, then, in case the invariant σ_κ lies outside the interval $0 \le \sigma_\kappa < 1$ for the polygon now attained with the generators V''_{a_κ}, V''_{b_κ}, for which σ_κ belongs to that interval. Thereby, yet notice, that $\sigma_\kappa = 1$ no longer need to be allowed, since for this value, we have the same groups as for $\sigma_\kappa = 0$. If, according to this, for the sake of uniformity, we introduce the p intervals $0 \le \sigma_\kappa < 1$ as p linear domains B'_κ, then insofar as the two endpoints are not to count as different, these are also to be considered as closely just as the two dimensional domains B_κ.

We now combine the $(1+2p)$ domains $\mathrm{B}_0, \mathrm{B}_\kappa, \mathrm{B}'_\kappa$, by conceiving, as points of a more comprehensive continuum, all possible combinations of $(1+2p)$ points from each of those domains. *This arises a* $(3n-\nu+6p-6)$*-dimensional continuum, which we will call* B'*, and in which, for each polygon of the signature* (p, n, ν)*, an equivalent one certainly finds itself. But there find themselves, not only one, but several such in general, so that the reduction, in fact, is to be described as an incomplete one.* Namely one wants to observe, that for the reductio, as we just effected it, those elementary transformations, which we designated in I on p. 331 ff. as such of the fourth kind[74], did not come into application at all yet. To the addition of the transformations still lacking will consequently correspond a subdivision of B', *and first in the individual partial domain* B *thus arising do we have the single-valued image of the* $(3n-\nu+6p-6)$*-dimensional group continuum of the signature* (p, n, ν) *before us.*

[74] Transformations, for which two neighboring polygons of the character $(1,1)$ were transformed into two new polygons of the character.

1.22 Report on the continua of the Riemann surfaces of the genus p

For the individual one of our principal-circle groups Γ, as soon, on the basis of the
[391] general existence theorem of automorphic functions (cf. p. 18 ff.), we pass over to the consideration of the associated functions $\varphi(\zeta)$, the question, whether the principal circle is a boundary circle or not, because of fundamental significance. Namely, if there is no boundary circle, then the polygon P (considered in the projective plane) has hyperbolic cusps, whose number we denote by μ (as above on p. 26 ff.). P will then be mapped by an associated function $z = \varphi(\zeta)$ onto a symmetric Riemann surface with μ transition or symmetry lines.

By putting aside the discussion of this case, we therefore turn ourselves, first of all, to the case of a *boundary-circle group of the character* (p, n) *resp. of the signature*[75] $(p, n; l_1, \ldots, l_n)$, whose polygon P we have mapped by a suitably chosen function $z = \varphi(\zeta)$ onto a *Riemann surface F of the genus p*. We mark on F the n places $e_1, e_2, \ldots, e_n$ and designate the surface thus outfitted as the *"signed" Riemann surface.*

Each automorphic function belonging to Γ leads to such a Riemann surface. But all these surfaces furnish, as we know, one and the same algebraic model, are therefore algebraically single-valuedly transformable into one another and are consequently to count as not essentially different for the considerations following next; we choose one of them as a representative of all.

The group Γ belongs to a $(2n+6p-6)$-dimensional continuum, whose character was investigated as penetratingly as possible in the foregoing part of the present chapter. On the other hand, our representing signed Riemann surface F belongs to the *total manifold of essentially different Riemann surfaces F of the genus p, each signed with n points* $e_1, \ldots, e_n$. This manifold has became the subject of many old and new investigations, which admittedly have not lead to an exhaustive conclusion. We have to report here briefly on the principal viewpoints of the theories coming into consideration.

We mention as the first principal theorem: *all (unsigned) Riemann surfaces of the genus p form a single continuum.* This fact is explicitly emphasized for the first time by Klein in section III of the pamphlet appearing in 1882 "*Über Riemann's Theorie algebraischen Functionen und ihrer Integrale*" (Leipzig, Teubner) and indeed, in the form,
[392] that the totality of all m-sheeted Riemann surfaces with ω two-sheeted branchpoints is set down as a continuum. If one chooses a number $m > 2p-2$, then in consequence of the Riemann-Roch theorem (cf. "M" I on p. 546), there are always m-valued functions, so that, for each algebraic model of genus p, we are able to find a Riemann surface of the chosen number m of sheets as a representative. Also, as is known, the assumption of only two-sheeted branchpoints is no restriction of generality; the number ω of these branchpoints, moreover, calculates itself, according to "M" I on p. 545, to be:

$$\omega = 2m + 2p - 2,$$

As far as the proof of the theorem stated goes, Klein refers himself l.c. to two investigations of Lüroth[76] and Clebsch[77], from which the theorem follows. These inves-

[75] Hereby, the values $l_i = \infty$, corresponding to the parabolic cusps, are also to be allowed.

[76] "*Note über Verzweigungsschnitte und Querschnitte in einer Riemann'schen Fläche*", Math. Ann. Vol. 4, p. 181 (1871).

[77] "*Zur theorie der Riemann'schen Flächen*", Math. Ann. Vol. 6, p. 216 (1872).

tigations are of fundamental significance of the doctrine of the form of a Riemann surface. The Riemann surface is initially presented in the fashion, that the ω branchpoints, in any arrangement, are bound from the first to the last branchpoint by any curve C not crossing over itself. All m sheets are then cut through along C, and the arising cut-boundaries between each two branchpoints following upon one another are bound in any manner, so that only two-sheeted boundary points arise. The problem is then, to determine, how for the surface so obtained, by subsequent modification of the arrangement of the branchpoints resp. the curve C, the arrangement of the sheets is modifiable. The penetrating and exhaustive solution of the problem leads to a surprisingly simple result, from which, in fact, the conclusion can be drawn, that each of our surfaces is transportable into each other by continuous wandering of the branchpoints.

In addition, it is to be remarked, that the above named pamphlet of Klein brings the theorem in question directly into expression at the very beginning. In place of the ordinary multiple-sheeted Riemann surface positioned over a plane, Klein works with arbitrary multiple-connected closed surfaces in space, on which associated algebraic functions are constructed on the basis of physical considerations. But that two even surfaces of the same degree of connectivity are equivalent in the sense of analysis situs, [393] i.e., are able to be continuously and single-valuedly mapped onto one another, has long been known[78]. From this, the continuous transportability of the one surface into the other, and with this, the continuous connectivity of all algebraic models of the same genus p, then yields itself easily.

By now further signing each of our Riemann surfaces of the genus p with n points $e_1, \dots, e_n$, we assign free movability on the entire surface to each individual one of these points independently of the remaining ones. *The Riemann surfaces of the genus p thus signed with the same number of points e will then also form a single continuum.*

For the determination of the number of dimensions of this continuum, the studies of Riemann's in Auf. 12 of his "*Theorie der Abelschen Functionen*"[79] are fundamental. If, for the present, we designate as "simply infinite" the value domain of a complex variable, *then the (unsigned) Riemann surfaces of the genus p form a* $(3p-3)$*-dimensional continuum.* Riemann express this thus, *that for the genus p, the algebraic models have* $(3p-3)$ *moduli (absolute invariants with respect to rational transformation).* In this form, the theorem holds, in case $p > 1$; for $p = 0$, there is no modulus, and for $p = 1$, as is known, one.

Klein develops a representation also including the cases $p = 0$ and $p = 1$ in section III of his pamphlet on Riemann. The result is, *that for the genus p, the number of the moduli is* $(3p-3+\rho)$, *whereby the number* ρ *gives how multiple infinitely often the individual Riemann surface is transformable into itself.* In this respect, however, as is known, the theorem holds: *for* $p = 0, \rho = 3$, *for* $p = 1, \rho = 1$, *but for* $p > 1$ *always* 0.

If we return to the signed surfaces, these transformations of the surface into itself become of significance. We remember beforehand, that for $p = 0$, the number n must be ≥ 3, but for $p = 1$, $n \geq 1$, since the combinations hereby excluded do not occur for

78 Cf. Moebius, "*Theorie der Elementarverwandtschaften*", Leipzig, Berichte Vol. 15 (1863) or Ges. Werke., Vol. 2 as well as C. Jordan "*Sur la déformation des surfaces*", Liouvilles Journal. Ser. 2, t.2, p. 105 (1866).

79 Crelle's Journal, Vol. 54 (1857) or Ges. Werke, p. 112 (First edition).

principal-circle groups. Then, in the case $p = 0$, we are able to bring three of the n points e, by a suitable transformation, to three prescribed places, so that only $(n-3)$ points are still freely movable on the surface. For $p = 1$, however, one can bring one
[394] point e to an arbitrarily prescribed place, and therefore, only $(n-1)$ points are still freely movable. By holding fast to the Riemann terminology, we are thus able to say, *that the signing of the surface brings with it* $(n-\rho)$ *further (complex) moduli.*

The total number of the moduli of the signed surface is therefore $(n+3p-3)$. Accordingly, if we again designate the value domain of a real variable as "simply infinite", then the theorem holds: *the continuum of the Riemann surfaces of genus p signed with n points e, is* $(2n+6p-6)$*-dimensional.* This number of dimensions agree exactly with that of the corresponding group continuum!

In order to be able to compare the two continua in question, the group continuum and the continuum of the Riemann surfaces, with definitive success, we had to bring the surface continuum into a surveyable representation *by actual production of the* $(n+3p-3)$ *complex resp.* $(2n+6p-6)$ *real moduli,* analogous to the representation of the group continuum by the $(2n+6p-6)$ independent invariants. Unfortunately, this representation of the continuum of Riemann surfaces has been brought to a satisfy conclusion only in the lowest cases.

In the case $p=0$, where we have to do with the one-sheeted z-plane, only the $(n-3)$ moduli originating from the signing come into consideration. We shall choose them, say, in the form of *the* $(n-3)$ *independent double ratios of the n points* $e_1, \ldots, e_n$; these double notions are complex and therefore represent $(2n-6)$ real moduli. If we put the double ratio of the four points e_1, e_2, e_3, e_k, to be denoted by (e_1, e_2, e_3, e_k), down in the form:

$$(e_1, e_2, e_3, e_k) = \frac{e_1 - e_k}{e_3 - e_k} : \frac{e_1 - e_2}{e_3 - e_2}$$

say, then, in case we use the linear transformation of the variable z still at our disposal in such a manner that $e_1 = 0$, $e_2 = 1$, $e_3 = \infty$ occurs, then for the fourth point e_k, the equation:

$$e_k = (e_1, e_2, e_3, e_k)$$

will occur. Accordingly, if we choose, as $(n-3)$ moduli, the double ratios, to be denoted for short by $\lambda_1, \lambda_2, \ldots, \lambda_{n-3}$:

$$\lambda_1 = (e_1, e_2, e_3, e_4), \lambda_2 = (e_1, e_2, e_3, e_5), \ldots, \lambda_{n-3} = (e_1, e_2, e_3, e_n),$$

then, as the values $e_4, e_5, \ldots, e_n$, these will be *unrestrictedly variable complex quantities, independent of one another.* The representation thereby obtained of the continuum of
[395] the signed Riemann surfaces of the genus $p=0$ is only then not single-valued, if there are limiting cases with coordinates of the points e_1, e_2, e_3; these cases would therefore still call for special means of representation.

For the genus $p = 1$, we choose the representing surfaces two-sheeted with four branchpoints. The one modulus of the surface not yet singled we are then able to *choose* either in *algebraic* form, say as *the double ratio* λ *of the four branchpoints resp. as the associated absolute invariant* J (cf. "M" I on p. 5 ff.) or too, in *transcendental* form, as the *period quotient* ω. In the latter case, which we favor, in order to obtain and single-valuedly represent the entire surface continuum, we have to restrict the "modulus" ω to the "*fundamental domain of the modular group*", known here from "M" I on p.

209 ff.. In order, further, to fix the moduli originating from the signing, we so choose the integral u of the first kind belonging to the surface, that a first point e_1 comes to lie at $u = 0$. We so determine a period parallelogram of the vertices $u = 0, \omega_1, \omega_2, \omega_1 + \omega_2$, that the quotient $\frac{\omega_1}{\omega_2}$ is equal to the "modulus" ω. Let the $(n-1)$ additional points $e_2, \ldots, e_n$ thereby carry themselves over into the places:

$$u_1 = p_1\omega_1 + q_1\omega_2, u_2 = p_2\omega_1 + q_2\omega_2, \ldots, u_{n-1} = p_{n-1}\omega_1 + q_{n-1}\omega_2.$$

as $(2n-2)$ *still lacking real moduli, we are then able to utilize the numbers* p_i, q_i *coming up here, which are all variable independently of one another in the intervals*:

$$0 \leq p_i < 1, \quad 0 \leq q_i < 1.$$

An algebraic model of the genus $p = 2$ is always hyperelliptic, so that the representing Riemann surface lets itself be arranged as a two-sheeted surface over the z-plane with six branchpoints situated, say, at $z = v_1, v_2, \ldots, v_6$. *The continuum of the unsigned surfaces is thus mapped single-valuedly onto the continuum of the surfaces with* $p = 0$ *signed with* 6 *points and therefore representable by three independent (complex) double ratios.* If a signing with n points $e_1, e_2, \ldots, e_n$ also supervenes, we shall add n further moduli, in the form of the double ratios (v_1, v_2, v_3, e_k) say, and in order to state in what sheet e_k is situated, we also have to assign, to the individual one of these double ratios, a determinate one of the two values:

$$\pm\sqrt{(e_k - v_1)(e_k - v_2)\cdots(e_k - v_6)}.$$

We describe the total continuum single-valuedly and completely, by not only carrying the three double ratios $(v_1, v_2, v_3, v_4), (v_1, v_2, v_3, v_5), (v_1, v_2, v_3, v_6)$, *but also the n further* [396]
ones (v_1, v_2, v_3, e_k), *and indeed the latter for each of the two adjoint roots, though all complex values independently of one another.* This representation breaks down and is to be completed only for those limiting cases, for which coincidences of the three branchpoints v_1, v_2, v_3 enter.

For the higher cases, with $p \geq 3$, the approaches developed so far for the actual production of the moduli appear only slightly workable. One can produce these either in an *algebraic* way, namely, *as absolute invariants of the "normal curve of the function* φ*" (cf. "M" I on p. 569) in the sense of the theory of linear invariants*, or in a *transcendental* way, by the *periods of the integrals of the first kind.* To these there would then still be added the moduli originating by the signing, in a representation form, which is suited to give n arbitrary movable points on the Riemann surface[80].

1.23 Report on the continua of the symmetric Riemann surfaces of the genus *p*.

If the projective polygon has $\mu > 0$ hyperbolic vertices, then we will (as already on p. 28) denote its character by (p', n'), in order to reserve the notations p, n for the corresponding Riemann surfaces. Among the fixed polygon vertices are then contained $\nu = n' - \mu$

[80] Further discussion of proofs concerning the moduli of the unsigned surfaces of the genus $p = 3$ find themselves in the paper of Klein "*Zur Theorie der Abel'schen Functionen*", Abschn. II, Math. Ann. Vol. 36 (1889) on p. 45.

elliptic and parabolic ones, so that the signature of the polygon is to be denoted by $(p', n'; l_1, l_2, \ldots, l_v)$.

The associated Riemann surface (we again imagine one chosen among all coming into consideration as a representative) is then, as we already determined before (p.391), *symmetric*, and exhibits μ transition or symmetry lines; its genus p calculates itself, according to (2) on p. 28, as:

$$p = 2p' + \mu - 1, \tag{1}$$

and it is signed with:

$$n = 2(n' - \mu) = 2v \tag{2}$$

pairwise symmetrically situated points e, which correspond to the elliptic and parabolic vertices of the polygon.

From (1) yields itself:

$$\mu = p + 1 - 2p'$$

[397] Accordingly, if for given p, we collect all symmetric surfaces corresponding to our polygons, then we have to substitute for p', one after the other, all whole numbers of the interval:

$$0 \leq p' < \frac{p+1}{2},$$

which gives, for the number of the symmetry lines, the values $\mu = p+1, p-1, p-3, \ldots$ down to 2 or 1.

Klein, in his pamphlet on Riemann, had now already taken up in 1882 the investigation of the symmetric surfaces, and thereby, above all, distinguished two essentially different cases, according as the surface, cut up along the μ symmetry lines, is still connected or falls into two separated symmetric halves. The surfaces of the latter kind Klein designates as "*orthosymmetric*"; we only have to do here with such surfaces, since our polygons of the ζ-plane (cf. figure 9 on p. 27) are cut into two separate pieces by the real ζ-axis. For the orthosymmetric surfaces of the genus p, however, Klein finds, that such with $(p+1)$ symmetry lines, such with $(p-1), (p-3)$, etc. come up, but no other ones. In each case, therefore, our polygons furnish examples for all kinds of orthosymmetric Riemann surfaces of the genus p.

On the suggestion of Klein, G. Weichold has subjected the symmetric surfaces to a penetrating investigation in his Leipzig dissertation of 1883 "*Über symmetrische Riemann'sche Flächen und die Periodicitätsmoduln der zugehörigen Abel'schen Normal integrale erser Gattung*"[81]. Furthermore, Klein himself has returned to the theory of our surfaces in very detailed fashion in the third part of his lecture "*Riemann'sche Flächen*"[82] from the summer semester of 1892.

If we gather together into one kind the orthosymmetric Riemann surfaces of the genus p with the same number μ of symmetry lines, then these will form a manifold, which is contained in the total continuum of the surfaces of genus p. Namely, the theorem subsists: *the orthosymmetric Riemann surfaces of the individual kind, given by*

[81] Cf. Zeitschr. f. Math. u. Phys. Vol. 28, p. 321.

[82] Appearing Handwritten and published on commission by B.G. Teubner, Leipzig. See also the paper "*Über Realitätsverhältnisse bei der einem beliebigen Geschlechte zugehörigen Normalcurve der φ*" Math. Ann. Vol. 42 (1892).

p and μ, likewise form, by themselves, a single continuum. This theorem is first stated in the dissertation of Weichold's just named, as well as proved in the lecture of Klein, similarly just cited. The proof is carried out on the basis of the general theorems on surfaces of Möbius and Jordan in the manner already indicated in the previous paragraph [398] for the general surfaces of the genus p[83]. But it remains still to be described, that an investigation, running parallel to the Lüroth-Clebsch developments, of the symmetric surfaces would be carried out directly for their form as actual multiple-sheeted Riemann surfaces over a plane.

The number of dimensions of the continuum of orthosymmetric surfaces, or, as we shall be able to say again, the number of the moduli of the individual surface, Klein determines in his cited lecture of the summer semester of 1892 on p. 136 ff. on the basis of the same principles as for the unsymmetric surfaces. He obtains as result: *the individual orthosymmetric surface of the genus p has $(3p-3+\sigma)$ real moduli, whereby the number σ gives how multiple infinitely often the surface is transformable into itself with retention of its symmetry lines.* Hereby, the value domain of a real variable counts as "simply infinite". *As is known, one then has $\sigma = 3$ for $p = 0$, $\sigma = 1$ for $p = 1$, and $\sigma = 0$ for $p > 1$.*

For $p = 0$, we choose as a representing surface the schlicht z-plane and make the real z-axis into the symmetry line. The ∞^3 transforms of the surface into itself are then the linear transformations of z with real coefficients. For $p = 1$, the symmetric models are investigated exhaustively in I on p. 224 ff. We find three kinds, which were represented l.c. by the figures 60, 62 and 63. But only the kind furnished by figure 62, with rectangular period parallelogram, gives an orthosymmetric surface, and indeed, one such with two symmetry lines, as it must be. We are able to so choose the integral u of the first kind, that of the two periods ω_1, ω_2, the first is purely imaginary and the second real, and that, besides this, one of the symmetry lines carries itself over onto the real u-axis. The ∞^1 transformations of the surface into itself, for which, at the same time, the symmetry lines remain maintained, are then, as is known, representable in the form $u' = \pm u + a$, where a is a real parameter.

These last statements become of importance, if we now undertake the signing of our individual surface with any $n = 2\nu$ pairwise symmetric points e. *Self-evidently, the orthosymmetric surfaces of the genus p with μ symmetric lines, signed in such a form with n points, will also form a single continuum.* In order, at the same time, to determine its number of dimensions, in the case of the genus $p = 0$, we arrange the $\frac{1}{2}n$ points e of the positive z-halfplane so that a first point e comes to lie at $z = i$, a second [399] one, say, on the imaginary z-axis above $z = i$.[84] This second point then furnishes *one* real modulus, while the remaining $(\frac{1}{2}n-2)$ points, which remain arbitrarily movable, furnish $(n-4)$ more additive real moduli. The total number of these is thus $(n-3)$. In the case $p = 1$, we are able to choose a first point e of one half of the surface, that its

[83] See the relevant footnote on p. 393.

[84] For $p = 0$, $p' = 0$, $\mu = 1$ holds, so that the number n' must be ≥ 3. From (2) on p. 396 thus follows:

$$\frac{1}{2}n = n' - 1 \geq 2.$$

image in the u-plane falls ont the imaginary u-axis[85]. This point furnishes only a real modulus, while the $(\frac{1}{2}n-1)$ remaining, freely movable points e furnish $(n-2)$ more real moduli. The total number of the moduli is consequently $(n-1)$. For $p>1$, we are not able to control any of the $1/2n$ points e of one half of the surfaces; rather, they all remain freely movable, and therefore furnish n real moduli. By conceiving all cases together, we are able to say, that the signing of the surface adds $(n-\sigma)$ real moduli to the previous $(3p-3+\sigma)$, whereby the number σ has the meaning noted above. By addition follows: *the individual signed orthosymmetric surface has* $(n+3p-3)$ *real moduli, or the continuum of the orthosymmetric Riemann surfaces of the genus p with μ symmetry lines, signed with n pairwise symmetric points, is* $(n+3p-3)$*-dimensional.*

From the relations (1) and (2) follows:

$$n+3p-3=2n'+\mu+6p'-6=3n'-\nu+6p'-6.$$

the continuum of the groups of the signature $(p',n';l_1,l_2,\ldots,l_\nu)$ was likewise $(3n'-\nu+6p'-6)$-dimensional. The corresponding continuum of the orthosymmetric Riemann surfaces therefore shows the same number of dimensions as the group continuum!

In order to be able to compare with success the two continua in question, the continuum of the principal-circle groups of the signature $(p',n';l_1,l_2,\ldots,l_\nu)$ and the continuum of the orthosymmetric Riemann surfaces of the genus p with μ symmetry lines, signed with n pairwise symmetric points e, we must bring the surface continuum into a surveyable representation by *actual production of the* $(n+3p-3)$ *real moduli*, analo-
[400] gous to the representation of the group continuum by the $(3n'-\nu+6p'-6)$ independent invariants. However, also here again, the representation of the surface continua is only brought to a satisfactory conclusion in the lowest cases.

In the case $p=0$, let the points $\overline{e_i},\overline{e_k}$ be symmetric to the points e_i,e_k of the positive z-halfplane. The double ratio λ of these four points, to be called (e_i,e_k) say, we may then put down as a real quantity belonging to the interval $0\le\lambda\le 1$.[86] *As* $(n-3)$ *moduli, we are then able to put down the* $(n-3)$ *double ratios.*

$$(e_1,e_2),(e_1,e_3),(e_2,e_3),(e_1,e_4),(e_2,e_4),\ldots,(e_1,e_{\frac{1}{2}n}),(e_2,e_{\frac{1}{2}n}),$$

which are then all real and lie between 0 *and* 1, *the limits included.* But the relations are here already more complicated, insofar as between each three double ratios (e_1,e_2), (e_1,e_i) (e_2,e_i) there subsists an inequality (intersection condition of two circles), and that upon the latter being fulfilled, after fixing the points e_1,e_2, the third point e_i is first ambiguously determined from the double ratios (intersection of two circles).

In the case $p=1$, we have, as a *first modulus, the absolute value* $|\omega|$ *of the purely imaginary period quotient* ω. Of the periods themselves, we wanted to choose ω_2 real and therefore ω_1 purely imaginary, but one half of the surface is to be mapped onto the half period parallelogram with the vertices $u=0,\frac{1}{2}\omega_1,\omega_2,\frac{1}{2}\omega_1+\omega_2$. We are then able to

[85] For $p=1$, $p'=0$, $u=2$ immediately follows from (1) on p. 396, so that, similarly, $n'\ge 3$ holds. From (2) on p. 396 now follows:

$$\frac{1}{2}n=n'-2\ge 1.$$

[86] Compare e.g. the developments concerning the double ratio of four points in "M" I on p. 5 ff.

displace a first point e_1 to $u_1 = \frac{1}{2}p_1\omega_1$, while the remaining ones, say, find their images at the places:

$$u_2 = \frac{1}{2}p_2\omega_1 + q_2\omega_2, u_3 = \frac{1}{2}p_3\omega_1 + q_3\omega_2, \ldots, u_{\frac{1}{2}n} = \frac{1}{2}p_{\frac{1}{2}n}\omega_1 + q_{\frac{1}{2}n}\omega_2.$$

The $(n-1)$ moduli originating from the signing are then the $(n-1)$ numbers p_1, p_2, q_2, p_3, q_3, $\cdots$, $p_{\frac{1}{2}n}$, $q_{\frac{1}{2}n}$, which are all variable independently of one another between the limits 0 and 1, the latter included.

Of the higher cases, those for $p = 2$, $n = 0$ still appear easily accessible, therefore the *unsigned hyperelliptic surfaces of the genus $p = 2$*. As a representative, we choose, in each case, a two-sheeted surface over the z-plane with six branchpoints $v_1, v_2, \ldots, v_6$, which are either real or complex conjugate in pairs. Accordingly, there are four cases initially imaginable, according as all six branchpoints v, or four, or two or finally, none are real. If, $\overline{z}$ is the complex value conjugate to z, then the transformation $z' = \overline{z}$ can always be carried out on the surface in two ways, according as we let an interchange of the sheets enter or not: *our surface therefore admits two symmetric transformations into itself, to be called S_1, S_2 for the moment, say.* If one sketches the surfaces for oneself, then one arrives at the following restricting result: *orthosymmetry is furnished, first of all, by the case of six real branchpoints, and indeed, for each of the two transformations S_1, S_2, whereby, both times, three transition lines come up, secondly, by the case of three pairs of complex conjugate branchpoints v with one symmetry line, the latter however, only for one of the two transformations S; the other transformation, as well as all remaining cases, furnish diasymmetry*[87]. [401]

The case $p = 3$ has also been much studied, where we have to distinguish the two cases $\mu = 4$ and $\mu = 2$. For obtaining the six real moduli in an algebraic way, one has to attach to *the singularity free plane curves of the fourth degree with four resp two real traces,* which were first[88], subjected to a detailed investigations and classification by Zeuthen[89]. However, it appears not to be easy, from the investigations present so far, to also arrive at a representation of the six moduli only for the unsigned surfaces, which permit the two surface continua coming into consideration here to be clearly surveyed.

1.24 Continuity of the mapping between the continuum of groups and the continuum of Riemann surfaces

[402]

The $(3n - \nu + 6p - 6)$-dimensional continuum of the groups Γ of a given signature $(p, n; l_1, l_2, \ldots, l_\nu)$ we had represented above by a domain B, which we will now denote more precisely by B_g. Let B_f correspondingly be the name of the equal -dimensional continuum of the Riemann surfaces F standing across from this. The individual group Γ or the individual surface F we then also designate as a "point" in B_g resp. B_f.

87 Compare, for this, the discussions of Klein in the lecture "*Riemann'schen Flächen*" cited on p. 397, on p. 124 ff. of the second part.

88 "*Sur les différentes formes des courbes du quotriéme ordre*" Math. Ann. Vol. 7, p. 410 (1874).

89 See also the relevant discussions of Klein in the oft-mentioned lecture, part I on p. 202 and part II on p. 156 ff.

The existence theorem of automorphic functions (p. 18 ff.) now informs us, in any case, that to each individual point in B_g uniquely corresponds a determinate point in B_f. We ask, *whether for continuous alteration of the point in* B_g, *perhaps the corresponding point in* B_f *also alters itself continuously.* We shall be able to answer this question affirmatively by a theorem, which we shall cite further on as the "*continuity theorem*".

In any case, the theorem subsists, first of all, *that we are always able to follow a continuous alteration of the invariants of the group by a similarly continuous alteration of the polygon P*. Namely, if, according to I on p. 335 ff., we calculate the generators $V_1, V_2, \ldots, V_{a_1}, V_{b_1}, \ldots$, for Γ from the (supernumerary in general) invariant system, then the coefficients of these substitutions are determinate rational functions of the invariants, which will consequently alter themselves continuously with respect to continuous alterations of the invariants j. If one now, according to I on p. 314, effects the construction of the polygon P by composition from one polygon of the character $(0, n+p)$ and p polygons of the character $(1,1)$, and choose these $(p+1)$ polygons to be composed as normal polygons of one and the same center, then in case this center remains fixed for the continuous alteration of the invariants or alters itself only continuously, then it is immediately evident, that the total polygon itself only alters itself continuously for these alterations.

We assert further: *with respect to each continuous alteration of the invariants, and therefore of the polygon, the functions of the polygon, the functions of the group* Γ *likewise go over continuously into the functions of the altered group.*

The theorem, standing at the center of the present study, has been proved in two ways. The first proof, originating from Poincaré, to which we return immediately in detail, was already developed above (p. 167 ff.). It attaches itself to the representation of
[403] the automorphic functions by the Poincaré series and rests on the proof of the uniform convergence of these series with respect to alteration of the invariants.

E. Ritter has furnished the second proof of our theorem in an investigation carried out with great thoroughness: "*Die Stetigkeit der automorphen Functionen bei stetiger Abänderung des Fundamentalbereiches*"[90]. Ritter attaches to the production of automorphic functions by means of the elementary potentials and integrals of the second kind of the polygon P. He introduces a numerical measure for the alteration of the polygon P, and pushes through and obtains an associated upper bound for the corresponding alteration of the automorphic functions, from which then emerges, that the alteration of the functions actually coverage uniformly to 0 with the measure of the alteration of the polygon P.

Ritter considers degenerations only in the case of the symmetric polygons P, and indeed, in §11 of the second paper just named. A degeneration of such a kind, as Ritter allows, is present, if a hyperbolic invariant becomes parabolic, i.e., assumes the boundary value 2. On the other hand, for our symmetric or unsymmetric polygons, still other limiting cases also come up, e.g., the one, for which a transversal invariant becomes parabolic. These cases are consequently not yet disposed of by Ritter.

[90] Tneil I "*Symmetrische Fundamentalbereiche*", Math. Ann. Vol. 45, p. 473(1894); Theil II "*Allgemeine Fundamentalbereiche*", Math. Ann. Vol. 46, p. 200 (1894).

The circumstantiality of the Ritter investigations, vis-a-vis the far more brief procedure of Poincaré, pays for itself moreover, in the fact that the Ritter considerations are independent of whether the polygon P is the discontinuity domain of a group Γ or upon reproduction leads to multiple covering of the ζ-plane resp. taken by itself already multiple covers part of the ζ-lane (cf. p. 51), while the Poincaré proof only has validity, to the extent that the use of the series named after him is permissible, i.e. only in the domain of the "single-valued" automorphic functions.

In this domain, alone coming into consideration here, the Poincaré proof already represented above (p. 167 ff.), as remarked, accomplished everything required. It rests on the above and uniform convergence of the Poincaré series of the dimensions $d < -2$, in case ζ is restricted to the domain denoted l.c. by B_0 and Γ to the "full neighborhood" of the first chosen group $\Gamma^{(0)}$ in the continuum B_g. But this convergence proceeds from the circumstance, that the series rearranged into the form (4) on p. 171 [404]
converges better than a geometrical series put down in (14) on p. 173 on the right side and itself convergent[91].

For the convergence proof in question, we have initially excluded above (cf. p. 168) the limiting cases of the continuum B_g, i.e., those groups Γ, for which a lowering of the character (p, n) of the polygon P, or a degeneration of this, enters. However, we only did this, since the character of these limiting cases was not yet sufficiently clarified at that time. For a degeneration of the polygon into a line (level curve of a hyperbolic substitution, which changes into the identity), the approach of the Poincaré series self-evidently breaks down. But we could carry out each passage to the limit (and indeed, in two ways in general) in such a manner, that as limiting case, *a non-degenerating polygon P of lower character* entered. For such a boundary polygon, however, the Poincaré convergence study represented on p. 167 ff. remains completely usable. In fact, for this it's only requisite, that the domain B_0 on p. 170 bounded about the point ζ_0 also remains in the interior of the polygon P upon alteration of the invariants, which is indeed always attainable for finitely extended P. All further steps of the development on p. 170 ff., however, then remain undisturbed. Thus our theorem of the continuous transition of the automorphic functions of a group into the functions of immediately neighboring groups *also remains in force under inclusion of the limiting cases in the continuum* B_g, *whose finitely extended polygons belong to lower characters*[92].

We now only need to remember, that two functions φ_1, φ_2 of an individual group are bound by an algebraic relation $G(\varphi_1, \varphi_2) = 0$, which, for its part, is suited to define

[91] Furthermore, it makes no difference, that the domain B_0, to which the convergence study refers itself, lies entirely in the interior of the principal circle (or entirely outside this), and therefore the convergence is not directly proved, say, for the points of P belonging to the principal circle itself. Namely, if, inside the *finitely* extended domain B_0, the series of a first group Γ, and therefore the automorphic functions to be represented by them, upon continuous alteration of the invariants, themselves go over *continuously* into those of the neighboring group, then the coefficients of each algebraic relation subsisting between two functions φ_1, φ_2 and with this, also the moduli of the Riemann surface, will thereby only be able to alter themselves, continuously.

[92] Poincaré himself has returned to the extension of the continuity studies to the limiting cases in the course of his paper "*Sur les groupes des équations linéaires*", Act. Math., Vol. 4, p. 236 ff. (1884).

[405] a representing Riemann surface F of the algebraic model. Upon continuous alteration of the group invariants, the functions φ, and with this, the coefficients of that relation $G(\varphi_1, \varphi_2) = 0$ alter themselves continuously, and therefore, the same thing holds for the representing Riemann surface. There thus arises, as an answer to the question of the "*continuity theorem*" raised at the beginning of the paragraph: *for any continuous movements of the points in the domain* B_g*, the corresponding point in the domain* B_f *likewise moves itself continuously without exception!.*

1.25 Single-valuedness of the mapping between the continuum of groups and the continuum of Riemann surfaces

To the individual point of the group continuum B_g uniquely corresponded a determinate point in the continuum B_f of the Riemann surfaces. We now ask, whether hereby two different points of B_g are able to furnish one and the same point of B_f, whether therefore two essentially different groups of the signature $(p, n; l_1, l_2, \ldots, l_\nu)$ are able to furnish one and the same signed Riemann surface F. The question is to be answered no, and therefore there subsists the following "*single-valuedness theorem*": *the mapping between the group continuum* B_g *and the continuum* B_f *of the surfaces is one such that to the individual point of* B_f*, there can never correspond more than a single point of* B_g.

This single-valuedness theorem is set up by Klein in the course of the paper "*Neue Beiträge zur Riemannian'schen Functiontheorie*"[93], not only for principal-circle group, but also such groups with infinitely many principal circles, to which one can attain the principal-circle groups by means of the composition process. In the case of the principal-circle groups, alone coming into consideration here, the single-valuedness theorem was first announced by Poincaré in October 17, 1881[94] and proved later on, as a "*lemma fundamental*" inside the paper "*Sur les groupes des équations linéaires*"[95]. The following representation reproduces this proof of Poincaré's.

For the first group Γ we choose a canonical polygon of $(2n+4p)$ sides in the projective plane, which let furnish the polygon P in the positive ζ-halfplane. If $\nu = n$, then P furnishes the entire surface F, but if $\nu < n$, and therefore we have the symmetric case with $\mu = n - \nu$ symmetry lines, then P furnishes only one of the two symmetric
[406] halves of F. The surface F resp. its half, corresponding to the boundary of P, carries a canonical cut system. We carry over the surface (surface half) thus cut up into a new positive halfplane of the variable ζ' onto a polygon P' of the second group Γ'. Then the two polygons P and P' of the halfplanes of ζ and ζ', by mediation of F, correspond to one another continuously, single-valuedly and *conformally*, whereby, in particular, the sides and vertices of P and P' correspond to one another single valuedly.

To the substitutions $1, V_1, V_2, \ldots$ of Γ let correspond the substitutions $1, V_1', V_2', \ldots$ of the isomorphic group Γ'. Let the substitution V_k carry P over into P_k and correspondingly, V_k' the polygon P' into P_k'. The two polygons P_k and P_k' are then similarly mapped

[93] Math. Ann. Vol. 21, p. 209 ff. (1882).
[94] Comptes Rendus, Vol. 93, p. 582.
[95] Act. math., Vol. 4, p. 231 (October 1883).

onto one another single-valuedly, continuously and conformally. As one sees, there arises from this a single-valued, continuous and conformal mapping of the two polygon nets N and N' in the halfplanes of ζ and ζ' to be produced from P and P', and indeed, the conformality of this relation does also not cease at the elliptic vertices lying in the interior of the halfplanes, since indeed, at two such corresponding polygon vertices, we always have the same angle $\frac{2\pi}{l}$.

Concerning the mapping of the two real axes of ζ and ζ' onto one another, we will initially draw no conclusion. On the other hand, our statements on the single-valued, continuous and conformal mapping of the two halfplanes onto one another concern all those portions of these, which have finite distance from the axes. In fact, each point $\zeta = \xi + i\eta$ and $\zeta' = \xi' + i\eta'$ with $\eta > 0$ resp. $\eta' > 0$, after running through finitely many polygons, is indeed drawn into the interior of the net N resp. N'.

The previous statements on the mapping of the two halfplanes of ζ and ζ' onto one another now already permit the conclusion to the drawn, *that the mapping in question necessarily is representable by a linear equation*:

$$\zeta' = \frac{a\zeta + b}{c\zeta + d}. \tag{1}$$

In order to show this, we understand by ζ_0 and ζ_0' the two values of the interior of the two halfplanes corresponding to one another. By understanding by $\overline{\zeta_0}$ and $\overline{\zeta_0}'$ are two conjugate complex values to ζ_0 and ζ_0', we introduce two new variables Z and Z' by:

$$Z = \frac{\zeta - \zeta_0}{\zeta - \overline{\zeta_0}}, \quad Z' = \frac{\zeta' - \zeta_0'}{\zeta' - \overline{\zeta_0}'}. \tag{2}$$

in the planes of these variables Z and Z', the two halfplanes considered so far map themselves onto two circular surfaces of the radii 1 about the origins $Z = 0, Z' = 0$. [407]
The mapping of these two circular surfaces onto one another is then again, without exception, single-valued, continuous and conformal in the interior of the two surfaces, whereby the two origins $Z = 0$, $Z' = 0$ correspond to one another. Further, however, the mapping is one such that, for $\lim|Z| = 1$ (resp. $\lim|Z'| = 1$), also $\lim|Z'| = 1$ (resp. $\lim|Z| = 1$) always takes place.

If, on the strength of this, we consider the quotient $\frac{Z'}{Z}$ in the interior of the unit circle of the Z-plane, then this represents here a finite, continuous and single-valued analytic function everywhere different from 0 without singular points. According to this, the logarithm of the quotient $\log(\frac{Z'}{Z})$, in the circle in question, furnishes, an everywhere finite, continuous and single-valued analytic function.

By writing $Z = X + iY$, with separation of the real and imaginary components, let the function last obtained furnish, for the same separation:

$$\log\left(\frac{Z'}{Z}\right) = U(X, Y) + iV(X, Y). \tag{3}$$

Then $U(X, Y)$ and $V(X, Y)$ are two functions of X and Y, conjugate to one another, which are everywhere single-valued, continuous and harmonic in the "interior" of the limit circle of the Z-plane.

Now we have, for the first of these two functions, the representation:

$$\log \frac{|Z'|}{|Z|} = U(X, Y). \tag{4}$$

If now, from the interior of the unit circle, we approach the periphery of this along any path, then $|Z'|$ also thereby uniformly approaches the value 1. Consequently, in the interior of the unit circle, one can assume a circle lying sufficiently close to this, and concentric with it, in such a manner, that along the periphery of this $|Z|$ and $|Z'|$ deviate "arbitrarily" little from 1, and therefore $U(X, Y)$ "arbitrarily" little from 0. On account of the harmonic behavior of $U(X, Y)$, it emerges from this, that $U(X, Y)$ can assume no value different from zero in the interior of that circle, and with this, anywhere in the interior of the unit circle at all.

But if $U(X, Y)$ is perceived as everywhere identical with 0 in the interior of the unit circle, then the conjugate function $V(X, Y)$ has an everywhere constant value there, which let be called ϑ. Equation (3) then furnishes, on the strength of this:

$$Z' = e^{i\vartheta} Z, \tag{5}$$

[408] and therefore, if we return to ζ and ζ' by means of (2), a linear relation (1), in fact, finds itself between ζ and ζ'.

With this, the group Γ goes over into Γ' by means of this substitution. The two groups are consequently not different from one another from the invariant standpoint, and furnish the same point in the domain B_g. The single-valuedness theorem is thus proved.

1.26 Generalities on the continuity proof of the fundamental theorem in the domain of the principal-circle groups

The relation of the group continuum B_g coming up for the signature $(p, n; l_1, l_2, \ldots, l_\nu)$ to the corresponding continuum B_f of the signed Riemann surfaces, as it stands now, is such a one, that to each point of B_g single-valuedly corresponds a point of B_f, which alters continuously, without exception, along with its original in B_g, as well as, conversely, to the individual point in B_f, at most one point in B_f can correspond. If we are able to show, that to the individual point in B_f here also always corresponds a point in B_g, then the two continua, B_g and B_f, are mapped onto one another mutually single-valuedly, and the two fundamental problems I and II on p. 45 ff., concerning the principal-circle groups, are answered by two corresponding *fundamental theorems*, the "*boundary circle theorem*" and the "*principal circle theorem*" in this manner, *that in fact, on each signed Riemann surface a polymorphic function* $\zeta = f(z)$ *exists, which accomplishes the required mapping os the surface onto a boundary-circle resp. principal-circle polygon.*

The proof of the theorem, that *to each* point of B_f a point of B_g must correspond, Klein and Poincaré have now attempted to present a continuity consideration (cf. p. 285) independently of one another and approximately simultaneously. The two detailed representations, already cited often, are named once more below[96], whereby it is to be remarked, that the paper of Poincaré only had to be dated a year later than that of Klein's, since Poincaré attached to the "group continua" and not (as Klein) to the "polygon continua", and thus had to first beforehand, carry through his extensive developments on polygon reduction (cf. p. 305).

The representation of Klein refers itself not only to the boundary-circle and principal-circle theorem, but, at the same time to the more general theorem, mentioned on p. 47, which comes up for groups, which come up by composition ("shuffling into one another") of arbitrary principal-circle groups. [409]

For the mapping of the continua B_g and B_f onto one another, Klein makes the assumption, indeed not proved, but appearing doubtless to him, that this mapping is an "*analytic*" one. The moduli of the Riemann surface, chosen as real quantities, are therefore to be $(2n+6p-6)$ real analytic functions of the group invariants.

From this, there next yields itself, for the relation between B_g and B_f in the neighborhood of individual places: if, at a particularly considered point of B_g, the fundamental determinant of those $(2n+6p-6)$ functions becomes neither 0 nor ∞, then the neighborhood of this point maps itself precisely onto the full neighborhood of the corresponding point in B_f. But that our mapping of B_g onto B_f must be of this nature everywhere, yields itself from the single-valuedness theorem. Namely, e.g., were the full neighborhood of a point of B_g mapped onto a portion of the neighborhood of the corresponding point in B_f, then on account of the continuity theorem, then there would necessarily have to be found points in B_g there, each one of which would furnish more than one point of B_f. Visualize the lowest case: if, upon continuously mapping one plane onto another, the full neighborhood of a point of the first plane draws itself together onto a part of the neighborhood of the corresponding point of the other plane, and therefore of an angle $< 2\pi$, then the points on the sides of the angle correspond in pairs to the same point of the first plane.

Further, Klein uses a known theorem of Weierstrass on the attaining of boundary values by analytic functions. If, in the continuum B_g, we describe a path to a certain point "Γ_0", then to this path will correspond a continuous image in B_f, which, corresponding to the limit Γ_0, will approach a certain limit point "F_0". From the Weierstrass theorem, Klein concludes that this limit is also actually attained.

In the further effectuation of the continuity proof, Klein now starts out from the assumption, first of all, that in the carrying-over of B_g into B_f, certain parts of domains still remain in B_f, none of whose points correspond to points in B_g. On the strength of this, Klein studies the boundary of such a domain and derives a contradiction from its assumed existence. First of all, namely, the individual boundary point, according to the argument last given, is always attainable by the mapping of B_g onto B_f. But then, it is even able to be passed beyond, in consequence of the principle of the mapping of the neighborhood of one place of B_g onto the full neighborhood of the corresponding place in B_f. By means of this argument, which already comes into application in a known proof of the fundamental theorem of algebra, thus follows the impossibility of [410]

[96] Klein, "*Neue Beiträge zur Riemann'schen Functiontheorie*" Math. Ann. Vol. 21 (p. 211), dated Oct. 2, 1882; Poincaré, "*Sue les groupes des équations linéaries*", Act. math. Vol.4 (pp. 233 and 276), dated Oct. 20, 1883.

partial domains remaining free in B_f.[97] Rather, B_g and B_f are mapped mutually single-valuedly onto one another, whereby the fundamental theorem would be proved.

Poincaré's continuity proof works basically with the same arguments. However, Poincaré raises objections to the representation of Klein's at a point, certainly very essential, concerning the shape of the two oft-named domains B. For the purpose of a better tie-in to the foregoing paragraphs, we have attached the report on the continuity proof of Klein's on the continuum of the *groups* B_g. Actually, as was already mentioned many times, not the continuum of the groups, but the continuum of the polygons P, to be called B_p say, underlay the representation of Klein's, to which corresponds the continuum B'_f of the cut-up Riemann surfaces. But the continuum B_p is now an *open* one; for we represented this, e.g. in the boundary-circle case of the character (p,n) above (p. 303 ff.), as a $(2n+6p-6)$-dimensional cube. Toward the boundary of the cube, the discontinuity domains of the associated modular group press themselves ever more densely together, and the polygons P themselves thereby assume a more and more distorted shape[98]. Also, for the continuum B'_f of the cut-up surfaces, toward the boundary of this, corresponding distortions of the crosscut system must enter. In the presence of the difficult relations coming into consideration here, to draw the conclusion, that the cube boundary of the B_p also actually carries itself over onto the boundary just mentioned of the domain B'_f, would certainly require considerable deliberation.

To have effected the transition from the continua B_p of the polygons to that B_g of the groups, is the great and essential step forward of Poincaré's. In an objective respect, it is hereby moreover to be remarked, that Poincaré restricts his detailed inves-
[411] tigation to the *boundary-circle case*[99]. This has the not inessential simplification as a consequence, that those limiting cases, which enter by hyperbolic vertex invariants becoming infinite, remain excluded. Secondly, in a formal respect, it is to be added, that Poincaré expresses the boundary-circle theorem in somewhat different form, to be immediately described more closely.

Namely, remember, that the individual polymorphic function ζ on a Riemann surface, according to equation (1) on p. 234, satisfies the differential equation of 3rd order:

$$[\zeta]_z = 2R(s,z), \tag{1}$$

whereby we imagine the Riemann surface over the z-plane given by the algebraic relation:

$$G(s,z) = 0 \tag{2}$$

between s and z. We are also able to let the linear different equation of the 2th order for the "polymorphic forms Z of zero dimension" to be formed from ζ according to (3) on

[97] In this argument, to be sure, the boundary points of the unclosed manifolds present for Klein do not receive sufficient consideration. See the objection of Poincaré's against the representation of the continuity proof furnished by Klein, to be immediately discussed in the text.

[98] Remember the relations of the modular group coming up in the theory of the elliptic functions, as well as the distortion coming up at the boundary of the triangle net for the period parallelograms.

[99] However, for the principal-circle theorem, see the assertion of Poincaré at the close of his paper l.c. on p. 311.

p. 118 take the place of the equation (1), which, in consequence of the calculation on p. 118 ff., represents itself in the form:

$$\frac{d^2 Z}{dz^2} + R(s,z)Z = 0. \tag{3}$$

Then the $(3p-3)$ moduli of the relation (2) and the n singular places of the differential equation (3) take the place of the $(n+3p-3)$ "complex" moduli of the signed Riemann surface. But the boundary-circle theorem amounts to this, that for arbitrarily given relation (2) of the genus p and n singular points arbitrarily fixed on the associated Riemann surface, after prescription of a "signature" $(p,n;l_1,\ldots,l_\nu)$ belonging to these singular points (as we are able to say briefly, using our language), there always exists one and only one differential equation (3), whose integral quotient ζ maps the suitably cut-up surface onto a boundary-circle polygon of that signature.

In the individual continuum B_g of boundary-circle groups, as was already mentioned many times above (e.g. p. 305). One intends to perceive a closed manifold. For the corresponding continuum B_f of the signed Riemann surfaces, the same properly would then have to be recognizable. Poincaré expresses himself in more detail concerning this l.c. on p. 277. We are able, first of all, to arrive at the boundary of the domain B_f, by two of the n points e coinciding, secondly, by a decrease of the genus p of the Riemann surface by one unit entering by the fusing of two branchpoints[100]. [412]
In the former case, the two coinciding points e yield a single such point of parabolic character (cf. p. 363), so that n decreases itself by one unit. But if we fashion the Riemann surface according to the principles of Lüroth and Clebsch developed on p. 392 and bring two branchpoints, which join these two sheets, into fusion, then the connection of the sheets will cease at the place of fusion, however, there in the two sheets, two new points e of parabolic character come up[101], so that, in the limiting case, the signature $(p-1,n+2;l_1,l_2,\ldots,l_{n+2})$ is present. In both cases, as one easily determines, the number of the dimensions of the boundary continuum lowers itself by *two* units with respect to that of B_f. From this, on the basis of the assumption, that for an m-dimensional unclosed manifold, the boundary must necessarily be $(m-1)$-dimensional, Poincaré then draws the conclusion, that the surface continuum B_f is a closed one[102]. On this foundation, Poincaré then develops the continuity proof essentially by means of the argument already sketched at the beginning of the paragraph.

Also, the Poincaré representation of the continuity proof is not free of doubts. First, as far as the function-theoretic character of the dependency of our two manifolds B_g and B_f on one another goes, the assumption obviously underlies the Poincaré argument, that this dependency is an *analytic* one. However, this assumption is never ex-

---

[100] Poincarè' himself, moreover, effects the lowering of the genus by introducing a "new double point" for the "curve" $G(s,z)=0$.

[101] Compare above on p. 333, the detailed representation of the relations for the polygons P of the genus $p=1$, which (in consequence of the principle of composition) possess general significance.

[102] Furthermore, Poincaré also applies the same method of argument for the group continua (cf. pp. 382 and 385).

plicitly proved or even just stated. What Poincaré develops l.c. in §§3 and 4, is not sufficient in general[103].

There is a further objection to the assumption, that the boundary of an m-dimensi-
[413] onal manifold is always $(m-1)$-dimensional. To be sure, this has been asserted on occasion by Riemann and Betti; however, the doctrine of sels shows, that the dimension concepts coming into question here already require clarification throughout[104]. In case one leaves aside the analytic character of the dependency between B_g and B_f, then unfortunately, in carrying B_g over onto B_f, one also has to base oneself on the general theorems of the doctrine of sets. It is a question thereby first and foremost, of the theorems concerning the *invariance of the number of dimensions*[105]. In the latter regard L.E.I Brouwer has recently set up three theorems[106], which signify a very essential step forward in the development of this portion of the doctrine of sets. In case the arguments of the continuity proof are to be carried to a conclusion free of objections apart from the question of the closedness of our manifolds and independently of the possibility of analytic dependency, then two more theorems would have to be added to the theorems mentioned below the text, the first one of which would correspond to the above-mentioned Weierstrass theorem, while the second would have to show, that the neighborhood of a place in B_g also actually carries itself over onto the *full* neighborhood of the corresponding place in B_f (cf. the argument sketched above on p. 409 for the two-dimensional manifolds).

If, for this state of affairs, the general effectuation of the continuity proof, in a fully satisfactory form, appear only slightly promising at this time, then we shall nevertheless have to attempt, at least in easily surveyable special cases of lower characters, to actually carry through the proof in question. In this undertaking, we will ignore the
[414] question, whether the mapping of the continua B_g and B_f onto one another is an ana-

[103] In §3 a linear differential equation of mth order with rational coefficients is considered and the proof carried out, that the invariants of the substitutions of an associated system of integrals are *integral functions* of the coefficients of the differential equation; §4 restricts itself to differential equations of the second order and shows the single-valuedness of the coefficients of this differential equation in certain invariants, more closely defined l.c., of the substitutions of a system of integrals. These developments refer themselves only to the boundary-circle case of the character $(0, n)$, so that only in this case is the analytic character of the dependency in question to be considered as proved by Poincaré.

[104] See, for this, the report of Schönflies, put out by the Deutschen Mathematiker-Vereinigung, "*Die Entwicklung der Lehre von Punktmannifaltigkeiten*", II Teil (1908), Kap. IV and V, especially kap. IV, §6.

[105] See Schönflies, l.c., in particular §6.

[106] The author thanks an announcement in a letter of Brouwer for the knowledge of these three theorem:

I. "An n-dimensional manifold can *not* contain the one-to-one continuous image of a manifold of *higher* dimension number."

II. "In an n-dimensional manifold, the one-to-one and continuous image of a manifold of the *same* dimension number fills up a certain domain *everywhere densely*."

III. "In an n-dimensional manifold, the one-to-one and continuous image of a manifold of *smaller* dimension is a *nowhere dense* point set."

Brouwer has proved these theorems in a work appearing soon in the Mathem. Annalen.

lytic one or not, and support ourselves solely on the continuity and single-valuedness theorems.

1.27 Effectuation of the continuity proof for the signature $(0,3;l_1,l_2)$

If one ignores the cyclic hyperbolic groups[107], (which one can, if one will, count as principal-circle groups), then there is only a single case with *one-dimensional* continua B_g and B_f, namely those of the signature $(0,3;l_1,l_2)$.

For the character $(0,3)$, the modular group reduces itself to the identical transformation, so that the group continuum B_g is identical with the polygon continuum. We represented the latter above (p. 290), with utilization of the form t of the hyperbolic vertex invariant, by the line of length 1, defined by means of the inequality $0 \leq t \leq 1$. The corresponding continuum B_f we represented, according to p. 398, by simple z-planes, in which the real axis, counted as a symmetry line, while the positive z-halfplane was signed with two arbitrary points e_1, e_2, the negative halfplane with the symmetric points $\overline{e_1}, \overline{e_2}$.

However, a somewhat different choice of the Riemann surface is expedient here. Each of our groups (as each group of the character $(0,3)$ generally) permits extension by reflections to a group of the second kind whose discontinuity domain is the "elementary triangle" of the signature $(0,3;l_1,l_2)$. If we also add the reflection in the real ζ-axis, then a $\overline{\Gamma}$ arises, whose discontinuity domain is the quadrangle representation of this quadrangle $\varepsilon_1,\varepsilon_2,\varepsilon_3,\varepsilon_4$, as figure 79 shows.

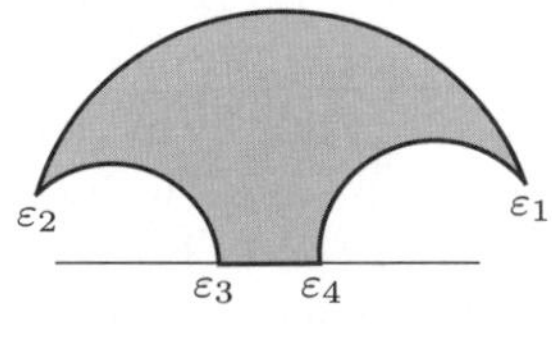

Fig. 79.

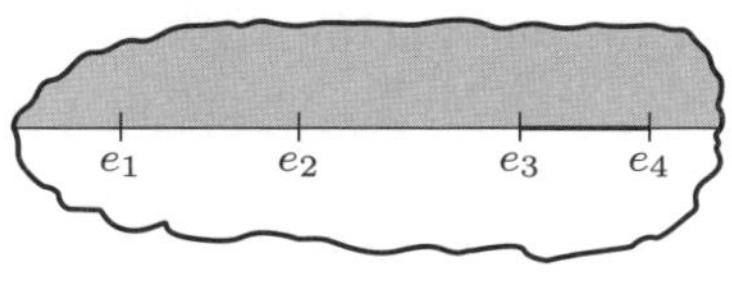

Fig. 80.

[107] In this case, the fundamental theorem asserts the mappability of each two-sheeted Riemann surface with four branchpoints onto a circular ring. The correctness of the theorem already emerges from the theory of the elliptic functions, but can also be effected by a continuity consideration in the manner of the text.

To the first of the two group extensions effected here corresponds the reflection
[415] of the z-plane considered previously in the circle through the four points $e_1, e_2, \overline{e_1}, \overline{e_2}$, but to the second extension, the reflection in the real z-axis. If, we introduce, as an automorphic function $z = \varphi(\zeta)$, one such, which maps the quadrangle of figure 79 onto the positive z-halfplane, then to the four vertices $\varepsilon_1, \varepsilon_2, \varepsilon_3, \varepsilon_4$ let correspond the points of the real z-axis (cf. figure 80), now to be denoted by e_1, e_2, e_3, e_4. The z-plane first considered then rests over the new one as a two-sheeted surface with the branchpoints e_3, e_4, and it is signed at the places e_1, e_2 in both sheets. By rather choosing this two-sheeted surface as a representative of the linear continuum B_f, we define, tying into the points e just defined, as the associated modulus, the double-ratio:

$$\lambda = \frac{(e_1 - e_2)(e_3 - e_4)}{(e_1 - e_3)(e_2 - e_4)}. \tag{1}$$

This modulus is real, and, in order to describe the entire continuum B_f, has to run through the interval $0 \le \lambda \le 1$.

Let the passage to $t = 0$ and $t = 1$ give the limiting values:

$$\lim_{t=0} \lambda = \lambda_0, \qquad \lim_{t=1} \lambda = \lambda_1.$$

The line B_g then carries itself over continuously and reversibly continuously (in consequence of our two oft-mentioned theorems) onto the segment of B_f delimited by λ_0 and λ_1. The fact that the limiting values λ_0, λ_1 belong to the image, as well as these values themselves, are, however, first to be determined more closely.

The limiting case $t = 0$ we could attain in three different ways, first by the fusing of ε_3 and ε_4 into the parabolic point, with decomposition of the polygon into two boundary-circle polygons, secondly, by the fusing of ε_1 and ε_2 into one parabolic cusp without decomposition of the polygon, thirdly by degeneration of the polygon into a line (simultaneous coincidence of ε_3 and ε_4, as well as of ε_1 and ε_2).

If we ignore the last case (on account of the breakdown of the Poincaré series), then in consequence of the continuity theorem, the function $z = \varphi(\zeta)$ will go over uniformly continuously into that of the limiting case, in each of the other two cases. In the one case, as the branchpoints e_3, e_4 more together, we get a decomposition of the surface into two separated sheets, in the other case, the two signed places e_1, e_2 come into coincidence in each sheet. Also, the equivalence of every two such surfaces, which are near these limiting cases (see the relevant investigations for the polygon on p. 361), is
[416] immediately recognizable. If, namely, e_3 has moved very close to e_4, then by means of a transformation:

$$\frac{z' - e_1}{z' - e_4} = \mu \frac{z - e_1}{z - e_4}$$

with sufficiently small μ, one can go over to a z'-plane, in which e_3 is again distant from e_4 while e_2 has now come close to the point e_1.

Each of the two limiting cases ($e_2 = e_1$ as well as $e_3 = e_4$) now furnishes, in consequence of (1), the value λ_0. *Moreover, as long as we are already able to completely determine, as here, the associated moduli of the Riemann surface by means of one of the possible limiting shapes of the polygon, we may restrict our consideration to this one limiting shape.*

Following this principle in the other limiting case $t=1$, we will prefer, among the possible (and consistently non-degenerating) boundary polygons, that entering for coincidence of ε_2 with ε_3; there is then a lowering of the group to a cyclic group of elliptic or parabolic type (according as l_1 is finite or not). For the limiting function $z=\varphi(\zeta)$, we have $e_2=e_3$, so that the boundary value is $\lambda_1=1$ and is similarly attained.

According to this, the two continua B_g and B_f are *mutually single-valuedly* mapped onto one another: *and therefore the fundamental theorem is proved for the signature* $(0,3;l_1,l_2)$.

1.28 Effectuation of the continuity proof for the signature $(0,3;l_1)$

There are, in all, three cases with *two-dimensional* continua B_g and B_f, corresponding to the signatures $(0,3;l_1)$, $(1,1;l_1)$ and $(0,4;l_1,l_2,l_3,l_4)$, of which the first leads to a principal-circle theorem, and each of the two others to a boundary-circle theorem.

Also, for the signature $(0,3;l_1)$, which we consider first, the group continuum B_g is identical with the polygon continuum and, as this, is represented by the surface of a square, by means of the two quantities t_1,t_2, corresponding to the hyperbolic vertex invariants j_2,j_3, which are subjected to the conditions:

$$0\le t_1\le 1,\qquad 0\le t_2\le 1. \tag{1}$$

For the Riemann surfaces, instead of the transcendental moduli $|\omega|$ and $\mu_1=\frac{1}{2}p_1\omega_1$ provided on p. 400, we are better able, in the special case here, to use *algebraic* moduli. Namely, the individual Riemannian surface belonging here admits *two* different sym- [417]
metric transformation into itself, each with two symmetry lines, as one wants to make this clear to oneself, in the plane of the integral u of the first kind, in the corresponding subdivided period rectangle of figure 81. Now introduce the function z, so that it carries the partial rectangle with the vertices $u=0$, $\frac{\omega_2}{2}$, $\frac{\omega_1+\omega_2}{2}$, $\frac{\omega_1}{2}$ over ont the positive z-halfplane. The representing Riemann surface then covers this z-plane doubly and has the four branchpoints denoted in figure 82 by $e_2,\dots,e_5$, while it is signed at e_1 in the two sheets.

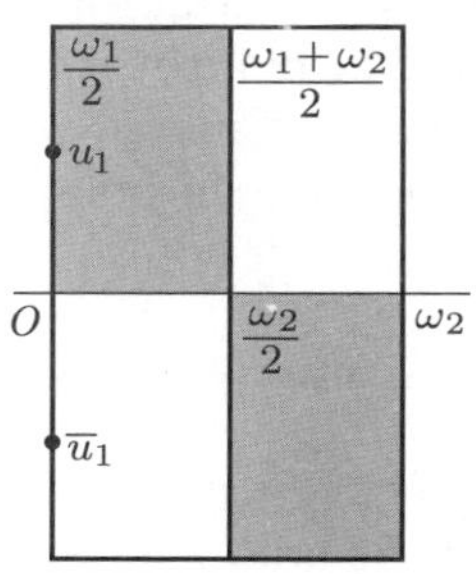

Fig. 81.

The individual one of our groups Γ, for its part, is correspondingly twice capable of extension by reflections, and indeed, it is a question here of the same extensions, which were already described for the groups of the previous paragraph (p. 414). The discontinuity domain of the $\overline{\Gamma}$ finally arising has the shape of the pentangle of figure 83, whose image furnished by $z = \varphi(\zeta)$ is then directly the positive z-halfplane of figure 82. The vertices $\varepsilon_1, \ldots, \varepsilon_5$ correspond to the points $e_1, \ldots, e_5$ of the real z-axis.

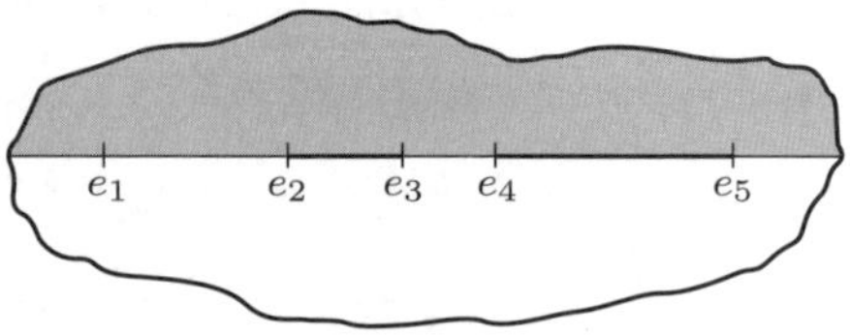

Fig. 82.

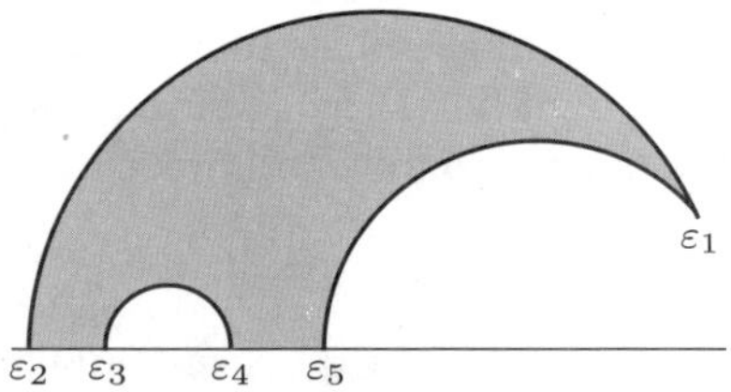

Fig. 83.

We now arrive at the most surveyable representation of the present continuum B_f, if we introduce the two *algebraic moduli* in the form of the double ratios:

$$\lambda_1 = \frac{(e_4 - e_1)(e_2 - e_3)}{(e_4 - e_2)(e_1 - e_3)}, \qquad \lambda_2 = \frac{(e_1 - e_3)(e_4 - e_5)}{(e_1 - e_4)(e_3 - e_5)}. \tag{2}$$

If we exclude the limiting cases initially, then λ_2 will be arbitrarily choosable in the interval $0 < \lambda_2 < 1$ and after choice of the quadruple e_1, e_3, e_4, e_5, essentially fixed (i.e. apart from linear transformation). The point e_2 is then still arbitrarily movable in the interval $e_1 < e_2 < e_3$, which has as consequence, that λ_1 has to describe the interval $0 < \lambda_1 < 1$ independently of λ_2. By bringing in the boundary cases λ_1 and λ_2 are therefore arbitrarily movable in the intervals:

$$0 < \lambda_1 \leq 1, \qquad 0 \leq \lambda_2 \leq 1, \tag{3}$$

[418] so that we also find the continuum B_f represented by the surface of a square.

For the mapping of the square B_g onto the square B_f, the boundary points of B_g are to be considered first of all.

Along the side $t_1 = 0$, the vertex invariant becomes $j_2 = 2$, therefore parabolic. According to the principle on p. 416 of the favoring of *one* limiting shape of the polygon (among the various possibilities), in case the favored shape already allows the moduli to be determined, we will fuse the vertices $\varepsilon_2, \varepsilon_3$ into a parabolic cusp, and thus arrive at the linear continuum of the signature $(0,3;l_1,\infty)$, which we treated in §27 on p. 416. On the Riemann surface, upon the coincidence of the branchpoints e_2 and e_3, the two sheets become disconnected here, by appearing signed at the coincidence place with two new points e. We obtain $\lambda_1 = 0$, and according to the result of §27 on p. 416, find the continuum $t_1 = 0$, $0 \leq t_2 \leq 1$ in question mapped single-valuedly and continuously onto the side $\lambda_1 = 0$, $0 \leq \lambda_2 \leq 1$ of the square B_f in such a manner, that $t_2 = 0$ furnishes the endpoint $\lambda_2 = 0$ (with the group of the signature $(0,3;l_1,\infty,\infty)$), and the point $t_2 = 1$ the endpoint $\lambda_2 = 1$. And indeed, we may and will choose, at the latter point, the cyclic parabolic group as limiting case, which arises for simultaneous coincidence of ε_2 with ε_3 and ε_1 with ε_5.

We arrange next to this, at the same time, the discussion of the side $t_2 = 1$, which requires the passage to the limit $\lim j_3 = \infty$. We effect this for the polygon of figure 83 in such a fashion, that we let the vertex ε_1 (naturally without altering the angle there) move up to the point ε_5. There arises the cyclic hyperbolic group of the invariant j_2 resp. t_1. The functions of the limiting case furnish an unsigned orthosymmetric surface of the genus $p = 1$, which arises from figure 82, by the point e_1 of the two sheets moving up to the branchpoint e_5. Thereby, λ_2 becomes $= 1$, and λ_1 is the modulus of this surface. But that the side $t_2 = 1$ of the square hereby maps itself single-valuedly and continuously onto the entire side $\lambda_2 = 1$ of the square (including the endpoints), is the content of the fundamental theorem proved at the beginning of §27 on p. 414, already mentioned in the theory of elliptic functions.

At the end $t_1 = 0$ of the line $t_2 = 1$, the parabolic limiting case of coincidental points $\varepsilon_2, \varepsilon_3$ is chooseable which we had already reached at the end $t_2 = 1$ of the line $t_1 = 0$. At this place, therefore, there is continuity of the chosen boundary polygons and similarly of the corresponding surfaces. We reach the other end $t_1 = 1$ of the line $t_2 = 1$, (after the coincidence of ε_1 with ε_5 already effected) by letting the points ε_3 and ε_4 coincide at [419]
any place of the segment $\varepsilon_2, \varepsilon_5$ of the real ζ-axis, say at the midpoint of this.

Precisely corresponding considerations hold for the sides $t_2 = 0$ (coincidence of ε_4 and ε_5) and $t_1 = 1$ (coincidence of ε_1 and ε_2). Notice thereby, in particular, the continuity of the chosen boundary polygons at the vertices $t_1 = 0$, $t_2 = 0$ (group of the signature $(0,3;l_1,\infty,\infty)$) and $t_2 = 0$, $t_1 = 1$ (simultaneous coincidence of ε_4 and ε_5 and ε_1 and ε_2). At the end $t_2 = 1$ of the $t_1 = 1$ of the square, ε_1 and ε_2 having already coincided, we let the points ε_3 and ε_4 coincide at the midpoint of the segment ε_2, ε_5 of the real ζ-axis. But this is not the boundary polygon, which we reached at the vertex in question along $t_2 = 1$: *here, therefore, the continuity of the boundary polygons is interrupted; also, it does not let itself be produced by other choices of the boundary polygons along the sides* $t_1 = 1$ *and* $t_2 = 1$.

For the closer investigation of the vertex in question, we will distinguish, in which direction we approach it from the square B_g, by fixing, say, the limiting value:

$$\lim \frac{j_3}{j_2} = \lim \frac{1-t_1}{1-t_2}, \tag{4}$$

lying between 0 and $+\infty$. In the limit itself, as just now, the points ε_3 and ε_4 are to coincide at the midpoint of the segment $\varepsilon_2, \varepsilon_5$. There then arises, as boundary polygon, the triangle $\varepsilon_1, \varepsilon_2, \varepsilon_5$ of figure 84, in which, after fixing the points $\varepsilon_2, \varepsilon_5$, the vertex ε_1 lies on a circle k through ε_2 and ε_5, which, in addition, is determined by the polygon angle at ε_1. The point ε_1 now describes this circle K from ε_2 to ε_5 continuously and completely. If the limiting value (4) runs through the continuum defined by:

$$0 \leq \lim \frac{j_3}{j_2} \leq +\infty. \tag{5}$$

The two endvalues 0 and $+\infty$, however, furnish jus the two boundary polygons met with before at the vertex $t_1 = 1,\ t_2 = 1$ (coincidence of ε_1 with ε_2 resp. ε_1 with ε_5). Between them, we now produce the continuity, by inserting the linear continuum of the triangles $\varepsilon_1, \varepsilon_2, \varepsilon_5$ just discussed.

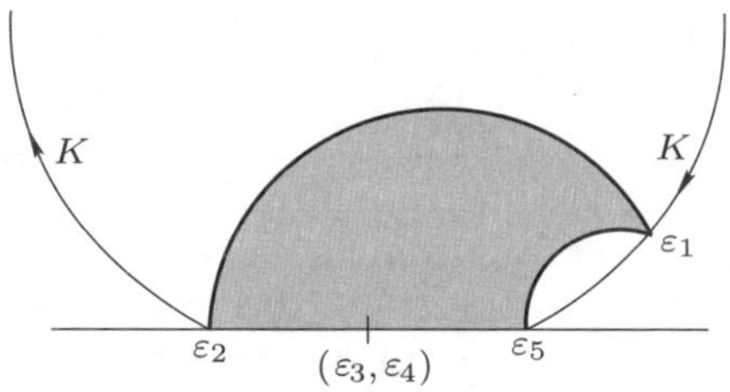

Fig. 84.

For the sake of visualizability, we set:

$$t_3 = \frac{1 - t_2}{2 - t_1 - t_2}, \tag{6}$$

[420] and interpret t_3 as a third rectangular coordinate besides t_1 and t_2. The function of t_1 and t_2 standing on the right in (6) is, apart from the vertex $t_1 = 1, t_2 = 1$, single-valued in the square, but becomes multiple-valued at that vertex in such a manner, that the interval of the limiting values (4) given in (5) is mapped single-valuedly and continuously onto the interval $1 \geqq t_3 \geqq 0$. The relation (6) we interpret in the space of the t as a "*paraboloid*"; to the square B_g corresponds, on the latter, a *pentangle* delimited by three lines and two hyperbolic portions, which let be called B'_g. B'_g is in general mapped single-valuedly onto the square, but infinitely multiple-valuedly onto the vertex $t_1 = 1,\ t_2 = 1$ in such a manner, that the various directions from the square to this vertex are single-valuedly and continuously coordinated to the pentangle side $t_1 = 1, t_2 = 1, 0 \leq t_3 \leq 1$. To the boundary of the pentangle B'_g we have now coordinated *a closed continuum of boundary polygons.*

How does this fashion itself with respect to the relations at the corresponding vertex $\lambda_1 = 1, \lambda_2 = 1$ of the square B_f?

We had here initially reached two limiting cases along the sides of the square, which we are briefly able to characterize by the immediately understandable figure 85. For the production of the continuity, we insert between them (by holding $e_3 = e_4$, say, fixed at the midpoint of the segment e_2, e_5) all the surfaces, which enter for continuous wandering of the place e_1 from e_2 to e_5 (through the infinite). For the representation of this continuum, we avail ourselves of the double-ratio:

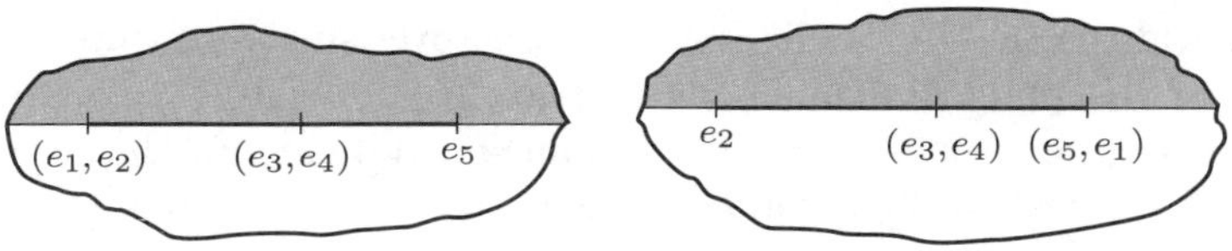

Fig. 85.

$$\lambda_3 = \frac{(e_1-e_2)(e_3-e_5)}{(e_1-e_3)(e_2-e_5)}, \tag{7}$$

which correspondingly has to grow from 0 to 1. This is connected with λ_1, λ_2 by:

$$\lambda_3 = \frac{\lambda_1 - 1}{\lambda_1\lambda_2 - 1} = \frac{1}{\lambda_2 + \dfrac{1-\lambda_2}{1-\lambda_1}}, \tag{8}$$

so that, here too, the continuum of the directions from the square B_f toward the vertex in question corresponds continuously and single-valuedly to the interval $0 \le \lambda_3 \le 1$. From (8) follows:

$$\lambda_1\lambda_2\lambda_3 - \lambda_1 - \lambda_3 + 1 = 0, \tag{9}$$

a relation, which under interpretation of λ_3 as a third rectangular coordinate, we recognize as a *surface of third degree.* On it, the quadrangle B_f again projects itself onto a *pentangle*, which now appears bounded by five lines. We call this B'_f and are able, on account of its mapping onto the square B_f, to repeat word for word the same thing which was stated above for B_g and B'_g. [421]

It follows now already, from the existence theorem of automorphic functions, without taking the aid of a fundamental theorem, *that the fifth side of the pentangle* B'_g *is single-valuedly and continuously coordinated to that of the pentangle* B'_f. Next transform the individual triangle $\varepsilon_1, \varepsilon_2, \varepsilon_5$ (cf. figure 84) so that ε_2 comes to lie at $\zeta = -1$, ε_5 at $\zeta = +1$ and ε_1 on the imaginary ζ-axis. The coincidence place of ε_3 and ε_4 then, in general, no longer lies at the midpoint of the segment ε_2, ε_5; rather, we shall obtain exactly the continuum represented by the fifth side of B'_g, by letting the place $\varepsilon_3 = \varepsilon_4$ describe the segment $\varepsilon_2, \varepsilon_5$ from ε_2 to ε_5.

The mapping of the triangle $\varepsilon_1, \varepsilon_2, \varepsilon_5$ onto the z-halfplane we correspondingly set up, so that $e_2 = -1$, $e_5 = +1$ and $e_1 = \infty$. The segment $\varepsilon_2, \varepsilon_5$ is the mapped single-valuedly and continuously onto the segment of the real z-axis from -1 to $+1$, and corresponding to the wandering place $\varepsilon_3 = \varepsilon_4$, its image $e_3 = e_4$ describes that segment from $z = -1$ to $z = +1$. The double ratio λ_3 hereby runs through its interval from 1 to 0, and with this, in fact, the fifth side of B'_f is described.

In consequence of the argument so far, *the closed boundary of the pentangle* B'_g *is mapped single-valuedly and continuously onto the closed boundary of* B'_f.

For disposing of the interior of the pentangle, we attach say, to that family of lines on the paraboloid portion B'_g, which is cut out by the places $t_3 =$const. We describe this family by letting its "parameter" t_3 grow from 0 to 1; thereby, the extreme values $t_3 = 0$ and $t_3 = 1$ furnish two sides of the pentangle B'_g itself.

The individual one of these lines carries itself over single-valuedly and continuously onto a curve of the domain B'_f, which joins two boundary points of B'_f and neither crosses over nor touches itself (single-valuedness theorem). The entire family of lines of B'_g furnishes, on B'_f, a family of curves, of which no two are able to have a point in common (single-valuedness theorem). The two extreme curves (for $t_3 = 0$ and $t_3 = 1$) are, as we already know, sides of the pentangle B'_f, which are homologous to the
[422] corresponding sides of B'_g. In the interior of B'_f, however, no gaps can remain uncovered by curves; for from the existence of such a gap, we would immediately conclude a contradiction to the continuity theorem: *with this, the domain* B'_g *and* B'_f *are mutually single-valuedly (with inclusion of their boundaries) mapped onto one another, and the fundamental theorem is proved for the signature* $(0,3;l_1)$.

1.29 Effectuation of the continuity proof for the signature $(1,1;l_1)$

The continuity proof fashions itself especially simply in the case of the signature $(1,1;l_1)$. The group continuum B_g was furnished here by the double triangle heavily outlined in figure 47 on p. 329, and by means of the coordination of the boundary curves of this double triangle, represented a continuum of the connectivity of the surface of the sphere, as long as we counted the group of the signature $(0,3;l_1,\infty,\infty)$, which belonged to one vertex of the double triangle, as part of the continuum.

As modulus of the associated Riemann surfaces, we wanted (cf. p. 395) to so choose the period quotient ω, that the present continuum B_f is representable, say, by the double triangle of the ω-halfplane heavily outlined in figure 86; the position of this domain in the halfplane is characterized by a few values ω.

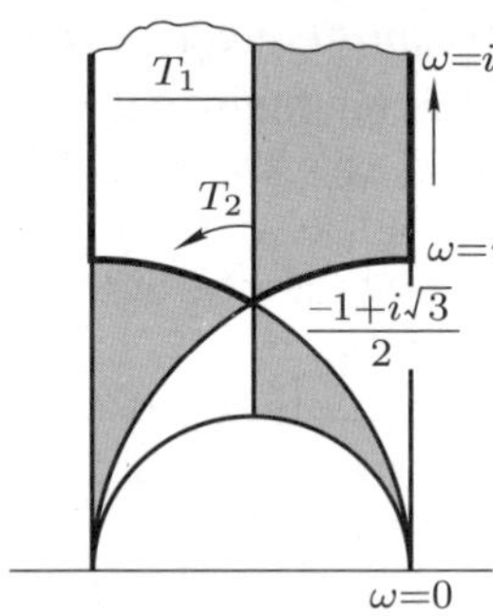

Fig. 86.

We now already drew attention on p. 330 to the fact, that the triangle net of the surface Φ_3 shows exactly the same structure as the triangle net of the ω-halfplane. The situation is now simply this, *that the double triangle* B_g *of figure 47 on p. 329 is mapped onto the double triangle* B_f *of figure 86 reversibly single-valuedly and continuously,* from which then, a corresponding mapping arises between the surface Φ_3 and the ω-halfplane.

In order to obtain this result, we have to introduce the periods ω_1, ω_2 of the integral of the first kind in such a manner, that the generators V_c and V_a of Γ furnish the following transformation of u:

$$(V_c) \qquad u' = u + \omega_1, \qquad (V_a) \qquad u' = u + \omega_2. \tag{1}$$

From the relation $V_a V_b V_c = 1$ or $V_b = (V_c V_a)^{-1}$, then follows, as the effect of V_b on u:

$$(V_b) \qquad u' = u - \omega_1 - \omega_2, \quad \text{or} \quad u' = u + \omega_3, \tag{2}$$

in case we define ω_3 by the equation $\omega_1 + \omega_2 + \omega_3 = 0$.

Next, we now carry over the boundary of the shaded elementary triangle of figure 47 onto the halfplane of the quotient ω of the period ω_1, ω_2. The vertex of that triangle cut out on the curve k_2 (cf. figure 47) by the symmetry line $x = z$ furnish, as discontinuity domain of the group Γ, a quadrangle with the generators V_a and V_c, which (in the language of p. 331) exhibits "diagonal" as well as "mid-line symmetry". Both these symmetries carry themselves over onto the period parallelogram of the u-plane, so that the latter is a square. With this, we have the "*harmonic*" case of the elliptic model (cf. "M" I on p. 10 ff.), and the orientation of the generators V_c and V_a to one another (see e.g., figure 48 on p. 331) yields the value $\omega = +i$ for the period quotient. [423]

From here, we pursue the side K_2 of the domain B_g up to the cusp situated on the boundary of the sheet Φ_3 at $y = z$. According to p. 328 (where, admittedly, the discussion refers itself, not to the curve K_2 but to K_1, all points of the curve K_2, whose equation was

$$xz - 2yt = 0,$$

furnish quadrangles with the generators V_a, V_c with mid-line symmetry. The associated period parallelograms are thus rectangles and ω becomes purely imaginary; and indeed, from the condition $x < z$ or $j_a < j_c$ taking place for B_g, follows $|\omega_2| < |\omega_1|$ for the period rectangle and therefore $|\omega| > 1$. If we have reached the cusp B_g situated on the boundary of Φ_3 along the side K_2, then the drop to the signature $(0,3;l_1,\infty,\infty)$ has entered, and therefore $\omega = +i\infty$ is reached. Our usual argument yields, that the side in question of the domain B_g carries itself over reversibly single-valuedly and continuously onto the imaginary ω-axis from $\omega = +i$ to $\omega = +i\infty$, i.e., therefore, onto the homologous side of the domain B_f.

One will very easily continue this consideration. If we proceed from the vertex of the domain B_g, already carried over onto $\omega = i$, along the symmetry line $x = z$, then we arrive at quadrangles with "diagonal symmetry"; $j_a = j_c$ thus has the consequence $|\omega_1| = |\omega_2|$, so that we move with ω on the unit circle of the halfplane. The endpoint $x = y = z$ of the side of B_g now described now furnishes a polygon for Γ, which as a quadrangle with the generators V_a, V_c, has diagonal symmetry, as too, as a quadrangle with the generators V_a, V_b and as a quadrangle with the generators V_b, V_c. If one puts in all symmetry lines, then there arises a decomposition, say, of the quadrangle of the generators V_a, V_c, always preferred here, into 12 alternately symmetric and congruent triangles of the angle $\frac{\pi}{2}, \frac{\pi}{3}, \frac{\pi}{6l_1}$, as this comes into expression in figure 87 on p. 423. The symmetry lines all carry themselves over onto the u-plane; there, therefore, we get the period parallelogram of figure 88 with the "*equianharmonic*" case, i.e., with

$$\omega = \frac{-1 + i\sqrt{3}}{2}.$$

the side of B_g represented by $x = z$ thus carries itself singlevaluedly and continuously onto the homologous side of B_f.

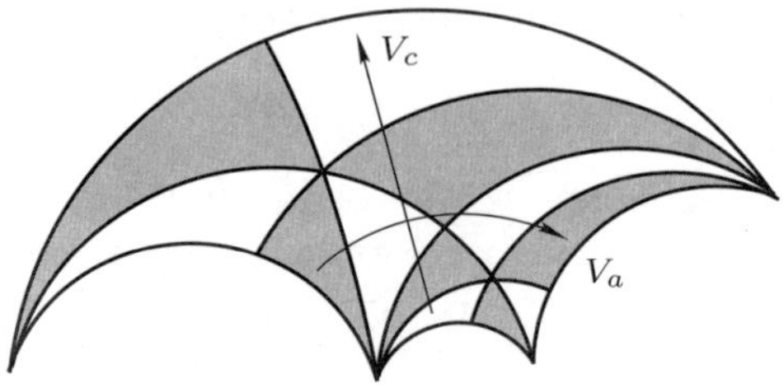

Fig. 87.

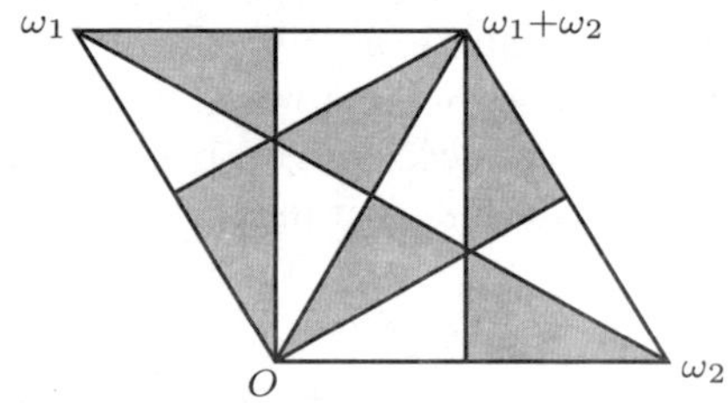

Fig. 88.

Finally, the third side of the shaded portion of B_g, given by $y = z$, also offers no difficulty. Here $|\omega_3| = |\omega_1|$ follows, from which yields itself, that the real component of ω is equal to $-\frac{1}{2}$. One will now further easily effect the mapping onto the homologous side of the shaded portion of B_f.

If we now call upon the coordination of the boundary curves of B_g and those of B_f by means of the transformations T_1 and T_2, then the remainder of the boundaries of our two domains dispose the themselves without further ado: *the closed boundary of the double triangle* B_g *carries itself over reversibly singlevaludedly and continuously onto that the double triangle* B_f.

In order, finally, to dispose of the interior B_g, we cover this double triangle simply and gaplessly with any suitably chosen family of curves. For this, say, we are able to choose the *curves of second degree*, which we cut out on the surface Φ_3 on p. 320 by means of the planes $z - 2\lambda t = 0$. One effects the transportation of this family onto B_f word for word, exactly as the corresponding measure in the case discussed on p. 421. *We find the two double triangles* B_g *and* B_f *reversibly singlevaluedly coordinated to one another, whereby the fundamental theorem is proved for the signature* $(1, 1; l_1)$.

We will not go more closely into the third case of *two-dimensional* domains B_g and B_f, belonging to the signature $(0, 4; l_1, l_2, l_3, l_4)$, since the consideration lets itself be fashioned exactly according to the prototype of the one just carried through to a conclusion. The domain B_g, as long as we ignore special cases of equal vertex invariants, lets itself be represented by the double triangle in figure 62 on p. 355. As modulus of the Riemannian surfaces we utilize either the double-ratio

$$\lambda = \frac{(e_1 - e_2)(e_3 - e_4)}{(e_1 - e_3)(e_2 - e_4)}$$

or, which perhaps recommends itself even more on account of the analogy to the foregoing, the period quotient ω for a two-sheeted surface over the z-plane, which is branched at the four signed places e_1, e_2, e_3, e_4. In the former case, the *plane of the complex variable* λ is the antitype of the continuum B_f, in the latter case, *the double triangle of the angle 0* in the ω-halfplane *belonging to the "modular function"* $\lambda(\omega)$ (cf. figure 67 in "M" I on p. 276). The reader will easily prove for himself, *that this double triangle is mapped one-to-one onto that of figure 62 on p. 355, where the fundamental theorem is proved in the case of the signature* $(0,4;l_1,l_2,l_3,l_4)$.[108] [425]

1.30 Effectuation of the continuity proof for the signature $(0,3)$

There are, in all, three signatures with three-dimensional continua B_g and B_f, namely $(0,3)$, $(1,1)$, $(0,4;l_1,l_2,l_3)$. In the first two cases, for which a signing with points e is not present, the genus of the Riemann surface is $p = 2$; it is a question of the two kinds of orthosymmetric surfaces of this genus, which we discussed on p. 401. In the third case, the genus p of the Riemann surface is $= 0$. All the three cases furnish principal-circle theorems.

For the signature $(0,3)$, the group continuum B_g is again identical with the polygon continuum and therefore, as the latter, by means of the vertex invariants j changed around into the form t, representable by the *cube which corresponds to the inequalities*:

$$0 \leq t_1 \leq 1, \quad 0 \leq t_2 \leq 1, \quad 0 \leq t_3 \leq 1. \tag{1}$$

The individual group, as always for the character $(0,3)$, admits two extension by reflections here, one of which effects the halving of the double triangle (in the projective plane), while the second is the reflection in the real ζ-axis. As discontinuity domain of the group of the second kind finally arising, we utilize the circular-arc polygon represented in figure 89.

For the orthosymmetric surfaces of the genus $p = 2$, it is a question of *those with six real branchpoints*, for which we already determined on p. 401, that the individual surface is orthosymmetric in two ways with three transition lines. These are the two symmetric transformations, which correspond to the two extensions of the group by reflections. To the symmetric subdivision of the discontinuity domain of the group corresponds the cutting-up of the Riemann surface (cf. figure 90) along the real axis (the six symmetry lines arranged here) into four halfsheets. We set the positive upper halfplane sheet in correspondence with the polygon of figure 89, whereby the points $e_1, e_2, \ldots$ are to correspond to the vertices $\varepsilon_1, \varepsilon_2, \ldots$. [426]

108 One finds a more penetrating representation in Part IV of the second work of the author cited on p. 286.

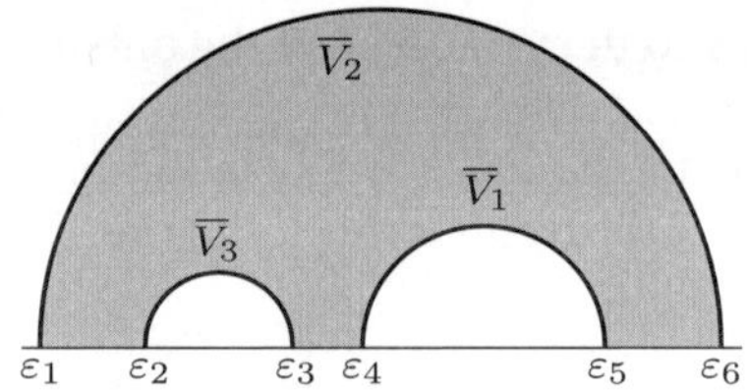

Fig. 89.

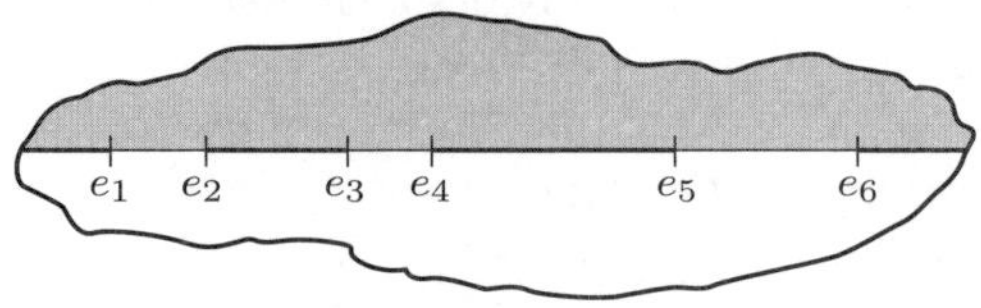

Fig. 90.

For the representation of the continuum B_f, we utilize as moduli the three double ratios:

$$\lambda_1 = \frac{(e_1-e_2)(e_3-e_6)}{(e_1-e_3)(e_2-e_6)}, \quad \lambda_2 = \frac{(e_3-e_4)(e_5-e_2)}{(e_3-e_5)(e_4-e_2)}, \quad \lambda_3 = \frac{(e_5-e_6)(e_1-e_4)}{(e_5-e_1)(e_6-e_4)}. \tag{2}$$

Each of these three values is restricted to the interval $0 \le \lambda \le 1$. At the same time, the three values λ are choosable independently of one another in this interval and, after a choice is made, furnish essentially (i.e., for the invariant standpoint) only one system of points $e_1, e_2, \dots, e_6$.

Namely, if we initially exclude the limiting cases $\lambda = 0$ and $\lambda = 1$, then it is permissible to set:

$$e_1 = 0, \quad e_4 = 1, \quad e_6 = \infty.$$

But then, from (1) follows:

$$e_2 = \lambda_1 e_3, \quad e_5 = \lambda_3^{-1}, \quad \lambda_2 = \frac{\lambda_1\lambda_3 e_3^2 - (\lambda_1\lambda_3+1)e_3+1}{\lambda_1\lambda_3 e_3^2 - (\lambda_1+\lambda_3)e_3+1}. \tag{3}$$

the second of these equations determines e_5; in consequence of the first, it is determined by the value λ_1, in what ratio the point e_2 divides the segment from e_1 to e_3. Finally, however, by addition of the value λ_2, the point e_3 is uniquely determined from the third equation (3), on account of its position between 0 and 1. If one finds this determination of e_3 unintuitive, we calculate this third equation around into:

$$1-\lambda_2 = \frac{(1-\lambda_1)(1-\lambda_3)e_3}{(1-\lambda_1 e_3)(1-\lambda_3 e_3)},$$

and convince oneself, that λ_2 decreases continuously and without reversal from 1 to 0, in case e_3 wanders continuously from $e_1 = 0$ to $e_4 = 1$. By bringing in the boundary
[427] values of the λ, we are able to determine *the continuum* B_f *of our orthosymmetric surfaces of the genus $p = 2$, with inclusion of the limiting cases, we are able to represent by means of the cube defined by the conditions*:

$$0 \le \lambda_1 \le 1, \quad 0 \le \lambda_2 \le 1, \quad 0 \le \lambda_3 \le 1. \tag{4}$$

The mapping of the cube B_g onto the cube B_f we begin with the discussion of the six boundary surfaces, of the former.

Let the three generating reflections, which the $\overline{\Gamma}$ belonging to the polygon in figure 89 has (besides $\zeta' = \overline{\zeta}$) be, as was already indicated in the figure itself, $\overline{V}_1, \overline{V}_2, \overline{V}_3$. From them, we obtain the generators of the original Γ in the form:

$$V_1 = \overline{V}_2\overline{V}_3, \quad V_2 = \overline{V}_3\overline{V}_1, \quad V_3 = \overline{V}_1\overline{V}_2. \tag{5}$$

The passage to the limit $\lim t_3 = 0$ for the invariant of V_3, we can and will effect by fusion of the points ε_5 and ε_6 into a parabolic cusp. On the Riemann surface, the branchpoints e_5 and e_6 come into coincidence; decrease to $p = 1$ enters, and points e lie in both sheets at the coincidence place, which correspond to the two parabolic cusps of the boundary polygon. Here we have arrived back at the signature $(0,3;\infty)$, treated in the paragraph before last (p. 416 ff.). The modulus λ_3 has become equal to 0; and one convinces oneself easily, that the coordinates t_1, t_2 in the square $t_3 = 0$, as well as the λ_1, λ_2 in the square $\lambda_3 = 0$, are exactly those with which we worked l.c. From the result at that time now follows immediately: *the square facial surface $t_3 = 0$ of the cube B_g is mapped onto the homologous surface $\lambda_3 = 0$ of B_f reversibly single-valuedly and continuously, whereby the vertices and sides of the one square correspond to the homologous vertices and sides of the other.*

Coordinated with $t_3 = 0$ are the boundary surfaces $t_2 = 0$ and $t_1 = 0$. *They too consequently appear mapped reversibly single-valuedly and continuously onto the homologous boundary surfaces of B_f.*

The plane $t_3 = 1$ we will approach perpendicularly and therefore, say, for constant t_1, t_2, whereby we will produce two different boundary polygons.

Namely, by understanding by V a hyperbolic substitution with the fixpoints $\varepsilon_2, \varepsilon_3$ (cf. figure 89), which displaces the ζ-plane from ε_2 toward ε_3, and denoting by σ a continuously variable parameter, we will first transform the symmetry circle of $\overline{V}_1$ by V^σ the circles of $\overline{V}_2$ and $\overline{V}_3$ remaining fixed. The new generating reflections are:

$$V^\sigma \overline{V}_1 V^{-\sigma}, \quad \overline{V}_2, \quad \overline{V}_3.$$

The substitution V_1 remains unaltered, V_2 retains its invariant t_2, since $\overline{V}_3 = V^\sigma \overline{V}_3 V^{-\sigma}$ [428]
holds, while t_3 becomes continuously variable, and for $\sigma = +\infty$ reaches the limit $t_3 = 1$.

The boundary polygon furnishes the cyclic hyperbolic group of the invariant t_1. The genus of the Riemann surface reduces itself to $p = 1$, with coincidence of the branchpoints e_3, e_4, e_5. The modulus of the surface is λ_1; on the other hand, $\lambda_3 = 1$, while λ_2 assumes an indeterminate form. According to the fundamental theorems concerning the cyclic hyperbolic groups, the linear continuum $0 \le t_1 \le 1$ is mapped onto that represented by $0 \le \lambda_1 \le 1$, reversibly single-valuedly and continuously; we imagine this mapping represented by the equation $\lambda_1 = \psi(t_1)$.

For the determination of the behavior of λ_2, we form a second shape for the boundary polygon. Now let the symmetry circles of $\overline{V}_1$ and $\overline{V}_3$ remains unaltered, while that of $\overline{V}_2$ is transformed by $V^{-\sigma}$; the generating reflections are then:

$$\overline{V}_1, \; V^{-\sigma}\overline{V}_2 V^\sigma, \; \overline{V}_3.$$

The boundary polygon now furnishes the cyclic hyperbolic group of the invariant t_2; and an analogous consideration to the one just now gives $\lambda_2 = \psi(t_2)$, understanding by ψ this dependency, as before. There has thus yielded itself: *the square boundary surface* $t_3 = 1$ *of* B_g *is mapped reversibly single-valuedly and continuously onto the homologous boundary surface* $\lambda_3 = 1$ *of* B_f*, whereby homologous sides and vertices correspond to one another. The mapping has the simple character coming to expression by:*

$$\lambda_1 = \psi(t_1), \qquad \lambda_2 = \psi(t_2),$$

in consequence of which, parallels to the sides of the one square furnish just such in the other.

One will immediately carry through an analogous consideration, with the corresponding result, for the coordinated facial surfaces $t_2 = 1$, $t_1 = 1$.

After disposal of the boundary surfaces, we lay the plane $t_3 = \tau$ through the interior of B_g, understanding by τ any value between 0 and 1. We will carry over the square intersection into the cube B_f, whereby, it is nothing more than a question of repetition of the continuity proof of the signature $(0,3;l_1)$. First of all, map the two sides $t_1 = 0$ and $t_2 = 0$ of our square given by $t_3 = \tau$ (according to the boundary surface coordination) onto two curves lying in the boundary surfaces $\lambda_1 = 0$ resp. $\lambda_2 = 0$ of B_f, which run out from a point in the edge $\lambda_1 = 0$, $\lambda_2 = 0$, corresponding to the polygon of the signature
[429] $(0,3;\infty,\infty)$ and the invariant $t_3 = \tau$, and end in the edges $\lambda_1 = 1, \lambda_2 = 0$ resp. $\lambda_1 = 0, \lambda_2 = 1$ at the same height $\lambda_3 = \psi(\tau)$. The other two sides of the square to be mapped simply carry themselves over, in the homologous boundary surfaces of B_f, onto two lines at the same height $\lambda_3 = \psi(\tau)$.

In carrying the square surface over, the same complications make their appearance toward the vertex $t_1 = 1$, $t_2 = 1$, which we met at the corresponding place on p. 419. Here too, we remove this, by distinguishing the various limiting values:

$$\lim \frac{j_2}{j_1} = \lim \frac{1-t_1}{1-t_2},$$

and with this, the various directions in which we are able to approach the vertex in question. We find, by the usual method of reasoning: *the square running through over cube* B_g *at the height* $t_3 = \tau$ *carries itself over reversibly single-valuedly and continuously onto the surface of an intersection through the cube* B_f*, which is spanned along the boundary of* B_f *by the quadrangle already described.*

Now nothing further remains, except to let the parameter τ grow from 0 to 1 and fill out in such a manner the entire B_g gaplessly and simply with a family of planes. In the transportation of these onto the cube B_f, one again has to support oneself on the two theorems of continuity and single-valuedness and in addition, only to notice, that the two limiting cases $\tau = 0$ and $\tau = 1$, as we already know, furnish the boundary surfaces $\lambda_3 = 0$ and $\lambda_3 = 1$ of B_f: *the two cubes* B_g *and* B_f *are mapped reversibly single-valuedly and continuously onto one another, with which is actually proved the fundamental theorem for the signature* $(0,3)$.

1.31 Representation of the three-dimensional continua B_g and B_f for the signature $(1,1)$

The three-dimensional group continuum B_g, which comes up for the signature $(1,1)$, we represent, according to the result obtained at the close of §12 on p. 330, by a *hexahedron*, characterized more closely there, which was bounded by *three planes, two surfaces of the second degree* (hyperboloids H_2 and H_3) and *one surface of third degree* (corresponding to the parabolic limiting case).

The corresponding continuum B_f comprises *the orthosymmetric surfaces of the genus $p=2$, which are representable as two-sheeted surfaces over the z-plane with six branchpoints lying symmetrically in pairs with respect to the real z-axis* (cf. p. 401). Let the branchpoints of the positive z-halfplane be situated at $z = v_1, v_2, v_3$. The function [430]
$z=\varphi(\zeta)$ present here one can also conceive as a principal function of the group of the genus $p=0$ of the signature $(0,4;2,2,2)$, whose polygon lying in the projective plane is given by figure 42 on p. 316. The places v_1, v_2, v_3 then correspond to those denoted l.c. by e'_1, e'_2, e'_3. B_f *is consequently simply the continuum of the halfplanes signed with three points v_1, v_2, v_3, whereby, all those signed halfplanes are to count as not distinct, which are congruent in the sense of our hyperbolic determination of measure, established in I on p. 26 ff.*

Accordingly, in order to introduce the *moduli* in invariant fashion, through each two of the three points v_i, we draw a semicircle orthogonal to the real z-axis and denote the feet of these semicircles lying on the real axis in the fashion given in figure 91. As moduli, we could hen utilize the three double-ratios:

$$\left\{\begin{aligned} \lambda_1 &= \frac{(u_1-v_2)(v_3-w_1)}{(u_1-v_3)(v_2-w_1)}, \\ \lambda_2 &= \frac{(u_2-v_3)(v_1-w_2)}{(u_2-v_1)(v_3-w_2)}, \\ \lambda_2 &= \frac{(u_3-v_1)(v_2-w_3)}{(u_3-v_2)(v_1-w_3)}, \end{aligned}\right. \tag{1}$$

which all belong to the interval $0 \le \lambda \le 1$, and whose logarithms, taken negatively, would give the distances of the three points v from one another in the sense of the hyperbolic determination of measure[109].

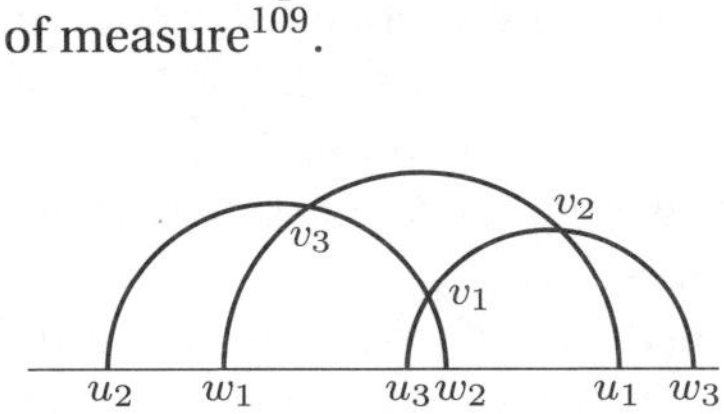

Fig. 91.

The comparison of the two continua B_g and B_f will let itself be fashioned yet somewhat more perspicuously if we let the three moduli:

[109] The representation of the continuum B_f chosen in the text breaks down only in the "parabolic" limiting case, to which we return later.

$$\lambda_a = \lambda_1 + \frac{1}{\lambda_1}, \quad \lambda_b = \lambda_2 + \frac{1}{\lambda_2}, \quad \lambda_c = \lambda_3 + \frac{1}{\lambda_3} \tag{2}$$

take the place of the $\lambda_1, \lambda_2, \lambda_3$, whose notation was chosen to tie into that of the group invariants j_a, j_b, j_c. The moduli (2), on account of $0 \le \lambda_i \le 1$, appear subjected to the conditions:

$$2 \le \lambda_a \le +\infty, \quad 2 \le \lambda_b \le +\infty, \quad 2 \le \lambda_c \le +\infty. \tag{3}$$

with the $\lambda_a, \lambda_b, \lambda_c$, moreover, the $\lambda_1, \lambda_2, \lambda_3$ are uniquely determined, on account of $0 \le \lambda_i \le 1$.

Upon giving λ_a, the "distance" of the points v_2 and v_3 (in the sense of the hyper-
[431] bolic determination of measure) is fixed. If we put v_2 at $z = i$ say, and v_3 between 0 and i on the imaginary z-axis, then v_3 is simply $= i\lambda_1$. Further, by giving λ_c, the point v_1 becomes restricted to the periphery of a "circle" about $v_2 = i$; one then develops the equation of this circle from the third equation (1) and finds, in case one writes $z = x + iy$:

$$x^2 + y^2 - \lambda_c y + 1 = 0.$$

Correspondingly, giving λ_b restricts our point v_1 to the periphery of the "circle" about v_3:

$$x^2 + y^2 - \lambda_1 \lambda_b y + \lambda_1^2 = 0.$$

In order that these two circles meet one another at all, the condition is to be required:

$$\lambda_a^2 + \lambda_b^2 + \lambda_c^2 - \lambda_a \lambda_b \lambda_c - 4 \le 0. \tag{4}$$

If the inequality (4) holds, then we have two points v_1, which lie symmetrically with respect to the circle through v_2 and v_3 orthogonal to the real z-axis (imaginary z-axis for the above choice of the places v_2, v_3). But if the equation (4) holds, then we have only one point v_1, which is situated on that circle through v_2 and v_3 itself. In this case, the Riemann surface shows an additional symmetry, insofar as it goes over into itself by the reflection in the circle through the points v_1, v_2, v_3.

Up until now, it has not yet been decided, in which order we will denote the three branchpoints of the positive z-halfplane by v_1, v_2, v_3. If we do not have the symmetric case just mentioned, the points v_1, v_2, v_3, as in figure 91, therefore furnish a proper circular-arc triangle in the z-halfplane, then let the notation of the vertices, say, be effected, so that the points v_1, v_2, v_3 always follow one another on the boundary of that triangle as in figure 91. In the case of the inequality (4), then of two points v_1, v_2 found before, one is always distinguished, so that a permissible value system $\lambda_a, \lambda_b, \lambda_c$ uniquely determines the points v_1, v_2, v_3.

Further, by also excluding the equality sign in (4) beforehand, a cyclic permutation of the notations is not still permissible. We arrange the latter, so that λ_a comes out not greater than λ_b, and also not greater than λ_c:

$$\lambda_a \le \lambda_b, \qquad \lambda_a \le \lambda_c, \tag{5}$$

whereby we remain in agreement with our condition set up for the group continuum, $j_a \le j_b, j_a \le j_c$.

An addition is now necessary only in the symmetric case, i.e. for the equality sign
[432] in (4), where the three points v_1, v_2, v_3 lie on a circle orthogonal to the real axis. The

requirement (5) to which we hold fast, says that the "distance" of the points v_2 and v_3 is not greater than that of the points v_1 and v_2, and also not greater than that of the points v_3 and v_1. As figure 92 shows, we are able to satisfy this requirement in two ways; and the transition from the one to the other way amounts to an interchange of the values λ_a and λ_c. In the case of the equality sign in (4) the two value systems $\lambda_a, \lambda_b, \lambda_c$ and $\lambda'_a = \lambda_a, \lambda'_b = \lambda_c, \lambda'_c = \lambda_b$ will furnish the same surface.

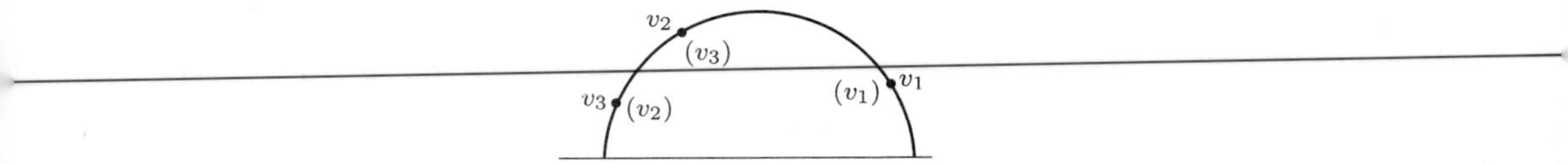

Fig. 92.

We clothe the result obtained in geometrical form, by interpreting the $\lambda_a, \lambda_b, \lambda_c$ as rectangular coordinates or better, instead of these, introducing, at the same time, the homogeneous coordinates ξ, η, ζ, τ by means of the proportion:

$$\lambda_a : \lambda_b : \lambda_c : 1 = \xi : \eta : \zeta : \tau. \tag{6}$$

In consequence of the coordinations (3), we then find ourselves in the tetrahedron bounded by:

$$\xi - 2\tau \geqq 0, \eta - 2\tau \geqq 0, \zeta - 2\tau \geqq 0, \tau \geqq 0. \tag{7}$$

the equation (4), however, furnishes us a *surface of third degree* in the form:

$$(\xi^2 + \eta^2 + \zeta^2)\tau - \xi\eta\zeta - 4\tau^3 = 0, \tag{8}$$

which we denote say, by F_3, in contrast to the Φ_3 of the boundary of B_g. One now easily represents the shape of the F_3 according to the arguments also followed for the Φ_3 (see e.g., p. 320 ff.). We make note: the F_3 has a knot-point at the tetrahedral point $\xi = 2\tau$, $\eta = 2\tau$, $\zeta = 2\tau$ and touches the tetrahedral surfaces running together here at the mid-lines $\eta = \zeta$ resp. $\zeta = \xi$ and $\xi = \eta$. Besides this, the F_3 appears tautly stretched along the plane $\tau = 0$ into the triangle $\xi\eta\zeta = 0$ and hangs stackly in the tetrahedron in question with a cusping toward the knot-point mentioned.

By now also remembering the requirements (5) and the argument attached to figure 92, we have as a result: *our surface continuum* B_f *finds, as antitype in the space of the* ξ, η, ζ, τ, *a tetrahedron, bounded by a sheet of the surface* F_3, *by the two planes* $\xi - \eta = 0$ *and* $\xi - \zeta = 0$ *as well, finally, by the plane* $\tau = 0$; *the two tetrahedral surfaces* $\xi - \eta = 0$ *and* $\xi - \zeta = 0$ *are mapped onto one another, insofar as homologous points of these furnish the same surface over the z-plane (cyclic permutation of the notations* v_i*); similarly, the two halves of the tetrahedral surface* F_3 *cut off by the symmetry line* $\eta = \zeta$ *are coordinated to one another (cf. figure 92), while only the fourth tetrahedron surface* $\tau = 0$ *remains uncovered.* The plane $\eta = \zeta$ is a symmetry plane of the tetrahedron B_f. Two points of B_f symmetric with respect to this yield two Riemann surfaces, of which the one goes over [433] into the other[110] by a reversal (say by the reflection $z' = -\bar{z}$ in the imaginary z-axis). We may not consider these two surfaces as equivalent; for we have not correspondingly done this for our groups Γ.

1.32 Effectuation of the continuity proof for the signature (1, 1)

The proof is now to be carried out, that the hexahedron B_g and the tetrahedron B_f are mapped onto one another reversibly single-valuedly. The domain B_g is decomposed by its symmetry plane $y = z$ into two symmetric halves, of which we denote the one characterized by $y \geqq z$, say, by $B_g^{(0)}$. Correspondingly, the tetrahedron B_f is decomposed by the symmetry plane $\eta = \zeta$ into two symmetric halves, one of which is to contains the image of $B_g^{(0)}$. We assume, that $B_f^{(0)}$ is described by $\eta \leq \zeta$; this is attainable, if need be, by undertaking a permutation of v_2 and v_3 in the distribution of the notations v_1, v_2, v_3 among the branchpoints. Our proof will now be able to restrict itself to showing, *that the pentahedron* $B_g^{(0)}$ *and the tetrahedron* $B_f^{(0)}$ *correspond to one another reversibly single-valuedly.* Since the same thing then follows immediately from symmetry considerations for the other halves of our domains B_g and B_f.

Now above (p. 334 ff.), we have coordinated, to the closed boundary of the pentahedron $B_g^{(0)}$, a consistently continuous closed system of boundary polygons. Hereby, in particular, at the edge $x = 0$, lying in the plane $t = 0$, we had to distinguish, in which direction we wanted to approach the individual edge-point from the domain $B_g^{(0)}$. Above and beyond this, at the midpoint $y = z$ of that edge, for the paths approaching tangentially to the plane $x = 2t$, it was also to be distinguished, in which of the infinitely many surfaces Φ_3 running through B_g we wanted to reach that point.

The transportation of the boundary points of the pentahedron $B_g^{(0)}$, thus provided with boundary polygons onto the the tetrahedron $B_f^{(0)}$ is indeed somewhat circumstan-
[434] tial, but finds no fundamental difficulty: *we find, as image of the pentahedron boundary, exactly the total boundary of the tetrahedron* $B_f^{(0)}$. In detail, the various boundary portions correspond to one another as the following table states:

Pentahedron $B_g^{(0)}$	tetrahedron $B_f^{(0)}$
Surface Φ_3	Vertex $\xi = 2\tau, \eta = 2\tau, \zeta = 2\tau$
Surface $xz - 2yt = 0$	Surface F_3
Plane $x = z$	Plane $\xi = \zeta$
Plane $y = z$	Plane $\eta = \zeta$
Plane $t = 0$	Plane $\tau = 0$

For each two boundary manifolds coordinated to one another in such a way, the continuity proof is to be carried through in detail, which never runs into difficulties. We will not represent everything here to the last detail, and restrict ourselves to the following remarks:

110 Hereby, according to the determination, an interchange of the notations v_2 and v_3 to be undertaken, which effects an interchange of λ_b and λ_c, and therefore, in fact, furnishes the transformation $\xi' = \xi, \eta' = \zeta, \zeta' = \eta, \tau' = \tau$.

First, the surface Φ_3 furnishes the the parabolic limiting cases, i.e., therefore, the two-dimensional continuum of the signature $(1,1;\infty)$. The moduli $\lambda_a, \lambda_b, \lambda_c$ break down here, as was already noticed on p. 396 [111]; they all assume the value 2. One could thereby aid oneself, by distinguishing the ∞^2 directions, in which one can approach the vertex $\lambda_a = 2$, $\lambda_b = 2, \lambda_c = 2$ (knot-point of the F_3) from the pentahedron $B_g^{(0)}$. In calculational terms, this is to be effected in such a manner, that one puts:

$$(\lambda_a - 2) : (\lambda_b - 2) : (\lambda_c - 2) = X : Y : Z,$$

say, and then, for the sake of visuality, interprets X, Y, Z as homogeneous coordinates in the plane. The condition (4) on p. 431 furnishes:

$$X^2 + Y^2 + Z^2 - 2YZ - 2ZX - 2XY \leq 0, \tag{1}$$

so that we are restricted to the surface of the "ellipse" represented by the "equation" (1). However, on account of (5) on p. 431, only the third of this surface described by $X \leq Y$, $X \leq Z$ comes into force, of which then, again, only the half cut off by the $Y = Z$ belongs to $B_f^{(0)}$. The validity of the fundamental theorem for the two-dimensional continuum present here, however, then already emerges from §29 on p. 422 ff., where indeed, the [435]
case of the signature $(1,1;\infty)$ was also included[112].

The surface $xz - 2yt = 0$ furnishes us, in the projective plane, the quadrangle with the generators V_a, V_c with midline symmetry, which, in the ζ-halfplane, gives the polygon shape characterized in figure 93. One half, say the polygon $d, b, e_1', b', d', e_2', e_3'$, yields a halfsheet of the Riemann surface, as figure 94 represents this. The symmetry remains maintained here, and therefore, in fact, we come to the surface F_3. To the inequality $j_a \leq j_c$ corresponds here $\xi \leq \zeta$. The equation holds thereby, i.e. if we find ourselves on the boundary $x = z$ of our surface $xz - 2yt = 0$, then our polygon, besides midline symmetry, also obtains diagonal symmetry, and in particular, the polygon d, b, e_1', e_2', e_3' becomes symmetric with itself with respect to a circle, which runs from e_2' to the "midpoint" of the segment d, b. We see, that the edge $x = z$, $xz - 2yt = 0$ furnishes the linear continuum of the signature $(0,3;2,4)$, for which our fundamental theorem was proved in §27 on p. 414 ff.

If we go beyond the edge in question to the plane $x = z$, then from now on only the diagonal symmetry remains subsisting, and our polygon, in the ζ-halfplane, obtains the type represented in figure 95. The half lying outside the symmetry circle d, e_2', b say, [436]
gives the halfsheet of the Riemannian surface, sketched in figure 96. The corresponding symmetry coming to light here yields $\lambda_a = \lambda_c$ and therefore, as we asserted, the plane $\xi = \zeta$. The supervening edge $x = y = z$ furnishes polygons with threefold diagonal symmetry; their type is given by figure 97, and the corresponding halfsheet of the surface is sketched in figure 98. We come to the linear continuum of the signature $(0,3;2,3)$ for which the fundamental theorem is likewise already proved in §27 on p. 414 ff.

---

[111] One has to choose the two-sheeted Riemannian surface over the z-plane here. So that, not the real z-axis, but a proper circle, represents the symmetry line. This circle is then to be drawn together into a point, whereby, besides the remaining v_1, v_2, v_3, it furnishes a fourth branch-point, which corresponds to the parabolic cusp of the boundary polygon.

[112] Compare the exhaustive representation in section III of the second work of the author cited on p. 286.

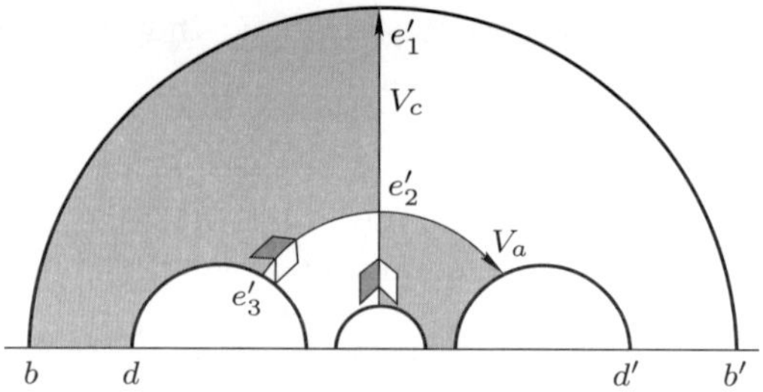

Fig. 93.

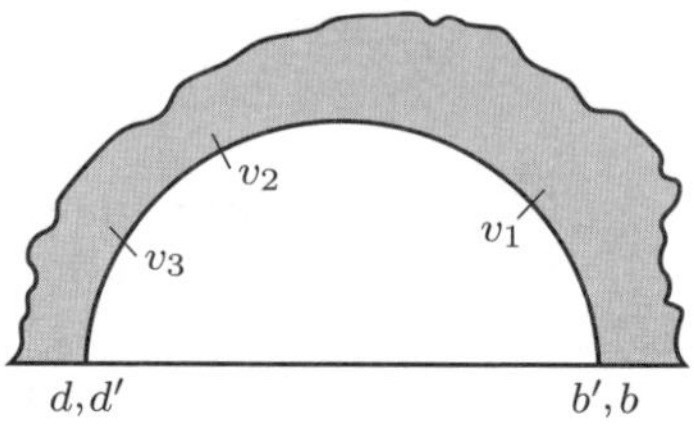

Fig. 94.

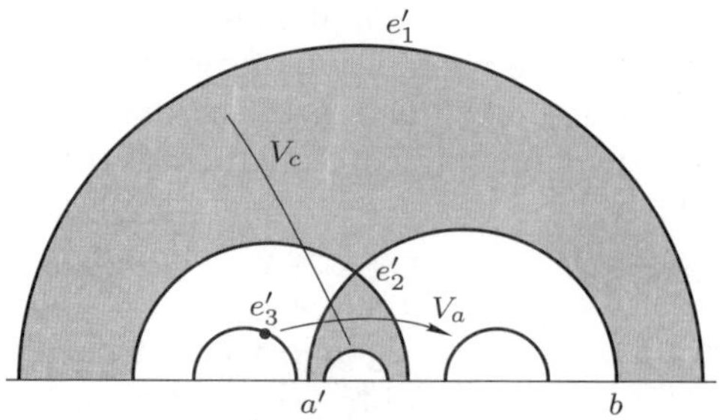

Fig. 95.

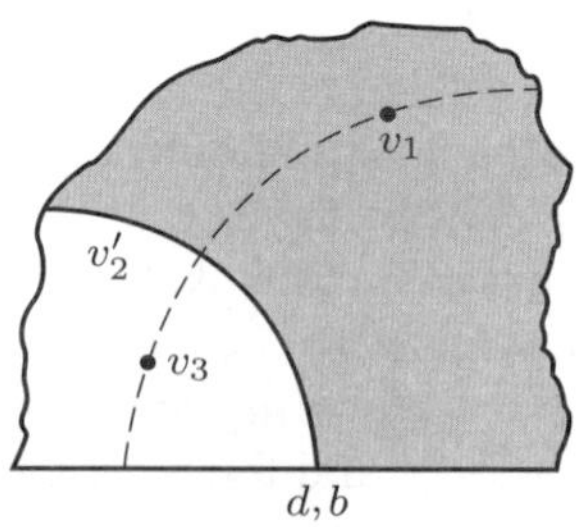

Fig. 96.

The supervening plane $y = z$ furnishes, in the projective plane, a plane quadrangle with the generating V_b, V_c with diagonal symmetry. figure 99 represents the polygon type entering here, which, just on account of the symmetry, furnish us $\lambda_b = \lambda_c$ and therefore $\eta = \zeta$. The linear continuum of the boundary models is especially interesting,

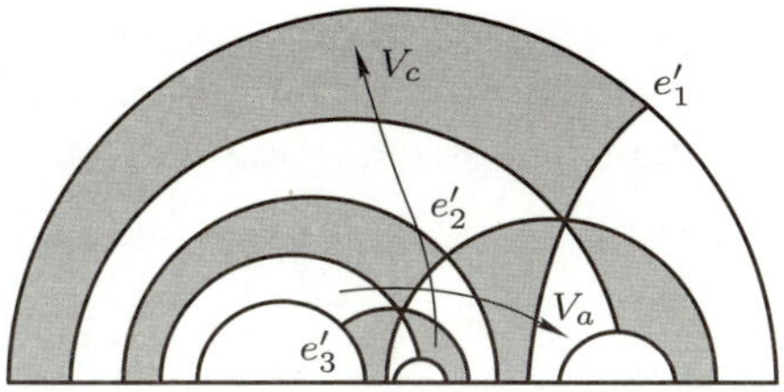

Fig. 97.

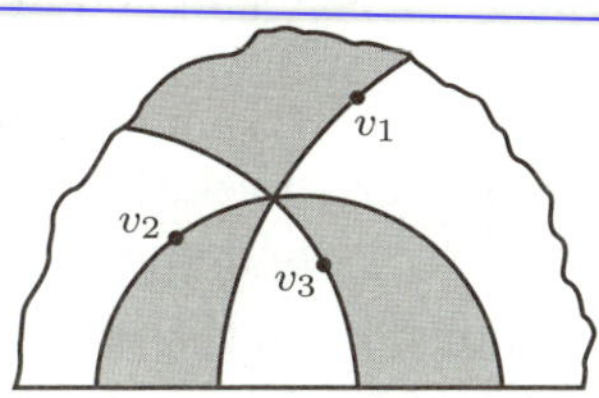

Fig. 98.

which corresponds to the transition to the points $x = 0,\ t = 0,\ y = z$. In this case, the segment of the real ζ-axis denoted in figure 99 by a, d draws itself together into a point, at which the points e'_2, e'_3 fuse into a parabolic cusp. We arrive at the linear continuum of the signature $(0,3;2,\infty)$, which we also reached from the surface $xz - 2yt = 0$, and therefore, from figure 93 by fusion of the points e'_2, e'_3 into a parabolic cusp. For the pentahedron $B_g^{(0)}$, for the representation of this continuum, we had to bring in the various surfaces Φ_3, which are tangential to the plane $x - 2t = 0$ at the plane in question. For the tetrahedron $B_f^{(0)}$, it is simply a question of the edge $\eta = \zeta$ of the surface F_3; on the Riemann surface, the branchpoints v_2 and v_3 have come into coincidence, have thereby canceled one another out and furnished one point e in each sheet. The fundamental theorem is also proved for this continuum already, in §27 on p. 414 ff.

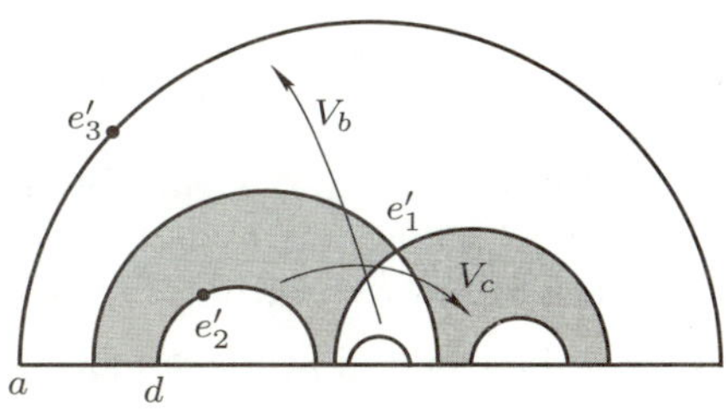

Fig. 99.

It would now remain, also to consider the planes $t = 0$ and $\tau = 0$, as well as their corresponding edges. But here we come to especially simple relations (cf. p. 335), and restrict ourselves to the following brief indications. If we approach the surface $t = 0$, say, by drawing together into points all four circles assigned to one another in figure 99 by V_b and V_c, then in the limiting case, the group Γ will only consist of the identical [437] substitution 1; for the polygon spans the entire ζ-plane. In the Riemann surface, the branchpoints v_2 and $\overline{v_2}$, as well, too, as v_3 and $\overline{v_3}$ have come into coincidence on the real axis. Our boundary function z, however, is simply given by

$$\frac{z-v_1}{z-\overline{v}} = \left(\frac{\zeta-e_1'}{\zeta-\overline{e_1'}}\right)^2.$$

We arrive at the plane $\tau = 0$, and are even able to recognize a single-valued correspondence at these places of elementary character, by continuing the investigation, an also considering the two other kinds of passage to the limit possible here, for which one of the points e_2', e_3' is always held in the "interior" of the ζ-halfplane.

Further, we now also have to full carry through the continuity study for the other boundary surfaces, considered before, of our domain $B_g^{(0)}$ and $B_f^{(0)}$. One will, hereby, in the usual manner, cover the individual boundary surface of $B_g^{(0)}$ by a family of curves simply and completely, and will always judiciously cut out such families by the families of our surfaces Φ_3 of third degree:

$$t(x^2+y^2+z^2) - xyz - (j_e+2)t^3 = 0,$$

where the parameter j_3 has to decrease from -2 to $-\infty$.

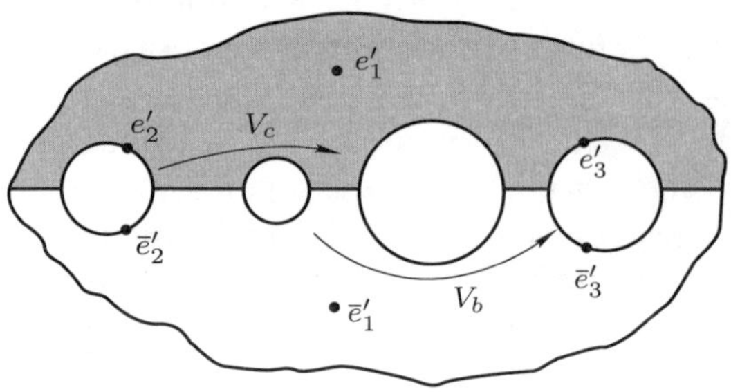

Fig. 100.

After finishing these individual considerations, one fills up the interior of the pentahedron simply and completely with the surfaces Φ_3 of this family and discusses the mapping onto the tetrahedron $B_f^{(0)}$, always utilizing the single-valuedness and continuity theorem. One arrives at the result, *that the domains* $B_g^{(0)}$ *and* $B_f^{(0)}$, *and consequently, also the hexahedron* B_g *and the tetrahedron* B_f *are reversibly single-valuedly and continuously mapped onto one another, with which the fundamental theorem for the signature* $(1,1)$ *is proved.*

[438] The third case of the continua B_g and B_f, which are three-dimensional, was that of the signature $(0,4;l_1,l_2,l_3)$. The theory of this fashions itself exactly at that of the case of the signature $(1,1)$ just carried to a conclusion; except this theory, at least if no two of the number, l_1, l_2, l_3 are equal to one another, is even more simple insofar as the inequalities $x \le y, x \le z$ and the corresponding ones $\xi \le \eta$, $\xi \le \zeta$ drop out. B_g, according

to p. 355 ff., put itself together from two pentahedra lying next to one another, one of which is bounded by a surface Φ_3 (corresponding to the parabolic limiting case), the three surfaces H_1, H_2, H_3 of p. 352 and the surface $t = 0$. B_f is simply the continuum of the halfplanes signed with e_1, e_2, e_3, whereby the notations of these points e_i are not permutable. Therefore B_f is the "dihedron", which is bounded by our surface F_3 and the plane $\tau = 0$. The effectuation of the continuity proof we omit here, since in fact, it is in no position to present anything new with respect to the foregoing development.

The developments of the preceding chapter verify the assertion made on p. 285, that the difficulty of the continuity proof is grounded in the circumstance, that one is not in a position to sufficiently survey the continua, of which something is to be proved. In the last considered special cases, for which we possessed a complete overview of the continua, the procedure of reasoning of the continuity proof leads relatively easily to our goal, while in the general case, the definitive effectuation of this proof runs aground, above all, on the circumstance, that we lack means for a perspicuous representation of the continuum B_f of the Riemann surfaces.

The state of affairs established the great significance of the direct methods of proof of our fundamental theorems, to be discussed in the next chapter. As soon as these theorems have found an objection-free proof in the general case as well, we shall be justified, *in also utilizing, in place of the moduli we are lacking, our invariants j of the polygons for the exhaustive representation of the continua of algebraic models.*

2

Proof of the principal-circle and the boundary-circle theorem

The two theorems which answered the fundamental problems I and II on p. 45 ff., we had already invested with the names "boundary-circle theorem" and "principal-circle theorem" in the preceding chapter. Problem III on p. 46 attached to an arbitrary Riemann surface with a genus $p > 1$, on which was drawn a system of p reentrant cuts, not meeting one another[1]. The associated fundamental theorem then asserts the mappability of the cut-up surface by a polymorphic function $\zeta = f(z)$ onto a domain bounded by $2p$ closed boundary curves of the kind described on p. 35 under β.[2] In order to also have a short and fitting name for this theorem, the name "*reentrant-cut theorem*" recommends itself. [439]

In the direct proofs of these fundamental theorems, to be developed further on, the first two, therefore the boundary-circle and the principal-circle theorems, are to be treated together, which is to be done in the present chapter. The proof of the reentrant-cut theorem, which is significantly more difficult and demands still further-going preparations, is to follow in the third chapter.

First of all, let us present here a few historical notices concerning the direct methods of the proof of all three theorems[3]. [440]

2.1 Historical information concerning the direct methods of proof of the fundamental theorems

The exist two direct appearances of the proof of the fundamental theorems, which rest on ideas of H.A. Schwartz, and of which, the first approach, by its nature, restricts itself to the cases with principal resp. boundary circle, and so far is only carried through for the boundary-circle theorem. We accordingly restrict the next following statements to

[1] For $p = 1$, problem III is already answered by a well-known theorem from the theory of doubly-periodic functions.

[2] It is a question of that kind of automorphic functions, to which the investigations of Schottky, cited in I on p. 104, are devoted.

[3] In contrast to the continuity methods of the previous chapter, the proofs in question are designated as "direct", insofar as they directly effect the existence of the postulated polymorphic function on the individual Riemann surface.

this theorem. The approach to the proof attaches to the *hyperbolic determination of measure*, which was taken as a point of departure in I on on p. 26 ff. for the surface with a boundary circle. In the sense of this determination of measure, the substitutions of the individual group Γ coming into consideration furnished congruent displacements of the "hyperbolic plane" into itself, whose "infinitely distant points" were, in known fashion, furnished by the periphery of that boundary circle.

We base represented the "hyperbolic arc-element" $d\sigma$ in formula (5) on p. 171, under the assumption that the boundary circle in the plane of the variable $\zeta = \xi + i\eta$ is the circle of radius 1 about $\zeta = 0$, in the form:

$$d\sigma = \frac{|d\zeta|}{1-\xi^2-\eta^2}. \tag{1}$$

If now, first of all, there is a polygon of the character (p, n), then we are able to conformally map this, by means of an associated automorphic function $z = \varphi(\zeta)$, onto a Riemann surface F over the plane of $z = x + iy$. If $ds = |dz|$ is then the differential on the surface F corresponding to $d\sigma$, then on account of the conformality of the mapping, we may set:

$$d\sigma = Q \cdot |dz|, \tag{2}$$

where Q only depends on the plane and not on the directions of the differentials of arc. If we consider $Q = Q(x, y)$ on the Riemann surface, then on account of the underlying automorphic character, this function, upon continuation over the surface, is *single-valued* on F. Thereby, this real function $Q(x, y)$ is, in general, finite, continuous and > 0 on F, exceptions from the easily determinate character only take place at the n places $e_1, e_2, \ldots, e_n$ of F, which originate from the fixed polygon vertices, at the point $z = \infty$ and at the branchpoints.

In the closer investigation of these exceptional places, it yields itself, that the function:

$$u(x, y) = 2\log Q(x, y), \tag{3}$$

[441] whereby the real value of the logarithm is meant, is also *single-valued on F and, in general, finite and continuous with exception only of those three kinds of points just named, at which u becomes logarithmically infinite in a manner easily given.* For this function $u(x, y)$, it is now possible to develop, from the relations (1) and (2) by means means of an intermediate calculation not to be given here, *the differential relation*:

$$\frac{\partial^2 u}{\partial x^2} + \frac{\partial^2 u}{\partial y^2} = 8e^u. \tag{4}$$

This latter fact is immediately understandable here from another aspect. In I on p. 30 ff, it was represented, that the relations of the hyperbolic determination of measure are elementarily realizable on *surface of constant negative curvatures*, whereby then, e.g., the various polygons of the net carried over onto that surface come out congruent in the elementary sense. In Bianchi's "*Vorlesungen über Differentialgeometrie*", one finds the mapping of the surface of the principal circle onto a so-called "*pseudospherical surface*" (surface of constant negative curvature) represented[4]. From the general formulas of the theory of curvature[5], here now yield themselves, for the conformal

[4] Cf. kap. XVI of the first edition of the German revision (Leipzig, 1899).

[5] Cf. Bianchi, l.c., §§ 35, 36 and 37 on p. 66 ff.

mapping of our Riemann surface F over the plane $z = x + iy$, the following rules: if $d\sigma$ is the arc-differential on the pseudospherical surface, then one has, first of all:

$$d\sigma = Q(x, y) \cdot \sqrt{dx^2 + dy^2}; \tag{5}$$

and on the other hand, the constant curvature K calculates itself from the relation:

$$K = -\frac{1}{Q^2}\left(\frac{\partial^2 \log Q}{\partial x^2} + \frac{\partial^2 \log Q}{\partial y^2}\right). \tag{6}$$

Now if we set $K = -4$, these are directly our formulas (2) and (4). *Accordingly, in our case, it is a question of a conformal mapping of our surface F onto a pseudospherical surface of the curvature $K = -4$.*

If we imagine F, say, transformed by a canonical cut system of the usual form (cf. I on p. 183) into a simply connected surface, then we know, that this furnishes a polygon of the known kind, i.e. therefore, a simply covered one, or as we will say briefly further on, a "*Schlicht*" domain on the pseudospherical surface. Also, the continuation over periodpaths of the un-cut-up surface or circuits about points e lead to congruent, adjacently positioned domains. The knowledge is now fundamental, *that these two properties of the mapping of our surface F onto the pseudospherical surface are also able to be directly deduced from the properties of the function $u(x, y)$ on the basis of the preceding formulas.* Upon this circumstance rests the idea of the methods of proof under [442]
discussion here of the boundary-circle theorem. If an *arbitrary Riemann surface F* is given, and if, on this, there are marked n places $e_1, e_2, \ldots, e_n$, to which n whole numbers $l_1, l_2, \ldots, l_n$ are assigned in known fashion, *then we ask for the existence of a solution u of the differential equation (4), single-valued on the surface F, which possesses the behavior of our function $u(x, y)$ considered above at the n points e_i, the branchpoints and the point $z = \infty$ and except for this, is everywhere finite and continuous on F.* If we then coordinate, to any first arc-differential $|dz|$ of F, an arbitrarily situated and directed arc-differential:

$$d\sigma = e^{\frac{u}{2}} \cdot |dz|$$

of the pseudospherical surface, then upon continuation from here, the entire mapping of the surface F onto the pseudosphere is uniquely determined, and has, as remarked, the principal properties which the boundary-circle theorem asserts.

The proof of the boundary-circle theorem is reduced in this fashion to the existence-proof of a solution u of the differential equation (4), provided with certain properties, on our arbitrarily chosen surface F. On the instigation of H. A. Schwartz, who furthermore had imparted the basic ideas of the preceding development only by word of mouth to various mathematicians, the Gottinger Gesellschaft der Wissenschaften, in December 1889, set as prize-problem for the year 1891, the presentation of an existence-proof of the solution u.

The further development has then been this, that this existence-proof of the solution u was actually furnished by E. Picard in a great series of papers[6]. In the basic ideas, the Picard arguments attach themselves to the known methods of Schwartz and [443]

[6] The first announcements are made in the Compt. Rend. of Sept. 23, 1889 and June 13, 1890; among the detailed representations, the two works are mainly to be named: "*De l'equation*

Neumann for the solution of the Laplacian differential equation $\Delta u = 0$. First of all, the "boundary-value problem" of the differential equation (4) is solved for domains chosen sufficiently small. Then transition to larger domains, as well as, finally, to the closed surface, is obtained by joining together the partial domains by means of the alternating procedure. For the last step, ie. for the transition to the closed surface, difficulties made their appearance, which were first to be overcome later.

In addition, in his investigation mentioned, Picard presupposes, that the n whole numbers $l_1, l_2, \ldots, l_n$ are all finite, which amounts to the fact, that the points $e_1, e_2, \ldots, e_n$ all have elliptic character. Poincaré, in a great work in the year 1898[7], has succeeded in extending the investigation also to the case, that the points e are arbitrarily of elliptic or parabolic character.

The second approach to the proof of the fundamental theorems is entirely different from that previously described, distinguishes itself in great measure by its capability of being generalized. For this approach too, one is indebted to word-of-mouth stimulation of Schwartz, who imparted his relevant ideas in Easter 1882 to Klein and a year later to Poincaré. In connection with the boundary-circle case we are able to sketch a first perception of the method of proof in question, say, in the following manner.

Let any boundary-circle polygon P be given, which we map onto a Riemann surface F by means of an associated automorphic function $z = \varphi(\zeta)$. Also, we can and will define the latter, by introducing a second suitably chosen algebraic function of this, which appears bound with z by the algebraic equation:

$$G(w, z) = 0, \tag{7}$$

whereby this equation leads from itself to the surface F and therefore can serve as the replacement of the surface.

We now bring in, further, the net N of the polygons belonging to P, covering the surface of the boundary circle schlichtly and map this by means of the function $z = \varphi(\zeta)$: *there arises a connected Riemann surface, consisting of infinitely many copies of the surface F layered over one another, which we will denoted by F_∞ and designated as a "covering surface" of F*. Thereby, we connect the various copies of the surface F

[444] over one another exactly as as the N originals, namely the polygons of the net N; for the covering surface F_∞ is indeed mapped reversibly single-valuedly, continuously and conformally onto the net N, and with this, onto the surface of the boundary circle. In particular, however, follows: *Corresponding to the structure of the net N, the covering surface represents a "simply connected" domain, inside which, each closed curve can be drawn together continuously and without tearing into a point.*

Now the following two ideas establish our new approach to the proof: *If any Riemann surface F, outfitted with n points e and associated whole numbers l, is given, then we are able to produce for this, in completely corresponding fashion, a uniquely determined covering surface F_∞*. The solution of this problem, which rests on considerations of "Analysic situs", poses no particular difficulty at all, as we shall soon see.

$\Delta u = Ke^u$ *sur une surface de Riemann fermée*", Journ. de. Math. (4e série), Vol. 9 (1893) and "*De l'integration de l'equation $\Delta u = e^u$ sur ua surface des Riemann fermée*", Crelle's Journ. Vol. 130, p. 243 (1904).

[7] "*Les fonctions fuchsiennes et l'èquation $\Delta u = e^u$*", Journ. de. Math. (5e série, Vol. 4, p. 137 ff.).

And now the second idea of Schwartz is: *that one is to attempt to map this "simply connected", infinitely-sheeted surface F_∞ conformally onto the Schlicht surface of a circle in the same manner as one is in the position of accomplishing this mapping for finitely-sheeted, simply connected surfaces with a boundary assumed analytic, say.* Afterwards, it is then yet to be proved, that the images, positioned next to one another, of the various copies of F are all circle-related, as well as, that the mapping can be effected essentially in only one way. Both of these, at least in the boundary-circle case, initially the sole one under consideration here, are not difficult.

Poincaré has now, first of all, taken up this Schwartzian impulse, and even generalized it in very remarkable fashion [8]. Namely, Poincaré has noticed, that it is possible, without increasing the actual difficulties of the investigation, to replace the algebraic relation (7) by an arbitrary *analytic* one, whereby then (as long as the algebraic case is not present), the surface F. lying over the z-plane, of the analytic function $w(z)$, is itself already infinitely sheeted. To be sure, the studies of Poincaré's at that time, which were to prove the possibility of mapping the covering-surface onto the Schlicht surface of a circle, were only first preliminary. From the aspect of presuppositions, restrictive assumptions underlay the development, and also, on the other hand, from the aspect of results, not everything desired was attained. In the latter respect, it was found, that the Schlicht image domain indeed lies in the interior of the boundary circle, but not [445]
that it necessarily fully covers this circle[9].

Not until 1900 did a few steps forward make their appearance, which partially concern themselves with isolated points of the general Poincaré argument of 1883, partially with special cases[10]. But then, in the year 1907, almost simultaneously and independently of one another, Poincaré himself and P. nKoebe have generally and completely solved[11] the problem, including our boundary-circle problem, referring itself to arbitrary analytic relations (7).

The designation "*uniformization*" of analytic functions or analytic curves, become curvent, brings to expression, that in the polymorphic function $\zeta(z)$, conversely, w as well as z become *single-valued* automorphic funcitons[12].

With this, therefore, at the same time, a second direct proof for our "boundary-circle theorem" was given which, as one can say, follows the "*method of the covering-*

[8] See the note "*Sur un théorème de la théorie générale des fonctions*", Bull. de la soc. mathém. de France of May 18, 1883, Vol. 11 (p. 112 ff.).

[9] Also compare the section "Uniformisierung analytischer Bezienungen mittelst automorpher Functionen" in Hilberts Paris address "*Mathematicshe Probleme*", Gött. Nachr. of 1900, p. 253 ff.

[10] See Osgood, "*on the existence of the Green's function for the most general simply connected plane region*", Amer. M. S Transact., Vol. I. p. 310 (1900), Brodén, "*Bemerkungen über die Uniformisierung analytischer Functionen*", Lund (Berlin printing office, 1905), as well as the works of Johansson "*Über die Uniformisierung Riemann'scher Flächen mit endlicher Anzahl von Windungspunkten*" Act. soc. Fenn. Vol. 33 (1906), " *Ein Satz über die conforme Abbildungeinfach zusammenhängender Flächen auf den Einheritskreis*", Math. Ann Vol. 62, p. 177 (1906).

[11] Poincaré "*Sur l'uniformisation des fonctions analytiques*", Act. Math, Vol. 31, p. 1 ff.; Koebe "*Über die Uniformisierung beliebiger analytischer Curven*" Gött. Nachr. of 1907, p. 191ff.

[12] Furthermore, concerning the significance of ζ as a "*uniformizing variable*" see the discussions on p. 40 ff.

surface", Already, a short time before, Koebe had succeeded, by means of the same method, in proving the "principal-circle theorem"[13]. In this connection, Koebe has treated both theorems in detail in the paper "*Über die Uniformisierung der algebraischen Curven, I*", which has been dedicated to F. Klein for his sixtieth birthday[14].

The method of the covering-surface, however, shows itself to be furthergoing. In [446] fact, Koebe has succeeded, in a certainly far more difficult investigation, by means of this method, in also proving the "reentrant cut theorem"[15]. For this theorem, Koebe has then also developed a second method, resting on an "*interacting procedure*", by whose deepening, he has even succeeded in proving that general fundamental theorem set up by Klein, which we mentioned on p. 47 (at the close of §11)[16]. In detailed fashion, Koebe has represented a first part of his investigations belonging here in the paper: "*Über die Uniformisierung der algebraisichen Curven, II*"[17].

On the basic works of Koebe's, so rich in consequences, we therefore possess direct proofs for the three off-mentioned theorems according to a unified method, namely that of the "covering surface". Therefore, for the disposal of the program taken on by setting up the three problems I, II and III on p. 45 ff., the most expedient thing for us will be to take as a point of departure for the further following detailed representation of the direct proof of our theorems, tying into Koebe, the "method of the covering surface". This is to be done in such a fashion, that we dispose of the principal-circle and the boundary-circle theorems in the present chapter, and in the following one, the reentrant cur theorem. In the next four paragraphs here follow, first of all, precatory developments of a more general kind.

2.2 Theorems on logarithmic potentials and Green's functions

In "M" I on p. 504 we have considered the "*logarithmic potentials*" (there called "potentials" for short) belonging to a Riemann surface. We have to utilize a series of general theorems from the theory of these potentials, which are to be collected together here beforehand for the avoidance of future interruptions. For the proofs, we mostly refer to the original papers, whereby, above all, the fundamental works of H. A. Schwartz[18]
[447] come into consideration, or also to the textbooks on function theory. In the latter re-

[13] "*Über die Uniformisierung reeller algebraischer Curven*", Gött. Nachr. of 1907, p. 171 ff.

[14] Math. Ann., Vol. 67, p. 143 (1909).

[15] "*Über die Uniformisierung der algebraischen Curven durch automorphe Funktionen mit imaginären Substitutionsgruppen*", Gött. Nachr. of 1909, p. 68 ff.

[16] "*Über die Uniformisierung der algebraischen Curven durch automorphe Funktionen mit imaginären Substitutionsgruppen*", (continuation and conclusion), Gött. Nachr. of 1910, p. 180 ff.

[17] Math. Ann. Vol. 69, p. 1 ff. (1910).

[18] "*Über die Integration der partiellen Differentialgleichung:*

$$\frac{\partial^2 u}{\partial x^2} + \frac{\partial^2 u}{\partial y^2} = 0$$

unter vorgeschriebenen Greaz - und Unstetigkeitsbedingungen", Berlin. Monatsber. of 1870, p. 767 ff.; "*Zur Integration der partiellen Differentialgleichung* $\Delta u = 0$", Crelle's Journ. Vol. 74, p.

spect, we prefer the multifaceted and surveyable representation of Osgood[19], who has treated the theory of the logarithmic potential in the 13th chapter of his work named below.

Let a connected domain B be spread out over the plane of the complex variable $z = x + iy$, which is to be finitely-sheeted, is to have only finitely many branchpoints and similarly many boundary curves which latter comprise the entire boundary R of B. In any case, the degree of the connectivity of this domain B is itself a finite one. Arbitrarily many of the sheets of B may stretch themselves into the infinite, or, as we will say, contain the point ∞. We will only assume for simplification, that the boundary R does not run through a point $z = \infty$.

By a "*logarithmic potential*" $u(x, y)$ of the domain B we understand, then, a *single-valued real function of position in B, which, apart from finitely many discontinuity points, behaves itself "harmonically" in the neighborhood of each place of B*. The latter property says, that the derivatives $\frac{\partial u}{\partial x}, \frac{\partial u}{\partial y}, \frac{\partial^2 u}{\partial x^2}, \frac{\partial^2 u}{\partial y^2}$ exist, that u together with these derivatives, as was already generally assumed, are single-valued as well as finite and continuous and that, finally, the Laplacian equation:

$$\frac{\partial^2 u}{\partial x^2} + \frac{\partial^2 u}{\partial y^2} = 0 \tag{1}$$

is fulfilled for the two derivatives of the second order. At each of the finitely many discontinuity places, however, a determinate function $\omega(x, y)$, becoming discontinuous there, is to exist, so that $u(x, y)$, in the neighborhood of that place, permits the representation:

$$u(x, y) = \omega(x, y) + \text{harm.Funct.} \tag{2}$$

Only the place $z = \infty$ and the branchpoints of the domain B hereby require a supplementary definition. We map the neighborhood of the point ∞ and that of a ν-sheeted branchpoint lying at z_0 by:

$$z' = \frac{1}{z}, \quad \text{resp.} \quad z' = \sqrt[\nu]{z - z_0}$$

onto the Schlicht neighborhood of the origin of the z'-plane. If u hereby goes over into $u'(x', y')$, then we say u behaves itself harmonically in the neighborhood of that special point of B, in case u' behaves itself harmonically in the neighborhood of the origin of the z'-plane. [448]

For simplification, we also assume, that none of the prescribed discontinuity places of $u(x, y)$ lie on the boundary R of the domain B. The boundary values of $u(x, y)$, which we will distinguish by the notation $U(x, y)$, will then represent an everywhere finite continuous sequence of values. Let the assumption hold, *that $u(x, y)$, from the interior of the domain B, goes over uniformly continuously into the boundary values U*. Also let

218 ff. (1870 and 71). Both papers are printed Vol. 2 of H.A. Schwartz "*Gesammelten mathematischen Abhandlungen*", (Berlin 1890) to which we shall refer further on for the sake of easy accessibility.

19 "*Lehrbuch der Funktionentheorie*", Vol.1 (Leipzig 1907).

the first derivatives $\frac{\partial u}{\partial x}, \frac{\partial u}{\partial y}$ still exist and be continuous on the boundary R; and similarly, let the interior values of these derivatives go over uniformly continuously into the boundary values.

Among the theorems on logarithmic potentials, we place the following at the forefront: *If, by means of any anylytic function $z' = f(z)$, we effect a conformal mapping of the domain B onto a domain B' over the z'-plane, which has all the properties of shape required of the domain B, then $u(x, y)$ hereby goes over again into a logarithmic potential $u'(x', y')$ of our kind for B'; this therefore obtains the character of invariance with respect to conformal mapping.* This is a basic property of the logarithmic potentials, which rests on the nature of the conformal mapping, and therefore, in calculational terms, on the Cauchy-Riemann differential relations[20].

A series of further properties of our potentials u yield themselves from the well-known Green's integral theorems. We next repeat the known theorem, also already utilized many times in "M" I on p. 504 ff.: *If B_0 is a partial domain of B outfitted with finitely many boundary curves (which may also be identical with B itself), in which u is free of discontinuities and consequently behaves itself harmonically everywhere, then u (apart from the case that u is everywhere constant in B_0 cannot assume its greatest or smallest value at any "interior" point of B_0; rather, each interior value of u lies between the maximum and minimum of the values u on the boundary R_0 of B_0, without being equal to one of these extreme values.* Namely, if u behaves itself harmonically in the neighborhood of a point z_0, then describe a circle in this neighborhood of a point z_0, then describe a circle in this neighborhood about z_0 as center; then the value of u of z_0 is the mean value of all values u entering on that circular periphery, from which then the given theorem follows easily[21].

[449] A consequence of the last theorem is, *that a logarithmic potential u of the domain B is uniquely determined by its boundary values U and the manner in which u becomes discontinuous in the domain B.* Namely, if there were a second potential u_1 with the same boundary values U and the same manner of becoming discontinuous at the prescribed points of B, then the difference $(u_1 - u)$ would be everywhere harmonic in B, and equal to zero along the entire boundary R of B, and therefore also zero everywhere in the interior of B.

To the logarithmic potentials of domain B belong, in particular, the "*Green's functions*" of this. If z_0 is an arbitrary point inside B, which we will briefly designate further on as "point O" of B, then we understand by r the absolute value of the difference $(z - z_0)$:

$$r = |z - z_0|.$$

We then designate as the Green's function $g(x, y)$ of the domain B belonging to the point O a logarithmic potential of B, which

1. in the neighborhood of the point O permits the representation:

$$g(x, y) = \log(r^{-1}) + \text{harman.Funct.}, \tag{3}$$

2. is otherwise everywhere harmonic in B, and

[20] Compare, for this, Osgood "*Lehrbuch usw.*" on p. 565.

[21] Cf. Osgood "*Lehrbuch usw.*", p. 544 ff.

3. vanishes everywhere along the boundary R of B.

If the Green's function of a domain B, with prescribed discontinuity place O, exists at all, *then, according to the theorem last stated, this is also uniquely determined.*

If one has the Green's function for an arbitrary point $z_0 = x_0 + iy_0$ of the domain B, *then one is in the position to solves, in general, the "boundary-value problem" (cf. "M" I on p. 510) for this domain.* Namely, if $u(x, y)$ is a potential, everywhere harmonic in B, then by means of Green's theorem, one exhibits the following equation for the value $u(x_0, y_0)$ of u at the place z_0 in question:

$$u(x_0, y_0) = \frac{1}{2\pi}\int_{(R)} U\frac{\partial g}{\partial n}ds, \tag{4}$$

whereby the integral is to be extended over the boundary R of B in the direction of circuit which is positive for this domain, U, as above, denotes the boundary values of u and by $\frac{\partial g}{\partial n}$ is meant the differential quotient of our Green's function belonging to the place z_0, taken along the inner normal of R.[22]

For the case that B, in particular, is a Schlicht circular disc, then we immediately succeed in forming the Green's function of B belonging to nay interior point O of this circular disc. If O' is symmetrically situated to O with respect to the circular periphery R, and if one denotes by ρ and ρ' the distances of any point $z = x + iy$ of the circular disc form O and O', then the quotient $\frac{\rho}{\rho'}$, as is known, is constant along the periphery R and has these the value k, say. *The Green's function of the circular disc belonging to the point O is then simply:* [450]

$$g(x, y) = -\log\rho + \log\rho' + \log k \tag{5}$$

and the substitution of this expression of the Green's function into the formula (4) furnishes the solution of the boundary-value problem for the circular disc by means of the Poisson integral[23].

The Possion integral, for its part, becomes the source of a series of further important theorems in potentials. E.g., we state the theorem: *A logarithmic potential $u(x, y)$ of our kind, in the domain B of its definition, has partial derivatives of all orders, each one of which behaves itself harmonically again*[24]. Above all, however, one can deduce, from the theorems on the Possion integral, with the aid of the theorem set up above on the maxima and minima of the potentials $u(x, y)$, the following convergence theorem, fundamental further on[25]:

If the infinite series:

---

[22] The definition of the integral in (4) on the right side poses no difficulties in the cases coming into consideration further on, where we always put the boundary R together from finitely many portions of analytic curves.

[23] For this, see especially the two papers of Schwartz, cited on p. 446, in the "*Gesammelten mathematischen Abhandlungen*" Vol. 2, p. 175 ff. as well as the supplement there on p. 360. Also, compare Osgood, "*Lehrbuch usw.*" on p. 554 and "M" I on p. 513.

[24] Cf. Osgood, "*Lehrbuch usw.*" on p. 560.

[25] See, for the detailed proof, Osgood, "*Lehrbuch usw.*" on p. 571.

$$u_1 + u_2 + u_3 + u_4 + \cdots \tag{6}$$

of potentials of the domain B converges uniformly in this domain, its boundary included, then the sum again represents a logarithmic potential of our kind there.

The potentials $u(x, y)$ stand in the relation to the *analytic functions* of the domain B exhibited in "M" on p. 505 ff. If, by means of the integral:

$$v(x, y) = \int \left(\frac{\partial u}{\partial x} dy - \frac{\partial u}{\partial y} dx \right), \tag{7}$$

carried from any initial point chosen fixed in B to the place $z = x + iy$, we form, from
[451] the logarithmic potential $u(x, y)$, the "*conjugate potential*" $v(x, y)$ to $u(x, y)$, *then, as is known*:

$$f(z) = u(x, y) + i v(x, y). \tag{8}$$

is an analytic function of z, defined in the domain B and uniquely continuable there, but which, naturally, in case B is multiple-connected, or if there is a discontinuity point, by no means needs itself be single-valued in B. From here, one can also carry over the theorems on analytic continuation backwards to the potentials $u(x, y)$; however, one can also independently develop a doctrine of "*harmonic continuation*" for the potentials $u(x, y)$, as this, e.g., is done in Osgood l.c., on p. 575 ff.

To the convergence theorems set up before corresponds, as we yet mention, the known theorem, *that a uniformly convergent series of analytic functions again furnishes an analytic function.*

2.3 More on the solution of the boundary-value problem

If one gives a continuous sequence of real values U on the boundary R of a domain B of our kind, then, as already remarked, there is at most one function $u(x, y)$, which is everywhere single-valued and harmonic in the domain B and goes over continuously into the proposed boundary values U. The "boundary-value problem" requires, the existence of this function $u(x, y)$ be proved and actually produced.

If we are in possession of the Green's function of the domain B for an arbitrary point O of this, then we could solve the boundary-value problem for this domain by means of the equation (4) on p. 449. Since the Green's function could be produced for the circular disc, the solution of boundary-value problem for the circular disc by means of the Possion integral attached itself to this.

Indirectly, with this, the boundary-value problem is also, at the same time, to be considered as solved for all those simply connected domains, which can be mapped by known analytic functions single-valuedly onto Schlicht circular discs. Here belong, e.g., multiple covered circular discs with an inner winding-point, as we had cut them out in "M" on p. 510 at the branchpoints of Riemann surfaces, also "circular discs" with the center $z = \infty$, ie. parts of the Schlicht z-plane running outside a very large circle, and finally, also the circular sectors utilized above (p. 9 ff.) in the existence proof of the automorphic functions.

Now, further, there intervenes the known "*combination method on the basis of an*
[452] *alternating procedure*" of Schwartz and Neumann, discussed in detail in "M" I on p. 514

ff. If one has two domains B_1 and B_2 which partially cover one another, and if one is in a position to solve the boundary-value problem for B_1 and B_2 individually, then that method yields the existence and method of production of the function $u(x, y)$, which solves the boundary-value problem for the domain arising from B_1 and B_2 by union.

These premisses were sufficient, in order in "M" I on p. 508 ff., to give the existence proof of algebraic functions on arbitrarily given Riemann surfaces, as well too, in the present volume on p. 9 ff., to give the existence proof of automorphic functions for an arbitrary discontinuity domain of the ζ-plane. The thing to do was simply cover the Riemann surface resp. the discontinuity domain with a finite number of Schlicht circular discs resp. by circular discs with winding points or by circular sectors bit by bit completely, and to exercise the combinational procedure described above upon addition of each new circular disc.

For our further purposes, it appears desirable, in the solution of the boundary-value problem, to go one step further. In fact, we will now so imagine the domain B taken as a basis in the previous paragraph, *that its boundary R consists of a finite number of portions of analytic curves, of which every two neighboring ones meet at a non-vanishing angle.* The solvability of the boundary-value problem for domains B of this general shape was already shown by Schwartz in the year 1870.[26]. We will briefly sketch his argument here.

If K is one of these curve-portions, then the assumption that K be analytic clothes itself in the form that the coordinates x and y of the points of K be representable by means of a real parameter by *two real analytic functions*:

$$x = \varphi(t), \qquad y = \psi(t) \tag{1}$$

of t. And indeed, let us describe K completely and simply, if we let the parameter t run through the interval

$$\alpha \leq t \leq \beta. \tag{2}$$

In addition (which will be fulfilled without further ado for the curves K to be utilized later), for each point t_0 of the interval (2), the derivatives of the functions (1) are also to fulfill the inequality: [452]

$$|\varphi'(t_0)| + |\psi'(t_0)| > 0. \tag{3}$$

If now, for the neighborhood of the point t_0 just considered, we put down the Taylor developments:

$$\begin{aligned}\varphi(t) &= a_0 + a_1(t - t_0) + a_2(t - t_0)^2 + \cdots,\\ \psi(t) &= b_0 + b_1(t - t_0) + b_2(t - t_0)^2 + \cdots\end{aligned}$$

of the functions $\varphi(t)$ and $\psi(t)$, then we obtain, in:

$$z = x + iy = (a_0 + ib_0) + (a_1 + ib_1)(t - t_0) + (a_2 + ib_2)(t - t_0)^2 + \cdots, \tag{4}$$

an analytic function z of t, which maps the real t-axis in the neighborhood of t_0 onto the corresponding portion of the curve K. By analytic continuation over the entire interval (2), we find the segment of the real t-axis bounded by this inequality (2) mapped onto the curve-portion K by means of the analytic function:

[26] Cf. Schwartz "*Ges. math. Abh.*", Vol. 2, p. 150 ff., as well as Osgood, "*Lehrbuch usw.*", p. 608ff.

$$z = \varphi(t) + i\psi(t). \tag{5}$$

Now Schwartz extends this mapping of the real t-axis furnished by the equation (5) *to complex values* of t. Since, for the point t_0 of our interval, the derivative $(\varphi'(t) + i\psi'(t))$ never vanishes, in consequence of (3), then a strip of the complex t-plane, chosen sufficiently narrow along that segment of the real t-axis, will furnish, by the analytic function (5), a consistently conformal image in the z-plane, which positions itself as a narrow strip about the image of the real t-axis, ie., the curve K. We will hereby assume the arrangement so effected, that the portion of the image belonging to the interior of the domain B originates from the portion of the strip lying in the "positive t-halfplane".

On this study, the solution of the boundary0-value problem for our domain B then bases itself. We construct, from one endpoint $t = \alpha$ to the other $t = \beta$, a circular arc, which runs through the positive t-halfplane and reaches the real t-axis at an angle η sufficiently small but different from zero. it bounds, with the oft-named segment of the real t-axis, a slender "circular crescent", for which the boundary-value problem is solv-
[454] able[27]. The mapping of this circular crescent by means of the analytic function (5) furnishes a consistently conformal image in the z-plane, which indeed has the boundary curve K in common with the domain B, and in addition, for angle η chosen sufficiently small, runs entirely inside B. For this domain, we are consequently in the position, to likewise solve the boundary-value problem.

The further development then bases itself again on the combination method of Schwartz and Neumann. If, for the two domains of the kind considered, meeting at a vertex of the boundary R, the angles η and η' are so large, that the two domains reaching this vertex already partially cover one another there, then we are also able to immediately fuse these. If, however, these two domains only first have the vertex in common, then we must avail ourselves of one more intermediary domain. We choose as such, say, a circular sector[28], cur out from a circle about the vertex in question as a center; apart from this vertex, the circular sector is to lie entirely in the interior of B, and have so large a central angle, that it reaches over, at the vertex on both sides into the domains to be fused here.

We now proceed in the manner just sketched with all individual analytic boundary portions of B. The gaps then still remaining open in the interior of domain B, we cover further with finitely many circular discs and finally arrive in this fashion at the following result: *For the domain B, taken as a basis on p. 447, we are able to solve the boundary-value problem, in case the boundary R of this consists of a finite number of analytic curve-portions, which form non-vanishing angles with one another at the vertices of this boundary*[29].

---

[27] In fact, one can easily map a circular crescent conformally onto the surface of a Schlicht full circle.

[28] A circular sector can easily be mapped onto a semicircle in such a manner, that the two radii bounding the sector go over into the diameter bounding the semicircle. The semicircle is then, as the circular crescent before, mappable onto a full circle, so that the solution of the boundary-value problem for the circular sector holds no difficulty.

[29] See the detailed representation of Osgood, "*Lehrbuch usw.*" on, p. 608 ff.

The preceding developments contain, at the same time, the Schwartzian "*principle of symmetry*" in its general form (cf. "M" I on p. 92), and with this, also the "analytic" resp. "harmonic continuation" of the functions of our domain B beyond an analytic boundary portion K of this[30], a subject, to which we return once more in the next paragraph.

2.4 The Green's function of a simply connected domain

[455]

Now, in particular, let the domain B be a *simply connected one*; the boundary R of B, which according to assumption is put together from a finite number of analytic portions with non-vanishing vertex angles, will consequently now consist of a *single line running back into itself.* In the interior of the domain, we mark an arbitrarily chosen point O, and understand by $g(x,y)$ the Green's function of the domain B belonging to this point as discontinuity point, whose existence and unique determinateness was proved in the foregoing paragraph. In the neighborhood of O, according to (3) on p.449, the representation holds:

$$g(x,y) = -\log r + \text{harman.funct.}, \tag{1}$$

except for this, $g(x,y)$ is everywhere harmonic inside B. The values of $g(x,y)$ inside B are everywhere > 0, and toward the boundary R, $g(x,y)$ goes over uniformly continuously into the value 0.

Now let c be any constant positive value. Then all points of the domain B at which $g(x,y)$ assumes this value c, form the "*level curve*" of our Green's function belonging to c and represented by

$$g(x,y) = c. \tag{2}$$

This curve is the bounding curve of that partial domain B_c of B, inside which the inequality $g(x,y) > c$ takes place. *This partial domain B_c necessarily consists of a single continuum, in whose interior the discontinuity point O is situated.* Namely, were this to consist of at least two separated portions, then O could only lie inside one of them, and therefore $g(x,y)$ would be everywhere harmonic in the interior of the other portion and constantly equal to c on its boundary. According to a theorem cited on p. 448, $g(x,y)$ would consequently also be equal to c everywhere inside this portion, and the harmonic continuation of our function would show, that this is generally equal to the constant c, which is excluded. [456]

On the basis of the same consideration, one shows further: *the boundary R_c of the domain B_c cannot fall into separated portions; it is, rather, connected, and accordingly bounds a "simply" connected partial domain B_c.* In each case, namely, B_c has an outer boundary curve, on the other side of which one arrives at the boundary R of B. If now an inner boundary curve, closed by itself, were to supervene, then this would again bound a domain not belonging to B_c, inside which $g(x,y)$ would be constantly equal to c.

30 See the details in Osgood, "*Lehrbuch usw.*" on, p. 580.

We ask further, whether it is possible, that the level curve R_c of our Green's function $g(x, y)$ exhibits *multiple points*. At such a point, simultaneous vanishing of the two derivatives $\frac{\partial g}{\partial x}$ and $\frac{\partial g}{\partial y}$ would then enter.

From the general behavior of a function of x and y in the neighborhood of a place, at which it is harmonic[31], would give, for the domain B_c, the shape sketched in figure 101 at a place of simultaneous vanishing of the derivatives $\frac{\partial g}{\partial x}, \frac{\partial g}{\partial y}$. The essential thing is, that the domain B_c, accentuated by shading, reaches the point in question with a number of cusps *in the same sheet*[32] of the total domain B. If it is a question of a v-fold point of the level curve, then the domain B_c runs up to this point with v vertices of the angles $\frac{\pi}{v}$ (in figure 101, v is taken equal to 3), and v intermediately situated quoins of the same angle $\frac{\pi}{v}$ remain free of the domain B_c. The dotted line in the interior of the domain B_c of figure 101 running close to the boundary shows, how one must run through the boundary, in order, here too, to still conceive it as a single line running back into itself.

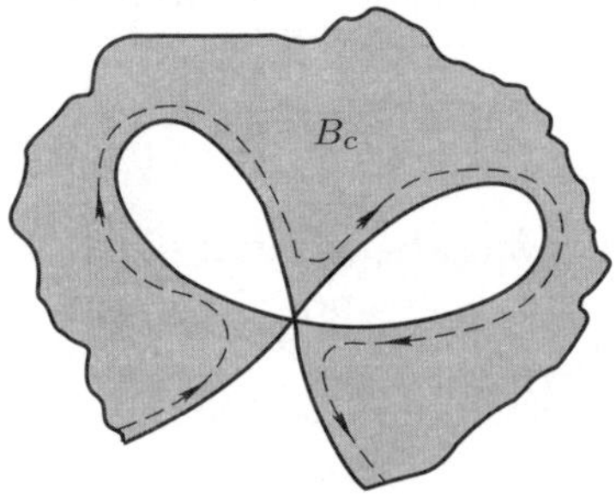

Fig. 101.

But now, as one sees, in the present case, R_c falls into at least v closed curves, of which v reach to the point in question with vertices. We pick out two of these latter curves and call them K_1 and K_2. The closed curve K_1, as belonging to the simply connected domain B, divides this domain B into two partial domains B_c' and B_c'', of which the former contains the point O and the domain B_c, while B_c'' is a ring-shaped domain
[457] remaining, with the outer boundary R. But now further, the second closed curve K_2 obviously belongs to the inner domain B_c', from which we immediately draw the conclusion as above, that $g(x, y)$ would have to be constant everywhere.

According to this, the occurrence of a multiple point for a level curve (2), or, as we are also able to say, the simultaneous vanishing of the derivatives $\frac{\partial g}{\partial x}, \frac{\partial g}{\partial y}$, is precluded in B. One can also express this fact, so that the differential quotient $\frac{\partial g}{\partial h}$ taken with respect to the inner curve normal of a level curve is always > 0: *The level curve (2) is a simply*

[31] See the details in Osgood, "*Lehrbuch usw*", p. 578, in particular, theorem 4 here.

[32] We assume, in the terminology of the text, that at the place where the multiple point of R_c lies, a branchpoint does not find itself. How the situation is to be understood in the latter case, yields itself according to known fundamental theorems, by one imagining the neighborhood of the branchpoint, for the moment mapped onto a Schlicht domain.

closed everywhere continuously curved curve, and the entire family of the level curves which enter, if we let c grow from 0 *to* ∞, *will cover our domain B simply and gaplessly.*

Now produce the potential conjugate to $g(x, y)$:

$$h(x, y) = \int \left(\frac{\partial g}{\partial x} dy - \frac{\partial g}{\partial y} dx \right), \tag{3}$$

by choosing the constant, hereby arbitrarily disposable, in any manner. *then*:

$$Z = e^{-g-ih} \tag{4}$$

is an analytic function of z, which maps the simply connected domain B conformally onto the Schlicht surface of the unit circle of the Z-plane. Thereby, to the point O corresponds the origin $Z = 0$, and to the neighborhood of O, the simply and completely covered neighborhood of $Z = 0$. Upon continuation of the mapping from here, according to the representations above, inner branchpoints in the image can in fact yield themselves[33]. On the other hand, upon mapping the domain boundary R, where g has continuously gone over into the value 0, we arrive at the periphery of the unit circle in the Z-plan. *The mapping is a consistently conformal one, and indeed, with inclusion of the analytic boundary R, apart form the fact, that the neighborhoods of inner winding points of the domain B furnish Schlichtly covered images in the Z-plane, as well as, that the vertices on the boundary R of B drop out of the image*[34].

According to the indications given at the close of §3 on p. 454, it is moreover possible, *to continue the Green's function* $g(x, y)$, *and with this, the function* $Z(z)$, *harmonically resp. analytically over and beyond the individual analytic boundary portion of the domain B.*[35] We will carry this out here yet a bit further. [458]

We denoted the individual analytic boundary portion by K on p. 452 and mapped the immediate neighborhood of this conformally onto the plane of a variable t, which was connected with z by the equation (5) on p. 453 the curve K itself hereby went over into the segment of the real t-axis defined by $\alpha \leqslant t \leqslant \beta$. In the portion of the neighborhood of this segment of the real t-axis belonging to the positive t-halfplane, including this segment itself, the complex function $(h - ig)$ proceeding from our Green's function $g(x, y)$, conceived in dependency on t, is now an analytic function, which itself assumes real values along the real t-axis. But then the principle of symmetry informs us: the analytic continuation of the function $(h - ig)$ into the negative t-halfplane is effected in such a manner, *that this function itself assumes conjugate complex values at points which are symmetric with respect to the real t-axis.*

But one can similarly carry out this consideration easily and directly for the curve portion K of the z-plane and the function $Z(z)$. Two points, lying close to the boundary K, inside and outside the domain B, which correspond to conjugate values, therefore

[33] This fact we conclude here from the shape of the level curves. One can also attach to the circumstance, that the image is always one-sheeted in the neighborhood of $Z = 0$, and that the boundary of the domain B maps itself onto the periphery of the unit circle. The impossibility of inner branchpoints lets itself be concluded from this by means of an argument to be given in more detail in §9.

[34] Compare, moreover, Osgood, " *Lehrbuch usw.*", p. 594 ff.

[35] Compare, for this, Osgood "*Lehrbuch usw*", p. 584ff.

to points of the t-plane symmetric with respect to the real t-axis one designates *as situated symmetrically with respect to the analytic curve portion K.*[36] But then, obviously, the theorem holds: *The continuation of the function $Z(z)$ over and beyond the analytic boundary portion K regulates itself in such a fashion, that at two points symmetric with respect to K, there occur values Z with equal h but opposite g, or that the one value Z' is reciprocal to the conjugate complex value $\overline{Z}$ of the other Z.*

2.5 Two theorems of Koebe

Besides the general known theorems developed so far, concerning logarithmic potentials and special Green's functions, Koebe still uses two special theorems in his proofs of the principal-circle and boundary-circle theorems, which he has to develop for the purpose of these proofs[37].

[459] Let the unit circle in the z-plane be denoted further on by K. Let B be a simply connected Schlicht domain, which belongs to the surface of the circle K and contains the origin of the z-plane, to be denoted by O, in its interior. The boundary R of B is a closed curve, which may also contain portions of the periphery K itself. Let the boundary point B lying closest to the center O have the distance d from O; we will designate it further on as the "*minimal distance*" of the boundary R from the center O.

The assumption is to hold for the domain B, *that the Green's function $g(x, y)$ belonging to the point O exist.* We will here explicitly introduce a development of this for the neighborhood of the point O.[38] If the number z in polar representation, is $z = re^{\vartheta i}$, then the series development holds for $g(x, y)$:

$$g(x,y) = \log\left(\frac{1}{r}\right) + a + \sum_{k=1}^{\infty} r^k (a^{(k)} \cos k\vartheta + b^{(k)} \sin k\vartheta), \tag{1}$$

which is always convergent inside a circle with the radius d about O.

If the minimal distance $d = 1$, then this says that the domain B is identical with the surface of the circle K. Then we simply have $g = \log(r^{-1})$, and therefore, in the development (1), the absolute term a as well as all further coefficients $a^{(k)}, b^{(k)}$ are equal to zero. For arbitrary minimal distance $d < 1$;

$$g(x,y) - \log\left(\frac{1}{r}\right) \tag{2}$$

is certainly everywhere harmonic in the domain B. On the boundary R of B, $g = 0$ and $r \leq 1$, whereby the equality sign (on account of $d < 1$) cannot, in any case, take place everywhere. The function (2) is therefore now certainly ≤ 0 along R, whereby too, the equality sign does not always hold here. With this, according to a theorem set up on p. 448:

$$g(x,y) - \log\left(\frac{1}{r}\right) < 0$$

[36] The symmetry with respect to a circle is a special case of this general definition.

[37] Cf. the works of Koebe's, cited above (p. 445), in the Gött. Nachr. of 1907, p. 203 ff and the Math. Ann. Vol. 67, pp. 208 and 215.

[38] Cf. Osgood, "*Lehrbuch usw.*", p. 573.

always holds in the interior of B. At the point O, which lies inside B, however, the function (2), in consequence of (1), has the value a. *With this, for the absolute term a of the development (1) of our Green's function $g(x, y)$, The inequality follows*:

$$a < 0, \tag{3}$$

As soon as the minimal distance $d < 1$.

The first theorem of Koebe now gives an essential sharpening of this result, and can be stated thus: *if one takes a number D in any manner from the interior $0 < D < 1$, and* [460] *prescribes the inequality $d \leq D$ for the minimal distance d, then however the domain B may be shaped, a determinate number $A < 0$ lets itself be given, so that the absolute term a of the development (1) of our Green's function $g(x, y)$ satisfies the inequality*:

$$a < A. \tag{4}$$

This theorem is provable by comparison of the function $g(x, y)$ with a second Green's function $G(x, y)$, whose domain B' contains B as a partial domain, and which, for its part, is only stipulated by the choices of the number D. The domain B' is two-sheeted and is obtained in the following fashion:

According to the assumption, there is at least one boundary point of B, which has the distance d from the origin O. From this point, we draw a line, not crossing over itself, which never enters the interior of B, initially up to the periphery of K and from there, outside K, say rectilinearly, up to $z = \infty$. Since $d \leq D$, then one can mark a point z_0 on this line, for which $|z_0| = D$ holds. Should there be several such points, then we understand by z_0 the last point of this kind run through in the drawing of this line. This point lies inside K; let that part of the line drawn, which runs from z_0 to $z = \infty$, be called L (cf. figure 102).

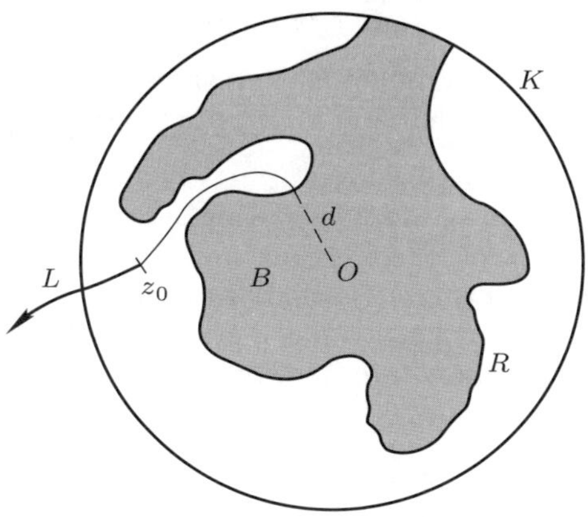

Fig. 102.

Now introduce the two-sheeted Riemann surface over the z-plane belonging to the function $\sqrt{z - z_0}$, here by utilizing the line L, which is permissible, as branch cut and

imagine, that the domain B is situated in the upper sheet. The Riemann surface is simply connected. The circle K furnishes a closed curve on it first after running through it twice. If we carry out a cut through two sheets along K and take away the part of the surface running outside K, then there remains a likewise simply connected domain with a boundary curve running back into itself. From the domain now, in the lower
[461] sheet, we take out all points characterized by $|z| < \frac{1}{2}D$, whereby a circular periphery of the radius $\frac{1}{2}D$ about 0 in the lower sheet[39] comes up as a second boundary curve. let the twofold connected domain thus arising be the domain B' to be constructed. *This domain B' is stipulated only by the chosen number D and the position of the point z_0 on the periphery of the circle of radius D about O'; in contrast to this, the functions of the domain B' are naturally independent of the particular effectuation of the branchcut L between z_0 and K.* Obviously, B is again a partial domain of B', which perhaps has part of the periphery K of the uppersheet in common with the boundary of B'.

According to the theorem of §3 on p. 451 ff., the Green's function $G(x,y)$ of the domain B' belonging to the point O exists and is represented, say, in the neighborhood of O by the series converging there:

$$G(x,y) = \log\left(\frac{1}{r}\right) + A + \sum_{k=1}^{\infty} r^k (A^{(k)} \cos k\vartheta + B^{(k)} \sin k\vartheta). \tag{5}$$

According to the statements just made, this function is indeed not only stipulated by the absolute value $|z_0| = D$, but also by the amplitude of z_0. But since two domains B' with the same $|z_0|$ and different amplitudes of z_0 are essentially transportable into one another simply by rotation about O, then upon alteration of the amplitude of z_0, only the coefficients $A^{(1)}, B^{(1)}, A^{(2)}, B^{(2)}, \ldots$ in (5) alter themselves, while the absolute term A remains unaltered: *the absolute term A in the development (5) of the Green's function $G(x,y)$ is a quantity depending solely on the chosen number D.*

The function:

$$G(x,y) + \log r \tag{6}$$

is everywhere harmonic in the domain B'. This vanishes on the boundary curve K of B', but on the second boundary curve of B', lying in the lower sheet, it has the constant negative value:

$$\log\left(\frac{D}{2}\right) < -\log 2.$$

Accordingly, the function (6) is everywhere negative in the interior of B' (cf. p. 448), and since, in particular, it assumes the value A at the point O, then follows: *the number A, depending only on D, satisfies the inequality*:

$$A < 0. \tag{7}$$

[462] We now effect the comparison between $g(x,y)$ and $G(x,y)$. The latter function is positive for each interior point of B'. Since B is a partial domain of B' different form the whole, then the values of $G(x,y)$ entering on the boundary R of B are necessarily ≥ 0, whereby the equality sign does not hold everywhere. The function defined for B:

[39] Notice, that no point of the branchcut L has a distance $< D$ from the center O.

$$G(x,y) - g(x,y), \tag{8}$$

which is everywhere harmonic in B, will accordingly be likewise ≥ 0 along the boundary R, excluding the constant validly of the equality sign. Accordingly, inside B:

$$G(x,y) - g(x,y) > 0,$$

holds everywhere, and we find, in particular, for the point O, the inequality:

$$A > a,$$

with which the theorem set up above is proved.

From the theorem proved, we draw a *conclusion*, which is utilized later. Let a number ε belonging to the interval $0 < \varepsilon < 1$ be chosen arbitrarily small but fixed. Let the number A belonging to $D = 1 - \varepsilon$ in the sense of the theorem proved be called α. Each domain B with a minimal distance $d \leq 1 - \varepsilon$ then has, as absolute term in the development (1) of the Green's function, a number a, for which $a < \alpha$ holds. Conversely, therefore, a domain for which that absolute term a belongs to the interval $\alpha \leq a < 0$, necessarily has a minimal distance $d > 1 - \varepsilon$. Therefore the conclusion holds: *If ε is taken from the interval $0 < \varepsilon < 1$ as an arbitrarily small but fixed number, then there is a number α, which is < 0, so that, for a domain B, for which the absolute term a in the development of the Green's function satisfies the inequality $a \geq \alpha$, then the minimal distance d necessarily satisfies the condition*:

$$1 - d < \varepsilon. \tag{9}$$

The preceding theorem comes into use for the principal-circle theorem. A theorem closely related to it, which is now to be derived, correspondingly serves for the boundary-circle theorem. Let the domain B be, as before, schlicht and simply connected; also let it again contain the origin O in its interior. The point $z = \infty$ is to belong neither to the interior nor the boundary R of B; B is therefore to be a finite domain. Thereby, it is very remarkable for the theorem to the set up, *that moreover, we need presuppose no upper bond for the "maximal distance" of the boundary R from the origin* [463]
O. The single requirement now still supervening is this, that the Green's function of B belonging to the point O is always to exist.

By d, we understand, as before, the "minimal distance" of the domain B, i.e. the shortest distance of a boundary point, from O. There then holds the following second theorem of Koebe: *if one chooses a number D, arbitrary but fixed, satisfying the condition $D > 0$, and prescribes the inequality $d \leq D$ for the minimal distance d of the domain B, then however B might be chosen according to prescription, there is a determinate finite number A, so that the absolute term a in the development (1) of the Green's function $g(x,y)$ of B satisfies the inequality*:

$$a < A. \tag{10}$$

The proof rests on the same principles as that of the first theorem. From a boundary point of the domain B, which has precisely the distance d from O, we draw a line, not crossing over itself and not penetrating into the interior of B, to the point $z = \infty$. Since $d \leq D$, there is, on this line, at least one point z_0 with the distance $|z_0| = D$ from the origin O. If several such points occur on the line, the let z_0 be the last of them run through upon description of the line. The portion of our line lying between this point z_0 and the endpoint $z = \infty$ we call L and again imagine this line L, as above, as a branchcut of the two-sheeted, simply connected Riemann surface belonging to the function $\sqrt{z - z_0}$.

We again imagine, that the domain B is situated in the upper sheet of this surface. But from the lower sheet, exactly as above, we cut out the circular disc defined by $|z| < \frac{1}{2}D$. The remaining domain B', which contains the point $z = \infty$, is simply connected, and has the circle of radius $\frac{1}{2}D$ about O in the lower sheet as boundary curve. The functions of the domain B' are only stipulated by the position of the point z_0, but will be independent of the possible effectuation of the branchcut L. If, in particular, we form the Green's function $G(x, y)$ of B' belonging to the point O, then we find as before, that the absolute term A in the development (5) of this only depends on the absolute value $|z_0| = D$, but is independent of the amplitude of the complex number z_0.

The domain B, is a portion of B' different from the whole and never reaches the boundary of B'. Consequently, the function defined for B:

$$G(x, y) - g(x, y), \tag{11}$$

[464] which is everywhere harmonic in this domain, is necessarily > 0 along the entire boundary R. With this, the function (11) will also certainly be > 0 at the interior point O (cf. p. 448), from which the inequality asserted in the theorem yields itself:

$$A > a.$$

As a corollary we note: if $D > 1$, *then for A, the inequality $A > 0$ holds.* Namely, if $D > 1$, then to the domain B also belong such, which contain the unit circle of the z-plane entirely in their interior. In a case of this kind, the function:

$$g(x, y) + \log r \tag{12}$$

is everywhere harmonic in the circle in question and always > 0 on its periphery. Therefore, for the value a of the function (12) at the origin $a > 0$ similarly holds, whereby, from (10), $A > 0$ follows all the more.

As above, we also attach here a *consequence* of the second Koebe theorem. If we choose a number ε from the interval $0 < \varepsilon < 1$, arbitrarily small but fixed, and set $D = \frac{1}{\varepsilon}$, then to this D, in the sense of the theorem, belongs a determinate number $A > 0$. We write $\frac{1}{A} = \alpha$, so that α too is a determinate finite number, satisfying the condition $\alpha > 0$ and fixed with ε. For *all* domains B and $d \le D$, $a < A$ holds, therefore a domain with $a \ge A$ cannot have $d \le D$. We thus obtain the consequence: *if ε is taken fixed from the interval $0 < \varepsilon < 1$ as an arbitrarily small number, then there is a determinate finite number α, satisfying the condition $\alpha > 0$, of such a kind, that for all domain B chosen finite as above, for which the absolute term a in the development (1) of the associated Green's function $g(x, y)$ satisfies the inequality $a \ge \frac{1}{\alpha}$, the minimal distance d necessarily satisfies the condition:*

$$d > \frac{1}{\varepsilon}. \tag{13}$$

2.6 Production of the covering surface F_∞ in the boundary-circle case

We now go to the establishment of the "method of the covering surface", and indeed, for the boundary-circle case first. Over the plane of a variable z, let any closed m-sheeted Riemann surface F of genus p be given, for which, according to a known rule, the number of the branchpoints is $\leq 2m+2p-2$. Let the surface be signed with n arbitrarily chosen, but fixed, points $e_1, e_2, \ldots, e_n$, to which we assign just as many whole [465]
numbers $l_1, l_2, \ldots, l_n$, which are to be ≥ 2 and also may be $= \infty$. We call the symbol $(p, n; l_1, l_2, \ldots, l_n)$, as usual, the "*signature*" of the surface. The number n can hereby also be equal to 0, where we then have the signature $(p, 0)$.

Notice, at the same time: *all those lower signatures are to be excluded, for which either no boundary-circle theorem comes into consideration, or for which this theorem holds from the very beginning.* These are, first of all, the signatures with $p = 0, n < 3$, but also all signatures $(0, 3; l_1, l_2, l_3)$. For the lowest of these, namely, we have elliptic or parabolic rotation groups; for the remaining ones, notice that here indeed, in the individual case (from the invariant standpoint), there is always only *one* surface F, such that the existence of the polymorphic function $\zeta(z)$ corresponding to our theorem already follows from that of the associated automorphic function $z(\zeta)$. For $p = 0$, we shall thus assume $n \geq 4$, and also exclude here the special case (0, 4; 2, 2, 2, 2), insofar as the latter indeed leads to a parabolic rotation group and therefore the correctness of the associated fundamental theorem for it always follows from the theory of the elliptic functions. From the same reason, for $p = 1$, we exclude the lowest case (1, 0). However, there are no further exclusions.

On F, introduce a canonical cut-system, say of that surveyable form which underlay the theorem in I on p. 319 Therefore, the $2p$ conjugate reentrant cuts a_i, b_i are to run out from one and the same point on the surface, and from the same point, the n cuts c_k are also to run out to the point e_k, whereby, let the sequence of these cuts correspond to the sequence of these points already utilized above. In carrying out these cuts, we will follow the following prescription: *each of the cuts, a_i, b_i, c_k are to build themselves up from finitely many analytic portions, e.g. from lines, which run together at the common source of the cuts at non-vanishing angles.* The fulfilling of this requirement meets with no difficulty.

The cut-up surface we will denote by F_1, its boundary by R_1. F_1 represents a simply connected domain, whose boundary consists of $(2n+4p)$ banks of the $(n+2p)$ cross-cuts. We will designate the individual such bank, reaching from the common source of the cuts up to a point e resp. from the common source back to this latter, as a "side" of the boundary R_1.

In the production of the covering surface F_∞ it is now a question, according to p. 443, of a layering over one another of infinitely many copies of the surface F_1, [466]
just obtained, and of joining these copies with one another exactly in the fashion, in which the infinitely many polygons P of a boundary-circle net N of the signature $(p, n; l_1, l_2, \ldots, l_n)$ appear bound to one another.

In order to now carry out this covering process in more detail, we pick out any non-degenerate boundary-circle polygon P_1, which has the signature $(p, n; l_1, l_2, \ldots, l_n)$ of our surface F. The fixed vertices of P_1 we call $\varepsilon_1, \varepsilon_2, \ldots, \varepsilon_n$ for the sake of distinction, and assume, that these present the same sequence on the boundary of P_1, as the points

$e_1, e_2, \ldots, e_n$ on the boundary of F_1 corresponding to it. For the sake of the visuality, we imagine P_1 situated in the projective plane and fashioned into a rectilinear polygon of $(2n+4p)$ sides and consistently concave or just non-convex angles.

We are now able to single-valuedly map the boundary of the polygon P_1 onto the similar boundary of the domain F_1, by above all letting the "vertices" $e_1, e_2, \ldots, e_n$ of the boundary of F_1 correspond to the vertices $\varepsilon_1, \varepsilon_2, \ldots, \varepsilon_n$ of P_1 and coordinating the intermediately situated sides of P_1 as well as the remaining sides of P_1, with preservation of the sequence, ie. therefore, to the homologous sides of the boundary of F_1. Since, further on, we are only concerned with considerations of "analysis situs", then it suffices, to coordinate two homologous sides to one another as units, without specializing this coordination for the individual points of the sides. In the some case, we coordinate the entire polygon P_1 and the surface F_1 to one another.

We now establish a process of of approximating the covering surface, by first carrying out the following construction in the projective plane. In the polygon net N belonging to P_1 we position all those polygons next to P_1, which have at least one parabolic boundary point in the common with P_1. Therefore, each polygon neighboring P_1 along a side is to be positioned, similarly, all polygons lying around an elliptic or accidental vertex of P_1, finally, however, at a parabolic vertex, always only the two neighboring polygons, which indeed each have a side in common with P_1. If we consider these two polygons as connected at a parabolic vertex, then it is a simply closed wreath of polygons, which we have laid out. Let these together with P_1, cover a domain, which we will denote by P_2.

Above all else, we determine, *that P_2 is again a simply connected domain.* Thereby, P_1 is a partial domain of P_2 different from the whole, and the boundary of P_1 runs
[467] entirely inside P_2, apart from the parabolic vertices possibly coming up, which the boundary of P_1 has in common with that of P_2.[40]

Further, we position next to P_2 all those polygons of the net N, which have at least one boundary point in common with this domain P_2, different from a parabolic point. By addition of those surrounding polygons, P_2 is extended to a domain P_3 of simple connectivity[41] composed of finitely many polygons. We continue in the same manner, and obtain *an infinite chain of simply connected domains $P_1, P_2, P_3, \ldots$, each individual one of which, P_ν, has a boundary built up from finitely many sides running back into itself.* Thereby holds, further: *for each finite ν, $P_{\nu-1}$ is a portion of P_ν different from the whole, since the boundary of $P_{\nu-1}$, apart form the parabolic vertices, at which $P_{\nu-1}$ always reaches to the boundary of P_ν, runs entirely int the interior of P_ν.* In addition,

[40] If one raises the objection, that for the polygon P_2, convex angles (sums of two polygon angles in the net N) are occasionally able to come up, and that consequently, the *simple* connectivity of P_2 is not directly evident, then by more penetrating angle considerations in the projective plane, we would be able to invalidate this objection. However, in order not to go too far into the theory of polygons, we are able to proceed in the following manner: P_2 always has an outer boundary, which appears put together from finitely many sides of the net N ; also, where two of these sides which are neighboring meet at a parabolic point, say, only finitely many polygons of the net N between them will reach up to this point. Now understand by P_2 simply the complex of the finitely many polygons of the net N included by this outer boundary. This domain P_2 is then certainly simply connected.

[41] See, if necessary, the previous footnote.

as is known, each interior point of the net N can be reached from the initial space P_1 after running through only finitely many polygons of the initial space P_1. Therefore follows, finally: *For each interior point of the net N or (in case we return to the* ζ*-plane) for each interior point of the surface of the boundary circle, there is a finite number* ν*, so that that point is situated in* P_ν*, and therefore also in* $P_{\nu+1}, P_{\nu+2}, \ldots$. In particular, this also holds for the accidental and elliptic vertices in the net N, *while the individual parabolic point, from a determinate* ν *on, becomes a boundary point of* P_ν *and then remains a boundary point of the following domains* $P_{\nu+1}, P_{\nu+2}, \ldots$. One can also express the present state of affairs thus, *that our domains* $P_1, P_2, P_3, \ldots$ *approach a limiting domain* P_∞ *uniformly, which is the Schlichtly covered surface of the ellipse of the projective plane resp. of the boundary circle in the* ζ*-plane*[42].

To the finitely many polygons, which comprise the complex P_ν, we now coordinate just a many copies of the surface F_1 and layer these over one another. We then join together each two of these surface-copies, which correspond to neighboring polygons of the complex, along the relevant (lying over one another) cut-boundaries ("sides" of the two copies F_1), and generate in this manner a connected surface-complex F_ν, which in consideration of the mutual positional relations of the various copies F_1 to one another, has the domain P_ν as a single-valued image. We consequently determine: *the domain* F_ν *is simply connected, covers the z-plane finite-sheetedly with finitely many branchpoints and has a boundary* R_ν *running back into itself, which appears composed from finitely many curve-portions with non-vanishing vertex angles.* The branchpoints originate partly form those of the original surface F, and they come up partly at the places e_k for layering the copies F_1 over one another. [468]

These statements hold for each finite number ν. But if we now produce the unbounded chain $F_1, F_2, F_3, \ldots$ of bounded surfaces, then we read off, from the statement above concerning the P_ν, concerning the natural relation of these: *for each finite* ν, $F_{\nu-1}$ *is a portion of the domain* F_ν *different from the whole, whereby the boundary* $R_{\nu-1}$ *of* $F_{\nu-1}$*, apart from the parabolic points* e_k*, at which* $F_{\nu-1}$ *always reaches up to the boundary of* F_ν*, runs entirely inside* F_ν. Each parabolic point e_k will accordingly remain situated on the boundary of our domains.

With increasing ν, the domain F_ν now approaches that model F_∞ more and more, which we will designate (cf. p. 443) as the *"covering surface" of the Riemann surface* F_1, to be utilized for the boundary-circle theorem. This covering surface, which finds its single-valued image in the entire polygon net $N = P_\infty$, is a completely determinate model, insofar as each of its sheets resp. each of its copies F_1 of the initial surface F_1 is to be reached by a determinate finite process. For each finite ν, F_ν is a component of the covering surface F_∞ different from the whole. On the other hand, for each interior point[43] and for each curve in the interior of F_∞, there is a finite number ν, so that the point resp. the curve belongs to the domain F_ν. Inside F_ν, therefore inside F_∞, [469]

---

[42] This statement presupposes the ordinary (not the projective) determination of measure in the plane of the ellipse resp. the ζ-plane.

[43] A point e_k of parabolic character, to be sure, from ta certain index on, lies on the *boundary of each* further domain F_ν, and therefore does not belong to the "inner" points the surface F_∞. However, for the argument of the text, such a point can also be counted as part of the inner points.

each closed curve is continuously contractible to a point without tearing: *the covering surface F_∞ accordingly represents a simply connected domain.*

2.7 Production of the covering surface in the principal-circle case

Now let an arbitrary *orthosymmetric* Riemann surface F be proposed, which lies m-sheetedly over the z-plane and has the genus p. Let the number of symmetry lines be called μ. The points e_k, with which we imagine F signed, are subjected here to be restriction, that none of them may lie on one of the μ symmetry lines[44]. The number of the points must consequently be even; $\frac{1}{2}n$ of them, $e_1, e_2, \ldots, e_{\frac{1}{2}n}$, lie in one half of surface divided off by the μ symmetry lines, while the rest of the point e are the points of the other half of the surface symmetrically situated with respect to them. To the former points we assign the whole number $l_1, l_2, \ldots, l_{\frac{1}{2}n}$, which are able to be ≥ 2 or also signify ∞; to the point e of the other half of the surface there belong resp. these same numbers l.

We have used here the same notations, which we introduced on p. 396 ff. in the discussion of the orthosymmetric surface. We must also again take up the notations p' and n' used at that time, but, to be sure, define them here directly by tying into our Riemann surface, without utilization of a principal-circle polygon. We take one of the halves of the surface cut off by the symmetry lines, which is therefore a Riemann surface provided with μ boundary curves, and imagine this again transformed into a closed surface, by hanging a simply connected domain into each boundary curve. Then p' is simply the genus of the closed surface thus arising. For the number n', however, according to (2) on p. 397, be simply have the definition:

$$n' = \frac{1}{2}n + \mu,$$

so that n' is the number of the points e lying in the surface-half increased by the num-
[470] ber of the boundary curves of this half. As "signature" of the surface-half, we will designate the collection $(p', n'; l_1, l_2, \ldots, l_{n'-\mu})$.

This signature is then exactly that one for principal-circle polygons (in the projective representation), which leads from itself to orthosymmetric surfaces of our kind. Further on, we shall always only have to operate with one surface-half. On this, we now start out form a canonical crosscut system, as it corresponds to the theorem in I on p. 319. We therefore have to lay out $2p$ reentrant cuts a_i, b_i pushing out from and returning back to one place of the surface-half, as well as to draw N' additional crosscuts c_k from the same place to the $\frac{1}{2}n$ points e and the μ boundary curves. We effect the arrangement of the cuts, so that we are again able to single-valuedly map the boundary of the cut-up surface-half, to be denoted by F_1, onto the boundary of an entirely arbitrarily chosen, non-degenerating polygon P_1 of the signature $(p', n'; l_1, l_2, \ldots, l_{n'-\mu})$. Thereby, we understand by P_1 a polygon situated in the projective plane, and indeed,

[44] Concerning the character of such degenerations of principal-circle polygons, for which an elliptic or parabolic vertex moves onto one of the μ principal-circle segments, see the developments of the previous chapter, e.g. on p. 416 or p. 420.

only the portion lying in the interior of the ellipse, so that, corresponding to the μ boundary curves of F_1 originating the symmetry lines, μ ellipse segments (principal-circle segments) take part in the boundary of P_1. The mapping of the boundaries of F_1 and P_1 onto one another is naturally understood in the same sense of "analysis situs", as the corresponding mapping in the previous paragraph (p. 466). In the same sense, the surface F_1 as a whole is also to correspond to the surface of the polygon P_1.

The preceding determinations still require two additions. First of all, namely, here too, the cuts a_i, b_i, c_k are not be fashioned entirely arbitrarily. Rather, we determine, as on p. 465, *that each of these is to be composed of finitely many analytic portions, e.g. from lines, and that thereby, at the entire boundary* F_1, *only vertex angles different from 0 may come up*. Here too, the fulfilling of this requirement meets with no difficulty. But as far as the μ boundary curves of F_1 originating from the symmetry lines go, *these are certainly analytic, indeed each is algebraic*; for one can introduce, on the original Riemann surface F over the z-plane, an algebraic function w in such a manner, that those μ symmetry lines map onto just as many real traces of the "algebraic curve"[45]:

$$f(w,z) = 0.$$

A second additional remark concerns the signatures of elementary character, to be excluded, say. In this regard, things lie more simply here than for the boundary-circle theorem; we have only to determine, *that for* $p' = 0$, *the inequality* $n' > 2$ *has to hold*, whereby the cases leading to cyclic groups are to be excluded. For them, the principal-circle theorem, as long as one will still speak of such, is either directly perspicuous (in the elliptic and parabolic cases), or (in the hyperbolic case) known from the theory of the elliptic functions. [471]

After these preparations, we are able to proceed to the production of the covering surface F_∞ exactly as in the previous paragraph. We first go into the polygon net N belonging to the polygon P_1, and first of all position next to P_1 all those neighboring polygons, which have at least one non-parabolic boundary point in common with P_1. There supervene, therefore, all polygons neighboring on the sides of P_1, then besides these, those polygons, which otherwise surround the vertices of P_1, lying in the interior of the ellipse (the principal circle). The domain P_2 enlarged in such a way, has *simple* connectivity[46].

We continue in the same way and get an infinite chain of more and more comprehensive simply connected domains $P_1, P_2, P_3, P_4, \ldots$, of which, with a single exception, all statements made above (p. 467) concerning the domain-chain at that time again hold. The new thing is just this, *that the boundary of* $P_{\nu-1}$ *not only reaches to the boundary of* P_ν *with its possible cusps, but always with all its ellipse segments on the boundary as well.* For these segments, obviously the same thing holds as for the parabolic cusps, namely, that they remain constantly on the boundary of the further

[45] See the reference to the literature on p. 396 ff.

[46] If need be, we shall agree in the footnote on p. 467. In order that that consideration remain valid here fully unaltered, we will, for the present, continue the individual polygons over to μ segments of the ellipse up to the hyperbolic vertices lying outside. For these vertices then, the consideration given above for the parabolic points also holds directly.

following domains. Also, the theorem[47] again holds here; *that the domain P, with increasing index v, uniformly approach a limiting domain P_∞, which is nothing other than the Schlichtly covered surface of the ellipse (the principal circle).*

To the finitely many polygons, which comprise the complex P_ν, we now again coordinate just as many copies of the surface F_1, any layer these over one another. The
[472] boundary portions of these copies lying over one another we then join to one another according to the precise prescription of the connectedness, which is present in the complex P_ν for the corresponding polygons. There arises a surface-complex F_ν, which finds its single-valued image in P_ν. In particular, there holds again: *the domain F_ν is simply connected, covers the z-plane finite-sheetedly with finite many branchpoints and has a boundary R_ν running back into itself, which is comprised of finitely many analytic curve-portions with non-vanishing vertex angles.*

Also, concerning the mutual relation of the domains $F_1, F_2, F_3, \ldots$ to one another, all statements on p. 468 mutually hold again with the sole exception, *that the boundary $R_{\nu-1}$ not only reaches the boundary of F_ν with possible parabolic points e_k, but also with all its segments originating from the symmetry lines.* These components consequently remain on the boundary of our domains.

Upon passage to the limit $\lim \nu = \infty$, we obtain, as image of the entire net $N = P_\infty$, *the "covering surface F_∞, coming into utilization for the principal-circle theorem, of the Riemann surface F_1"*, in which, as on p. 468. We recognize a completely determinate model. The arguments at the close of §6 also carry themselves over immediately; we note, in particular, the theorem, *that the covering surface F_∞ represents a simply connected domain.*

2.8 The Green's functions of the domain F_ν and their convergence in the principal-circle case

In the domain F_ν, and indeed, in the interior of the initial space F_1, we choose an arbitrary place z_0 to be denoted by O, which however, is not to be a point e_k, nor a branchpoint, and is always to be situated in the finite part of the plane. The Green's function of the domain F_ν belonging to the point O exists, according to the theorem on p. 454, and is denoted by u_ν, say[48]. If we set $z - z_0 = re^{\vartheta i}$, then for u_ν, there holds the development, convergent in the neighborhood of O:

$$u_\nu = \log\left(\frac{1}{r}\right) + a_\nu + \sum_{k=1}^{\infty} r^k (a_\nu^{(k)} \cos k\vartheta + b_\nu^{(k)} \sin k\vartheta) \tag{1}$$

since we have to give developments of this kind many times further on, let it be permitted, to utilize the abbreviation:

$$u_\nu = \log\left(\frac{1}{r}\right) + a_\nu + \sum_{k=1}^{\infty}, \tag{2}$$

[47] Naturally, as above, meant in the sense of the elementary (not the projective) determination of measure.

[48] Let the notation g remain reserved for another Green's function, to be defined more closely soon.

so that, therefore, the summation symbol comprises such terms, which vanish at 0. If one forms the difference $(u_{\nu+1} - u_\nu)$, then one obtains a function defined for the domain F_ν and everywhere harmonic in this. The boundary points of F_ν, apart from possible parabolic points e_k, and the boundary segments coming up in the principal-circle case furnished by the symmetry lines, are consistently inner points of $F_{\nu+1}$, at which $u_{\nu+1} > 0$ thus holds. Therefore, $(u_{\nu+1} - u_\nu)$ will never be negative on the boundary R_ν, and not be everywhere equal to 0. For each inner point[49] of F_ν, therefore, the inequality $u_{\nu+1} - u_\nu > 0$ holds (cf. p. 448) this argument, valid for all indices ν, yields *the chain of inequalities*: [473]

$$u_1 < u_2 < u_3 < u_4 < \cdots, \tag{3}$$

as valid for each inner point of the covering surface F_∞, whereby, however, those initial terms of this infinite chain are to be left out, which contain functions not defined for that point[50].

At the point O, in consequence of (2), the value of the difference $(u_{\nu+1} - u_\nu)$ is equal to $(a_{\nu+1} - a_\nu)$, and since this point is an inner point of F_ν for each index ν, then $a_{\nu+1} - a_\nu > 0$ always holds. We thus obtain the theorem: *for the absolute terms a_ν in the developments (2) of our Green's functions, the chain of inequalities holds*:

$$a_1 < a_2 < a_3 < a_4 < \cdots, \tag{4}$$

We next pursue the inequality (4) and ask, whether the numbers a_ν with increasing index ν approach a *finite* upper limit or not. In the boundary-circle as well as the principal-circle case, it lets itself be shown, that such a finite limit exists. But the argument leading to this result is extremely protracted in the boundary-circle case, while in the other case it is just as short. Accordingly, we *initially let the restriction to the principal-circle case* enter.

In this case, namely, we have the advantage, that for each finite index ν, we are able to give a so-called "*invariant*" for u_ν in the domain F_ν, ie. in our case, a determinate logarithmic potential, independent of ν, for which it is established, that in the interior of the entire domain F_ν, it is greater than u_ν, In contrast to this, such a majorant does not let itself be found in general in the boundary-circle case. For the definition of the majorant, we go back to the originally proposed Riemann surface, draw the μ symmetry lines, cut up along these and retain only our one surface-half, without, however, laying out the canonical cut system on this. We therefore have a surface provided with μ boundary curves, on which we mark our place O, chosen before. *The Green's function $g(x, y)$ of the surface-half belonging to this place O is the desired majorant and has, say, in the neighborhood of O, the convergent development*: [474]

$$g = \log\left(\frac{1}{r}\right) + A + \sum_{k=1}^{\infty}. \tag{5}$$

By now adding the cuts $a_i, b_i, c_k, g > 0$ will obviously always hold along these cuts, apart from the end points of the cuts c_k running along the symmetry lines. If we accordingly transplant g onto all copies F_1, which comprise F_ν, then we obtain in g a potential

[49] The parabolic place e_k, in contrast to p. 468, are not counted here as inner points.

[50] If the point in question lies in the interior of F_ν or on the boundary of this domain, without already being a point of $F_{\nu-1}$, then one simply has to begin the chain (3) with u_ν.

single-valued in the domain F_ν, which vanishes along the segments of the boundary R_ν furnished by symmetry lines, but is >0 at all remaining points of R_ν, which further becomes discontinuous as $\log(\frac{1}{r})$ at the point O, as well as at the homologous points of the remaining copies F_1 comprising F_ν, and that, finally, F_ν is otherwise everywhere harmonic in F_ν. As one sees, this function g is independent of ν.

We now compare u_ν with the majorant g, and, for this purpose, form the difference $(g-u_\nu)$. Along R_ν this difference is never negative, and, apart from the symmetry line segments, is also certainly >0. Of the discontinuity points, the one at O has dropped out for the difference $(g-u_\nu)$, while for the remaining points homologous with O, the discontinuities remain subsisting as $\log(r^{-1})$. Therefore, for all further inner pints, $g-u_\nu$ is necessarily >0, and we draw from this, in particular, for the point O, the conclusion $A-a_\nu>0$. The number a_ν thus remains beneath the absolute term A of the development (5) for each index ν, so that $\lim_{\nu=\infty} a_\nu=\infty$ is perceived as impossible: *in the principal-circle case, the existence of a finite determinate upper limit:*

$$a=\lim_{\nu=\infty} a_\nu \tag{6}$$

for the absolute terms a_ν in the developments (2) of our Green's functions u_ν is proved.

Equipped with this result, we now also go onto the chain of the inequalities (3), and
[475] raise the question, whether the series of functions $u_1, u_2, u_3, \dots$ converges to a limit function or not. In order to decide concerning this, the difference $(u_{\nu+\nu'}-u_\nu)$ is to be estimated for any two indices ν and ν'. However, we carry out this investigation, in expedient fashion, not in the domain F_ν, for which $(u_{\nu+\nu'}-u_\nu)$ is initially defined, but in a Schlicht domain, to be obtained by means of the Green's $u_{\nu+\nu'}$.

Namely, by understanding, as remarked, by ν and ν' any two positive whole numbers, we will initially bring in only the Green's function $u_{\nu+\nu'}$ and form the associated conjugate $v_{\nu+\nu'}$, with arbitrary choice of the disposable constants. According to §4 on p. 455 ff., the analytic function:

$$e^{-u_{\nu+\nu'}-iv_{\nu+\nu'}}=Z=X+iY \tag{7}$$

of z maps the simply connected domain $F_{\nu+\nu'}$ onto the Schlicht surface of the unit circle of the Z-plane, to be called K further on. Let the component F_ν of $F_{\nu+\nu'}$ thereby furnish the partial domain B of the circular disc K, different from the whole, whereby, however, the boundary R of B has a finite number of segments, as well as, in certain cases, also of points, in common with the periphery K.

We will now transplant the functions $u_{\nu+\nu'}$ and u_ν onto the circular surface K resp. the domain B, and denote these, in dependency on Z resp. on the real variables X and Y, by:

$$u_{\nu+\nu'}=U_{\nu+\nu'}(X,Y), \quad u_\nu=U_\nu(X,Y). \tag{8}$$

For the first of them, there finds itself immediately from (7):

$$U_{\nu+\nu'}=\log\frac{1}{|Z|}, \tag{9}$$

i.e. $U_{\nu+\nu'}$ in the Green's function of the circular surface K. Since, further:

$$U_{\nu+\nu'}-U_\nu=u_{\nu+\nu'}-u_\nu \tag{10}$$

remains harmonic at the origin $Z = 0$ originating from the point O, the U_ν itself similarly becomes discontinuous here as $-\log|z|$, and we obtain, as representation of U_ν in the neighborhood of $Z = 0$:

$$U_\nu = \log\frac{1}{|Z|} + a'_\nu + \sum_{k=1}^{\infty}. \tag{11}$$

Since, besides this, U_ν vanishes along the boundary R, image of R_ν, then *we recognize in U_ν the Green's function of the domain B belonging to $Z = 0$*, as this furthermore could [476]
also have been deduced from the invariance of our potentials with respect to conformal mapping.

We will specialize the equation (10) for the point $Z = 0$ resp. the point O, by going back to the developments (9) and (11) for the calculation of the left side, to the associated series (2) for that of the right side. Obviously, it follows:

$$-a'_\nu = a_{\nu+\nu'} - a_\nu, \tag{12}$$

so that, from (4), we deduce the further inequality:

$$a'_\nu < 0. \tag{13}$$

The convergence study to be presented now bases itself on the consequence from the first theorem of Koebe given on p. 462, whereby one wants first of all to convince oneself, that the domains K and B present here, as well, too, as the boundary R of the latter, correspond to the assumptions of the Koebe theorem cited. Accordingly, if we imagine a number ε of the interval $0 < \varepsilon < 1$ arbitrarily small, but taken fixed, then there is an associated determinate number α, which is < 0, for which the statements in that "*consequence*" are valid.

Now, however, from the inequalities (4) and the finiteness of the limit (6), there follows *the existence of a determinate finite index ν_0*, so that for each index $\nu \geq \nu_0$ and ν' remaining arbitrary, the inequality:

$$a_{\nu+\nu'} - a_\nu < |\alpha| \tag{14}$$

holds, understanding by $|\alpha|$ the absolute value of that number α determined with ε. Accordingly, from (12) follows, with consideration of the fact that a'_ν and α have negative numerical values:

$$\alpha < a'_\nu < 0. \tag{15}$$

On the strength of this, however, the "consequence" informs us immediately: *if ν is chosen $\geq \nu_0$, then the inequality $1-d < \varepsilon$ holds for the minimal distance d of the domain boundary R from the origin $Z = 0$ and therefore*:

$$d > 1 - \varepsilon. \tag{16}$$

We now effect the estimation of $(u_{\nu+\nu'} - u_\nu)$, not in F_ν, but in more suitable fashion in B, where $(U_{\nu+\nu'} - U_\nu)$ is everywhere harmonic. $U_\nu = 0$ on the boundary R, while [477]
$U_{\nu+\nu'}$ itself, as the Green's function (9) of the circular surface K, is never negative there.

Since, however, according to the result (16) just obtained, $|Z| > 1-\varepsilon$ is always fulfilled along R, then follows, as valid for each point of boundary R:

$$0 \le U_{\nu+\nu'} - U_\nu = U_{\nu+\nu'} < \log\left(\frac{1}{1-\varepsilon}\right).$$

Since now, as we know, the function $(U_{\nu+\nu'} - U_\nu)$ behaves itself harmonically everywhere inside B, then follows (cf. p. 448) the inequality, valid for all points of the domain B:

$$0 \le U_{\nu+\nu'} - U_\nu < \log\left(\frac{1}{1-\varepsilon}\right).$$

By returning to the domain F_ν, we therefore obtain the result: *after the choice of an arbitrarily small number ε of the interval $0 < \varepsilon < 1$, there is an associated finite index ν_0, so that for each $\nu \ge \nu_0$ and arbitrary ν', the inequality holds in the entire domain F_ν, including the boundary R_ν:*

$$0 \le u_{\nu+\nu'} - u_\nu < \log\left(\frac{1}{1-\varepsilon}\right). \tag{17}$$

Now *the uniform convergence of the series of functions $u_1, u_2, u_3, \ldots$ to a determinate limit-function*:

$$\lim_{\nu=\infty} u_\nu = u, \tag{18}$$

existing on the covering surface F_∞, is easily perceivable. If, namely, B is any domain lying in the interior of the covering surface, which may also contain parabolic points e as well as boundary segments originating from symmetry lines in finite number on its boundary, then (cf. p. 467) there is a finite whole number λ, such that B is a component of F_λ. We then simply have to define the function u for domain B by the infinite series:

$$u = u_\lambda + (u_{\lambda+1} - u_\lambda) + (u_{\lambda+2} - u_{\lambda+1}) + (u_{\lambda+3} - u_{\lambda+2}) + \cdots. \tag{19}$$

this series is uniformly convergent in the domain B; for, in consequence of (17), we are able to keep the value

$$\begin{aligned}&(u_{\lambda+\mu+1} - u_{\lambda+\mu}) + (u_{\lambda+\mu+2} - u_{\lambda+\mu+1}) + \cdots + (u_{\lambda+\mu+\nu'} u_{\lambda+\mu+\nu'-1})\\ &= (u_{\lambda+\mu+\nu'} - u_{\lambda+\mu})\end{aligned}$$

below each positive value, chosen ever so small, at each point of B, for finite μ chosen sufficiently large and arbitrary ν'. But according to the theorem on p. 450, this uniformly convergent series again represents a potential of B, which naturally, at the point O, possess the discontinuity $\log(r^{-1})$ of the first term u_λ and, except for this, is everywhere harmonic. *The existence of the limit function u on the covering surface F_∞ is thus proved.*

The inequality (17) also holds on the boundary R_ν, where $u_\nu = 0$, and indeed, for arbitrary ν'. Since the functions $u_{\nu+\nu'}$, for growing ν', approach the limit function u uni-
[478] formly along the entire boundary R_ν, then $\lim u_{\nu+\nu'} = u$, along R_ν, cannot be greater than the value standing on the right in (17). If we set the latter $= \eta$ for short, then the theorem arises: *after choice of arbitrarily small under $\eta > 0$, a finite index ν_0 is givable,*

so that, for $\nu \geq \nu_0$*, the limit-function u, along the entire boundary* R_ν*, satisfies the inequality*:

$$0 \leq u \leq \eta. \tag{20}$$

We are able to simply say, *our limit-function u, along the boundary* R_ν*, uniformly approaches the value 0 for increasing* ν.

Finally, we will also refer here to the possibility of the harmonic continuation of our Green's functions on the basis of the principle of symmetry (cf. pp. 454 and 458). If we pick out any boundary portion of F_ν furnished by symmetry lines of the original surface, then we are able, initially, to continue the domain F_ν over and beyond this, by hanging on[51] a domain F'_ν, symmetric with respect to that symmetry line.

At points, which lie symmetrically with respect to the common transition line between F_ν *and* F'_ν*, the function* u_ν*, upon harmonic continuation, will simply assume opposite values.*

Since, in addition, the boundary portion in question also remains such for $F_{\nu+1}$, $F_{\nu+2}, \ldots$, as well, finally, as for F_∞, *then the same rule holds for the continuation of the functions* $u_{\nu+1}, u_{\nu+2}, \ldots$*, indeed, also for that of the limit function u.* We come back to this once more in the next paragraph.

2.9 Mapping of the covering surface onto a circular disc. Proof of the principal-circle theorem

The results obtained now lead us easily to the close of our studies on the principal-circle theorem. In fact, we need only investigate the image of the domain $F_{\nu+\nu'}$ laid out by the function (7) on p. 475 the Schlicht surface of the circle denoted here by K for $\lim \nu' = \infty$, n order to arrive at the mapping of the covering surface underlying the principal-circle theorem. However, we prefer to attach ourselves directly to the limit-function u, in order to arrive at our goal, on the basis of the statements of the previous paragraph, with another passage to the limit.

For this purpose, we understand by v the potential conjugate to u, by choosing the constants hereby disposable in any manner. Because of the simple connectivity of the covering surface, v is also single-valued on this, and therefore, we possess, in: [479]

$$\zeta = e^{-u-iv}, \tag{1}$$

a function single-valued and regular analytic everywhere on the F_∞, which the function Z given in (7) on p. 475 would approach for $\lim \nu' = \infty$ and for correct choice of the arbitrary constants contained in $v_{\nu+\nu'}$.

By now mapping the F_∞ by means of the function $\zeta(z)$, let the entire covering surface furnish the conformal image P_∞, and the partial domain F_ν the component P_ν. Naturally, these images P_∞ and P_ν initially have nothing to do with the polygon complexes so denoted in §6 ff.. Our present models P_∞ and P_ν are now to be investigated more closely.

---

[51] This would be built up from copies F'_1 of the half of the original surface not utilized above (p. 469 ff.), and indeed, in precisely symmetric fashion to the construction of the domain F_ν from the copies F_1.

Since, for u, the inequality $u \geq 0$ always holds on the covering surface, then $|\zeta|$ is also always ≤ 1. It follows *the image P_∞ and therefore, also its component P_ν, never reach out beyond the unit circle of the ζ-plane, to be denoted again by K*, say.

The boundary of P_ν is the image of the boundary R_ν of the partial domain F_ν in the ζ-plane. For this boundary F_ν however, the inequality (20) on p. 478 holds. After the choice of an arbitrary small number $\eta > 0$, we therefore have a finite ν_0, such that for $\nu \geq \nu_0$, the inequality:

$$0 \leq \log \frac{1}{|\zeta|} \leq \eta$$

holds for all points of R_ν, which by returning to the quantity ε in place of η, we have to calculate this around to:

$$1 - \varepsilon \leq |\zeta| \leq 1. \tag{2}$$

We are able to say: *after choice of an arbitrarily small number $\varepsilon > 0$, here is a finite ν_0, so that for $\nu \geq \nu_0$, the minimal distance d of the domain boundary P_ν from the center $\zeta = 0$, satisfies the inequality*:

$$1 - \varepsilon \leq d \leq 1. \tag{3}$$

Next, now, the conclusion of schlichtness of the image P_∞ lets itself be drawn. The point O of the covering surface, at which u becomes infinite, furnishes, as image, the point $\zeta = 0$. The neighborhood of O is mapped onto the simply and completely covered neighborhood of $\zeta = 0$. Since, except for this, u remains everywhere harmonic and, in particular, therefore, finite, then no further portion of P_∞ can reach the point $\zeta = 0$. In the immediate neighborhood of $\zeta = 0$, therefore, the image P_∞ only covers the ζ-plane *one-sheetedly*.

[480] From the single-valuedness and regularity of the function $\zeta(z)$ in the domain F_∞, it now by no means follows directly, that the image P_∞ could not also partially cover the unit circle K multiple-sheetedly. Let us consequently assume, that we are able to produce a point ζ_0 with $|\zeta_0| < 1$, which is covered by at leat two sheets of the image P_∞. The two places of P_∞ lying over one another at ζ_0 will furnish two inner points z_0' and z_0'' of the covering surface F_∞. We are able to give a determinate finite index ν_0', so that for all $\nu \geq \nu_0'$, the two points z_0' and z_0'' are already inner points of F_ν.

On the other hand, choose any ε of the interval $0 < \varepsilon < (1 - |\zeta_0|)$, determine the associated finite number ν_0 and find from (3), that for $\nu \geq \nu_0$, the minimal distance d of the boundary of the domain P_ν satisfies the inequality:

$$d \geq 1 - \varepsilon > |\zeta_0|. \tag{4}$$

From this, a contradiction yields itself. Namely, if we imagine $\nu \geq \nu_0$ and also $\nu \geq \nu_0'$ chosen in any manner, then the domain P_ν, as a component of the domain P_∞, only covers the neighborhood of $\zeta = 0$ simply, but that of ζ_0, as we assumed above, at least doubly. If we accordingly picture the concentric circles about $\zeta = 0$ with radii $\leq |\zeta_0|$, then the smaller of these circles of P_ν are only simply covered, but those with radii, which are almost equal to $|\zeta_0|$, at least partially twofoldly. There is consequently a greatest of these circles, whose surface of P_ν is still covered simply everywhere. Its periphery will therefore exactly reach to the boundary of the domain P_ν, and its radius furnishes the minimal distance d of the boundary of domain P_ν. The inequality $d \leq |\zeta_0|$, holding for this radius, stands in contradiction to (4), *so that assumption of multiple covering of the ζ-plane by P_∞ it not to be held.*

We now need to allude only once more to the inequality (3), and conclude from this, that the boundaries of the domains P_ν, for $\lim \nu = \infty$, uniformly approach the periphery of the unit circle K. With this arises the result: *the function $\zeta(z)$ maps the covering surface F_∞ conformally onto the schlichtly covered surface of the unit circle of the ζ-plane.*

In order to recognize in $\zeta(z)$ that polymorphic function, whose existence the principal-circle theorem asserts, we now have to present the following additional consideration.

The covering surface F_∞ was composed of infinitely many copies of the surface F_1, which we now will distinguish by the notations $F_1, F_1', F_1'', \ldots$. In order to survey the positional relations between these copies in one schlicht picture, we had, in §7, attached to an arbitrarily chosen net N of principal-circle polygons of the signature coming into consideration. In order to avoid confusions with the conformal image furnished by the function $\zeta(z)$, we will call in §7 the net utilized $\overline{N}$ and denote its polygons, as they correspond to the copies $F_1, F_1', F_1'', \ldots$ by $\overline{P}_1, \overline{P}_1', \overline{P}_1'', \ldots$. These polygons have exactly the same positional relations to one another as the copies $F_1^{(i)}$ corresponding to them. The relation between F_∞ and $\overline{N}$ was, in addition, only this, that the copy $F_1^{(i)}$ was set in correspondence as a whole with the polygon $\overline{P}_1^{(i)}$, and that the individual "side" of the boundary of the boundary of $F_1^{(i)}$ again corresponded as a whole to a polygon side of $\overline{P}_1^{(i)}$. [481]

Now the net $\overline{N}$ is a regular one, to which a group Γ of linear substitution belongs. In particular, let the substitution $\overline{V_i}$ carry the initial polygon $\overline{P}_1$ over into the arbitrarily chosen other one $\overline{P}_1^{(i)}$. $\overline{V_i}$ represents a reversibly single-valued continuous transformation of the net $\overline{N}$ into itself, for which the individual polygon always goes over again into a polygon of $\overline{N}$. The possibility of such a transformation is stipulated by the regularity of the arrangement of the polygons in the net $\overline{N}$.

Now this regularity also directly passes over to the arrangement of the copies $F_1, F_1', F_1'', \ldots$ of the covering surface F_∞. According to this, there is a reversibly singe-valued mapping of the covering surface F_∞ onto itself, which corresponds to the substitution $\overline{V_i}$, and which we have to define thus: the copy F_1 corresponds as a whole to the copy $F_1^{(i)}$, and from here, the mapping to be understood in the same sense is to be continued, by copying neighboring on homologous sides always being set into correspondence as well. But we are able to immediately specialize this mapping of the surface F_∞ onto itself for the *points* of this, by simply letting correspond, to the points of the individual copy $F_1^{(k)}$, the homologous points of the coordinated copy $F_1^{(l)}$. In this manner, the F_∞ appears mapped *conformally* onto itself, and therefore we have found, corresponding to the individual substitution $\overline{V_i}$, *a reversibly single-valued analytic transformation of the covering surface F_∞ into itself.*

We now return to the analytic function $\zeta(z)$, and with this, to be image P_∞ of F_∞. The infinitely many copies $F_1, F_1', F_1'', \ldots$ furnish, in the image, infinitely many partial domains $P_1, P_1', P_1'', \ldots$ positioned *next to one another*, which, in their totality, yields a net N schlichtly covering the surface of the unit circle K. The transformation obtained [482]
furnishes, *a reversibly single-valued analytic transformation V_i of the net N, and with*

this, of the surface of the circle K, into itself, whereby, in particular, the initial space P_1 goes over into $P_1^{(i)}$. According to the discussions on p. 406 ff., it yields from this, *that the transformation* V_i *is a linear substitution of* ζ:

$$V_i(\zeta) = \frac{\alpha_i\zeta + \beta_i}{\gamma_i\zeta + \delta_i}, \tag{5}$$

whereby we have, in fact, recognized, in $\zeta(z)$, *the linearly-polymorphic function, whose existence proof was to be carried out.* The "*single-valuedness theorem*" or, as one also says, the "*unicity theorem*", that in each case, no more than a single such polymorphic function $\zeta(z)$ can exist on our original Riemann surface, we have already proved on p. 405 ff.

In conclusion, it is perhaps not superfluous, to allude to the significance of the *principle of symmetry* for the preceding study. If we enlarge the first copy F_1 by addition of its mirror image along the μ symmetry lines of the originally proposed Riemann surface F (which then, to be sure exhibits, in its two halves, cuts a, b, c running symmetrically), then, as one immediately deduces from the discussions at the close of the previous paragraph (p. 478), the analytic function $\zeta(z)$ is continuable into the newly attached surface half according to the principle of symmetry. *As image of this half there then arises simply the mirror image of* P_1 *with respect to the "principal circle"* K. Self-evidently, we then obtain further, upon continuation over the cuts a, b, c of the new surface half, a net situated outside the principal circle K, which is simply the mirror image of the net contained inside K. *The principal circle theorem is thus proved in its full extent.*

From the inequalities (4) on p. 473 on our studies only first referred themselves to the principal-circle theorem; in particular, it was initially not possible for us to prove the subsisting of equation (6) on p. 474 for the boundary-circle case. As soon as we have succeeded in doing this, *we are able to carry the preceding study over to the boundary-circle case in all essential points,* as one wants to convince oneself in detail. The sole distinction is this, that all the remarks drop out, which are connected with the μ symmetry lines and the boundary portions of the domains F_ν, F_∞ resp. P_ν, and P_∞, originating form them. With this, naturally, the discussions concerning analytic continuation of the polymorphic function $\zeta(z)$ over and beyond the circle K also drop out; in fact, the
[483] latter is now the "boundary circle". *In this situation, the boundary-circle theorem will also be proved in its full extent, as soon as we have succeeded in showing the theorem attached to the equation (6) on p. 474 in the boundary-circle case.*

2.10 Introduction of new series of functions in the boundary-circle case

We return, in the boundary-circle case, to series of the Green's functions:

$$u_\nu = \log\left(\frac{1}{r}\right) + a_\nu + \sum_{k=1}^{\infty} \tag{1}$$

of our simply-connected domains F_ν, for which the inequalities (3) and (4) on p. 473, resp, the theorems attached to these, are valid. The proof of the existence of a finite upper bound for the series $a_1, a_2, a_3, \ldots$, of the absolute terms in (1) succeeds by a rather circumstantial indirect proof-procedure, in which we therefore start out from the *assumption, that* $\lim_{\nu=\infty} a_\nu = \infty$.

Also there again (as above on p. 475), the mapping of the simply connected domain $F_{\nu+\nu'}$ onto the schlicht surface of a circle K accomplishes a most essential service for us. However, vis-a-vis (7) on p. 475, we will alter the scheme of the mapping function $Z(z)$ a bit, by writing:

$$e^{-(u_{\nu+\nu'}-a_{\nu+\nu'})-i v_{\nu+\nu'}} = Z = X + iY, \tag{2}$$

and for the unique determination of the conjugate potential $v_{\nu+\nu'}$, making the stipulation, that in the development of this valid for the neighborhood of O, the absolute term vanishes. Furthermore, for abbreviating further following calculations, we will assume the point O situated at $z = 0$.[52]

The circle K, whose surface is the image of the domain $F_{\nu+\nu'}$, is now, to be sure, no longer the unit circle of the Z-plane. *Rather, the radius of this circle K, to be denoted, say, by* $\rho_{\nu+\nu'}$*, is obviously given by:*

$$\rho_{\nu+\nu'} = e^{a_{\nu+\nu'}}; \tag{3}$$

consequently, *with increasing index, this radius, according to assumption, grows beyond all bounds.* But we have the advantage, that the mapping modulus is equal to 1 at the point O. In consideration of the stipulation concerning the constant contained in the potential $v_{\nu+\nu'}$, we find, for the exponent in (2) on the left side a development of the form: [484]

$$-(u_{\nu+\nu'} - a_{\nu+\nu'}) - i\,v_{\nu+\nu'} = \log r + i\vartheta + c_1 z + c_2 z^2 + \cdots,$$

which we are also able to write:

$$\log Z = \log z + c_1 z + c_2 z^2 + \cdots.$$

Conversely, therefore, there exists for $\log(z^{-1})$, in the Z-plane in the vicinity of $Z = 0$ convergent development of the form:

$$\log\left(\frac{1}{z}\right) = \log\left(\frac{1}{Z}\right) + C_1 Z + C_2 Z^2 + \cdots.$$

We deduce from this further[53]:

$$\log\left(\frac{1}{r}\right) = \log\frac{1}{|Z|} + \sum_{k=1}^{\infty}{}'. \tag{4}$$

Let the partial domain of the circular surface K corresponding to the component F_ν of $F_{\nu+\nu'}$ be called B again, as on p. 475. We transplant the values of the potential u_ν onto B and write these as a function of X and Y:

[52] Should all places $z = 0$ of our Riemann surface be branchpoints or points e, then we shall first of all linearly transform z in suitable fashion.

[53] We remember, that we always wanted to include terms in summation symbols, as in (4) on the right, which vanish at O resp. $Z = 0$. We distinguish different sums of this kind, as e.g. in (1) and (4), if need be, by indices on the summation signs.

$$u_\nu = U_\nu(X,Y). \tag{5}$$

By substituting the expression (4) for the logarithmic term in (1) on the right, there follows:

$$U_\nu = \log\frac{1}{|Z|} + a_\nu + \sum_{k=1}^{\infty} + \sum_{k=1}^{\infty}{}',$$

and since the two sums on the right side represent harmonic functions, which vanish at $Z=0$, then we are able to bring therm together into a single summation symbol:

$$U_\nu = \log\frac{1}{|Z|} + a_\nu + \sum_{k=1}^{\infty}{}''. \tag{6}$$

We note as a result: *the function $U_\nu(X,Y)$, which is the Green's function of the domain B belonging to the origin, has a development (6), convergent for the neighborhood of $Z=0$, with the same absolute term a_ν, which comes up in (1).*

Now the assumptions of the second Koebe theorem and its "consequence" are cut
[485] out for the relations present here. One only wants to notice, that the radius $\rho_{\nu+\nu'}$ for each finite index is itself *finite*. Consequently, B is also finite; but, as was already emphasized on p. 462, a determinate, finite "maximal distance" of the domain boundary from the origin by no means belonged to the presuppositions of that theorem. Now choose, in the sense of the "consequence", a number ε of the interval $0<\varepsilon<1$ arbitrarily small, and imagine the associated finite number determined, satisfying the condition $\alpha>0$. Since $\lim_{\nu=\infty} a_\nu = \infty$ holds according to assumption, then we are able to give a finite index ν_0, so that for all $\nu \geq \nu_0$, the condition $a_\nu \geq \frac{1}{\alpha}$ is fulfilled. The "consequence" then furnishes the result: *after choice of an arbitrarily small number ε, taken from the internal $0<\varepsilon<1$, there is always a finite index ν_0, such that for all $\nu \geq \nu_0$, the minimal distance d of the boundary belonging to B from the origin $Z=0$ satisfies the inequality*:

$$d > \frac{1}{\varepsilon}. \tag{7}$$

We shall soon have to apply this inequality.

It has now been an especially effective idea of Koebe's, besides the Greens' functions, to yet introduce two further similarly defined series of functions $u_1', u_2', u_3', \ldots$ and $u_1'', u_2'', u_3'', \ldots$ on the covering surface F_∞. If μ is any index, then the potentials u_μ' and u_μ'' are to be defined for the domain F_μ, and indeed, by the following determinations: *at the point O, the potentials u_μ' and u_μ'' are to become discontinuous as the real, resp. the imaginary component freed from i, of z^{-1}, and except for this, they are to be everywhere harmonic in F_μ and vanish on the boundary R_μ as the Green's function u_μ*. The existence of these potentials, according to §3 on p. 451 ff., is not in question.

Transplanted onto the Z-plane, let the u_μ', u_μ'' furnish the potentials:

$$u_\mu' = U_\mu'(X,Y), \quad u_\mu'' = U_\mu''(X,Y). \tag{8}$$

These are, in any case, defined for the neighborhood of $Z=0$, and have here, if we write $Z=\rho e^{\theta i}$, convergent developments of the form:

$$U_\mu' = \frac{\cos\theta}{\rho} + A_\mu' + \sum_{k=1}^{\infty}{}', \quad U_\mu'' = -\frac{\sin\theta}{\rho} + A_\mu'' + \sum_{k=1}^{\infty}{}'', \tag{9}$$

where the sums coming up on the right here are naturally not to be the ones so denoted before. in particular, the functions $U'_{\nu+\nu'}$ and $U''_{\nu+\nu'}$ are defined for our circular surface K and vanish on its periphery; similarly, the functions U'_ν and U''_ν are defined for the domain B and vanish on the boundary R of this.

Under the assumptions now guiding us, we are now able to draw the conclusion, that the functions of the two series $u'_1, u'_2, \ldots$ and $u''_1, u''_2, \ldots$ converge uniformly to two limit functions: [486]

$$u' = \lim_{\mu=\infty} u'_\mu, \quad u'' = \lim_{\mu=\infty} u''_\mu. \tag{10}$$

The passage to the limit. The argument works with the same reasoning in both cases; we carry it out for the series $u'_1, u'_2, u'_3, \ldots$.

The difference

$$u'_{\nu+\nu'} - u'_\nu = U'_{\nu+\nu'} - U'_\nu \tag{11}$$

is also harmonic at $Z = 0$, and therefore in the entire domain B. If we write:

$$U'_{\nu+\nu'} - U'_\nu = \left(U'_{\nu+\nu'} - \frac{\cos\theta}{\rho}\right) - \left(U'_\nu - \frac{\cos\theta}{\rho}\right),$$

Then there follows, for the absolute value of the difference (11):

$$|U'_{\nu+\nu'} - U'_\nu| \le \left|U'_{\nu+\nu'} - \frac{\cos\theta}{\rho}\right| + \left|U'_\nu - \frac{\cos\theta}{\rho}\right|. \tag{12}$$

Now $(U'_{\nu+\nu'} - \frac{\cos\theta}{\rho})$ is everywhere harmonic inside K and equal to $-\frac{\cos\theta}{\rho_{\nu+\nu'}}$ along the periphery (where $U'_{\nu+\nu'}$ vanishes), so that $|U'_{\nu+\nu'} - \frac{\cos\theta}{\rho}| \le \frac{1}{\rho_{\nu+\nu'}}$, along the periphery and therefore is $< \frac{1}{d}$, since the domain B with the minimal distance d is a partial domain, different from the whole, of the circular surface K of the radius $\rho_{\nu+\nu'}$. According to an oft-applied theorem on p. 448, the inequality:

$$\left|U'_{\nu+\nu'} - \frac{\cos\theta}{\rho}\right| < \frac{1}{d} \tag{13}$$

will consequently also hold inside K and therefore along the boundary R of B. Along the boundary $U'_\nu = 0$, and therefore:

$$\left|U'_\nu - \frac{\cos\theta}{\rho}\right| = \left|\frac{\cos\theta}{\rho}\right| \le \frac{1}{d},$$

so that again, according to (12) and (13):

$$|U'_{\nu+\nu'} - U'_\nu| < \frac{2}{d}$$

holds along R. This same inequality will accordingly subsist for the entire domain B. By returning to the covering surface and utilizing the theorem attached to formula (7), we have: *after choice of an arbitrarily small number ε taken from the interval $0 < \varepsilon < 1$, there is a finite index ν_0, so that for each $\nu \ge \nu_0$ and arbitrary ν', the inequalities:* [487]

$$|u'_{\nu+\nu'} - u'_\nu| < 2\varepsilon, \qquad |u''_{\nu+\nu'} - u''_\nu| < 2\varepsilon \tag{14}$$

hold in the entire domain F_ν, including the boundary R_ν. We have also stated the result immediately for the series $u_1'', u_2'', \ldots$; in fact, the argument carries itself over directly to this series, in case one will only replace $\cos\theta$ everywhere by $-\sin\theta$.

To this attaches itself the known procedure of proof, which we have described above on p. 477 in detail, in connection with the inequality (17) there: *the uniform convergence of our series of functions to two limit functions, u' and u'', defined for the entire interior of the covering surface, is therefore actually proved.*

In order, at the same time, to yet undertake an estimate of the limit functions u' and u'', we go back once more to the inequalities (14) and remember, that u_ν' as well as u_ν'' vanishes along R_ν. Along R_ν, therefore, the inequalities $|u'_{\nu+\nu'}| < 2\varepsilon$ and $|u''_{\nu+\nu'}| < 2\varepsilon$ hold, and indeed for arbitrary ν'. Since the functions $u'_{\nu+\nu'}$ and $u''_{\nu+\nu'}$, with ν' increasing more and more, now uniformly approach the limit functions u' and u'', then the latter, taken absolutely, cannot, in any case, be greater than 2ε along R_ν: *Along the boundary R_ν, with only $\nu \geq \nu_0$, our limit functions satisfy the conditions.*

$$|u'| \leq 2\varepsilon, \quad |u''| \leq 2\varepsilon. \tag{15}$$

we are able to say briefly: *our limit functions u' and u'', along R_ν, for ν growing more and more, uniformly approach the value* 0.

We emphasize once more, that all these consequences rest on the assumption $\lim_{\nu=\infty} a_\nu = \infty$. By now investigating a certain mapping of the covering surface F_∞ effected by the limit functions u' and u'', we shall arrive at a result, whose impossibility lets itself be shown easily. Conversely, we shall draw the conclusion from this, that our assumption $\lim_{\nu=\infty} a_\nu = \infty$ was false, whereby then, according to the discussions on p. 482 ff., the boundary-circle theorem will be proved.

2.11 Connection of the limit functions u', u'' with one another and with Green's functions u_μ

After obtaining the limit functions u', u'', we are able to make wide-ranging statements concerning the Green's functions u_μ of our domains F_μ, which again admit conclusions to be drawn concerning those limit functions themselves.

[488] Let B be any domain lying in the interior of the covering surface. We are then able, first of all, to choose μ_0' so great, that for each finite index $\mu \geq \mu_0'$, the domain B is a component of F_μ. The analytic function:

$$Z_\mu(z) = e^{-(u_\mu - a_\mu) - i v_\mu}, \tag{1}$$

defined for the domain F_μ, maps F_μ onto the surface of a circle of radius $\rho_\mu = e^{a_\mu}$, now to be denoted more precisely by K_μ. The image B_μ of the domain B in the Z_μ-plane is a partial domain of this circular surface.

We will now denote the reciprocal value of the function (1), in dependency on z, by $\varphi_\mu(z)$:

$$\varphi_\mu(z) = \frac{1}{Z_\mu} = e^{(u_\mu - a_\mu) + i v_\mu}, \tag{2}$$

And undertake an estimate of this function in the domain B. For this purpose, we go back to the functions (8) on p. 485, and determine for these the inequalities (cf. p. 486), valid along the periphery K_μ:

$$\left|U'_\mu - \frac{\cos\theta}{\rho}\right| \leq \frac{1}{\rho_\mu}, \quad \left|U''_\mu + \frac{\sin\theta}{\rho}\right| \leq \frac{1}{\rho_\mu} \tag{3}$$

which, in consequence of known methods of reasoning, also hold in the interior of K_μ and therefore, in particular, in entire domain B_μ.

Now obviously:

$$\left|(U'_\mu + iU''_\mu) - \frac{\cos\theta - i\sin\theta}{\rho}\right| \leq \left|U'_\mu - \frac{\cos\theta}{\rho}\right| + \left|U''_\mu + \frac{\sin\theta}{\rho}\right|$$

Accordingly, if we remember that $\rho e^{\theta i}$ is the polar representation of the complex variable Z_μ, and that:

$$\frac{\cos\theta - i\sin\theta}{\rho} = \frac{1}{\rho e^{\theta i}} = \frac{1}{Z_\mu} = \varphi_\mu(z)$$

therefore holds, then by returning the covering surface F_∞, we find, from the last equation, with consideration of (3):

$$|(u'_\mu + iu''_\mu) - \varphi_\mu(z)| \leq \frac{2}{\rho_\mu}, \tag{4}$$

valid for all $\mu \geq \mu'_0$ *in the total domain* B.

Now we remember, that the functions u'_μ and u''_μ, with increasing index, approach the limit functions u' and u'' uniformly inside B. After choice of a positive but arbitrarily small η different from 0, there is an associated finite index μ''_0, so that for all μ, which do not lie below μ'_0 and μ''_0, the inequality holds everywhere inside B: [489]

$$|(u' + iu'') - (u'_\mu + iu''_\mu)| < \eta. \tag{5}$$

Since now again:

$$|(u' + iu'') - \varphi_\mu(z)| \leq |(u' + iu'') - (u'_\mu + iu''_\mu)| + |(u'_\mu + iu''_\mu) - \varphi_\mu(z)|$$

is correct, then we conclude from (4) and (5):

$$|(u' + iu'') - \varphi_\mu(z)| < \eta + \frac{2}{\rho_\mu}, \tag{6}$$

valid in the entire domain B *and for all* μ, *which do not lie below* μ'_0 *and* μ''_0.

In order to shorten somewhat this latter stipulation concerning the permissible indices μ, we choose an arbitrarily small positive number ε different from zero and then set:

$$\eta = \frac{\varepsilon}{6}, \qquad \frac{1}{\rho_\mu} \leq \frac{\varepsilon}{6} \tag{7}$$

To this η then belongs a determinate finite number μ''_0; on the other hand, the second requirement (7) is fulfilled on account of $\lim \rho_\mu = \infty$, as soon as μ does not lie below a determinate index μ_0'''. Now, finally, choose μ_0 as a finite number, so that none of the numbers $\mu_0', \mu_0'', \mu_0'''$ is greater than μ_0. *Then, in the entire domain* B, *be inequality*:

$$|(u' + iu'') - \varphi_\mu(z)| < \frac{1}{2}\varepsilon \tag{8}$$

holds for all indices μ, *which are* $\geq \mu_0$.

From this, one may easily conclude the uniform convergence of the series of functions $\varphi_1(z), \varphi_2(z), \ldots$ in the domain B, whereby, naturally, the initial terms are again to be excluded, which are not, or only partially, defined for B. *Namely, if* $\mu \geq \mu_0$, *and* μ' *signifies an entirely arbitrarily choosable index*, then from:

$$\varphi_{\mu+\mu'}(z) - \varphi_\mu(z) = ((u' + iu'') - \varphi_u(z)) - ((u' + iu'') - \varphi_{\mu+\mu'}(z))$$

follows the inequality:

$$|\varphi_{\mu+\mu'}(z) - \varphi_\mu(z)| \leq |(u' + iu'') - \varphi_\mu(z)| + |(u' + iu'') - \varphi_{\mu+\mu'}(z)|$$

and therefore, we draw from (3) the conclusion:

$$|\varphi_{\mu+\mu'}(z) - \varphi_\mu(z)| < \varepsilon. \tag{9}$$

From this, with known method of reasoning, it follows the asserted uniform convergence of our series of functions in the domain B to a limit function, which, in conse-
[490] quence of (8), is none other than our function $(u' + iu'')$. Now since B is an arbitrary domain in the interior of the covering surface, we note as a result: *our analytic functions, defined for the domains* $F_1, F_2, F_3, \ldots$

$$\varphi_1(z), \varphi_2(z), \varphi_3(z), \ldots \tag{10}$$

converge uniformly to an analytic limit function, to be considered as defined for the entire interior of the covering surface F_∞, *namely to*:

$$\lim_{\mu=\infty} \varphi_\mu(z) = u' + iu'', \tag{11}$$

understanding by u' *and* u'' *our limit functions obtained in the previous paragraph.*

The new knowledge, which we obtain here concerning these limit functions u', u'' is this, *that* $(u' + iu'')$ *is an analytic function, or expressed otherwise, that* u'' *is the conjugate potential of* u'.[54] We will externally characterize this result, by letting the notation v' take the place of u'' from now on.

---

[54] Koebe, in his first announcement concerning the boundary-circle theorem "*Über die Uniformisiarung beliebiger analytischer Kurven*" Gött. Nachr. of 1907, p. 206 ff., in place of the functions $u_1'', u_2'', u_3'', \ldots$, works directly with the conjugate potentials $v_1', v_2', v_3', \ldots$ of the $u_1', u_2', u_3', \ldots$. In the detailed representation in the Math. Ann. Vol. 67, p. 219 ff., which we have followed here, Koebe, for the sake of symmetry, has introduced, besides the u_μ', the coordinated functions u_μ'', whereby, then, to be sure, the proof was still to be carried out, that the sum $(u_\mu' + iu_\mu'')$, for $\lim \mu = \infty$, also actually yielded an analytic function.

2.12 Mapping of the covering surface by means of the function $(u' + iv')$. Proof of the boundary-circle theorem

By means of the analytic function $(u' + iv')$, which we will denote by:

$$\zeta(z) = u' + iv', \tag{1}$$

and which is uniquely defined for the entire interior of the covering surface F_∞, we will now map this surface onto the ζ-plane. Let this image be called P_∞, and the partial domain F_ν furnish the component P_ν; naturally, these domains P_∞ and P_ν initially have nothing to do with the polygon complexes so denoted in §6.

If it is now a question of determining the character of the conformal image P_∞ of the F_∞, then the approximation function $\varphi_\mu(z)$ of the function $\zeta(z)$ already shows us the way. The function $\varphi_\mu(z)$ furnishes, as a conformal image of F_μ, that schlichtly cov- [491]
ered portion of the φ_μ-plane, which runs outside the circle with the radius $\frac{1}{\rho_\mu}$ about the origin of this plane. However, we are also able, without these approximation functions, to base the study solely on the statements, which were made at the close of §10 on p. 487 concerning the limit functions u' and v'.

The partial domain F_ν with the boundary R_ν furnishes, mapped by the function $\zeta(z)$, the domain P_ν, whose boundary therefore corresponds to the boundary R_ν. If we again take up the quantity of measure ε mentioned in (15) on p. 487, and choose the index ν not smaller than the associated finite number ν_0, then the inequality (15) on p.487 holds on the boundary R_ν, from which we draw the conclusion:

$$|\zeta| \le |u'| + |v'| < 4\varepsilon \tag{2}$$

therefore, the image of the boundary R_ν in the ζ-plane will certainly lie in the interior of the circle of radius 4ε about the origin.

From this, by a consideration entirely analogous to p. 480, the conclusion of the schlichtness of the image P_∞ lets itself be drawn. The point O of F_∞, at which $z = 0$, is the sole discontinuity point of the function ζ, and indeed, ζ becomes infinite as z^{-1}. The point O accordingly carries itself over onto the point $\zeta = \infty$< and the neighborhood of O thereby furnishes the schlichtly covered neighborhood of the image point $\zeta = \infty$. At all remaining points of the covering surface, ζ has finite values, *In any case, therefore, the image P_∞ will only cover the neighborhood of $\zeta = \infty$ one-sheetedly.*

The single-valuedness and regularity of $\zeta(z)$ inside F_∞ now by no means yet stipulates the schlichtness of the image P_∞. We will consequently test, whether a place ζ_0 different form the origin $\zeta = 0$ can be covered by at least two sheets of the image P_∞. To the two places lying over one another at ζ_0 will correspond two inner points z_0' and z_0'' of the covering the surface F_∞. We are also then already able to give a determined finite index ν_0', so that for all $\nu \ge \nu_0'$, the two points z_0', z_0'' simultaneously become inner points of F_ν.

On the other hand, we choose an arbitrarily small ε of the interval $0 < \varepsilon < \frac{1}{4}|\zeta_0|$ and determine the associated number ν_0, so that for $\nu \ge \nu_0$, the inequality (2) holds on the boundary of F_ν. Then the assumption of the multiple-sheetedness of P_∞ at ζ_0 easily leads to a contradiction.

Namely, if we choose ν neither smaller than ν_0 nor smaller than ν_0', the domain P_ν, [492] as a component of P_∞, covers the neighborhood of $\zeta = \infty$ simply, but that of ζ_0 at least doubly. If we therefore picture the concentric circles about $\zeta = 0$ with all radii $> |\zeta_0|$, then the exterior of these circles, for sufficiently large radii, are only covered simply by P_∞, but for radii, which lie sufficiently close to $|\zeta_0|$, at least doubly. Accordingly, there is a smallest one among those circles, whose exterior was still simply covered by P_∞. Its periphery will therefore just reach to the boundary of the domain P_ν, and P_ν consequently had boundary points with a distance $> |\zeta_0| > 4\varepsilon$ from the origin, which, in fact, stands in contradiction to (2). *Therefore, the image P_∞ covers the ζ-plane multiple at no place different from the origin.*

We now need return only once more to the inequality (2) and conclude from this, that the boundaries of the domains P_ν, for $\lim \nu = \infty$, draw themselves together uniformly about the zero of the ζ-plane. With this arises the result: *the function $\zeta(z)$ maps the entire interior of the covering surface F_∞ conformally onto the schlichtly covered ζ-plane, excepting only the origin $\zeta = 0$ of this.* Furthermore, for the sake of convenience, we will let the reciprocal value $\zeta' = \frac{1}{\zeta}$ take the place of ζ. *Then, as conformal image of the entire interior of the covering surface F_∞, these appears the schlichtly covered ζ'-plane, excepting the single point $\zeta' = \infty$ of this.*

Now we are able to carry over the considerations on p. 481, which base themselves on the regular construction of the covering surface F_∞ from infinitely many copies of the original surface F_1, word for word to the present relations. To those infinitely many copies correspond, in the ζ'-plane, infinitely many domains $\overline{P}_1, \overline{P}'_1, \overline{P}''_1, \ldots$, positioned next to one another, which join themselves together into a regular net $\overline{N}$, schlichtly covering the entire ζ'-plane except for the point $\zeta' = \infty$. Also, the discussions concerning the reversibly single-valued analytic transformations of the F_∞, and with this, of the net $\overline{N}$ into itself, remain subsisting unaltered. In particular, we draw the conclusion of the existence of an analytic, reversibly single-valued transformation $V_i(\zeta')$ of the net $\overline{N}$ into itself, for which the initial domain $\overline{P}_1$ goes over into an arbitrarily prescribed domain $\overline{P}_i^{(i)}$ of $\overline{N}$.

But now, from the principles of the theory of entire functions, it follows very easily, that a one-to-one mapping of the ζ'-plane into itself, which is everywhere conformal in the finite part of the plane, for which the point $\zeta = \infty$ corresponds to itself, is necessarily a *linear integral transformation.* The transformation V_i of the ζ'-plane into itself
[493] therefore has the form:

$$V_i(\zeta') = \alpha_i \zeta' + \beta_i. \tag{3}$$

If we collect all these transformations for $i = 0, 1, 2, \ldots$, *then there arises a group* Γ *with the discontinuity domain $\overline{P}_1$ and with the single boundary point $\zeta = \infty$.* In consequence of I on p. 222 ff., Γ is consequently a *parabolic rotation group,* and therefore, $\overline{P}_1$, and with this, our original surface F_1, has one of those small signatures belonging to these groups.

But we have explicitly excluded these signatures above (p. 465), as that belonging to the boundary-circle theorem the contradiction, which we already indicated on p.483, is consequently uncovered, and with this, our assumption made on p. 483, that $\lim_{\nu=\infty} a_\nu = \infty$, has proved to be untenable. *Also in the boundary-circle case, the absolute terms a_ν in the developments (1) on p. 472 of the Green's functions u_ν of our domains*

F_V approach a "finite" limit, so that, according to the discussions at the close of §9 *on p. 482, the boundary-circle theorem is now also proved.*

The extraordinary significance, which accrues to the results thereby obtained, we have already described in the epilogue to the preceding chapter on p. 438. The boundary-circle theorem is by no means of the most general "functional theorem"; for it refers itself in the form, as we have brought it into representation here, not to the general case of analytic models, but only to algebraic models, and it is surpassed in this domain by that general fundamental theorem of Klein's cited on p. 47, which contains the boundary-circle theorem and reentrant cut theorem only as simplest special cases. But the boundary-circle theorem has a very characteristic advantage over the reentrant cut theorem and over the general fundamental theorem just mentioned, *i.e. the character of invariance with respect to an arbitrary transformation of the canonical cut system.* In all remaining cases of the general theorem of Klein, the polymorphic function $\zeta(z)$ depends to a certain degree on the choice of the cut system: *for the boundary-circle theorem, it is independent of this choice*[55]*. An alteration of the crosscut system stipulates an alteration of the polygon to the mapped* for unaltered total net N, *as we considered it in detail in I on p. 320 as "transformation of the canonical polygons".* The polymorphic function $\zeta(z)$ of the boundary-circle theorem belongs to the uncut surface and therefore to the proposed algebraic model as the single quantity of its kind. *In it is found that element, which for an unsigned surface of genus $p = 1$ is the associated elliptic integral of the first kind.* With the discovery of the polymorphic function $\zeta(z)$ of the boundary-circle theorem, the theory of the algebraic models has accordingly found its natural conclusion, whereby we must certainly admit, that the development of this theory in the general case stands far behind the special case of the unsigned surfaces of the genus $p = 1$, since in the general case, we have no analytic representations of the functions of the model in ζ, which could measure up in usefulness to Jacobian theta series of the case $p = 1$. [494]

[55] Naturally, also for the principal-circle theorem; however, we do not pursue separately in the text, since this theorem only refers itself to a special species of Riemann surfaces, namely the orthosymmetric ones.

3

Proof of the reentrant cut theorem

We now turn ourselves to the "reentrant cur theorem", which answers to the third of the fundamental problems formulated on p. 46. This theorem, set up by Klein in the year 1882, asserts the following: *if an arbitrary finite-sheeted closed Riemann surface F with finitely many branchpoints and a genus $p > 0$ is given, and if this surface is cut through along any p reentrant cuts not meeting one another and not dividing the surface into separate pieces, then there is one and essentially also only one polymorphic function, which maps the surface onto a discontinuity domain of kind named under β on p. 35 with 2p boundary curves mapped pairwise linearly onto one another.* [495]

For $p = 1$, we come back to a known theorem from the theory of elliptic functions, according to which, each surface of this genus provided with a reentrant cut lets itself be carried over conformally onto a circular ring, whose two boundary curves correspond to the banks of the cut. Namely, if one adds the conjugate reentrant cut and maps the surface thus cut up onto a period parallelogram by means of the integral u of the first kind, then let these vertices lie at $u = 0$, $\omega_1, \omega_2, \omega_1 + \omega_2$, and in particular, let the side from 0 to ω_2 correspond to one bank of the initially drawn reentrant out. For this assumption, the function:

$$\zeta = e^{\frac{2i\pi u}{\omega_2}}$$

furnishes, in fact, as the image of the period parallelogram on the ζ-plane, a circular ring, whose boundary curves are mapped onto one another by the loxodromic or (in special cases, e.g., the one of a purely imaginary period quotient) hyperbolic substitution:

$$\zeta' = e^{\frac{2i\pi\omega_1}{\omega_2}} \cdot \zeta.$$

Corresponding to this state of affairs, we shall be able to presuppose $p > 1$ from now on. The general reentrant cut theorem remained unproved until 1908. First in February of the year named, Koebe has given a proof of the reentrant cur theorem to the mathematical society at Göttingen[1], which rests on the method of the covering surface. The note in the Gött. Nachr. of 1909, p.68 ff, as well as the paper in Vol. 69 of the [496]

[1] See the relevant notice in the report on the proceedings of the Göttingen mathematical society in the Jahresbericht der Deutschen Mathematiker-Vereinigung of 1908 on p.50.

Math. Annalen cited there gives the representations of the proof. The following representation of the proof of the reentrant cut theorem attaches itself to the latter work of Koebe's in the general train of thought, but deviates from this often in the detailed argument, especially so far the the mapping of the partial domains F_n of the covering surface F_∞ onto schlicht domains. Also, the general theorems on conformal mapping in §1 through 4 appear here in a somewhat more precise form, insofar as the calculations are carried through to numerical statements concerning the estimating constants coming into consideration.

A short time after Koebe, R. Courant, in the second chapter of his dissertation[2], has developed a second proof of the reentrant cut theorem. This appears as a special application of mapping theorems, which Courant carried out in the first chapter of his work, tying into the theory of Hilbert on the treatment of the most general problems of conformal mapping on the basis of the Dirichlet Principle[3].

3.1 Theorems on schlicht infinite images of a circular surface

We understand by K_r the periphery of the circle of radius r about the origin of the plane of a complex variable z. In the interior of K_1, ie., for $|z| < 1$, let a single-valued analytic function $f(z)$ be given by a series converging there:

$$f(z) = \frac{1}{z} + a_1 z + a_2 z^2 + \cdots \tag{1}$$

[497] *with vanishing absolute term.* At the origin itself, therefore, $f(z)$ is to become infinite as z^{-1}; in addition, we assume, *that $f(z)$, inside K_1, assumes no value more then once.*

This function $f(z)$ will map the inner surface of the circle K_1 onto a schlicht, ie. never covering the image plane multiple, domain, which contains the point ∞ of the image plane, corresponding to the origin $z = 0$, in its interior. The images of the circular peripheries K_r for $r < 1$ will cover that domain simply and gaplessly and indeed, in such a manner, that for $r_1 > r_2$, the image of K_{r_2} surrounds that of K_{r_1}. If we accordingly denote by M_r the maximum of the absolute value $|f(z)|$ along K_r, then:

$$M_{r_1} < M_{r_2} \tag{2}$$

will certainly hold for $r_1 > r_2$.

The next consideration has the goal of calculating an upper bound for the value M_r. In consequence of the assumptions on $f(z)$, the two functions:

$$\varphi(z) = f(z) - \frac{1}{2}, \qquad \psi(z) = \frac{1}{2}\varphi(z), \tag{3}$$

inside K_1 are regular and single-valued everywhere, ie. also including the origin. If we now denote by M'_r and M''_r, the maxima of the absolute values $|\varphi(z)|$ and $|\psi(z)|$ along K_r, then from:

---

[2] "*Über die Anwendung des Dirichlet'schen Principes auf die Probleme der conformen Abbildung*", Gottingen, 1910; printed in Vol. 71 of Math. Annalen.

[3] See, concerning this, the note of Hilbert's "*Zur Theorie der conformen Abbildung*" in the Gött. Nachr. of July 17, 1909.

$$f(z) = z^{-1} + \varphi(z), \qquad \varphi(z) = z\psi(z).$$

There yield themselves the consequences:

$$M_r \leq \frac{1}{r} + M'_r, \qquad M'_r = rM''_r. \tag{4}$$

For any $r_1 < 1$, $\psi(z)$ is now a finite, single-valued and regular function inside and on K_{r_1}. This will map the surface of the circle K_{r_1} onto a simply connected, one-or several-sheeted *finite* domain of the ψ-plane. Since this image, as emphasized, does not contain the point ∞ of the ψ-plane, then for each interior point of the image, however this might be shaped, a boundary point lets itself be given, which stands further from the origin $\psi = 0$ of the image place than that interior point. If, accordingly, $r_2 < r_1$, then:

$$M''_{r_2} < M''_{r_1}$$

certainly holds, and therefore, from the second formula (4) follows:

$$M'_{r_2} = r_2 M''_{r_2} < r_2 M''_{r_1} = \frac{r_2}{r_1} M'_{r_1}. \tag{5}$$

From (2), (4) and (5) here consequently yields itself: [498]

$$M_{r_1} < M_{r_2} \leq \frac{1}{r_2} + M'_{r_2} < \frac{1}{r_2} + \frac{r_2}{r_1} M'_{r_1};$$

and since, in consequence of the first equation (3):

$$M'_{r_1} \leq M_{r_1} + \frac{1}{r_1}$$

is correct, then follows, further:

$$M_{r_1} < \frac{1}{r_2} + \frac{r_2}{r_1}\left(M_{r_1} + \frac{1}{r_1}\right)$$

and therefore:

$$M_{r_1} < \frac{r_1^2 + r_2^2}{r_1 r_2 (r_1 - r_2)}. \tag{6}$$

For r_2 one may substitute here any value of the interval $0 < r_2 < r_1$. If, in particular, we set $r_2 = r_1(-1 + \sqrt{2})$, then the right side of (6) becomes minimal for given r_1. We leave out the index 1 and note as a result: *if $f(z)$ is given inside K_1 by the convergent series (1), and if this function assumes no value more than once there, then for any $r < 1$, the maximum M_r of $|f(z)|$ along K_r satisfies the inequality*:

$$M_r < \frac{2 + 2\sqrt{2}}{r} \tag{7}$$

independently of whatever particular function $f(z)$ might be present.

We now let the following development:

$$f(z) = \frac{1}{z} + a_0 + a_1 z + a_2 z^2 + \cdots \tag{8}$$

take the place of (1) and hold fast, except for this, to all assumptions concerning $f(z)$.

The function:

$$f_1(z) = f(z) - a_0$$

then fulfills the preconditions, without exception, of the theorem just stated, so that the inequality (7) holds for the maximum of $|f_1(z)|$ along K_r.

We designate as the "*oscillation*" S_r of the function $f(z)$ along the periphery K_r the maximum of the absolute value:

$$|f(z) - f(z')|,$$

which enters for any two places z and z' of the periphery K_r. Since obviously:

$$|f(z) - f(z')| = |f_1(z) - f_1(z')| \leq |f_1(z)| + |f_1(z')|$$

holds, and since, besides this, the two amounts standing on the right, according to be
[499] remark just made, do not exceed the quantity M_r mentioned in (7), there follows: *if $f(z)$ is given inside K_1 by the series (8) convergent there, and if this function takes on no value there more than once, then for an $r < 1$, the oscillation S_r of the function along K_r satisfies the inequality*:

$$S_r < \frac{4 + 4\sqrt{2}}{r} \tag{9}$$

independently of which particular function might be present.

3.2 Theorems on schlicht finite models of a circular surface

Now let a function be given by:

$$f(z) = z + b_2 z^2 + b_3 z^3 + \cdots, \tag{1}$$

which is regular in the interior *and on the boundary* of K_1 and so constituted, that it assumes no value more than once there. If we set $w = f(z)$, then by this function, the surface of the circle K_1, the boundary included, is mapped onto a schlicht finite portion of the w-plane, and indeed, in such a manner, that the origin $z = 0$ again furnishes the origin of the image plane, and so that the mapping modulus (the magnification ratio of the mapping) is equal to 1 at this place.

For any $r \leq 1$, let L_r be the image of the circular periphery K_r in the w-plane. This line is then a closed line, not cutting itself, running once around the origin. For the distance, which the individual points of the line L_r have from the origin of the w-plane, a lower as well as an upper bound is givable, which depends on r and not on the special function present.

The following consideration leads to the obtention of the lower bound for the "minimal distance" of the line L_r from the origin. Set:

$$w_1 = f_1(z) = \frac{1}{f(z)} = \frac{1}{2} - b_2 + c_1 z + c_2 z^2 + \cdots \tag{2}$$

and obtain thus a function $f_1(z)$, which satisfies the preconditions of the theorem stated at the close of §1. Let the image of K_r, and therefore of L_r, in the w_1-plane be called $L_r^{(1)}$; this will represent a closed line running once around the origin $w_1 = 0$. Since now, the oscillation of $f_1(z)$ along K_r is smaller than $\frac{4+4\sqrt{2}}{r}$ and this oscillation has the meaning in the w_1-plane of the greatest chord of the line $L_r^{(1)}$, then no point of the line $L_r^{(1)}$ will able to reach. a distance from the origin included by the line, which [500]
would be $\geq \frac{4+4\sqrt{2}}{r}$. Therefore, we find, by return to the w-plane, that all points of L_r have a distance $> \frac{-1+\sqrt{2}}{4} r$ from the origin. The inequality (9) in the previous paragraph was indeed only stated for $r < 1$. But it is immediately self evident from the present mapping of the circular surface K_1 into the w-plane, that the result just obtained on the minimal distance of L_r also remains subsisting for $r = 1$, in case it holds for all $r < 1$: *if the function $w = f(z)$, given by the series (1), regular inside and on K_1, takes on no value more than once there, then for all $r \leq 1$, the minimal distance d_r of the line L_r, which corresponds to the periphery K_r, from the origin of the image place, satisfies the inequality*:

$$d_r > \frac{-1+\sqrt{2}}{4} r, \tag{3}$$

whichever particular $f(z)$ might be present. In particular, therefore, there holds for the minimal distance d_1 of the boundary L_1:

$$d_1 > \frac{-1+\sqrt{2}}{4}. \tag{4}$$

In addition, we are also able to immediately deduce the more general theorem (3) from the special inequality (4). Notice namely, that $z^{-1} \cdot f(z)$ has the value 1 at the origin $z = 0$ and is therefore a function regular everywhere inside and on K_1 and different from 0, thus:

$$\log\left|\frac{f(z)}{z}\right| \tag{5}$$

is a harmonic potential everywhere inside and on K_1, for which, in consequence of (4), the inequality:

$$\log\left|\frac{f(z)}{z}\right| > \log\left(\frac{-1+\sqrt{2}}{4}\right)$$

holds along the boundary. This inequality will therefore also hold inside K_1, so that, in particular, for each point of K_r, the inequality:

$$\log|f(z)| > \log r + \log\left(\frac{-1+\sqrt{2}}{4}\right)$$

holds, which furnishes the rule (3).[4]

[501] The number standing on the right in (4) is a lower "bound" for d_1 and by no means a lower "limit", to which the distance d_1 could come arbitrarily close for suitable function $f(z)$ of our kind. If, in particular, w is a linear function of z then according to assumption, it has the form:

$$w = \frac{z}{az+1}$$

and therefore represents a parabolic substitution. If we choose z so that the real z-axis belongs to the orbit curves of this substitution, and so that the individual positive real value z is transformed into a smaller value w, then this amounts to a being real and $a > 0$. Since, in addition, w is to be regular in and on K_1, then $a < 1$ follows. The absolutely smallest value of w along the line L_1, circular in this case, originates from $z = 1$, and is $w = \frac{1}{a+1} > \frac{1}{2}$. For linear functions $f(z)$, therefore:

$$d_1 > \frac{1}{2} \tag{6}$$

holds, and indeed the bound has here, at the same time, the meaning of a limit (not attachable).

One obtains the greatest value of w along the circle K_1 at $z = -1$, where:

$$|w| = \frac{1}{1-a} \tag{7}$$

holds. Since a is indeed < 1, but may come arbitrarily close to the value 1, then for the value (1) of the our function, ∞ is an upper limit (not attainable). From this yields itself, *that also in general, ie. for arbitrary function $f(z)$ of our kind, there cannot exist a finite upper limit for the "maximal distance" of the line L_1 corresponding to the circle K_1.*

The result obtained does not preclude, *that for each $r < 1$ a finite upper bound is givable depending only on r and independent of the special function $f(z)$, beneath which the maximal distance D_r of the line L_r is obtained in each case, a bound, which then must, to be sure, be itself infinite for* $\lim r = 1$. The existence of such a bound is
[502] actually proved by Koebe, and indeed, by means of a consideration, which in its character, ties into his developments for the proof of the theorems on p. 459 ff. We shall here even press through to an explicit expression for this upper bound in r.

The potential (5), harmonic everywhere in and on K_1, has the value 0 at the point $z = 0$. Should this constant be equal to 0, then $|f(z)| = |z|$ and therefore $D_r = r$, from which our assertion becomes self-evident. If the potential is not constant (which we accordingly only have to consider), then this takes on positive as well as negative values along K_1, and therefore, one can, in particular, produce points on K_1, at which $|f(z)| < 1$

[4] The Green's function of the domain bounded by L_1, belonging to the point $w = 0$, which accomplishes the mapping of this domain backwards onto the surface of the surface of the circle K_1, has a vanishing absolute term in its development considered for the neighborhood of $w = 0$. The theorem (4) of the text, as far as its character goes, thus attaches itself as a special theorem of the general consequence, which we deduced above on p. 464 from the second Koebe theorem at that time, whereby, to be sure, that "consequence" is cut out for other presuppositions. The present special theorem is more precise, insofar as in (4), on the right side sits, a completely determinate number.

occurs. The image L_1 of the circle K_1 will consequently run segmentwise inside the unit circle of the w-plane, or (which says the same thing) the unit circle of the w-plane will always run segmentwise outside the image domain B_1 of the circular surface K_1, bounded by L_1.

Now fix any point with $|w| = 1$ outside L_1 and turn the w-plane so that the point fixed arrives at $w = 1$. The maximal distance D_r is not altered thereby; whatever holds for it for the new position of the domain B_1, also held for the original position, so that the assumption is permitted that the point $w = 1$ lies outside L_1.

Now draw a cut Q (cf. figure 103) running outside L_1 from $w = 1$ to the point $w = \infty$. Only for the sake of easier visualizability, arrange this cut, so that it comes in tangentially to the real w-axis from the right toward its initial point $w = 1$, that it further coincides with the positive real w-axis in its last part (which indeed certainly emerges definitely from the "finite" domain B_1 for a finite w), as well as, that, finally, the line [503]
Q, without becoming pushed over the origin, by holding fast to its initial and its end point, is continuously deformable (cf. figure 103) into the portion of the real positive w-axis reaching from 1 to ∞.

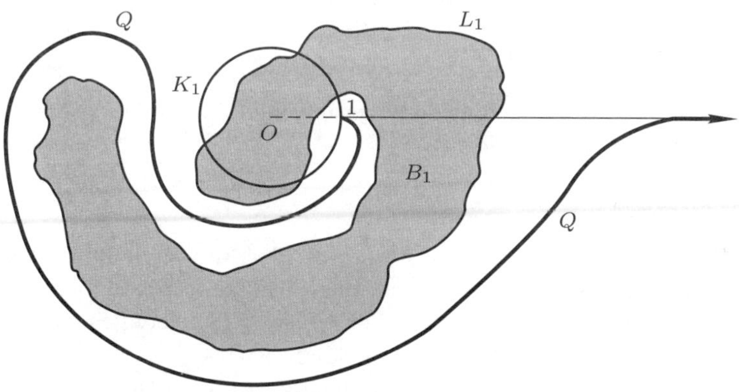

Fig. 103.

These determinations have the purpose, of fashioning, in the most surveyable possible form, the mapping of the w-plane by means of the function:

$$t = \log(w - 1). \tag{8}$$

If we define this function $t(w)$ single-valuedly in the w-plane cut along Q, simply by prescribing the function $t(0) = \pi i$ for $w = 0$, then the cut w-plane carries itself over onto a stripe bounded by two curves, running parallel at a distance of 2π, whose lower (corresponding to one bank of Q) boundary approaches the real t-axis asymtatically on the left, and on the right even joins with the real t-axis (cf. figure 104). To the dotted portion of the real w-axis theorem 0 and 1 in figure 103, there corresponds thereby the parallel to the negative real t-axis from $t = \pi i$ to $t = \infty$, dotted in figure 104. *This line runs entirely in the interior of our image stripe.* The domain B_1 finds its conformal image is a *schlicht domain B_1' similarly situated entirely in the interior of the image stripe*, which always contains the point $t = \pi i$ in its interior[5].

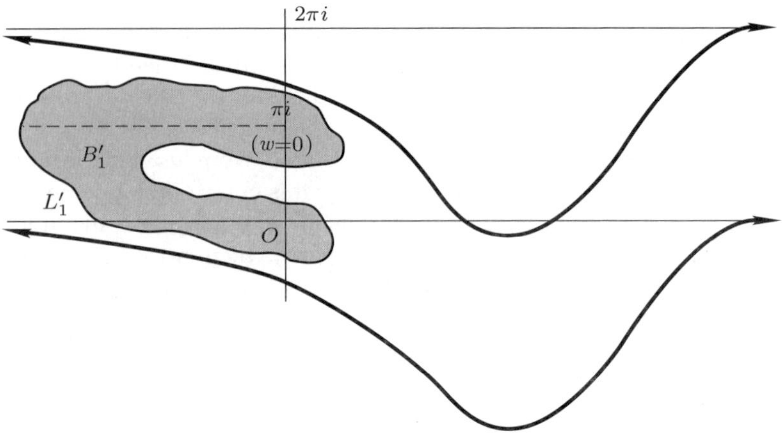

Fig. 104.

[504] The strip obtained can be considered as an initial domain for the cyclic group to be generated from the substitution:

$$t' = t + 2i\pi, \tag{9}$$

and, in fact, we will imagine the entire t-plane divided into mutually equivalent strips. Then, conversely, the t-plane furnishes, as image over the w-plane, an infinite-sheeted Riemann surface of simple connectivity, for which Q can serve as "branchcut". The principal function of this surface is naturally $t(w)$.

After we have set the entire t-plane into relation with w in such manner, we must effect one more mapping of the t-plane, for its part. We set:

$$s = \sqrt{t - 3\pi i}, \tag{10}$$

and must, corresponding to this function, position a second sheet over the previous t-plane, which is to be branched with the first at the two points $t = 3\pi i$ and $t = \infty$. As branchcut, we utilize the line from $3\pi i$ to ∞ running parallel to the negative real t-axis, *which is situated entirely in that strip of the t-plane, which proceeds from the initial strip from the substitution (9)*. The function (10), however, we will define single-valuedly on the two-sheeted t-surface, by stipulating the function value:

$$s = \sqrt{\pi}(-1 + i), \tag{11}$$

with root $\sqrt{\pi}$ taken positively, in the lower sheet of our surface carrying the domain B_1', at $t = \pi i$ (corresponding to the value $z = 0$). To the lower sheet of the t-surface then corresponds the halfplane lying on the left of the imaginary s-axis, and it is, as image

[5] The drawing given in figure 104 is naturally to be understand only schematically.

of B_1', *a finite schlicht domain* B_1'' again finds itself, which never reaches the boundary of the halfplane, ie. The imaginary s-axis.

Now, finally, we map the repeatedly mentioned s-half plane by means of the function:

$$Z = \frac{s + \sqrt{\pi}(1-i)}{s - \sqrt{\pi}(1+i)} \tag{12}$$

in such a manner onto the schlicht surface of the Z-plane, that to the values s given in (11), and therefore to the origin of the original z-plane, corresponds the point $Z = 0$. *To the domain B_1'', and therefore, to the surface of the circle K_1 of the z-plane, thereby corresponds, without exception, single-valuedly and conformally, a schlicht domain B_1'''* [505]
of the Z-plane, which also lies with its boundary entirely inside the unit circle of the Z-plane.

The function $Z(z)$ accomplishing this mapping is consequently single-valued and regular in and on K_1. The same thing also holds even for the quotient $\frac{Z}{z}$, since to the value $z = 0$ corresponds the value $Z = 0$ again. But this quotient, as we just found is absolutely < 1 everywhere along the periphery K_1 itself; accordingly, the potential:

$$\log\left|\frac{Z}{z}\right|, \tag{13}$$

harmonic everywhere in and on K_1, will be < 0 everywhere along K_1 and therefore also always have negative values inside K_1. From this yields itself, *that for each $r < 1$, $|Z| < r$ holds along the periphery K_r of the z-plane.*

But now, as follows from (8),(10) and (12), w as a function of Z, is represented by:

$$w = 1 - e^{-2i\pi\left(\frac{1+iZ}{1-Z}\right)^2} \tag{14}$$

According to this, the inequality will hold for $|w|$ along K_r:

$$|w| \le 1 + \text{ upper limit of } \left|e^{-2i\pi\left(\frac{1+iZ}{1-Z}\right)^2}\right|, \tag{15}$$

whereby the upper limit refers itself to all values Z with $|Z| < r$.

An estimate of this upper limit is easy to effect. The circular surface of the radius r about $Z = 0$ recovers itself, in the plane of the variable:

$$Z' = \frac{1+iZ}{1-Z},$$

as a schlicht finite circular surface of the center $Z' = \frac{1+ir^2}{1-r^2}$, which contains the origin $Z' = 0$ neither in its interior nor on its periphery, and consequently lies entirely inside the circle of radius $\frac{2\sqrt{1+r^4}}{1-r^2}$. If we accordingly go on to the plane of the variable $Z'' = Z'^2$, then as the image as the circular surface just mentioned, which corresponds to all values Z with $|Z| < r$, it lies inside the circle of radius $\frac{4(1+r^4)}{(1-r^2)^2}$ about $Z'' = 0$.

If we set $Z'' = X'' + iY''$, then follows: [506]

$$\left|e^{-2\pi i\left(\frac{1+iZ}{1-Z}\right)^2}\right| = \left|e^{-2\pi i Z''}\right| = \left|e^{-2\pi i(X''+iY'')}\right| = e^{2\pi Y''},$$

and since, as we just saw, $Y'' < \frac{4(1+r^4)}{(1-r^2)^2}$ for $|Z| < r$, then , with this, we have effected an estimate for the upper limit intended in (15):

$$|w| = |f(z)| < 1 + e^{8\pi \frac{1+r^4}{(1-r^2)^2}}. \tag{16}$$

This inequality, valid along the periphery K_r, now exhibits on the right, in fact, an expression, which is finite for each $r < 1$, but, to be sure, as was already stated before, becomes infinite for $\lim r = 1$ *itself.*

The two bounds obtained for $|w|$ along K_r we will denote for short as follows:

$$g(r) = \frac{-1+\sqrt{2}}{4} r, \qquad G(r) = 1 + e^{8\pi \frac{1+r^4}{(1-r^2)^2}} \tag{17}$$

The result obtained we will then immediately express in a somewhat more general form. First, let the two points z_0 and w_0 take the place of the two origins, which are situated arbitrarily, but in the finite part of our two planes. Secondly, we assume, that the mapping modulus for the mapping of the z-plane onto the w-plane furnished by $w = f(z)$, has any finite value M_0 at the place z_0. Then, what was proved for w before, holds for $\mathrm{M}_0^{-1} \cdot w$; and therefore the bounds of $|w|$ along K_r are now simply $\mathrm{M}_0 g(r)$ and $\mathrm{M}_0 G(r)$. Thirdly, we will assume, that a circle K_R with arbitrary finite R takes the place of the circle K_1 about z_0. By the transformation:

$$z' = R^{-1} \cdot (z - z_0),$$

This is changed into the circle K_1' of the z'-plane. If M_0 is the modulus of the mapping of the z-plane onto the w-plane at z_0, then obviously $\mathrm{M}_0 \cdot R$ is the modulus of the mapping of the z'-plane onto the w-plane at $z' = 0$. Furthermore, the circle K_r about z_0 obviously furnishes the circle $K'_{\frac{r}{R}}$ about $z' = 0$, so that we get, as bounds for $|w|$ along K_r:

$$\mathrm{M}_0 R g\left(\frac{r}{R}\right), \qquad \mathrm{M}_0 R G\left(\frac{r}{R}\right)$$

Finally, we are able to yet restrict the assumptions, so that the requirements to be placed on the function $f(z)$ are no longer necessarily fulfilled along the periphery K_R itself, but only hold everywhere inside this periphery. Namely, if we choose a positive number δ different form zero and arbitrarily small, then $f(z)$ will fulfill all previous requirements in the interior and on the periphery $K_{R-\delta}$. Now for each determinate $r < R - \delta$, the expressions:

$$\mathrm{M}_0 \left(R-\delta\right) g\left(\frac{r}{R-\delta}\right), \qquad \mathrm{M}_0 (R-\delta) G\left(\frac{r}{R-\delta}\right), \tag{18}$$

[507] are *continuously* dependent on δ for $\lim \delta = 0$, including the limit. In (18), however, for $|w|$, it is a question of "boundary" and not of "limits" say, to which $|w|$ could care arbitrarily case. Since now, on the other hand, δ was choosable arbitrarily small, we may, on account of the continuity of the expressions (18) emphasized, also take $\delta = 0$ right away.

By collecting all individual results together, we have obtained the following fundamental theorem: *if a function $w = f(z)$ is given inside the circle K_R of finite radius R*

about the point z_0 of the z-plane, which is everywhere single-valued and regular there, and which assumes no value there more than once, and if M_0 *is the mapping modulus at the point* z_0, *which furnishes the point* w_0 *say, the absolute value of* $w - w_0 = f(z) - f(z_0)$, *along the periphery* K_r *with an* $r < R$, *satisfies the inequalities:*

$$\mathrm{M}_0 R g\left(\frac{r}{R}\right) < |f(z) - f(z_0)| < \mathrm{M}_0 R G\left(\frac{r}{R}\right), \tag{19}$$

independently of which particular function might be present.

One can also clothe the result in geometrical form, by returning to the images L_r of the circular peripheries K_r. For each individual $r < R$, L_r will be a curve surrounding the image point w_0 of z_0 once, for whose minimal distance d_r and maximal distance D_r from the point w_0, the inequalities hold:

$$d_r > \mathrm{M}_0 R g\left(\frac{r}{R}\right), \qquad D_r < \mathrm{M}_0 R G\left(\frac{r}{R}\right), \tag{20}$$

understanding by g and G the functions given in (17).

3.3 The distortion theorem for circular domains

For the mapping of the surface of the circle K_R onto the w-plane by means of our function $w = f(z)$, now let $\mathrm{M} = |f'(z)|$ be the mapping modulus at an arbitrary place z. If again, as in formula (19), §2, the mapping modulus $\mathrm{M}_0 = |f'(z_0)|$ is given at the center z_0 of K_R, then we are able, on the basis of the theorem set up at the close of §2, to show, that for each $r < R$, an upper as well as a lower bound of the values $\mathrm{M} = |f'(z)|$ inside K_r is givable independently of $f(z)$.

In order, first of all, to obtain the upper bound, we inset, between the peripheries K_r and K_R, the periphery K_ρ, with $\rho = \frac{1}{2}(r + R)$, and are then able to put down, for an arbitrary place z inside or on K_r, by means of the Cauchy theorem:

$$f'(z) = -\frac{1}{2i\pi}\int_{(K_\rho)} \frac{f(\zeta)}{(\zeta - z)^2}\, d\zeta, \tag{1}$$

where the integration variable ζ has to describe the circle K_ρ, as is indicated by the lower index on the integral sign in the positive sense of circuit. [508]

If $f(z)$ were constantly equal to $f(z_0)$, then $f'(z)$ would be $= 0$. In this case, equation (1) reads:

$$0 = -\frac{1}{2i\pi}\int_{(K_\rho)} \frac{f(z_0)}{(\zeta - z)^2}\, d\zeta.$$

If we subtract this formula from (1), then follows:

$$f'(z) = -\frac{1}{2i\pi}\int_{(K_\rho)} \frac{f(\zeta) - f(z_0)}{(\zeta - z)^2}\, d\zeta, \tag{2}$$

From (2), we find, for the absolute value of the derivative standing on the left:

$$\mathrm{M} = |f'(z)| \le \frac{1}{2\pi}\int_{(K_\rho)} \frac{|f(\zeta) - f(z_0)|}{|\zeta - z|^2}\, |d\zeta|, \tag{3}$$

But now, in consequence of (19), §2, here holds along K_ρ:

$$|f(\zeta)-f(z_0)|<\mathrm{M}_0 RG\left(\frac{\rho}{R}\right).$$

On the other hand, obviously:

$$|\zeta-z|>\rho-r \text{ and therefore } \frac{1}{|\zeta-z|^2}<\frac{1}{(\rho-r)^2},$$

while the integral of all $|d\zeta|$ furnishes the length $2\rho\pi$ of the periphery K_ρ. With this follows from (3):

$$\mathrm{M}=|f'(z)|<\frac{1}{2\pi}\mathrm{M}_0 RG\left(\frac{\rho}{R}\right)\cdot\frac{1}{(\rho-r)^2}\cdot 2\rho\pi.$$

By collecting everything and substituting the expression of ρ in r and R, it follows:

$$\mathrm{M}=|f'(z)|<\frac{2\mathrm{M}_0 R(R+r)}{(R-r)^2}G\left(\frac{1}{2}+\frac{r}{2R}\right) \tag{4}$$

for each place z inside or on K_r.

From the upper bound for $|f'(z)|$ thus obtained, we are easily able to derive a lower bound. In order to simplify the calculations, we imagine, if necessary, the z-plane so
[509] displaced into itself, that $z_0=0$, and that the smallest value of $\mathrm{M}=|f'(z)|$ along K_r, to be denoted by M' say, occurs at the place $z=-r$. By the hyperbolic substitution:

$$z'=\frac{R^2z+rR^2}{rz+R^2}, \tag{5}$$

the surface of the circle K_R becomes so transformed into itself, that the points $z=-r$ and $z=0$ go over into $z'=0$ resp. $z'=+r$. At $z=0$, the modulus μ of this mapping (5) easily calculates itself as:

$$\mu=\left(\frac{dz'}{dz}\right)_{z=0}=\frac{R^2-r^2}{R^2}, \tag{6}$$

and at $z=-r$ is reciprocal to this.

If we now set

$$f(z)=f\left(\frac{R^2z'-rR^2}{-rz'+R^2}\right)=f_1(z').$$

Then the function $f_1(z')$, inside the circle K'_R of the z'-plane, satisfies all the requirements, which we placed on the function $f(z)$ inside K_R. In particular, however, for the mapping of the surface of that circle by means of the function $w=f_1(z')$ onto the w-plane, we have the mapping moduli $\mu\mathrm{M}'$ resp. $\mu^{-1}\mathrm{M}_0$ at the places $z'=0$ and $z'=r$. For the latter moduli, as holding on the periphery K'_r of the z'-plane, the inequality (4) holds, in case one substitutes for M_0 on the right side the modulus $\mu\mathrm{M}'$, now occurring at the center:

$$\mu^{-1}\mathrm{M}_0<\frac{2\mu\mathrm{M}'R(R+r)}{(R-r)^2}G\left(\frac{1}{2}+\frac{r}{2R}\right).$$

Now substitute for μ its expression (6), and remember, that M' was to be the smallest value of $\mathrm{M}=|f'(z)|$ along K_r. It follows, as valid along K_r:

$$\mathrm{M} = |f'(z)| > \frac{\mathrm{M}_0 R^3}{2(R+r)^3}\left(G\left(\frac{1}{2}+\frac{r}{2R}\right)\right)^{-1}. \tag{7}$$

Since the bound standing here on the right increases with decreasing r, then the inequality (7) also holds for $|z| < r$.

By collecting the results (4) and (7), we arrive at the following theorem: *if $f(z)$ is single-valued and regular in the interior of the circle K_R with center z_0, and if this function takes on no value there more than once, then for each determinate $r < R$, the inequality holds at all places z inside and on K_r:*

$$\frac{\mathrm{M}_0 R^3}{2(R+r)^3}\left(G\left(\frac{1}{2}+\frac{r}{2R}\right)\right)^{-1} < |f'(z)| < \frac{2\mathrm{M}_0 R(R+r)}{(R-r)^2} G\left(\frac{1}{2}+\frac{r}{2R}\right), \tag{8}$$

understanding by M_0 the mapping modulus at the point z_0, whichever particular function $f(z)$ might be present. [510]

The ratio of the two moduli $\mathrm{M}_1 = |f'(z_1)|$ and $\mathrm{M}_2 = |f'(z_2)|$ to one another gives us the degree of the "*distortion*" which enters for the two places z_1, z_2, upon going over to the image in the w-plane. We may, in fact, consider the quotient of two arc-differentials $|dw_1|$ and $|dw_2|$ as a measure of the distortion, which correspond to two equal arc-differentials $|dz|$ drawn at z_1 and z_2. For abbreviation, we set the expression, dependent only on the quotient $r : R$:

$$\frac{4(R+r)^4}{R^2(R-r)^2}\left(G\left(\frac{1}{2}+\frac{r}{2R}\right)\right)^2 = H\left(\frac{r}{R}\right). \tag{9}$$

From this theorem just stated then easily follows further: *Holding fast to the previous assumptions, it follows, for any two places z_1, z_2 inside or on K_r, for the measure of the distortion, the inequality*:

$$\left(H\left(\frac{r}{R}\right)\right)^{-1} < \left|\frac{f'(z_1)}{f'(z_2)}\right| < H\left(\frac{r}{R}\right). \tag{10}$$

Koebe designates the theorem hereby stated as the "*distortion theorem*". If, in particular, we set $z_2 = z_0$, then from (10) follows the inequality, analogous to (8):

$$\mathrm{M}_0\left(H\left(\frac{r}{R}\right)\right)^{-1} < |f'(z)| < \mathrm{M}_0 H\left(\frac{r}{R}\right). \tag{11}$$

However, as one sees, the bounds are drawn more narrowly in the inequality (8). We shall also apply the name "*distortion theorem*" to the theorem coming into expression by (8).

As an example, we again consider a linear function:

$$w = f(z) = \frac{\alpha z + \beta}{\gamma z + \delta}, \qquad \alpha\delta - \beta\gamma = 1, \tag{12}$$

and assume $z_0 = 0$, which signifies no restriction of the generality. The elementary case $\gamma = 0$, where $f'(z)$ is constant, will subsume itself under the formulas to be set up and count as excluded for the argument. The pole of $f(z)$ situated at $z_1 = -\frac{\delta}{\gamma}$ may not belong to the interior of K_R:

$$|z_1| = \left|\frac{\delta}{\gamma}\right| \geq R. \tag{13}$$

Now in the present case, we have:

$$M = |f'(z)| = \frac{1}{|\gamma z + \delta|^2} = \frac{1}{|\gamma|^2} \cdot \frac{1}{|z_1 - z|^2}, \quad M_0 = \frac{1}{|\delta|^2}.$$

[511] The maximum M for all places with $|z| \leq r$ will accordingly occur on the periphery of K_r itself at the point lying closest to the place z_1, where

$$|z_1 - z| = |z_1| - r,$$

and the minimum of the opposite point of K_r, where

$$|z_1 - z| = |z_1| + r,$$

then we have:

$$\frac{1}{|\gamma|^2} \cdot \frac{1}{(|z_1| + r)^2} \leq |f'(z)| \leq \frac{1}{|\gamma|^2} \cdot \frac{1}{(|z_1| - r)^2}.$$

By absorbing the mapping modulus M_0, we are able to write for this:

$$\frac{M_0}{\left(1 + \frac{r}{|z_1|}\right)^2} \leq |f'(z)| \leq \frac{M_0}{\left(1 - \frac{r}{|z_1|}\right)^2}$$

In the limiting case $|z_1| = R$, the smallest value on the left and simultaneously, the greatest value on the right enters, *With this, we get, as a special theorem for linear functions*:

$$\frac{M_0 R^2}{(R + r)^2} \leq |f'(z)| \leq \frac{M_0 R^2}{(R - r)^2} \tag{14}$$

or also, as a consequence of the corresponding inequality (10):

$$\left(\frac{R - r}{R + r}\right)^2 \leq \frac{|f'(z_1)|}{|f'(z_2)|} \leq \left(\frac{R + r}{R - r}\right)^2 \tag{15}$$

for any points z, z_1, z_2 lying inside or on K_r. In both cases, the given bounds are, at the same time attainable limits.

3.4 The distortion theorem for arbitrary domains

Now let an arbitrary finite, schlicht and connected domain be given in the z-plane, which we call B, and whose boundary let be denoted by R.[6] The latter can be composed of one or also of a finite number of curves running into themselves, so that we only assume, concerning the degree of the connectivity of B, that this is finite. In order however, not to allow for each possibility, we will assume, that each of the boundary curves of B is composed of finitely many analytic portions.

[512] Further, let any function $w = f(z)$ be given, which is everywhere single-valued and

[6] Further on, we only need R in this meaning and no longer as the radius of a circle K_R.

regular in the interior of B, and which assumes no value more than once there. At a determinate point z_0 chosen inside B, let the mapping modulus belonging to this function have finite value $\mathrm{M}_0 = |f'(z_0)|$.

In order now to carry over the theorems of the previous paragraph to the more general relations present here, we understand by B_0 any partial domain of B, whose boundary R_0 is everywhere finitely distant from the boundary R of the domain B. Therefore, a determinate number $d > 0$ is to let itself be given for the "minimal distance" of the boundary R_0 from the boundary R. The domain B_0, in each case, is also to contain the point z_0. Under these assumptions, we shall succeed in giving a lower and an upper bound for the values of the moduli $\mathrm{M} = |f'(z)|$ in the interior and on the boundary of B_0, which are both finite and independent of the particular function $f(z)$.

If, in B we draw parallel curves to the boundary curves at the distance d (which obviously themselves again appear built up from finitely many analytic portions), then these will bound a partial domain B_0', which can consist of several pieces. But certainly the connected piece B_0'' containing z_0, whose boundary let be called R_0'', will contain the domain B_0 as a component. From this, the following is clear: we are able to draw a line from z_0 to each arbitrarily prescribed point z inside B_0 or on the boundary of B_0, whose minimal distance from the boundary R is not $< d$, and whose length lies beneath a fixed givable finite number l. In fact, we are able to avail ourselves, e.g., of the rectilinear line joining z_0 to z, except that everywhere this exists over the boundary R_0'' out of B_0'', we go around the boundary itself up to the reentrance place. The line described, on the strength of this, puts itself together from sequents of a straight line, whose length remains beneath the maximal distance of two boundary points of the finitely extended domain B, as well as from portions of the boundary R_0'', which has a finite total length; the length of the line therefore remains beneath a givable finite number l.

The line described from z_0 to z we will now cover with a chain of circles of the radius d, according to following principle: let the first circle $K^{(1)}$ have the center z_0; let the second $K^{(2)}$ have that point z_1 as center, where the line in question finally exits from the circle of radius $\frac{1}{2}d$ concentric with $K^{(1)}$; let the third circle $K^{(3)}$ have the place z_2 as center, where the line finally exits from the circle of radius $\frac{1}{2}d$ concentric with $K^{(2)}$ etc. The length of the line described from z_0 to the center of the circle $K^{(2)}$ is $\geq \frac{1}{2}(\nu-1)d$. [513]
Accordingly, if we set:

$$\nu = 1 + E\left(\frac{2l}{d}\right), \tag{1}$$

understanding by $E\left(\frac{2l}{d}\right)$ the greatest number contained in $\frac{2l}{d}$, then the circle $K^{(\nu)}$ at the latest, if not an earlier one, will be so situated, that the end point z of the line from the center of that circle has the distance $< \frac{1}{2}d$. There therefore exists an index $\nu_0 \leq \nu$, such *that the endpoint z of the line has a distance $< \frac{1}{2}d$ from the center of the circle $K^{(\nu_0))}$*.

Now we apply the inequalities (8) on p. 509 to our circles, whereby we have to set $r = \frac{1}{2}d$, $R = d$, and put, for abbreviation, the finite constant:

$$G\left(\frac{3}{4}\right) = 1 + e^{\frac{2696\pi}{49}} = G. \tag{2}$$

For the mapping modulus M_1 of our function $f(z)$ at the center of $K^{(2)}$ follows:

$$\frac{4}{27}\mathrm{M}_0 G^{-1} < \mathrm{M}_1 < 12\mathrm{M}_0 G.$$

With utilization of this result and renewed application of (8) on p. 509, there follows, for the mapping modulus M_2 at the center of $K^{(3)}$:

$$\left(\frac{4}{27}\right)^2 \mathrm{M}_0 G^{-2} < \mathrm{M}_2 < 12^2 \mathrm{M}_0 G^2.$$

By continuation of the same procedure of reasoning, we find, for the mapping modulus M_{ν_0-1} at the center of the circle $K^{(\nu_0)}$ the inequalities:

$$\left(\frac{4}{27}\right)^{\nu_0-1} \mathrm{M}_0 G^{-\nu_0+1} < \mathrm{M}_{\nu_0-1} < 12^{\nu_0-1}\mathrm{M}_0 G^{\nu_0-1}.$$

But now, the end point z of the line, with the mapping modulus $\mathrm{M} = |f'(z)|$, lies at a distance $< \frac{1}{2}d$ from the center of the circle $K^{(\nu_0)}$. There thus yields itself:

$$\left(\frac{4}{27}\right)^{\nu_0} \mathrm{M}_0 G^{-\nu_0} < \mathrm{M} < 12^{\nu_0}\mathrm{M}_0 G^{\nu_0},$$

and therefore, since $\nu_0 \leq \nu$ holds:

$$\left(\frac{4}{27}\right)^{\nu} \mathrm{M}_0 G^{-\nu} < \mathrm{M} < 12^{\nu}\mathrm{M}_0 G^{\nu}. \tag{3}$$

[514] In this fashion, we have arrived at the general *"distortion theorem": in the finite, schlicht and connected domain B, whose boundary R is built up from finitely many analytic portions, and which may be simply or multiple connected let a function $f(z)$ be given, which is everywhere single-valued and regular inside B, and which assumes no value there more than once. Then for each point z in the interior and on the boundary of a partial domain B_0, whose boundary R_0 has an everywhere finite distance from the boundary of the domain B, the inequality subsists*:

$$\left(\frac{4}{27}\right)^{\nu} G^{-\nu}\mathrm{M}_0 < |f'(z)| < 12^{\nu} G^{\nu}\mathrm{M}_0, \tag{4}$$

understanding by M_0 the value $f'(z_0)$, and there subsists further, for each pair of points z_1, z_2 inside or on the boundary of B_0, the inequality:

$$3^{-4\nu}G^{-2\nu} < \frac{|f'(z_1)|}{|f'(z_2)|} < 3^{4\nu}G^{2\nu}, \tag{5}$$

whatever particular function may be present. Therefore, v is a finite whole number, characteristic of the domain B_0 and given by (1), and G signifies the finite constant given in (2).

3.5 Consequences of the distortion theorem

We hold fast to the assumptions on the domain B and the function $f(z)$ and assume, beside this, *that $f(z)$ has its required properties, not only in B, but also on the boundary*

R of B, as well, too, as even outside B directly at the boundary. The latter requirement we will clothe more precisely thus: we imagine a parallel curve of the distance d drawn to each boundary curve of B outside this domain, understanding by d a positive number, chosen sufficiently small[7], but different from 0. These parallel curves bound a domain $\overline{B}$, which arises from B by positioning plane bands of breadth d along the boundary curves. Our assumption is accordingly, *that for the $d > 0$, chosen fixed independently of $f(z)$, the function $f(z)$ also has its properties in the enlarged domain $\overline{B}$.*

Let the number of the boundary curves of B be n, and let $n \geq 2$ hold. One of these bounds B form the outside, but the remaining $(n-1)$ are inner boundary curves. [515]

If we now go to the w-plane by means of the function $w = f(z)$, then B will furnish a schlicht image domain $\overline{B}'$, and B a finite and schlicht image domain B'. One of the boundary curves (not necessarily that one which corresponds to the outer curve of B) will bound B' form the outside, and the remaining $(n-1)$ boundary curves bound B' from within. Each of these latter curves includes a simply-connected domain, which appears cut our of the domain B'. We will denote by $J_1, J_2, \ldots, J_{n-1}$ the surface content of these $(n-1)$ domains included by the domain B'. On the other hand, let J_n denote the surface content of the simply connected domain included by the outer boundary curve of the domain B', so that J_n is representable as the sum of the content numbers $J_1, J_2, \ldots, J_{n-1}$ and the content of B'.

For the content numbers just named, we are now able to prove the inequality:

$$J_1 + J_2 + \cdots + J_{n-1} < qJ_n, \tag{1}$$

in which q is a number of the interval $0 < q < 1$ is dependent only on the domain B, but not of the particular function $f(z)$ present.

For the proof of the inequality (1), we choose on arbitrary point z_0 inside B, whose minimal distance from the boundary R let be denoted by r. We may assume, that the mapping modulus $\mathrm{M}_0 = |f'(z_0)| = 1$; for this is attainable, if necessary, by a similarly transformation to be exercised on the w-plane, for which all content numbers J assume the same factor, and therefore the inequality (1) remains subsisting without alteration of q.

The circle K_r of radius r about z_0 belongs entirely to the domain B, so that we are able to apply to this circle the theorem of §2 on p. 500. In consequence of this latter theorem, the image of the circle K_r in the w-plane, has a surface content F_r, for which the number:

$$m = \frac{3-2\sqrt{2}}{16}\pi r^2 \tag{2}$$

gives a lower bound:

$$F_r > m. \tag{3}$$

[7] The individual boundary curve of B is indeed considered as finitely extended, and also the domain situated outside B surrounded by them is not to vanish. Also, we assume, that each finite portion of a boundary curve not only bounds B, but also the portions of the plane situated outside B. For d chosen sufficiently small, it is then always possible to position the plane band of the breadth d described in the text, smoothly at the boundary outside B and to thus produce the domain $\overline{B}$.

If, on the other hand, one applies the distortion theorem (4) on p. 514 to the domain $\overline{B}$, then it follows, along the boundary R:

$$|f'(z)| < 12^{\nu} G^{\nu}. \tag{4}$$

[516] Now let L_n be the given finite length of the outer boundary of B, and correspondingly, let L'_n be the length of the outer boundary of B', which bounded the surface of the content J_n before. Then we have:

$$L'_n = \int |dw| = \int |f'(z)||dz|,$$

where the first integral is to be extended over the outer boundary of B', and correspondingly, the second over that of B. Since along the latter, the inequality (4) holds, there follows:

$$L'_n < 12^{\nu} G^{\nu} \int |dz|, \quad L'_n < 12^{\nu} G^{\nu} L_n.$$

The content J_n of the domain included by this line now satisfies the inequality:

$$J_n \leq \frac{L'^2_n}{4\pi}.$$

For the surface number J_n, accordingly, the finite value:

$$M = \frac{12^{2\nu}}{4\pi} G^{2\nu} L_n^2 \tag{5}$$

will give an upper bound:

$$J_n < M. \tag{6}$$

From (3) and (6) follows immediately:

$$\frac{F_r}{J_n} > \frac{m}{M}, \tag{7}$$

where on the right side stands a number lying between 0 and 1, dependent on our domain B and the number d, but independent of $f(z)$.

Since the image of the circle K_r is a partial domain of B' and the content of B' is given by $J_n - (J_1 + J_2 + \cdots + J_{n-1})$, then we have:

$$F_r \leq J_n - (J_1 + J_2 + \cdots + J_{n-1}),$$

from which we further deduce (with utilization of (7)):

$$\frac{J_n}{J_1 + J_2 + \cdots + J_{n-1}} \geq \frac{J_n}{J_n - F_r} = \frac{1}{1 - \frac{F_r}{J_n}} > \frac{1}{1 - \frac{m}{M}}.$$

Consequently, if we write:

$$1 - \frac{m}{M} = q, \tag{8}$$

then $0 < q < 1$, and we have:

$$J_1 + J_2 + \cdots + J_{n-1} < qJ_n. \tag{9}$$

We have accordingly obtained the theorem: *if $f(z)$ is single-valued and regular in the interior of the domain $\overline{B}$, produced from our n-fold connected domain B, and if $f(z)$ assumes no value more than once there, then the total surface content included by $(n-1)$ inner boundary curves of the image domain is smaller that the content J_n multiplied by q, which is bounded by the outer boundary circle of the image B'; and indeed, q is thereby a number of the interval $0 < q < 1$, dependent only on B and the strip width d, but independent of the particular function $f(z)$.* [517]

For setting up a second consequence, we draw *any curve C of finite length* in the domain B. Let the image of this in B' be the curve C', whose length we denote by L. On the other hand, let L_0 be the length of C.

For obtaining an inequality for the curve length L, we attach to the representation:

$$L = \int_{(C')} |dw| = \int_{(C)} |f'(z)| \cdot |dz|,$$

where the first integral is to be extended along C' and the second, correspondingly, along the original curve C. Let the maximum of $|f'(z)|$ along C be denoted by M_C; then there holds, in any case:

$$L \leq M_C \int_{(C)} |dz| = M_c L_0.$$

by squaring, we obtain:

$$L^2 \leq M_C^2 L_0^2. \tag{10}$$

If we put in for z_1, in the inequality (5) on p. 514, that place of C, at which the maximum M_C enters, while z_2 is any arbitrary place z of B, then follows:

$$M_C < 3^{4v} G^{2v} |f'(z)|,$$

and therefore, again, by squaring:

$$M_C^2 < 3^{8v} G^{4v} |f'(z)|^2. \tag{11}$$

We denote by $d\omega$ a surface element of the domain B at the place z and by $d\omega'$ its image in the domain B'. One then has: $d\omega' = |f'(z)|^2 d\omega$ and therefore follows, form the inequality (11), by multiplication by $d\omega$:

$$M_C^2 d\omega < 3^{8v} G^{4v} d\omega'.$$

Integrate the differential standing on the left over B and therefore the one standing on the right, correspondingly over B': [518]

$$M_C^2 \int_{(B)} d\omega < 3^{8v} G^{4v} \int_{(B')} d\omega'.$$

If we denote by J the content of B' and, correspondingly, by J_0 that of B, then we have:

$$M_C^2 J_0 < 3^{8v} G^{4v} J.$$

With utilization of (10) there accordingly follows, further:

$$L^2 < \left(\frac{3^{8\nu} G^{4\nu} L_0^2}{J_0}\right) \cdot J.$$

For abbreviation, we write:

$$Q = \frac{3^{8\nu} G^{4\nu} L_0^2}{J_0}, \tag{12}$$

and have to state as a result: *if the presuppositions of the last theorem set up hold, and if a curve C of finite length is drawn in the domain B, then there is a positive finite number Q, independent of the particular function $f(z)$, of such a kind, that the square of the length L of the image curve of C in the domain B' is smaller than the product of that number Q and the content J of the image domain B'*:

$$L^2 < QJ. \tag{13}$$

With this, the general preparations for the proof of the reentrant cut theorem are ended.

3.6 Production of the covering surface F_∞ for a Riemann surface provided with p reentrant cuts

The proof of the reentrant cut theorem, as was already remarked on p. 446, was to be carried out by means of the method of the covering surface. We now have to produce the foundation for this.

Let an arbitrary Riemann surface F with a genus $p > 1$ be given. We imagine drawn on F, in one of the infinitely many possible ways, a system of p reentrant cuts, running separately from one another and not decomposing the surface into separate pieces, and call the surface thus cut out F_1. The cut-up surface F_1 is p-fold connected and has the $2p$ crosscut banks as just as many boundary curves, each running back into itself.

Across from the cut-up surface F_1, we set any arbitrary discontinuity domain P_1' of the kind named under P on p. 35, with $2p$ boundary curves, coordinated pairwise
[519] to one another (cf. figure 12 on p. 35). Between F_1 and P_1' we are able to produce a reversibly single-valued continuous mapping, understood in the sense of "analysis situs", whereby, in particular, the p pairs of crosscut banks of F_1 correspond to the p pairs of coordinated boundary curves of P_1'.

For the production of this mapping, we imagine the p reentrant cuts drawn on F_1 as cuts b in the sense of figure 40 in I on p.183, add p conjugate cuts according to the prescription of the this figure, as well as p cuts c running from a point of the F_1 to the cuts a. On the other hand, in P_1', we join each two boundary curves, coordinated to one another, by a crosscut corresponding to the individual reentrant cut a and then draw, from any inner point of the domain P_1', corresponding to the c, p further crosscuts after the cuts just drawn before. F_1 and P_1' are fashioned in this manner into simply connected domains, both of whose boundary curves we are able to immediately map single-valuedly onto one another, in such a fashion, that the p cuts b are in fact coordinated to the p pairs of the boundary curves of P_1'. The fashioning of the mapping for the inner points also, of these two simply connected domains then hold no more difficulty.

By now imagining the cuts a and c, as well as the cuts in P_1' corresponding to them, taken away again, we reproduce P_1' by means of the generators of the associated group, and produce in this fashion a polygon net N', which covers the entire plane of the polygons schlichtly, apart from infinitely many discretely situated boundary points.

We will now undertake the generation of this net in individual steps, and indeed, we first of all position at the $2p$ boundary curves of P_1', just as many new domains equivalent with P_1', each individual one of which accordingly exhibits $(2p-1)$ more uncovered boundary curves. All together, with this, we have obtained $(2p+1)$ domains, connected to one another, with $2p(2p-1)$ boundary curves still remaining uncovered; we will denote this complex by P_2'. At each of the $2p(2p-1)$ boundary curves we now position a further domain equivalent with P_1', and obtain a complex of $(4p^2+1)$ domains equivalent to P_1' to be denoted by P_3', which still exhibits $2p(2p-1)^2$ uncovered boundary curves.

We continue in the same way, and find, after $(n-1)$ steps from P_1', a complex of:

$$\frac{p(2p-1)^{n-1}-1}{p-1} \tag{1}$$

domains equivalent with P_1', to be denoted by P_n', which still exhibits: [520]

$$2p(2p-1)^{n-1} \tag{2}$$

uncovered boundary curves, and by no means covers the plane simply. As limiting model P_∞' for $\lim n = \infty$, we obtain the polygon net N' already mentioned, which covers the plane schlichtly up to the isolatedly situated boundary points.

On the basis of the single-valued continuous mapping of the domain P_1' onto the cut-up Riemann surface F_1, we will now repeat the reproduction process of the domain P_1' as a covering process of the surface F_1 in precisely corresponding fashion. We therefore, first of all, cover F_1 with $2p$ further copies of this surface, by hanging one further copy along each of the $2p$ boundary curves of F_1. The $(2p+1)$ copies F_1 so arranged together furnish a Riemann surface with $2p(2p-1)$ boundary curves to be denoted by F_2, of which the domain denoted above by P_2' furnishes a single-valued continuous image.

Along each of the $2p(2p-1)$ boundary curves, we next position one further copy F_1 and arrive at the surface F_3, which corresponds to the complex P_3'. After $(n-1)$ steps, we get a surface to be denoted by F_n, which appears built up by covering by the number of copies of the surface F_1 given in (1), and still contains uncovered boundary curves in the number (2). These boundary curves are crosscut banks of the copies F_1 situated at the boundary of F_n. It is self-evident, that F_n again finds its single-valued continuous image in the schlicht domain P_n'.

If we now again effect the passage to the limit $\lim n = \infty$, there then arises, as the antitype of the net $N' = P_\infty'$, *the "covering surface" F_∞ of the surface F_1, put together from infinitely many copies F_1 and to be utilized for the reentrant cut theorem.* This surface F_∞ is a completely determined model, insofar to each copy F_1 belonging to it is reachable from the first one by a uniquely determined chain of surface copies F_1.

As in the boundary-circle case lacking parabolic points e_k, the domain F_{n-1} here never reaches up to the boundary of the next domain F_n. On the other hand, essential

distinctions naturally subsist between the covering surface at that time and the F_∞ here now. Notice, e.g., the following state of affairs: if one removes the partial domain F_n from the covering surface of the boundary-circle case without parabolic e_k, then one single connected infinitely-sheeted surface remains, which has, in the finite part
[521] of the plane, one boundary curve running back into itself; if one removes the partial domain F_n from the F_∞ here, then $2p(p-1)^{n-1}$ separate pieces remain, each of which is infinitely sheeted and exhibits one closed boundary curve in the finite part of the plane.

3.7 Mapping of the surface F_n onto a schlicht domain for special reentrant cuts

The mapping of the bounded surface F_n onto the schlicht domain P_n' was indeed single-valued and continuous, but not conformal. This mapping only served the purpose of obtaining a clear picture for the mutual positional relations of the copies F_1 comparing the surfaces F_n to one another.

The goal of the next developments is, *for any finite n, to map the bounded surface F_n conformally onto a schilicht plane domain P_n*, which then, to be sure, initially has nothing to do with a group of linear substitutions. We will first all carry through this mapping for a special choice of the P reentrant cuts, which we have immediately to describe more closely, and begin with the case $n = 1$, i.e. with the bounded surface F_1 itself.

We may assume, that the given Riemann surface has only two-sheeted branchpoints, which, if necessary, are attachable by rational transformation of the surface. If this is positioned μ-sheetedly over the plane of the complex variable z, then the number w of the branchpoints is given by:

$$w = 2\mu + 2p - 2. \tag{1}$$

According to the investigations of the Lurӧth and Clebsch cited on p. 392, we are able, by suitable arrangement of the branchcuts, to give the surface the following form: the two uppermost sheets are joined to one another in the manner of a two-sheeted (hyperelliptic) surface of the genus p with $(2p+2)$ branchpoints, which we call $e_1, e_2, \ldots, e_{2p+2}$, and between then run $p+1$ branchcuts from e_1 to e_2, from e_3 to e_4 etc., finally from e_{2p+1} to e_{2p+2}(cf. figure 105 on p. 522). The third sheet, following below the second, is joined with the second by two further branchpoints and a branchcut running between them, similarly, the fourth sheet with the third etc. finally the μth with the $(\mu-1)^{\text{st}}$. Thereby, the total number:

$$2p + 2 + 2(\mu - 2) = w$$

of the branchpoints is extended.

[522] We lay all the p reentrant cuts in the uppermost sheet, and indeed, the first is to run around the branchcut joining e_1 and e_2, the second around the one going from e_3

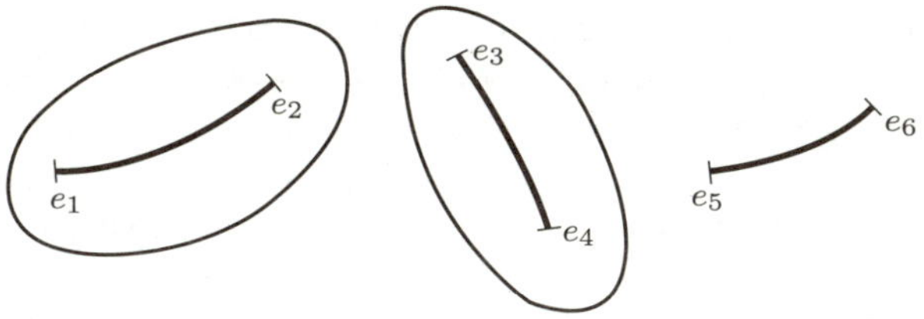

Fig. 105.

to e_4 etc. (cf. figure 105, where $p = 2$ holds)[8]. We now again undertake a completion of our cut-up surface F_1 into a closed surface, by adding, corresponding to each individual one of the p reentrant cuts, an individual full sheet of the z-plane. The latter we imagine cut up by just this reentrant cut into two parts, and indeed, we hang the inner part onto the outer bank of the reentrant cut of F_1 in question, but the outer part onto the inner cut bank. The bounded surface F_1, in this manner, by addition of the p full sheets, becomes a closed $(\mu + p)$-sheeted surface F_1', whose number of branchpoints is the previous one given in (1). According to a known rule (cf. "M" I on p.494), the genus p' of F_1' calculates itself as:

$$p' = -(\mu + p) + 1 + \frac{1}{2}w = 0. \tag{2}$$

One also obtains the same result from the single-valued continuous image P_1' of the surface F_1. If one imagines P_1' so chosen in the ζ-plane, that the point $\zeta = \infty$ lies in the interior of P_1', then this domain has $2p$ inner boundary curves, which include just as many simply-connected portions of the ζ-plane. If we add these portions to the domain P_1', then there arises a simply connected closed domain, namely the full ζ-plane.

On the surface F_1' of the genus $p' = 0$, according to the existence of algebraic functions on Riemann surfaces (cf. "M" I on p.508), there is now a *principal function*, which we will call $\eta_1(z)$, and which we are able to fix uniquely by three stipulations. To this end, we will stipulate, that $\eta_1(z)$ becomes infinite at a place z_0 in the interior of F_1, [523]
arbitrary, but chosen fixed once and for all, and permits the development:

$$\eta_1(z) = \frac{1}{z - z_0} + c_1(z - z_0) + c_2(z - z_0)^2 + \cdots. \tag{3}$$

In the neighborhood of z_0, with wanting absolute term-$\eta_1(z)$ is completely determined by this. In fact, two conditions for $\eta_1(z)$ are given by the requirement, that this function become infinite at z_0 as $(z - z_0)^{-1}$, and $\eta_1(z)$ is hereby determined up to an additive constant. The latter is fixed by the requirement, that in (3), no absolute term different from 0 is to come up.

[8] As one sees, we have to do here, not with an arbitrary, but with a very special system of reentrant cuts. The consideration of the text consequently has only a preliminary and orienting significance, while in the next paragraph, the generally valid investigation will follow. If, moreover, one does not want to reach back to the developments of Lüroth and Clebsch cited above, then in this situation, it is permissible, to directly attach the investigation to the particular Riemann surface described in the text.

The function $\eta_1(z)$ maps the closed Riemann surface F_1' conformally onto the schlichtly covered η_1-plane. *Thereby, in particular, the partial domain F_1 of F_1' will find its conformal image in a schlicht domain P_1 of the η_1-plane, which, corresponding to the crosscut banks of F_1, exhibits $2p$ closed boundary curves, and in addition, contains the point $\eta_1=\infty$, corresponding to the point z_0, in its interior.* This is the mapping of F_1 which we wanted to achieve.

In carrying over this development to F_n, the need makes itself felt, to have shortened notations for the numbers given in (1) and (2) on p. 519 ff. We write:

$$s_n=\frac{p(2p-1)^{n-1}-1}{p-1},\quad p_n=p(2p-1)^{n-1}; \tag{4}$$

S_n is therefore the number of the copies F_1 which comprise F_n, while $2p_n$ signifies the number of the boundary curves of F_n (crosscut banks, remaining uncovered, of the copies F_1 lying at the boundary of F_n). In all, we have $2p$ different kinds of crosscut banks. One wants to clarify for oneself, by an inductive argument applied to $P_1', P_2', \cdots$, *that at the boundary of F_n, the individual kind of reentrant-cut bank is represented $(2p-1)^{n-1}$ times, therefore each kind equally often.*

In consequence of this circumstance, there are just as many inner banks as outer banks of the individual kind of reentrant cut. Accordingly, we are able to effect the completion of the surface F_n to a closed Riemann surface F_n' in the same fashion as in the case $n=1$. First of all, $(2p-1)^{n-1}$ copies of the z-plane are each divided into two parts by the first reentrant cut, whereupon we hang the inner parts onto the boundary curves of F_n originating from the outer banks of the first crosscut, but the outer parts onto the boundary curves originating from the inner banks. With $(2p-1)^{n-1}$ further copies of the z-plane, we proceed exactly thus for the boundary curves originating from the second crosscut etc.

[524] One arrives thus at a closed (μs_n+p_n)-sheeted Riemann surface F_n' with ws_n two-sheeted branchpoints. The genus p_n' of this surface, corresponding to the relations (2), calculates itself, with utilization of the expressions (4) as

$$p_n'=-(\mu s_n+p_n)+1+\frac{1}{2}ws_n=0. \tag{5}$$

Naturally, one can also read this result off, as above in the case $n=1$, from the shape of the single-valued image P_n' of the surface F_n.

The closed Riemann surface F_n' of the genus $p_n'=0$ has a uniquely determined principal function $\eta_n(z)$, which, in the neighborhood of the place z_0 chosen above in the initial copy F_1 of F_n, admits the development:

$$\eta_n(z)=\frac{1}{z-z_0}+c_1^{(n)}(z-z_0)+c_2^{(n)}(z-z_0)^2+\cdots. \tag{6}$$

This principal function $\eta_n(z)$ maps the closed surface F_n' onto the schlicht full plane η_n. *Thereby, in particular, the bounded surface F_n, as a partial domain of F_n', finds its conformal image in a schlicht domain P_n of the η_n-plane, which has $2p_n$ closed boundary curves and, moreover, contains in its interior the point $\eta_n=\infty$, corresponding to the place z_0.* This domain P_n is the conformal image of F_n which we wanted to produce.

3.8 Mapping of the surface F_n onto a schlicht domain for arbitrary reentrant cuts

Also for an entirely arbitrarily chosen system of p reentrant cuts, with retention of the essential viewpoints of the previous argument, we are able to accomplish the mapping of the surface F_n onto a schlicht domain P_n by means of an analytic function $\eta_n(z)$.

We begin again with the case $n = 1$, and imagine the given surface as a "*Riemann special surface*", positioned over the "z-sphere". Concerning the choice of the p reentrant cuts, we presuppose nothing special. The shape of the individual cuts, however, we may imagine so regulated, that these appear built up from *a finite number of segments of great special circles*.

Along the individual reentrant cut, which we will call S, we draw two closed curves S' and S'' on both sides close to S, but at a never vanishing distance, which we again build up-from finitely many segments of great circles. We will carry out the following construction: from the vertices, at which the circular arcs of the curve S' meet, we draw radii to the center of the sphere, and produce the plane circular sectors, which are bounded by each two radii following one another, and the intermediately situated [525] circular arcs of the curve S'. These circular sectors form, in following one another, a cyclic closed chain, and furnish, in such a manner, the conical surface of a pyramid with the vertex at the center of the sphere and of the boundary S'. This conical surface represents a simply connected domain; and the connectivity also then remains a simple one, even if we add to the conical surface the surface band running between S and S'. We imagine the same construction carried out for the curve S'', as well as, in corresponding fashion, for all $(p-1)$ remaining reentrant cuts.

With the aid of the $2p$ simply connected domains thus produced, we now effect the fashioning of the bounded surface F_1 into a closed surface. Along the boundary of F_1 furnished by the individual crosscut bank, we add the surface band lying opposite with its pyramidal surface. By proceeding thus along all $2p$ boundary curves of F_1, there arises a *closed* surface F_1', which, as one perceives immediately by returning to the image P_1' of F_1, is again *simply* connected.

The surface F_1' is a polyhedral surface, which is composed of finitely many plane and spherical portions. The methods of Schwartz and Newman, developed in "M" I on p.508 ff. for the production of functions, can be directly carried over to such surfaces, as this was already stated by Schwartz in the year 1870[9]. Namely, the imbricate covering of the polyhedral surface succeeds with finitely many domains, the individual one of which is conformally mappable onto the surface of a schlicht circle.

Since now, the surface F_1' obtained has the genus $p_1' = 0$, then there exists on it a principal function $\eta_1(z)$, which we again fix by the requirement, that for this, in the neighborhood of the point z_0, chosen fixed once and for all in the interior of F_1, the

[9] Cf. §14 of the paper of Schwartz, "*Uber die integration der partiellen Differentialgleichung*:

$$\frac{\partial^2 u}{\partial x^2} + \frac{\partial^2 u}{\partial y^2} = 0$$

unter vorgeschriebenen Grenz - und Unstetigkeitgkbestimmungen". Berl. Monatsber. of 1870, p.767 ff., Ges. math Abn., Vol. 2.

development holds:

$$\eta_1(z) = \frac{1}{z - z_0} + c_1(z - z_0) + c_2(z - z_0)^2 + \cdots. \tag{1}$$

This principal function maps the polyhedral surface F_1' onto the schlichtly covered
[526] η_1-plane. *There, in particular, the bounded surface F_1 is mapped conformally onto a schlicht domain P_1 of the η_1-plane, which has $2p$ closed boundary curves corresponding to the $2p$ cut banks, and in addition, contains in its interior the point $\eta_1 = \infty$, corresponding to the place z_0.*

For the surface F_n, which is now a bounded Riemann spherical surface, the enumerations carried through in the previous paragraph remain subsisting completely. Accordingly, the $2p_n$ boundary curves of F_n again fall into $2p$ systems of $(2p-1)^{n-1}$ similar reentrant cut banks. at the $(2p-1)^{n-1}$ boundary curves of the individual kind, we shall now again produce the closing of the surface F_n by means of just as many simply connected domains, each consisting of a plane band and a pyramidal surface. There arises, finally, a closed polyhedral surface F_n', consisting of finitely many plane and spherical portions, whose connectivity, as it yields itself by return to the image domain P_n', is simple. The associated principal function $\eta_n(z)$, which we define uniquely by the prescription:

$$\eta_n(z) = \frac{1}{z - z_0} + c_1^{(n)}(z - z_0) + c_2^{(n)}(z - z_0)^2 + \cdots, \tag{2}$$

again leads us to our goal: the principal function $\eta_n(z)$ maps the closed polyhedral surface F_n' conformally onto the simply and completely covered η_n-plane. *Accordingly, the bounded surface F_n, as a partial domain of F_n', is mapped by $\eta_n(z)$ onto a schlicht domain P_n of the η_n-plane, which has $2p_n$ boundary curves, and contains, in its interior, the point $\eta = \infty$, corresponding to the place z_0.*

Each of the cuts S, which, to the extent that they are uncovered on the F_n, furnish the boundary curves of our schlicht domain P_n, we put together above from finitely many arcs of great circles. Each such circle is mapped by the analytic function $\eta_n(z)$ onto an analytic curve of the η_n-plane. *Each of the $2p_n$ boundary curves of P_n therefore appear composed of finitely many portions of analytic curves.*

If we imagine the surface F_n as a partial domain of the covering surface F_∞, then we are able to consider the surface band, added along the individual boundary curve of F_n in the production of the closed surface F_n', as situated on the F_∞, and indeed, in the copy F_1 neighboring with F_n along that boundary curve. But this surface band, by means of the function (2), maps itself, in the η_n-plane, onto a ring-shaped domain, attaching itself to the boundary curve in question, which positions itself schlichtly in
[527] the interior of the simply connected domain surrounded by that boundary curve. *Accordingly, while the function $\eta_n(z)$ is initially defined on the F_∞ only for the partial domain F_n, $\eta_n(z)$, upon analytic continuation over the individual boundary curve of F_n, shows itself, all the same, to be single-valued and regular along a strip of never vanishing breadth attaching itself outside to this boundary curve; also, the mapping of the surface F_n, enlarged by the $2p_n$ strips, still gives, in the η_n-plane, a schlicht domain, again provided with $2p_n$ boundary curves.*

As one will have noticed, the theorems of §5 are calculated for these relations; in fact, we shall have to apply the latter below to our domain p_n.

3.9 Introduction of a system of analytic transformations belonging to the domain P_n

The surface F_n was built up from s_n copies of the surface F_1. To these s_n copies, positioned over one another correspond to just as many partial domains of P_n, positioned next to one another, which we will denote by $P_n^{(0)}, P_n^{(1)}, \ldots, P_n^{(s_n-1)}$, and which, in their totality, exactly exhaust the domain P_n. Since each of these domains is mapped single-valuedly and conformally onto one and the same bounded surface F_1 (for those s_n copies F_1 are all congruent to one another), then each two of the domains $P_n^{(0)}$, $P_n^{(1)}$, $\cdots, P_n^{(s_n-1)}$ are single-valuedly and conformally mappable onto one another, just by mediation of F_1 itself. We are able, in particular, to state the theorem: *these are s_n analytic transformations $T_n^{(0)}$, $T_n^{(1)}, \ldots, T_n^{(s_n-1)}$, of which $T_n^{(i)}$ so transforms the partial domain $P_n^{(0)}$ into $P_n^{(i)}$, that two corresponding points η_n and $T_n^{(i)}(\eta_n)$ correspond to one and the same point of the surface F_1.*

The domains $P_n^{(0)}, P_n^{(1)}, \ldots, P_n^{(s_n-1)}$ have exactly the same positional relations to one another, as those P_n' of the domain N', put together with domains equivalent to P_1' (cf. p. 519), which indeed are single-valuedly coordinated to those s_n domains $P_n^{(i)}$ by mediation of F_n. These partial domains of P_n' are obtained from the initial domain P_1' by application of linear substitutions $V_0, V_1, \ldots, V_{s_n-1}$. Obviously the analytic transformation $T_n^{(i)}$ and the linear substitution V_i correspond to one another hereby, and in particular, V_0 as well as $T_n^{(0)}$ is the identical operation.

The analogy goes yet somewhat further, insofar *as the transformations $T_n^{(i)}$, exactly as the substitutions V_i, let themselves generated by repetition and combination, in known fashion, from p of them, with which the p inverse transformations are to be considered as given.*

Let a first cut S of F_1 furnish, as images of its two banks, two boundary curves of $P_n^{(0)}$ coordinated to one another, of which we denote the one by R_0, the other by R_0'. [528]
Along R_0', let $P_n^{(0)}$ be neighboring with $P_n^{(1)}$, a domain, which let arise form $P_n^{(0)}$ by the transformation $T_n^{(1)}$. Then, in particular, $R_0' = T_n^{(1)}(R_0)$, and therefore the transformation inverse to $T_n^{(1)}$ carries R_0' over into R_0. But now, among the $T_n^{(i)}$, these is a transformation, to be denoted by $T_n^{(p+1)}$, which carries R_0' over into R_0 in exactly the same fashion, namely, that transformation, which carries $P_n^{(0)}$ over into the domain neighboring along R_0. Now notice, that an analytic transformation (function) is completely determined, if we given the image of a finitely extended curve portion laid out by it. Therefore follows, *that $T_n^{(p+1)}$ is identical with the transformation inverse to $T_n^{(1)}$.*

In this fashion, let the p boundary curve pairs of $P_n^{(0)}$ furnish the p transformations:

$$T_n^{(1)}, T_n^{(2)}, \ldots, T_n^{(p)}, \tag{1}$$

and the p transformations $T_n^{(p+1)}, \cdots$ inverse to them. By considering the latter p transformations a given along with the former ones, *we shall, in fact, be able to generate all s_n transformations $T_n^{(i)}$ from the p transformations (1).*

In order to now prove this in general, we carry $P_n^{(0)}$ over into any partial domain $P_n^{(i)}$ of P_n by means of the transformation $T_n^{(i)}$. Let the boundary curves of $P_n^{(l)}$ homologous to R_0 and R_0' be calledR_i and R_i':

$$R_i = T_n^{(i)}(R_0), \quad R_i' = T_n^{(i)}(R_0'). \tag{2}$$

Now also let the partial domain $P_n^{(i+1)}$, neighboring with $P_n^{(i)}$ along R_i', belong to the total domain P_n and proceed form $P_n^{(0)}$ by $T_n^{(i+1)}$. Then, insofar as R_i' is the boundary curve of $P_n^{(i+1)}$ homologous with R_0, the equation:

$$R_i' = T_n^{(i+1)}(R_0)$$

with hold, while on the other hand, by substitution of $R_0' = T_n^{(1)}(R_0)$ into the second equation (2), it follows

$$R_i' = T_n^{(i)} T_n^{(1)}(R_0).$$

From the principles already utilized, of the determinacy of analytic functions, there consequently yields itself the identity of $T_n^{(i+1)}$ and $T_n^{(i)} T_n^{(1)}$. By repeated application of this procedure of reasoning, one generates, as asserted, all s_n transformations $T_n^{(i)}$ from the p partial transformations (1) and their inverse transformations.

[529] Naturally one must guard oneself, from applying further principles, say, of our groups of *linear* substitutions V, to the transformation present here. We are indeed able, without further ado, to speak of an entire group of analytic transformations, which have the p transformations (1) as generators. But there is no speaking of the fact, that $P_n^{(0)}$, say, is a discontinuity domain of this group, since, after all, nothing at all is stated further concerning the behavior of the analytic functions $T_n^{(i)}(\eta_n)$ in the portions of the η_n-plane remaining free from the domain p_n, than that which lets itself be deduced from the single-valuedness and regularity of $\eta_n(z)$ on F_∞ right at the boundary outside F_n.

The preceding statements hold for each finite n. Now, however, our goal is, to present a convergence study referring to the limiting case $\lim n = \infty$. Hereby, we shall be able to determine the existence of a limit function:

$$\zeta(z) = \lim_{n=\infty} \eta_n(z), \tag{3}$$

which will be the polymorphic function of the reentrant out theorem sought by us, and for which, then, the group to be generated from the $T_\infty^{(1)}, T_\infty^{(2)}, \ldots, T_\infty^{(p)}$ will obtain the known significance.

3.10 Application of the distortion theorem to the domain P_n

The point z_0, which we utilized for the production of the function $\eta_n(z)$, was to be chosen in the interior of the initial surface F_1; also, let this not be a branchpoint of our surface. Consequently, a circle will let itself be described about z_0 as center with a finite radius, which likewise runs entirely inside F_1 and in its interior, has neither a branchpoint of the surface, nor runs through one such with its periphery. The surface of this circle will accordingly be a schlicht circular surface situated in F_1. We may also assume the radius of this circle equal to 1; for this is immediately attainable, if necessary, by exercise of a similarly transformation. We therefore call the circle K_1 and also, introduce, at the same time, the circle $K_{\frac{1}{2}}$ of the radius $\frac{1}{2}$ concentric with K_1.

Let the image of the periphery $K_{\frac{1}{2}}$ in the η_n-plane be the curve C_n; the schlicht portion of the η_n-plane situated outside C_n is the conformal image of the surface of the circle $K_{\frac{1}{2}}$. Since, in the interior of K_1, the development holds for $\eta_n(z)$:

$$\eta_n(z) = \frac{1}{z - z_0} + c_1^{(n)}(z - z_0) + c_2^{(n)}(z - z_0)^2 + \cdots, \tag{1}$$

and the function $\eta_n(z)$ can assume no value there more than once, then according to the theorem on p. 498, the maximum of $|\eta_n(z)|$ is smaller than $(4 + 4\sqrt{2})$ along $K_{\frac{1}{2}}$. We accordingly note as a result: *the curve C_n, which is the image of the circle $K_{\frac{1}{2}}$ in the η_n-plane, runs entirely in the portion of this plane characterized by $|\eta_n| < 4 + 4\sqrt{2}$.* [530]

On the other hand, we are able to prove, that the curve C_n, for n remaining arbitrary, cannot shrink together below a certain degree of smallness. For this purpose, we call the value η_n at the point $(z_0 + 1)$ situated on the periphery K_1:

$$\eta_n(z_0 + 1) = \alpha_n, \tag{2}$$

where α_n is a number with a value, $|\alpha_n| < 4 + 4\sqrt{2}$. If we map the η_n-plane by:

$$\eta'_n = \frac{1}{\eta_n - \alpha_n}. \tag{3}$$

onto the η'_n-plane, then let C_n furnish the curve C'_n in the η'_n-plane. The simply connected domain included by C'_n is the conformal image of the circle $K_{\frac{1}{2}}$. Thereby, to the center z_0 corresponds the the origin $\eta'_n = 0$, and indeed, the mapping modulus at this place is equal to 1. $\eta'_n(z)$ is also single-valued and regular in the interior of K_1 and assumes no value more than once there. We are accordingly able to apply theorem 19 on p. 507, where we have to put in $M_0 = 1$, $R = 1$ and $r = \frac{1}{2}$; we find: *the curve C'_n runs inside that ring-shaped domain of the η'_n-plane, which is fixed by*:

$$\frac{-1 + \sqrt{2}}{8} < |\eta'_n| < 1 + e^{\frac{136\pi}{9}} < e^{16\pi}; \tag{4}$$

And therefore, the curve C_n runs inside the ring-domain of the η_n-plane characterized by:

$$e^{-16\pi} < |\eta_n - \alpha_n| < 8 + 8\sqrt{2}. \tag{5}$$

Since it now runs around the place α_n of the η_n-plane, it is self-evident, that for no finite n can it sink below the degree of smallness determined by the left side of (5) and independent of n.

To the partial domains $F_1, F_2, \ldots, F_{n-1}$ of F_n, let correspond the partial domains $P_{n,1}, P_{n,2}, \ldots, P_{n,n-1}$, of which, naturally, each is also a partial domain of the next following. Obviously, namely $P_{n,k+1}$ proceeds from $P_{n,k}$, by adding, to each of the $2p_k$ boundary curves, a domain $P_n^{(i)}$, which corresponds to a surface copy F_1, $P_{n,1}$ is naturally identical with the initial domain on p. 527 denoted by $P_n^{(0)}$. If we take away from P_n the portion of the η_n-plane running outside the curve C_n, then a domain remains provided with $(2p_n + 1)$ boundary curves, which let be called P'_n; correspondingly, the [531] domain $P'_{n,k}$ proceeds from $P_{n,k}$ by leaving out the port of the η_n-plane running outside C_n, which has $(2p_k+1)$ boundary curves, namely, the outer boundary curve C_n and the $2p_k$ inner boundary curves.

While n is a finite, but arbitrarily large number, we understand by l a finite index chosen fixed, which always lies under n. P'_{l+1} will be that domain, which emerges from P_{l+1} by removal of the part of the η_{l+1}-plane running outside the curve C_{l+1}. In corresponding fashion, $P'_{l+1,l}$ arises from $P_{l+1,l}$. Since it is a question of a determinate finite number l, then we have before us, in $P'_{l+1,l}$ and P'_{l+1}, fixed determinate domains of our structure, of which the latter emerges from the former by addition of $2p_l$ domains $P^{(i)}_{l+1}$. We will emphasize, in particular, that the various boundary curves of this domain $P^{(i)}_{l+1}$ are everywhere finitely distant from one another.

In the η_n-plane, we have, in $P'_{n,l}$ and $P'_{n,l+1}$, conformal images of $P'_{l+1,l}$ laid out by the function $\eta_n = f(\eta_{l+1})$. For the images, the presuppositions of the general distortion theorem at the close of §4 on p. 514 hold in their full extent. Namely, our boundaries are schlicht throughout, and their boundary is composed of finitely may analytic portions, which indeed arise by means of the analytic functions $\eta_{l+1}(z)$ and $\eta_n(z)$ from portions of great circles of the original Riemann z-sphere. But we may also apply the inequality (5) on p. 514 of the distortion theorem including the boundaries of our domains. In fact, along each boundary curve of $P'_{l+1,l}$ resp. P'_{l+1}, we are able to add one more strip of non-vanishing breadth lying outside, without the enlarged domain ceasing to satisfy the conditions of the distortion theorem. For the curve C_{l+1}, and also for the inner boundary curves of $P'_{l+1,l}$. This is self-evident, insofar as the latter domain is a partial domain of P'_{l+1}. For the inner boundary curves of P'_{l+1}, one wants to avail oneself of those strips, which correspond to the surface bands laid along the boundary curves of F_{l+1} on the Riemann sphere.

The length of the curve C_n is given by:

$$\int_{(C_n)} |d\eta_n| = \int_{(C_{l+1})} |f'(\eta_{l+1})||d\eta_{l+1}|, \tag{6}$$

[532] where the integral standing on the right, as indicated, is to be extended along the curve C_{l+1} in the η_{l+1}-plane. If we understand by M the maximum of $f'(\eta_{l+1})$ along C_{l+1} and by λ the finite length of the curve C_{l+1}, then there obviously holds:

$$\int_{(C_{l+1})} |f'(\eta_{l+1})||d\eta_{l+1}| \leq M\lambda.$$

Since, on the other hand, in consequence of (5), the length of the curve C_n is certainly greater than the length of the periphery of a circle with the amount standing on the left in (5) as radius, there follows:

$$M > \frac{2\pi}{\lambda} e^{-16\pi}. \tag{7}$$

Therefore, there are certainly values $|f'(\eta_{l+1})|$ along C_{l+1}, which are greater than the amount given in (7) on the right.

Now since the inequality (5) on p. 514 of the distortion theorem holds for P'_{l+1} including the boundary, we are able to substitute for the denominator $|f'(z_2)|$ such a value exceeding the amount (7) on the right. Then, for each other place z_1, or, as we shall say here η_{l+1}, inside and on the boundary of the domain P'_{l+1} the inequality holds:

$$|f'(\eta_{l+1})| > 3^{-4v} G^{-2v} \cdot \frac{2\pi}{\lambda} e^{-16\pi}. \tag{8}$$

But the constant G defined in (2) on p. 513 is $> e^{16\pi}$. By decreasing the right side of (8), we are accordingly also able to write:

$$|f'(\eta_{l+1})| > \frac{2\pi}{\lambda \cdot 3^{4v} \cdot G^{2v+1}}. \tag{9}$$

with this, however great the finite number n might be chosen, is obtained a lower bound, depending only on l, but independently of n, for the mapping modulus $|f'(\eta_{l+1})|$ inside and on the boundary of P'_{l+1}.

We will now set our sights on that inner boundary point of $P'_{n,l+1}$, which has the smallest distance m from the boundary of the foregoing domain $P'_{n,l}$. Let the line from that point to the boundary point of $P'_{n,l}$ situated closest be called Λ; it obviously belongs, except for its two endpoints, to the interior of a domain $P_n^{(i)}$ situated at the boundary $P_{n,l+1}$. If the straight line Λ maps itself onto the curve Λ' of the η_{l+1}-plane, then, for the length m of the line Λ, we have the representation:

$$m = \int_{(\Lambda)} |d\eta_n| = \int_{(\Lambda')} |f'(\eta_{l+1})||d\eta_{l+1}|.$$

If therefore, the length of the curve Λ', is denoted by λ', say, then it follows from the last equation that: [533]

$$m > \frac{2\pi\lambda'}{\lambda \cdot 3^{4v} \cdot G^{2v+1}}.$$

Now Λ' joins an inner boundary point of P'_{l+1} with an inner boundary point of $P'_{l+1,l}$. Therefore, the length λ' of Λ' is not smaller than the minimal distance d of the inner boundary of P'_{l+1} from the inner boundary of the domain $P'_{l+1,l}$, na amount d, which has a value different from 0 independently of n. Corresponding to this, the inequality holds for the minimal distance m:

$$m > \frac{2\pi d}{\lambda \cdot 3^{4v} \cdot G^{2v+1}}.$$

Finally, $P'_{n,l+1}$ is a partial domain of P'_n, which would only be identical with P'_n in the lowest case $n = l+1$. The inner boundary points of P'_n are therefore able to lie no more closely to those of $P'_{n,l}$, then the inner boundary points of $P'_{n,l+1}$. By returning to the domains P_n and $P_{n,l}$, we are consequently able to bring the following result to expression: *if l is a fixed, and n > l an arbitrary finite index, then there is a lower bound*:

$$s = \frac{2\pi d}{\lambda \cdot 3^{4v} \cdot G^{2v+1}}, \tag{10}$$

for the minimal distance of the boundary P_n from the boundary of the domain $P_{n,l}$, which is independent of n and different from 0.

3.11 Application of the consequences of the distortion theorem to the domain P_n

The consequences of the distortion theorem, set up in §5 on p. 514 ff., are now to be brought into application to the domain P_n, whereby we hold fast to the notations $P_n, P_{n,k}$ as well too, as $P'_n, P'_{n,k}$, in the meaning so far. By $P_{n,1} (= p_n^{(0)})$ we therefore have to understand that part of P_n, which corresponds to the initial copy F_1. The curve C_n, corresponding to the circle $K_{\frac{1}{2}}$ of the surface copy (cf. p. 530) is situated entirely in the portion of the η_n-plane characterized by $|\eta_n| < 4 + 4\sqrt{2}$. On the other hand, the $2p$ parts of the η_n-plane still remaining free of the domain $P_{n,1}$, we will say "the $2p$ gaps of the domain $P_{n,1}$", all find themselves inside C_n. Accordingly, the total surface content of these gaps, to be denoted by $J_{n,1}$ say, will be smaller than the surface of the circle of the radius $(4 + 4\sqrt{2})$:

$$J_{n,1} < (48 + 32\sqrt{2})\pi. \tag{1}$$

[534] We now form, for further following utilization, the domain P'_1, which arises from P_1 by leaving out the portion of the η_1-plane situated outside C_1. On the other hand, in a first gap of $P_{n,1}$, we position the neighboring domain $P_n^{(1)}$, from which we still take out the part corresponding to the surface of the circle $K_{\frac{1}{2}}$. The remainder, which now again contains $2p$ inner groups, let be called $P_n^{(1)\prime}$; this is a schlicht finite domain, which is mapped single-valuedly and conformally onto the similarly schlicht finite domain P'_1. The mapping, without compromising its character, also lets itself be carried over onto strips of never vanishing breadth, which one positions along each boundary curve outside our domain. For the boundary curves originating from the periphery $K_{\frac{1}{2}}$, this is self-evident from the mapping of $P_n^{(1)}$ onto P_1; for the remaining boundary curves, again utilize the strips, which we positioned along the boundary curves of F_1 on the surface of the Riemann sphere.

For our domain P'_1 and its image $P_n^{(1)\prime}$, therefore, the theorem stated in connection with (9) on p. 516 holds. In consequence of this, the total content of the $2p$ inner gaps of $P_n^{(1)\prime}$ and therefore, all the more, the total content of the $(2p-1)$ inner gaps of $P_n^{(1)}$. Thereby, q is a number of the interval $0 < q < 1$, which is independent of the mapping function, and is therefore stipulated only by the domain P'_1, ie. by the surface F_1, and by the choice of the point z_0.

We apply the same consideration to all remaining domains of $P_n^{(2)}, P_n^{(3)}, \cdots$, to the positioned in the groups of $P_{n,1}$. By means of these positionings, however, we arrive at the domain $P_{n,2}$ with $2p_2$ gaps. Therefore, as one finds by addition, the total content $J_{n,2}$ of these gaps will be smaller than the total content $J_{n,1}$, multiplied by q, of all gaps $P_{n,1}$. With utilization of (1) we thus find:

$$J_{n,2} < (48 + 32\sqrt{2})\pi q. \tag{2}$$

We repeat the same consideration for the further domains $P_n^{(k)}$, to be arranged in the gaps of $P_{n,2}$, which indeed, are all mapped single-valuedly and conformally onto P_1. It follows:

$$J_{n,3} < (48 + 32\sqrt{2})\pi q^2. \tag{3}$$

By continuation of the consideration, the general theorem arises: *the total surface content $J_{n,k}$ of the $2p_k$ gaps of the domain $P_{n,k}$ satisfies the inequality*:

$$J_{n,k} < (48+32\sqrt{2})\pi q^{k-1}. \tag{4}$$

whereby q is a number belonging to the interval $0<q<1$, which depends solely on the [535]
underlying surface F_1 and the choice of the place z_0. Since $P_{n,n}$ is identical to P_n, we shall simply write J_n for $J_{n,n}$. *For the total surface content J_n of the $2p_n$ gaps of P_n, we thus have the inequality*:

$$J_n < (48+32\sqrt{2})\pi q^{n-1}. \tag{5}$$

The second consequence of the distortion theorem coming to expression by the inequality (13) on p. 518 is also cut out for the present relations. The $2p_k$ boundary curves of the domain $P_{n,k}$ have, since they run in the finite part of the plane and are each composed of finitely many analytic pieces, finite lengths, which we will denote by $L_{n,k}^{(1)}, L_{n,k}^{(2)}, \ldots, L_{n,k}^{(2p_k)}$. For $n=1$, we come back to the $2p$ boundary curves of P_1. In order to be able to suitably apply the theorem (13) on p. 518, we want to remind ourselves of the two lines S' and S'' on p. 524, which we had drawn on the given Riemann surface next to the individual cuts. We carry the band of surface bounded by S' and S'' over onto the η_1-plane by means of the function $\eta_1(z)$, and recover this in a plane band, in whose interior runs the boundary curve of P_1, just coming into consideration. But in order not to come right up to the boundary of this plane band with the domain to be considered, we will imagine a narrower strip drawn in the interior of the band, which indeed also still contains the boundary curve of P_1 in its interior, but except for this, holds itself finitely distant everywhere from the boundary of the plane band just mentioned. Such strips we imagine adduced for each boundary curve of P_1 and call these $St_1^{(1)}, St_1^{(2)}, \ldots, St_1^{(2p)}$. Self-evidently, these run separately from one another in the η_1-plane, without reaching over one another anywhere.

Let the first boundary curve of $P_{n,k}$, whose length we called $L_{n,k}^{(1)}$, as boundary curve of a domain $P_n^{(i)}$, correspond to that boundary curve of P_1, about which we delimited the strip $St_1^{(1)}$. The preparations are so made, that all presuppositions of the theorem attached to (13) on p. 518 are fulfilled by the strip $St_1^{(1)}$. There is therefore a finite number Q_1, depending only on the domain $St_1^{(1)}$, so that the square of the length $L_{n,k}^{(1)}$ fulfills the inequality:

$$(L_{n,k}^{(1)})^2 < Q_1 J^{(1)}, \tag{6}$$

understanding by $J^{(1)}$ the content of the image domain of $St_1^{(1)}$ surrounding the boundary curve of $P_{n,k}$ considered.

In entirely corresponding fashion, one finds for the boundary curves of $P_{n,k}$ gener- [536]
ally:

$$(L_{n,k}^{(i)})^2 < Q_i J^{(i)}, \quad i=1,2,\ldots,2p_k. \tag{7}$$

Among the numbers Q_i, we naturally have only $2p$ different ones, which belong to the $2p$ strips $St_1^{(1)}, St_1^{(2)}, \ldots, St_1^{(2p)}$. If we call the greatest of them Q, then obviously:

$$(L_{n,k}^{(i)})^2 < Q J^{(i)}$$

holds, from which we obtain by addition:

$$\sum_{i=1}^{2p_k} (L_{n,k}^{(i)})^2 < Q \sum_{i=1}^{2p_k} J^{(i)}. \tag{8}$$

Into each individual gap of $P_{n,k-1}$, a domain conformal with P_1 now fits as a subordinate component here of $P_{n,k}$. On the $(2p-1)$ interior boundary curves of this component there then find themselves $(2p-1)$ image strips, which no more collide with one another, than the strips $St_1^{(1)}, \dots, St_1^{(2p)}$ of the domain P_1. From this yields itself, that the sum standing one the right in (8) is smaller than the total surface content of the $2p_{k-1}$ gaps of the domain $P_{n,k-1}$. This total content was denoted by $J_{n,k-1}$ and obeys the inequality:

$$J_{n,k-1} < (48 + 32\sqrt{2})\pi q^{k-2}.$$

Therefore, from (8) arises the following result: *the sum of the squares of the arc-lengths of all $2p_k$ boundary curves of $P_{n,k}$ satisfy the inequality*:

$$\sum_{i=1}^{2p_k} (L_{n,k}^{(i)})^2 < (48 + 32\sqrt{2})\pi Q q^{k-2}, \tag{9}$$

where q has the same significance as in (4) and also, Q is a finite positive number independent of n and k, which only depends on our surface F_1 and the strips $St_1^{(1)}, \dots, St_1^{(2p)}$, chosen once and for all.

In particular, for the the domain $P_{n,n} = P_n$ itself, we will call the lengths of the boundary curves $L_n^{(1)}, \dots, L_n^{(2p_n)}$. For these, therefore, it follows from (9):

$$\sum_{i=1}^{2p_n} (L_n^{(i)})^2 < (48 + 32\sqrt{2})\pi Q q^{k-2}. \tag{10}$$

3.12 Effectuation of the convergence proof of the functions $\eta_n(z)$

[537]

We have now assembled all means, of proving the convergence of the functions:

$$\eta_1(z), \eta_2(z), \dots, \eta_n(z), \dots, \tag{1}$$

to a limit function:

$$\lim_{n=\infty} \eta_n(z) = \zeta(z), \tag{2}$$

defined for all interior points of the covering surface F_∞.

We understand by B any domain chosen fixed, situated in the “interior” of the covering surface. There is then a determinate finite number l, so that B is a partial domain of F_l. While l therefore lies fixed, let $n > l$ signify an arbitrary finite index exceeding the number l. We consider the difference:

$$\eta_{n+m}(z) - \eta_n(z), \tag{3}$$

and are able to show, that for n chosen sufficiently large and m remaining arbitrary, this can be made arbitrarily small in the entire domain B.

We shift the discussion into the η_n-plane and write, for abbreviation, η instead of η_n. B furnishes here a schlicht domain B_η, which is partial domain of $P_{n,l}$ and therefore, all the more, of P_n. Let the difference (3), in dependency on η, be written, for abbreviation:

$$\eta_{n+m}(z) - \eta_n(z) = \Delta(\eta). \tag{4}$$

$\Delta(\eta)$ represents a function, single-valued and regular not only in B_η, but in the entire domain P_n, without the place $\eta = \infty$, which corresponds to the point z_0, being assumed. In fact, indeed, in consequence of (6) on p. 524, the following development holds in the neighborhood of z_0:

$$\Delta = (c_1^{(n+m)} - c_1^{(n)})(z - z_0) + (c_2^{(n+m)} - c_2^{(n)})(z - z_0)^2 + \cdots, \tag{5}$$

so that Δ exhibits a zero of first order at the place $\eta = \infty$.

For this state of affairs, we need only estimate $\Delta(\eta)$ *for any place η_0 of the domain B_η situated in the finite part of the plane.* This can be proved on the basis of the Cauchy integral theorem. In order to put this down, we describe about η_0 as center a circle K_r with radius so large, that all boundary curves of P_n, to be denoted by $R_1, R_2, \ldots, R_{2p_n}$, say, the inside K_r. If we take away the part of the η-plane running outside K_r from the domain P_n, the domain P'_n remains, provided with $(2p_n + 1)$ boundary curves. The function $\Delta(\eta)$ is single-valued and regular in this domain P'_n, including all boundary [538]
curves. In consequence of the Cauchy integral theorem, we are therefore able to set:

$$\Delta(\eta_0) = \frac{1}{2i\pi}\int_{(K_r)} \frac{\Delta(\eta)}{\eta - \eta_0}\,d\eta + \frac{1}{2i\pi}\sum_{k=1}^{2p_n}\int_{(R_k)} \frac{\Delta(\eta)}{\eta - \eta_0}\,d\eta, \tag{6}$$

where the integrals, as indicated, are to be extended over the boundary curves of P'_n.

Since, for $\lim r = \infty$, in consequence of (5), the value $|\Delta(\eta)|$ uniformly approaches the value 0 along K_r, then the value of the first integral in (6), for r chosen sufficiently great, can be made arbitrarily small, without the remaining components of the equation (6) altering themselves. This integral therefore vanishes by itself, so that we find:

$$\Delta(\eta_0) = \frac{1}{2i\pi}\sum_{k=1}^{2p_n}\int_{(R_k)} \frac{\Delta(\eta)}{\eta - \eta_0}\,d\eta. \tag{7}$$

If we understand by $\eta^{(k)}$ any place of the boundary curve R_k chosen fixed, then, in each case:

$$\int_{(R_k)} \frac{\Delta(\eta^{(k)})}{\eta - \eta_0}\,d\eta = \Delta(\eta^k)\int_{(R_k)} \frac{d\eta}{\eta - \eta_0} = 0$$

holds, since the function under the last integral sign is regular everywhere on and inside the integral curve R_k. With utilization of this circumstance, we may rewrite equation (7) into the form:

$$\Delta(\eta_0) = \frac{1}{2i\pi}\sum_{k=1}^{2p_n}\int_{(R_k)} \frac{\Delta(\eta) - \Delta(\eta^{(k)})}{\eta - \eta_0}\,d\eta. \tag{8}$$

We are now easily able to estimate the absolute value of the individual integral standing on the right. Since, first of all, η_0 lies in the interior of $P_{n,l}$, and η, in (8), only signifies boundary points of P_n, then according to the theorem attached to the equation (1) on p. 533, $|\eta-\eta_0|$ is always $> s$, and therefore:

$$\left|\frac{1}{\eta-\eta_0}\right| < \frac{1}{s}, \tag{9}$$

understanding by s the number n given there, different from 0 and independent of η. If, further, we denote the oscillation (cf. p. 498) of the function $\Delta(\eta)$ along the curve R_k by $S_k(\Delta(\eta))$, then self-evidently, for the points η of just this curve R_k:

$$|\Delta(\eta)-\Delta(\eta^{(k)})| \le S_k(\Delta(\eta)). \tag{10}$$

[539] Finally, however:

$$\int_{(R_k)} |d\eta| = L_n^{(k)} \tag{11}$$

holds, where on the right, as in the previous paragraph, is meant the length of the curve R_k.

If we accordingly denote the kth integral of the sum (8) by $\int_{(k)}$ for short, then we find for its absolute value:

$$\left|\int_{(k)}\right| \le \int_{(R_k)} \frac{|\Delta(\eta)-\Delta(\eta^{(k)})|}{|\eta-\eta_0|}|d\eta| < \frac{1}{s} S_k L_n^{(k)},$$

whereby S_k is written is abbreviation for $S_k(\Delta(\eta))$. Since now, $2SL \le S^2+L^2$ always holds, we are also able to write:

$$\left|\int_{(k)}\right| < \frac{1}{2s}(S_k^2+(L_n^{(k)})^2). \tag{12}$$

Now the oscillation of $\Delta(\eta)=\eta_{n+m}-\eta_n$ along R_k is no greater than than the sum of the oscillations of η_{n+m} and η_n there. But the oscillation of $\eta_n=\eta$ is obviously smaller than the curve length $L_n^{(k)}$, and that of η_{n+m} is smaller than the length $L_{n+m,n}^{(k)}$ of the corresponding boundary curve of the domain $P_{n+m,n}$ in the η_{n+m}-plane. With this holds:

$$S_k < L_n^{(k)} + L_{n+m,n}^{(k)} \tag{13}$$

from which we immediately draw the conclusion:

$$\begin{aligned} S_k^2 &< (L_n^{(k)})^2+(L_{n+m,n}^{(k)})^2+2L_n^{(k)}L_{n+m,n}^{(k)}, \\ S_k^2 &< 2(L_n^{(k)})^2+2(L_{n+m,n}^{(k)})^2. \end{aligned}$$

The inequality (12) consequently goes over into:

$$\left|\int_{(k)}\right| < \frac{1}{2s}\left(3(L_n^{(k)})^2+2(L_{n+m,n}^{(k)})^2\right),$$

and therefore, from (8) follows:

$$|\Delta(\eta_0)| < \frac{1}{4\pi s}\left\{3\sum_{k=1}^{2p_n}(L_n^{(k)})^2+2\sum_{k=1}^{2p_n}(L_{n+m,n}^{(k)})^2\right\}. \tag{14}$$

Now the inequality (1) on p. 536, as well as the inequality (9) there, comes into utilization, in which latter substitute now, for the indices n and k, $n+m$ and n. It follows,

$$|\Delta(\eta_0)| < \frac{1}{s}(60+40\sqrt{2})Qq^{n-2}. \tag{15}$$

With this, we have reached our goal. Since Q and q are independent of n and q belongs to the interval $0 < q < 1$, we are able, for n chosen sufficiently large, to suppress the value $|\Delta(\eta_0)|$ beneath each givable positive quantity different from 0. *But through this is proved the uniform convergence of the infinite series of analytic functions*: [540]

$$\eta_l + (\eta_{l+1} - \eta_l) + (\eta_{l+2} - \eta_{l+1}) + \cdots, \tag{16}$$

in the domain B; *the sum is the analytic limit function*:

$$\zeta(z) = \lim_{n=\infty} \eta_n(z), \tag{17}$$

which is defined for the entire interior of the covering surface F_∞.

In it, we already possess the polymorphic function of the reentrant cut theorem. But in order to see this definitively, we yet have to determine two facts, first, namely, this, that $\zeta(z)$ is "*linearly*"-polymorphic, and secondly, this, that $\zeta(z)$ is essentially *uniquely* determined.

3.13 Proof of the linearity theorem

Concerning the mapping, which the limit function $\zeta(z)$ accomplishes from the entire interior of the covering surface F_∞, several statements emerge immediately from the developments, so for. The image of the F_∞ in the ζ-plane is to be denoted by P_∞, and the partial domain F_n of F_∞ will yield the partial domain $P_{\infty,n}$. The latter domain lies schlichtly on the ζ-plane and has $2p_n$ boundary curves, which bound the $2p_n$ inner gaps of $P_{\infty,n}$.

The inequalities (4) on p. 534 and (9) on p. 536 hold for fixed k for each n, therefore, on account of the uniform convergence of the $\eta_n(z)$ to $\zeta(z)$ including possible equalities, also for $\lim n=\infty$. If, after the passage to the limit $\lim n=\infty$, we write the notation n instead of k in those inequalities, then we obtain the theorem concerning the image of the F_∞ in the ζ-plane: *there holds, for the total content* $J_{\infty,n}$ *of the* $2p_n$ *gaps of the schlicht domain* $P_{\infty,n}$:

$$J_{\infty,n} \le (48+32\sqrt{2})\pi q^{n-1}; \tag{1}$$

similarly, we find, for the lengths $L^{(k)}_{\infty,n}$ *of the* $2p_n$ *boundary curves of the domain* $P_{\infty,n}$:

$$\sum_{k=1}^{2p_n} (L^{(k)}_{\infty,n})^2 \le (48+32\sqrt{2})\pi Q q^{n-2}. \tag{2}$$

The developments of §10 also carry themselves over to the case $n=\infty$. First of all, the condition (5) on p. 530 remain remains subsisting for $\lim n=\infty$, ie. for the curve C_∞. To this attaches itself the application of the distortion theorem developed [541] there, and therefore, we note: *for each finite l, for the distance of a boundary point of the domain $P_{\infty,l+1}$ from an interior point or a boundary point of the domain $P_{\infty,l}$, there is a positive lower bound s_l, different from 0. This bound will subsist all the more for the distance of a boundary point of any domain $P_{\infty,n}$ with $n > l$ from an interior or boundary point of the domain $P_{\infty,l+1}$.*

Since p is assumed > 1, the number:

$$2p_n = 2p(2p-1)^{n-1}$$

of the gaps of $P_{\infty,n}$, itself grows beyond all limits as $\lim n = \infty$. In contrast to this, there arises from (1):

$$\lim_{n=\infty} J_{\infty,n} = 0. \tag{3}$$

it follows: *the image P_∞ of the covering surface F_∞ covers the entire ζ-plane schlichtly up to an infinite set of boundary points, which we will call M_g, and which, in consequence of (3), has no content.*

If one fixed an individual one of these boundary points, i.e. therefore, any place of the ζ-plane not belonging to the interior of P_∞, then for each n, this will be situated in a uniquely determined one of the $2p_n$ gaps. If one then further marks any two boundary points different from one another, then these will have a determinate finite distance from one another. If, on the strength of this, we choose the number n so large, that the amount standing one the right in (2) comes out $< 4e^2$, then, all the more, the $2p_n$ lengths $L^{(k)}_{\infty,n}$ will be smaller than $2e$. From this yields itself directly, that the two points must lie in two of the $2p_n$ gaps of $P_{\infty,n}$ different from one another, and that therefore *each* joining-line of the two places must press through the interior of $P_{\infty,n}$. It follows: *for any two different points of the set M_g, there is no joining line, which would consist only of boundary points; M_g therefore represents a set of discrete points.*

We further carry over the theorem of §9 on p. 527 to the limiting case $n = \infty$. To the infinitely many copies F_1, positioned over one another, which form the F_∞, correspond infinitely many domains $P^{(0)}_\infty, P^{(1)}_\infty, P^{(2)}_\infty, \cdots$, positioned next to one another, which, in their totality, cover the domain P_∞ gaplessly and simply. Let $P^{(0)}_\infty$ be identical with $P_{\infty,1}, P_{\infty,2}$ consists of $P^{(0)}_\infty$ and the $2p$ neighboring domains $P^{(i)}_\infty$; similarly, $P_{\infty,3}$ consists of $P^{(0)}_\infty$ and all domains $P^{(k)}_\infty$, separated from $P^{(0)}_\infty$ by no more than one
[542] domain $P^{(i)}_\infty$. Continuing in the same fashion, we are able to say: *$P_{\infty,n}$ consists of $P^{(0)}_\infty$ and all $P^{(k)}_\infty$ separated from this initial space by at most $(n-2)$ domains $P^{(i)}_\infty$.* One wants to hold fast to this definition for an argument to be presented further below.

Above all, we determine: the covering surface F_∞ is regular, in the sense that it is structured about each of its copies F_1, exactly as about any others; it accordingly admits infinitely many reversibly single-valued transformations into itself, whose individual one is determined by giving that surface copy F_1, into which the initial copy is carried over by means of that transformation. These infinitely many transformations for F_∞ into itself obviously form a group.

If we go to the ζ-plane, then we arrive at the transformation of the result of §9 on p. 527 to the limiting case $n = \infty$. *We obtain a group of analytic functions*:

$$\zeta' = T^{(0)}_\infty(\zeta),\ \zeta' = T^{(1)}_\infty(\zeta),\ \zeta' = T^{(2)}_\infty(\zeta), \ldots, \tag{4}$$

each one of which transformations the net P_∞ of the domains $P_\infty^{(0)}, P_\infty^{(1)}, P_\infty^{(2)}, \cdots$, reversibly uniquely into itself, in such a manner, that upon exercise of $T_\infty^{(i)}$, the initial space $P_\infty^{(0)}$ is carried over into $P_\infty^{(i)}$. The group is producible from p generators (cf. p. 523), which carry $P_\infty^{(0)}$ over into p neighboring $P_\infty^{(i)}$.

The developments just presented now permit us to show, that we are concerned with a group of linear ζ-substitutions. The analytic character of the individual substitution $T_\infty^{(i)}$ refers, to start with, only to the interior points of P_∞; for $T_\infty^{(i)}(\zeta)$ is only defined for these so far. But now we are able to immediately define $T_\infty^{(i)}(\zeta)$ also for the points of the set M_g as well. Namely, to the individual boundary point corresponds a determinate chain of domains $P_\infty^{(k)}$, characteristic only for it, which lead from the initial space $P_\infty^{(0)}$ to that boundary point. The chain, transformed by the individual $T_\infty^{(i)}$, then leads to that boundary point, which corresponds to that first one by means of the transformation $T_\infty^{(i)}$. *After this extension, the individual transformation $T_\infty^{(i)}$ furnishes a single-valued continuous function in the entire ζ-plane, apart from the fact, that it exhibits a pole of the first order at that place situated inside P_∞, which proceeds from the point $\zeta = \infty$, lying inside the initial domain $P_\infty^{(0)}$, by the transformation inverse to $T_\infty^{(i)}$.*

The question now poses itself, whether $T_\infty^{(i)}(\zeta)$ is analytic at the individual boundary [543]
point. We need to prove this only for the p generators. Consequently, let $T(\zeta)$ be one of these, which transforms $P_\infty^{(0)}$ into the neighboring domain $P_\infty^{(1)}$. We will then, similarly, as on p. 538, apply the Cauchy theorem, in order to represent the function $T(\zeta)$ for an arbitrary interior point of the domain P_∞, finite and different from the pole $T^{-1}(\infty)$. If we imagine ζ_0 chosen fixed, then we are able to so choose a determinate finite number l, that ζ_0 is also already an interior point of $P_{\infty,l}$. Then if n is an arbitrary index exceeding the number l, then ζ_0 is also situated in $P_{\infty,n}$ and moreover, $T(\zeta)$ is analytic and single-valued everywhere inside and on the boundary of $P_{\infty,n}$. One now wants to add to the $2p_n$ inner boundary curves of $P_{\infty,n}$, one more closed finite outer boundary curve, which is to include ζ_0 and all inner boundary curves, that so that the pole $T^{-1}(\infty)$ of $T(\infty)$ lies outside this curve[10]. We call this new curve C and will remove the part of the ζ-plane running outside C from the domain $P_{\infty,n}$. In the truncated domain, which let be called $P'_{\infty,n}$, the function $T(\zeta)$ will then still be everywhere continuous.

We refer the Cauchy theorem to this domain $P'_{\infty,n}$, and find $T(\zeta_0)$ represented by:

$$T(\zeta_0) = \frac{1}{2i\pi}\int_{(C)} \frac{T(\zeta)}{\zeta-\zeta_0}d\zeta + \frac{1}{2i\pi}\sum_{k=1}^{2p_n}\int_{(R_k)} \frac{T(\zeta)}{\zeta-\zeta_0}d\zeta \tag{5}$$

where the integrals refer themselves to the $(2p_n+1)$ boundary curves of $P'_{\infty,n}$, whose $2p_n$ interiors we denote, as on p. 538, by $R_1, R_2, \ldots$.

The individual integral of the sum on the right in (5), which let be denoted by $\int_{(k)}$ for abbreviation, we change around according to the principle already utilized on p. 538. Let be understood by ζ_k any point of the kth boundary curve R_k, which let be chosen fixed. Then the equation (cf. p. 538) holds:

$$\int_{(R_k)} \frac{T(\zeta_k)}{\zeta-\zeta_0}d\zeta = T(\zeta_k)\int_{(R_k)} \frac{d\zeta}{\zeta-\zeta_0} = 0,$$

so that we may use, for the kth integral of the sum (5), the following manner of writing:

[10] This pole find itself in a domain $P_\infty^{(i)}$, which is neighboring with $P_\infty^{(0)}$.

$$\int_{(k)} = \int_{(R_k)} \frac{T(\zeta) - T(\zeta_k)}{\zeta - \zeta_0} d\zeta. \tag{6}$$

For the absolute value of the integral follows:

$$\left|\int_{(k)}\right| = \int_{(R_k)} \frac{|T(\zeta) - T(\zeta_k)|}{|\zeta - \zeta_0|} d\zeta. \tag{7}$$

[544] But now:

$$|T(\zeta) - T(\zeta_k)| \leq S_k(T(\zeta)),$$

where on the right is meant the oscillation of $T(\zeta)$ along R_k. Further, $|\zeta - \zeta_0| \geq s_l$ holds, and therefore:

$$\frac{1}{|\zeta - \zeta_0|} \leq \frac{1}{s_l},$$

understanding by s_l the positive number different from 0 so denoted above (p. 541), since ζ_0 signifies an interior point of $P_{\infty,l}$ and ζ a boundary point of $P_{\infty,n}$. Finally:

$$\int_{(R_k)} |d\zeta| = L_{\infty,n}^{(k)}$$

is the length of the kth boundary curve R_k of $P_{\infty,n}$. From (7) thus follows:

$$\left|\int_{(k)}\right| \leq \frac{1}{s_l} S_k L_{\infty,n}^{(k)} \leq \frac{1}{2s_l}(S_k^2 + (L_{\infty,n}^{(k)})^2), \tag{8}$$

where S_k is put in abbreviation for $S_k(T(\zeta))$.

Now go back to the above definition of $P_{\infty,n}$ (p. 542). Since T was to be a generator, then $P_\infty^{(1)} = T(P_\infty^{(0)})$ was neighboring with $P_\infty^{(0)}$. Correspondingly, $T(P_{\infty,n})$ consists of $P_\infty^{(1)}$ and all domains $P_\infty^{(i)}$, which are separated from $P_\infty^{(1)}$ by no more than $(n-2)$ such domains. But a domain $P_\infty^{(i)}$, which is separated from $P_\infty^{(1)}$ by at most $(n-2)$ domains, will be separated from the neighboring domain $P_\infty^{(0)}$ by at most $(n-1)$ domains, *so that $T(P_{\infty,n})$ is a component of $P_{\infty,n+1}$*. On the other hand, a domain of the complex $P_{\infty,n-1}$, which is separated from $P_\infty^{(0)}$ by at most $(n-3)$ domains, will be separated from $P_\infty^{(1)}$ by at most $(n-2)$ domains, and therefore belong to $T(P_{\infty,n})$. *Accordingly, $P_{\infty,n-1}$, for its part, is a component of $T(P_{\infty,n})$.*

The boundary curve R_k, transformed by T, which is to be called $T(R_k)$, consequently belongs, as boundary curve, either to the domain $P_{\infty,n-1}$ or $P_{\infty,n}$ or finally $P_{\infty,n+1}$. But since $S_k(T(\zeta))$ is obviously smaller than the length of this boundary curve, then there arises:

$$\sum_{k=1}^{2p_n} S_k^2 < \sum_{k=1}^{2p_{n-1}} (L_{\infty,n-1}^{(k)})^2 + \sum_{k=1}^{2p_n} (L_{\infty,n}^{(k)})^2 + \sum_{k=1}^{2p_{n+1}} (L_{\infty,n+1}^{(k)})^2.$$

The condition (2) accordingly furnishes:

$$\sum_{k=1}^{2p_n} S_k^2 < (48 + 32\sqrt{2})\pi Q(q^{n-3} + q^{n-2} + q^{n-1}).$$

[545] And therefore, all the more:

$$\sum_{k=1}^{2p_n} S_k^2 < 3(48+32\sqrt{2})\pi Q q^{n-3}.$$

Since $q<1$, it follows from (2), further:

$$\sum_{k=1}^{2p_n} (L_{\infty,n}^{(k)})^2 < (48+32\sqrt{2})\pi Q q^{n-3},$$

And therefore, from (8) follows:

$$\left|\sum_{k=1}^{2p_n}\int_{(k)}\right| \le \sum_{k=1}^{2p_n}\left|\int_{(k)}\right| < \frac{1}{s_l}(96+64\sqrt{2})\pi Q q^{n-3}.$$

Now go back to equation (5), whose first term on the right side let be taken the left. By passage to the absolute values on the right and left follows:

$$\left|T(\zeta_0)-\frac{1}{2i\pi}\int_{(C)}\frac{T(\zeta)d\zeta}{\zeta-\zeta_0}\right| < \frac{1}{s_l}(48+32\sqrt{2})Qq^{n-3}. \tag{9}$$

If, for changing choice of n, we also hold fast to the curve C first chosen, then in (9) on the left stands a *determinate* value *independent of* n. In contrast to this, the right side of the inequality (9), by n chosen sufficiently large, can be brought arbitrarily close to the value 0. *Consequently, the value in (9) on the left is identical to* 0, *and therefore the equation subsists*:

$$T(\zeta_0)=\frac{1}{2i\pi}\int_{(C)}\frac{T(\zeta)d\zeta}{\zeta-\zeta_0}. \tag{10}$$

The integral standing on the right here now furnishes a single-valued and analytic function, not only for the points ζ_0 in the interior of P_∞, but, according to known theorems for the total domain included by C, therefore also for the points of the set M_g. This function, in consequence of (10), agrees with our function $T(\zeta)$ at all points ζ_0 in the interior of P_∞. The agreement also remains subsisting at the boundary points, just because we know, that $T(\zeta)$ is also continuous in its interior: *the function $T(\zeta)$ is a reversibly single-valued analytic function, not only in the interior of P_∞, but also at the points of the set M_g, and therefore, in the entire ζ-plane.*

Such a function, according to known fundamental theorems of function theory, is necessarily a *linear one*: [546]

$$T(\zeta)=\frac{\alpha\zeta+p}{\gamma\zeta+\delta}. \tag{11}$$

Now since this result holds for each of the p generators, then it holds, at the same time, for all the transformations $T_\infty^{(i)}(\zeta)$ of our group. With this, we have proved the *"linearity theorem": the group of the single-valued transformations of the covering surface F_∞ into itself represents itself, in the limit function $\zeta(z)$, as a group of "linear substitutions" with $P_\infty^{(0)}$ as "discontinuity domain", so that $\zeta(z)$ is, in fact, a linearly-polymorphic function of the desired kine on our given Riemann surface.* There only remains now to yet decide the question, whether $\zeta(z)$ is the sole function of this kind, or whether there are several of them.

3.14 Proof of the unicity theorem. Proof of the reentrant cut theorem

The fact, that the polymorphic function $\zeta(z)$ of the surface F_1 obtained is essentially, ie. apart form linear transformation, unique, we prove indirectly, by starting from the existence of a second linearly-polymorphic function $\overline{\zeta}(z)$, which again maps F_1 onto a discontinuity domain of our kind. We will call the latter $\overline{P}_\infty^{(0)}$, the equivalent domains arranging themselves next to this by $\overline{P}_\infty^{(1)}$, $\overline{P}_\infty^{(2)}$, $\cdots$, and denote the entire net by $\overline{P}_\infty$. The domain $\overline{P}_\infty$ is mapped single-valuedly and conformally onto the covering surface F_∞; thereby again let the partial domain $\overline{P}_{\infty,n}$ of $\overline{P}_\infty$ correspond to the partial domain F_n of F_∞. For the sake of simplicity, we also stipulate, that the place z_0 of original surface copy F_1 correspond to the place $\overline{\zeta}=\infty$, which is attainable if need be, by linear transformation of $\overline{\zeta}$.

The arguments of paragraphs 10 and 11, which concerned the applications of the distortion theorem to the net P_∞ of the domain $P_\infty^{(i)}$, carry themselves over directly to the net $\overline{P}_\infty$. Thereby, the numbers q and Q are even the same as at that time, since, indeed, our present domain $\overline{P}_\infty^{(0)}$ is mapped conformally onto $P_\infty^{(0)}$ by mediation of F_1. We note, in particular, for the lengths $L_{\infty,n}^{(k)}$ of the $2p_n$ boundary curves of $\overline{P}_{\infty,n}$, the inequality corresponding to the condition (2) on p. 540:

$$\sum_{k=1}^{2p_n}(\overline{L}_{\infty,n}^{(k)})^2 \leq (48+32\sqrt{2})\pi Q q^{n-2}. \tag{1}$$

[547] By mediation of the covering surface F_∞, the total interior of the domain $\overline{P}_\infty$ is mapped reversibly single-valuedly and conformally onto that of P_∞, whereby, according to our stipulation, the places $\overline{\zeta}=\infty$ and $\zeta=\infty$ correspond to one another. But we are also able to immediately extend the single-valuedness of the correspondence to the two sets $\overline{M}_g$ and M_g of the boundary points, since homologous domain chains of the two nets $\overline{P}_\infty$ and P_∞ furnish boundary points which are coordinated to one another. The mapping between ζ and $\overline{\zeta}$ is then, in particular, also a continuous one at all boundary points.

The mapping determined we will now represent by the equation:

$$\overline{\zeta}=f(\zeta). \tag{2}$$

This function $f(\zeta)$ is then reversibly single-valued for all values ζ, also including the value $\zeta=\infty$, and continuous for all finite ζ; in addition, it is always analytic for all places ζ not belonging to the set M_g. But that the analytic character also remains preserved at the boundary points, one perceives by repetition of the argument attached to the Cauchy theorem on p. 543. If again, ζ_0 is an arbitrary finite point chosen inside P_∞, then arrange the indices l and n as earlier (p. 543). Also, we again go from $P_{\infty,n}$ to a finite domain $P'_{\infty,n}$, by addition of an outer boundary curve C, which includes ζ_0 and all $2p_n$ inner boundary curves of $P_{\infty,n}$. The Cauchy theorem then furnishes:

$$f(\zeta_0)=\frac{1}{2i\pi}\int_{(C)}\frac{f(\zeta)d\zeta}{\zeta-\zeta_0}+\frac{1}{2i\pi}\sum_{k=1}^{2p_n}\int_{(R_k)}\frac{f(\zeta)d\zeta}{\zeta-\zeta_0}. \tag{3}$$

For this equation, the argument attached to (5) on p. 543 lets itself be applied anew, whereby the single unessential distinction subsists, that for the estimation of the oscillation of $f(\zeta)$ along R_k, our present inequality (1) is to be utilized in place of the inequality (2) on p. 540. We again to come to the result, that the $2p_n$-termed sum, standing on the right in (3), vanishes by itself, from which follows the representation:

$$f(\zeta_0) = \frac{1}{2i\pi}\int_{(C)} \frac{f(\zeta)d\zeta}{\zeta - \zeta_0}. \tag{4}$$

Carrying over the argument attached to (10) on p. 545 shows, on the strength of this, that the analytic character of $f(\zeta)$ also remains preserved at the points of the set M_g. As soon as this is established, it is self-evident, that $\bar{\zeta}$ is a linear function of ζ; and [548]
indeed, we have:

$$\bar{\zeta} = \alpha\zeta + \beta,$$

since to the value $\zeta = \infty$ again corresponds the value $\bar{\zeta} = \infty$. With this, we obtain the *"unicity theorem": there is essentially, i.e., apart form a linear transformation, only one single linearly-polymorphic function $\zeta(z)$ of the desired kind on the cut-up Riemann surface F_1.*

By collecting all results, there arises the *"reentrant cut theorem": on each Riemann surface of a genus $p > 0$, which we have provided with any system of p reentrant cuts not decomposing this into separate pieces, these exists one and essentially also only one linearly-polymorphic function $\zeta(z)$, which carries the cut-up surface conformally over onto a schlicht domain provided with $2p$ closed boundary curves, pairwise coordinated to one another linearly (discontinuity domain).*

3.15 Koebe's proof of the general Kleinian fundamental theorem

Of the furthergoing investigations of Koebe's[11], the two notes "*Über die Uniformisierung der algebraischen Curven durch automorphe Functionen mit imaginären Substitutions Grunppen*", which were already cited on p. 446, come especially into consideration here, and indeed, because Koebe has succeeded here in giving a proof of the most general fundamental theorem of Klein[12], mentioned on p. 47, which refers to domains, which arise by composition from v polygons with boundary-circles[13] of the signatures[14]:

$$(p_i; n_i; l_1^{(i)}, l_2^{(i)}, \dots, l_{n_i}^{(i)}), \quad i = 1, 2, \dots, v. \tag{1}$$

Here, at least the principal viewpoints of this proof of Koebe's are to be described briefly.

Besides the methods of investigation so far, one more supervenes as essentially [549]

[11] Besides the works cited above (p. 445 ff.), we also cite Koebe's note "*Über die Hilbert'sche Uniformisierungsmethode*", Gött. Nachr. of 1910, p.59, which attaches itself to the note of Hilbert's, cited on p.496, "*Zur Theorie der conformen Abbildung*", Gött. Nachr. of 1909, p.314.

[12] Stated in 1882 in Vol. 20 of the Math. Ann., p.206 ff.

[13] Even elliptic or parabolic rotation groups are also able to take part in the composition; however, for the sake of brevity, we ignore these.

[14] Koebe will give a detailed representation of his proof in the work "*Uber die Uniformisierung der algebraischen Curven III*", which is to appear shortly in the Mathematische Annalen.

new, which Koebe designates as an "*iteration procedure*". We are able to best describe the character of this procedure by application to the reentrant cut theorem.

The surface F_1, cut through along the p reentrant cuts, we had, on p. 525 ff., mapped onto a schlicht domain P_1 of the η_1-plane, provided with $2p$ boundary curves. The boundary curves were coordinated to one another in pairs and indeed, this coordination (which we did not yet pursue especially at that time for P_1) is an *analytic* one, as is directly provable from the consideration of the surface bands adduced above (p.524) on the Riemann surface along the reentrant cuts.

The iteration procedure now has, as its goal the carrying of this domain P_1 conformally over onto a new schlicht domain with $2p$ boundary curves, for which, however, the analytic coordination of the boundary curves is, in particular, a *linear* one. This happens in the following fashion: we retain only a single pairs of boundary curves mapped onto one another and therefore imagine the $(2p-2)$ remaining gaps of our domain P_1 us filled schlichtly. Because of the analytic coordination of its two boundary curves still on hand, we are able to map the domain so obtained conformally onto a closed Riemann surface of the genus $p = 1$. Let the integral n of the first kind of this surface have that period path, which arises from the boundary curve of P_1, corresponding to the period ω_2, while oppositely situated points on the bank of the reentrant cut concerned, n differs by ω_1. Then, by means of the function:

$$v_2 = e^{\frac{2i\pi u}{\omega_2}} \tag{2}$$

we map the Riemann surface, and with this, our domain of the η_1-plane, onto the v_2-plane, where we again get a domain provided with two boundary curves. But the latter are now mapped onto one another by the *linear* substitution:

$$v_2' = e^{\frac{2i\pi\omega_1}{\omega_2}} v_2. \tag{3}$$

To the domain P_1 corresponds a partial domain, to be called $P^{(2)}$ say, of the domain just obtained of the v_2-plane. The advantage obtained is, that for *one* of the p pairs of boundary curves, we have a *linear* mapping, while, each of the two curves of the $(p-1)$ remaining pairs, now as before, are mapped analytically onto one another. As soon as we have linear mapping of the boundary curves, then the concept of the discontinuity domain of the group generated by the mapping comes into force. We are therefore able,
[550] at least in each of the two gaps belonging to the favored boundary curves, to situate an infinite chain of equivalent domains, which position themselves smoothly next to another, and each of which naturally still offers $(2p-2)$ inner gaps (as P_2).

We now effect this same operation on the second pair of the boundary curves and arrive at a schlicht domain $P^{(3)}$, situated in a v_3-plane, for which the boundary curves of that second pair are connected linearly. Thereby, to be sure, for the analytic character of the connection of the boundary curves of the first pair, the linear character is lost again. But for the first pair, something essential remains maintained: *Namely, the associated analytic transformation, applied to* $P^{(3)}$, *also furnishes a smooth arrangement of the equivalent domains next to one another in the two gaps preferred first.* Herein lies the audacity of the entire consideration: while, in general, for non-linear analytic transformations, one is able to say nothing at the very beginning concerning the continuation into the domain gaps with regard to the schilichtness, the transformation belonging to the first pair behaves itself, now as before, in $P^{(3)}$, as a linear substitution with respect to the arrangement of the domains. This fact is a consequence of of the circumstance, that the conformal mapping of the v_2-plane onto that of v_3, extends over the full plane, always from the two gaps belonging to the second boundary curve pair, we are consequently able to immediately refer the reproduction of the second boundary curve pair to the simply infinite chain of domains, which we had already laid out on the first boundary curve pair. If the first two transformations for $P^{(3)}$ are called T_1 and T_2 (of which, therefore, T_2 is presently linear), then all transformations $T_2^m T_1^n$ (formed for all positive, negative and vanishing integral m, n) will already effect smooth arrangement of the domains next to one another.

One now treats, according to the same method, the third, the fourth etc. boundary curve pairs, and, after one has disposed of the pth pair, one has to begin again with the first. The iteration procedure consists in the unbounded repetition of the same procedure, whereby, bit by bit, smooth arrangement of the domains next to one another enters for *all* further transformations to be produced from the p generating transformations.

There remains to show, hereby, the *convergence to a limit function* $\lim_{n=\infty} v_n$, as well as the *linearity* and unicity in the limiting case. As Koebe has shown this l.c., this succeeds on the basis of the same principles, which underlie the method of the covering surface, so that, with this, first of all, is obtained a second procedure of proof for the reentrant cut theorem.

As Koebe has now proved in the second of the notes cited above, exactly the same [551]
method of investigation also preserves its traction in the proof of the general Kleinian fundamental theorem cited before. The principal viewpoints of this argument are the following:

Let an arbitrary closed Riemann surface of genus p, signed with n points e, be proposed. On this is to be drawn any crosscut system of the kind discussed on p. 47, as it arises by the composition of ν boundary-circle polygons. The total system is therefore to fall into ν separate crosscut systems, each of canonical character (cf. figure 40 in I on p.183). If the signature of the ith of these systems is given by (1), then we have:

$$p = p_1 + p_2 + \cdots + p_\nu, \quad n = n_1 + n_2 + \cdots + n_\nu.$$

The surface thus cut up, as was already determined on p. 47, is still ν-fold connected and has ν boundary curves built up form crosscut banks.

This surface Koebe next maps onto a schlicht ν-fold connected domain with analytic boundary curve coordination. The coordination thereby refers itself to the different portions of one and the same of the ν closed boundaries, the sequence of the coordinated sides is the same, as it present at the boundary of t canonical limit-circle polygons.

Now the iteration procedure begins. Only the first boundary, that of the signature $(p_1, n_1; l_1^{(1)}, \ldots)$, will be retained, while the $(\nu-1)$ remaining gaps are filled out schlichtly. The enlarged domain can be mapped onto a Riemann surface of genus p_1 and the signature $(p_1, n_1; l_1^{(1)}, \ldots)$, then this, on the basis of the boundary-circle theorem proved mapped, backwards onto a boundary-circle polygon of the same signature. After effectuation of the mapping, we again put in the images of the $(\nu-1)$ remaining gaps and arrive back at a domain provided with ν boundaries, which has "*boundary-circle character*" with resect to its first boundary, at least. Therefore, smooth arrangement of the domains is attained with respect to this boundary circle.

Before carrying the process further, Koebe reflects the net obtained so far into the interior of the boundary circle, and applies the continuation simultaneously to the outer and inner net. This continuation consists in the transformation to boundary-circle character now being effected with subsequent reflection to a second, a third etc. boundary, and that, after disposal of the νth boundary, again beginning withe the first.

[552] Here too, on the basis of the consequences of his distortion theorem, Koebe succeeds in the showing the convergence of the processes, as well as the linearity and unicity of the limit function, with which, in fact, is proved the general theorem of Klein's mentioned.

For all the importance, which accrues to the investigations hereby carried to an end, we are nevertheless not able to omit referring here once more to the special position of the boundary-circle theorem (cf. p. 493). The cut system of the Riemann surface,which underlies the reentrant cut theorem, is choosable in infinitely many essentially different ways; and the same thing obviously holds for the cut system, falling into ν separate systems, of the general theorem considered last.

Corresponding to this, there are, in these cases, always infinitely many essentially different polymorphic functions $\zeta(z)$.

On the other hand, the polymorphic function of the boundary-circle theorem, which exists one each signed surface, was invariant with respect to a transformation of the canonical cut system. *The function $\zeta(z)$ of the boundary-circle theorem theorem has, in this regard, a unique position, and it appears as a uniquely determined attribute of the uncut surface.*

The polymorphic function $\zeta(z)$ of the boundary-circle theorem then also accomplished the most with regard to its "*uniformizing power*". If we first draw a canonical cut system of the boundary-circle kind on a surface outfitted with n points e_i and associated whole numbers l_i; and then a cut system of the kind of the general theorem, falling into ν separate systems, then the polymorphic function belonging to the last system will be single-valued in the function $\zeta(z)$ of the boundary-circle theorem (as it belongings to the cut system chosen first); but by no means is this function of the boundary-circle theorem also conversely single-valued in the polymorphic function of the general theorem.

In this state of affairs one nay designate the discovery of the polymorphic function $\zeta(z)$, of the boundary-circle case, on a signed surface, by Klein and Poincaré, as well as the definitive existence proof of this function by Koebe, Picard and Poincarê as a high point, which the Riemann function theory has reached in this direction. The development would be able to even be designated as a complete one, if, unfortunately, there did not yet remain, to this time, a desideratum (cf. p. 279) in this regard, namely, that we still possess no actual usable analytic representation law of the remaining functions of the surface in ζ.

A

An addition to the transformation theory of automorphic functions

In the theory of the elliptic modular functions, the treatment of the problem of the transformation of nth degree takes up quite a lot of space. The developments in "M" II are devoted essentially to this transformation theory and its manifold applications of an algebraic and arithmetic nature. The subject has, besides the material, also a historical interest, insofar as the thing to do there was to subsume part of the results of the transformation theory of elliptic functions, known for a very long time, under the theory of the modular functions of "M" I, built up on a group-theoretic-geometric basis, and to make them understandable in a new fashion. [553]

Such a tradition, now, to be sure, does not adhere to the transformation theory of other single-valued automorphic functions. In the year 1887, Poincaré has first referred, in his work "*Les fonctions fuchsiennes et l'arithmétique*"[1], to the possibility of a transformation theory for such automorphic functions, whose groups were derived, in the work named (cf. also I on p. 501), from the theory of the indefinite integral ternary quadratic forms. Poincaré did not make an attempt to pursue this idea further; also, such an attempt, apart from a pair of slight starts[2], was not undertaken until recently from another aspect. All the same, developments lie in this direction, which, as concerns their character, subsume themselves most beautifully under the general character of the present book. We will do this here, in closing, as least in one case, by treating the *transformation of third degree of the triangle function belonging to the signature* $(0,3;2,4,5)$. We become acquainted with a development here, which will appear strikingly related to the treatment of the transformation of the 7th degree of the elliptic functions by Klein[3], even if the theory to be developed here will, in a certain respect, [554]
fashion itself in a fashion even richer in relations. We first present a few general remarks concerning the transformation of higher degree of automorphic functions.

---

[1] Journ. de. Math., 4th Ser., Vol. 3.

[2] See, concerning this, in particular, Kap. III "*Congruenzgruppen und Transformationstheorie*" of the paper of the author "*Weitere Untersuchungen über automorphen Gruppen et.*", Math. Ann. Vol. 39, p. 62 ff. (1891).

[3] Math. Ann. Vol. 14, p. 428 (1878); cf. also "M" I, p. 692.

A.1 General approach to the transformation of single-valued automorphic functions

If $J(\omega)$ is the well-known modular function of first level thus denoted, and if a, b, c, d are four whole numbers of positive determinant:

$$ad - bc = n > 0,$$

then one says (cf. "M" II on p. 44), the function:

$$J'(\omega) = J\Big(\frac{a\omega + b}{c\omega + d}\Big)$$

arises by transformation of the nth degree from $J(\omega)$. If $n = 1$, then $J'(\omega) = J(\omega)$. If $n > 1$, then an algebraic equation subsists between $J'(\omega)$ and $J(\omega)$:

$$f(J', J) = 0,$$

which is designated as a "modular equation of first level for transformation of nth degree", and is the subject of extensive algebraic and arithmetic developments (cf. "M" II on p. 54 ff.).

By attempting to generalize this approach, we understand by Γ any of our groups of genus $p = 0$ and by $J(\zeta)$ any suitably chosen principal functions of Γ. Let the substitutions of Γ, as usual, be called $V_0 = 1, V_1, V_2, V_3, \ldots$, and, to the extent that we need their developed form, we write them in the form:

$$\zeta_i = \frac{\alpha_i\zeta + \beta_i}{\gamma_i\zeta + \delta_i}, \quad \alpha_i\delta_i - \beta_i\gamma_i = 1, \tag{1}$$

i.e., therefore, "*unimodularly*". Let the fundamental domain of the group Γ, which is mapped single-valuedly and conformally onto the J-plane, be called P_0, and go over by V_i into the domain P_i. Let the entire net of the domains $P_0, P_1, P_2, \ldots$ be called N, as usual.

Now understand by T a linear substitution of ζ not contained in Γ:

$$\zeta' = T(\zeta) = \frac{a\zeta + b}{c\zeta + d}, \tag{2}$$

with non-vanishing determinant $(ad - bc)$, and ask, *whether, in particular, there can be*
[555] *such substitutions T, for which the transformed principal functions:*

$$J'(\zeta) = J(T(\zeta)) = J\Big(\frac{a\zeta + b}{c\zeta + d}\Big) \tag{3}$$

is bound to the original one $J(\zeta)$ by an algebraic relation:

$$f(J', J) = 0. \tag{4}$$

The transformed function $J'(\zeta)$ is the principal function of the group $\Gamma' = T^{-1}\Gamma T$ of all substitutions:

$$V_i' = T^{-1} V_i T, \tag{5}$$

which proceed from V_i by transformation by means of T. The polygon $P_0' = T^{-1}(P_0)$ is to be chosen as the fundamental domain of Γ' in the ζ-plane; this goes over by the substitution V_i' into $P_i' = V_i'(P_0')$.

Since, in consequence of (4), the two single-valued functions $J(\zeta)$ and $J'(\zeta)$ are to be connected algebraically, then they have equal domains of definition: *The total set of the domains $P_0', P_1', P_2', \ldots$ coincides precisely with the net N of the domains $P_0, P_1, P_2, \ldots$, so that the domain N is carried over into itself by the transformation T.*

We may suppose equation (4) irreducible, and assume, that it has the degree μ in J'; to it then belongs a certain μ-sheeted connected Riemann surface F_μ over the J-plane. We will now determine, how the substitutions of the group Γ work on the single-valued function $J'(\zeta)$, and for this purpose, go from a place ζ of the initial domain P_0, where there are the function values $J(\zeta)$ and $J'(\zeta)$, in the interior of the net N to the equivalent place $V_i(\zeta)$. The function J reproduces itself for each such path, and therefore, our paths carry themselves over in the J-plane onto closed paths, but in the surface F_μ, onto paths from the place corresponding to the initial place resp. from the associated value pair J, J', to any of the μ places lying over one another here. Such paths lead from the "branch", just denoted in particular by J', of our algebraic function J' of J defined by (4) to any of the μ branches of this function, to be distinguished, say, by the notations $J', J'', J''', \ldots, J^{(\mu)}$. Conversely, by carrying a suitable path of the Riemann surface F_μ over onto the net N in the manner mentioned, one can also arrive at each branch of our algebraic function: *By exercise of the substitutions $V_0, V_1, V_2, \ldots$ of Γ on $J'(\zeta)$, one obtains all the μ branches $J'(\zeta), J''(\zeta), \ldots, J^{(\mu)}(\zeta)$ of our algebraic function and only these.* In particular, we are able to imagine the first μ substitutions $V_0 = 1, V_1, V_2, \ldots, V_{\mu-1}$ so [556]
chosen, that they give the μ branches:

$$J'(\zeta) = J'(V_0(\zeta)),\ J''(\zeta) = J'(V_1(\zeta)),\ \ldots,\ J^{(\mu)}(\zeta) = J'(V_{\mu-1}(\zeta)), \tag{6}$$

completely.

If now, any substitution V_k of Γ similarly furnishes the branch $J^{(i+1)}$, which we already obtained by V_i, then:

$$J'(V_k(\zeta)) = J'(V_i(\zeta))$$

holds, and therefore:

$$V = V_k V_i^{-1}$$

will be a substitution of Γ, which leaves $J'(\zeta)$ invariant. If, conversely, V is any substitution of Γ, which transforms $J'(\zeta)$ into itself, then:

$$V_k = V V_i$$

will lead to the came branch $J^{(i+1)}$ as V_i. From this, by means of known methods of reasoning, follows: *All the substitutions of Γ, which leave $J'(\zeta)$ unaltered, form a subgroup Γ_μ of finite index μ in Γ, for which the substitutions $V_0 = 1, V_1, V_2, \ldots, V_{\mu-1}$ represent a "system of representatives".*

Now the transformed group $\Gamma' = T^{-1}\Gamma T$ included *all* the substitutions of the net N into itself, which reproduce the associated "principal function" $J'(\zeta)$. Therefore Γ_μ is also contained as a subgroup $\Gamma'_{\mu'}$ in Γ'. Hereby, since the preceding consideration can be attached exactly thus to the group Γ' and the function J' (in place of J), the index μ' is equal to the degree, which our relation (4) attains in J.

Two groups, which have a subgroup in common, which is of finite index in each of the two groups, are called "*commensurable*". *Our two groups* Γ *and* $\Gamma' = T^{-1}\Gamma T$ *must, accordingly, necessarily be commensurable.* With this, however, we have, at the same time, obtained the sufficient condition for the existence of a transformation T of the desired kind, since, conversely, from the commensurability of the groups Γ and Γ', the existence of an algebraic relation (4) follows immediately: *If the group* Γ *of the principal function* $J(\zeta)$ *is carried over by transformation by means of* T *into a group* $\Gamma' = T^{-1}\Gamma T$ *"commensurable" with* Γ, *then there exists, between the original function* $J(\zeta)$ *and the transformed one* $J'(\zeta) = J(T(\zeta))$, *an algebraic relation* (4), *which is to be designated as the associated "transformation equation".*

With T, all infinitely many substitutions $T_k = TV_k$ will also accomplish such a transformation. But as we already saw, we arrive in this way to any μ different functions:

$$J'(\zeta) = J(T(\zeta)),\ J''(\zeta) = J(TV_1(\zeta)),\ \ldots,\ J^{(\mu)}(\zeta) = J(TV_{\mu-1}(\zeta)), \tag{7}$$

so that we have to designate the μ transformations:

$$T,\ T_1 = TV_1,\ T_2 = TV_2,\ \ldots,\ T_{\mu-1} = TV_{\mu-1} \tag{8}$$

[557] as the "*representatives*" for all the transformations TV_k present here.

One can also conceive the μ functions $J'(\zeta), J''(\zeta), \ldots, J^{(\mu)}(\zeta)$ as the μ solutions of the transformation equation (4). With respect to the substitutions of the group Γ, these μ functions undergo a group of permutations, which represents *the Galois group (monodromy group) of the transformation equation* (4). To these μ functions belong μ conjugate subgroups:

$$\Gamma_\mu,\ V_1^{-1}\Gamma_\mu V_1,\ V_2^{-1}\Gamma_\mu V_2,\ \ldots,\ V_{\mu-1}^{-1}\Gamma_\mu V_{\mu-1}. \tag{9}$$

The substitutions common to all these groups form, inside Γ, *A distinguished subgroup of finite index, whose fundamental domain, carried over onto the* J*-plane, furnishes the Riemann surface of the Galois resolvent of the transformation equation.*

All these statements are the transfers of well-known theorems from the transformation theory of the principal moduli in the theory of the modular functions. It is also clear, that in case the proposed group has a genus $p > 0$, a "*transformation correspondence*" with analogous group-theoretic and algebraic properties takes the place of the transformation equation. However, we do not pursue this further.

The principal question will now be, under which conditions a group Γ actually goes over by a transformation T into a group $\Gamma' = T^{-1}\Gamma T$ commensurable with it. Here too, the relations of the modular group briefly touched upon above are typical. This group was a simplest example of a great series of arithmetically defined groups, which were investigated in I on p. 446 from various viewpoints. The definition of an individual such group Γ presupposes a certain real number field in such a manner, that the substitution coefficients α, β, γ, δ of Γ belong as *whole numbers* to this field. Thereby, in order to initially have only the simplest case here before our eyes, we suppose the substitutions as *unimodular*, i.e. we assume that the determinant of the individual substitution is

$\alpha\delta - \beta\gamma = 1$.[4] For the modular group we are concerned with all integral unimodular substitutions of this kind in the field of the rational numbers. If there is a number field of second or higher degree, then a certain *structure of the substitutions* supervenes, which remains maintained upon combination of these. [558]

The substitutions:

$$T(\zeta) = \frac{a\zeta + b}{c\zeta + d},$$

for which a transformation theory of the kind discussed is to be possible, then has to follow the same law of formation as the substitutions V of Γ with the sole exception, *that the determinant of T:*

$$ad - bc = n$$

is any whole positive numbers of the underlying field. This is the direct transfer of the approach known from the theory of the modular functions. It will also be permissible to speak of a "*transformation of degree n*", in the case that n is not a rational whole number[5].

In order to immediately again write the substitution:

$$V' = T^{-1}VT = \begin{pmatrix} \alpha' & \beta' \\ \gamma' & \delta' \end{pmatrix},$$

arising from $V = \left(\begin{smallmatrix} \alpha & \beta \\ \gamma & \delta \end{smallmatrix}\right)$ by transformation by means of T, *unimodular*, we have to set

$$\begin{cases} n\alpha' = ad\alpha - bc\delta + cd\beta - ab\gamma, \\ n\beta' = bd(\alpha - \delta) + d^2\beta - b^2\gamma, \\ n\gamma' = ac(\delta - \alpha) - c^2\beta + \alpha^2\gamma, \\ n\delta' = ad\delta - bc\alpha - cd\beta + ab\gamma. \end{cases} \tag{10}$$

Also, secondly, V' again has *the structure* of the substitutions V of Γ, since we presupposed this for the substitution T. Therefor, thirdly, as long as the transformed substitution V' is to again belong to the subgroup $\Gamma'_{\mu'}$ and therefore, to the group Γ, only the *integrality* still remains to be required. Bur in consequence of (10), the requirement [559]
clothes itself in *congruences with respect to the modulus n*, which the four coefficients $\alpha, \beta, \gamma, \delta$ must fulfill, and which, e.g., are always fulfilled, if the conditions:

$$\alpha \equiv \delta,\ \beta \equiv 0,\ \gamma \equiv 0 \pmod{n} \tag{11}$$

---

[4] In the case of the group of the signature $(0,3;2,4,5)$, obtaining detailed treatment below, the relations are arithmetically somewhat more complicated, insofar as the requirement of only unimodular substitutions cannot be maintained.

[5] If we apply this approach to the groups treated in I on p. 533, whose underlying number fields are quadratic or arise by combination of quadratic fields, then we arrive at the case of the transformation first mentioned by Poincaré, named in the introduction (p. 551). Tieing into the associated indefinite ternary forms, from which Poincaré derived the groups Γ in question, the approach to the transformation already lets itself be formulated in the field of the rational numbers: All the "*unimodular*" integral ternary substitutions of the form into itself furnish the group Γ; a transformation T is furnished by just such a substitution of "*arbitrary non-vanishing determinant*".

are met.

As one sees, the relations fashion themselves precisely as for the modular group. Through the congruences (11), a distinguished subgroup of finite index inside Γ is defined, which we will call *the principal congruence group of level n*; this furnishes us, by means of its fundamental domain, the *Galois resolvent of the transformation equation for the degree n*. The most comprehensive subgroup Γ_μ belonging commonly to Γ and $\Gamma' = T^{-1}\Gamma T$ is a *congruence group of level n*, containing that principal congruence group in itself, whose fundamental domain gives the function-theoretic foundation for setting up the transformation equation itself. If the principal congruence group inside Γ is a subgroup Γ_{M} of index M, then the total group Γ will reduce itself, with respect to the modulus *n, to a finite group* G_{M} *of order* M, which gives the *Galois group of the transformation equation* in new representation form.

All these statements correspond to well-known theorems from the transformation theory of the modular functions. But since we will carry our approaches through for the group of the signature $(0,3;2,4,5)$, and indeed, in the lowest case thereby coming into consideration of the degree $n = 3$, we shall beforehand have to remember, somewhat more penetratingly, the arithmetic structure of the group Γ in question.

A.2 The arithmetic character of the group of the signature $(0,3;2,4,5)$

The group Γ of the signature $(0,3;2,4,5)$ is one of the arithmetically simplest groups actually incommensurable with respect to the modular group. The arithmetic theory of this is completely contained in the developments of I on p. 586 ff.. But since these developments are worded very generally, then it is necessary to specialize them here for our group under discussion[6].

[560] By j we understand from now on the real whole algebraic number:

$$j = \frac{-1+\sqrt{5}}{2} \tag{1}$$

satisfying the quadratic equation:

$$j^2 + j - 1 = 0. \tag{2}$$

The associated real quadratic number field can be represented by means of the bases $[1, j]$; in particular, we arrive precisely at all whole numbers of this field, if we form the sums $(a + bj)$, be means of all pairs of rational whole numbers a, b.

This number field underlies the representation of our group Γ of the signature $(0,3;2,4,5)$. If, namely, in the positive halfplane of the variable $\zeta = \xi + i\eta$, we draw the three semicircles with the equations:

---

[6] The following statements deviate a bit (in consequence of another choice of the initial triangle) from I on p. 586 ff., but attach themselves immediately to the works of the author "*Entwicklungen zur Transformation fünfter und siebenter Ordnung einiger specieller automorpher Functionen*", Act. math. Vol. 17, p. 345 (1893) and "*Über eine einfache Gruppe von 360 Operationen*", Gött. Nachr. Von 1896, Heft 3.

$$\begin{cases} \xi^2+\eta^2+2\xi-1=0, \\ (1+j-\sqrt{j})(\xi^2+\eta^2)-(1+j+\sqrt{j})=0, \\ \xi=0, \end{cases} \tag{3}$$

then these just include a circular-arc triangle with the desired angles $\frac{\pi}{2}$, $\frac{\pi}{4}$, $\frac{\pi}{5}$. The reflections belonging to these circles as symmetry circles:

$$\begin{cases} (\overline{V}_1) \qquad \zeta'=\dfrac{-\overline{\zeta}+1}{\overline{\zeta}+1}, \\ (\overline{V}_2) \qquad \zeta'=\dfrac{1+j+\sqrt{j}}{(1+j-\sqrt{j})\overline{\zeta}}, \\ (\overline{V}_3) \qquad \zeta'=-\overline{\zeta}, \end{cases} \tag{4}$$

where by $\overline{\zeta}$ is the complex value conjugate to ζ. The generators, emerging from these, of the group Γ of the signature $(0,3;2,4,5)$ itself are:

$$\begin{cases} (V_1=\overline{V}_2\overline{V}_3) \qquad \zeta'=\dfrac{1+j+\sqrt{j}}{(-1-j+\sqrt{j})\zeta}, \\ (V_2=\overline{V}_3\overline{V}_1) \qquad \zeta'=\dfrac{\zeta-1}{\zeta+1}, \\ (V_3=\overline{V}_1\overline{V}_2) \qquad \zeta'=\dfrac{(1+j-\sqrt{j})\zeta+(-1-j-\sqrt{j})}{(1+j-\sqrt{j})\zeta+(1+j+\sqrt{j})}. \end{cases} \tag{5}$$

As it must be, these substitutions satisfy the conditions:

$$V_1^2=1,\ V_2^4=1,\ V_3^5=1,\ V_1V_2V_3=1.$$

Following the substitutions type emerging in (5), we form, by means of any whole numbers A, B, C, D of our quadratic field, generally, the substitutions: [561]

$$\zeta'=\frac{(A+B\sqrt{j})\zeta+(C+D\sqrt{j})}{(-C+D\sqrt{j})\zeta+(A-B\sqrt{j})}, \tag{6}$$

and, for abbreviation, will denote the individual such substitution V, by giving its four whole numbers A, B, C, D, by the symbol

$$V=(A,B,C,D). \tag{7}$$

The type of these substitutions V reproduces itself upon combination of two of them. But since all these substitutions (5) subsume themselves under this type, then all substitutions of our group Γ will always possess the type under discussion.

However, there is one more deviation here from the general statements of the previous paragraph. According to those statements, we would unite all *unimodular* substitutions (6) into a group, and apply to this the transformation T. But with this, we do not yet meet our desired group of the signature $(0,3;2,4,5)$; for among the three generators (5), the first two have the determinant 2 and the third the determinant 4; and, in case the substitutions are written unimodularly, the integrality of the coefficients would be lost[7].

How the indeterminacy emerging here is to be overcome, is answered by the general developments of I on p. 589 ff.. We imagine formed *all dimodular and quadrimodular substitutions* (6), i.e., all substitutions of the determinants 2 and 4, whereby, among the quadrimodular ones, all the unimodular substitutions multiplied by 2 will also find themselves.

If we then, first of all, consider the quadrimodular substitutions, for which:

$$A^2 - jB^2 + C^2 - jD^2 = 4$$

therefore holds, then there yields itself from this:

$$A^2 + C^2 \equiv j(B^2 + D^2), \quad (\text{mod } 4). \tag{8}$$

Now since there are four incongruent numbers modulo 2 and numbers congruent modulo 4 furnish squares congruent modulo 2, then there are, in all, in our quadratic field modulo 4, only the four quadratic residues:

$$0,\ 1,\ 1+3j,\ 2+j, \tag{9}$$

as one determines immediately by squaring the numbers incongruent modulo 2. Accordingly, $(B^2 + D^2)$ is congruent modulo 4 either with one of the four numbers:

$$0,\ 2,\ 2+2j,\ 2j \tag{10}$$

[562] or with one of the six numbers:

$$1,\ 1+3j,\ 2+j,\ 2+3j,\ 3+j,\ 3. \tag{11}$$

To the six residues (11) correspond the congruences:

$$j(B^2 + D^2) \equiv j,\ 3+2j,\ 1+j,\ 3+3j,\ 1+2j,\ 3j, \quad (\text{mod } 4), \tag{12}$$

where here on the right stand the six residues modulo 4 not yet represented in (10) and (11). Since these, as we saw, are not able to be congruent with the sum of two square $(A^2 + C^2)$ with respect to the modulus 4, then the congruence (8) requires:

$$A^2 + C^2 \equiv j(B^2 + D^2) \equiv 0,\ 2,\ 2+2j,\ 2j, \quad (\text{mod } 4).$$

But from this yields itself, that, for the quadrimodular substitutions (6):

$$A \equiv C,\ B \equiv D, \quad (\text{mod } 2) \tag{13}$$

necessarily holds.

If we now combine the two substitutions:

---

[7] Only the substitution V_1 would remain integral, since one easily recognizes in $\frac{1+j+\sqrt{j}}{\sqrt{2}}$ a whole algebraic number.

$$V = (A, B, C, D), \; V' = (A', B', C', D')$$

and write:

$$VV' = V'' = (A'', B'', C'', D''),$$

then (written in detail) the four equations hold:

$$\begin{cases} A'' = AA' + jBB' - CC' + jDD', \\ B'' = AB' + BA' + CD' - DC', \\ C'' = AC' + jBD' + CA' - jDB', \\ D'' = AD' + BC' - CB' + DA'. \end{cases} \tag{14}$$

If now, V and V' are quadrimodular, therefore the congruences (13) hold for these two substitutions, then one immediately recognizes, in A'', B'', C'', D'', numbers divisible by 2. After removal of the common factor 2 from the four coefficients of V'', there therefore remains again a quadrimodular substitution of the type (6) with integral coefficients: *All the quadrimodular substitutions* (6) *with whole numbers A, B, C, D of our quadratic field, form a group by themselves.*

One disposes of the dimodular substitutions in the simplest manner on the basis of the fact, that the substitution V_2, given under (5), which is dimodular, transforms the group of the quadrimodular substitutions into itself. In fact, we have, for an arbitrary substitution $V = (A, B, C, D)$:

$$V' = V_2^{-1} V V_2 = (A, D, C, -B),$$

so that V' is again quadrimodular, if V was. The group of the quadrimodular substi- [563]
tutions is therefore capable of extension by addition of V_2 to a more comprehensive group, in which the former is distinguished subgroup of index 2. The substitutions $V' = VV_2$ supervene, where V has to run through all quadrimodular substitutions. But we have:

$$V' = VV_2 = (A + C, B + D, -A + C, -B + D). \tag{15}$$

Now since the congruences (13) hold for V, we are able to remove the common factor 2 from the coefficients of the substitutions V' obtained in (15), and consequently again recognize one of our *dimodular* substitutions. But we also thus arrive at all dimodular substitutions; for if V' is any one of them, then $V'V_2^{-1} = V$ is quadrimodular and therefore $V' = VV_2$ is representable in the form (15): *The group of all quadrimodular substitutions extended by V_2 is identical with the group of all dimodular and quadrimodular substitutions, whose existence we have thus proved.*

For the position of the initial triangle (3) chosen above, the arithmetically defined group always contains the group of the signature (0,3;2,4,5), and, on the other hand, is properly discontinuous (cf. I on p. 597 ff.). From this, then, with consideration of the shape of the circular-arc triangle $\frac{\pi}{2}, \frac{\pi}{4}, \frac{\pi}{5}$, follows very easily on the basis of the argument developed in I on p. 617 ff., *that the group of the signature* (0,3;2,4,5), *for the choice of the initial triangle made in* (3), *is straight away identical with the group of all dimodular and quadrimodular substitutions* (6).

A.3 Introduction of the transformation of third degree

Since the group Γ contains the dimodular substitutions (6) on p. 561 themselves, and moreover, the number 3 is a prime in the underlying quadratic number field, 3 comes into consideration here as the lowest transformation degree. An individual transformation of the third degree, to which we will attach, is:

$$\zeta' = T(\zeta) = \frac{(1+j)\zeta - j}{j\zeta + (1+j)}; \tag{1}$$

in fact, this has the type (6) on p. 561 of the substitutions of our group Γ, except that the determinant of the substitution (1) is equal to 3.

We begin by looking for the congruence group Γ_μ of third level contained com-
[564] monly in Γ and the transformed group $\Gamma' = T^{-1}\Gamma T$. If $V = (A, B, C, D)$ is a substitution of Γ and $V' = T^{-1}VT = (A', B', C', D')$ the corresponding substitution of Γ', then as long as one again assigns V' the determinant of V, one find easily:

$$\begin{cases} A' = A, & B' = \frac{1}{3}(B(1+2j)+2D), \\ C' = C, & D' = \frac{1}{3}(-2B + D(1+2j)). \end{cases} \tag{2}$$

One sees immediately, that V' is always and only then again integral and therefore contained in Γ, if V fulfills the congruence:

$$D \equiv B(1-j), \quad (\text{mod } 3). \tag{3}$$

The solution of the equations (2) for A, B, C, D yields:

$$\begin{cases} A = A', & B = \dfrac{1}{3}(B'(1+2j) - 2D'), \\ C = C', & D = \dfrac{1}{3}(2B' + D'(1+2j)). \end{cases} \tag{4}$$

The transformed substitutions V' again contained in Γ thus fulfill the congruence:

$$D' = B'(-1+j), \quad (\text{mod } 3); \tag{5}$$

and one also sees, conversely, that *each* substitution of Γ, which fulfills the congruence (5), is obtained from another substitution $V = (A, B, C, D)$, which satisfies the congruence (3), by transformation by means of T: *The common subgroup Γ_μ of the group Γ and the transformed group $\Gamma' = T^{-1}\Gamma T$ is the congruence group of third level defined by the condition:*

$$D \equiv B(-1+j), \quad (\text{mod } 3). \tag{6}$$

That the congruence (6) defines a subgroup, one also easily sees directly. Namely, if $V_1 = (A_1, B_1, C_1, D_1)$ and $V_2 = (A_2, B_2, C_2, D_2)$ are only two substitutions of Γ, and if one obtains, by their combination, the substitution $V_3 = V_1V_2 = (A_3, B_3, C_3, D_3)$ of Γ, then according to (14) on p. 562, one has:

$$A_1B_2 + B_1A_2 + C_1D_2 - D_1C_2 = \sigma B_3,$$
$$A_1D_2 + B_1C_2 - C_1B_2 + D_1A_2 = \sigma D_3,$$

where $\sigma = 1$ or 2, according as both substitutions V_1, V_2 were dimodular or not. If the congruence (6) now holds for V_1 and V_2, then there follows, as valid modulo 3:

$$A_1B_2 + B_1A_2 + (-1+j)C_1B_2 + (1-j)B_1C_2 \equiv \sigma B_3,$$
$$(-1+j)A_1B_2 + B_1C_2 - C_1B_2 + (-1+j)B_1A_2 \equiv \sigma D_3.$$

With consideration of $(-1+j)^2 \equiv -1$, (mod 3), the validity of the congruence (6) also follows for V_3. [565]

All the substitutions V of Γ, which, for the transformation by T, furnished the subgroup Γ_μ, and which, therefore, satisfy the congruence (3), also self-evidently form a subgroup to be considered as defined by (3). It receives the same index as Γ_μ, and let it be called Γ'_μ on this account; for there holds:

$$\Gamma'_\mu = T\Gamma_\mu T^{-1}, \tag{7}$$

so that the discontinuity domain of Γ'_μ proceeds from that of Γ_μ by the transformation T and therefore contains the same number of circular-arc triangles as that of Γ_μ.

For obtaining the discontinuity domain of Γ_μ, the circumstance is important, that we are also able to produce Γ'_μ by yet another transformation of Γ_μ. Namely, with the substitution (A,B,C,D), Γ_μ always also contains the substitution $(A,B,-C,D)$. If we go from this substitution to:

$$(A',B',C',D') = (A,B,-C,-D), \tag{8}$$

by also changing the fourth whole number in sign, then the substitution (8) thus obtained will also be contained in Γ, and indeed, it satisfies the congruence (3), and is therefore contained in Γ'_μ. Also, we obviously obtain, in this fashion, from Γ_μ, the entire Γ'_μ. But now, as one easily calculates sign change of C and D is achieved by transformation by means of the reflection $\overline{V}_3$ in the imaginary ζ-axis:

$$\overline{V}_3^{-1}(A,B,C,D)\overline{V}_3 = (A,B,-C,-D).$$

One can thus also produce Γ'_μ from Γ_μ by transformation with $\overline{V}_3$:

$$\Gamma'_\mu = \overline{V}_3^{-1}\Gamma_\mu\overline{V}_3.$$

In conjugation with (7), from this follows further, immediately:

$$(\overline{V}_3T)\Gamma_\mu(\overline{V}_3T)^{-1} = \Gamma_\mu,$$

with which the theorem is obtained: *The congruence group* Γ_μ *of third level is carried over into itself by transformation by means of the substitutions of the second kind:*

$$\overline{V}_3T(\zeta) = \frac{-(1+j)\overline{\zeta}+j}{j\overline{\zeta}+(1+j)}; \tag{9}$$

this substitution (9) *is a reflection, and its symmetry circle is given by:*

$$(\xi+2+j)^2+\eta^2=6+3j. \tag{10}$$

[566] Let the addition of this reflection extend Γ_μ to the "group of the second kind" $\overline{\Gamma}_\mu$.

Indeed, notice hereby, *that the reflection obtained is by no means contained in the group* Γ *extended by* $\overline{V}_3$. The known symmetry is consequently a property of the subgroup Γ_μ and not of the total group. The symmetry circle (10) will therefore run diagonally through the triangles of our net belonging to $\overline{\Gamma}$, and indeed, in a manner, which we soon have to determine more closely.

Beforehand, let us draw attention to the fact, *that our group* Γ_μ *is transformed into itself by a second, essentially new substitution, namely by:*

$$T'(\zeta)=\frac{(3+2j)\sqrt{j}\zeta+(1+j)(3+2\sqrt{j})}{(1+j)(-3+2\sqrt{j})\zeta-(3+2j)\sqrt{j}}. \tag{11}$$

Namely, if one writes:

$$T'^{-1}(A,B,C,D)T'=(A',B',C',D'),$$

then one easily determines the relations:

$$\begin{cases} A'=A,\\ B'=-\frac{1}{3}(11+5j)B+(5+3j)C-\frac{2}{3}(3+2j)D,\\ C'=-(3+2j)B+(5+3j)C-2(1+j)D,\\ D'=-\frac{2}{3}(3+2j)B+(4+2j)C-\frac{1}{3}(7+4j)D. \end{cases} \tag{12}$$

Since B and D satisfy the congruence (6), one again recognizes, in B' and D', whole numbers. On the other hand, from (12) follows:

$$D'+(1-j)B'=-(4+j)B+3(2+j)C-(3+2j)D,$$

and therefore:

$$D'+(1-j)B'\equiv j(D+(1-j)B)\equiv 0,\quad (\text{mod } 3).$$

Since, finally, the transformed substitution has the same determinant as the original, our assertion is proved.

T' is an elliptic substitution of period 2. The addition of this substitution to $\overline{\Gamma}_\mu$ will extend this group to a group $\overline{\Gamma}_{\frac{1}{2}\mu}$, in which $\overline{\Gamma}_\mu$ is a distinguished subgroup of index 2. Naturally, $\overline{\Gamma}_{\frac{1}{2}\mu}$ is no more contained in $\overline{\Gamma}$, as $\overline{\Gamma}_\mu$. *However, there is another subgroup* $\overline{\Gamma}'_\mu$ *of the second kind contained in* $\overline{\Gamma}_{\frac{1}{2}\mu}$, *which is contained in the extended* $\overline{\Gamma}$. Namely, we find, in:

$$T'T^{-1}=\begin{pmatrix} -1+(1+j)\sqrt{j}, & (2+j)(1+\sqrt{j})\\ (2+j)(-1+\sqrt{j}), & -1-(1+j)\sqrt{j}\end{pmatrix}, \tag{13}$$

[567] a substitution contained in Γ. From this emerges, first, *that* $J(T'(\zeta))=J(T(\zeta))$ *holds, and that* T *and* T' *are therefore two transformations, of which each can be represented by the other.* On the other hand, we obtain, in:

$$T'T^{-1}\overline{V}_3(\zeta) = T'(\overline{V}_3T)^{-1}(\zeta) = \frac{(1-(1+j)\sqrt{j})\overline{\zeta} + (2+j)(1+\sqrt{j})}{(2+j)(1-\sqrt{j})\overline{\zeta} - 1 - (1+j)\sqrt{j}}, \tag{14}$$

a substitution contained in $\overline{\Gamma}$, *which transforms* Γ_μ *into itself.* The addition of this substitution to Γ_μ furnishes the group denoted above by $\overline{\Gamma}'_\mu$.

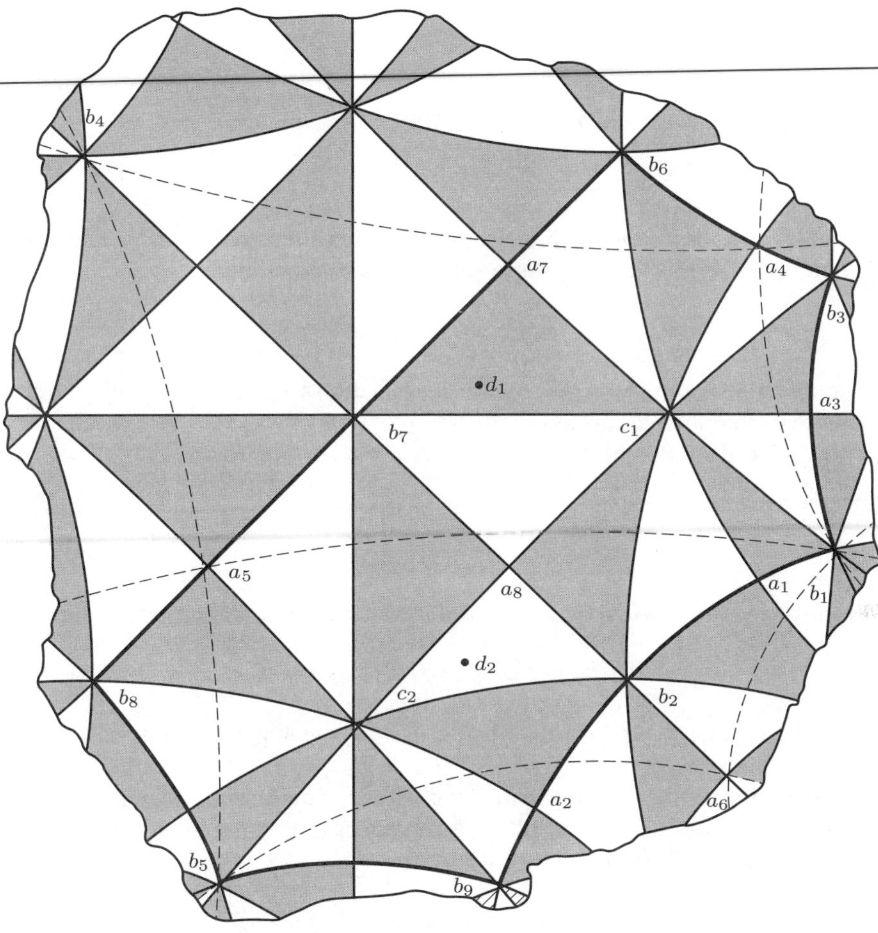

Fig. 106.

The statements so far are sufficient, in order to construct the discontinuity domain of the group Γ_μ. In the appended figure 106, those vertices of the triangle division of $\overline{\Gamma}$, which are surrounded by 4 triangles, are generally denoted by a; correspondingly, points b and c are such vertices, which are surrounded by 8 resp. 10 triangles. Conceive the triangle $a_3b_1c_1$ as initial space of $\overline{\Gamma}$, so that b_1 lies at $\zeta = i$, the circle $b_1a_1b_2$ represents the unit circle, upon which a_1 and b_2 are situated as the places: [568]

$$\zeta = -j(\sqrt{j} + i\sqrt{2}),\ \zeta = -\sqrt{j} + ij,$$

ans finally, the circle $b_1 a_3 b_3$ furnishes the imaginary ζ-axis, on which the points a_3 and b_3 resp. carry the values:

$$\zeta = \frac{i}{\sqrt{2}}(1+j+\sqrt{j}),\ \zeta = i(1+j)(1+\sqrt{j})$$

(the first of them is the fixpoint of V_1 [cf. (5) on p. 560]). To the point denoted in the figure by a_4 belongs the substitution of period 2:

$$V_4 = V_3^{-1} V_1 V_3 = (0, 1, 2+j, 2+j), \tag{15}$$

so that the point a_4 is the image of the fixpoint of V_4:

$$\zeta = (1+\sqrt{j})(-\sqrt{j}+i\sqrt{2}). \tag{16}$$

The substitutions V_2 and V_4 satisfy the congruence (6) on p. 564 and are accordingly contained in Γ_μ. The fixpoints of these substitutions, situated at:

$$\xi = 0,\ \eta = 1 \text{ resp. } \xi = -j-\sqrt{j},\ \eta = (1+\sqrt{j})\sqrt{2},$$

are points of the symmetry circle (10) of the reflection $\overline{V}_3 T$. This symmetry circle (10) is therefore the circle $b_1 a_4$, dotted in the figure[8].

With $\overline{V}_3 T$ and V_2, the operation:

$$\overline{V}_3 T V_2(\zeta) = \frac{-\overline{\zeta}+(1+2j)}{(1+2j)\overline{\zeta}+1}$$

is also contained in $\overline{\Gamma}_\mu$, which is again a reflection, namely that of the symmetry circle:

$$(1+2j)(\xi^2+\eta^2)+2\xi-(1+2j) = 0. \tag{17}$$

This cuts the symmetry circle (10) at the point b_1 at the angle $\frac{\pi}{4}$ and is the circle $b_1 a_5$, dotted in the figure.

Now we bring in the elliptic substitution T' of period 2, whose fixpoint is denoted
[569] by d_1. On the basis of the relation (12), one proves easily, that the substitution V_4, transformed by T', is the following:

$$V_5 = T'^{-1} V_4 T' = (0, 4+3j, 4+2j, 2+j).$$

If we combine this with the reflection $\overline{V}_3 T V_2$, then there again arises a reflection in:

$$V_5 \overline{V}_3 T V_2(\zeta) = \frac{(4+3j)\overline{\zeta}+2+j+3(2+j)\sqrt{j}}{(2+j-3(2+j)\sqrt{j})\overline{\zeta}-(4+3j)}.$$

Therefore, in fact, the fixpoint a_5 of V_5 lies on the symmetry circle (17), and the latter is crossed at a right angle by the symmetry circle:

---

[8] These two points suffice for the determination of the circle in question, since, indeed, this must run perpendicularly to the orthogonal circle of our triangle net itself.

$$(2+j-3(2+j)\sqrt{j})(\xi^2+\eta^2)-2(4+3j)\xi-(2+j+3(2+j)\sqrt{j})=0 \tag{18}$$

of the reflection just obtained.

With this, the discontinuity domain for the most comprehensive group $\Gamma_{\frac{1}{2}\mu}$ has already closed itself off. We have, as a result: *A discontinuity domain of the group* $\overline{\Gamma}_{\frac{1}{2}\mu}$ *is furnished by the triangle $b_1a_4a_5$; of the three sides of these, two, namely b_1a_4 and b_1a_5, are the symmetry circles of the generating reflections $\overline{V}_3T$ and $\overline{V}_3TV_2$; the third side is halved at d_1, and the two halves are mapped onto one another by the third generator T'.*

The obtention of the discontinuity domains for the remaining subgroups $\overline{\Gamma}_\mu$, $\overline{\Gamma}'_\mu$, Γ_μ contained in $\overline{\Gamma}_{\frac{1}{2}\mu}$ now holds no more difficulty. The triangle $b_1a_4a_5$ is transformed by T' into the triangle $b_4a_5a_4$ of the figure. *By addition of the second to the first triangle, we obtain, in the circular-arc quadrangle $b_1a_4b_4a_5$, a discontinuity domain for the group $\overline{\Gamma}_\mu$, which, with this, is generatable from the four reflections $\overline{V}_3T$, $\overline{V}_3TV_4$, $V_5\overline{V}_3TV_2$, $\overline{V}_3TV_2$.* The substitution T' indeed transforms this quadrangle into itself, but thereby carries the triangle net of the group Γ into that of the transformed group $T'^{-1}\Gamma T'$.

If we add, to the quadrangle $b_1a_4b_4a_5$, its mirror image along b_1a_5, then there arises, *in the double quadrangle $b_1a_4b_4a_5b_5a_6$, finally, a discontinuity domain for the group Γ_μ, in which we consequently recognize a group of the genus $p=0$, which is generatable from the two elliptic substitutions V_2, V_5 and the hyperbolic substitution $\overline{V}_3TV_4\overline{V}_3TV_2$.*

Moreover, it is easy, to so transform the domain obtained by allowed alterations, that it appears put together from a number of triangles not divided up into separate pieces. One will transform the triangles $b_1b_2a_6$ by V_2 etc. By suitable continuation, one obtains *the discontinuity domain of Γ_μ in the shape of the domain $b_1b_3b_6b_8b_5b_9$, built up from two regular rectangular pentangles and outlined heavily in the figure, whereby we give the associatedness of the boundary curves, in immediately understandable fashion, by the table:* [570]

$$(b_1b_2)\text{–}(b_1b_3),\ (a_4b_3)\text{–}(a_4b_6),\ (b_6b_7)\text{–}(b_2b_9),$$
$$(a_5b_7)\text{–}(a_5b_8),\ (b_5b_8)\text{–}(b_5b_9).$$

We note, above everything else, the result: *Our group Γ_μ has index $\mu=10$, so that it is to be denoted as Γ_{10} from now on.*

Finally, we also dispose, without trouble, of the group $\overline{\Gamma}'_{10}$, contained in the group $\overline{\Gamma}$, which we generated above by the addition of the substitution (14) to Γ_{10}. Instead of this substitution (14), we could also utilize the substitution composed of the elliptic substitution T' and the reflection $\overline{V}_3TV_2$ in the circle b_1a_5:

$$T'\overline{V}_3TV_2(\zeta)=\frac{(3+j+\sqrt{j})\overline{\zeta}+1+j+(3+2j)\sqrt{j}}{(1+j-(3+2j)\sqrt{j})\overline{\zeta}-3-j+\sqrt{j}}, \tag{19}$$

whose effect on the discontinuity domain of the Γ_{10} is directly determinable. By exercise of the reflection and subsequent application of T', namely, the quadrangle $b_1a_5b_5a_6$ is flipped over into the quadrangle $b_1\,a_4b_4a_5$ in such a fashion, that the points b_5, a_5 resp. arrive at b_1, a_4. One reads off immediately from the figure, that the regular pentangle $b_7b_8b_5b_9b_2$ hereby comes to lie on the pentangle $b_6b_3b_1b_2b_7$. But from this yields itself immediately: *A discontinuity domain for the extended group* $\overline{\Gamma}_{10}$ *is furnished by the regular pentangle* $b_1 b_3 b_6 b_7 b_2$*; this group is consequently generatable from three substitutions, namely* V_2, V_4 *and the substitution* (19) *of the second kind, which transforms the pentangle side* $b_7 b_2$ *into* $b_6 b_7$.

A.4 Setting up the transformation equation of tenth degree

In order now to actually obtain the transformation equation of tenth degree between $J'(\zeta) = J(T(\zeta))$ and $J(\zeta)$ corresponding to the transformation T of third degree, we avail ourselves, by bringing in the method followed by Klein[9] for the modular equations of the elliptic functions, of a *principal function* $\varphi(\zeta)$ of the group Γ_{10} belonging to the genus $p = 0$.

If, in the determination of this principal functions $\varphi(\zeta)$, one would only first take
[571] the Γ_{10} alone into consideration and not yet the total group Γ, and with this, the filling up of the discontinuity domain of the Γ_{10} by circular-arc triangles, then it would e.g. be natural, to so choose $\varphi(\zeta)$, that this function vanishes at the point d_1, (cf. figure 106 on p. 567) and maps the circular-arc quadrangle $b_1 a_4 b_4 a_5$ onto the unit circle of the φ-plane. To the substitution T', the reflection $\overline{V}_3 T V_2$ and the substitution $T'\overline{V}_3 T V_2$ of the second kind then correspond resp. to the substitutions:

$$\varphi' = -\varphi,\ \varphi' = \frac{1}{\overline{\varphi}},\ \varphi' = -\frac{1}{\overline{\varphi}}, \tag{1}$$

where as usual, $\overline{\varphi}$ is the complex value conjugate to φ. The third of these substitutions (1), which therefore corresponds to the substitution (19) on p. 570 contained in the group $\overline{\Gamma}$, represents the transformation by so-called "*diametral symmetry*". If we imagine the values of φ so positioned on the "φ-sphere", that the real axis, the imaginary axis and the unit circle are three great circles crossing one another orthogonally, then the diametral symmetry simply consists in each point of the sphere going over into the one lying diametrically opposite.

We are now able, without the diametral symmetry losing its representation form (1), to undertake one more arbitrary rotation of the φ-sphere about its center, which, according to I on p. 42, amounts to introducing, instead of φ, any principal function:

$$\varphi' = \frac{(a+ib)\varphi + (c+id)}{(-c+id)\varphi + (a-ib)}$$

with the meaning of the real coefficients given there. We will make use of this, by fixing φ in a manner better adopted to the total group Γ and its triangle net.

We will, first of all, so rotate, that the two diametral points $\varphi = 0$ and $\varphi = \infty$ correspond to the two points of the discontinuity domain denoted in the figure (cf. p. 567) by c_1 and c_2. Thereby, let it at first still remain undecided, whether $\varphi = 0$ takes place at the point c_1 or at c_2, so that hereby is introduced a first ambiguity in the choice of φ. We will then, further, rotate about the diameter of the sphere running from $\varphi = 0$ to $\varphi = \infty$

9 "*Über die Transformation die elliptischen Functionen und dre Auflösung der Gleichungen fünften Grades*" Math. Ann. Vol. 14, p. 141 (1878). Compare also "M" I on p. 636.

so far, that any third value φ, e.g. that taking place at the vertex b_3, becomes purely imaginary. Hereby is introduced a second ambiguity; for the value in question can be as well chosen positive or negative imaginary. *We accordingly have, in all, four more admissible choices of* $\varphi(\zeta)$*; however, these four principal functions are connected most simply, since, in one of them* $\varphi(\zeta)$*, the three others represent themselves in the forms:*

$$\varphi' = -\varphi, \; \varphi' = \frac{1}{\varphi}, \; \varphi' = -\frac{1}{\varphi}. \tag{2}$$

We reserve, for the present, the decision, concerning which of these four functions is to be our principal function $\varphi(\zeta)$. [572]

In order to explore the connection of the principal function $\varphi(\zeta)$ with the principal function $J(\zeta)$ of the total group Γ, we position the φ-plane, and therefore, the discontinuity domain of the Γ_{10}, in known fashion, over the J-plane as a ten-sheeted closed Riemann surface. One immediately reads off the branching of this surface from figure 106. *At* $J = 0$, *two sheets run isolated (points* a_4, a_5*), the remaining sheets hang together in pairs at four branchpoints; at* $J = 1$, *similarly, two sheets run isolated (points* b_1, b_5*), the remaining sheets hang together in fours at each of two branchpoints; at* $J = \infty$, *finally, the sheets hang together in fives at each of two branchpoints, at which our determination regarding the values* $\varphi = 0$ *and* $\varphi = \infty$ *hold.*

For the representation of J as a rational function of tenth degree of φ, we deduce from these statements the schema:

$$\begin{aligned} J : J-1 : 1 = \; & (\varphi^2 + a\varphi + b)(\varphi^4 + c\varphi^3 + d\varphi^2 + e\varphi + f)^2 \\ & : (\varphi^2 + \alpha\varphi + \beta)(\varphi^2 + \gamma\varphi + \delta)^4 : \varepsilon\varphi^5. \end{aligned} \tag{3}$$

Our next task consists in bringing into discovery the values of the eleven coefficients still unknown here.

Now this equation must be invariant with respect to the above mentioned diametral symmetry in such a manner, that upon exercise of the third substitution (1), J goes over into $\overline{J}$; for this substitution corresponds to the operation (19) on p. 570, contained in the extended group $\overline{\Gamma}$. The right side of equation (3) will consequently also go over into itself in all components of the equation upon exercise of the third substitution (1) and subsequent sign change of i. By observing, that β and γ certainly do not vanish, we first of all apply this requirement of invariance to the middle term of the two trinomials standing on the right side of (3). It yields itself, that the two equations:

$$\varphi^2 - \frac{\overline{\alpha}}{\overline{\beta}}\varphi + \frac{1}{\overline{\beta}} = 0, \; \varphi^2 + \alpha\varphi + \beta = 0$$

must coincide, and similarly, the two equations:

$$\varphi^2 - \frac{\overline{\gamma}}{\overline{\delta}}\varphi + \frac{1}{\overline{\delta}} = 0, \; \varphi^2 + \gamma\varphi + \delta = 0.$$

Thus we have: [573]

$$\alpha = -\frac{\overline{\alpha}}{\overline{\beta}}, \; \beta = \frac{1}{\overline{\beta}}, \; \gamma = -\frac{\overline{\gamma}}{\overline{\delta}}, \; \delta = \frac{1}{\overline{\delta}}. \tag{4}$$

The two solutions of $\varphi^2+\gamma\varphi+\delta=0$ are the values of the function $\varphi(\zeta)$ at the points b_3 and b_8 of the discontinuity domain of Γ_{10}. One of these values was to be purely imaginary, and therefore this also holds for the second, since it emerges from the first by the third substitution (1). For a purely imaginary value φ, $\overline{\varphi}=-\varphi$ holds, so that the third equation (1) furnishes $\varphi'=\frac{1}{\varphi}$ for that second root. With this $\delta=\varphi\varphi'=1$ follows, so that we have to write the third and fourth equation (4) more precisely thus:

$$\overline{\gamma}=-\gamma,\ \delta=1. \tag{5}$$

For obtaining further conditions for the unknown coefficients, we utilize (cf. Klein l.c. or "M" I on p. 636) the circumstance, that the functional determinant of the two functions:

$$(\varphi^2+\alpha\varphi+\beta)(\varphi^2+\gamma\varphi+\delta)^4,\ \varphi^5,$$

because of the identity, necessarily subsisting:

$$(\varphi^2+a\varphi+b)(\varphi^4+c\varphi^3+d\varphi^2+e\varphi+f)^2=(\varphi^2+\alpha\varphi+\beta)(\varphi^2+\gamma\varphi+\delta)^4+\varepsilon\varphi^5, \tag{6}$$

must contain the factor $(\varphi^4+c\varphi^3+d\varphi^2+e\varphi+f)$. Carrying through the calculation of this schema furnishes, for the coefficients c, d, e, f, the following expressions in α, β, γ:

$$\begin{cases} c=\frac{4}{5}\alpha+\frac{1}{5}\gamma, \\ d=\frac{3}{5}\beta-\frac{3}{5}, \\ e=-\frac{4}{5}\alpha-\frac{1}{5}\beta\gamma, \\ f=-\beta. \end{cases} \tag{7}$$

In order to obtain further relations, in particular, also such for a and b, we utilize the identity (6) itself and compare the coefficients of φ^9, φ and the absolute term on the left and right. In this way, we find:

$$\begin{cases} a+2c=\alpha+4\gamma, \\ af^2+2bef=\alpha+4\beta\gamma, \\ bf^2=\beta. \end{cases} \tag{8}$$

[574] With utilization of (7), the first and third of these equations yield:

$$a=-\frac{3}{5}\alpha+\frac{18}{5}\gamma,\ b=\frac{1}{\beta}, \tag{9}$$

but the second equation (8) furnishes:

$$\alpha(\beta^2-1)=6\beta\gamma(\beta-1). \tag{10}$$

If now, $\beta=1$, then $\overline{\alpha}$ would be $=-\alpha$, and the equation $\varphi^2+\alpha\varphi+\beta=0$ would also have purely imaginary solutions, which we could denote by $\varphi(b_1)$ and $\varphi(b_5)$, tying in with figure 106. Also, b would be $=1$ and a purely imaginary, so that $\varphi(a_4)$ and $\varphi(a_5)$

would be purely imaginary. Since, at the two points c, the values $\varphi = 0$ and $\varphi = \infty$ occur, then in the φ-plane, the six points $\varphi(a_4)$, $\varphi(a_5)$, $\varphi(b_1)$, $\varphi(b_5) = \varphi(b_4)$, $\varphi(c_1)$, $\varphi(c_2)$ would lie on the imaginary axis, and therefor, for each other choice of the principal functions φ, lie on the periphery of one and the same circle. But if we choose $\varphi(\zeta)$ in the manner mentioned on p. 571 (above), then the points $\varphi(b_1)$, $\varphi(a_4)$, $\varphi(b_4)$, $\varphi(a_5)$ lie on the unit circle. But then $\varphi(c_1)$ does not belong to this periphery. Therefore β is different from 1, so that from equation (10), we deduce the relation:

$$\alpha(\beta+1) = 6\beta\gamma. \tag{11}$$

Further, were $\alpha = 0$, then, since β does not vanish, γ would necessarily have to be = 0, and therefore, in consequence of (9) and (7), a, c and e would vanish. Consequently, in the identity (6), apart from the single term $\varepsilon\varphi^5$, only even powers of φ would come up, so that ε would be = 0. But this is excluded, and therefore α is different from 0. The first equation (4), as well as (11), now furnishes:

$$\beta = -\frac{\alpha}{\overline{\alpha}}, \ \gamma = \frac{1}{6}(\alpha - \overline{\alpha}), \tag{12}$$

where upon, in (9) and (7), one also gets unique expressions for $a, b, \ldots, f$ in α.

For obtaining α itself, however, the conditions set up so far do not yet suffice. We yet being in the fact, that the functional determinant of the two functions:

$$(\varphi^2 + a\varphi + b)(\varphi^4 + c\varphi^3 + d\varphi^2 + e\varphi + f)^2, \ \varphi^5$$

must contain the factor $(\varphi^2 + \gamma\varphi + 1)$ in the cube. However, it is sufficient, if from this schema, we only deduce the one relation:

$$2ac + 3b + d = 15 + 15\gamma^2,$$

which, with the aid of the expressions of a, b, c, d, γ in α and $\overline{\alpha}$ already obtained, [575]
calculates itself around into the form:

$$-\frac{1}{3}\alpha\overline{\alpha} + \frac{1}{75}\overline{\alpha}^2 - \frac{\overline{\alpha}}{\alpha} - \frac{1}{5}\frac{\alpha}{\overline{\alpha}} = \frac{26}{5} + \frac{5}{36}(\alpha - \overline{\alpha})^2. \tag{13}$$

Since the numerical coefficients in this equation are all real, then the interchange of α and $\overline{\alpha}$ is admissible:

$$-\frac{1}{3}\alpha\overline{\alpha} + \frac{1}{75}\alpha^2 - \frac{\alpha}{\overline{\alpha}} - \frac{1}{5}\frac{\overline{\alpha}}{\alpha} = \frac{26}{5} + \frac{5}{36}(\alpha - \overline{\alpha})^2. \tag{14}$$

By subtraction of these two equations from one another follows:

$$\frac{1}{75}(\alpha^2 - \overline{\alpha}^2) = \frac{4}{5}\frac{\alpha^2 - \overline{\alpha}^2}{\alpha\overline{\alpha}}. \tag{15}$$

Now were α real, then from (12) would follow $\beta = -1$, $\gamma = 0$, and therefore, in consequence of (9), b would also be $= -1$ and a real. With this, all six values $\varphi(a_4)$, $\varphi(a_5)$, $\varphi(b_1)$, $\varphi(b_4)$, $\varphi(c_1)$, $\varphi(c_2)$ would be real, which, according to an argument already carried out above, is impossible. Secondly, were α purely imaginary, then, by a corresponding method of reasoning, one would find those six values φ to be purely imaginary, which is likewise impossible. Accordingly, $(\alpha^2 - \overline{\alpha}^2)$ is different from 0, and therefore follows from (15):

$$\alpha\overline{\alpha} = 60. \tag{16}$$

By utilizing this result, the addition of equations (13) and (14), after brief intermediate calculation, furnishes:

$$(\alpha - \overline{\alpha})^2 = -180.$$

With this, the four values yield themselves for α:

$$\alpha = \sqrt{15} + 3i\sqrt{5},\ \sqrt{15} - 3i\sqrt{5},\ -\sqrt{15} + 3i\sqrt{5}, -\sqrt{15} - 3i\sqrt{5}.$$

These correspond to the four choices of the principal function $\varphi(\zeta)$ still possible. We are able to effect a verification of this result, by actually calculating out the four forms of the function:

$$\frac{(\varphi^2 + \alpha\varphi + \beta)(\varphi^2 + \gamma\varphi + \delta)^4}{\varphi^5}:$$

$$(\varphi^2 + (\sqrt{15} \pm 3i\sqrt{5})\varphi + \frac{1 \mp i\sqrt{3}}{2})(\varphi^2 \pm i\sqrt{5}\varphi + 1)^4 \cdot \varphi^{-5},$$

$$(\varphi^2 - (\sqrt{15} \pm 3i\sqrt{5})\varphi + \frac{1 \mp i\sqrt{3}}{2})(\varphi^2 \mp i\sqrt{5}\varphi + 1)^4 \cdot \varphi^{-5}.$$

From one of these expressions, the three others must proceed, up to a constant factor[10], by the substitutions (2). Testing verifies this.

[576] We now determine the principal function $\varphi(\zeta)$ uniquely, by choosing for α the first of the given values, $\sqrt{15}(1 + i\sqrt{3})$. The coefficients $\beta, \gamma, a, b, \ldots, f$, as already remarked, are uniquely determined with this; finally, ε also determines itself uniquely from the identity (6). We obtain as a result: *The principal function $J(\zeta)$, as a rational function of the tenth degree of the principal function $\varphi(\zeta)$ of the subgroup Γ_{10}, is represented by:*

$$\begin{aligned} &J : J - 1 : 1 \\ &= \left(\varphi^2 - \frac{3\sqrt{3}}{\sqrt{5}}(1 - i\sqrt{3})\varphi + \frac{1}{2}(1 + i\sqrt{3})\right)\left(\varphi^4 + \frac{1}{\sqrt{5}}(4\sqrt{3} + 13i)\varphi^3\right. \\ &\quad \left. - \frac{3}{10}(1 + i\sqrt{3})\varphi^2 - \frac{1}{2\sqrt{5}}(9\sqrt{3} + 25i)\varphi - \frac{1}{2}(1 - i\sqrt{3})\right)^2 \\ &\quad : (\varphi^2 + \sqrt{15}(1 + i\sqrt{3})\varphi + \frac{1}{2}(1 - i\sqrt{3}))(\varphi^2 + i\sqrt{5}\varphi + 1)^4 \\ &\quad : \frac{2^7 \cdot 3^4\sqrt{3}}{5^2\sqrt{5}}(1 + i\sqrt{3})\varphi^5. \end{aligned} \tag{17}$$

In consideration of later developments, we must make even more precise statements concerning the distribution of values of the principal function $\varphi(\zeta)$ in the discontinuity domain of the Γ_{10} (cf. figure 106, p. 567). First of all, the two values, to be denoted by $\varphi(a_4)$ and $\varphi(a_5)$, are the solutions of the quadratic equation:

[10] Notice, that the coefficient ε is not yet determined.

$$\varphi^2 - \frac{3\sqrt{3}}{\sqrt{5}}(1 - i\sqrt{3})\varphi + \frac{1}{2}(1 + i\sqrt{3}) = 0,$$

whereby, it only initially still remains undecided, which root belongs to the point a_4 and which to a_5:

$$\varphi(a_4),\ \varphi(a_5) = \frac{1 - i\sqrt{3}}{2} \cdot \frac{3\sqrt{3} \pm 4\sqrt{2}}{\sqrt{5}}.$$

In the same sense, one finds:

$$\varphi(b_1),\ \varphi(b_5) = -\frac{1 + i\sqrt{3}}{2}(\pm 4 + \sqrt{15}),$$

$$\varphi(b_3),\ \varphi(b_7) = -i\frac{\pm 3 + \sqrt{5}}{2}.$$

The values $\varphi(c_1)$ and $\varphi(c_2)$, as we remember, are equal 0 and ∞, whereby, it is again undecided, whether $\varphi(c_1)$ is equal to 0 or equal to ∞. Two of the four values $\varphi(a)$ still lacking are:

$$\varphi(a) = \frac{-1 + i\sqrt{3}}{2} \cdot \frac{3 + \sqrt{5}}{2} \cdot \frac{(-3 \mp \sqrt{-3 + 8\sqrt{6}})\sqrt{3} + i(5\sqrt{5} \pm \sqrt{3}\sqrt{3 + 8\sqrt{6}})}{4\sqrt{5}},$$

and the two others proceed from these values by the third substitution (1) on p. 571 (diametral symmetry). [577]

In order to eliminate the indeterminacies still present here, we will, according to a method exercised often in "M" I, mechanically carry the the discontinuity domain of the Γ_{10}, by bending the boundary curves, assigned to one another, together, over into the φ-plane, in which we imagine marked beforehand the points corresponding to the values $\varphi(a_4)$, $\varphi(a_5)$, ..., for easier orientation. The conformal image of the discontinuity domain on the φ-plane is, to be sure, not very usable, since a few of the points laid out are too near to the origin, while others lie far away. But one can remove this annoyance by going over to a circle-related figure. This is done adjacently in figure 107, in which one finds half of the discontinuity domain carried over, namely, the regular pentangle with the center c_1. The imaginary axis has here still remained a line, while the real φ-axis has become a circle with finite radius. [578]

In the figure, we have indicated a distribution of values of φ, by marking the three points $\varphi = \pm 1$, $\varphi = -i$, which involves the assumption $\varphi(c_1) = 0$. However, the indeterminacy still persists, that we do not know, whether this distribution of values is the correct one, or whether the distribution of values actually taking place first proceeds from the one just considered by one of the three transformations (2) on p. 572. But the transformation $\varphi' = -\varphi$ cannot occur, since the places lying near the origin:

$$\frac{-1 + i\sqrt{3}}{2} \cdot \frac{4\sqrt{2} - 3\sqrt{3}}{\sqrt{5}},\quad \frac{1 + i\sqrt{3}}{2}(4 - \sqrt{15}),\quad i\frac{3 - \sqrt{5}}{2}$$

have positive imaginary components. Were $\varphi' = -\varphi^{-1}$ the transformation leading to the correct distribution of values, then, as figure 107 informs us, the numbers $\varphi(b_1)$ and $\varphi(a_4)$ would have absolute values > 1 and positive imaginary components. But the only numbers admissible because of the requirement $|\varphi(b_1)| > 1$, $|\varphi(a_4)| > 1$:

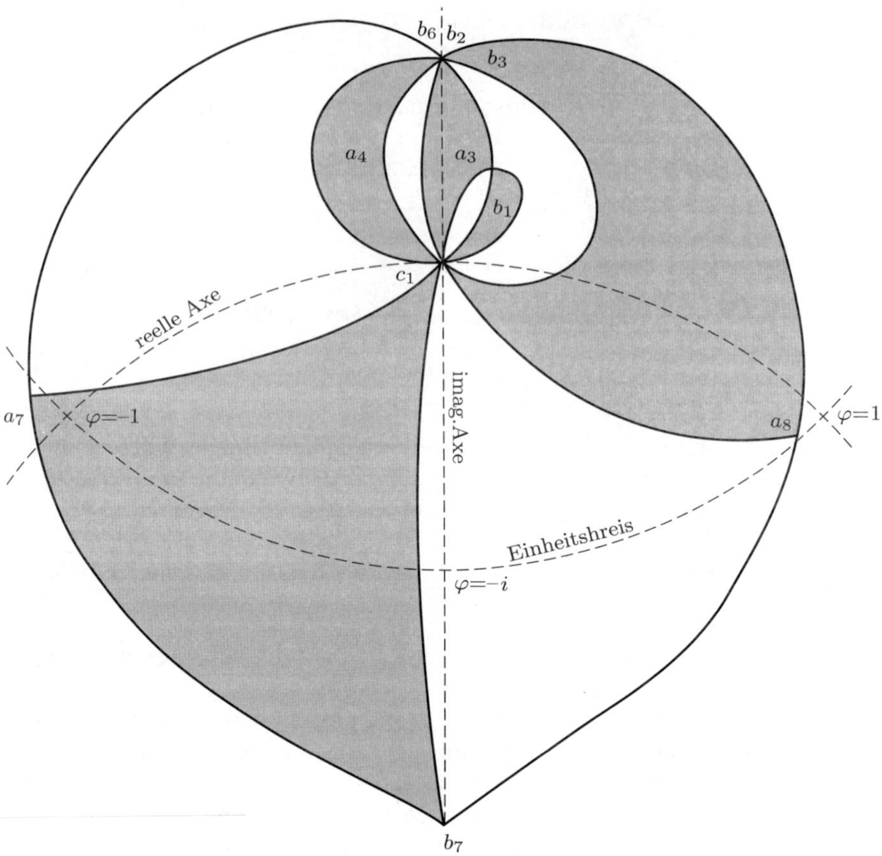

Fig. 107.

$$-\frac{1+i\sqrt{3}}{2}(4+\sqrt{15}),\ \frac{1-i\sqrt{3}}{2}\cdot\frac{3\sqrt{3}+4\sqrt{2}}{\sqrt{5}},$$

have negative imaginary components. Finally, if $\varphi' = \varphi^{-1}$ would hold, then again, in consequence of figure 107, $|\varphi(b_1)| > 1$, $|\varphi(a_4)| > 1$ would necessarily hold, and $\varphi(b_1)$ would have to have a positive, but $\varphi(a_4)$ a negative real component. But this does not take place. *According to this, we note, as a result:*

$$\begin{cases} \varphi(c_1)=0,\ \varphi(c_2)=\infty,\ \varphi(b_1)=\dfrac{1+i\sqrt{3}}{2}(4-\sqrt{15}), \\ \varphi(b_2)=\varphi(b_3)=\varphi(b_6)=i\dfrac{3-\sqrt{5}}{2},\ \varphi(a_4)=\dfrac{-1+i\sqrt{3}}{2}\cdot\dfrac{4\sqrt{2}-3\sqrt{3}}{\sqrt{5}},\ \cdots. \end{cases} \tag{18}$$

The transformation T' of period 2, given in (11) on p. 566, furnished, as we determined above, a transformation of the group Γ_μ into itself. Consequently:

$$\varphi'(\zeta)=\varphi(T'(\zeta)) \tag{19}$$

is also a principal function of Γ_μ, and therefore is connected linearly with $\varphi(\zeta)$. In addition, since, as we remarked, $T'^2 = 1$, the linear substitution of φ into φ' will itself be of period 2. Thereby, in particular, the value $\varphi = \varphi(b_1)$ must go over into $\varphi' = \varphi(b_4)$ and similarly, the value $\varphi = \varphi(a_4)$ into $\varphi' = \varphi(a_5)$. After we have uniquely determined the values $\varphi(b_1), \ldots$, the linear relation between φ and φ' is now consequently directly giv- [579]
able. *The original principal function* $\varphi(\zeta)$ *and the one* $\varphi'(\zeta)$, *transformed by* T', *stand in the linear relation:*

$$\varphi' = \frac{i\sqrt{5}\varphi - 8 + 2i\sqrt{3}}{(8+2i\sqrt{3})\varphi - i\sqrt{5}}. \tag{20}$$

For obtaining the transformation equation of the tenth degree:

$$f(J', J) = 0, \tag{21}$$

all means now lie ready. Since, according to p. 567, we can set:

$$J'(\zeta) = J(T(\zeta)) = J(T'(\zeta)),$$

then one finds, after putting in $T'(\zeta)$ for ζ in (17), the relation between $J'(\zeta)$ and $\varphi'(\zeta)$, reading the same as (17):

$$\begin{aligned} &J' : J' - 1 : 1 \\ =\; &\left(\varphi'^2 - \frac{3\sqrt{3}}{\sqrt{5}}(1 - i\sqrt{3})\varphi' + \cdots\right)(\varphi'^4 + \cdots)^2 \\ &: (\varphi'^2 + \sqrt{15}(1+i\sqrt{3})\varphi' + \cdots)(\varphi'^2 + i\sqrt{5}\varphi' + 1)^4 \\ &: \frac{2^7 \cdot 3^4\sqrt{3}}{5^2\sqrt{5}}(1+i\sqrt{3})\varphi'^5. \end{aligned} \tag{22}$$

The transformation equation of the tenth degree (21) *now yields itself by elimination of* φ *and* φ' *from the equations* (17), (20) *and* (22); *in the usual manner*[11], *we consider these three equations as a replacement of the transformation equation.*

A.5 The Galois group of the transformation equation and its cyclic subgroups

According to p. 559, the Galois group (monodromy group) of the transformation equation of the tenth degree is isomorphic with that finite group G_M, to which our group Γ of the signature $(0, 3; 2, 4, 5)$ reduces itself modulo 3. In this reduction, the whole algebraic number j takes over the role of a "*Galois imaginary*", which satisfies the irreducible congruence:

$$j^2 + j + 2 \equiv 0, \quad (\text{mod } 3). \tag{1}$$

Correspondingly, the A, B, C, D become whole imaginary numbers of the form $(a+bj)$, where a and b are rational whole numbers, only to be distinguished modulo 3.

The determinant of the individual substitution is initially $\equiv \pm 1 \pmod 3$. But in case the determinant is $\equiv -1$, we are able to multiple the four coefficients of the substitution [580]

[11] See the reference on p. 570.

with the whole number $(j-1)$, which satisfies the congruence:

$$(j-1)^2 \equiv -1, \quad (\text{mod } 3),$$

whereupon, without disturbing the structure of our substitutions, the determinant becomes $\equiv 1$ in all cases: *The Galois group of the transformation equation is isomorphic with the group* G_M *of all "unimodular" substitutions, incongruent modulo* 3*:*

$$V \equiv (A, B, C, D) \tag{2}$$

for which, therefore:

$$A^2 + C^2 - j(B^2 + D^2) \equiv 1, \quad (mod\ 3) \tag{3}$$

holds. Thereby, however, two substitutions, which are transportable into one another by sign change of the coefficients, are to be considered as not distinct.

In order, at the same time, to determine the periods of the individual substitutions, we derive, with consideration of (3), from the rule (14) on p. 562, for the square and the fourth power of the substitution (2):

$$V^2 \equiv (A^2+1, AB, AC, AD),$$
$$V^4 \equiv (A^4 - A^2 - 1, A^3B + AB, A^3C + AC, A^3D + AD).$$

Since simultaneous sign change of the A, B, C, D does not alter the substitution, then we are able to reduce A to one of the five residues 0, 1, $1+2j$, j, $1+j$ (mod 3). If, first of all, $A \equiv 0$, then, as one sees, $V^2 \equiv 1$, without V being $\equiv 1$; therefore V is of the period *two*. If secondly, $A \equiv 1$, then there follows:

$$V^2 \equiv (A, -B, -C, -D) \equiv V^{-1};$$

therefore $V^3 \equiv 1$, i.e. V is either the identity $(1,0,0,0)$ belonging here or of period *three*. If, thirdly, $A \equiv 1+2j$, then we have:

$$A^2 + 1 \equiv j^2 + j + 2 \equiv 0;$$

accordingly, V^2 is of period two and V of period *four*. For $A \equiv j$, it follows easily from (1) that:

$$A^4 - A^2 - 1 \equiv A, \ A^3 + A \equiv -1;$$

with this, we now get:

$$V^4 \equiv (A, -B, -C, -D) \equiv V^{-1},$$

so that $V^5 \equiv 1$, and therefore, since the identity $V \equiv 1$ is excluded, V has period *five*. If, last of all, $A \equiv 1+j$, then:

$$A^4 - A^2 - 1 \equiv -A, \ A^3 + A \equiv 1$$

holds, whereupon one again finds the period *five* for V: *The group* G_M *contains, besides*
[581] *the identity* $V \equiv 1$, *substitutions of period two, namely for* $A \equiv 0$, *substitutions of period three, namely for* $A \equiv \pm 1$, *substitutions of period four, namely for* $A \equiv \pm(1+2j)$, *and substitutions of period five, namely for* $A \equiv \pm j$ *and* $A \equiv \pm(1+j)$.

In order to determine the order M of the group, observe, that the five quadratic residues 0, 1, $1+2j$, $2+j$, 2 find themselves among the 9 numbers $a+bj$, incongruent modulo 3. By addition of each two of these residues, one determines easily: Of the 81 incongruent number pairs B, D, 17 pairs furnish the sum $B^2+D^2\equiv 0$, while the individual one of the 8 numbers not congruent to 0 are each represented by 8 pairs B, D in the form (B^2+D^2).

If now, $A\equiv 0$, then we may restrict C to the numbers 0, 1, $1+2j$, j, $1+j$. For $C\equiv 0$, according to the preceding remarks, the congruence (3) furnishes 8 admissible number pairs B, D, of which, however, twos are transportable into one another by simultaneous sign change of B, D. These 8 pairs, therefore, furnish only 4 different substitutions. For $C\equiv 1$, the congruence (3) furnishes, in all, 17 pairs B, D, but for the remaining three values C, 8 pairs each. Therefore, in all, one has $4+17+3\cdot 8=45$ substitutions with $A\equiv 0$. If, secondly, $A\equiv 1$, then for $C\equiv 0$, we have, in all, 17 pairs B, D, but for the remaining 8 numbers C, 8 each. The identity is contained among the $17+8\cdot 8=81$ substitutions. For $A\equiv 1+2j$, we have, for each of the two numbers $C\equiv 1+2j$ and $C\equiv 2+j$, as incongruent solutions of (3), in all, 17 pairs B, D each, but for the remaining 7 number C, 8 pairs B, D each, which gives $2\cdot 17+7\cdot 8=90$ operations. Finally, for $A\equiv j$ and $A\equiv 1+j$, we find no number C, for which A^2+C^2 would be $\equiv 1$; in both cases, therefore, for all 9 incongruent numbers C, we have 8 admissible pairs B, D each, which gives $2\cdot 9\cdot 8=144$ operations: *The group G_M of the substitutions* (2), *incongruent mod* 3, *contains, besides the identity,* 45 *substitutions of period two,* 80 *substitutions of period three,* 90 *substitutions of period four, and* 144 *substitutions of period five; since addition substitutions do not come up, the order of our group* $M=360$, *so that we shall denote this group from now on as the* G_{360}.

The substitution $(0,0,1,0)$ also belongs to the 45 substitutions of period two. In order to determine, with now many substitutions this is conjugate, we look up all those substitutions $V\equiv(A,B,C,D)$, which transform $(0,0,1,0)$ into itself, or, what amounts to the same thing, which are permutable with $(0,0,1,0)$. This requirement:

$$(0,0,1,0)\cdot V\equiv V\cdot(0,0,1,0),$$

however, on the basis of the rule (14) on p. 562, one clothes in the more detailed form: [582]

$$(C,-D,-A,B)\equiv(C,D,-A,-B).$$

Therefore, either $B\equiv D\equiv 0$ or $A\equiv C\equiv 0$ must take place, so that we are led to the 8 substitutions:

$$\begin{cases}(1-j,0,\pm(1-j),0),\ (0,0,1,0),\ (1,0,0,0),\\ (0,1,0,\pm j),\ (0,j,0,\pm 1).\end{cases} \tag{4}$$

According to this, the substitution $(0,0,1,0)$ *is one of* $360:8=45$ *conjugates, i.e. all* 45 *substitutions of period two are conjugate.*

The first substitution (4), namely $(1-j,0,1-j,0)$, is of period four. It can at most be permutable with the substitutions V, which are also permutable with its square $(0,0,1,0)$. But the calculation shows, that $(1-j,0,1-j,0)$ is only permutable with the substitutions standing in the first row of (4), while the four substitutions standing in the second row of (4) transform $(1-j,0,1-j,0)$ into its inverse substitution

$(1-j,0,-1+j,0)$. *Thus,* $(1-j,0,1-j,0)$ *is one of* $360/4=90$ *conjugates, i.e. all* 90 *substitutions of period four are conjugate.*

Of the 80 substitutions of period three, we pick out the two particular ones:

$$(1,1,0,-1+j),\ (1,1,0,1-j). \tag{5}$$

If $V \equiv (A,B,C,D)$ is to be permutable with the first of them:

$$(1,1,0,-1+j)\cdot V \equiv V\cdot(1,1,0,-1+j),$$

then one finds, on the basis of the rule (14) on p. 562:

$$\begin{aligned}&(A+jB+(1+j)D, A+B+(1-j)C, -(1+j)B+C+jD, -(1-j)A+C+D)\\ \equiv&(A+jB+(1+j)D, A+B-(1-j)C, (1+j)B+C-jD, -(1-j)A-C+D).\end{aligned}$$

For the subsisting of the congruence, as one easily shows,

$$C\equiv 0,\ D\equiv(-1+j)B$$

is sufficient and necessary. If these two conditions are fulfilled, then the congruence (3) reduces itself to $A^2\equiv 1$, so that the first substitution (5) shows itself to be permutable, all together, with the 9 substitutions:

$$(1,B,0,(-1+j)B). \tag{6}$$

[583] Similarly, one finds, that the second substitution (5) is permutable with the 9 substitutions:

$$(1,B,0,(1-j)B).$$

On the other hand, the calculation of the scheme:

$$(1,1,0,-1+j)\cdot V \equiv V\cdot(1,1,0,1-j)$$

to be carried out on the basis of rule (14) on p. 562, shows that there is no substitution V in our G_{360}, which transforms the one substitution (5) into the other. *Each of the substitutions* (5) *is accordingly one of* $360/9=40$ *conjugates, so that the* 80 *substitutions of period three fall into two systems of* 40 *conjugate substitutions each, represented by the substitutions* (5).

Here the question yet remains to be decided, whether the individual substitution of period three is conjugate with its inverse or not. However, one now shows immediately from (14) on p. 562:

$$(1,1,0,-1+j)(0,1,1,-1+j)\equiv(0,1,1,-1+j)(1,-1,0,1-j),$$

so that the substitution $(1,1,0,-1+j)$ is transformable into its inverse by $(0,1,1,-1+j)$. The corresponding thing naturally holds for the second substitution (5). *The individual substitution of period three and its inverse always belong to one and the same system of* 40 *conjugate substitutions.*

The five substitutions:

$$(j, 1-j, 0, 0),\ (1+j, 1+j, 0, 0),\ (1+j, -1-j, 0, 0),\ (j, -1+j, 0, 0),\ (1, 0, 0, 0) \tag{7}$$

are the different powers of the first of them, which has period give. At the same time, these are the only substitutions contained in the G_{360} with $C \equiv 0$, $D \equiv 0$; for, as one easily calculates the congruence $A^2 - jB^2 \equiv 1$ furnishes exactly the five substitutions (7). If $(j, 1-j, 0, 0)$ is to be permutable with (A, B, C, D), then, in consequence of (14) on p. 562:

$$\begin{aligned} &(jA-(1+j)B, (1-j)A+jB, jC-(1+j)D, (1-j)C+jD) \\ &\equiv (jA-(1+j)B, (1-j)A+jB, jC+(1+j)D, -(1-j)C+jD) \end{aligned}$$

must hold. But this congruence is only fulfilled for $C \equiv 0$, $D \equiv 0$, so that we come back exclusively to the 5 powers (7) of the substitution $(j, 1-j, 0, 0)$. The substitution $(j, 1-j, 0, 0)$, according to this, is permutable only with its own powers. The same holds for $(1+j, 1+j, 0, 0)$, since this substitution is the second power of $(j, 1-j, 0, 0)$ and, on the other hand, $(j, 1-j, 0, 0)$ is representable as the third power of $(1+j, 1+j, 0, 0)$. *Accordingly, the substitution* $(j, 1-j, 0, 0)$ *and, similarly, also the substitution* $(1+j, 1+j, 0, 0)$, *is one of* $360/5 = 72$ *conjugate substitutions; i.e. all* 72 *substitutions of period five with* $A \equiv \pm j$ *are conjugate, and similarly, all* 72 *substitutions of period five with* $A \equiv \pm(1+j)$. [584]

The cyclic subgroups of the G_{360} are hereby completely disposed of: *The* G_{360} *contains, besides the* G_1 *consisting of the identity alone, as far as cyclic subgroups go,* 45 *conjugate* G_4 *and in these, just as many conjugate* G_2*, two systems of* 20 *conjugate* G_3 *each and, finally,* 36 *conjugate cyclic* G_5.

On the basis of these statements, the theorem lets itself be easily proved: *The* G_{360} *is a simple group.* Namely, a distinguished subgroup G_μ contained in the G_{360} contains, besides the identity, either no G_2 or all 45 substitutions of period two, which we are able to collect in the statement, that G_μ necessarily contains $45 \cdot \varepsilon_1$ substitutions of period two, where ε_1 is either $= 0$ or $= 1$. If $\varepsilon_1 = 1$, this G_μ will contain either none or all conjugate G_4 and therefore contain $90 \cdot \varepsilon_1 \varepsilon_2$ substitutions of period four, where ε_2 is also $= 0$ or $= 1$. As one determines by continuing, the G_μ will contain $(40 \cdot \varepsilon_3 + 40 \cdot \varepsilon_4)$ substitutions of period three and $144 \cdot \varepsilon_5$ substitutions of period five, where each of these three latter numbers ε is also $= 0$ or $= 1$. The order μ of each distinguished subgroup is thus representable in the form:

$$\mu = 1 + 45\varepsilon_1 + 90\varepsilon_1\varepsilon_2 + 40\varepsilon_3 + 40\varepsilon_4 + 144\varepsilon_5.$$

But μ must now be a divisor of 360, whereby we are able to ignore $\mu = 1$ and $\mu = 360$ itself. If ε_5 were $= 1$, then μ would be $= 180$, which is, however, not attainable by any permissible choice of the remaining ε. Therefore $\varepsilon_5 = 0$ and with this, μ relatively prime to 5. The number μ is therefore already a divisor of 72, so that $\varepsilon_1\varepsilon_2 = 0$. But obviously, the sum $(1 + 45\varepsilon_1 + 40\varepsilon_3 + 40\varepsilon_4)$ is not to be made into a divisor of 72 different from 1 for any admissible choice of ε_1, ε_3, ε_4. *Accordingly, a distinguished subgroup* G_μ *different* *from G_1 and G_{360} does not exist, so that our G_{360} is, in fact, a simple group.*[12]

A.6 The non-cyclic subgroups of the G_{360} and the extended $\overline{G}_{720}$

[585] We will set up the non-cyclic subgroups of the G_{360} by tying into the cyclic G_2, G_3, G_4, G_5 already known by means of a method of reasoning, which we first describe in general. If G_μ is any subgroup of the G_{360}, and we know, that this is one of ν conjugate groups, then G_μ is permutable, all together, with $360\nu^{-1} = \mu'$ substitutions; the latter form a $G_{\mu'}$, which represents the most comprehensive subgroup, in which G_μ is contained distinguishably. We then pass from G_μ to $G_{\mu'}$ and repeat the same argument for $G_{\mu'}$.

This method of reasoning is not continuable further, as soon as $\mu\nu = 360$, i.e. as soon as G_μ is only permutable with its own substitutions. If, in this case, we call the ν conjugate groups:

$$G_\mu,\ G'_\mu,\ \ldots,\ G_\mu^{(\nu-1)}, \tag{1}$$

then we obtain, after transformation by means of any substitution V of the G_{360}, in:

$$V^{-1}G_\mu V,\ V^{-1}G'_\mu V,\ \ldots,\ V^{-1}G_\mu^{(\nu-1)}V, \tag{2}$$

a permutation of the groups (1). If V and V' hereby give the same permutation, then, as one easily perceives, the substitutions $V'V^{-1}$ transforms each group (1) into itself. But all the substitutions, which transform each individual group (1) into itself, form a distinguished subgroup inside this total group different form the G_{360}[13], which is therefore necessarily the G_1. No two different substitutions V, V', consequently, are able to furnish the same permutation (2).

To the 360 substitutions V correspond, in this fashion, just as many permutations of the ν groups (1), which form a permutation group isomorphic with the G_{360}. Since this is contained in the total group of all $\nu!$ permutation of the ν groups (1), 360 is a divisor of $\nu!$, and therefore $\nu \geqslant 6$, and with this, $\mu \leqslant 60$. *Each subgroup G_μ, different from the total group G_{360}, has an order $\mu \leqslant 60$.* Should the highest value $\mu = 60$ be reached, then G_{360} is isomorphic with a permutation group of 6 things. But the only such group, which has the order 360, is the group of the even permutations of 6 things. We shall recognize our G_{360} as isomorphic with this group below.

[586] The individual G_2, as one of 45 conjugate groups, is contained distinguishably in a group G_8, e.g. the G_2 arising from $(0,0,1,0)$ in the G_8 of the substitutions:

[12] The G_{360} already comes up as a group of linear substitutions in a form closely related to the text in E. H. Moore; see his note "*A doubly-infinite system of simple groups*", Bull. of the New York Math. Soc., Vol. 3, No. 3 (1893). L. E. Dickson gives a more detailed representation in his book "*Linear groups with an exposition of the Galois field theory*", (Leipzig, 1901). There (p. 264), the form of the substitutions present in the text is also already alluded to; these are designated as "hyperorthogonal substitutions".

[13] We are not able to reach the case $\mu = 360$, $\nu = 1$ using the method of reasoning of the text, since the G_{360} is a simple group.

$$\begin{cases} (1-j,0,1-j,0), & (0,1,0,j), \\ (0,0,1,0), & (0,j,0,1), \\ (1-j,0,-1+j,0), & (0,j,0,-1), \\ (1,0,0,0), & (0,1,0,-j), \end{cases} \tag{3}$$

which we already placed together in (4) on p. 582. This G_8 has the dihedral type and is, at the same time, the most comprehensive group, in which the cyclic G_4 to be generated from $(1-j,0,1-j,0)$ is contained distinguishably. Corresponding to the 45 G_2 resp. G_4, we find: *There are, in the G_{360}, 45 conjugate dihedral groups G_8; the individual G_2 always takes part in five such G_8, but in only one of these five as a distinguished subgroup.*

The individual G_8 is already permutable only with its own substitutions, so that, from these dihedral groups G_8, we are not able to arrive at any further subgroups G_μ. But now, in the dihedral group G_8, two four-groups G_4 are contained distinguishably; for the group (3), we have, as one four-group:

$$(0,0,1,0),\ (0,1,0,j),\ (0,j,0,-1),\ (1,0,0,0), \tag{4}$$

and as the other:

$$(0,0,1,0),\ (0,j,0,1),\ (0,1,0,-j),\ (1,0,0,0). \tag{5}$$

We notice, at the same time: The substitution $(0,0,1,0)$ cannot take part in four-groups other than these two. Namely, the four substitutions of such n four-group G_4 are all permutable with one another; but only the 8 substitutions (3) of the dihedral-G_8 are permutable with $(0,0,1,0)$. Since now, the individual four-group contains three substitutions of period two, we obtain, in all, $2 \cdot 45/3 = 30$ four-groups.

One also easily determines this number from the fact, that the individual four group is always furnished by three of the 45 conjugate dihedral groups. Thus, the group (4) is furnished by those three G_8, in which the three substitutions $(0,0,1,0)$, $(0,1,0,j)$, $(0,j,0,-1)$ are contained distinguishably. Now these 30 four-groups fall into two systems of 15 conjugate G_4 each, represented by the groups (4) and (5). Namely, were all 30 G_4 conjugate, then the individual one would be conjugate in a more comprehensive group G_{12}, which would have to contain a dihedral group G_8, which is a contradiction. With this, we have, as a result: *As the only four-groups contained in the G_{360}, we have two systems of* 15 *conjugate G_4 each, which are represented by the groups* (4) *and* (5). [587]

As one of 15 conjugates, the individual four-G_4 is contained distinguishably in a G_{24}, which, as is known, has the octahedral type. With this, we find, further: *In the G_{360} are contained two systems of* 15 *conjugate octahedral groups G_{24} each, which, for their part, again contain two systems of* 15 *conjugate tetrahedral groups each; at the same time, since the four-groups were single, these are also the only octahedral and tetrahedral groups contained in the G_{360}.* As an example, we note that octahedral-G_{24}, which is generatable from the substitutions:

$$S_0 \equiv (1-j,0,1-j,0),\ T_0 \equiv (0,1+j,1-j,1-j),\ U_0 \equiv (1,1-j,1,1+j). \tag{6}$$

Namely, these latter satisfy the congruences:

$$S_0^4 \equiv 1,\ T_0^2 \equiv 1,\ U_0^3 \equiv 1,\ S_0 T_0 U_0 \equiv 1,$$

and, with this, according to "M" I on p. 456[14] actually generate an octahedral group.

Since the individual G_{24} is only permutable with its own substitutions, our procedure of reasoning arrives at an end for them. Accordingly, we now consider the two systems of 20 conjugate cyclic groups G_3 each. If we first of all go back to the substitutions of period three themselves, then the individual one of them, as one of 40 conjugates, is contained distinguishably in a G_9, e.g. the first substitution (5) on p. 582 in the G_9 of the substitutions:

$$(1, B, 0, (-1+j)B), \tag{7}$$

which, as one sees, contains, besides the identity 8 substitutions of period three. Since each of these substitutions is permutable with each other one[15], then each of them, by means of our method of reasoning, leads from itself to the same G_9, so that the individual substitution of period three finds itself in only one G_9.

If now, all 8 substitutions (7) belonged to only one system of 40 conjugate substitutions of period three, then, according to the result just obtained, we would obtain only $40/8 = 5$ conjugate G_9. But then, the individual one would be contained distinguish-
[588] ably in a G_{72}, while, however, such a group cannot exist. Therefore, the individual G_9 contains substitutions of period three from both systems. Since now, in all, we have 80 substitutions of period three, we obtain, as a result: *The G_{360} contains 10 conjugate groups G_9, whose individual one consists of exclusively permutable substitutions, and each contains two cyclic G_3 from the individual one of the two systems.*

The individual one of the G_9, as one of 10 conjugates groups, is contained distinguishably in a G_{36}. In particular, the G_9 of the substitutions (7) furnishes a G_{36}, which further contains the 9 substitutions of period two:

$$(0, B, 1, (-1+j)B) \tag{8}$$

and the 18 substitutions of period four:

$$(1-j, B, \pm(1-j), (-1+j)B). \tag{9}$$

That these 36 substitutions form a group, is self-evident; *for we have to do here with that G_{36}, which furnishes the congruence group Γ_{10} underlying the transformation equation of tenth degree and defined by* (6) *on p. 564.* But the G_9 is necessarily contained distinguishably in the G_{36}, since, besides the identity, it contains only substitutions of period three, and the remaining substitutions (8) and (9) of the G_{36} do not belong to this period.

Moreover, the substitutions (7) and (8), taken by themselves, form a G_{18}; and this is the most comprehensive group, in which the individual G_3 contained in (7) is distinguished. *The G_{360} contains ten conjugate G_{36} and, in these, ten conjugate G_{18}, each one of which includes four cyclic G_3 as distinguished subgroups.* In addition to this, in the

---

[14] Also see Dyck "*Gruppentheoretische Studien*", Math. Ann. Vol. 20, p. 1 (1882).

[15] This proceeds immediately from the easily provable congruence:

$$(1, B, 0, (-1+j)B)(1, B', 0, (-1+j)B') \equiv (1, B+B', 0, (-1+j)(B+B')).$$

G_{18}, dihedral groups G_6 also occur in a certain easily determinable number; however, these G_6 do not come further into consideration.

The individual G_5, as one of 36 conjugate groups, is contained distinguishably in a dihedral G_{10}. *The* G_{360} *will contain* 36 *conjugate dihedral group* G_{10} *of this kind.* However, our approach does not furnish further groups.

But now there exist still further especially important subgroups in the G_{360}, namely, two systems of 6 conjugate icosahedral-G_{60} each, which therefore have the maximum the order μ, which we determined above. It is self-evident, that, by means of our procedure of reasoning, we are not able to arrive at these groups; for the icosahedral group is a simple group. However, it is easy, to define the icosahedral group from generators, as this was effected before by the substitutions (6) for an octahedral-G_{24}. We set, first of all: [589]

$$S_1 \equiv (j, 1-j, 0, 0), \; T_1 \equiv (0, j, 0, 1), \; U_1 \equiv (1, 1-j, -1-j, j), \tag{10}$$

and easily prove for these substitutions:

$$S_1^5 \equiv 1, \; T_1^2 \equiv 1, \; U_1^3 \equiv 1, \; S_1 T_1 U_1 \equiv 1.$$

If one sets, secondly:

$$S_2 \equiv (j, 1-j, 0, 0), \; T_2 \equiv (0, j, 0, -1), \; U_2 \equiv (1, 1-j, 1+j, -j), \tag{11}$$

then, correspondingly, the congruences hold

$$S_2^5 \equiv 1, \; T_2^2 \equiv 1, \; U_2^3 \equiv 1, \; S_2 T_2 U_2 \equiv 1,$$

and, in addition, $S_1 = S_2$. According to "M" I on p. 456, the substitutions (10), as well as the substitutions (11), generate an icosahedral-G_{60}.

These two groups are certainly not identical. Namely, were they, then the substitution:

$$T_1 T_2 \equiv (1-j, 0, -1+j, 0)$$

of period four would also be contained in the G_{60}, which is not the case. Both G_{60} have the G_5 to be generated from $S_1 = S_2$ in common. From this lets itself be deduced, that these two groups are not able to be conjugate with one another. Namely, the individual G_{60} is one of six conjugate groups. Were those two G_{60} contained in this system of six G_{60}, then the individual G_5 would always be contained in at least two G_{60}. Now since the individual G_{60} contains 6 G_5, then we would have at most $6 \cdot 6 : 2 = 18$ conjugate G_5, while we nevertheless have 36 of them (cf. p. 584). With this, we note: *Inside the* G_{360} *are contained two systems of six conjugate icosahedral groups* G_{60} *each, which are able to be represented by the two groups to be generated from the substitutions* (10) *resp.* (11). From the existence of the G_{60} follows further, at the same time, according to the remarks presented at the beginning of the paragraph, as a principal theorem: *The* G_{360} *is isomorphic with the group of the even permutations of* 6 *things.*

The individual G_{60} contains five conjugate tetrahedral-G_{12}, and indeed, as the sole subgroups of this type. We find: *The individual one of the two systems of the* 15 *conjugate* G_{12} *is coordinated to the individual system of the* 6 *groups* G_{60}, *and indeed, in such a manner, that the individual tetrahedral group* G_{12} *is always contained in two icosahedral groups of the coordinated system.*

Finally, one can easily show, *that, besides the* 12 *icosahedral groups named, no further groups of this structure are contained in the* G_{360}. Namely, if G_{60} is any icosahedral group contained in the G_{360}, there is a group conjugate with the G_{60}, which
[590] contains the substitution $S \equiv (j, 1-j, 0, 0)$ of period five, and besides this, a yet to be determined substitution $T \equiv (0, B, C, D)$ of period two, which, combined with S into $ST \equiv ((1+j)B, \ldots)$ furnishes a substitution of period three. With this, $B \equiv j$ and therefore, in consequence of (3) on p. 580:

$$C^2 - jD^2 \equiv -j$$

holds. The solution of this congruence restricts T to the following ten substitutions:

$$\begin{aligned} T \equiv\ & (0, j, 0, \pm 1),\ (0, j, 1, \pm(1+j)),\ (0, j, \pm(1+j), j), \\ & (0, j, -1, \pm(1+j)),\ (0, j, \pm(1+j), -j). \end{aligned}$$

But since the five substitutions:

$$T,\ S^{-1}TS,\ S^{-2}TS^2,\ \ldots,\ S^{-4}TS^4$$

obviously furnish the same G_{60}, we obtain only two different G_{60} from the ten T calculated which are already identical with the icosahedral groups already discussed.

The group Γ was capable of extension by the reflection $\overline{V}_3(\zeta) = -\bar{\zeta}$ in the imaginary axis to a group $\overline{\Gamma}$ of the second kind, which, reduced modulo 3, furnishes *a group* $\overline{G}_{720}$ *of order* 720:

$$\overline{G}_{720} = G_{360} + G_{360}\overline{V}_3.$$

In order to determine how many substitutions conjugate with the reflection $\overline{V}_3$ are contained in $\overline{G}_{720}$, we enumerate the substitutions $V \equiv (A, B, C, D)$ permutable with $\overline{V}_3$. But the congruence $\overline{V}_3 V = V\overline{V}_3$ reads, explicitly:

$$(A, B, C, D) \equiv (A, B, -C, -D),$$

and requires either $C \equiv 0$, $D \equiv 0$ or $A \equiv 0$, $B \equiv 0$. We therefore obtain here the ten substitutions of the dihedral-G_{10} belonging to $(j, 1-j, 0, 0)$. From this follows: *The reflection* $\overline{V}_3$*, inside the* $\overline{G}_{720}$*, is one of* $360 : 10 = 36$ *conjugate substitutions.*

If, further, in:

$$G_\mu,\ G'_\mu,\ G''_\mu,\ \ldots,\ G_\mu^{(\nu-1)}, \tag{12}$$

we have a system of ν conjugate subgroups of the G_{360}, and we assume, at the same time, the extreme case $\mu\nu = 360$, then the following alternative holds. Either these ν groups are *only* conjugate among themselves also in the $\overline{G}_{720}$, where then, the individual G_μ, as one of ν conjugates is contained distinguishably in a group $\overline{G}_{2\mu}$ of the second kind of the order $720 : \nu = 2\mu$. Or the transformation, say, by $\overline{V}_3$, furnishes a system different from (12):

$$\overline{V}_3 G_\mu \overline{V}_3,\ \overline{V}_3 G'_\mu \overline{V}_3,\ \ldots,\ \overline{V}_3 G_\mu^{(\nu-1)} \overline{V}_3, \tag{13}$$

[591] which is first conjugate in the $\overline{G}_{720}$, but not yet in the G_{360}, with the system (12). In the latter case, the G_μ are *not* capable of extension to groups $\overline{G}_{2\mu}$ inside the $\overline{G}_{720}$. Naturally,

the second case can only then take place, if there are, in the G_{360}, two non-conjugate systems of the same structure.

The application of this general consideration to our subgroups G_{μ} discovered above in the G_{360} offers not the slightest difficulty. We find: *The* 10 *conjugate groups* G_{36} *are capable of extension inside the* G_{720} *to* 10 *conjugate* $\overline{G}_{72}$ *(which we already know here from p. 566); the two systems of* 15 *octahedral-*G_{24} *each and, likewise, the two systems of* 6 *icosahedral-*G_{60}*, become conjugate inside the* $\overline{G}_{720}$.

With this, we have obtained sufficient group-theoretic foundations for the following developments.

A.7 The two resolvents of sixth degree of the transformation equation

The decomposition of the Galois group G_{360} of the transformation equation permits us to become acquainted with the *resolvents of the transformation equation of the* 10th *degree.* The most important of these are *two resolvents of* 6th *degree,* which correspond to the two systems of icosahedral groups G_{60}, *two resolvents of* 15th *degree,* which correspond to the two systems of octahedral groups G_{24}, and finally, *the Galois resolvent of* 360th *degree,* corresponding to the entire G_{360} itself. The goal of the further investigations will have to be, to set up these resolvents and carry through the solution of the resolvents and the transformation equation itself by the Galois resolvents.

The two systems of the G_{60} furnish *two systems of six conjugate congruence groups* Γ_6 *of third level of index* 6 *each*; inside the extended $\overline{\Gamma}$, both systems are transformable into one another. The discontinuity domain of the individual Γ_6 consists of six double triangles of the net N of the group Γ. A trial shows, that there is essentially only one single possibility, corresponding to the branching theorem (cf. "M" I on p. 346) belonging to our net N, of building an *unsymmetrical* domain of this kind. figure 108 gives the domain in one of the two arrangements, symmetrical to one another; the reflection in the midline $c_0 b_2$ furnishes the other domain. By identifying, one after the other, the [592]
individual one of the six double triangles of these two domains with the initial triangle of the net N fixed above (p. 560), we obtain the discontinuity domain of all $2 \cdot 6$ groups Γ_6. As one sees, we have again taken up the notations a, b, c for the vertices; and indeed, the vertices belonging together into one cycle in this specific case carry the same notations.

Our groups Γ_6 now belong to the genus $p = 0$. We define a principal function $\psi(\zeta)$, belonging to the discontinuity domain of figure 108, uniquely by the determination:

$$\psi(c_0) = 0, \ \psi(b_1) = 1, \ \psi(c_1) = \infty. \tag{1}$$

If we map the plane of the principal function $\psi(\zeta)$, and therefore, the domain of figure 108, onto the J-plane, there arises *a six-sheeted Riemann surface over this plane,* which has the following branching: *At* $J = 0$, *there are two branchpoints of two sheets each, while two sheets run isolated; at* $J = 1$, *there finds itself one branchpoint of two sheets, and one further one, which carries the value* $\psi = 1$, *of four; at* $J = \infty$, *that sheet, which carries the value* $\psi = 0$ *there, runs isolated, while the five remaining sheets are connected at a branchpoint carrying the value* $\psi = \infty$.

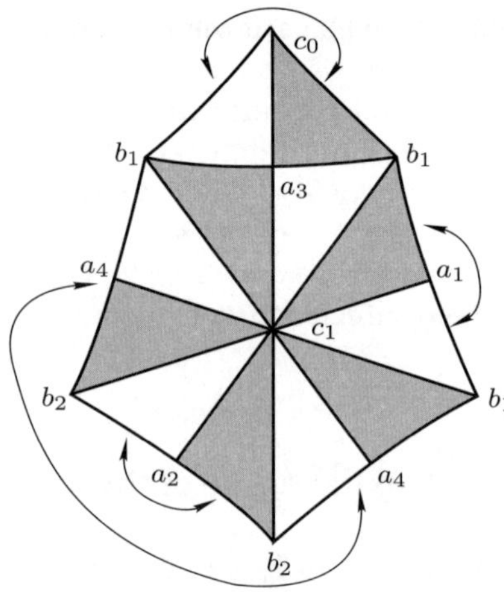

Fig. 108.

For the representation of J as a rational function of the sixth degree of ψ, we deduce from this the schema:

$$J : J-1 : 1 = (\psi^2 + a\psi + b)(\psi^2 + c\psi + d)^2 : (\psi + e)^2(\psi - 1)^4 : f\psi.$$

The calculation of the the coefficients $a, b, c, \dots, f$, still unknown here, offers no difficulty. First of all, we utilize the circumstance, that the functional determinant of the two functions $(\psi+e)^2(\psi-1)^4$ and ψ must contain the factor $(\psi^2+c\psi+d)$, then we find (as above on p. 573), by comparison of coefficients, the following representation of c and d in e:

$$5c = 3e - 1,\ 5d = e. \tag{2}$$

[593] Since, on the other hand, the functional determinant of:

$$(\psi^2 + a\psi + b)(\psi^2 + c\psi + d)^2,\ \psi$$

must exhibit the factor $(\psi+e)(\psi-1)^3$, one correspondingly finds the further relations:

$$\begin{cases} 4a + 3c = 5e - 15, \\ 2ac + 3b + d = 15 - 15e, \\ bd = 5e. \end{cases} \tag{3}$$

The last two equations (2) and (3), since e certainly does not vanish, immediately give $b = 25$. The first equation (3), in consequence of (2), gives:

$$5a = 4e - 18.$$

If we put $b = 25$ and the representations of a, c and d by e into the second equation (3), there yields itself a quadratic equation for e, whose solutions are:

$$e = -\frac{11 \pm 3i\sqrt{15}}{2}.$$

One now immediately determines the associated values a, c, d and calculates the coefficient f still lacking from the identical equation:

$$(\psi^2 + a\psi + b)(\psi^2 + c\psi + d)^2 - (\psi + e)^2(\psi - 1)^4 = f\psi,$$

by substituting $\psi = 1$ say. *The representation of J as a rational function of sixth degree of the principal function ψ consequently has the form:*

$$\begin{aligned} J : J - 1 : 1 = &\left(\psi^2 - \frac{8\sqrt{5} \pm 6i\sqrt{3}}{\sqrt{5}}\psi + 25\right)\left(\psi^2 - \frac{7\sqrt{5} \pm 9i\sqrt{3}}{2\sqrt{5}}\psi - \frac{11 \pm 3i\sqrt{15}}{10}\right)^2 \\ &: \left(\psi - \frac{11 \pm 3i\sqrt{15}}{2}\right)^2 (\psi - 1)^4 : \pm \frac{2^8 \cdot 3^4 i\sqrt{3}}{5^2 \cdot \sqrt{5}}\psi; \end{aligned} \tag{4}$$

these relations, by conceiving them, for given J, as equations of the sixth degree for ψ, directly represent the two resolvents of the sixth degree sought. The $2 \cdot 6$ solutions ψ of these resolvents furnish the principal functions of the twelve groups Γ_6. With respect to the substitutions of the total group Γ, the six functions of the individual system of the Γ_6 will experience the 360 even permutations. We therefore have to do here with two equations of the sixth degree, whose Galois group (monodromy group) is the group of the even permutations of its roots.

With consideration of further developments, we yet have to determine, which of [594]
the two equations (4) belongs to the discontinuity domain drawn in figure 108. It is easy to decide concerning this. We being in the equation (4) belonging to the upper sign, and remember, that the real values J take place along the sides of the triangle division of our net N. But in figure 108, one determines, that all the triangle sides carry themselves over onto a single curve of the ψ-plane running back into itself. We are easily able to give the equation of this curve for the resolvent (4). Namely, in order that J become real, it is necessary and sufficient, that:

$$(2\psi - 11 - 3i\sqrt{15})^2(\psi - 1)^4\psi^{-1}$$

come out purely imaginary. If we accordingly set $\psi = x + iy$, and understand by $\mathfrak{R}(A)$ the real component of the complex number A, then there yields itself, as the carrier of all real values J, the curve of seventh degree:

$$\mathfrak{R}((2x - 11 + i(2y - 3\sqrt{15}))^2(x - 1 + iy)^4(x - iy)) = 0. \tag{5}$$

In order to draw this curve in the x, y-plane, calculate from (4), first of all, the values:

$$\psi(a_1), \psi(a_2) = \frac{1}{\sqrt{5}}\left(4\sqrt{5} + 3i\sqrt{3} \pm 2\sqrt{3}\left(\sqrt{-3 + 2\sqrt{6}} + i\sqrt{3 + 2\sqrt{6}}\right)\right),$$
$$\psi(a_3), \psi(a_4) = \frac{1}{4\sqrt{5}}\left(7\sqrt{5} + 9i\sqrt{3} \pm \sqrt{15}\left(\sqrt{3 + 8\sqrt{6}} + i\sqrt{-3 + 8\sqrt{6}}\right)\right),$$

mark the points concerned in the x, y-plane approximately by: [595]

$$6.14 + 6.66i,\ 1.86 - 2.01i; \ldots$$

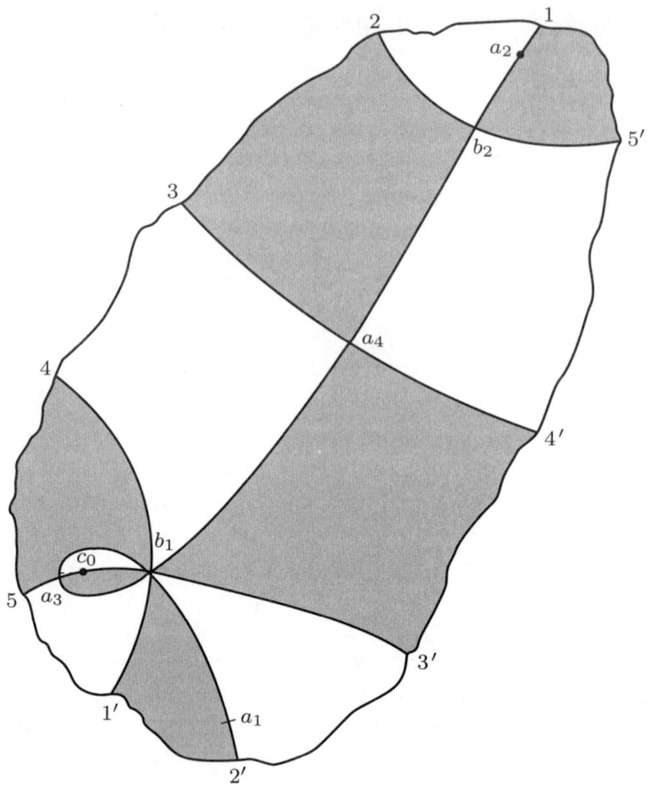

Fig. 109.

and also add the points

$$\psi(c_0) = 0,\ \psi(b_1) = 0,\ \psi(b_2) = \frac{1}{2}(11 + 3i\sqrt{15})$$

the remaining course of the curve is sketched in figure 109. The curve cuts the x-axis, indicated dottedly in the figure, four-fold at $x = 1$, simply at $x = 0$, and besides this, simply at each of the points $x = \frac{1}{2}(11 \pm 3\sqrt{15})$; it further has five asymptotes of equal difference of direction $\frac{\pi}{5}$, which correspond to the branches denotes by $1, 1', 2, 2', \ldots$.

A glance at the figures 108 and 109 now suffices, in order to recognize, that we have to do here with the polygon of figure 108 bent together and not with the polygon symmetric to this figure. We thus note, as a result: *The resolvent* (4) *with the upper sign belongs to the polygon chosen in figure 108.*

A.8 The discontinuity domains of the Γ_{15} and Γ_{30} belonging to the octahedral and tetrahedral groups

In the G_{360}, we found (p. 587) two systems of 15 octahedral groups G_{24} each. To them correspond *two systems of* 15 *conjugate congruence groups* Γ_{15} *of third level of index* 15 *each*, which are conjugate inside the extended $\overline{\Gamma}$, but not yet inside the original group Γ. It is sufficient, to produce the discontinuity domains for an individual Γ_{15}; among the octahedral groups G_{24}, we favor that one, say, which we produced above (p. 587) from the substitutions:

$$S \equiv (1-j,0,1-j,0),\ T \equiv (0,1+j,1-j,1-j),\ U \equiv (1,1-j,1,1+j). \tag{1}$$

We remember, at the same time, that this G_{24} contains in itself the dihedral-G_8 of the substitutions:

$$\begin{cases} (1-j,0,1-j,0),\ (0,0,1,0),\ (1-j,0,-1+j,0),\ (1,0,0,0), \\ (0,1,0,j),\ (0,j,0,1),\ (0,j,0,-1),\ (0,1,0,-j) \end{cases} \tag{2}$$

(cf. p. 586) and is consequently representable in the form:

$$G_{24} = G_8 + UG_8 + U^2G_8. \tag{3}$$

A discontinuity domain of the Γ_{15} belonging to this G_{24} is represented in figure 110. In order to characterize the positioning of this in the triangle net N, the initial triangle, chosen above (p. 560), of the total group Γ is heavily outlined in the figure, and the generators of the group Γ are appended. The vertex b_1 therefore lies at $\zeta = i$, the circular arc $b_1 b_3$ represents the part of the imaginary ζ-axis with $|\zeta| > 1$ obtaining itself, and the circular arc $b_1 b_3'''$, a portion of the unit circle. For the vertices, we have again consistently taken up the notations a, b, c; vertices, which belong together into one cycle, have the same lower index, and are distinguished by upper indices. We give the associatedness of the boundary curves tabularly, in an immediately understandable notation: [596]

$$\begin{cases} (b_1 b_3) \longrightarrow (b_1 b_3'''), & S = V_2^{-1} \equiv (1-j,0,1-j,0), \\ (a_3 b_3) \longleftrightarrow (a_3 b_3'), & V_4 \equiv (0,-1+j,1-j,1+j), \\ (a_4 b_3') \longleftrightarrow (a_4 b_3''), & V_5 \equiv (0,1,0,j), \\ (a_7 b_4') \longleftrightarrow (a_7 b_4''), & V_6 \equiv (0,j,0,1), \\ (b_5 b_4) \longleftrightarrow (b_5 b_4'), & V_7 \equiv (0,1-j,1-j,-1-j), \\ (b_3''' b_4'') \longleftrightarrow (b_3'' b_4), & V_8 \equiv (1,1-j,-1,1+j), \end{cases} \tag{4}$$

whereby, next to each pair of coordinated boundary curves is noted the associated substitution.

One can verify these assertions in the following manner: First of all, convince oneself, that the coordination of the boundary curves is one such, which corresponds to the branching theorem (cf. "M" I on p. 346) belonging to our triangle net. The discontinuity domain therefore defines a subgroup of the group Γ, and indeed, obviously, a Γ_{15} of index 15, since the domain mentioned is built up of 15 double triangles. This Γ_{15} is clearly of the genus $p = 0$; and the fixed vertices a_3, a_4, b_5, a_7 and b_1 lie on the boundary of the domain, whose generators belong to the periods 2, 2, 2, 2 and 4. *The group* Γ_{15} *defined by the given domain therefore has the signature* $(0,5;2,2,2,2,4)$.[16] [597]

[16] This Γ_{15} is capable of extension neither by a reflection of $\overline{\Gamma}$ nor by a reflection foreign to this group.

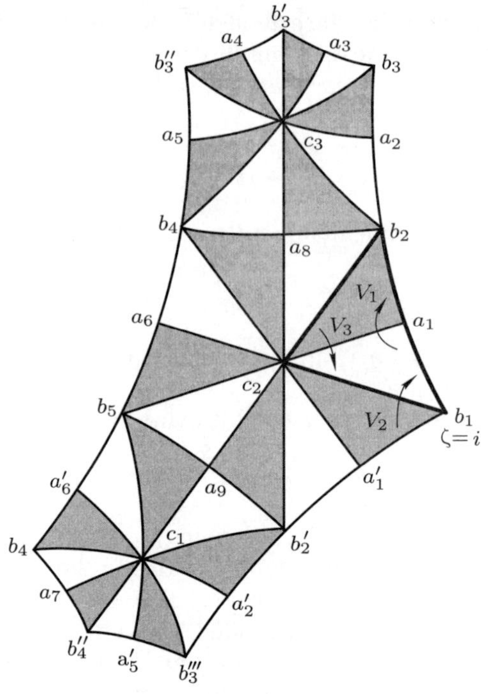

Fig. 110.

Secondly, one wants to represent the defined substitutions $V_4, V_5, \ldots, V_8$ in the generators V_1, V_2, V_3 of the total group and will, in each case, find the congruence given in (4) verified. The substitutions $S = V_2^{-1}, V_5, V_6$ belong to the G_8 given in (2), and therefore, to the G_{24} chosen. But the same thing also holds for V_4 and V_7; for the congruence hold:

$$V_4 \equiv U^2 \cdot (0, j, 0, -1), \ V_7 \equiv U^2 S^{-1} \equiv U^2 V_2.$$

Finally $V_8 = V_7 V_6$, so that V_8 also belongs to the G_{24}. All generators of the Γ_{15} defined by the domain of figure 110, and therefore, this group itself, is contained in the Γ_{15} belonging to our G_{24}; both groups G_{15}, as groups of the same index, are therefore identical: *The* Γ_{15} *belonging to the octahedral group* G_{24} *is the group of the signature* $(0,5;2,2,2,2,4)$ *defined by the discontinuity domain of figure 110.*

In the G_{24} is contained a distinguished tetrahedral group. This group G_{12} is composed, first, of the four-group, also contained distinguishably in the G_{24}:

$$V_2^2 \equiv (0,0,1,0), \ V_5 \equiv (0,1,0,j), (0,j,0,-1), \ V_0 \equiv (1,0,0,0), \tag{5}$$

and besides this, of the eight substitutions of period three contained in the G_{24}. One will produce a discontinuity domain of the Γ_{30} belonging to the G_{12} without trouble from the domain of figure 110. Exercise the substitution V_2 on this domain, which, without itself belonging to the G_{12}, transforms it into itself, and hang the new domain arising onto the former one. One thus obtains, as a discontinuity domain of the Γ_{30}, the

polygon represented in figure 111. For the coordination of the boundary curves, the following argument holds: Those generators V of the Γ_{15}, which belong to the four-group (5) or are of period three, remain maintained as generators of the Γ_{30}; for the second added half of the polygon, then, to a generator V of this kind, there arranges itself the further one $V_2VV_2^{-1}$. The remainder of the generators do not yet belong to the Γ_{30}; one of them, V, then gives, in V_2V, a generator of Γ_{30}. To the first category belong only the generators V_5 and V_8 of the Γ_{15}, next to which, therefore, we get the substitutions $V_2V_5V_2^{-1}$, $V_2V_8V_2^{-1}$. V_2, V_4, V_6, V_7 form the second category, which therefore furnish [598]
the generators V_2^2, V_2V_4, V_2V_6, V_2V_7 of the Γ_{30}. We again place the coordination of the boundary curves of figure 111 next to the associated substitutions together tabularly:

$$\begin{cases} (b_1b_3'') \longleftrightarrow (b_1b_3'''), & V_2^2, \\ (b_2b_3) \longrightarrow (b_2'b_3'''), & V_2V_4, \\ (a_1b_3) \longleftrightarrow (a_1b_3'), & V_5, \\ (b_3''b_4') \longrightarrow (b_3'b_4), & V_8, \\ (b_4b_5') \longrightarrow (b_4''b_5''), & V_2V_7, \\ (b_4'b_5') \longrightarrow (b_4''b_5), & V_2V_6, \\ (a_2b_2') \longleftrightarrow (a_2b_2''), & V_2V_5V_2^{-1}, \\ (b_2b_5) \longrightarrow (b_2''b_5''), & V_2V_8V_2^{-1}. \end{cases} \tag{6}$$

From the coordination of the boundary curves given with this, one immediately determines, that we have to do here with a polygon of the genus $p = 1$. Of fixed edges, which are fixpoints of the generators of Γ_{30}, there find themselves, on the boundary of the polygon, the three a_1, a_2, b_1, which are fixpoints of the substitutions of period two V_2^2, V_5, $V_2V_5V_2^{-1}$ of the Γ_{30}. We consequently note as a result: *The* Γ_{30} *belonging to the chosen tetrahedral group* G_{12} *is a group of the signature* $(1,3;2,2,2)$, *whose discontinuity domain is given by the polygon of figure 111 and the boundary curve coordination* (6). [599]

Each of the six conjugate icosahedral groups G_{60} of the individual system contains five conjugate tetrahedral groups G_{12}, so that the individual G_{12} always finds itself in two conjugate icosahedral groups G_{60} and G_{60}', and indeed, obviously, as the greatest common component of these two groups. The Γ_{30} belonging to the domain of figure 111 is accordingly the greatest common component of two of our conjugate groups Γ_6 and Γ_6' of one system.

This, in fact, verifies itself in figure 111. If, by means of allowed modification, one arranges the discontinuity domain of Γ_{30} cyclically about the point denoted by c_2 in figure 111, then there arises, as a new form of the discontinuity domain of the Γ_{30}, the one given in figure 112, whereby the points of the old arrangement denoted by c_2, b_1 $(\zeta = i)$ and b_5'' carry these same notations in the new one. We again put the coordination of the boundary curves for figure 112 together tabularly:

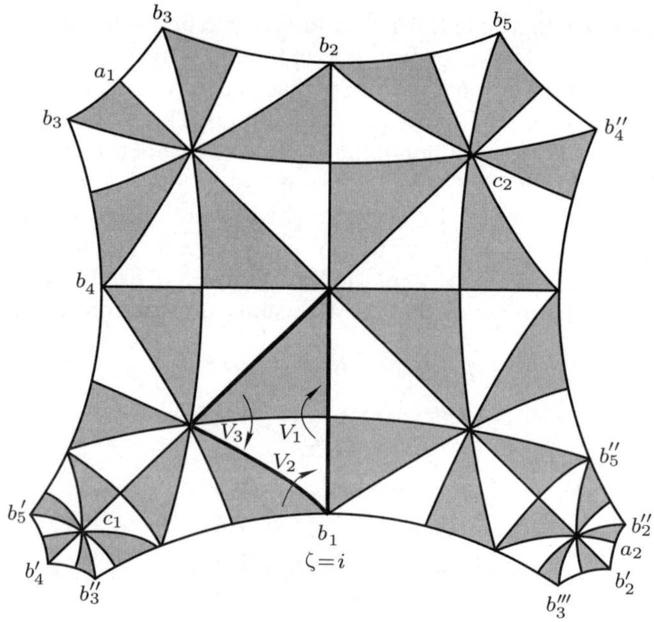

Fig. 111.

$$\begin{cases} (b_1 b_2 b_3 b_4) \longrightarrow (b_1' b_2' b_3' b_4'), \\ (a_1 b_4) \longrightarrow (a_1 b_4'''), \\ (b_4'' b_6') \longrightarrow (b_4''' b_6), \\ (a_2 b_6) \longrightarrow (a_2 b_6'), \\ (b_7 b_5 b_8 b_7''') \longrightarrow (b_7' b_5'' b_8' b_7''), \\ (b_4'' b_7) \longrightarrow (b_4' b_7'''). \end{cases} \tag{7}$$

[600] If we now add, to the Γ_{30}, the substitution, to be denoted by W, of the period five with the fixpoint c_2, then we obtain, in:

$$\Gamma_6 = \Gamma_{30} + W\Gamma_{30} + W^2\Gamma_{30} + W^3\Gamma_{30} + W^4\Gamma_{30},$$

a group, whose discontinuity domain, as one shows by reduction of the coordination of the boundary curves of figure 112 by means of the rotation W about c_2, is the partial polygon with six double triangles, heavily outlined in figure 112. The coordination of the boundary curves of this partial polygon is given in its interior. As one sees, we have arrived back here at the discontinuity domain of figure 108, through which, in fact, we were able to define one system of six conjugate Γ_6.

The second group Γ_6' conjugate with the Γ_6 lets itself be obtained in the same manner, by cyclically arranging the domain of figure 111, by allowed modification, about the point c_1, which furnishes the similarly denoted point of figure 112. Namely, the group Γ_6' is simply given by $\Gamma_6' = V_2^{-1}\Gamma_6 V_2$, as one concludes from the circumstance, that indeed, the Γ_{30}, but not the Γ_6 is transformed into itself by V_2. We close with the theorem: *If one combines the 6 groups Γ_6 of the individual system in pairs, which gives,*

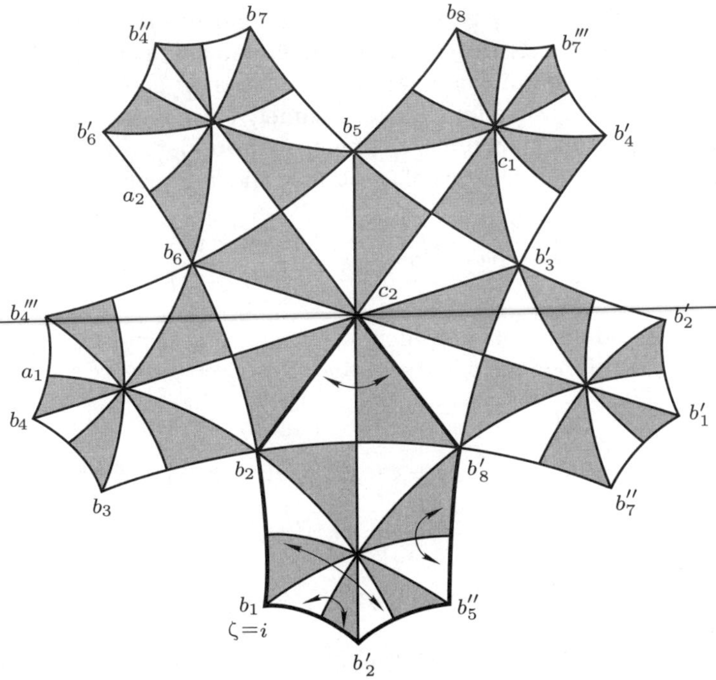

Fig. 112.

in all, 15 *pairs* Γ_6, Γ'_6, *then the greatest common component of the groups* Γ_6, Γ'_6 *of the individual pair is one of our* 15 *conjugate* Γ_{30} *of the signature* (1,3;2,2,2).

A.9 The two resolvents of the 15th degree of the transformation equation

The preceding group-theoretic-geometric considerations lay the basis for obtaining the *two resolvents of* 15th *degree of the transformation equation* corresponding to the Γ_{15}. To be sure, the relations are somewhat difficult here; the direct approach of a principal function of the chosen Γ_{15} and the attempt to determine the relation between it and the principal function $J(\zeta)$ of the total group from the branching of the polygon of figure 110 over the J-plane according to the functional determinant method, is without prospect. Rather, we must here take a somewhat wider detour. [601]

According to the concluding theorem of the precious paragraph, the Γ_{30} of the discontinuity domain of figure 111 is the greatest common component of two groups Γ_6 and $\Gamma'_6 = V_2^{-1}\Gamma_6 V_2$, whose principal functions:

$$\psi(\zeta),\ \psi'(\zeta) = \psi(V_2(\zeta)), \tag{1}$$

are to be chosen according to the prescription given on p. 592. *We have, for our* Γ_{30} *of the genus* $p = 1$, *in* $\psi(\zeta)$ *and* $\psi'(\zeta)$, *directly a function system, which in this specific case, assumes the individual pair of associated values at only one point of the discontinuity domain.* Since $\psi(\zeta)$ and $\psi'(\zeta)$ are five-valued in this domain, these functions enter into an algebraic relations with one another, which, in each of them, ascends to the fifth degree. This relation is givable immediately. Namely, since ψ and ψ' are two different solutions of the equation (4) on p. 593 belonging to the upper sign (cf. p. 595), then one finds, by setting the two expressions of $(J-1)$ in ψ and ψ' equal:

$$\left(\psi-\frac{11+3i\sqrt{15}}{2}\right)^2(\psi-1)^4\psi'-\left(\psi'-\frac{11+3i\sqrt{15}}{2}\right)^2(\psi'-1)^4\psi=0.$$

If we free the left side from its non-identically vanishing factor $(\psi-\psi')$, there then follows, as an *algebraic relation of the genus* $p=1$ *between* ψ *and* ψ':

$$\begin{aligned}(\psi+\psi')^4\psi\psi'-3(\psi+\psi')^2(\psi\psi')^2+(\psi\psi')^3\\ -3(5+i\sqrt{15})((\psi+\psi')^3\psi\psi'-2(\psi+\psi')(\psi\psi')^2)\\ +3\frac{31+19i\sqrt{15}}{2}((\psi+\psi')^2\psi\psi'-(\psi\psi')^2)\\ -2^2\cdot 7\cdot(2+3i\sqrt{15})(\psi+\psi')\psi\psi'+3(8+37i\sqrt{15})\psi\psi'+\frac{1}{2}(7-33i\sqrt{15})=0.\end{aligned} \tag{2}$$

This relation is symmetric in ψ and ψ'. The interchange of ψ and ψ' represents a transformation into itself of period two of the algebraic model furnished by (2); and indeed, it is a question thereby of that transformation, which belongs to the transformation V_2 of the Γ_{30} into itself.

Before we set up functions of lesser valuedness for the Γ_{30}, we go to Γ_{15}, which arises from the Γ_{30} by addition of V_2. The symmetric functions:

$$t(\zeta)=\psi(\zeta)\cdot\psi'(\zeta),\ u(\zeta)=\psi(\zeta)+\psi'(\zeta) \tag{3}$$

of the ψ, ψ', which, in consequence of (2), are bound to one another by the relation:

$$\begin{aligned}tu^4-3t^2u^2+t^3-3(5+i\sqrt{15})(tu^3-2t^2u)+\frac{3}{2}(31+19i\sqrt{15})(tu^2-t^2)\\ -2^2\cdot 7(2+3i\sqrt{15})tu+3(8+37i\sqrt{15})t+\frac{1}{2}(7-33i\sqrt{15})=0,\end{aligned} \tag{4}$$

[602] furnish us functions of the Γ_{15}. Now it is essential, that not only $t(\zeta)$ but also $\sqrt{t(\zeta)}$ is a single-valued automorphic function of the Γ_{15}. Namely, the functions ψ and ψ' have their zeros and poles, in the polygon of figure 111, exclusively at the points c; and indeed, $\psi(\zeta)$ becomes 0 fivefold at c_2 and simply ∞ at each of the remaining points c, while $\psi'(\zeta)$ becomes 0 fivefold at c_1 and simply infinite at each of the remaining points c. For the left half of the polygon mentioned, i.e. for the discontinuity domain of the Γ_{15} given in figure 110, it accordingly follows, that $t=\psi\cdot\psi'$ becomes 0 fourfold at the point c_1 of this domain and twofold ∞ at each of the two points c_2 and c_3. Since other zeros and poles are lacking, and since, on the other hand, the Γ_{15} is of genus $p=0$, it follows, *that:*

$$v(\zeta)=\sqrt{t(\zeta)}=\sqrt{\psi(\zeta)\psi'(\zeta)} \tag{5}$$

is also a single-valued function of the Γ_{15}, *which is* 0 *twofold at* c_1 *and simply* ∞ *at* c_2 *and* c_3. By (5), the function v is first determined up to sign; we reserve the choice of this.

The algebraic relation subsisting between the functions $u(\zeta)$ and $v(\zeta)$ we are able to derive from (4). Namely, if we substitute $t = v^2$ here, then the equation arising:

$$u^4v^2 - 3u^2v^4 + v^6 - (15+3i\sqrt{15})(u^3v^2 - 2uv^4) + \frac{1}{2}(93+57i\sqrt{15})(u^2v^2 - v^4)$$
$$-(56+84i\sqrt{15})uv^2 + (24+111i\sqrt{15})v^2 + \frac{1}{2}(7-33i\sqrt{15}) = 0,$$

shows itself to be reducible. In fact, the left side of this equation lets itself be decomposed into the two factors:

$$\left(u^2v - \frac{1}{2}(15+3i\sqrt{15})\right)uv - v^3 + (12+3i\sqrt{15}v)$$
$$\pm\left(uv^2 - \frac{1}{2}(15+3i\sqrt{15})v^2 + \frac{1}{2}(11+3i\sqrt{15})\right),$$

which interchange themselves for sign change of v. Since we are now still able to arrange the sign of $v(\zeta)$, we determine this in such a manner, *that the two functions* $u(\zeta)$ *and* $v(\zeta)$ *of the group* Γ_{15} *of the genus* $p = 0$ *enter into the algebraic relation:*

$$u^2v + uv^2 - v^3 - \frac{1}{2}(15+3i\sqrt{15})uv - \frac{1}{2}(15+3i\sqrt{15})v^2 + (12+3i\sqrt{15})v + \frac{1}{2}(11+3i\sqrt{15}) = 0 \tag{6}$$

with one another.

This relation verifies, that $v(\zeta)$ is a two-valued function in the polygon of the Γ_{15}, and shows, further, that $u(\zeta)$ is three-valued there. But we are now also easily able, and [603]
indeed in two ways, to arrive at a principal function of the Γ_{15}. If we first conceive the equation (6) as a quadratic equation for u, then this belongs to a two-sheeted Riemann surface of the genus $p = 0$ over the v-plane. At the branchpoints of this surface, the discriminant of the quadratic equation (6) vanishes:

$$v\left(5v^3 + (15+3i\sqrt{15})v^2 - \frac{1}{2}(51-21i\sqrt{15})v - (22+6i\sqrt{15})\right). \tag{7}$$

One of the branchpoints lies at $v = 0$; and since, on account of the genus $p = 0$, only *one* further branchpoint comes up, then the cubic factor in (7) must necessarily contain the square of a linear factor. This linear factor also causes the derivative of the cubic factor to vanish:

$$15v^2 + (30+6i\sqrt{15})v - \frac{1}{2}(51-21i\sqrt{15}) = 0,$$

and can easily be found from this. On the strength of this, one decomposes the discriminant (7) into the factors:

$$v(v+4)\left(v\sqrt{5} - \frac{1}{2}(\sqrt{5}-3i\sqrt{3})\right)^2, \tag{8}$$

so that the second branchpoint of the Riemann surface in question is situated at $v = -4$. As solution of the equation (6) for u, we note:

$$u = -\frac{1}{2}v + \frac{1}{4}(15+3i\sqrt{15}) \pm \frac{1}{2}\left(v\sqrt{5} - \frac{1}{2}(\sqrt{5}-3i\sqrt{3})\right)\sqrt{\frac{v+4}{v}}. \tag{9}$$

We are now able to obtain *a principal function* $\chi(\zeta)$ *of the* Γ_{15} in the form:

$$\chi(\zeta) = \sqrt{\frac{v(\zeta)+4}{v(\zeta)}} \tag{10}$$

and will so determine the sign still disposable here, *that* $\chi(\zeta)$ *enters into the relation, following from* (9), *with the uniquely defined functions u and v:*

$$u = -\frac{1}{2}v + \frac{1}{4}(15+3i\sqrt{15}) - \frac{1}{2}\left(v\sqrt{5} - \frac{1}{2}(\sqrt{5}-3i\sqrt{3})\right)\cdot\chi. \tag{11}$$

The second manner of defining a principal functions of the Γ_{15} attaches to the interpretation of u and v as rectangular coordinates in a plane. Equation (6) then represents a curve of third degree, which, as belonging to the genus $p = 0$, must have a double point. The desired principal function is then cut out on this curve by a pencil of lines with the center at the double point. *But this representation of the principal function* $\chi(\zeta)$ *on the curve of third degree emerges immediately from the equation* (11) *in the form:*
[604]

$$\chi = -\frac{4u+2v-3(5+i\sqrt{15})}{2\sqrt{5}v-\sqrt{5}+3i\sqrt{3}}. \tag{12}$$

Simultaneous vanishing of numerator and denominator of the right side of this equation enters at the point with the coordinates:

$$u_0 = \frac{7\sqrt{5}+9i\sqrt{3}}{2\sqrt{5}},\quad v_0 = \frac{\sqrt{5}-3i\sqrt{3}}{2\sqrt{5}}, \tag{13}$$

whereby, as one easily shows, a double point of our curve of third degree is in fact obtained.

One now finds the rational expressions of the third and second degree of u and v in the principal function $\chi(\zeta)$ from (10) and (11) very easily:

$$\begin{cases} u = \psi + \psi' = \dfrac{(\sqrt{5}-3i\sqrt{3})\chi^3 + 3(5+i\sqrt{15})\chi^2 - 3(3\sqrt{5}-i\sqrt{3})\chi - (23+3i\sqrt{15})}{2^2(\chi^2-1)}, \\ v = \sqrt{t} = \sqrt{\psi\psi'} = \dfrac{2^2}{\chi^2-1}. \end{cases} \tag{14}$$

If we now conceive $\psi(\zeta)$ *and* $\chi(\zeta)$ *as a system of functions of the group* Γ_{30}, *which are* 5- *resp.* 2*-valued in the associated discontinuity domain, then, according to* (14), *these appear bound by the algebraic relation*:

$$\begin{aligned} 4\psi^2(\chi^2-1)^2 - \psi(\chi^2-1)((\sqrt{5}-3i\sqrt{3})\chi^3 + 3(5+i\sqrt{15})\chi^2 \\ -3(3\sqrt{5}-i\sqrt{3})\chi - (23+3i\sqrt{15})) + 64 = 0. \end{aligned} \tag{15}$$

The resolvent of the 15th degree of the transformation equation, whose discovery is the goal of the present investigation, can now be obtained by elimination of ψ from this equation (15) and the equation (4) on p. 593 (for the upper sign). However, this calculation too, on account of its extent, shows itself to be unable to be carried through. Accordingly, we take another path, by explicitly determining *the distribution of values of the principal function χ in the discontinuity domain of the* Γ_{15}.

Since, at the points c_1, c_2, c_3 of the polygon of the Γ_{15}, we had the values $v(c_1) = 0$, $v(c_2) = \infty$, $v(c_3) = \infty$, then we first of all determine:

$$\chi(c_1) = \infty, \ \chi(c_2), \chi(c_3) = \pm 1. \tag{16}$$

To the double point of the curve (6) correspond two places of the polygon of the Γ_{15}, at which, in consequence of (13):

$$\psi + \psi' = \frac{7\sqrt{5} + 9i\sqrt{3}}{2\sqrt{5}}, \ \psi\psi' = -\frac{11 + 3i\sqrt{15}}{10}.$$

The values ψ occurring at these two place are therefore the solutions of the quadratic equation: [605]

$$\psi^2 - \frac{7\sqrt{5} + 9i\sqrt{3}}{2\sqrt{5}}\psi - \frac{11 + 3i\sqrt{15}}{10} = 0.$$

But here on the left stands the basis of the quadratic factor in the first term on the right of equation (4) on p. 593. In the relations of figure 108 on p. 592, therefore, the two value systems ψ, ψ' coming into consideration here are the following:

$$\psi = \psi(a_3), \ \psi' = \psi(a_4) \text{ and } \psi = \psi(a_4), \ \psi' = \psi(a_3).$$

If one now goes into the positioning of the polygons of the groups Γ_6 and Γ_6' belonging to ψ and ψ' into the polygon of figure 111, then one easily determines, that in the polygon of figure 110, the two places $a_5 = a_5'$ and a_9 carry the two value systems ψ, ψ' just given. *To the double point of the curve of third degree thus correspond the two places $a_5 = a_5'$ and a_9 of our domain.* If we again take up the notations u_0, v_0 for the coordinates of the double point, then from (12) and (13) follows:

$$\frac{u - u_0}{v - v_0} = -\frac{\sqrt{5}}{2}\chi - \frac{1}{2}.$$

On the other hand, one calculates from the equation of the curve of third degree, according to elementary rules, that the two limiting values for the left side of the last equation occurring at the double point are the solutions of the quadratic equation:

$$\left(\frac{u - u_0}{v - v_0}\right)^2 + \left(\frac{u - u_0}{v - v_0}\right) - \frac{41 + 15i\sqrt{15}}{16} = 0.$$

For the two associated values χ we therefore find the equation:

$$\chi^2 - \frac{1}{4}(9 + 3i\sqrt{15}) = 0.$$

We therefore note, as a result:

$$\chi(a_5), \chi(a_9) = \pm\frac{1}{2}\sqrt{9+3i\sqrt{15}}. \tag{17}$$

For further calculations, we present the following general argument: If one knows the values ψ and ψ' at an individual vertex of the polygon of the Γ_{15}, then with this are also known the values u and v^2, as well as the value v, in consequence of the relation, yielding itself from (6):

$$v = -\frac{2uv^2 - (15+3i\sqrt{15})v^2 + 11 + 3i\sqrt{15}}{2u^2 - 2v^2 - (15+3i\sqrt{15})u + 24 + 6i\sqrt{15}}. \tag{18}$$

But then, we find the associated $\chi(\zeta)$ from (12).

[606] Thus, first of all, at the vertex b_1 ($\zeta = i$):

$$\psi = \psi' = \frac{1}{2}(11+3i\sqrt{15}),$$

from which one calculates, in the fashion described:

$$u = 11+3i\sqrt{15},\ v = -\frac{1}{2}(11+3i\sqrt{15}).$$

With this, equation (12) furnishes:

$$\chi(b_1) = \frac{1}{8}(8\sqrt{5}+i\sqrt{3}). \tag{19}$$

At the points b_2 and b_3, ψ and ψ' have the values:

$$\psi = 1,\ \psi' = \frac{1}{2}(11+3i\sqrt{15}),$$

from which we calculate, for u and v:

$$u = \frac{1}{2}(13+3i\sqrt{15}),\ v = \pm\sqrt{\frac{1}{2}(11+3i\sqrt{15})} = \pm\frac{1}{2}(3\sqrt{3}+i\sqrt{5}).$$

Both values v are to be applied, since we must obtain two different values χ.[17] Equation (12) yields:

$$\chi(b_2), \chi(b_3) = -\frac{1}{4}(\sqrt{5}-i\sqrt{3}\pm(1-i\sqrt{15})i), \tag{20}$$

two values, which satisfy the quadratic equation:

$$\chi^2 + \frac{1}{2}(\sqrt{5}-i\sqrt{3})\chi - \frac{1}{4}(3+i\sqrt{15}) = 0. \tag{21}$$

There follow the vertices b_4 and b_5, which carry the value system $\psi = 1$, $\psi' = 1$ and therefore $u = 2$, $v = \pm 1$. Here too, both values v are to be utilized, since we must get two different values χ.[18] These two values calculate themselves from (12) to be:

---

[17] Notice, that for the values u, v in question, the right side of (18) appears in the form $\frac{0}{0}$

[18] Also for $u = 2$, $v^2 = 1$, the right side of (18) assumes the form $\frac{0}{0}$.

$$\chi(b_4),\ \chi(b_5) = \sqrt{5},\ -i\sqrt{3}. \tag{22}$$

Here, however, a new difficulty enters. The two points b_4 and b_5 are of different kinds, insofar as, on the polygon of the Γ_{15}, considered closed, the point b_4 is surrounded by four, but b_5 only two, double triangles. For setting up the resolvent of the 15th degree, we must accordingly know, which of the two values given on the right in (22) is $\chi(b_4)$ and which is $\chi(b_5)$.

For this purpose, by comparison of the coordination of edges in figures 110 and 111, we will make the following determination: If we position the polygon of the Γ_{30} given [607]
in figure 111 over the χ-plane (the polygon of figure 110 considered closed), then there arises *a two-sheeted Riemann surface of the genus* $p = 1$ *with four branchpoints,* which originate from the vertices a_3, a_7, b_1, b_5 of figure 110. Now the relation (15) belongs to this surface. If we therefore conceive this as a quadratic equation for ψ, then the discriminant, ascending to the 10th degree in χ, must be composed of the square of a cubic factor and a biquadratic factor with the four roots $\chi(a_3)$, $\chi(a_7)$, $\chi(b_1)$, $\chi(b_5)$. The actual calculation of the discriminant of (15), in fact, shows that this contains the square of:

$$\chi^3 + i\sqrt{3}\chi^2 - \chi - i\sqrt{3},$$

but besides this, the biquadratic factor:

$$(\chi - \sqrt{5})\left(\chi - \frac{1}{8}(3\sqrt{5} + i\sqrt{3})\right)\left(\chi^2 + \frac{5}{8}(\sqrt{5} + 3i\sqrt{3})\chi + \frac{1}{8}(7 + 9i\sqrt{15})\right).$$

With this, we set:

$$\chi(b_5) = \sqrt{5},\ \chi(b_4) = -i\sqrt{3}, \tag{23}$$

while the two values $\chi(a_3)$, $\chi(a_7)$ are the solutions of the quadratic equation:

$$(\chi - \chi(a_3))(\chi - \chi(a_7)) = \chi^2 + \frac{5}{8}(\sqrt{5} + 3i\sqrt{3})\chi + \frac{1}{8}(7 + 9i\sqrt{15}). \tag{24}$$

At the point a_4, the solutions of the quadratic equation:

$$\psi^2 - \frac{8\sqrt{5} + 6i\sqrt{3}}{\sqrt{5}}\psi + 25 = 0, \tag{25}$$

which one takes from the first term of the right side of (4) on p. 593, are the values ψ, ψ' occurring. With this, it follows:

$$u = \frac{8\sqrt{5} + 6i\sqrt{3}}{\sqrt{5}},\quad v = \pm 5,$$

and the relation (18) gives, as the appropriate value, $v = 5$. With this, however, (12) furnishes immediately:

$$\chi(a_4) = -\frac{3}{\sqrt{5}}. \tag{26}$$

For the determination of the values $\chi(a_1)$, $\chi(a_2)$, $\chi(a_6)$, $\chi(a_8)$ yet lacking, one has to combine each solution of (25) with each of:

$$\psi^2 - \frac{7\sqrt{5}+9i\sqrt{3}}{2\sqrt{5}}\psi - \frac{11+3i\sqrt{5}}{10} = 0.$$

Here, however, the calculations fashion themselves not substantially new, so that we
[608] forego the individual values $\chi(a_1),\ldots,$ and will only produce the biquadratic expression, which vanishes for them. For this purpose, observe, that all the values $\chi(b)$ and $\chi(c)$ are known. We are already able to build up the resolvent of the 15th degree, to the extent, that we are able to represent $(J-1)$ in closed form up to a numerical factor in χ. The lacking biquadratic factor then simply yields itself, according to the functional determinant method, in the form:

$$\chi^4 - \frac{4}{\sqrt{5}}\chi^3 + \frac{1+15i\sqrt{15}}{20}\chi^2 + \frac{3\sqrt{5}-15i\sqrt{3}}{10}\chi + \frac{9+3i\sqrt{15}}{20}. \tag{27}$$

Finally, the numerical factor just mentioned in the resolvent of the 15th degree is yet to be determined. For this purpose, we put down an approximate calculation, valid for the neighborhood of the point c_1. There, J, ψ and χ simultaneously become infinite; and indeed, the highest terms of the relations (4) on p. 593 and (15) on p. 604 furnish the relations:

$$J = \frac{5^2\sqrt{5}}{2^8\cdot 3^4 i\sqrt{3}}\cdot\psi^5,\ \psi = \frac{1}{4}(\sqrt{5}-3i\sqrt{3})\chi.$$

It follows from this, further:

$$\chi^5 = \frac{3^4 i\sqrt{3}}{2^7\cdot 5^2\sqrt{5}}(\sqrt{5}+3i\sqrt{3})^5\cdot J,$$

from which the numerical factor in question can be read off.

By collecting all the results of our calculations, we find, *as the expression of one resolvent of the* 15th *degree of our transformation equation of the* 10th *degree:*

$$\begin{aligned}
&J : J-1 : 1\\
&= \left(\chi + \frac{3}{\sqrt{5}}\right)\left(\chi^2 + \frac{5\sqrt{5}+15i\sqrt{3}}{8}\chi + \frac{7+9i\sqrt{15}}{8}\right)\left\{\left(\chi^2 - \frac{9+3i\sqrt{15}}{4}\right)\right.\\
&\quad\left.\cdot\left(\chi^4 - \frac{4}{\sqrt{5}}\chi^3 + \frac{1+15i\sqrt{15}}{20}\chi^2 + \frac{3-3i\sqrt{15}}{2\sqrt{5}}\chi + \frac{9+3i\sqrt{15}}{20}\right)\right\}^2\\
&\quad : \left(\chi - \frac{3\sqrt{5}+i\sqrt{3}}{8}\right)(\chi-\sqrt{5})^2\cdot\left\{(\chi+i\sqrt{3})\left(\chi^2 + \frac{\sqrt{5}-i\sqrt{3}}{2}\chi - \frac{3+i\sqrt{15}}{4}\right)\right\}^4\\
&\quad : \frac{3^4 i\sqrt{3}}{2^7\cdot 5^2\sqrt{5}}(\sqrt{5}+3i\sqrt{3})^5(\chi^2-1)^5,
\end{aligned} \tag{28}$$

from which the second resolvent of the 15th *degree proceeds simply by sign change of* i*.*

A.10 Note on the grups Γ_{20} belonging to the ten conjugate G_{18}

[609]

We determined on p. 588, that a distinguished subgroup G_{18} of index 2 is contained in the individual one of the ten conjugate G_{36}, which contains the substitutions of the periods two and three but not those of the period four of the G_{36}. To these G_{18} correspond *ten conjugate congruence groups* Γ_{20} *of the third level, which also belong to the genus* $p = 0$. Namely, if we choose as an example the Γ_{10} belonging to the double quadrangle of figure 106, then, according to a remark just made, the substitutions of period two belonging to the vertices a_4, a_5, a_6 are indeed contained in the associated subgroup Γ_{20}, but, of the substitutions of the period four belonging to the vertices b_1, b_4 and b_5, only the second powers. If we accordingly position the discontinuity domain of the Γ_{20} as a two-sheeted surface over the φ-plane (the discontinuity domain of the Γ_{10} considered closed), then here yield themselves thereby, two and only two branch-points, which correspond to the vertices b_1 and b_4 (or b_5). The Γ_{20} is therefore actually of the genus $p = 0$.

The production of principal functions of this Γ_{20} is possible without further ado. Since, namely, in consequence of the transformation equation (17) on p. 576, the values of the principal function $\varphi(\zeta)$ at the vertices b_1 and b_4 are the roots of the quadratic equation:

$$\varphi^2 + \sqrt{15}(1 + i\sqrt{3})\varphi + \frac{1}{2}(1 - i\sqrt{3}) = 0,$$

then follows immediately: *From the ten principal functions* $\varphi(\zeta)$, *principal functions for the ten conjugate* Γ_{20} *let themselves be obtained in the form:*

$$\sqrt{\frac{\varphi - \frac{1}{2}(4 - \sqrt{15})(1 + i\sqrt{3})}{\varphi + \frac{1}{2}(4 + \sqrt{15})(1 + i\sqrt{3})}}. \tag{1}$$

The preceding consideration attains its proper significance further below. Namely, the Galois resolvent of the transformation equation, as we shall soon see, has a Riemann surface of the genus $p = 10$. The ten linearly independent functions φ,[19] producable from the integrals of the first kind, furnish in their quotients, as in known, functions of a valuedness $\leqslant 2p-2$, i.e. therefore $\leqslant 18$. But now, our principal functions (1) of the Γ_{20} yield precisely 18-valued functions for the "principal congruence group" Γ_{360} of the third level belonging to the subgroup G_1 of the G_{360}. The suspicion is therefore natural, *that we have before us, in those ten principal functions* (1), *quotients of our ten linearly independent functions* φ. We shall find this directly verified below, and thereby obtain from the approaches developed here, the means of actually carrying through [610]

[19] Cf. "M" I on p. 543. We want not to confuse these ten functions φ with the ten principal functions φ of the Γ_{10}.

the solution of the transformation equation by its Galois resolvent.[20]

A.11 The Riemann surface of the Galois resolvent of the transformation equation

The subgroup G_1 of the G_{360}, consisting of the identity alone, furnishes, as the corresponding subgroup in Γ, *the distinguished principal congruence group* Γ_{360} *of index* 360. The associated discontinuity domain, upon mapping by $J(\zeta)$, yields *a regular-symmetric* 360*-sheeted Riemann surface* over the J-plane, which, according to p. 557 ff., is *The Riemann surface of the Galois resolvent of our transformation equation.* We call this surface F_{360} and, for the sake of easier perception, may also imagine it as a closed surface situated in space, arising by bending together the coordinated boundary curves of the discontinuity domain. *This surface* F_{360} *permits* 360 *single-valued transformations of the first kind into itself and just as many of the second kind;* the former furnish us a direct geometric interpretation of our group G_{360}, while the addition of the 360 transformations of the second kind yield the interpretation of the group $\bar{G}_{720}$ of the
[611] second kind.

For the regular-symmetric net N of $2\cdot 360$ triangles, which spans the F_{360} closed in space, we will retain the earlier notations for the vertices. *The net will therefore exhibit* 180 *points a, corresponding to the* 180 *two-sheeted branchpoints situated at* $J=0$, *further,* 90 *vertices b, which belong to the* 90 *four-sheeted branchpoints at* $J=1$, *and finally,* 72 *vertices c, which stand opposite just as many five-sheeted branchpoints as* $J=\infty$.

We attach here, at the same time, a few direct consequences from the structure of the G_{360} and the enumerations just carried through. Since we have 36 conjugate cyclic G_5 and 72 points c on the F_{360}, there follows: *The individual cyclic* G_5 *has two fixpoints c.* Similarly, there yields itself: *The individual one of the* 45 *cyclic* G_4, *which are conjugate inside the* G_{360}, *has two fixpoints b; each of the* 45 *groups* G_2 *always have, besides these two points b, four more points a as fixpoints.*

As far as the *symmetry lines* of our surface F_{360} go, a glance at our triangle net (see e.g. figure 111) shows that we have *two* kinds of such lines. On a line of the one kind lie

[20] The following procedure of reasoning also offers itself: In each Γ_{20} there is again contained a distinguished subgroup Γ_{40} of index 2; it is a question of those ten conjugate Γ_{40}, which correspond to the ten groups G_9. The individual Γ_{20} (as one easily deduce from the doubling of the double quadrangle of figure 106) was now a group of the signature $(0,6;2,2,\ldots,2)$. Since the G_9 now no longer contains a single substitution of period two, the discontinuity domain over the plane of the principal function (1) belonging to Γ_{40} furnishes a two-sheeted surface with *six* branchpoints, which, with this, has the genus $p=2$. But, on the hyperelliptic surface, the two-valued function is always representable as the quotient of the two linearly independent differentials of the first kind (cf. "M" I on p. 571). In our case, we are also able to conceive those two differentials of the first kind as such of the Γ_{360}, from which, then the suspicion stated in the text emerges as valid.

If, moreover, we do not pursue further in the text the direct relation obtained between the ten conjugate Γ_{20} and the ten linearly independent functions φ of the group Γ_{360}, then this has its basis in the fact, that for this Γ_{360}, functions of an even far smaller valuedness exist (special functions), whose very interesting theory is now to be developed.

only points a and b, in alternating sequence; on a line of the other kind lie points a, b and c, and indeed, in the arrangement $a, c, b, c, a, c, \ldots$. *That two symmetry lines of the same kind are conjugate,* is self-evident.

In order to enumerate the number of the symmetry lines, we notice that an individual such line admits a cyclic G_μ of displacements into itself. The order μ of this cyclic group yields itself from the number of the triangle sides of the net N, which build up the symmetry line. If we have a symmetry line of the first kind (one such without points c), then this consists of 2μ sides, while a line of the second kind must have 4μ sides. Now pursue, e.g. in figure 112, which represents the discontinuity domain of a Γ_{30}, on the basis of the coordination of the boundary curves given l.c., the symmetry line running through the vertices b_5'', b_8'. It contains the pairs of sides $(b_5'' b_8')$, $(b_8' b_3')$, $(b_3' b_4')$; next to these $(b_4'' b_6')$ arranges itself and then $(b_6 b_5)$, whereupon, then, the pair $(b_5 b_8)$, equivalent with the first, attaches itself. The symmetry line in question contains five pairs of sides, following one another an inequivalent with respect to the Γ_{30} and therefore, with respect to the Γ_{360}. Now since cyclic G_μ with $\mu > 5$ do not come up in the G_{360}, then for each symmetry line of the first kind, $\mu = 5$ holds. As an example of a symmetry line of the second kind, bring in the one running through the points c_1, c_2 of figure 112. One will easily find, that they close themselves with respect to the Γ_{30} and accordingly, also with respect to the Γ_{360}, after running through 20 triangle sides. Therefore $4\mu = 20$, [612]
and with this, $\mu = 5$ again. But now, we have at our disposal, all together, 360 sides (ab) for the symmetry lines of the first kind and $2 \cdot 360$ sides (ac) and (bc) for those of the second kind. We consequently obtain the result: *The regular-symmetric net of the F_{360} exhibits two systems of* 36 *conjugate symmetry lines each.*

The 36 symmetry lines of the individual system are, as we saw, uniquely coordinated to the 36 conjugate G_5, in such a manner, that the individual symmetry line is displaced into itself by the five substitutions of its G_5. Conversely, therefore, to the individual G_5 belongs one symmetry line from each of the two systems. If $\bar{V}$ and $\bar{V}'$ are the two symmetric transformations of the F_{360} in these two symmetry lines, and if V is the generator of the G_5, then V is permutable with $\bar{V}$ as well as with $\bar{V}'$:

$$V\bar{V} = \bar{V}V, \; V\bar{V}' = \bar{V}'V.$$

By addition of $\bar{V}$ to G_5 consequently arises a subgroup $\bar{G}_{10} = G_5 + G_5\bar{V}$ of the $\bar{G}_{720}$, in which V is contained distinguishably; and similarly, V will be contained distinguishably in the subgroup $\bar{G}'_{10} = G_5 + G_5\bar{V}'$ of $\bar{G}_{720}$. But now V, in the $\bar{G}_{720}$, is one of 72 conjugate substitutions. There are therefore, in the $\bar{G}_{720}$, only $720/72 = 10$ substitutions, which which V is permutable. The two groups $\bar{G}_{10}$ and $\bar{G}'_{10}$ are thus identical, so that $\bar{V}'$ is one of the five substitutions $V^\nu \bar{V}$. Now since:

$$\bar{V}'^2 = V^\nu\bar{V} \cdot V^\nu\bar{V} = V^{2\nu} \cdot \bar{V}^2 = V^{2\nu} = 1$$

must hold, then $\nu = 0$, and therefore $\bar{V}' = \bar{V}$. There accordingly yields itself: *Among the* 360 *transformations of the second kind of the surface F_{360} into itself, there are* 36 *and only* 36 *symmetric transformations; these are uniquely coordinated to the* 36 *conjugate G_5, and each of them has two symmetry lines, one from each individual system.*

By means of the symmetry lines, one can develop numerous additional geometric discussions concerning the structure of the G_{360}. With consideration of later investigations, however, we only draw attention to the fact, that five points a and, diametrically

opposite (on the symmetry line), five points b always lie on the individual symmetry line. They are the fixpoints of substitutions of period two, which are permutable with the G_5. We arrive back at the 36 dihedral groups G_{10} and above (p. 588), and are able to state the theorem: *The individual pair of associated symmetry lines, as one of* 36 *con-*
[613] *jugate pairs, is carried over into itself, all together, by ten transformations of the* G_{360}, *forming a dihedral group.* Each of the individual substitutions of period two adding itself to the G_5 has one fixpoint a on the individual symmetry line, lying opposite this a (on the symmetry lines) its two fixpoints b, while the two fixpoints a still lacking are on the remaining symmetry lines through these points b diametral to these latter.

Now as far as the *function-theoretic character* of our F_{360} goes, we first of all determine the genus p. As a 360-sheeted surface over the J-plane, our F_{360} has the 180 points a, 90 points b and 72 points c, already named above, as branchpoints, with two, resp. four and five cyclically connected sheets. Further branchpoints, however, do not come up. In consequent of a known rule (cf. (2) in "M" I on p. 494), we accordingly calculate, *that our Riemann surface* F_{360} *belongs to the genus* $p = 10$.

Further, we immediately determine, *that the* F_{360} *cannot be hyperelliptic.* Namely, were it hyperelliptic, then there would exist two-valued functions on it, which, with respect to the 360 single-valued transformations of the surface into itself, would experience a G_{360} or a G_{180} of linear substitutions; but such groups do not exist. In this situation, the ten linearly independent functions φ arising from the integrals of the first kind, according to "M" I on p. 569, map our surface onto a "normal curve" of the 18th degree situated properly in the space R_9 of nine dimensions, which admits 360 collineations into itself, corresponding to the 360 transformations of the surface into itself.

We have already mentioned these functions before (p. 609 ff.), and discussed their relation to the ten conjugate Γ_{10}. But we already indicated there, that we would have to work here with special functions of lesser valuedness. As soon as such functions exist, we shall always prefer them to the p functions φ. We have here an opportunity, to bring into application the developments on special functions, which we attached to the Riemann-Roch theorem in "M" I on p. 546. In fact, by means of the general method of reasoning given there, we arrive at the simplest function system existing on our surface F_{360}.

We are already able to deduce a considerably more convenient system of functions from the developments on p. 601 ff. If one forms the quotient of the two functions denoted there by $\psi(\zeta)$ and $\psi'(\zeta)$, then, in the polygon of the Γ_{30} (figure 112), this represents a six-valued function, whose six zeros coincide at c_2, while the six poles coincide
[614] at the point c_1. In addition, in consequence of (14) on p. 604:

$$\sqrt{\frac{\psi}{\psi'}} = \frac{4}{\psi'(\chi^2 - 1)}, \tag{1}$$

so that even the root of our quotient is a single-valued function of the Γ_{30}, which has a 0 of order 3 at c_2, but an ∞ of order 3 at c_1. The third root of the expression (1):

$$\sqrt[6]{\frac{\psi}{\psi'}} = \frac{\sqrt[3]{4}}{\sqrt[3]{\psi'(\chi^2 - 1)}} \tag{2}$$

is consequently likewise still unbranched, but can no longer be single-valued, since it would be a single-valued function of the Γ_{30} of the genus $p = 1$. However, one can show[21], *that it is a single-valued function of the* Γ_{360}, *so that we have obtained here a twelve-valued function of the* F_{360}.

We now satisfy symmetry, by immediately placing next to one another the six solutions of one of our two resolvents of the sixth degree. Let these be distinguished from one another by the notations $\psi_1, \psi_2, \ldots, \psi_6$. We then define six ratio-magnitudes $\omega_1, \omega_2, \ldots, \omega_6$ by the proportion:

$$\omega_1 : \omega_2 : \cdots : \omega_6 = \sqrt[6]{\psi_1} : \sqrt[6]{\psi_2} : \cdots : \sqrt[6]{\psi_6}, \tag{3}$$

and form from them at the same time, the sixfold infinite family:

$$\alpha_1\omega_1 + \alpha_2\omega_2 + \cdots + \alpha_6\omega_6, \tag{4}$$

by means of arbitrarily choosable constants α. The quotient of any two quantities of the family (4) is a twelve-valued function of the F_{360}. Moreover, we shall later see directly, and could also already prove with the means standing at our disposal here, that the six quantities ω are linearly-independent.

If we consequently bring into application here the developments of "M" I on p. 551 ff., and interpret, in the spirit of these, the ω as homogeneous coordinates of a space R_5 of five dimensions, then the theorem arises: *By means of the quantities* (3), *we map the surface* F_{360} *onto a "normal curve" of twelfth degree, situated properly in the space* R_5 *of five dimensions; this obviously experiences, as the normal curve of the* φ, 360 *collineations into itself with respect to the* G_{360}. In fact, with respect to the operations of the G_{360}, the ω permute themselves, apart from supervening multiplicative sixth roots of unity.

It is self-evident, that one can also attach the same approach to the second resolvent of sixth degree, whose solutions let be $\psi'_1, \psi'_2, \ldots, \psi'_6$. However, we shall perceive further below, that the family to be obtained from these: [615]

$$\alpha'_1\omega'_1 + \alpha'_2\omega'_2 + \cdots + \alpha'_6\omega'_6$$

is identical with the family (4), from which, for the roots of the one resolvent of sixth degree, by means of 36 particular constants β, there yield themselves representations of the form:

$$\sqrt[6]{\psi'_k} = \beta_{k1}\sqrt[6]{\psi_1} + \beta_{k2}\sqrt[6]{\psi_2} + \cdots + \beta_{k6}\sqrt[6]{\psi_6}, \ (k = 1, 2, \ldots, 6) \tag{5}$$

in the roots of the other resolvent.[22] For the invariant standpoint, therefore, we obtain, from the two resolvents of sixth degree, only one family (4) and, consequently, only *one* curve C_{12} in the R_5.

Now apply to these results the "reciprocity theorem of Brill and Noether" divulged in "M" I on p. 553. In the second equation given in (5) l.c., one has to substitute, corresponding to the sixfold infinite family of twelve-valued ratio-magnitudes:

[21] The developments at the close of the present paragraph are only to have the significance of an orienting sketch, so that we pass over the proof for the sake of brevity.

[22] E.g., the quotient $\psi_1 : \psi'_1$ would also be easily accessible to the direct function-theoretic study.

$$\tau' = 6,\ m = 12,\ p = 10.$$

The theorem named furnishes, as the "reciprocal" or "complementary" family to the family (4), one such wish:

$$m' = 2p - 2 - m = 6,\ \tau = \tau' - m + p - 1 = 3,$$

i.e. therefore, a threefold infinite family of six-valued ratio-magnitudes, whose quotients are *six-valued "special functions"* of the F_{360}. We will denote this family by:

$$\gamma_1 x_1 + \gamma_2 x_2 + \gamma_3 x_3. \tag{6}$$

This furnishes, as image of the F_{360}, a curve C_6 of sixth degree situated in the R_2, i.e. plane, whose equation, tying into the naming of the F_{360}, let be denoted by:

$$F(x_1, x_2, x_3) = 0, \tag{7}$$

and which, corresponding to the G_{360}, again admits 360 *collineations into itself.* This plane curve C_6 has the genus $p = 10$ and is therefore *singularity-free.*

In the family (6), we have now actually obtained the simplest functions of the F_{360}, and, for understanding the situation, will, at the same time, also divulge the theorem: *The sixfold infinite family* (4) *arises from the quadratic combinations:*

$$x_1^2,\ x_2^2,\ x_3^2,\ x_2x_3,\ x_3x_1,\ x_1x_2, \tag{8}$$

[616] *the tenfold infinite family of the functions φ from the cubic ones:*

$$x_1^3,\ x_2^3,\ x_3^3,\ x_1^2x_2,\ x_1^2x_3,\ x_2^2x_1,\ x_2^2x_3,\ x_3^2x_1,\ x_3^2x_2,\ x_1x_2x_3. \tag{9}$$

The six quantities (8) as well as the 10 quantities (9) are linearly independent; for the relation (7) of 6th degree is irreducible. In addition, this theorem will be immediately perspicuous below.

It would, in itself, be possible, to investigate the curve C_6 obtained more closely by means of the purely function-theoretic method of reasoning solely applied up until now. However, in consequence of the 360 collineations of the C_6 into itself, we obtain, in the *ternary invariant theory,* a new and very valuable means of investigation. We shall all the more bring this in here, since the *ternary group of the* 360 *collineations* has already been known for a long time, and already explored penetratingly "invariant-theoretically".

C. Jordan already attempted to enumerate, more than 30 years ago, all existing ternary groups of finite order[23]. However, the G_{360}, exactly as the ternary G_{168}, coming up for the transformation of 7th degree of the elliptic functions, was overlooked in this enumeration. One is indebted to G. Valentiner[24] for the discovery of the ternary

23 "*Mémoire sur les équations différentielles linéaires à intégrale algébrique*", Crelle's Journ. Vol. 84 (1878), and "*Sur la détermination des groupes d'ordre fini contenus dans le groupe linéaire*", Atti della R. Accademia di Napoli (1880).

24 "*De endelige Transformations-Gruppers-Theori*", Abh. der Dänischen Akademie, 6te Reihe, Vol. 5, p. 64 (1889).

group G_{360}, who also completely gave the cyclic subgroups contained in the G_{360}, but moreover, only noticed the occurrence of subgroups G_{60} of the icosahedral type. A detailed invariant-theoretic investigation of the G_{360} of plane collineations is carried out by A. Wiman[25]. In the following developments for setting up the equation of the C_6, we follow the Wiman representation narrowly in many points.

For obtaining the equation of the C_6, we have three approaches at our disposal, which attach themselves to the three kinds of most comprehensive subgroups G_{24}, G_{60} and G_{36} of our G_{360}. Let the train of thought to be embarked upon be described briefly by attaching to the G_{24}.

The choice of three special ratio-magnitudes x_1, x_2, x_3 for the representation of the family (6) amounts to the choice of a coordinate system in the plane of the C_6. [617]
Now, first of all, bring in an individual octahedral group G_{24}, and remember the (to be reproduced immediately) figure 7 in I on p. 71, which represents the "octahedral division" in the projective (elliptic) plane. For a suitable choice of the coordinates, which we shall denote from now on as the "*oktaedrischen Coordinaten*", the "ternary form" of the G_{24}, which carries the division mentioned over into itself, assumes a very simple form, which, in particular, permits us to easily give the "full form system" of the G_{24}. Among the forms of the 6th degree to be composed of these ground-forms, there must then be a special form $F(x_1, x_2, x_3)$, which, besides the substitutions of the G_{24}, admits still further substitutions into itself, namely those of the G_{360}. Set equal to 0, this form $F(x_1, x_2, x_3)$ furnishes our curve C_6. In this way, we obtain the curve C_6 in the "*oktaedrischen Coordinaten system*" or also, as a "*Covariante der Oktae dergruppe*".

It is self-evident, that we are able to attach a corresponding consideration to an individual icosahedral group G_{60} and therefore represent our C_6 in the "*ikosaedrischen Coordinaten system*", or as a "*Covariante der Ikosaedergruppe*". Finally, the individual ternary G_{36} has a well-known meaning: Namely, as we shall explain below in more detail, it represents the collineations of a plane curve C_3 of third degree into itself of the so-called "*harmonic double-ratio*". We will denote this curve itself as the "*harmonic C_3*" for short, and a system chosen in connection with it as a "*harmonic coordinate system*". Our third approach will furnish the C_6 in the "*harmonic coordinate system*" or as a "*covariant of the harmonic curve C_3*".[26]

The ternary group G_{360} has given rise to a whole series of further investigations, which reach up to the most recent time. We refer here, first of all, to the developments generally, be reserving particular reference for such places, at which the following representation touches more closely with the investigations in question.

A short time after appearance of the Wiman work, Gerbaldi applied himself to the [618]

[25] "*Über eine einfache Gruppe von* 360 *ebenen Collineationen*", Math. Ann. Vol. 47, p. 531 (1895).

[26] If one wants to raise the objection to the representation in the text, that the statements of the present paragraph are furnished partially without proof, then conceive the immediately following developments independently, in such a manner, that one considers the ternary group G_{360} as given directly. In fact, the developments of § A.12 are independent of the preceding function-theoretic investigations and establish, for their part, the existence and form of the ternary G_{360}. One then obtains the connection to the transformation theory of the principal function belonging to the group Γ of the signature $(0,3;2,4,5)$ by means of a method of reasoning based on the limit-circle theorem, which is given below (p. 653) beneath the text.

continuation of the Wiman investigation. In a first work[27] he has calculated the resolvent of the 6th degree on the foundation of the ternary approach, which was obtained one year earlier, say, by the author by means of the function-theoretic method of reasoning followed above (p. 591 ff.) (cf. Note on p. 559). Later on, in a great investigation spreading itself out over four years[28], Gerbaldi has explored the ternary group G_{360} invariant-theoretically and geometrically in all directions, and treated the various resolvents. He has, in this manner, exhaustively worked over the program[29] initially laid out by Wiman, and thereby, to be sure, (as already suspected by Wiman) has been able to arrive at his goal only by a very considerable expenditure of calculations.[30]

Similarly tying into Wiman, Lachtin[31] has calculated the resolvent of the 6th degree and given a linear differential equation of the 3rd order with three singular points, in whose integrals the roots of that resolvent are rationally representable. It is a question hereby of that "differential resolvent" of the ternary G_{360}, which for just this group, corresponds precisely to the differential equation of the third order of the ternary group G_{168}, given in "M" I p. 706 under (9) in connection with the investigations of Halphen and Hurwitz. In a later work[32], Lachtin generalizes his investigation, to the extent, that he no longer refers only to the curve C_6, but generally to the plane of the coordinates x_1, x_2, x_3, which underlie the ternary G_{360}. The "differential resolvent" initially clothes itself here in the form of three linear partial differential equations of the 2nd order. In place of the special resolvent of the 6th degree of the precious study, we then get an algebraic equation of 6th degree of general character; Lachtin expresses himself, in the introduction to the last-named work, concerning the significance, which he ascribes
[619] to his investigations for the general theory of the equations of the 6th degree.

This significance of the ternary group G_{360} for the general theory of the equations of the 6th degree stands at the center of the investigations of Klein and Gordan to be named now. Klein[33] utilizes the "Galois problem" of the G_{360}, to be set up below (in § 17), in the same way as the "normal problem" for the solution of the general equation of the 6th degree, as, in the second part of his "Lectures on the Icosahedral", he reduces the solution of the general equation of the fifth degree to the "icosahedral equation" as normal equation. Before the discovery of the ternary G_{360}, Klein[34] had constructed

27 "*Sul gruppo semplice di* 360 *collineazioni piane*", Math. Ann. Vol. 50, p. 473 (1897).

28 "*Sul gruppo semplice di* 360 *collineazioni piane*". Rend. di Palermo Vols 12, 13, 14, 16 (1898–1902).

29 See the closing words of his oft-cited work, Math. Ann. Vol. 47, p. 556.

30 For the resolvents of the 6th, 10th and 15th degrees, the function-theoretic method based on our transcendental approach above (p. 570 ff.) is the natural and simplest one.

31 "*Die Differentialresolvente einer algebraischen Gleichung* 6*ten* *Grades mit einer Gruppe* 360*ter* *Ordnung*", Math. Ann. Vol. 51, p. 463 (1898).

32 "*Die Differentialresolvente einer algebraischen Gleichung* 6*ten* *Grades allgemeiner Art*", Math. Ann. Vol. 56, p. 445 (1902).

33 "*Sulla risoluzione delle equazioni di sesto grado*", Acc. dei Linc. Vol. 8 (1899); "*Über die Auflösung der allgemeinen Gleichungen* 5*ten* *und* 6*ten* *Grades*", Crelle's Journal Vol. 129, p. 151 (1905), printed in Vol. 61 of the Math. Ann. on p. 50 ff.

34 "*Zur Theorie der allgemeinen Gleichungen* 6*ten* *und* 7*ten* *Grades*", Math. Ann. Vol. 28, p. 499 (1886).

himself a quaternary collineation group [35] G_{360} for the theory of the equations of the 6th degree. Two extended investigations of Gordan[36] also belong essentially to this domain. In the first of these works, Gordan gives very extensive invariant theoretic calculations, which have as their goal, to explicitly produce and establish in more detail the differential equations already considered by Lachtin. In the second work, Gordan gives the continuation of his calculations in connection with the last two works of Klein cited. Most recently, Coble[37] has succeeded in considerably simplifying and fashioning more perspicuously the Gordan investigations.

A.12 The curve C_6 in the octahedral coordinate system

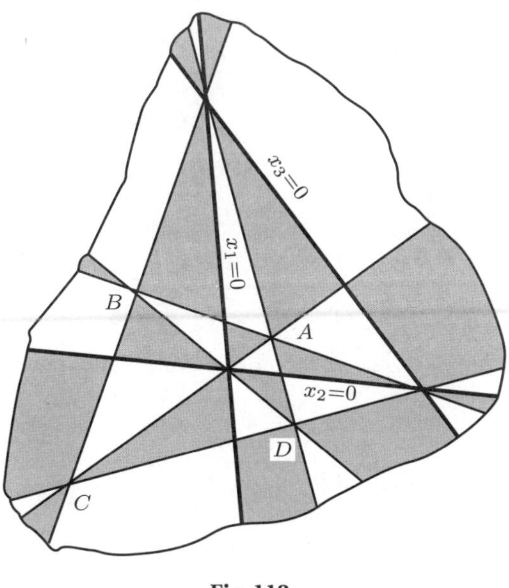

Fig. 113.

In figure 113, the octahedral division of the projective plane, mentioned before, is reproduced. The nine lines of this division are the *perspectivity axes* of the nine collineations of period two contained in the G_{24}, which represent "*harmonic perspectivities*". As is known, they fall into two systems of 6 resp. 3 operations, conjugate inside the G_{24}. The three latter perspectivities, with the identity, form the four group G_4, dis- [620]

[35] Also see the paper of Klein "*Über die Auflösung gewisser Gleichungen vom 7ten und 8ten Grade*", Math. Ann. Vol. 15 (1879).

[36] "*Die partiellen Differentialgleichungen des Valentinerproblems*", Math. Ann. Vol. 61, p. 453 (1905); "*Über eine Klein'sche Bilinearform. (Ein Beitrag zur Auflösung der Gleichungen 6ten Grades)*", Math. Ann. Vol. 68, p. 1 (1910).

[37] "*The reduction of the sextic equation to the Valentiner formproblem*", Math. Ann. Vol. 70, p. 337 (1910).

tinguished in the G_{24}. We choose the axes of these three operations as coordinate axes $x_1 = 0$, $x_2 = 0$, $x_3 = 0$. The octahedral coordinates x_1, x_2, x_3 are, by this, determined up to 3 constant factors; we so choose the latter, that the intersection point of three of the six conjugate perspectivity axes denoted in the figure by A, receive the coordinates $x_1 = 1$, $x_2 = 1$, $x_3 = 1$.

In the coordinate system thus chosen, the G_{24} represents itself simply in this fashion, that we combine the six permutations of the x with the eight transformations obtainable by sign change of the x. Each two of the 48 transformations thus arising, which go over into one another by simultaneous sign change of the original x_1, x_2, x_3, furnish the same collineation; in this specific situation, say, we choose that one of two such transformations, which is unimodular as a ternary substitution.

As generators of the G_{24}, we note:

$$\begin{aligned} &(S) \quad x_1' = -x_2,\ x_2' = +x_1,\ x_3' = +x_3, \\ &(T) \quad x_1' = +x_3,\ x_2' = -x_2,\ x_3' = +x_1, \\ &(U) \quad x_1' = +x_2,\ x_2' = +x_3,\ x_3' = +x_1, \end{aligned}$$

whereby (with insignificant deviation from (6) on p. 587) $ST = U$ holds now, and not $= U^{-1}$. The distinguished G_4 contains, besides the identity, the three perspectivities:

$$\begin{aligned} (TS^2T) &\quad x_1' = +x_1,\ x_2' = -x_2,\ x_3' = -x_3, \\ (U^2S^2U) &\quad x_1' = -x_1,\ x_2' = +x_2,\ x_3' = -x_3, \\ (S^2) &\quad x_1' = -x_1,\ x_2' = -x_2,\ x_3' = +x_3. \end{aligned}$$

The collineations of the G_{24} carry the conic section given by:

$$x_1^2 + x_2^2 + x_3^2 = 0 \tag{1}$$

[621] over into itself; we will designate this as the "octahedral conic section" belonging to the G_{24}. In the sense of the developments in I on p. 15 ff., it is a question here of the absolute conic section of the elliptic determination of measure, to which the division of figure 113 refers itself (cf. (2) in I on p. 17).

In (1) we have obtained the single covariant of 2nd degree of the G_{24}. In addition, the relations here even lie so simply, that we do not need to present a separate consideration on the full form system of the G_{24}. It is to be perceived immediately from the form of the ternary G_{24}, that each covariant of the 6th degree must be a linear combination of the three forms:

$$x_1^6 + x_2^6 + x_3^6,\ x_1^4x_2^2 + x_1^4x_3^2 + x_2^4x_1^2 + x_2^4x_3^2 + x_3^4x_1^2 + x_3^4x_2^2,\ x_1^2x_2^2x_3^2.$$

We therefore have, as a schema: *Our curve C_6 must let itself be represented, in the octahedral coordinate system, by means of certain constants a, b, c, in the form:*

$$a(x_1^6 + x_2^6 + x_3^6) + b(x_1^4x_2^2 + x_1^4x_3^2 + \cdots + x_3^4x_2^2) + cx_1^2x_2^2x_3^2 = 0. \tag{2}$$

In order to determine the coefficients a, b, c, one wants to consider beforehand, that, corresponding to the $2 \cdot 15$ G_{24}, one can initially choose the octahedral coordinate system in 30 ways; and since two conjugate systems are producable from one another by collineations of the G_{360}, for which the left side of (2) is invariant, then there are *only two essentially different choices of octahedral coordinates* x_1, x_2, x_3, belonging to two groups not conjugate inside the G_{360}. These coordinate systems are called essentially different, in the sense that they lead to *different representations* (2) *of our curve* C_6. For the transition from one system to the other is indeed effected by a linear substitution of the x, but this substitution does not belong to the G_{360}, so that it does not leave the left side of (2) invariant. Nevertheless, *a transformation of the second kind*, corresponding, say, to a reflection of the $\bar{G}_{720}$, lets itself be given, *which carries the one system into the other and, thereby, the* C_6 *into itself.*

Accordingly, as long as it is not decided, which of the two essentially different coordinate systems is intended by the x_1, x_2, x_3, each manner of determining the coefficients a, b, c in (2), will necessarily have to lead to *two different value systems.* The choice of one of them then amounts to a unique determination of the coordinate system. In order to briefly characterize the situation found, we could designate the x_1, x_2, x_3 as the "*unsymmetrical octahedral coordinates*"; we shall soon make a choice of coordinates satisfying the requirement of symmetry and standing in closer relation to the x_1, x_2, x_3, for which, just on account of this relation, it is permissible to speak of "*symmetric octahedral coordinates*". [622]

Before we set up the coordinate transformation, which leads from one of the unsymmetric systems to the other, we must consider, how the ternary form of the G_{24} alters itself for a certain alteration of the coordinate system, to be described immediately. We hold fast to the stipulation, that the coordinate axes are to be the three axes of the perspectivities contained in the distinguished four-group G_4. In contrast to this, there is to be the alteration, that be fixpoint A of the generating collineation U may now receive any three finite *real or complex* coordinates α_1, α_2, α_3 different from 0. The new form of the G_{24} in the x is to be characterized in this way, that upon calculation in the coordinates x_1', x_2', x_3', by means of:

$$x_1 = \alpha_1 x_1', \; x_2 = \alpha_2 x_2', \; x_3 = \alpha_3 x_3', \tag{3}$$

the G_{24}, in these x_1', x_2', x_3', again obtains the simple form given above.

The role, which the real octahedral division of figure 113 plays here, becomes clear from the following consideration: If the quotient $\alpha_1 : \alpha_2$ is real, while $\alpha_1 : \alpha_3$ is not real, then, in the real plane of the x_1, x_2, x_3, besides the sides of the coordinate triangle, only the two perspectivity axes:

$$\alpha_1 x_2 \pm \alpha_2 x_1 = 0, \tag{4}$$

running through the vertex $x_1 = 0$, $x_2 = 0$, are real (corresponding to the axes AC and BD of figure 113). In the real coordinate plane, accordingly, only the dihedral division of that subgroup G_8 of the G_{24} is still visible, which comprises the collineations of the G_{24} which are now still real. If both quotients of the α are complex, then, in the G_{24}, only the four collineations of the distinguished G_4 are still real, which even retain their forms $x_1' = x_1$, $x_2' = -x_2$, $x_3' = -x_3$ etc. given above unaltered. Correspondingly, in figure 113, only the three sides of the coordinate triangle remain subsisting in the real part of the plane, representing the "four-division".

We now come back to our two unsymmetrical octahedral coordinate systems, in order to determine their mutual relation. According to p. 586, the individual dihedral

group G_8 is always contained in two octahedral groups G_{24}, G'_{24}, not conjugate inside the G_{360}. We choose, say, the dihedral-G_8 belonging to the collineation S and remember, that in such a G_8, are always contained *two* four-groups, not conjugate inside the
[623] G_{360}. In our G_8, therefore, besides the group denoted so far by G_4, we have a second group G'_4, which has the substitutions 1 and S^2 in common with the first. The G'_4 then, for its part, is distinguished inside an octahedral group G'_{24}, not conjugate with the G_{24}; and we are able to deduce, from this G'_{24}, the second of the unsymmetrical coordinate systems.

This second G'_{24}, in the original coordinate system of the x_1, x_2, x_3, can contain no real collineations besides the G_8. Otherwise, it and, consequently, also the G_{360}, generatable by combination of the G_{24}, G'_{24}, would be entirely real; but there is no real ternary group G_{360} of such a kind (cf. I on p. 15 ff. and p. 69 ff.). If we accordingly introduce the axes belonging to the three perspectivities of the G'_4:

$$x''_1 = x_1 + x_2 = 0,\ x''_2 = x_1 - x_2 = 0,\ x''_3 = x_3 = 0 \tag{5}$$

(lines BD, AC and $x_3 = 0$ of figure 113), as the coordinate axes belonging to the G'_{24}, then the point A' with the coordinates α'_1, α'_2, α'_3, for our G'_{24}, corresponding to the point A of figure 113, can no longer be real. Rather, as an imaginary point, this will be situated on the perspectivity axis:

$$\alpha'_1 x''_2 - \alpha'_2 x''_1 = \alpha'_1(x_1 - x_2) - \alpha'_2(x_1 + x_2) = (\alpha'_1 - \alpha'_2)x_1 - (\alpha'_1 + \alpha'_2)x_2 = 0.$$

But now, the latter axis is the axis $x_2 = 0$ belonging to the real dihedral-G_8; thus $\alpha'_1 = \alpha'_2$, and we may therefore set both coordinates α'_1, α'_2 equal to 1. If we accordingly write:

$$x''_1 = x'_1,\ x''_2 = x'_2,\ x''_3 = \alpha'_3 x'_3,$$

or (by denoting the reciprocal value of α'_3 by α):

$$x'_1 = x_1 + x_2,\ x'_2 = x_1 - x_2,\ x'_3 = \alpha x_3, \tag{6}$$

then the G'_{24}, calculated in the coordinates x'_1, x'_2, x'_3, must again obtain the simple form taken as a basis above for the G_{24}. *There is thus a non-real constant α, so that the transformation* (6) *leads to the second unsymmetric octahedral-coordinate system x'_1, x'_2, x'_3.*

From the form of the G'_{24}, our cuve C_6, corresponding to the schema (2), must be representable in the form:

$$a'(x'^6_1 + x'^6_2 + x'^6_3) + b'(x'^4_1 x'^2_2 + \cdots + x'^4_3 x'^2_2) + c' x'^2_1 x'^2_2 x'^2_3 = 0. \tag{7}$$

For the determination of the unknown constants in (7) and (2) we are now able to uti-
[624] lizes the fact, that the equation (7) must be transformed into the equation (2) by the substitution (6). But if we put the expressions (6) of the x' in the x into the equation (7), then there yields itself, after brief intermediate calculation:

$$\begin{aligned}2(a'+b')(x_1^6+x_2^6) + a'\alpha^6 x_3^6 + (30a' - 2b')(x_1^4x_2^2 + x_2^4x_1^2) + (2b'+c')\alpha^2(x_1^4x_3^2 + x_2^4x_3^2)\\ +2b'\alpha^4(x_3^4x_1^2 + x_3^4x_2^2) + (12b' - 2c')\alpha^2 x_1^2x_2^2x_3^2 = 0.\end{aligned} \tag{8}$$

But this equation always and only then subsumes itself under the schema (2) if the conditions take place:

$$\begin{cases} 2a' + 2b' = a'\alpha^6, \\ 15a' - b' = b'\alpha^4, \\ 2b' + c' = 2b'\alpha^4. \end{cases} \tag{9}$$

If now a' were $= 0$, then b' would also be $= 0$ and $c' = 0$; and therefore a' is not equal to 0, and we may consequently set $a' = 1$. The elimination of b' from the first two equations (9) yields:

$$\alpha^{10} + \alpha^6 - 2\alpha^4 - 32 = 0,$$

an equation of the fifth degree for α^2, with the five solutions:

$$\alpha^2 = 2,\ \alpha^2 = \frac{-3 \pm i\sqrt{7}}{2},\ \alpha^2 = \frac{1 \pm i\sqrt{15}}{2}. \tag{10}$$

The first solution, as real, is unusable in any case. The second equation (10) leads to the ternary group G_{168}, which comes up for the transformation of the 7th degree of the elliptic functions[38]. Namely, remember that, in this G_{168} too, two systems of octahedral groups, not conjugate among one another, come up, and that, there too, each two non-conjugate G_{24}, G'_{24} have a dihedral-G_8 in common. Our schema utilized here must therefore, without any alteration, also lead to that covariant of the 6th degree of the ternary G_{168}, which, in "M" I on p. 733, is denoted by X (z_1, z_2, z_3), in the coordinate system z_1, z_2, z_3 at that time.

For the G_{360}, there accordingly remain only the two values:

$$\alpha^2 = \frac{1 \pm i\sqrt{15}}{2} = \left(\frac{\sqrt{5} \pm i\sqrt{3}}{2}\right)^2, \tag{11}$$

which, put into (9), furnishes the two coefficient systems:

$$a' = 1,\ b' = -\frac{15 \pm 3i\sqrt{15}}{4},\ c' = 15 \mp 3i\sqrt{15}. \tag{12}$$

By putting the calculated values into (8), and leaving out the common factor α^6 of all terms, we must obtain the relation (2) with $a = 1$. Therefore, we find, correspondingly, [625]
for the coefficients a, b, c in (2):

$$a = 1,\ b = -\frac{15 \mp 3i\sqrt{15}}{4},\ c = 15 \pm 3i\sqrt{15}. \tag{13}$$

As we foresaw, we have obtained here two value systems of the coefficients a, b, c. We now choose the octahedral coordinate system of the x uniquely, by setting[39]:

---

[38] Cf. Klein "*Über Transformation* 7ter *Ordnung der elliptischen Functionen*", Math. Ann. Vol. 14 (1878) or "M" I on p. 692 ff.

[39] A sign change of α amounts to an exercise of the perspectivity S^2, and, since this belongs to the G_{24}, is without significance.

$$\alpha = \frac{\sqrt{5}+i\sqrt{3}}{2}, \tag{14}$$

and obtain as a result: *the curve* C_6 *of the sixth degree represents itself, in the octahedral coordinate system* x_1, x_2, x_3 *finally chosen, by the following equation:*

$$x_1^6+x_2^6+x_3^6-\frac{15-3i\sqrt{15}}{4}(x_1^4x_2^2+x_1^4x_3^2+x_2^4x_1^2+x_2^4x_3^2+x_3^4x_1^2+x_3^4x_2^2)$$
$$+(15+3i\sqrt{15})x_1^2x_2^2x_3^2=0. \tag{15}$$

In order to be able to generate all the 360 collineations of this C_6 into itself, we must know at least one further collineation outside the G_{24}. We favor, say, that operation U', which, corresponding to our collineation U above (cf. p. 620), represents the cyclic interchange of the coordinates in the second system, belonging to the G'_{24}. This collineation, belonging to the G'_{24} and therefore to the G_{360}, writes itself, in consequence of (6), in the original coordinates x:

$$x_1'+x_2'=x_1-x_2,\ x_1'-x_2'=\frac{\sqrt{5}+i\sqrt{3}}{2}x_3,\ \frac{\sqrt{5}+i\sqrt{3}}{2}x_3'=x_1+x_2,$$

and from here, is easily calculated around into a unimodular ternary substitution of the x_1, x_2, x_3. Since, for the third generator of the octahedral-G_{24}, the representation $U=ST$ holds, then we have to note, as a result: *the* G_{360} *of the* 360 *collineations of the curve* (15) *into itself is genaratable from the three collineations:*

$$(S)\quad x_1'=-x_2,\ x_2'=x_1,\ x_3'=x_3,$$
$$(T)\quad x_1'=x_3,\ x_2'=-x_2,\ x_3'=x_1,$$
$$(U')\quad \begin{cases} x_1'=\frac{1}{2}x_1-\frac{1}{2}x_2+\frac{\sqrt{5}+i\sqrt{3}}{4}x_3,\\ x_2'=\frac{1}{2}x_1-\frac{1}{2}x_2-\frac{\sqrt{5}+i\sqrt{3}}{4}x_3,\\ x_3'=\frac{\sqrt{5}-i\sqrt{3}}{4}x_1+\frac{\sqrt{5}-i\sqrt{3}}{4}x_2.\end{cases}$$

[626] In the second of our two unsymmetric coordinate systems, as we saw, the curve (15) is given by:

$$x_1'^6+x_2'^6+x_3'^6-\frac{15+3i\sqrt{15}}{4}(x_1'^4x_2'^2+\cdots+x_3'^4x_2'^2)+(15-3i\sqrt{15})x_1'^2x_2'^2x_3'^2=0. \tag{16}$$

Correspondingly, in these x', the G_{360} has the form, which is obtained from the form of the collineation group producably by means of the generators S, T, U' simply by sign change of i in the substitution coefficients.

If one moreover denotes, by $\bar{x}_1, \bar{x}_2, \bar{x}_3$, the complex values conjugate to x_1, x_2, x_3, then the equation (15) of our C_6 can also be clothed in the form:

$$\bar{x}_1^6+\bar{x}_2^6+\bar{x}_3^6-\frac{15+3i\sqrt{15}}{4}(\bar{x}_1^4\bar{x}_2^2+\cdots+\bar{x}_3^4\bar{x}_2^2)+(15-3i\sqrt{15})\bar{x}_1^2\bar{x}_2^2\bar{x}_3^2=0.$$

As one sees, the two equations (15) and (16) are consequently interchanged by the "*ternary substitution of the second kind*":

$$x_1' = \bar{x}_1,\ x_2' = \bar{x}_2,\ x_3' = \bar{x}_3. \tag{17}$$

If we first go from the C_6 given by (15) to the equation (16) by the transformation (6), better clothed in the form:

$$x_1' = \frac{x_1 + x_2}{\sqrt{2}},\ x_2' = \frac{x_1 - x_2}{\sqrt{2}},\ x_3' = \frac{\sqrt{5}+i\sqrt{3}}{2\sqrt{2}} x_3,$$

and afterwards exercise the transformation (17), then we arrive back at the equation (15). *Accordingly, in:*

$$x_1' = \frac{\bar{x}_1 + \bar{x}_2}{\sqrt{2}},\ x_2' = \frac{\bar{x}_1 - \bar{x}_2}{\sqrt{2}},\ x_3' = \frac{\sqrt{5}-i\sqrt{3}}{2\sqrt{2}} \bar{x}_3, \tag{18}$$

we obtain a "collineation of the second kind", which carries our curve C_6 given by (15) *over into itself, and which, consequently, added to the G_{360}, generates, with it, the "collineation group of the second kind" $\bar{G}_{720}$.* If we denote the substitution (18) by $\bar{V}$, then one shows immediately, that $\bar{V}^2 = 1$; it is therefore here a question of one of the 36 *reflections* of the $\bar{G}_{720}$. Further, one calculates easily:

$$\bar{V}S\bar{V} = S^{-1},\ \bar{V}U\bar{V} = U',$$

S, U and U' used in the meaning thus far. The G_{24} is generatable from S and U, the G_{24}' from S and U'. *In* (18), *we have therefore obtained that reflection, which transforms the groups G_{24}, G_{24}' into one another, not yet conjugate inside the G_{360}.* [627]

There yet remains the discussion of the "*symmetric octahedral coordinate system*". Since the operation S is transformed by the reflection $\bar{V}$ into its inverse S^{-1}, then a coordinate triangle, whose vertices are the three fixpoints of the collineation S, is transformed into itself by $\bar{V}$. Such a triangle is choosable in 45 ways; however, these 45 triangles, as the 45 cyclic subgroups G_4 of the G_{360}, are all conjugate, and therefore not essentially different for our purpose.

Now the fixpoints of S are, first, the previous vertex $x_1 = 0$, $x_2 = 0$, and further, the two intersection points of the lines $x_1 \pm ix_2 = 0$ with the previous third axis $x_3 = 0$, which accordingly remains subsisting as such. For the transition to the symmetrical coordinates, which we call X, Y, Z, say, we thus have the schema:

$$X = \alpha(x_1 + ix_2),\ Y = \beta(x_1 - ix_2),\ Z = \gamma x_3,$$

and must so determine the still disposable constants α, β, γ, that not only the coordinate triangle, but the coordinates X, Y, Z are themselves invariant with respect to the reflection $\bar{V}$. First of all, we calculate, as the expression of the reflection (18) in the X, Y, Z:

$$X' = \frac{\alpha}{\bar{\alpha}} \cdot \frac{1+i}{\sqrt{2}} \bar{X},\ Y' = \frac{\beta}{\bar{\beta}} \cdot \frac{1-i}{\sqrt{2}} \bar{Y},\ Z' = \frac{\gamma}{\bar{\gamma}} \cdot \frac{\sqrt{5}-i\sqrt{3}}{2\sqrt{2}} \bar{Z}.$$

If we consequently choose the α, β, γ, say, in the form:

$$\alpha = e^{-\frac{\pi i}{8}},\ \beta = e^{\frac{\pi i}{8}},\ \gamma = \sqrt{5+i\sqrt{15}},$$

then we obtain, as coordinate transformation:

$$X = e^{-\frac{\pi i}{8}}(x_1 + i x_2),\ Y = e^{\frac{\pi i}{8}}(x_1 - i x_2),\ Z = \sqrt{5 + i\sqrt{15}}x_3, \tag{19}$$

and the reflection $\bar{V}$ obtains the simple form:

$$X' = \bar{X},\ Y' = \bar{Y},\ Z' = \bar{Z}. \tag{20}$$

The equation of the C_6, since it is transformed into itself by the reflection (20), will consequently now receive consistently real coefficients, and the expression of the collineation group G_{360} in the X, Y, Z is conjugate to itself; i.e. upon replacement of all substitution coefficients by their conjugate complex values, the group goes over into
[628] itself. The calculation verifies this; we find: *The curve C_6 of the sixth degree, in the symmetric octahedral coordinate system, receives the real equation:*

$$15\sqrt{15}(X^4 - Y^4)(2XY - Z^2) + 100X^3Y^3 + 150X^2Y^2Z^2 + 30XYZ^4 - Z^6 = 0. \tag{21}$$

The remark concerning the associated form of the G_{360} is, with this, self-evident.

A.13 The curve C_6 in the icosahedral coordinate system

The train of thought of the development in setting up the equation of the C_6 in the icosahedral coordinate system is analogous in all essential points to the investigation just concluded. In the ternary representation of an individual icosahedral group G_{60}, we abstain, to be sure, from the real form furnished by figure 16 in I on p. 72; rather, with Wiman (cf. l.c. p. 548), we take as a point of departure that form of the ternary G_{60}, collinear with that real one, which Klein utilizes in "Icos." on p. 211 ff. in the "problem of the A", since, in this way, we are able to directly make use of the form-theoretic developments given there.

Instead of the coordinates A_1, A_2, A_0, we use, tying in better with that so far, the notations y_1, y_2, y_3. As generators of the G_{60} of the ternary substitutions, we then have (cf. "Icos." on p. 213), with use of the abbreviation $e^{\frac{2i\pi}{5}} = \varepsilon$:

$$(S)\quad y_1' = \varepsilon^4 y_1,\ y_2' = \varepsilon y_2,\ y_3' = y_3,$$

$$(T)\quad \begin{cases} \sqrt{5}y_1' = (\varepsilon^2 + \varepsilon^3)y_1 + (\varepsilon + \varepsilon^4)y_2 + 2y_3, \\ \sqrt{5}y_2' = (\varepsilon + \varepsilon^4)y_1 + (\varepsilon^2 + \varepsilon^3)y_2 + 2y_3, \\ \sqrt{5}y_3' = y_1 + y_2 + y_3. \end{cases}$$

Only four substitutions, forming one of the five four-groups, to which T also belongs, are real for this representation. The absolute conic section, which, for the icosahedral division of figure 16 in I on p. 72, is not real, comes out real in the coordinate system of the y chosen here; this has, according to "Icos." on p. 214, (6), the equation:

$$y_1 y_2 + y_3^2 = 0, \tag{1}$$

and will be designated from now on as the "icosahedral conic section" belonging to the G_{60}.

Of independent covariants of the sixth degree, we have, in consequence of "Icos." on p. 214, besides the third power of $(y_1y_2+y_3^2)$, only one more single form, which we bring in in the form B, given l.c. on p. 218 in (14): [629]

$$y_1^3y_2^3-(y_1^5+y_2^5)y_3-2y_1^2y_2^2y_3^2+8y_1y_2y_3^4. \tag{2}$$

Since the C_6 is irreducible, then we immediately obtain the schema: *Our curve C_6, in the icosahedral coordinate y_1, y_2, y_3, must necessarily represent itself in the form:*

$$y_1^3y_2^3-(y_1^5+y_2^5)y_3-2y_1^2y_2^2y_3^2+8y_1y_2y_3^4+a(y_1y_2+y_3^2)^3=0,$$

or, better arranged:

$$(1+a)y_1^3y_2^3-(y_1^5+y_2^5)y_3+(3a-2)y_1^2y_2^2y_3^2+(3a+8)y_1y_2y_3^4+ay_3^6=0, \tag{3}$$

with a constant a yet to be determined.

For the calculation of a, we argue exactly as above (p. 622 ff.). Since we have two systems of icosahedral groups G_{60}, not conjugate inside the G_{360}, we are again able to choose the icosahedral coordinates y_1, y_2, y_3 in two essentially different (i.e. leading to two different equation forms (3)) ways, and shall have to decide to favor one of these two kinds below. Now the individual cyclic G_5, e.g. that arising from the present collineation S, was contained in two non-conjugate icosahedral groups G_{60}, G'_{60}. The fixpoints of this collineation S furnish the vertices of the coordinate triangle of the y. If we consequently choose the second of the two "unsymmetric icosahedral coordinate systems" y'_1, y'_2, y'_3, tying into the G'_{60}, by favoring the collineation S, also contained in it, then the vertices of the coordinate triangle of the y remain maintained, *so that the transition from one coordinate system to the other is to be effected by a transformation of the form:*

$$y'_1=\alpha_1y_1,\ y'_2=\alpha_2y_2,\ y'_3=\alpha_3y_3 \tag{4}$$

with three finite α different from 0.

Now the G'_{60}, in the second coordinate system of the y'_1, y'_2, y'_3, again has the same form, as the G_{60} in the first. We conclude from this immediately, that our C_6 must let itself be represented by means of the y' in the form:

$$(1+a')y_1'^3y_2'^3-(y_1'^5+y_2'^5)y'_3+(3a'-2)y_1'^2y_2'^2y_3'^2+(3a'+8)y'_1y'_2y_3'^4+a'y_3'^6=0.$$

The equation thus obtained must now be carried back, by the transformation (4), into the equation (3) of our C_6, multiplied by a factor b different from 0, in the first system of the y_1, y_2, y_3. But one determines immediately, that this always and only then take place, if the following sixequations hold: [630]

$$\begin{cases}(1+a')\alpha_1^3\alpha_2^3=b(1+a),\\ \alpha_1^5\alpha_3=\alpha_2^5\alpha_3=b,\\ (3a'-2)\alpha_1^2\alpha_2^2\alpha_3^2=b(3a-2),\\ (3a'+8)\alpha_1\alpha_2\alpha_3^4=b(3a+8),\\ a'\alpha_3^6=ba.\end{cases} \tag{5}$$

Now were one of the quantities a, a' equal to 0, then, in consequence of the last equation (5), the other would also vanish; and then, the elimination of b from the remaining equations would yield:

$$\alpha_1^3\alpha_2^3 = \alpha_1^2\alpha_2^2\alpha_3^2 = \alpha_1\alpha_2\alpha_3^4 = \alpha_1^5\alpha_3 = \alpha_2^5\alpha_3.$$

If, therefore, which is permissible, one sets $\alpha_3 = 1$, then follows:

$$\alpha_1\alpha_2 = 1,\ \alpha_1^5 = \alpha_2^5 = 1,\ \alpha_1 = \varepsilon^{\nu}, \alpha_2 = \varepsilon^{-\nu},$$

where ν is a whole number. The substitution (4) would therefore be a power of S, which, however, cannot effect the transition from the y to the y'. Therefore a and a' are certainly not equal to 0.

The elimination of b from the equations (5) yields, further:

$$\begin{cases} a(1+a')\alpha_1^3\alpha_2^3 = a'(1+a)\alpha_3^6, \\ a(3a'-2)\alpha_1^2\alpha_2^2 = a'(3a-2)\alpha_3^4, \\ a(3a'+8)\alpha_1\alpha_2 = a'(3a+8)\alpha_3^2, \\ a\alpha_1^5 = a\alpha_2^5 = a'\alpha_3^5. \end{cases} \tag{6}$$

The first three equations (6), with utilization of the abbreviation:

$$\alpha_1\alpha_2\alpha_3^{-2} = \beta, \tag{7}$$

may be written as three linear homogeneous equations for the three non-vanishing quantities aa', a, a':

$$\begin{cases} aa'(\beta^3-1) + a\beta^3 - a' = 0, \\ 3aa'(\beta^2-1) - 2a\beta^2 + 2a' = 0, \\ 3aa'(\beta-1) + 8a\beta - 8a' = 0. \end{cases} \tag{8}$$

Therefore, the equation of fifth degree holds for β:

$$\begin{vmatrix} \beta^3-1 & \beta^3 & -1 \\ 3(\beta^2-1) & -2\beta^2 & 2 \\ 3(\beta-1) & 8\beta & -8 \end{vmatrix} = 0, \tag{9}$$

[631] and we find from (8), by solution for a and a', the representation of the quantities in β:

$$a = \frac{8}{3(5\beta+4)},\quad a' = \frac{8\beta}{3(4\beta+5)}. \tag{10}$$

Of the solutions of the equation (9), $\beta = 0$ shows itself to be unusable without further ado, similarly, the double root $\beta = 1$, which, as one easily shows, again makes the substitution (4) into a power of S. The two solutions coming into consideration for us now remaining are only:

$$\beta = \frac{-7 \pm i\sqrt{15}}{8}.$$

By choosing, from these two roots, that one, say, with the upper sign:

$$\beta = \frac{-7 + i\sqrt{15}}{8}, \tag{11}$$

we have, with this, definitively chosen the unsymmetrical icosahedral coordinate system of the y_1, y_2, y_3. The equations (10) furnish, as associated values a and a':

$$a = -\frac{\sqrt{3} + 5i\sqrt{5}}{6\sqrt{3}}, \; a' = -\frac{\sqrt{3} - 5i\sqrt{5}}{6\sqrt{3}}. \tag{12}$$

The last equation (6) yields, on the strength of this:

$$\alpha_1^5 = \alpha_2^5 = \frac{-61 - 5i\sqrt{15}}{64}\alpha_3^5 = -\left(\frac{1 + i\sqrt{15}}{4}\right)^5 \alpha_3^5.$$

If we consequently set, which is permissible:

$$\alpha_3 = -\frac{1 - i\sqrt{15}}{4},$$

then follows, further:

$$\alpha_1^5 = \alpha_2^5 = 1, \; \alpha_1 = \varepsilon^\mu, \; \alpha_2 = \varepsilon^\nu.$$

The equation (7) now yields:

$$\varepsilon^{\mu+\nu} = \beta\alpha_3^2 = \frac{-7 + i\sqrt{15}}{8} \cdot \frac{-7 - i\sqrt{15}}{8} = 1,$$

so that $\alpha_1 = \varepsilon^\mu$, $\alpha_2 = \varepsilon^{-\mu}$. Since we may now still exercise an arbitrary power of S on the coordinate system of the y, without essentially altering this, then it is allowed to set the whole number just denoted by μ equal to 0.

By putting in the value a found into the schema (3), there yields itself as a principal result: *The equation of our curve of sixth degree in the chosen unsymmetric icosahedral coordinate system is:* [632]

$$\begin{aligned}5(\sqrt{3} - i\sqrt{5})y_1^3 y_2^3 - 6\sqrt{3}(y_1^5 + y_2^5)y_3 - 15(\sqrt{3} + i\sqrt{5})y_1^2 y_2^2 y_3^2 \\ +15(3\sqrt{3} - i\sqrt{5})y_1 y_2 y_3^4 - (\sqrt{3} + 5i\sqrt{5})y_3^6 = 0;\end{aligned} \tag{13}$$

this coordinate system, by the transformation:

$$y_1' = y_1, \; y_2' = y_2, \; y_3' = \left(\frac{\sqrt{3} + i\sqrt{5}}{2\sqrt{2}}\right)^2 y_3, \tag{14}$$

goes over into the second icosahedral coordinate system of the y'*, in which our curve* C_6 *is represented by:*

$$\begin{aligned}5(\sqrt{3} + i\sqrt{5})y_1'^3 y_2'^3 - 6\sqrt{3}(y_1'^5 + y_2'^5)y_3' - 15(\sqrt{3} - i\sqrt{5})y_1'^2 y_2'^2 y_3'^2 \\ +15(3\sqrt{3} + i\sqrt{5})y_1' y_2' y_3'^4 - (\sqrt{3} - 5i\sqrt{5})y_3'^6 = 0.\end{aligned} \tag{15}$$

One again sees immediately, that the transformation of the second kind:

$$y_1' = \bar{y}_1,\ y_2' = \bar{y}_2,\ y_3' = \bar{y}_3,$$

interchanges the two equations (13) and (15), and the same consideration attaches to this, which we attached on p. 626 to the transformation (17) at that time. We note, as a theorem: *The curve C_6 given by* (13) *is transformed into itself by the "collineation of the second kind":*

$$(\bar{V}) \qquad y_1' = \bar{y}_1,\ y_2' = \bar{y}_2,\ y_3' = \left(\frac{\sqrt{3}-i\sqrt{5}}{2\sqrt{2}}\right)^2 \bar{y}_3; \tag{16}$$

$\bar{V}$ is a reflection of the extended group $\bar{G}_{720}$, which transforms the two icosahedral groups G_{60} and G'_{60} into one another, and which consequently, with the collineations S and T, permits the entire $\bar{G}_{720}$ to be generated.

Next to the two unsymmetric systems, we now arrange, as a third, the "*symmetric icosahedral coordinate system*" x, y, z, which we have to so choose, that the reflection $\bar{V}$ represents itself in the form:

$$x' = \bar{x},\ y' = \bar{y},\ z' = \bar{z}. \tag{17}$$

Then, namely, the collineation group G_{360} is conjugate to itself, and the equation of the C_6 will obtain real coefficients. In order that the reflection (16) obtain the form just prescribed, we have to set

$$x = \alpha y_1,\ y = \beta y_2,\ z = \gamma\frac{\sqrt{3}+i\sqrt{5}}{2\sqrt{2}} y_3,$$

where the α, β, γ are three arbitrarily choosable *real* constants. By making the choice $\alpha = \beta = \sqrt{3}$, $\gamma = \sqrt{2}$, the coefficients of the curve equation not only become real, but even consistently rational. *In the "symmetric" icosahedral coordinate system defined by:*

$$x = \sqrt{3}y_1,\ y = \sqrt{3}y_2,\ z = \frac{\sqrt{3}+i\sqrt{5}}{2} y_3, \tag{18}$$

the curve C_6 of sixth degree represents itself in the, not only real, but also rational, equation form:

$$10x^3y^3 - 9(x^5+y^5)z - 45x^2y^2z^2 - 135xyz^4 + 27z^6 = 0. \tag{19}$$

[633] We still want to determine here, how the coordinates x, y, z let themselves be transformed into the symmetric octahedral coordinates X, Y, Z, to which belonged the equation (21) on p. 628 for our C_6. In the latter system, $Z = 0$ was the axis of a perspectivity, whose center lay at the opposite vertex $X = 0$, $Y = 0$. In our icosahedral coordinates, the perspectivity U, given in "Icos." on p. 213 in (4), has the simplest form; it writes itself, in the symmetric coordinates:

$$x' = -y,\ y' = -x, z' = -z,$$

and consequently has the axis $x - y = 0$, while the center is situated at the intersection point of the two lines $x + y = 0$, $z = 0$. Therefore let that axis be the side $Z = 0$ of the new coordinate triangle, while the sides $X = 0$, $Y = 0$ are to intersect at the given perspectivity center. In order to obtain the transition, we insert a mediating system X', Y', Z by:

$$X' = z,\ 2Y' = x + y,\ 2Z = x - y,$$

or, conversely:

$$x = Y' + Z,\ y = Y' - Z,\ z = X', \tag{20}$$

and then obtain the definitive $X,\ Y$ by means of the binary transformation:

$$X' = aX + bY,\ Y' = cX + dY.$$

Now put the expressions (20) for $x,\ y,\ z$ into the equation (19) and arrange according to powers of Z:

$$\begin{aligned}-10Z^6 - 15Z^4(3X'^2 + 6X'Y' - 2Y'^2) + 15Z^2(9X'^4 + 6X'^2Y'^2 - 12X'Y'^3 - 2Y'^4)\\ +27(X'^6 - 135X'^4Y'^2 - 45X'^2Y'^4 - 18X'Y'^5 + 10Y'^6) = 0.\end{aligned} \tag{21}$$

The corresponding ordered (multiplied by 10) equation (21) on p. 628 is:

$$\begin{aligned}-10Z^6 + 300Z^4XY + 15Z^2(-10\sqrt{15}X^4 + 100X^2Y^2 + 10\sqrt{15}Y^4)\\ +(300\sqrt{15}X^4 + 1000X^2Y^2 - 300\sqrt{15}Y^4)XY = 0.\end{aligned} \tag{22}$$

The comparison shows, first of all, that the component of the equation (21) free of Z must exhibit the factor $(3X'^2 + 6X'Y' - 2Y'^2)$; in fact, it is equal to the product of this factor with the expression: [634]

$$9X'^4 - 18X'^3Y' - 3X'^2Y'^2 - 6X'Y'^3 - 5Y'^4.$$

The further comparison of (21) and (22) yields:

$$\begin{aligned}-20XY &= 3X'^2 + 6X'Y' - 2Y'^2,\\ -10\sqrt{15}X^4 + 100X^2Y^2 + 10\sqrt{15}Y^4 &= 9X'^4 + 6X'^2Y'^2 - 12X'Y'^3 - 2Y'^4,\\ -15\sqrt{15}X^4 - 50X^2Y^2 + 15\sqrt{15}Y^4 &= 9X'^4 - 18X'^3Y' - 3X'^2Y'^2 - 6X'Y'^3 - 5Y'^4\end{aligned}$$

as necessary and sufficient conditions for the transformability of the equation (21) into (22). These relations are not independent of one another, since, by combination of the second and third, one easily generates the square of the first. As the conditions to be satisfied, we therefore take the first (whose right side we decompose into its linear factors), and the equation arising by elimination of X^2Y^2 from the second and third

$$\begin{aligned}-20XY &= (\sqrt{3}X' + (\sqrt{3} + \sqrt{5})Y')(\sqrt{3}X' + (\sqrt{3} - \sqrt{5})Y'),\\ -40\sqrt{15}(X^4 - Y^4) &= 27X'^4 - 36X'^3Y' - 24X'Y'^3 - 12Y'^4.\end{aligned}$$

It is no restriction of the generality of our argument[40], if we deduce, from the first of these two relations, the schema:

---

[40] Notice, that the relation (21) on p. 628 admits the collineation:

$$X' = e^{\frac{\pi i}{4}}Y,\ Y' = e^{-\frac{\pi i}{4}}X,\ Z' = Z,$$

which interchanges the sides $X = 0,\ Y = 0$.

$$\begin{cases} \alpha X = \sqrt{3}X' + (\sqrt{3}+\sqrt{5})Y', \\ \beta Y = \sqrt{3}X' + (\sqrt{3}-\sqrt{5})Y', \end{cases} \qquad (\alpha\beta = -20). \tag{23}$$

With consideration of $\alpha\beta = -20$, it follows from this:

$$20^4(X^4 - Y^4) = \beta^4(\sqrt{3}X' + (\sqrt{3}+\sqrt{5})Y')^4 - \alpha^4(\sqrt{3}X' + (\sqrt{3}-\sqrt{5})Y')^4.$$

In order that this be the second of the relations, the two equations:

$$\alpha^4 = 400(4+\sqrt{15}),\ \beta^4 = 400(4-\sqrt{15})$$

[635] are sufficient and necessary. Simultaneous alterations of α and β by the factors i^{ν} and $i^{-\nu}$ are inessential.

Consequently, α and β determine themselves to be:

$$\alpha = 2\sqrt{5}\sqrt{\frac{\sqrt{5}+\sqrt{3}}{\sqrt{2}}},\ \beta = -2\sqrt{5}\sqrt{\frac{\sqrt{5}-\sqrt{3}}{\sqrt{2}}}.$$

The substitution of these values into (23) and solution for X' and Y' yields:

$$\begin{cases} X' = \sqrt{\frac{2}{3}}\left(\sqrt{\frac{\sqrt{5}-\sqrt{3}}{\sqrt{2}}}X - \sqrt{\frac{\sqrt{5}+\sqrt{3}}{\sqrt{2}}}Y\right), \\ Y' = \sqrt{\frac{\sqrt{5}+\sqrt{3}}{\sqrt{2}}}X + \sqrt{\frac{\sqrt{5}-\sqrt{3}}{\sqrt{2}}}Y. \end{cases} \tag{24}$$

From (20) and (24) seises, finally, the theorem concluding our study: *The icosahedral coordinate system x, y, z is connected with the octahedral one X, Y, Z in the following manner:*

$$\begin{cases} x = \sqrt{\frac{\sqrt{5}+\sqrt{3}}{\sqrt{2}}}X + \sqrt{\frac{\sqrt{5}-\sqrt{3}}{\sqrt{2}}}Y + Z, \\ y = \sqrt{\frac{\sqrt{5}+\sqrt{3}}{\sqrt{2}}}X + \sqrt{\frac{\sqrt{5}-\sqrt{3}}{\sqrt{2}}}Y - Z, \\ z = \sqrt{\frac{2}{3}}\sqrt{\frac{\sqrt{5}-\sqrt{3}}{\sqrt{2}}}X - \sqrt{\frac{2}{3}}\sqrt{\frac{\sqrt{5}+\sqrt{3}}{\sqrt{2}}}Y. \end{cases} \tag{25}$$

A.14 The curve C_6 in the harmonic coordinate system

The third of the approaches discussed on p. 617 assumes the shortest approach to the curve C_6, for which we consider the C_6 as a "covariant of a harmonic curve of third degree", i.e. a C_3 with the harmonic double-ratio. According to "M" II on p. 242, a plane curve C_3 of third degree of the genus $p = 1$ admits 18 *collineations into itself,* which naturally form a group G_{18}. This G_{18} lets itself be represented, in the associated integral

u of the first kind with the pair of primitive periods ω_1, ω_2, in the especially accessible form:

$$u' \equiv \pm u + \frac{\lambda\omega_1 + \mu\omega_2}{3}, \text{ (mod } \omega_1, \omega_2) \tag{1}$$

(cf. l.c. equation (12)), whereby, λ, μ are therefore to be reduced to the nine number pairs incongruent mod 3. If the harmonic case is present, i.e. if the Aronhold invariant $T = \frac{64}{27} g_3$ of the curve of third degree vanishes, then the number of the collineations of the C_3 into itself raises itself to 36. In the transcendent representation, this G_{36} of the collineations writes itself: [636]

$$u' \equiv i^\nu u + \frac{\lambda\omega_1 + \mu\omega_2}{3}, \text{ (mod } \omega_1, \omega_2), \tag{2}$$

where i is the imaginary unit and therefore ν must run through a residue system mod 4. Thereby, if one avails oneself of a reduced period-pair, $\omega_1 = i\omega_2$ will hold.

The equation of the curve C_3, if we restrict ourselves, say, at the same time, to the harmonic case coming into consideration for us, can, according to "M" II on p. 246, be clothed in the form:

$$x_1^2 x_3 - 4x_2^3 + g_2 x_2 x_3^2 = 0, \tag{3}$$

whereby, as a coordinate system, one designated l.c. as "canonical", is taken as a basis. The geometrical character of this system is (cf. l.c. p. 248) this, that $x_3 = 0$ is an inflection tangent, whose tangency point lies at the vertex $x_2 = 0$, $x_3 = 0$, that, further, $x_1 = 0$ is the harmonic polar of the inflection point, and that, finally, $x_2 = 0$ represents the linear polar of the inflection tangent with respect to the triple of the remaining tangents running from the inflection point to the C_3. From this emerges, as was already determined l.c., that, corresponding to the 9 inflection points, we have, in all, *nine canonical coordinate triangles* at our disposal. The individual triangle is therefore always transformed into itself by *four* collineations of the G_{36}, forming a G_4; these G_4 are cyclic, since, e.g. the G_4 belonging to the coordinate system of the equation given in (3) arising from the collineation:

$$x_1' = x_1, \; x_2' = i x_2, \; x_3' = -i x_3 \tag{4}$$

is of period 4.

One now determines immediately, that the individual one of our 10 conjugate subgroups G_{36} of the G_{360} is isomorphic with our collineation group considered here, given by (2) in transcendental form. But this collineation-G_{36} is (apart from transformations of the coordinate system) unique[41]; therefore, we obtain the theorem: *The individual* [637]

[41] One perceives this most simply, by attaching to a collineation-G_{18} contained in the G_{360} and remembering, that, in this are contained four cyclic G_3, distinguished in the G_{18} (cf. p. 588). The fixpoints of the individual G_3 accordingly furnish a triangle, which is transformed into itself by all 18 collineations of the G_{18}. If x_1, x_2, x_3 are the coordinates referred to one of these four triangles, then the associated G_3 necessarily arises from:

$$x_1' = x_1, \; x_2' = \rho x_2, \; x_3' = \rho^2 x_3 \quad (\rho = e^{\frac{2i\pi}{3}}). \tag{5}$$

one of our collineation groups G_{18} has a harmonic curve C_3 of third degree as a lowest covariant.

In order to obtain our curve C_6 from here, we will, for the simplification of the calculations, introduce new coordinates z_1, z_2, z_3, in place of the canonical coordinates x_1, x_2, x_3, by the transformation:

$$x_1 = z_3 \cdot \sqrt{2i\sqrt{3}g_2\sqrt{g_2}},\ x_2 = z_1 \cdot \sqrt{g_2},\ x_3 = z_2 \cdot 2i\sqrt{3},$$

whereby, the equation (3) of our C_3 goes over into:

$$z_1^3 + 3z_1 z_2^2 + 3z_2 z_3^2 = 0. \tag{6}$$

From now on, as intended, we will designate the new coordinates as the "*harmonic coordinates*". The canonical triangle has remained subsisting as coordinate triangle for the transformation exercised. Now since the vertices of this triangle were fixpoints of a cyclic G_4 (cf. (4)), then, in case we choose the X, Y, Z, introduced on p. 627, tieing into this G_4, the theorem follows: *The harmonic coordinate system of the z_1, z_2, z_3 has the coordinate triangle in common with the symmetric octahedral coordinate system X, Y, Z.* In addition, we shall find, that, exactly as in the earlier cases, the harmonic system z_1, z_2, z_3 is *choosable in two ways,* which are interchangable by an operation first contained in the $\bar{G}_{720}$; we return to this soon.

[638] A few theorems from the invariant theory of the ternary cubic forms[42] suffices for obtaining a schema for the curve C_6. If $f(z_1, z_2, z_3)$, or $f(z)$ for short, is any cubic form, whose first and second derivatives let be denoted by f_i resp. f_{i_k}, then one has, first of all, as a cubic covariant, the Hessian determinant, to be denoted, say, by $h(z)$:

---

[41] In addition, the "sides" of the coordinate triangle are permuted by the operations of the G_{18}. As one easily proves, one can so choose the factors still disposable in the x_1, x_2, x_3, that the G_{18} arises by combination of the collineations (5) with the six permutations of the coordinates x_1, x_2, x_3 themselves. With this, however, we have, in fact, obtained the collineation-G_{18} of a plane C_3, referred to a coordinate system designated as "*singualr*" in "M" II on p. 251. With this, the theorem set up in the text on the existence of a covariant C_3 for the individual collineation-G_{18} of the G_{360} is established.

For understanding the present relations, we yet note, how the three other singular triangles ("inflection triangles" of the curve of 3rd degree) represent themselves in the singular coordinate system chosen. A second such triangle is furnished by the three sides:

$$x_1 + \rho^\nu x_2 + \rho^{2\nu} x_3 = 0,\ (\nu = 0, 1, 2),$$

and the latter two by:

$$\rho^{\pm 1} x_1 + x_2 + x_3 = 0,$$
$$x_1 + \rho^{\pm 1} x_2 + x_3 = 0,$$
$$x_1 + x_2 + \rho^{\pm 1} x_3 = 0.$$

The invariance of these triangles with respect to the G_{18} above is immediately obvious.

[42] Cf. e.g. Clebsch-Lindemann, "*Vorlesungen über Geometrie*", 5. Abth., or Salmon-Fiedler, "*Höhere ebene Curven*", Cap. 5 Abschn. 5.

$$\begin{vmatrix} f_{11} & f_{12} & f_{13} \\ f_{21} & f_{22} & f_{23} \\ f_{31} & f_{32} & f_{33} \end{vmatrix},$$

which is of 3rd dimension in the coefficients of the form f. Next to this, here arranges itself, as a covariant of 6th degree, the bordered Hessian determinant:

$$\begin{vmatrix} f_{11} & f_{12} & f_{13} & h_1 \\ f_{21} & f_{22} & f_{23} & h_2 \\ f_{31} & f_{32} & f_{33} & h_3 \\ h_1 & h_2 & h_3 & 0 \end{vmatrix},$$

where h_1, h_2, h_3 are the partial derivatives of the first order of $h(z)$. This covariant, to be called $g(z)$ say, as one sees, is of the 8th dimension in the coefficients of the form f.

But $g(z)$ is not the only covariant of 6th degree and 8th dimension. Since, namely, the Aronhold invariant S has the dimension 4, then we obtain, in $Sf(z)h(z)$, a second covariant of 6th degree and 8th dimension. But then the theory of invariants shows, that each further covariant of 6th degree, and therefore, in particular, the left side $F(z_1, z_2, z_3)$ of the equation of our C_6, must be represented by means of constant factors λ, μ in the form:

$$F(z) = \lambda g(z) + \mu S f(z) h(z).$$

Since, furthermore, $F(z)$ is irreducible, then λ cannot vanish and may accordingly also be set = 1.

We now calculate, in our case (6), by leaving out numerical factors:

$$f = z_1^3 + 3z_1 z_2^2 + 3z_2 z_3^2,\ h = z_1^2 z_2 - z_1 z_3^2 - z_2^3,$$
$$g = z_3^6 - 9z_3^4 z_1 z_2 - z_3^2(8z_1^4 - 30z_1^2 z_2^2 - 6z_2^4) - z_1 z_2(z_1^4 - 6z_1^2 z_2^2 + 21z_2^4).$$

We consequently have the following schema: *The equation of the curve C_6 of 6th degree is necessarily representable in the form:*

$$z_3^6 + (3\mu - 9)z_3^4 z_1 z_2 + 30z_3^2 z_1^2 z_2^2 + (6 - 2\mu)z_1^3 z_2^3 + z_3^2((\mu - 8)z_1^4$$
$$+(3\mu + 6)z_2^4) - z_1 z_2((\mu + 1)z_1^4 + (21 - 3\mu)z_2^4) = 0. \tag{7}$$

[639]

For the determination of μ, it suffices to allude to the fact, that the collineation of the G_{36} yielding itself from (4):

$$z_1' = iz_1,\ z_2' = -iz_2,\ z_3' = z_3$$

of period 4 is transformed into its inverse collineation by four harmonic perspectivities belonging to the G_{360}. Each of these perspectivities thus has the form:

$$z_1' = \alpha z_2,\ z_2' = \alpha^{-1} z_1,\ z_3' = z_3. \tag{8}$$

This operation must transform the equation (1) into itself, so that the two biquadratic forms:

$$(\mu - 8)z_1^4 + 3(\mu + 2)z_2^4,\ (1 + \mu)z_1^4 + 3(7 - \mu)z_2^4$$

must be absolutely invariant with respect to (8). There thus holds:

$$(\mu-8)\alpha^4 = 3(\mu+2),\ (1+\mu)\alpha^4 = 3(7-\mu),$$

from which, by elimination of α, the quadratic equation for μ arises:

$$(\mu+1)(\mu+2)+(\mu-7)(\mu-8) = 0.$$

We thus find for μ the two values:

$$\mu = 3 \pm 2i\sqrt{5}. \tag{9}$$

It consequently verifies itself, *that we are able to choose the harmonic coordinate system in two essentially different ways; the one choice furnishes, as equation of the curve* C_6:

$$\begin{aligned} z_3^6 + 6i\sqrt{5}z_3^4 z_1 z_2 + 30 z_3^2 z_1^2 z_2^2 - 4i\sqrt{5} z_1^3 z_2^3 + z_3^2((-5+2i\sqrt{5})z_1^4 + 3(5+2i\sqrt{5})z_2^4) \\ - z_1 z_2((4+2i\sqrt{5})z_1^4 + 3(4-2i\sqrt{5})z_2^4) = 0, \end{aligned} \tag{10}$$

and the other:

$$\begin{aligned} z_3^6 - 6i\sqrt{5}z_3^4 z_1 z_2 + 30 z_3^2 z_1^2 z_2^2 + 4i\sqrt{5} z_1^3 z_2^3 + z_3^2((-5-2i\sqrt{5})z_1^4 + 3(5-2i\sqrt{5})z_2^4) \\ - z_1 z_2((4-2i\sqrt{5})z_1^4 + 3(4+2i\sqrt{5})z_2^4) = 0. \end{aligned} \tag{11}$$

The relation of the two coordinate system to one another is easy to give. If we call the left sides of the two equations (10) and (11) $F(z)$ and $\bar{F}(z)$ for short, then it reveals itself, that these two forms $F(z)$ and $\bar{F}(z)$ are carried over into one another by the transformation:

$$z_1' = e^{\frac{\pi i}{4}} \sqrt[4]{3} z_2,\ z_2' = e^{\frac{3\pi i}{4}} \frac{1}{\sqrt[4]{3}} z_1,\ z_3' = z_3. \tag{12}$$

The transformation (12) *accordingly leads from one harmonic coordinate system to the*
[640] *other.* Furthermore, the application of the transformation (12) to our forms $f(z)$ and $h(z)$ yields:

$$f(z') = e^{\frac{7\pi i}{4}} \sqrt[4]{27} h(z),\ h(z') = e^{\frac{5\pi i}{4}} \frac{1}{\sqrt[4]{27}} f(z), \tag{13}$$

so that, apart from constant factors, the two cubic forms $f(z)$ and $h(z)$ are carried over into one another by the transformation (12).

The latter fact has a remarkable consequence. We perceive, that the collineation group G_{36}, transformed from the z to the z', is the collineation group of the forms $h(z')$ and $f(z')$, i.e. again the original G_{36}. This group is consequently carried over into itself by the transformation (12); and since, moreover, the square of (12) belongs to the G_{36}, *then the* G_{36}, *by addition of the transformation* (12), *is extended to a* G_{72}, *in which the* G_{36} *is contained distinguishably.*

We have already become acquainted with this group extension on p. 566 in another form. In fact, it is a question here of that extension of the group denoted l.c. by $\Gamma_\mu(\Gamma_{10})$, which we attained by the transformation T' of third degree given in (11) on p. 566. This operation T' carried our original group Γ over into the transformed group Γ', which had the Γ_{10} in common with Γ, which corresponds to our subgroup G_{36} of the G_{360}. The proper significance of the two equation forms (10) and (11), resp. the two associated coordinate systems is therefore this, *that the two harmonic coordinate systems go over into one another by a collineation* (12), *corresponding to the transformation T' of third degree.*

One obtains a curve of 6th degree invariant with respect to the G_{72} in $\frac{1}{2}(F(z)+\bar{F}(z))=0$, or in detail:

$$z_3^6+30z_3^2z_1^2z_2^2-5z_3^2(z_1^4-3z_2^4)-4z_1z_2(z_1^4+3z_2^4)=0. \tag{14}$$

This C_6, with 72 collineations into itself, has a real equation, and also possesses, as one easily shows, at least one real trace; this is consequently transformed into itself by the collineation of the 2nd kind:

$$z_1'=\bar{z}_1,\ z_2'=\bar{z}_2,\ z_3'=\bar{z}_3 \tag{15}$$

in such a way, that the real trace thereby corresponds to itself point for point. *The G_{72} is consequently capable of extension by the "reflection"* (15) *to a group $\bar{G}_{144}$ of the second kind.* As one immediately surveys, it is a question of the extended group denoted on p. 566 by $\overline{\Gamma}_{\frac{1}{2}\mu}$, so that the operation (15) corresponds to the reflection (9) on p. 565, denoted at that time by $\bar{V}_3T=T^{-1}\bar{V}_3$, and the real component of the curve (14) is the image of the symmetry line dotted in figure 106 on p. 567. [641]

Now we found, in (14) on p. 567, that the combination of the transformation T' of the 3rd degree with the reflection $\bar{V}_3T$ furnished that substitution $\bar{V}$ contained in the extended $\overline{\Gamma}$, whose addition to the Γ_{10} yielded the $\overline{\Gamma}_{10}$ with the rectangular circular-arc pentangle, described on p. 570, as discontinuity domain. If we therefore combine the collineations (12) and (15) into:

$$z_1'=e^{\frac{\pi i}{4}}\sqrt[4]{3}\,\bar{z}_2,\ z_2'=e^{\frac{3\pi i}{4}}\frac{1}{\sqrt[4]{3}}\bar{z}_1,\ z_3'=\bar{z}_3, \tag{16}$$

then we have obtained here a collineation of the second kind contained in the $\bar{G}_{720}$, which extends the G_{36} to a $\bar{G}_{72}$. Upon the exercise of this transformation (16), the equations (10) and (11) then also actually go over into themselves individually, as immediately follows from:

$$F(z')=\bar{F}(\bar{z}),\ \bar{F}(z')=F(\bar{z}).$$

If one will again designate the two harmonic coordinate systems as "unsymmetric", then between then a *symmetric* system will arrange itself, which, however, must be simply identical with the symmetric octahedral coordinate system of the X, Y, Z, referred to the same coordinate triangle. We note, *as the transition from the first harmonic coordinate system (that of the equation* (10)*) to the octahedral system of the X, Y, Z:*

$$z_1=e^{\frac{\pi i}{8}}\sqrt[4]{\frac{5}{2\sqrt{3}}}\sqrt{1+i\sqrt{5}}X,\ z_2=e^{\frac{3\pi i}{8}}\sqrt[4]{\frac{5}{6\sqrt{3}}}\sqrt{1-i\sqrt{5}}Y,\ z_3=Z. \tag{17}$$

Equation (10), after sign change, goes over by this transformation directly into the real equation (21) on p. 628 of our C_6. By combination of (12) and (17), one obtains the connection between the second harmonic system of the z_1', z_2', z_3' and that of the X, Y, Z.

A.15 The real traces of the C_6 and the character of the points a, b, c

According to p. 612, we are able to select a coordinate system in 36 ways, in which the curve of 6th degree receives *two real traces*, which correspond to a pair of associated symmetry lines of the two kinds on the Riemann surface F_{360} (cf. p. 611 ff.). As these symmetry lines, those two real traces must also go over into themselves by 10 real transformations forming a dihedral group. A coordinate system of the kind in question is to be so chosen, that one individual dihedral subgroup G_{10} of the group of our 360
[642] collineations, and only this G_{10} comes out real.

If we first transform the icosahedral coordinates y_1, y_2, y_3; introduced on p. 628, by:

$$y_1 = x_1 + ix_2,\ y_2 = x_1 - ix_2,\ y_3 = x_3,$$

to new coordinates x_1, x_2, x_3, then the "icosahedral conic section" (1) on p. 628 assumes the equation form:

$$x_1^2 + x_2^2 + x_3^2 = 0,$$

and the ternary icosahedral group G_{60} obtains the real form, which underlies the icosahedral division given in I on p. 72 in figure 16. The generators S, T named on p. 628 and the operation U defined in (4) on p. 213 in "Icos.", in the coordinates x_1, x_2, x_3 so defined, have consistently real coefficients:

$$(S)\quad x_1' = x_1 \cos\frac{2\pi}{5} + x_2 \sin\frac{2\pi}{5},\ x_2' = -x_1 \sin\frac{2\pi}{5} + x_2 \cos\frac{2\pi}{5},\ x_3' = x_3,$$
$$(T)\quad x_1' = -\frac{1}{\sqrt{5}}x_1 + \frac{2}{\sqrt{5}}x_3,\ x_2' = -x_2, x_3' = \frac{2}{\sqrt{5}}x_1 + \frac{1}{\sqrt{5}}x_3,$$
$$(U)\quad x_1' = -x_1,\ x_2' = x_2,\ x_3' = -x_3.$$

If we then carry out a second transformation to coordinates ξ, η, ζ by[43]:

$$x_1 = \xi,\ x_2 = \eta,\ \frac{\sqrt{3} + i\sqrt{5}}{2\sqrt{3}} x_3 = \zeta,$$

then S and U still remain real, while T receives complex coefficients. But since the dihedral G_{10} to be generated from S and U is contained in no subgroup of the icosahedral group different from the G_{10} and G_{60} themselves, the collineations of the G_{10}, as it has to be, are now the only real collineations of the G_{60}.

In consequence of (18) on p. 633, the ξ, η, ζ are connected with the symmetric icosahedral coordinate system of the x, y, z thus:

$$x = \sqrt{3}(\xi + i\eta),\ y = \sqrt{3}(\xi - i\eta),\ z = \sqrt{3}\zeta. \tag{1}$$

Our curve C_6 accordingly (cf. (19) on p. 633) represents itself in the coordinates ξ, η, ζ as follows:

$$10(\xi^2 + \eta^2)^3 - 18\xi\zeta(\xi^4 - 10\xi^2\eta^2 + 5\eta^4) - 45(\xi^2 + \eta^2)^2\zeta^2 - 135(\xi^2 + \eta^2)\zeta^4 + 27\zeta^6 = 0.$$

[643] If we also set $\zeta = 1$, and interpret ξ and η as rectangular coordinates, then we obtain the

43 A confusion of the coordinate ζ, utilized here in passing, with the complex argument ζ of the automorphic functions, is not be feared.

equation of our curve C_6 in the form:

$$10(\xi^2+\eta^2)^2 - 18\xi(\xi^4 - 10\xi^2\eta^2 + 5\eta^4) - 45(\xi^2+\eta^2)^2 - 135(\xi^2+\eta^2) + 27 = 0. \qquad (2)$$

And the G_{10} of the real collineations of (2) into itself is generated from:

$$(S)\quad \xi' = \xi\cos\frac{2\pi}{5} + \eta\sin\frac{2\pi}{5},\ \eta' = -\xi\sin\frac{2\pi}{5} + \eta\cos\frac{2\pi}{5},$$

$$(U)\quad \xi' = \xi,\ \eta' = -\eta.$$

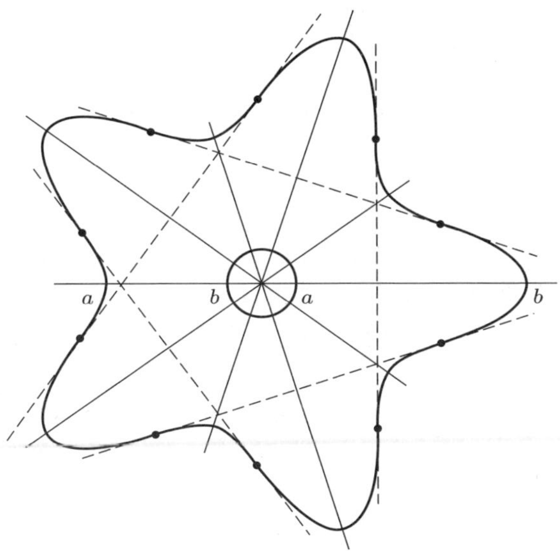

Fig. 114.

The curve (2) now, in the $\xi\eta$-plane, actually has two real traces, which are represented in figure 114. This real curve goes over into itself by the rotation S of the period 5 about the origin, and has the five lines drawn through the origin as symmetry axes (perspectivity axes). It consists of an inner, almost circular trace, and an outer branch surrounding this; the former corresponds to the symmetry line, which only contains the points a and b, the latter to the symmetry line, which exhibits the points a, b and c. Thereby, the points a and b (fixpoints of the perspectivities) are cut out on the individual trace by the symmetry axes mentioned in such a way, that a point a always lies opposite a point b (cf. p. 612). If we take the ξ-axis as an example of one of these symmetry lines, then one finds its intersection point with the C_6 by solution of the equation of 6th degree:

$$\xi^6 - \frac{9}{5}\xi^5 - \frac{9}{2}\xi^4 - \frac{27}{2}\xi^2 + \frac{27}{10} = 0,$$

whose left side lets itself be split up in the following manner:

$$\left(\xi^2 - 3\xi - \frac{3}{2}\right)\left(\xi^2 + \frac{3+2\sqrt{6}}{5}\xi - \frac{3-3\sqrt{6}}{5}\right)\left(\xi^2 + \frac{3-2\sqrt{6}}{5}\xi + \frac{3+3\sqrt{6}}{2}\right) = 0.$$

[644] All together, the individual perspectivity has 6 fixpoints, namely 4 points a and 2 points b. There is no doubt (and lets itself be verified by means of formulas to be set up later), that in the case of the ξ-axis, the two points b are furnished by the first factor in the last equation in the form:

$$\xi = \frac{3 \pm \sqrt{15}}{2}, \quad (\xi = 3.436\ldots \text{ resp. } = -0.436\ldots).$$

Of the fixpoints a, only two are real in the $\xi\eta$-plane, and are furnished by the second factor of the equation before last in the form:

$$\xi = \frac{-3 - 2\sqrt{6} \pm 3\sqrt{-3 + 8\sqrt{6}}}{10}, \quad (\xi = 0.432\ldots \text{ resp. } = -2.012\ldots).$$

For $\xi = \frac{3}{2}$, the equation (2) of our C_6 assumes the form:

$$10\left(\eta^2 - \frac{15}{4}\right)^3 = 0.$$

We immediately read off from this, that the line $2\xi - 3 = 0$ is a *double inflection tangent of the curve* C_6 with the two inflection points with the coordinates $\xi = \frac{3}{2}$, $\eta = \pm\frac{\sqrt{15}}{2}$. The five real double inflection tangents, dotted in the figure, cut out 10 points on the outer curve trace, which represent all inflection points. Now, according to the Plücker formulas, our singularity-free curve C_6 has, in all 72 inflection points; and these points permute themselves with respect to the 360 collineations only among one another. According to p. 611, we are here led to the points c: *The points c furnish the* 72 *inflection points on the curve* C_6 *of sixth degree; and, in particular, therefore, the intersection points, marked in the figure, of the outer branch with the five double inflection tangents are the* 10 *points c, which this branch carries.*

Tying into this, we are also able to clarify the geometrical character of the points a and b. A determinate conic section always runs through five consecutive points of our C_6. If we imagine this conic section formed for a vertex of one of the real traces (intersection point with a symmetry axis), then the conic section will itself be symmetric with respect to this axis, and therefore cut the curve in an even number of points, therefore in yet a sixth point: *The vertex a or b is thus a "sextactic" point of the curve.*

These sextactic points, too, have invariant character with respect to collineations,
[645] so that all (real or imaginary) points a and b of the C_6 are necessarily sextactic points. But now, according to a rule set up by Cayley (cf. "M" I on p. 697), a singularity-free curve of the nth degree has $n(12n - 27)$ sextactic points, so that our C_6 receives 270 such points in all. Since these 270 points are are exactly exhausted by the points a and b, there follows: *The geometric character of the* 180 *points a and* 90 *points b of our curve* C_6 *is this, that these represent all the sextactic points of the curve.*

A.16 Further geometrical theorems on the collineation group G_{360}

Since the various coordinate systems used so far are investigated in their connection, then it is immaterial in itself, which system we will utilize further on. However, the calculations, for the most part, fashion themselves most accessibly for the use of the octahedral coordinates x_1, x_2, x_3, since, in them, the associated octahedral group G_{24} (cf. p. 620) represents itself in the simplest fashion in the x_i. We shall, because of this, mostly favor this coordinate system.

Now, first of all, a few theorems concerning points and lines of the plane are to be set up, which are invariant with respect to subgroups of the G_{360}. Hereby, numerous geometrical relations between these models make their appearance, which, however, are mostly only translations of theorems derived above (p. 580 ff.) in cyclic groups and dihedral groups in the G_{360}.[44]

To the 45 conjugate G_2 correspond 45 *perspectivity axes and just as many perspectivity centers.* A first one of these axes is given by $x_1 = 0$; one obtains the remaining ones from it by repeated application of the generators S, T, U' of the G_{360}, noted on p. 625. Four axes, inequivalent with respect to the above G_{24} are given by:

$$\begin{cases} x_1 = 0, \ x_2 + x_3 = 0, \ x_1 + x_2 + \dfrac{\sqrt{5} + i\sqrt{3}}{2} x_3 = 0, \\ \dfrac{1 + i\sqrt{3}}{2} x_1 + \dfrac{1 + \sqrt{5}}{2} x_2 + \dfrac{-1 + i\sqrt{3}}{2} \cdot \dfrac{-1 + \sqrt{5}}{2} x_3 = 0. \end{cases} \tag{1}$$

By means of permutations and sign-change of the x (operations of the G_{24}), the first equation furnishes three axes all together, the second six, the third twelve and the fourth 24, with which their total number is reached. Analogously, from the center $x_1 = 1$, $x_2 = 0$, $x_3 = 0$ of the first perspectivity, one can calculate the remaining 44 centers.

We mention, secondly, *the* 36 *double inflection tangents*, which are coordinated to the conjugate G_5. Two, inequivalent with respect to the G_{24}, are: [646]

$$\frac{-1 + \sqrt{5}}{2} x_1 + \frac{1 + i\sqrt{3}}{2} x_2 = 0, \ x_1 + x_2 + \left(\frac{1 + i\sqrt{3}}{2} - \frac{-1 + \sqrt{5}}{2} \right) x_3 = 0. \tag{2}$$

The G_{24} generates, from the first one, 12 in all, and from the second one 24, with which all 36 are obtained.

We mention, further, *two systems of* 60 *lines each, corresponding to the cyclic* G_3, which arrange themselves into two systems of 20 triangles each. Meant are the three lines joining the fixpoints of each individual G_3. We note, say, only for one G_3 contained in the G_{24}, the three lines:

$$x_1 + x_2 + x_3 = 0, \ \rho^{\pm 1} x_1 + \rho^{\mp 1} x_2 + x_3 = 0, \tag{3}$$

whereby ρ is set as an abbreviation for the third root of unity $\frac{-1+i\sqrt{3}}{2}$.

The 36 cyclic G_5 correspondingly furnish $3 \cdot 36$ lines, which fall into two systems, namely, first the 36 double inflection tangents and secondly, *a system of* 72 *conjugate lines*, which one easily obtains from the inflection tangents by means of the icosahedral conic section to be discussed below. Finally, we obtain from the 45 cyclic G_4 thrice

44 Compare the detailed representations of Gerbaldi in the work in the Rend. di Palermo, Vol. 12 ff., cited on p. 618.

45 lines, falling into the system of the 45 perspectivity axes and *one system of* 90 *conjugate lines,* which one obtains most easily from the axes first named by means of the octahedral conic section.

The intersection point of two perspectivity axes is always the fixpoint of a cyclic group. The $45 \cdot 22$ intersection points furnished by the 45 axes coincide in the following manner: First, these points coincide *in 10's at* 36 *conjugate places, which are fixpoints of the* G_5. Through the individual such fixpoint, in fact, always run 5 axes; the remainder of the 72 additional fixpoints of the G_5, through which none of the perspectivity axes run, lie on the curve C_6 and furnish the *system of* 72 *points c*. Secondly, the intersection points of the perspectivity axes coincide *in 6's at* 45 *conjugate fixpoints of the cyclic* G_4, which are identical with the 45 *perspectivity centers.* Through each such center, namely, run four axes. The remainder of the fixpoints of these G_4 furnish the 90 *points b situated on* C_6. Finally the intersection points of the perspectivity axes coincide *in threes at the* $2 \cdot 60$ *fixpoints of the* G_3, each individual one of which is run through by three axes; these points do not lie on the C_6. With this, all $36 \cdot 10 + 45 \cdot 6 + 2 \cdot 60 \cdot 3 = 45 \cdot 22$ intersection
[647] points of the axes are exhausted.

We also arrive here immediately at the interpretation of the *dihedral groups* G_6, G_8 and G_{10}, by interpreting the preceding relations dualistically. For the $45 \cdot 22$ lines joining the perspectivity centers, centain coincidences take place. First, these lines coincide 36 *times into tens,* and thereby furnish the 36 double inflection tangents, whose individual one is invariant with respect to the dihedral-G_{10}, generatable from the five associated G_2. Secondly, the lines in question coincide 45 *times in 6's,* and furnish thus the 45 perspectivity axes, whose individual one is invariant with respect to the dihedral-G_8 to be generatable from four associated G_2. Thirdly, the lines joining the centers coincide *twice* 60 *times into 3's* and yield thus the $2 \cdot 60$ lines, whose individual one is invariant with respect to the G_6 to be generated from the three associated G_2.

Of special interest are the conic section systems belonging to our collineation group G_{360}. We have first to name the *two systems of* 15 *octahedral conic sections each,* belonging to the G_{24}. If we start out from the conic section (1) on p. 620 of the favored G_{24}, and apply the group generators S, T, U' (cf. p. 625) to this, then we obtain, for the first system, the three conic sections, inequivalent with respect to the G_{24}:

$$\begin{cases} x_1^2 + x_2^2 + x_3^2 = 0, \\ \dfrac{-\sqrt{5} + 3i\sqrt{3}}{2} x_1^2 + \sqrt{5}(x_2^2 + x_3^2) + 2i\sqrt{3} x_2 x_3 = 0, \\ \rho x_1^2 + \rho^2 x_2^2 + x_3^2 + \dfrac{i\sqrt{3}(\sqrt{5} + i\sqrt{3})}{2} (\rho x_2 x_3 + \rho^2 x_3 x_1 + x_1 x_2) = 0. \end{cases} \tag{4}$$

By exercise of the permutations and sign change of the x_1, x_2, x_3, the first equation furnishes one, the second equation six and the third eight conic sections. By putting down these same equations in the second unsymmetric octahedral coordinate system, and then transforming back to the first, one arrives at the second system of the 15 octahedral conic sections. Two octahedral conic sections of this system, inequivalent with respect to the G_{24}, are given by:

$$\begin{cases} x_1^2+x_2^2+\dfrac{1+i\sqrt{15}}{4}x_3^2=0, \\ x_1^2+x_2^2-\sqrt{5}\dfrac{\sqrt{5}-i\sqrt{3}}{4}x_3^2-3\dfrac{1-i\sqrt{15}}{4}x_1x_2+i\sqrt{3}\dfrac{1-i\sqrt{15}}{4}(x_1+x_2)x_3=0. \end{cases} \tag{5}$$

From the first of these conic sections, by means of the G_{24}, one generates, in all, three, from the second the remaining twelve octahedral conic sections of the second system. [648]

Especially remarkable, on account of their slight number, are the *two systems of six icosahedral conic sections each.* The tetrahedral group G_{12}, contained in our favored G_{24}, is, for its part, contained in two conjugate icosahedral groups G_{60}. There must therefore be two conic sections, which are invariant with respect to the G_{12} and are permuted by the collineations of the G_{24} not belonging to the G_{12}. The only conic sections, for which this holds, are given by:

$$\rho^{\pm 1}x_1^2+\rho^{\mp 1}x_2^2+x_3^2=0; \tag{6}$$

in them, we therefore have before us two icosahedral conic sections of the one system. The other four icosahedral conic sections of this system one obtains from these by means of the collineations S, T, U':

$$\begin{cases} x_1^2+x_2^2+x_3^2-i\sqrt{3}\dfrac{\sqrt{5}-i\sqrt{3}}{2}(x_2x_3+x_3x_1+x_1x_2)=0, \\ x_1^2+x_2^2+x_3^2+i\sqrt{3}\dfrac{\sqrt{5}-i\sqrt{3}}{2}(-x_2x_3+x_3x_1+x_1x_2)=0, \\ x_1^2+x_2^2+x_3^2+i\sqrt{3}\dfrac{\sqrt{5}-i\sqrt{3}}{2}(x_2x_3-x_3x_1+x_1x_2)=0, \\ x_1^2+x_2^2+x_3^2+i\sqrt{3}\dfrac{\sqrt{5}-i\sqrt{3}}{2}(x_2x_3+x_3x_1-x_1x_2)=0. \end{cases} \tag{7}$$

The equations of the six conic sections of the second system fashion themselves even more surveyably:

$$\begin{cases} x_2^2+x_3^2-\dfrac{1+i\sqrt{15}}{2}x_1^2\pm 2i\sqrt{3}x_2x_3=0, \\ x_3^2+x_1^2-\dfrac{1+i\sqrt{15}}{2}x_2^2\pm 2i\sqrt{3}x_3x_1=0, \\ x_1^2+x_2^2-\dfrac{1+i\sqrt{15}}{2}x_3^2\pm 2i\sqrt{3}x_1x_2=0. \end{cases} \tag{8}$$

If, with consideration of the relation of the icosahedral conic sections to the principal functions $\varphi(\zeta)$, we call the left sides of the equations (8), say, in order, $\psi_1(x_1,x_2,x_3)$, $\ldots$, $\psi_6(x_1,x_2,x_3)$, then these expressions $\psi_1,\ldots,\psi_6$, upon exercise of the generators S, T, U' of the collineation group, apart from cubic roots of unity as factors, will permute themselves simply. In fact, one easily finds:

$$\begin{aligned} (S)\quad & (\psi_1',\ldots,\psi_6')=(\psi_3,\psi_4,\psi_2,\psi_1,\psi_6,\psi_5), \\ (T)\quad & (\psi_1',\ldots,\psi_6')=(\psi_6,\psi_5,\psi_3,\psi_4,\psi_2,\psi_1), \\ (U)\quad & (\psi_1',\ldots,\psi_6')=(\psi_4,\psi_1,\psi_3,\psi_2,\rho\psi_5,\rho^2\psi_6). \end{aligned}$$

[649] Numerous additional theorems concerning our collineation group yield themselves from the closer study of these systems of conic sections. One thus finds immediately, e.g.: *The individual octahedral conic section of the one system always finds itself in double tangency with three octahedral conic sections of the other system.* The three tangency chords are three perspectivity axes, which cut out on that first conic section three (situated on the C_6) point-pairs b. The conic section tangents at the points b of the individual pair furnish, as intersection point, the (not situated on the C_6) associated perspectivity center. For the icosahedral conic section, the theorem holds: *Each icosahedral conic section finds itself in double tangency with the six conic sections of the other system, whereby, the six tangency chords are double inflection tangents of the C_6.*[45] The conic section cuts the C_6 in twelve inflection points c; and the poles belonging to the six inflection tangents (with respect to the icosahedral conic section) are the six (not lying on the C_6) fixpoints of the cyclic groups G_5 contained in the corresponding G_{60}. These six points, in the language of Clebsch[46], form a "*tenfold Brianchon hexangle*", while its polars, i.e. the six double inflection tangents, correspondingly furnish a "*tenfold Pascal hexagon*". Hereby, in case one gives the ternary G_{60} the real form, it is naturally a question of none other than the rectilinear icosahedral division of figure 16 in I on p. 72. One will also easily put in here the six likewise real double inflection tangents, of which the individual one always runs through five "edge midpoints" of the icosahedron.[47]

Next to the conic sections considered there would arrange themselves, as the next higher models, *the ten conjugate curves of third degree* of the G_{36} and their *ten Hessian covariants of the same degree.* However, since the calculations fashion themselves somewhat circumstantially in the octahedral coordinate system, we will go no more into their discussion.

A.17 The Galois resolvent of the transformation equation

[650]

We now again take up the train of thought interrupted at the end of § A.9, by going to the setting-up of the Galois resolvent of 360th degree of the transformation equation. We find the expression of this resolvent, by building up *the principal function J of the group Γ* as a covariant of the collineation group G_{360} *in the form of a rational function on the curve C_6.*[48]

[45] The six icosahedral conic sections of the individual system also stand in remarkable relation to one another. Namely, for each two conic sections of the system, the two simultaneous invariants vanish; and one can show, that this property suffices, in order to define the entire system of the six conic sections projectively. Gerbaldi has occupied himself since 1882 with this conic section system, without, to be sure, suspecting, at that time, the important relation of this to the collineation group G_{360}; see Gerbaldi's work "*Sui gruppi di sei coniche in involuzione*", Act. d. Turiner Acad. Vol. XVII, p. 566 ff.

[46] See the reference in "Icos.", p. 218 or also in I on p. 74.

[47] One finds a few further discussions concerning the conic sections in Wiman l.c.. In addition, see the comprehensive representation by Gerbaldi in the paper cited on p. 618 in the second note.

[48] See the corresponding relations for the collineation group G_{168} in "M" I on p. 732 ff.

If, similarly, as for the G_{168}, we attempt to put down the covariants of the form $F(x_1, x_2, x_3)$ of sixth degree standing on the left of the equation (15) on p. 625 of the C_6, then we have, first of all, *the Hessian covariant of the twelfth degree*:

$$H'(x_1, x_2, x_3) = \begin{vmatrix} F_{11} & F_{12} & F_{13} \\ F_{21} & F_{22} & F_{23} \\ F_{31} & F_{32} & F_{33} \end{vmatrix},$$

understanding by $F_{11}, F_{12}, \ldots$ the partial second derivatives of F (cf. p. 638). Set equal to 0, this covariant cuts out *the 72 inflection points c* on the C_6. This is accomplished, naturally, by each curve of the family:

$$\lambda H'(x) + \mu F(x)^2 = 0$$

with a λ different from 0.

Among these curves, there finds itself one, which decomposes into the first system of the six icosahedral conic sections, and a second, which, correspondingly, furnishes the other system of the conic sections of this kind. We therefore have no need to accomplish the somewhat circumstantial calculation of the covariant H', but utilize, *for cutting out the* 72 *points c on the* C_6, *with preference of the section* (8) *on p. 648, the covariant of twelfth degree:*

$$H(x) = \prod\left\{\left(x_1^2 + x_2^2 - \frac{1 + i\sqrt{15}}{2}x_3^2\right)^2 + 12x_1^2x_2^2\right\}, \tag{1}$$

whereby the product sign refers itself to the three cyclic permutations of the indices of the x.

As second covariant, we put down *the Hessian determinant of F bordered with the first derivatives of H*:

$$G'(x_1, x_2, x_3) = \begin{vmatrix} F_{11} & F_{12} & F_{13} & H_1 \\ F_{21} & F_{22} & F_{23} & H_2 \\ F_{31} & F_{32} & F_{33} & H_3 \\ H_1 & H_2 & H_3 & 0 \end{vmatrix}.$$

this is *of degree* 30 and therefore cuts out on the C_6 an invariant system of 180 points, [651]
as long as G', say, was not to contain the factor F. Now there are, on the C_6, two invariant systems of 180 points, namely, first 180 points a and secondly, the doubly counted systems of the 90 points b. The assumption, that $G'(x) = 0$ cuts out the 180 points a, however, leads to a contradiction. Namely, an individual point a always lies on a perspectivity axis as well. This axis cannot be a component of the curve $G'(x) = 0$, since otherwise, the covariant $G'(x) = 0$ of the 30th degree would contain all 45 perspectivity axes. Now since the curve $G'(x) = 0$, as a covariant of F, it is similarly transformed into itself by the perspectivity in question, then this curve must either cut the perspectivity axis perpendicularly at the place a and then touch the C_6 there, or it has a double point at a. In both cases, a counts doubly as an intersection point of $G'(x) = 0$ and $F(x) = 0$. But this, too, is impossible, since these two curves furnish 180 and not $2 \cdot 180$ intersection points. Accordingly, in case $G'(x)$ does not vanish identically along the curve C_6, then, by the curve $G'(x) = 0$, the system of the 90 points b counted doubly is cut

out. But, here too, we need not calculate out the covariant $G'(x)$; for this intersection of the points b is already accomplished *by each of the two systems of the octahedral conic sections,* whose individual one (in consequence of the developments given on p. 649) touches the C_6 at the six points b. By favoring the conic sections (5) on p. 647, we avail ourselves, *for cutting out the doubly counted system of the* 90 *points b on the* C_6*, the covariant of* 30th *degree:*

$$G(x) = \prod\left(x_1^2 + x_2^2 + \frac{1+i\sqrt{15}}{4}x_3^2\right)$$
$$\cdot\prod\left\{x_1^2 + x_2^2 - \sqrt{5}\frac{\sqrt{5}-i\sqrt{3}}{4}x_3^2 - 3\frac{1-i\sqrt{15}}{4}x_1x_2 + i\sqrt{3}\frac{1-i\sqrt{15}}{4}(x_1+x_2)x_3\right\}, \quad (2)$$

where the first product sign refers itself to cyclic permutations of the indices of the x, the second to twelve suitably chosen operations of our favored octahedral group G_{24}.

In the third place, we form *the functional determinant of F, G and H:*

$$K'(x_1, x_2, x_3) = \begin{vmatrix} F_1 & F_2 & F_3 \\ G_1 & G_2 & G_3 \\ H_1 & H_2 & H_3 \end{vmatrix},$$

which belongs *to the degree* 45. If K' is not identically 0 along the C_6, then $K' = 0$ cuts
[652] out on the C_6 an invariant system of 270 points, therefore either the points b counted thrice or the points a and b, each counted simply. But one sees, by means of the argument just attached to $G'(x)$, *that* $K'(x) = 0$ *must necessarily represent the* 45 *perspectivity axes.* Thus we also do not need to calculate K' out, since we are able to immediately form this covariant up to a numerical factor from the equation (1) on p. 645. *We accomplish the cutting out of the points a and b on the* C_6*, each counted simply, by setting to zero the covariant of* 45th *degree:*

$$K(x) = x_1x_2x_3(x_2^2 - x_3^2)(x_3^2 - x_1^2)(x_1^2 - x_2^2)\prod\left(x_1 + x_2 + \frac{\sqrt{5}+i\sqrt{3}}{2}x_3\right)$$
$$\cdot\prod\left(\frac{1+i\sqrt{3}}{2}x_1 + \frac{1+\sqrt{5}}{2}x_2 + \frac{-1+i\sqrt{3}}{2}\cdot\frac{-1+\sqrt{5}}{2}x_3\right), \quad (3)$$

where the first product sign refers itself to 12 particular, and the second to all 24 operations of the favored G_{24}.

If we will now represent the principal function $J(\zeta)$ of the group Γ as a 360-valued rational function on the curve C_6, then we observe, that the 360 zeros of J coincide in pairs at the 180 places a, while the 360 poles coincide in fives as the 72 places c. Consequently, the expression $(JGH^2 : K^2)$, which only still depends on the quotients of the x_1, x_2, x_3, is a function free of poles and zeros, i.e. a constant. Similarly, the 360 zeros of $J-1$ coincide in fours at the 90 places b, and therefore, one also recognizes a constant in $((J-1)H^5 : G^2)$. For the representation of J we thus obtain the schema:

$$J : J-1 : 1 = \alpha K^2 : \beta G^3 : \gamma GH^5, \quad (4)$$

where α, β, γ are three constants, which only come into force in their quotients.

For the determination of these constants, one can first of all utilize the circumstance, that $(\alpha K^2 - \beta G^3)$ must vanish for any point c, whose coordinates we are able to calculate, while, secondly, $(\beta G^2 + \gamma H^5)$ must become zero for any point a, whose coordinates would be calculated. But these calculations, on account of the high degrees of the expressions G, H and K, fashion themselves downright circumstantially, so that we will satisfy ourselves with the representation of J in the unclosed form (4). *In this equation* (4), *in conjunction with the equation:*

$$F(x_1, x_2, x_3) = 0 \tag{5}$$

of our curve C_6, we have before us now the Galois resolvent of the transformation equation. The algebraic problem would consist in calculating, for given J, by solution of the equations (4) and (5), a first associated place x_1, x_2, x_3, as a "root of the Galois resolvent", whereupon, then, the Galois character of the problem thereby comes into force, since in that first solution system x_1, x_2, x_3, all 360 solution systems are representable linearly with number-theoretically simple substitution coefficients, namely by the substitutions of our G_{360}.[49] [653]

After disposal of this problem, the roots of the resolvents of the 6th and 15th degrees, as well as the roots of the transformation equation itself, are rationally calculatable. By therefore now turning ourselves to our last problem, namely, to *setting up that resolvent and those of the transformation equation,* we shall have to conceive this task in this sense, that we must represent the principal functions $\psi(\zeta)$, $\chi(\zeta)$, $\varphi(\zeta)$ of the groups Γ_6, Γ_{15} and Γ_{10} as rational functions on the curve C_6.

A.18 The solution of the resolvents of 6th and 15th degree

For representing the solutions ψ of our two resolvents (4) on p. 593 of 6th degree in a solution of the Galois resolvent, i.e. therefore, for representing the principal functions $\psi(\zeta)$ of the groups Γ_6 as rational functions on the curve C_6, we have to avail ourselves of the icosahedral conic sections. If we attach the consideration, say, to the conic section:

$$x_2^2 + x_3^2 - \frac{1+i\sqrt{15}}{2} x_1^2 + 2i\sqrt{3} x_2 x_3 = 0, \tag{1}$$

[49] If, as is done by Wiman, one develops the theory of the collineation group G_{360} indepently, i.e. independently of the transcendental foundation presented here, stemming from the theory of the automorphic functions, then one can obtain the tie-in to the transcendental theory at the place now reached in the representation of the text. Namely, if one maps the curve C_6 onto the J-plane by means of the function defined in (4), then one obtains, as image, a regular-symmetric Riemann surface F_{360} of 360 sheets over the J-plane, which has the branching described on p. 611. In consequence of the "*boundary-circle theorem*", there exists on this F_{360} a polymorphic function $\zeta(J)$ of boundary-circle character, which, corresponding to the single-valued transformations of the F_{360} into itself, itself undergoes linear substitutions of the first resp. second kinds. In particular, to the symmetric transformations of the surface into itself correspond *reflections* of ζ; and it is immediately clear from this, that the individual halfsheet of the Riemann surface is mapped onto a triangle of the angles $\frac{\pi}{2}$, $\frac{\pi}{4}$, $\frac{\pi}{5}$ in the ζ-plane, bounded by symmetry circles. With this, we have again obtained the variable ζ, which was placed at the forefront above as the argument of the associated automorphic functions.

[654] then this cuts out on the curve C_6 a system of twelve points c, equivalent with respect to the associated icosahedral-G_{60}. Now a glance at figure 108 informs us, that the 72 points c, with respect to the G_{60}, fall into one system of twelve equivalent points, represented by the point c_0 of that figure, and a remaining system of 60 equivalent points, represented by c_1. That first point system is therefore cut out by the conic section (1), and the second system of the 60 points c by the five other icosahedral conic sections conjugate with (1).

Now for the principal function chosen on p. 592, $\psi(c_0) = 0$, $\psi(c_1) = \infty$ held; the 60 zeros of ψ on the C_6 thus coincide in fives at the twelve places c of the first system, while the 60 poles are the points c of the second system. By means of the function-theoretic argument already applied for the Galois resolvent, we thus find, that ψ, up to a constant factor, is represented by the quotient of the 5th power of the form standing in the left in (1) and the product of the five conjugate forms, or (which is shorter) by the quotient of the 6th powers of the form (1) and our covariant $H(x)$:

$$\psi = C\frac{\left(x_2^2 + x_3^2 - \frac{1+i\sqrt{15}}{2}x_1^2 + 2i\sqrt{3}x_2x_3\right)^6}{H(x)}.$$

For the determination of the constant C, one has only to observe, that the collineation $x_1' = -x_1$, $x_2' = -x_2$, $x_3' = x_3$ of period two, which has the point b with the coordinate $x_1 = i$, $x_2 = 1$, $x_3 = 0$ as fixpoint, does *not* transform the function ψ into itself. This point b is accordingly equivalent, with respect to the G_{60}, with the point denoted by b_1 in figure 108, since only this vertex b_1 (but not b_2) is surrounded by four double triangles in the discontinuity domain considered closed. But now $\psi(b_1)$ was $= 1$, so that the equation set down for $\psi = 1$, $x_1 = i$, $x_2 = 1$, $x_3 = 0$, must be correct. From this follows immediately:

$$C = \frac{1+i\sqrt{15}}{2},$$

so that we obtain the final representation for ψ:

$$\psi = \frac{1+i\sqrt{15}}{2}\frac{\left(x_2^2 + x_3^2 - \frac{1+i\sqrt{15}}{2}x_1^2 + 2i\sqrt{3}x_2x_3\right)^6}{H(x)}. \qquad (2)$$

In order to determine, to which of the two resolvents (4) on p. 593 this root ψ be-
[655] longs, put the coordinates $x_2 = i$, $x_3 = 1$, $x_1 = 0$ into (2); these furnish a point b, of which one easily perceives, that it is equivalent, with respect to the G_{60}, to the point b_2 of the discontinuity domain of figure 108. The calculation furnishes, at this place:

$$\psi(b_2) = \frac{11+3i\sqrt{15}}{2},$$

so that we have to do with the resolvent (4) on p. 593 belonging to the *upper* sign.

Now since the left sides of the equations (8) on p. 648, for which we will again take up the notations $\Psi_k(x_1, x_2, x_3)$ or $\Psi_k(x)$ for short, permute themselves simply with respect to the ternary substitutions of our G_{360}, apart from third roots of unity as factors, in consequence of p. 648, then the representation (2) carries itself over immediately to the five remaining roots of our resolvent. If we denote these roots, as on p. 614, by

$\psi_1, \psi_2, \ldots$, then these arises, as a result: *The six roots of the resolvent* (4) *on p. 593 of* 6th *degree, with the upper signs, represent themselves as rational functions on the curve* C_6 *thus:*

$$\psi_k = \frac{1+i\sqrt{15}}{2} \cdot \frac{\Psi_k(x)^6}{H(x)}, \ (k = 1, 2, \ldots, 6). \tag{3}$$

For the second resolvent of 6th degree, entirely analogous arguments hold. We satisfy ourselves with divulging, for the root ψ', belonging to the first equation (7) on p. 648, of the second resolvent, the representation:

$$\psi' = -\frac{1+i\sqrt{15}}{2} \frac{\left(x_1^2 + x_2^2 + x_3^2 - i\sqrt{3}\frac{\sqrt{5}-i\sqrt{3}}{2}(x_2 x_3 + x_3 x_1 + x_1 x_2)\right)^6}{H(x)}, \tag{4}$$

whose correctness one will easily verify.

In consequence of (3) we are able to represent the quotients of the 6th roots of the ψ_k, in which we already recognized, on p. 614, single-valued functions of the group Γ_{360}, by the proportion:

$$\sqrt[6]{\psi_1} : \sqrt[6]{\psi_2} : \cdots : \sqrt[6]{\psi_6} = \Psi_1(x) : \Psi_2(x) : \cdots : \Psi_6(x). \tag{5}$$

In the $\Psi_k(x)$ we therefore have before us those six ratio-magnitudes, which we introduced on p. 614 with the notation $\omega_1, \omega_2, \ldots, \omega_6$. Also, the remaining statements at that time are now clear. *The* $\Psi_k(x)$ *form six linearly independent quadratic forms of the* x_1, x_2, x_3*;* for one succeeds easily in representing the six expressions $x_1^2, x_2^2, \ldots, x_1 x_2$ linearly in the $\Psi_k(x)$:

$$\begin{cases} x_1^2 = \dfrac{-1+i\sqrt{15}}{24}(\Psi_1 + \Psi_2) + \dfrac{1}{12}(\Psi_3 + \Psi_4 + \Psi_5 + \Psi_6), \\ x_2^2 = \dfrac{-1+i\sqrt{15}}{24}(\Psi_3 + \Psi_4) + \dfrac{1}{12}(\Psi_1 + \Psi_2 + \Psi_5 + \Psi_6), \\ x_3^2 = \dfrac{-1+i\sqrt{15}}{24}(\Psi_5 + \Psi_6) + \dfrac{1}{12}(\Psi_1 + \Psi_2 + \Psi_3 + \Psi_4), \\ x_2 x_3 = \dfrac{1}{4i\sqrt{3}}(\Psi_1 - \Psi_2), \\ x_3 x_1 = \dfrac{1}{4i\sqrt{3}}(\Psi_3 - \Psi_4), \\ x_1 x_2 = \dfrac{1}{4i\sqrt{3}}(\Psi_5 - \Psi_6). \end{cases} \tag{6}$$

We perceive, that the *transition from the principal functions* $\psi(\zeta)$ *to the threefold infi-* [656]
nite family of the six-valued special functions, whose existence we had found on p. 615, is a very simple one:

$$x_1 : x_2 : x_3 = \frac{1}{\sqrt[6]{\psi_1} - \sqrt[6]{\psi_2}} : \frac{1}{\sqrt[6]{\psi_3} - \sqrt[6]{\psi_4}} : \frac{1}{\sqrt[6]{\psi_5} - \sqrt[6]{\psi_6}}. \tag{7}$$

At the same time, there now yield themselves, explicitly, the *linear representations of the* 6th *roots of the solutions* ψ *of the one resolvent of the* 6th *degree in the* 6th *roots of the other resolvent*, put down in (5) on p. 615; one thus finds, e.g., for the 6th root of the solution ψ' of the second resolvent, represented in (4), on the basis of the relations (5) and (6):

$$\sqrt[6]{\psi'} = \frac{\sqrt{5}-i\sqrt{3}}{4\sqrt{3}}\left(\rho^2\left(\sqrt[6]{\psi_1}+\sqrt[6]{\psi_3}+\sqrt[6]{\psi_5}\right)+\rho\left(\sqrt[6]{\psi_2}+\sqrt[6]{\psi_4}+\sqrt[6]{\psi_6}\right)\right). \tag{8}$$

In order to represent the roots of the two resolvents of the 15th degree as rational functions on the curve C_6, we remember, that the octahedral groups G_{24} of the individual system were uniquely coordinated to the pairs of icosahedral groups G_{60} of the associated system of the G_{60}. Accordingly, let $\Psi = 0$ and $\Psi' = 0$ be any two conjugate icosahedral conic sections, and $\Omega = 0$ the octahedral conic section coordinated to this pair.

This conic section $\Omega = 0$ touches the curve C_6 at six points b, equivalent with respect to the associated G_{24}. The discontinuity domain of figure 110 shows, that there is only one simple system of *six* equivalent points b for the G_{24}, which are represented by the vertex b_1 of the figure mentioned. For this point b_1, in consequence of (19) on p. 606, there holds:

$$\chi(b_1) = \frac{3\sqrt{5}\pm i\sqrt{3}}{8}. \tag{9}$$

[657] The product $\Psi\cdot\Psi'$ is a covariant of the G_{24}, so that, by $\Psi\cdot\Psi' = 0$, we cut out on the C_6 a system of 24 points c equivalent with respect to the G_{24}. In consequence of (5) on p. 602 and (10) on p. 603, these points are the 24 poles of the function χ on the C_6. Our repeated applied function-theoretic method of argument thus furnishes the theorem: *The roots of the resolvent of* 15th *degree represent themselves on the curve* C_6 *in the form:*

$$\chi - \frac{3\sqrt{5}\pm i\sqrt{3}}{8} = C\frac{(\Omega(x))^2}{\Psi(x)\Psi'(x)}, \tag{10}$$

where C is a constant in each case.

The determination of C offers no difficulty. If we choose $\Psi = \rho x_1^2 + \rho^2 x_2^2 + x_3^2$, $\Psi' = \rho^2 x_1^2 + \rho x_2^2 + x_3^2$ as an example, then $\Omega = x_1^2 + x_2^2 + x_3^2$ belongs to this pair. For the corresponding resolvents of 6th degree (4) on p. 593, the lower sign holds (as we know from the discussion of these resolvents just ended), and therefore for this example, in the schema (10), the lower sign likewise occurs on the left:

$$\chi - \frac{3\sqrt{5}-i\sqrt{3}}{8} = C\frac{(x_1^2+x_2^2+x_3^2)^2}{x_1^4+x_2^4+x_3^4-x_2^2x_3^2-x_3^2x_1^2-x_1^2x_2^2}. \tag{11}$$

Now bring in the collineation of period 4:

$$\begin{aligned}
x_1' &= \frac{1}{2}x_1 - \frac{1}{2}x_2 + \frac{\sqrt{5}+i\sqrt{3}}{4}x_3,\\
x_2' &= -\frac{1}{2}x_1 + \frac{1}{2}x_2 + \frac{\sqrt{5}+i\sqrt{3}}{4}x_3,\\
x_3' &= -\frac{\sqrt{5}-i\sqrt{3}}{4}(x_1+x_2)
\end{aligned}$$

with the fixpoint with the coordinate [50]:

$$x_1 = 1,\ x_2 = 1,\ x_3 = \frac{\sqrt{3}+i\sqrt{5}}{2}. \tag{12}$$

Since

$$\Omega(x') = \frac{\sqrt{5}-i\sqrt{3}}{8}\left(\frac{-\sqrt{5}+3i\sqrt{3}}{2}x_3^2 + \sqrt{5}(x_1^2+x_2^2) - 2i\sqrt{3}x_1x_2\right)$$

holds, then χ is not transformed into itself by the operation in question. Nevertheless, the second power $x_1' = -x_2,\ x_2' = -x_1,\ x_3' = -x_3$ of this operation carries χ over into itself. The point with the coordinate (12) is accordingly equivalent to the vertex b_5 of figure 110, at which, in consequence of (23) on p. 607: [658]

$$\chi(b_5) = \sqrt{5} \tag{13}$$

takes place. The substitution of the values (12) and (13) into the schema (11) furnishes, in the usual manner, the value of the constant C. *The solution χ, chosen here as an example, of our resolvent of* 15*th degree, consequently represents itself thus:*

$$\chi = \frac{3\sqrt{5}-i\sqrt{3}}{8} - \frac{\sqrt{5}-i\sqrt{3}}{2}\frac{(x_1^2+x_2^2+x_3^2)^2}{x_1^4+x_2^4+x_3^4-x_2^2x_3^2-x_3^2x_1^2-x_1^2x_2^2}. \tag{14}$$

A.19 Solution of the transformation equation of 10th degree

For the solution of the transformation equation (17) on p. 576, i.e. for the representation of the principal functions $\varphi(\zeta)$ of the group Γ_{10} as rational functions on the curve C_6, we have to bring in the ten curves of third degree $f(z_1, z_2, z_3) = 0$ and their Hessian curves $h(z_1, z_2, z_3) = 0$ from p. 638 ff. We will hereby hold fast to the harmonic coordinate system of the z_1, z_2, z_3, since (as was already remarked on p. 649) the equations of the $2\cdot 10$ curves C_3 come out somewhat unsurveyably in the octahedral coordinate system of the x_1, x_2, x_3.

For a first Γ_{10} resp. G_{36}, we have the two C_3:

$$f_1(z) = z_1^3 + 3z_1z_2^2 + 3z_2z_3^2 = 0,\ h_1(z) = z_1^2z_2 - z_1z_3^2 - z_2^3 = 0. \tag{1}$$

The coordinate triangle of the z is furnished by the fixpoints of a cyclic G_4 contained in the G_{36}. Since the ten G_{36} each contain nine of the 45 cyclic G_4, then each G_4 occurs in *two* groups G_{36}. These two G_{36} are transformed into one another by one of the four perspectivities written in (8) on p. 639, say:

$$z_1' = \frac{1+i\sqrt{5}}{\sqrt{2\sqrt{3}}}z_2,\ z_2' = \frac{1-i\sqrt{5}}{\sqrt{6\sqrt{3}}}z_1,\ z_3' = z_3.$$

[50] It is simply a question here of that substitution, which writes itself $x_1' = x_3$, $x_2' = x_2$, $x_3' = -x_1$ in the second of the two unsymmetric octahedral coordinate systems.

The exercise of this transformation on (1) furnishes, as the two curves C_3 belonging to the second G_{36}:

$$\begin{cases} f_2(z) = z_1^2 z_2 + z_1 z_3^2 + \left(\dfrac{2-i\sqrt{5}}{3}\right)^2 z_2^3 = 0, \\ h_2(z) = \left(\dfrac{2+i\sqrt{5}}{3}\right)^2 z_1^3 - 3z_1 z_2^2 + 3z_2 z_3^2 = 0. \end{cases} \tag{2}$$

The remaining $2\cdot 8$ equations $f_3(z)=0$, $h_3(z)=0$, ..., also in the harmonic coordinate system of the z_1, z_2, z_3, do not come out entirely simply.

[659] Now let an arbitrary one of the ten group G_{36} be proposed, whose curves C_3 let be given by $f(z)=0$, $h(z)=0$. Each of these two curves C_3 intersects our curve C_6 in a system of 18 points, equivalent with respect to the G_{36}. In consequence of the discontinuity domain of the Γ_{10} given by figure 106, there are, in all, four systems of 18 points equivalent with respect to the G_{36}: first, two systems of 18 points a each, represented by the vertices a_4 and a_5 in figure 106, second, two systems of nine points b each, to be counted doubly, represented by the vertices b_1 and b_5 of the figure mentioned. It is now clear without further ado, that the two systems of the points b, named secondly, are each cut out by our curve C_3 on the C_6, each counted doubly. Thus, e.g., the curve $f_1(z)=0$ touches the C_6 at the point b furnished by the vertex $z_1=0$, $z_3=0$ of the coordinate triangle; both curves the C_3 and the C_6 go over into themselves by the perspectivity belonging to the axis $z_3=0$, and consequently intersect that axis perpendicularly. The same state of affairs is present for the intersection of the curve $h_1(z)=0$ with the C_6 at the vertex $z_2=0$, $z_3=0$: *The curves $f(z)=0$ and $h(z)=0$ touch the curve C_6 at the nine points b, equivalent with respect to the G_{36}, which are represented by the polygon vertices b_1 and b_5 of figure 106.*

In consequence of (17) on p. 576, the values $\varphi(b_1)$ and $\varphi(b_5)$ are the roots of the quadratic equation:

$$\varphi^2 + \sqrt{15}(1+i\sqrt{3})\varphi + \frac{1-i\sqrt{3}}{2} = 0,$$

and indeed, more precisely, according to (18) on p. 578, there holds:

$$\varphi(b_1) = \rho^2(\sqrt{15}-4),\ \varphi(b_5) = \rho^2(\sqrt{15}+4). \tag{3}$$

If we accordingly form the quotient:

$$\frac{\rho\varphi - \sqrt{15}+4}{\rho\varphi - \sqrt{15}-4}, \tag{4}$$

then the 36 zeros of this will coincide in fours at the nine tangency points of one of the two curves $f(z)=0$, $h(z)=0$, and the 36 poles in fours at the nine tangency points of the other C_3 with the C_6.

Now the poles and zeros of the function (4) are exactly interchanged by the transformation denoted on p. 566 by T'. Namely, the effect of this transformation on φ is
[660] that given in (20) on p. 579, a substitution, which, for our present purpose, we better clothe in the form:

$$\frac{\rho\varphi'-\sqrt{15}-4}{\rho\varphi'-\sqrt{15}+4}=-\left(\frac{\sqrt{5}+\sqrt{3}}{\sqrt{5}-\sqrt{3}}\right)^2\cdot\left(\frac{1+i\sqrt{5}}{1-i\sqrt{5}}\right)^4\cdot\frac{\rho\varphi-\sqrt{15}+4}{\rho\varphi-\sqrt{15}-4}. \tag{5}$$

If, consequently, in order to avoid further going calculations, we assume, that the zeros of the function (4) are identical with the tangency points of the C_6 and the curve $f=0$, then we must, to be sure, let it remain to be seen, whether, in the functions (4), are meant the roots φ of our transformation equation (17) on p. 576, or the roots of the other form (22) on p. 579 of our transformation equation, connected with them by the relation (5). With this reservation, we are able to state the theorem: *The representation of the roots φ of our transformation equation as rational functions on the curve C_6 is accomplished by:*

$$\frac{\rho\varphi-\sqrt{15}+4}{\rho\varphi-\sqrt{15}-4}=C\left(\frac{f(z)}{h(z)}\right)^2, \tag{6}$$

where C is a constant.

The calculation of of C, to be sure, only up to a sign, whose unique determination would require further going calculations, one can carry through in the following way. The transformation given in (12) on p. 639, which interchanges the two harmonic coordinate systems used there, certainly does not correspond directly to the transformation of third degree denoted by T' on p. 566, but was nevertheless equivalent to T' with respect to the G_{36}. Accordingly, to the transformation (12) on p. 639 will correspond the carrying of φ over into the function φ' written on the left in (5):

$$\frac{\rho\varphi'-\sqrt{15}+4}{\rho\varphi'-\sqrt{15}-4}=C\left(\frac{f(z')}{h(z')}\right)^2.$$

Now, in consequence of (13) on p. 640, there holds, for the transformation applied here:

$$\frac{f(z')}{h(z')}=i\sqrt{27}\frac{h(z)}{f(z)},$$

so that, with consideration of (5), from the equation before last follows:

$$\left(\frac{\sqrt{5}-\sqrt{3}}{\sqrt{5}+\sqrt{3}}\right)^2\left(\frac{1-i\sqrt{5}}{1+i\sqrt{5}}\right)^4\frac{\rho\varphi-\sqrt{15}-4}{\rho\varphi-\sqrt{15}+4}=27C\left(\frac{h(z)}{f(z)}\right)^2.$$

By multiplication with (16), there yields itself:

$$\left(\frac{\sqrt{5}-\sqrt{3}}{\sqrt{5}+\sqrt{3}}\right)^2\left(\frac{1-i\sqrt{5}}{1+i\sqrt{5}}\right)^4=27C^2,$$

from which the value of C^2 follows uniquely. *The representation of φ is thus (up to the sign, remaining undecided) accomplished by:* [661]

$$\frac{\rho\varphi-\sqrt{15}+4}{\rho\varphi-\sqrt{15}-4}=+\frac{(4-\sqrt{15})(1-4i\sqrt{15})}{27\sqrt{3}}\left(\frac{f(z)}{h(z)}\right)^2. \tag{7}$$

The transformation (16) on p. 641 of the second kind furnishes a verification of this result, which carries the G_{36} over into itself. It furnished, for φ, the transformation of the "diametral symmetry":

$$\varphi' = -\frac{1}{\bar{\varphi}}, \tag{8}$$

while it works on $f(z)$ and $h(z)$ as follows:

$$f(z') = e^{\frac{7\pi i}{4}} \sqrt[4]{27} h(\bar{z}),\ h(z') = e^{\frac{5\pi i}{4}} \frac{1}{\sqrt[4]{27}} f(\bar{z}). \tag{9}$$

The application of the transformation in question to equation (6) thus furnishes, after brief intermediate calculation:

$$\frac{4-\sqrt{15}}{4+\sqrt{15}} \cdot \frac{\rho^2\bar{\varphi} - \sqrt{15} - 4}{\rho^2\bar{\varphi} - \sqrt{15} + 4} = 27C\left(\frac{h(\bar{z})}{f(\bar{z})}\right)^2.$$

If, in this equation, we alter the sign of the imaginary component everywhere, then there follows:

$$\frac{4-\sqrt{15}}{4+\sqrt{15}} \cdot \frac{\rho\varphi - \sqrt{15} - 4}{\rho\varphi - \sqrt{15} + 4} = 27\bar{C}\left(\frac{h(z)}{f(z)}\right)^2,$$

an equation, whose multiplication with (6) gives:

$$(4-\sqrt{15})^2 = 27C\bar{C}.$$

This condition, in fact, corresponds to the value of C calculated before.

Last of all, we yet allude to the fact, *that for the principal functions, given in* (1) *on p. 609, of the ten conjugate groups* Γ_{20}, *rational representations on the* C_6 *now also yield themselves:*

$$\sqrt{\frac{\rho\varphi - \sqrt{15} + 4}{\rho\varphi - \sqrt{15} - 4}} = \frac{(\sqrt{5} - \sqrt{3})(2 + i\sqrt{5})}{3\sqrt{\pm 6\sqrt{3}}} \cdot \frac{f(z)}{h(z)}.$$

They represent themselves, in this fashion, directly as quotients of "functions φ" of the Riemann surface F_{360}; for we indeed have ten linearly independent functions of this kind for our F_{360} of the genus $p = 10$, as "ratio-magnitudes" in the ten cubic expressions
[662] $z_1^3, z_2^3, \ldots, z_1 z_2 z_3$.

Our representation now ended has shown, in any case, that fruitful and remarkable developments are also to be met with in the domain of the transformation of the automorphic functions. The chosen example may even be able to vie with those beautiful developments, which Klein has based on his theory of modular functions concerning the transformation of seventh degree of the elliptic functions (Math. Ann. Vol. 14 on p. 428 (1878) and "M" I on p. 692 ff.). Thus, let our representation come to a close with this small slice from the transformation theory of the automorphic functions; in it, the character and the beauty of that direction of mathematics produced be Klein comes into expression once more, which strives to set into lively interplay the various domains of mathematical research.

Commentaries

1

Commentary by Richard Borcherds on Elliptic Modular Functions

Introduction

Why read two large volumes (*Lectures on the Theory of Elliptic Modular Functions*) about some obscure functions that most mathematicians have barely heard of? Klein's elliptic modular function $j = 1728J$ may hold some sort of record for its number of bizarre properties:

- The simplest explicit formula for j looks at first glance like a strange joke:

$$j(\tau) = \frac{(1+240\sum_{m,n>0} m^3 q^{mn})^3}{q\prod_{n>0}(1-q^n)^{24}} = q^{-1} + 744 + 196884q + 21493760q^2 + \cdots \qquad (1)$$

  (where $q = e^{2\pi i\tau}$).
- j is the simplest non-constant holomorphic function on the upper half plane invariant under the group $SL_2(Z)$ (so that $j((a\tau+b)/(c\tau+d)) = j(\tau)$ for all integers a, b, c, d with $ad - bc = 1$).
- j is the j-invariant for complex elliptic curves; in other words it gives an isomorphism from the moduli space of elliptic curves $C/\langle 1,\tau\rangle$ to the complex plane C.
- Complex multiplication [5]: $j(\tau)$ for imaginary quadratic τ is an algebraic integer generating the Hilbert class field of the corresponding order. These values are called singular moduli. (The term "singular" does not refer to a singularity in the sense of singularity theory; it is an old term for "special in some way".) For example

$$j((1+\sqrt{-163})/2) = -262537412640768000 \qquad (2)$$

  is an integer as $Z[(1+\sqrt{-163})/2]$ has class number 1, which implies that

$$e^{\sqrt{163}} = 262537412640768743.99999999999925\cdots \qquad (3)$$

  is very close to an integer. Extending these properties to other number fields is Hilbert's 12th problem, party solved by Shimura.
- There is a very short proof of Picard's theorem (that an entire function cannot omit two values) using j: if f is an entire function omitting the values 0 and 1728 (where j^{-1} has branch points) then $\exp(ij^{-1}(f(z))$ is a bounded holomorphic function and must therefore be constant.

- Petersson and Rademacher [4] found a convergent asymptotic infinite series involving Bessel functions for the coefficients of j, similar to the Hardy–Ramanujan series for the partition function. This series converges astonishingly fast: for example, the first term of the series for the coefficient $c(1) = 196884$ of $j(\tau)$ is $2\pi I_1(4\pi) = 196550 \cdot 665\ldots$.
- As a byproduct of their work on the Gross–Zagier formula for the derivative of the L-series of an elliptic curve, Gross and Zagier [7] found a description of the prime factorization of a product of differences of two values of j: for example, one special case of their result is

$$j((\sqrt{-163}+1)/2) = -2^{18} \cdot 3^3 \cdot 5^3 \cdot 23^3 \cdot 29^3 \tag{4}$$

- John Mckay discovered an extraordinary relation between the monster simple group and j: the smallest representation of the monster has dimension 196883, 1 less than the coefficient 196884 of j. Thompson, Conway, and Norton [1] extended this observation to the moonshine conjectures, relating the monster to elliptic modular functions of higher level.
- j satisfies the product formula, discovered independently by Koike, Norton, and Zagier:

$$j(\sigma) - j(\tau) = p^{-1} \prod_{m>0, n\in\mathbb{Z}} (1-p^m q^n)^{c(mn)} \tag{5}$$

  where the numbers $c(n)$ are the Fourier coefficients of $j(\tau) - 744 = \sum c(n) q^n$ and $q = e^{2\pi i\tau}$ and $p = e^{2\pi i\sigma}$. This product formula is the Weyl–Kac denominator formula for the infinite dimensional monster Lie algebra.
- Another product formula for j is

$$j(\tau) = q^{-1} \prod_n (1-q^n)^{a(n^2)} \tag{6}$$

  where the numbers $a(n)$ are the Fourier coefficients of the unique weight 1/2 modular form $3q^{-3} + O(q)$ for $\Gamma_0(4)$ satisfying the Kohnen plus-space condition $a(n) = 0$ if n is 2 or 3 mod 4. Zagier [7] interpreted this in terms of a duality between forms of weight 1/2 and 3/2, sometimes called Zagier duality.
- The modularity theorem, used by Wiles to prove Fermat's last theorem, can be rephrased as saying that any rational elliptic curve can be parametrized by elliptic modular functions of higher level.
- Elliptic modular functions are closely related to Apéry's proof that $\zeta(3)$ is irrational. See [8] for more details.
- The inverse of the modular function satisfies a third order non-linear differential equation involving the Schwarzian derivative: see formula (4) of section 2.16 of KL.

One wonders what further surprises about elliptic modular functions await discovery.

Klein and Fricke's monographs are unsurpassed for they coverage of the classical theory of elliptic modular functions, and are still almost the only place where one can find many examples of high-level modular functions worked out in detail.

We now give a brief guide to some of the changes in notation and terminology.

I.1 KF introduce the elliptic modular function J as an invariant related to either the binary quartic form

$$ax^4+4bx^3y+6cx^2y^2+4dxy^3+ey^4 \tag{7}$$

or the elliptic integral

$$\int \frac{dx}{\sqrt{x(x-1)(x-\lambda)}}. \tag{8}$$

These days J is more often thought of as an invariant of the corresponding elliptic curves

$$y^2 = ax^4+4bx^3+6cx^2+4dx+e \tag{9}$$

or

$$y^2 = x(x-1)(x-\lambda) \tag{10}$$

or as a function of τ in the upper half plane, in which case the corresponding elliptic curve is $C/\langle 1,\tau\rangle$. The fundamental property of J is that two complex elliptic curves are isomorphic if and only if they have the same J-invariant. This means that J is almost an isomorphism from the moduli space of elliptic curves to the complex plane, except for some problems at the two orbifold points where J is 0 or 1, corresponding to the two complex elliptic curves with extra automorphisms.

The invariant λ above is the Hauptmodul for the group $\Gamma(2)$ in the same way that J is the Hauptmodul for the group $\Gamma(1)$, where $\Gamma(N)$ is the subgroup of $SL_2(Z)$ of matrices congruent to 1 mod N, and a Hauptmodul is a roughly a generator for the field of meromorphic functions invariant under the group.

The function μ is a modular function of level 4.

Notation:

- Instead of J used in KF, it is now common to use $j = 1728J$ as this has better integrality properties.
- There are now about a dozen different notations for various multiples of the Eisenstein series g_2 and g_3, with 2 and 3 sometimes replaced by 4 and 6, and the letter g replaced by $c, e, g, E, G, \ldots$.
- The symbol Δ is used by KF for $(2\pi)^{12}q\prod(1-q^n)^{24}$, but these days it is often rescaled by a factor of $(2\pi)^{12}$) and used for $q\prod(1-q^n)^{24}$.
- The function λ is an invariant for elliptic curves with the points of order 2 marked, and is related to j by $j = 2^8(\lambda^2-\lambda+1)^3/(\lambda(1-\lambda))^2$. It can also be regarded as a level 2 modular function.

I.2 The periods of an elliptic integral studied here form the period lattice L in the complex plane C, and the corresponding elliptic curve is C/L.

The periodquotient w in the upper half plane is now usually denoted by τ (while KL later use τ for a modular function of level 5). KL regard w and the modular functions μ and λ as multiple-valued functions of J, with branch-points at $J = 0, 1, \infty$, whereas now one usually regards J as a single-valued function of w, μ, or λ. This use of multivalued inverse functions is rather as if one studied trigonometry using the multi-valued function arcsin instead of its single-valued inverse function sin. Treating w as a multiple-valued function of J does have some advantages: for example, w satisfies a 3rd-order differential equation given at the end of this section.

I.3 The s-functions studied here are multiple valued functions on the Riemann sphere with branch points of various orders at 3 points. They are said to be of the first, second, or third kind if the corresponding Riemann surface is isomorphic to a sphere, the plane, or the unit disc. As before, it is more usual these days to work with the inverse functions, which are essentially rational functions on the sphere invariant under some finite group, or elliptic functions on the plane, or functions on the unit disk invariant under a group with fundamental domain some triangle (as in the well-known Escher woodcuts). Some but not all elliptic modular functions are s-functions.

I.4 The functions J, λ, ξ, μ, ζ are modular functions of levels 1, 2, 3, 4, 5; in fact they generate the field of modular functions for $\Gamma(N)$, $N = 1,2,3,4,5$. This section defines modular forms and functions. What KL call the dimension of a modular form is now usually called minus the weight k.

I.5 This section covers standard properties of the Weierstrass elliptic function and related functions. Not much has changed since KF was written, though the formulas might look more familiar if ω in the upper half plane is replaced by by τ, and $r = e^{2\pi i\omega}$ by $q = e^{2\pi i\tau}$.

II.1 The elements of the group $PSL_2(C)$ of automorphisms of the Riemann sphere are classified into elliptic, parabolic, hyperbolic, and loxodromic elements.

II.2 The fundamental domain of the modular group $PSL_2(Z)$ acting on the non-real complex numbers consists of the points z with $|z| \geq 1$, $|\Re z| \leq 1/2$. The standard generators $S(z) = z+1$ and $T(z) = -1/z$ are now usually denoted in reverse order by T and S. The extended modular group is the group $PGL_2(Z)$ containing the modular group with index 2 and acting on the Riemann sphere. If we identify each complex number with its complex conjugate we get an action of the extended modular group on the upper half plane. The extended modular group is somewhat out of fashion these days, possibly because half its elements act as anti-holomorphic transformations, but it has the advantage that it is a reflection group (or Coxeter group). This makes its fundamental domain better behaved than that of the modular group. In this case the fundamental domain is a $(2,3,\infty)$ hyperbolic triangle. (This notation means that the angles of the triangle are $\pi/3$, $\pi/3$, and π/∞).

II.3 This chapter discusses the reduction theory of integral binary quadratic forms $ax^2 + 2bxy + cy^2$, mostly following Gauss's book Disquisitiones arithmeticae. If one maps such a form to the roots of $ax^2 + 2bx + c$ then the action of $SL_2(Z)$ on these forms corresponds to the action of $SL_2(Z)$ on the complex plane. In particular one can classify the equivalence classes of binary forms with given negative discriminant by classifying the corresponding complex numbers in the fundamental domain of $SL_2(Z)$, which shows that the number of such classes is finite.

II.4 The group of matrices of $SL_2(Z)$ congruent to I mod 2, now usually denoted by $\Gamma(2)$, is a normal subgroup of index 6 in $SL_2(Z)$, with quotient the nonabelian group $SL_2(F_2)$ of order 6. KF describe generators and fundamental domains for $\Gamma(2)$ and some related groups.

II.5 KL discuss the fundamental domain for general discrete subgroups and give a formula for the Euler characteristic of the quotient of the upper half plane by the subgroup. The formula looks a bit simpler in terms of Thurston's orbifold Euler characteristic χ (see Thurston's notes [6]), when it just says that $\chi(S/G) = \chi(S)/|G|$ for a finite group G acting on a surface S. Of course one then has to convert the orbifold Euler characteristic into a topological Euler characteristic; for example, an elliptic point of order n contributes 1 to the topological Euler characteristic but only $1/n$ to the orbifold Euler characteristic.

II.6 The group $\Gamma_{\{n\}}$ is the normal subgroup of the modular group generated by $\tau \mapsto \tau + n$ and its conjugates. It is contained in the subgroup $\Gamma(n)$, but is not equal to it except when $n < 6$, and in fact for $n \geq 6$ it has infinite index so is not a congruence subgroup or even an arithmetic subgroup. The case $n = 6$ is a sort of borderline case: for $n < 6$, $n = 6$, and $n > 6$ the compactifications of the quotient Riemann surfaces have positive, zero, or negative curvature, corresponding to the three simply connected Riemann surfaces (the sphere, the plane, and the upper half plane). Klein was fascinated by the correspondence of the cases for $n = 3, 4, 5$ with the symmetry groups of the Platonic solids Don't miss the spectacular picture of the fundamental domain of $\Gamma(7)$ (Fig. 86). KF obtain this by a 2-step process. They first take a quotient of the upper half plane (plus cusps) by the group $\Gamma_{\{n\}}$, which (for $n > 6$) is a copy of the hyperbolic plane, here represented as a unit disk. Taking the quotient by $\Gamma_{\{n\}}$ has the effect of changing the $(2,3,\infty)$ triangles (fundamental domains of the extended modular group) into the $(2,3,7)$ triangles of fig. 86. This unit disk is now acted on by $\Gamma(7)/\Gamma_{\{7\}}$, and Fig 86 is a picture of the fundamental domain of this group. If one identifies the edges as in the figure, one gets a model of the celebrated Klein quartic, a curve so extraordinary that there is an entire book about it [3]. It is the unique genus $g = 3$ curve with the maximum possible number $168 = 84(g-1)$ of automorphisms. This construction makes it obvious that the curve is acted on by the group $PSL_2(7)$ of order 168.

The surface $H/\Gamma(N)$ now often regarded as the moduli space of elliptic curves with full level N structure.

II.7 Notation: KL call a subgroup of class n if it contains $\Gamma_{\{n\}}$, and of level n if it contains $\Gamma(n)$ (which they denote by $\Gamma_{\{\mu(n)\}}$), so for $n < 6$ class n and level n are equivalent, but for $n \geq 6$ they are not. (The standard error in this subject is to assume that $\Gamma_{\{n\}} = \Gamma(n)$.)

The group $G_{\mu(n)}$ in KL is the projective special linear group $PSL_2(Z/nZ)$ over the integers mod n.

II.8, 9 KL classify the subgroups of the groups $SL_2(Z/pZ)$ and $PSL_2(Z/pZ)$ for p prime, and show that the groups $PSL_2(Z/pZ)$ are simple for $p > 3$. The finite groups G of rotations in R^3 are cyclic, dihedral, tetrahedral (order 12), octahedral (order 24) and icosahedral (order 60), which can be classified most easily by finding the corresponding orbifolds S^2/G. These have a strong tendency to turn up as subgroups of $PSL_2(Z/pZ)$. The reason for this is that the group S^3 of unit quaternions is a double cover of $SO_3(R)$, so any group of rotations has a double cover in S^3. On the other hand, the quaternions can be regarded as a 2-dimensional complex vector space, so all these double covers have 2-dimensional complex representations. One can realize these complex representations over over some algebraic number field and reduce modulo some prime to get a representation of the finite rotation group into the special linear group over some finite field.

III.1, 2 Terminology: An integral of the first kind on a Riemann surface is essentially what is now called a holomorphic 1-form, and an integral of the second kind is essentially a meromorphic 1-form whose residues at the poles are all zero.

These sections give a standard description of Riemann surfaces of curves. The main change since KL is that the construction of Riemann surfaces as branched covers of the complex plane using scissors and sticky tape has been mostly replaced by the use of manifolds and sheaves.

III.3, 4, 5 In this sections KL use the Riemann existence theorem to construct modular functions. For example, if the quotient of the upper half plane by a group has genus 0, then we automatically get a modular function giving an isomorphism from this quotient to the Riemann sphere with some points removed.

III.6, 7 These sections give a detailed discussion of the relation of the Klein quartic to modular functions of level 7. There are 3 modular weight 2 forms z_1, z_2, z_4 of level 7. They satisfy the relation $z_1^3z_4 + z_2^3z_1 + z_4^3z_2$ and give an isomorphism of the quotient $H/\Gamma(7)$ to the corresponding projective quartic curve (the Klein quartic). From the point of view of the Klein quartic, z_1, z_2, z_4 are a basis of the space of holomorphic 1-forms, so this is just the canonical embedding.

Reference

1. Conway, J. H.; Norton, S. P. Monstrous moonshine. Bull. London Math. Soc. 11 (1979), no. 3, 308–39.
2. Gross, Benedict H.; Zagier, Don B. On singular moduli. J. Reine Angew. Math. 355 (1985), 191–220.
3. Levy, Paul. The eightfold way. The beauty of Klein's quartic curve. Edited by Silvio Levy. Mathematical Sciences Research Institute Publications, 35. Cambridge University Press, Cambridge, 1999. x+331 pp. ISBN: 0-521-66066-1
4. Petersson, and Rademacher, H. The Fourier Coefficients of the Modular Invariant $J(\tau)$ Hans Rademacher American Journal of Mathematics Vol. 60, No. 2 (Apr., 1938), pp. 501–512
5. Shimura, Goro. Introduction to the arithmetic theory of automorphic functions. KanÃt' Memorial Lectures, No. 1. Publications of the Mathematical Society of Japan, No. 11. Iwanami Shoten, Publishers, Tokyo; Princeton University Press, Princeton, N.J., 1971
6. William Thurston, The Geometry and Topology of Three-Manifolds (Chapter 13), Princeton University lecture notes (1978–1981)
7. Zagier, Don B. Traces of singular moduli. Motives, polylogarithms and Hodge theory, Part I (Irvine, CA, 1998), 211–244, Int. Press Lect. Ser., 3, I, Int. Press, Somerville, MA, 2002.
8. Zagier, Don B. Elliptic modular forms and their applications. The 1-2-3 of modular forms, 1–03, Universitext, Springer, Berlin, 2008.

University of California, Berkeley
Email: borcherds@berkeley.edu

2

Commentary by Jeremy Gray

A short historical introduction

The four volumes by Robert Fricke, under the guidance of Felix Klein, that are now presented in English are one of the largest neglected monuments in the history of mathematics. Many mathematicians today, one suspects, are content to read only the modern theory, and console themselves with the thought that surely these tomes lack sufficient rigour to be useful a century or more after they were written. The first three rest, after all, between the inspiring discoveries of Poincaré and Klein in the early 1880s and the first rigorous attempts on the uniformisation theorem by Poincaré and Koebe in 1907.

The comparable neglect by historians of mathematics is the mirror image. The real difficulty of the mathematics, and its daunting extent – more than two thousand pages – likewise mixes with the concern about what are the important historical questions that a study of these books would illuminate.

On reading these books, each group of scholars may also come to the same conclusion, but now a more positive one. The books offer a series of partial answers to the question of what a theory of modular and automorphic functions is about. First, the theory of elliptic functions, one of the central themes in 19th-century mathematics, is presented as an indication of what can be done. This extensive theory is then generalised, with the aim of producing a theory for all algebraic integrals along the lines of a geometric function theory that goes back to Riemann and that would culminate in a theory of automorphic functions in general.

The most original feature in *Die elliptischen Funktionen und ihre Anwendungen*, Volume 1, is, however, the role assigned to group theory. It is not true that after Camille Jordan's *Traité des Substitutions et des Équations Algébriques* of 1870 the importance of group theory was steadily accepted. Even Poincaré, for all his identification of geometry with the study of suitable groups, and his intuitive grasp of the role of groups in the study of Riemann surfaces, did not study specific groups in the detail given here. Elsewhere, voices could be found that were unpersuaded of the merits of group theory: Kronecker's was not the only one.

Klein's famous book of 1884 *Vorlesungen über das Ikosaeder und die Auflösung der Gleichungen vom fünften Grade* is not simply a survey of the topic of the quintic equation. It was a long polemic in favour of group theory that sought to cast invariant theory (as practised by Hermite, Brioschi, and Kronecker) into a subservient position. Klein argued that the logic of groups and their subgroups was the appropriate framework for the study of polynomial equations, and that it led one naturally to look for the useful results that invariant theory came to only after mysterious and lengthy calculations. Klein completely, and explicitly, embraced the view of Jordan, and before him Galois, that questions about polynomial equations and their roots should be transferred to the study of the Galois group associated to the equation. In so doing he helped create the modern interpretation of Galois theory, and at the time his view was highly contentious.

Group theory, and more precisely Galois theory, are, however, central in the two volumes on elliptic functions, in accordance with Klein's vision of the subject. Volume 1 ends with the 'next' group after $PSL(2,5)$ on this approach, which is the group $PSL(2,7)$ of order 168 that Klein had studied in his (1878/79) to such good effect before his life was transformed by the arrival of Poincaré. It is studied in the way Klein had treated the icosahedral group, and it hints at Klein's ambition to study all polynomial equation in the spirit that the quintic had been analysed. Volume 2 broadens the approach and embraces a wide variety of material.

The two volumes on automorphic functions have a somewhat different character from the earlier ones and from each other. The first was written in 1897, but the second in 1912, after Poincaré and Koebe had returned to the subject. In these volumes, and especially in the second, the attempt is to show how much of the original continuity method of Klein for proving the uniformisation theorem can be made to work, and to build upon it a unified, systematic theory. But even here there remains the hope, which Klein also expressed at the Karlsruhe meeting of 1912, that what had been done for elliptic functions could be done for automorphic functions in general, and in this way a complete theory of functions with symmetry properties be developed.

Riemannian geometric function theory and Galois theory are the core of the Göttingen vision of mathematics. Galois theory was a major aspect of Hilbert's influential *Zahlbericht*, also published in 1897. The richness of the material in these four volumes shows how intimately Klein drew on the work of Galois and Riemann to create a programme for unifying much of mathematics and taking it in new directions.

It was both an ambitious programme with a potential for further research and an attempt to rescue Klein's past from the collapse of his health in 1882. As Klein slowly recovered from the collision with Poincaré that derailed his plans for the future, he realised that he would never be able to conduct research with the same intensity again. His own hope of becoming Riemann's intellectual successor, manifest in many of his papers in the 1870s, had to be redirected and the future of the theory of function theory on Riemann surfaces passed to Poincaré, only to be abandoned by him a few years later. The related theory of abelian functions lay in Weierstrass's hands, but in the 1880s it must have seemed to Klein that it too was stagnating. Fricke and Klein saw in each other the opportunity to recover what Klein had feared he had lost.

Robert Fricke had been a student of Klein's in the early 1880s and became a mathematics teacher in Braunschweig in 1885, which left him enough time to do his own research. Over the years he published several papers, a textbook on analysis, and an influential *Lehrbuch der Algebra* (1924) which was arguably the last of its kind before the turn to modern algebra initiated by van der Waerden in 1931. He was one of the editors of Klein's *Gesammelte Mathematische Abhandlungen* in the 1920s and Dedekind's *Gesammelte Mathematische Werke*, which were published in 1930, the year in which he died at the age of 68.

The four volumes Fricke produced amplify Klein's hope that group theory, more precisely the theory of transformation groups, could penetrate and unify the theories of analytic functions and geometry. It is a sad irony that, just at the time when Fricke tells us that he worried that the project might end unfinished, it was again transformed from outside by Poincaré and Koebe. But the vision has persisted, through many incarnations from Teichmüller to Thurston and on to today, and the riches of these four volumes now have a fresh chance to speak to mathematicians and historians of mathematics.

References

1. Dedekind, R. 1930. *Gesammelte Mathematische Werke*, R. Fricke, E. Noether, O. Ore (eds.), Vieweg & Sohn.
2. Fricke, R. 1924. *Lehrbuch der Algebra*, Vieweg & Sohn.
3. Hilbert, D. 1897. Die Theorie der algebraischen Zahlkörper (Zahlbericht), *Jahresbericht der Deutschen Mathematiker-Vereinigung* 4, 175–546. In *Gesammelte Abhandlungen* 1, 63–363, tr. I. Adamson, *The theory of algebraic number fields*, Berlin, Springer, 1998.
4. Jordan, C. 1870. *Traité des Substitutions et des Équations Algébriques*, Gauthier-Villars, Paris, repr. Gabay 1989.
5. Klein, C.F. 1878–79. Über die Transformation siebenter Ordnung der elliptischen Functionen, *Mathematische Annalen* 14, 428–471, repr. *Gesammelte Mathematische Abhandlungen* 3, 90–135
6. Klein, C.F. 1884. *Vorlesungen über das Ikosaeder und die Auflösung der Gleichungen vom fünften Grade*, Teubner, repr. Birkhäuser 1993, transl. G. G. Morrice *Lectures on the Icosahedron*, London, 1888, Dover 1956.
7. Klein, C.F. 1921–23. *Gesammelte Mathematische Abhandlungen*, R. Fricke, A. Ostrowski, H. Vermeil, E. Bessel-Hagen (eds.), Springer.

Open University
Email: jeremy.gray@open.ac.uk

3

Commentary by William Harvey on Automorphic Functions

The project has the highly laudable goal of providing the English-speaking world with a readable translation of the classic Fricke-Klein textbook on Fuchsian and Kleinian groups. This pair of volumes (F-K3, F-K4) by R. Fricke and F. Klein provided the graduate students of the early 20th century with an encyclopaedic study of the discrete groups arising from the theory of Riemann surfaces by way of the uniformisation theorem of Koebe and Poincaré. Apart from the little book *Automorphic Functions* by L.R. Ford and the often-overlooked second volume of Appell-Goursat's textbook *Théorie des Fonctions Algébriques,* (due to P. Fatou, as the authors state), there was no competing text before the 1960s, but the diifficulties posed by the arcane writing style and lack of a suitably organised format of Lemma, Theorem and Proof, in addition to the language barrier, meant that the subject was inaccessible to students and was sidelined for many years. It remains difficult to read today because of a definite verbosity and lack of clarity in the exposition: one has to struggle as much with working out what is to be proved as in understanding the proof.

This was unfortunate because there is much of great value in F-K even today. It contains the first comprehensive account of the combination theorems for Fuchsian groups, based on the free product construction for discrete groups in the plane discovered by Klein, which were then deployed (presumably by Fricke) for the purpose of setting up explicit trace parameters (moduli) for hyperbolic surfaces of given topological type, as well as for the more general case of orbifold and punctured surfaces. Their geometric methods were taken up by Fenchel and Nielsen in their famous unpublished manuscript (now in print), before being greatly extended by Bernard Maskit in the 1960s as a tool in developing the theory of Kleinian groups. These parameter spaces, early precursors of Teichmüller's moduli spaces, were later called *Fricke spaces* by Bers and Gardiner (see Lipman Bers' collected works) in their study of the Teichmüler modular group action. With the introduction of more systematic approaches to moduli space using metric methods, the study of hyperbolic surfaces seen through the same perspectives as K-F now presents a formidable body of theory and applications of their construction method, thanks to the work of many, including lots of recent Fields medallists. *Topological recursion,* as it is now known, has become an essential ingredient in current approaches to a wide swathe of problems in combinatorics, complex geometry and theoretical physics.

As a second key issue, F-K addressed the highly significant issue of how to test finitely generated groups for discreteness, using the patterns of algebraic conditions known as *trace relations*. This is still a topic of much current research today. And a simple scroll through the table of contents of the volumes yields further important matters addressed there, at least 50 years before they were taken up again in earnest.

It would be interesting to examine how these difficult textbooks were received by students and research workers of the time, if only we had the sort of usage data available nowadays. My impression is that F-K suffered in comparison with the more popular research topic of modular forms and functions which had been the subject of the first treatise (K-F1, K-F2) by these authors. No doubt the main advantage of the earlier books was the accessibility and the wealth of examples and applications offered by the modular group and modular forms in number theory and geometry. The main drawback of the material in F-K lies in the elaborate style and slow pace, together with a lack of good examples and applications outside the analysis of arithmetically generated lattices. Good applications to the study of Kleinian groups and 3-manifolds only emerged after the modern methods of deformation theory were introduced by the school of Ahlfors and Bers after 1960, followed by the remarkable contributions of W.P.Thurston in 3 dimensional geometry and topology.

Outline of the books:

An introductory section on hyperbolic plane geometry provides essential background for the ground-breaking geometric analysis of hyperbolic surfaces and orbifolds – a term unknown at the time– although the treatment here is very slow and laboured, it does produce a consistent approach to dissection of hyperbolic surfacesi into constituent parts using the notion of *signature*.

King's College University of London
Email: bill.harvey@kcl.ac.uk

4

Commentary by Barry Mazur

Generations of mathematicians have been introduced to geometry, analysis, algebra, and number theory by reading Felix Klein's accounts of these subjects. Each of Klein's expository masterpieces has a startlingly broad scope, such as his vital Erlangen Program (*Vergleichende Betrachtung über neuere geometrische Forschungen* 1872) which initiated the viewpoint that the concept of 'groups of symmetries' underpins (and defines) the classical geometries.

Klein carried through this 'Erlangen program' with Sophus Lie in the context of continuous geometries, but he also pressed for a modern group theoretic approach more broadly—an approach we take as second nature today; e.g., for finite groups, as in his 1884 treatise on regular solids, *Lectures on the Icosahedron and the Solution of the Fifth Degree:*

> it is not the figures themselves...which will form the subject of our study, but rather...those elementary geometrical operations by which those figures coincide with themselves.

—or, for discrete group actions, as in the introduction to volume 1 of Fricke-Klein's *Lectures on Automorphic Functions:*

> [T]he theory of groups stands in the foreground, obtaining more and more validity for itself in almost all parts of higher mathematics. Namely, certain function-theoretic formations of ideas, which go back to Riemann in their original source and whose additional development is a principal task of the present work, prove themselves easily approachable to group-theoretic methods. In this situation it will not be surprising if the first volume of the lectures on the so-called automorphic functions is solely dedicated to the group-theoretic foundations of the future investigations.

The sweep of any of Klein's mathematical historical narratives is always grand. For example, his 1894 "Lectures on Mathematics," delivered at Northwestern University begins with the declaration:

> It will be the object of our Colloquia to pass in review some of the principal phases of the most recent development of mathematical thought in Germany.

And consider Felix Klein's monumental *Seminar-Protokolle* which is a detailed account of 40 years of seminar lectures by him, his colleagues and students, and distinguished visitors, and which spans over 8000 pages in 29 volumes.

In the two volumes of *Elliptic Functions* we are shown a strikingly clear and very explicit investigation of the fundamentals some of the standard modular groups and the geometry corresponding to them, as well as the classical theory of binary quadratic forms, and connections with the subject-matter of Gauss's *Disquistiones Arithmeticae.* The exposition is made vivid by fully worked-out examples of fundamental domains and modular curves for many important (low level) cases; for example, the modular curve known as the *Klein quartic* with its neat uniformization given by a tiling of the unit disc by $(2,3,7)$ hyperbolic triangles. The construction of more general specific modular equations and their interaction with one another has been a treasure trove for students as well as for seasoned practitioners.

The volumes *Lectures on the Theory of Automorphic Functions* offer a full-fledged exploration of the theory of actions of discrete groups on the open unit disc, and a study of the Riemann surfaces connected to these actions, and their corresponding moduli.

All these volumes are interlaced with personal comments that unify the discussion, such as the footnote:

> In autumn 1873 I had myself a lively conversation with the deceased Clifford on this, that one must consider as a task of modern mathematics, to place into lively interplay those mathematical disciplines which come to us as separated but adjacent; we agreed, that this may be the most difficult for synthetic geometry and number theory.

Or consider the somewhat pointed comment:

> Poincaré's developments at this point always appear incomplete, since they do not adequately place in the foreground the nature of the manifold of the boundary points.

The publication of this English translation will benefit a world-wide readership, for the substance of these volumes is presented in such a strikingly explicit way as to be both useful and illuminating; and the expository style puts us intimately in touch with a vitally important moment in the development of mathematics.

Harvard University
Email: mazur@math.harvard.edu

5

Commentary by Series-Mumford-Wright

We are honored to introduce this English translation of Fricke and Klein's magnum opus on Automorphic Functions. It is a tribute to the enduring nature of the material that the reader will still find so much worth perusing, and we should all be grateful both to the translator Arthur DuPre and to the editors for their efforts which will make the contents so much more easily accessible to the English speaking world.

Our main interest has been the third volume, published in 1897, in which Fricke, following the plan sketched by his "highly esteemed teacher and friend" Klein, lays out the theory of 'Discontinuous Groups of Linear Substitutions of One Variable'—now known as Kleinian groups. Following the examples of the groups of the regular solids and the modular and Picard groups, Fricke introduces many basic notions such as fundamental domains (in both two and three dimensions), their associated tessellations (Netzen), and the limit set (Grenzegebilde). (Somewhat confusingly for the modern reader, the term Discontinuitätsbereich is used to mean a fundamental domain for a group acting in a given region of discontinuity.) He describes the connection of the fundamental polygon to generators and relations, leading to a discussion of 'canonical polygons' from which one reads off moduli which classify the associated groups and surfaces, in essence Fricke-Klein coordinates for Teichmüller space. There follows detailed study of 'rotation groups' (those with a single fixed point, which we would nowadays classify as the elliptic and parabolic elementary groups), 'hyperbolic groups' with two fixed points, Fuchsian groups (those with an invariant 'Hauptkreise' or 'principal circle'), culminating in Chapter 3 Part II with a remarkably far-sighted study of groups made by pairing circles in various configurations, which we would now call generalised Schottky groups.

As Fricke rightly says, this last class of group "is much more difficult and we can only give examples". He describes what we now call quasifuchsian groups (whose discontinuity set consists of two simply connected 'Netzen'), and others whose limit sets are totally disconnected. The illustrations in this chapter are quite astonishing for their complication and accuracy. No explanation is given as to how they were made: perhaps variously by Klein and Fricke (both of whom had substantial drafting skills) or perhaps by their students with help from the publishers, but we can only speculate. Other similar figures beautifully drawn with amazing precision are to be found in the archives of the Mathematische Institut, Göttingen, one of which is reproduced in color on the cover of the AMS Notices Volume 54, Number 8, September 2007, together with an explanatory note by Bill Casselman.

Until the 1970s and the advent of modern computer graphic facilities, the Klein-Fricke illustrations remained definitive. Almost all similar drawings in subsequent texts, for example those by Magnus and Lehner, are reproductions from this source. The book Indra's Pearls (Cambridge University Press 2002, 2015) describes a series of computer experiments, starting in the late 1970s and directly inspired by the illustrations in Chapter 3, investigating Kleinian group limit sets. One of us (DW) went to considerable lengths to create computer renditions of Figures 145 and 156, and it is astonishing to see just how precise the originals are.

We have neither space nor expertise to discuss all the vast array of other topics in these remarkable four volumes, but are confident that not only the historian but also the practising mathematician will find a wealth of fascinating material. It is our great pleasure to commend them to the modern reader.

Caroline Series
University of Warwick
David Mumford
Harvard University and Brown University
David Wright
Oklahoma State University
Email: C.M.Series@warwick.ac.uk, dbmumford@gmail.com, djwright921@gmail.com

6

Commentary by Domingo Toledo

It is a great pleasure to see that an English translation of the classical books of Fricke and Klein is finally available.

The historical significance of these books is enormous. The two volumes on Automorphic Functions (F-K I and F-K II) develop the rich connections between real hyperbolic geometry in dimensions two and three and Riemann surfaces. These connections, in general terms well-known today, come through actions of discrete groups of isometries. The quotient of the hyperbolic plane by such a group is a Riemann surface, while the uniformization theorem says essentially that all Riemann surfaces are obtained this way. Meromorphic functions on a Riemann surface then correspond to meromorphic functions on the hyperbolic plane invariant under the discrete group, called *automorphic functions*. These books present the knowledge at that time, resulting from the work of many mathematicians, including Fricke, Fuchs, Klein, Picard, Poincaré, Riemann, Schwarz.

The interest in these books does not stop with their historical significance. I would say that careful reading of these volumes is still one of the best ways to thoroughly understand these connections.

The last statement is not obvious and needs some justification. After all, there are modern expositions that take into account work that has been done since the publication of F-K. For example, the uniformization theorem is much better understood now, with more rigorous proofs than the continuity proof in F-K II. So, why go back to the past?

One answer to this question, that applies equally well to classical works in many other areas of mathematics, is that modern expositions tend to be very polished, showing the end result of maybe a century of working over the material. This polishing process, in order to be efficient, often omits some of the intermediate steps and this can result in obscuring the original motivation. To the modern beginner, particularly to a student, the subject may look like a miracle beyond human understanding, and consequently be discouraged about the possibility of ever doing original work in the subject The outcome could be radically different for a beginner who could also see the natural development of the subject as arising from concrete and understandable problems. Going back to original sources often provides this background. I believe that this is the real value of reading the old classics.

These books give a clear idea of the concrete origins of the subject. Plenty of motivation is presented, much attention is given to detail. Calculations are concrete, clear, concise. Some of my favorite parts are: (1) Constructions of fundamental domains in Part I of F-K I, (2) The arithmetic construction of groups acting on the projective model of the hyperbolic plane, Part III of F-K I. The numerous diagrams help the reader visualize many objects. They also convey very clearly to the reader the concrete visual understanding that the authors had developed.

There are many potential difficulties in reading old books, I will make some comments on F-K I, which also apply to reading other classics.

These books are not written in the present definition - theorem - proof style. To the modern reader they may look more like a narrative of a walk through the woods than a formal presentation of mathematics. Theorems are usually not stated as such, but are italicized statements that describe something interesting that the author sees, points out, and refers to it by page number when needed. These statements are often repeated, perhaps with small variations, in a manner that seems inefficient to the modern reader. I find that this inefficiency helps me to gradually get used to the terminology and the way of thinking used in the book.

In reading classics, it also helps to have some perspective on the common knowledge at the time the book was written. For example, function theory was very extensively developed at the time, as was projective geometry and, to a somewhat lesser extent, the differential geometry of constant curvature spaces. Group theory was at a reasonably mature stage of development, both abstract groups and concrete realizations as transformation groups (which here are called “groups of substitutions"). In fact, it is well-known that transformation groups were central to Klein’s view of mathematics. But topology was at a relatively primitive stage. This can be seen, for instance, in the definition of properly discontinuous group (Part I, 1.2), and in the theorem (in Part I, 2.2) that a group of isometries of the hyperbolic plane is properly discontinuous if and only if it contains no infinitesimal transformations. This last statement seems nonsensical to us, but it and its proof could be translated into our present language by careful and patient reading.

My advice: on first reading, when running into such statements that do not seem to make sense, ignore them and substitute our present understanding of the topological concepts. But concentrate on the details of the more geometrical aspects, such as: construction of fundamental domains; constructions, geometric or arithmetic, of concrete groups; explicit matrix representations. Understand as many examples as possible. And, above all, enjoy the the beautiful mathematics!

University of Utah
Email: toledo@math.utah.edu

7

Commentaries by Other Mathematicians

A modern, not historically inclined, working mathematician may wonder about the usefulness of the reedition of a monograph by Klein and Fricke in his field of research which is almost a hundred years old. The immense progress in the development of the field seems offers only one way to learn the subject by taking a highway allowing one to move fast to acquire the present knowledge and proceed with contributing to it with new results. The reedition of the book will allow a way to look back at the missed beauty of country roads and hiking trails and to acquire more wholesome picture of your favorite subject that may greatly deepen its knowledge in both historical, factual and mathematically unified context. There is also a practical use, specific to this remarkable book, that it offers a wealth of profound results and concrete examples that even from distance of hundred years are still in the mainstream of current research.

Igor Dolgachev
University of Michigan
Email: idolga@umich.edu

The four volumes written by Robert Fricke over one hundred years ago are now available in a translation by Arthur Du Pre. The first two volumes are based on notes of lectures by Felix Klein, and contain a full treatment of the theory of Elliptic modular functions as it had developed at that time. The second two are based on lectures given by Fricke, and treat Automorphic Functions. Together, these four volumes give a comprehensive treatment of the foundations of these fields as they were first developed.

Although there are many much more modern treatments of these subjects, it is important for a serious scholar to understand how they developed and these volumes contain not only an historical perspective, but many of the ideas that are an integral part of the theory today. Moreover, it is interesting to realize the differences between what was considered a proof then and now. Although the mathematics was not nearly as rigorous as it is now, the ideas were at least as deep and insightful. Unfortunately, the books are difficult to read because the exposition is unwieldy and the German is difficult, even for a native German speaker. Once one understands what Fricke is trying to do and the context in which he is doing it, however, one can appreciate how beautiful many of the ideas are.

Du Pre's translation is an attempt to make these books accessible to an English speaking audience. Since most mathematics these days is written in English, most mathematicians are able to read English and so, having these volumes in English makes them available to a broad audience. The translation is quite literal. In fact, some of the sentences are word by word translations so that the reader has to unscramble the German sentence structure to make sense of them. This means that while the book is now in English, it is still a very difficult read. Nevertheless, the perseverant reader will find great satisfaction in working through these books.

Linda Keen
Lehman College, The City University of New York
Email: LKeen@gc.cuny.edu

For many, even most, contemporary mathematicians, the mathematical past is largely inaccessible. This ignorance is unfortunate. Humans have not been on the earth for such a long time that it is appropriate to forget our history, or that of our accomplishments, mathematics among them. The indifference to the mathematical past and the distance from it will, I suppose, nevertheless become more pronounced if or when the Asian nations assume a major role in mathematics and even English ceases to be the principal, or even an adequate, medium of communication. The possibility of reducing mathematics to a trivial pursuit, a struggle for a large number citations, for a prize, or just for a tenured position, is also there. It is difficult not to be pessimistic!

Robert Langlands
Institute for Advanced Study
Email: rpl@ias.edu

The theory of elliptic and automorphic functions of one variable was created in the XIXth century. During the next decades it served as a blueprint for the theory of moduli spaces of algebraic curves of higher genera, Jacobian and Abelian varieties, and their further generalisations: automorphic functions for discrete subgroups of algebraic and more general Lie groups etc.

Moreover, this theory played a key role in such achievements as

1. Proving the Fermat theorem,
2. Classifying the instanton solutions of Einstein's general relativity equations,
3. Conception of Grothendieck's "dessins d'enfant" program for understanding the Galois group of the field of all algebraic numbers,

to name only a few. Reading the great classical treatment by Fricke and Klein from this perspective can be a very rewarding experience, both for a younger student intending to start a research program in this domain, and for a teacher lecturing and/or preparing exercises for her/his classes.

Yu.I. Manin
Max-Planck-Institut für Mathematik
Email: manin@mpim-bonn.mpg.de

Graduate students and seasoned researchers will enjoy reading these classic works on elliptic and automorphic functions. It pays to read the seminal works by the greats, and with these translations these classic works will be reintroduced to generations of mathematicians.

Ken Ono
Emory University
Email: ono@mathcs.emory.edu